AF323155

PROGRESS IN MOLECULAR AND ENVIRONMENTAL BIOENGINEERING – FROM ANALYSIS AND MODELING TO TECHNOLOGY APPLICATIONS

Edited by **Angelo Carpi**

INTECHOPEN.COM

**Progress in Molecular and Environmental Bioengineering
– From Analysis and Modeling to Technology Applications**
Edited by Angelo Carpi

CBS, Edition **2017**

Published by InTech

Janeza Trdine 9, 51000 Rijeka, Croatia

Notice
Statements and opinions expressed in the chapters are these of the individual contributors and not necessarily those of the editors or publisher. No responsibility is accepted for the accuracy of information contained in the published chapters. The publisher assumes no responsibility for any damage or injury to persons or property arising out of the use of any materials, instructions, methods or ideas contained in the book.

Publishing Process Manager Ana Pantar

Technical Editor Teodora Smiljanic

Cover Designer InTech Design Team

Image Copyright Denis Vrublevski, 2010. Used under license from Shutterstock.com

Additional hard copies can be obtained from orders@intechopen.com

Progress in Molecular and Environmental Bioengineering
– From Analysis and Modeling to Technology Applications, Edited by Angelo Carpi
p. cm.
ISBN 978-953-307-268-5

Contents

Contents

Contents

Contents

Preface

This book is an example of a successful and rapid expansion of bioengineering within the scientific world. In fact, it consists of two parts: one dedicated to molecular and cellular engineering and the other to environmental bioengineering.

The content classification mainly reflects the increasing number of studies on genetically modified microrganisms (GMO) directed towards non-biomedical industry. An important application field of these studies is the ecosystem as indicated by the chapters included in the part on environmental bioengineering.

Indeed, because some molecules from GMO are expected to provide either a biomedical use or an industrial application in a different field (see the chapters on Hydrogels and on Diacylglycerol Acyltransferase), the inclusion of the correspondent chapter in the industrial or biomedical part of the book is arbitrary.

This uncertainty and option in the classification of some topics occurs also because the biological component of a bioengineering study can consist in the methodology used or in the aim. An example of the first instance is the genetic manipulation of a microrganism for a specific molecular production, while the second case can be exemplified by the production of a biocompatible material or device with a non-biological methodology.

This reflects the more general characteristics of bioengineering which include either the manipulation of biology or a living being to obtain and use a specific biotechnology for a non-biomedical purpose or the use of disciplines typical of engineering (mathematics, physics, mechanics, chemistry, electromagnetism...) to solve a biomedical problem. These characteristics of bioengineering are partially responsible for the apparent heterogeneity of the topics included in the book.

Indeed, the core of the content crossing all the book sections is molecular or cellular engineering aimed at the production of GMO or specific molecules (usually proteins) for biomedical, industrial or environmental use. This core consists of thirteen chapters describing results obtained with up to date biotechnologies which include:

- insertion of DNA sequences from a different microorganism species (chimeric genes),

- biochemical change of the existing microorganism gene sequences (conjugative plasmids or transposoms),
- DNA and RNA transfer into embryos,
- particular bioprocessing, optimization and enrichment of medium culture.

Of these thirteen core studies, four include genetic engineering methodology: one optimizes the production of clean energy from cellulolytic material (plants and wood) degradation by chimeric β-glucosidases, one describes the GMO use for environmental bioremediation (pollutant removal), one represents a model of combinational bioengineering in embryos and the last one deals with the production of diacylglycerol acyltransferases, an enzyme which appears very promising for research on adipose tissue, for the management of obesity along with related diseases as well as for food industry.

Of the further nine studies characterized by bioprocess engineering with optimized cultures, seven principally aim at obtaining new important products for the use in ecosystem, one attempts to improve drug delivery and one optimizes the production of the principal biocatalysts which account for about 40% of total worldwide enzyme sales.

Two other groups of chapters include:

- design and modeling in molecular, tissue and enviromental bioengineering,
- production and important applications of biomaterials in the biomedical field as well as in other fields like agriculture and electronics.

Hence, this book includes a core of studies on bioengineering technology applications so important that its progress is expected to improve human health and ecosystem. These studies provide an important update on technology and achievements in molecular and cellular engineering as well as in the relatively new field of environmental bioengineering.

Moreover, because 'knowledge is a complex process which requires integration of the simple disciplinary dimension within a wider and more complex structure' this book will hopefully attract the interest of not only the bioengineers, researchers or professionals, but also of everyone who appreciates life and enviromental sciences.

Finally, I consider that each of the Authors has provided their extraordinary competence and leadership in the specific field and that the Publisher, with its enterprise and expertise, has enabled this project which includes various nations and continents.

Thanks to them I have the honour to be the editor of this book.

Dr. Angelo Carpi
Clinical Professor of Medicine and Director of the Division of Male Infertility at the
Department of Reproduction and Aging in the Pisa University Medical School, Pisa,
Italy

Part 1

Molecular and Cellular Engineering: Modeling and Analysis

Fractional Kinetics Compartmental Models

Davide Verotta
Department of Bioengineering and Therapeutic Sciences
Department of Biostatistics
University of California, San Francisco,
USA

1. Introduction

Dynamic models of many processes in the physical and biological sciences give rise to systems of differential equations called compartmental systems. These assume that state variables are continuous and describe the movement of material from compartment to compartment as continuous flows. Together with the mass balance requirements of compartmental systems, these assumptions lead to highly constrained systems of ordinary differential equations, which satisfy certain physical and/or physiological constraints. In this chapter we deal with equivalent structures represented using systems of differential equations of fractional order, that is fractional compartmental systems. The calculus of fractional integrals and derivatives is almost as old as calculus itself going back as early as 1695, to a correspondence between Gottfried von Leibnitz and Guillaume de l'Hôpital. Until a few decades ago, however, expressions involving fractional derivatives, integrals and differential equations were mostly restricted to the realm of mathematics. The first modern examples of applications can be found in the classic papers by Caputo (Caputo) and Caputo and Mainardi (Caputo and Mainardi) (dealing with the modeling of viscoelastic materials), but it is only in recent years that it has turned out that many phenomena can be described successfully by models using fractional calculus. In physics fractional derivatives and integrals have been applied to fractional modifications of the commonly used diffusion and Fokker–Planck equations, to describe sub-diffusive (slower relaxation) processes as well as super-diffusion (Sokolov, Klafter et al.). Other examples are of applications are in diffusion processes (Oldham and Spanier), signal processing (Marks and Hall), diffusion problems (Olmstead and Handelsman). More recent applications are in mainly in physics: finite element implementation of viscoelastic models (Chern), mechanical systems subject to damping (Gaul, Klein et al.), relaxation and reaction kinetics of polymers (Glockle and Nonnenmacher), so-called ultraslow processes (Gorenflo and Rutman), relaxation in filled polymer networks (Metzler, Schick et al.), viscoelastic materials (Bagley and Torvik), although there are recent applications in splines and wavelets (Unser and Blu ; Forster, Blu et al.), control theory (Podlubny ; Xin and Fawang), and biology (El-Sayed, Rida et al.) (bacterial chemotaxis), pharmacokinetics (Dokoumetzidis and Macheras ; Popovic, Atanackovic et al. ; Verotta), and pharmacodynamics (Verotta). Surveys with collections of applications can also be found in Matignon and Montseny , Nonnenmacher and Metzler (Nonnenmacher and Metzler), and Podlubny (Podlubny). A brief history of the development of fractional calculus can be found in Miller and Ross (Miller and Ross).

In this chapter we discuss and show results related to a number of issues related to the definition and use of fractional differential equations to define compartmental systems, in particular we: (1) review ordinary compartmental systems, (2) review fractional calculus, with particular regard to the mathematical objects needed to deal with fractional differential equations ;(3) define commensurate fractional differential equation (linear kinetics) compartmental models; (4) discuss and describe the conditions that allow the formulation of non-commensurate fractional differential equations to represent compartmental systems; (5) show relatively simple analytical solutions (based on the use of Mittag-Leffler functions) for the input-output response functions corresponding to commensurate and non-commensurate fractional (linear kinetics) compartmental models; (6) demonstrate the use of non-linear regression to estimate the parameters of fractional kinetics compartmental models from data available from (simulated) experiments; (7) describe general formulations for fractional order non-linear kinetics compartmental models.

2. Compartmental models

A compartment is fundamentally an idealized store of a substance. If a substance is present in a biological system in several forms or locations, then all the substance in a particular form or all the substance in a particular location, or all the substance in a particular form and location are said to constitute a compartment. Thus, for instance, erythrocytes, white blood cells, and platelets blood, can each be considered as a compartment. The function of the compartment as a store can be described by mass balance equations. The general form of the mass balance equation for a compartment is as follows. If x_i is the quantity of substance in compartment i that interchanges matter with other compartments constituting its environment, then the mass balance takes the form

$$\sum R_{ij} - \sum R_{ji} \tag{1}$$

where $\sum R_{ij}$ represents the summation of the rates of mass transfer into i from relevant compartments or the external environment, and $-\sum R_{ji}$ the summation of the rates of mass transfer from i to other compartments of the system or into the environment. The transfer of material between compartments takes place either by physical transport from one location to another or by chemical reactions. The treatment of a compartment as a single store is an idealization, since a compartment is a complex entity. For example, the concentration of erythrocytes in blood is generally not uniform and one could devise detailed models to describe their distribution. However, in general a compartment is characterized by the idealized average concentration in a compartment. In the rate of mass transfer to other compartments is thus generally of the form

$$R_{ij} = R_{ij}(x_j) \tag{2}$$

where x_j is the quantity of substance in compartment j. Mathematically, the process of aggregation involved in a lumped representation leads to ordinary differential equations as opposed to the partial differential equations that would be required to describe distributed effects. In the formulation of a model of chemical and material transfer processes in a biological system, the system is first divided into (n) relevant and convenient compartments.

The mathematical model then consists of mass balance equations for each compartment and relations describing the rate of material transfer between compartments. The general form of equation defining the dynamics of the i-th compartment is given by

$$\cdot\frac{dx_i}{dt} = R_{io} + \sum_{\substack{j=1 \\ j\neq i}}^{n} R_{ij}(x_j) - \sum_{\substack{j=1 \\ j\neq i}}^{n} R_{ji}(x_i) + R_{oi}(x_i) \tag{3}$$

where now R_{oi} indicates the flux of material from compartment i into the external environment, and R_{io} the flux of material into compartment i from external environment. The second stage requires specifying the functional dependences of each flux, which may be linear or nonlinear. Two commonly occurring types of functional dependence are the linear dependence and the threshold/saturation dependence, which includes the Michaelis-Menten form and the Hill equation sigmoid form. The linear and Michaelis-Menten dependences can be described mathematically in the form

$$R_{ij} = k_{ij} x_j \tag{4}$$

where. k_{ij} is a constant defining the fractional rate of transfer of material into compartment i from compartment j, and

$$R_{ij} = \frac{a_{ij} x_j}{b_{ij} + x_j} \tag{5}$$

where a_{ij} is the saturation value of flux R_{ij} and b_{ij} is the value of x_j at which R_{ij} is equal to half its maximal value. In many instances, the adoption of a linear time- invariant dynamic model for a metabolic system is adequate, at least within certain ranges of exogenous inputs and endogenous production rates. For a linear compartmental linear the state variables, x_j, appear in linear combinations only, and as a consequence the superposition theorem applies: the total response to several inputs is the sum of the responses to the individual inputs. In particular a linear (time-invariant) compartmental model can be written as

$$\frac{d}{dt}\begin{pmatrix} x_1(t) \\ \cdots \\ x_m(t) \end{pmatrix} = \begin{pmatrix} k_{11} & \cdots & k_{1m} \\ \cdots & \cdots & \cdots \\ k_{m1} & \cdots & k_{mm} \end{pmatrix}\begin{pmatrix} x_1(t) \\ \cdots \\ x_m(t) \end{pmatrix} + \begin{pmatrix} f_1(t) \\ \cdots \\ f_m(t) \end{pmatrix} = \mathbf{A}\mathbf{x}(t) + \mathbf{f}(t) \tag{6}$$

$$\mathbf{Y}(t) = \mathbf{B}\mathbf{x}(t)$$

with initial conditions $\mathbf{x}(0) = \mathbf{x}_0$, where now $\mathbf{f}(t)$ is the (vector valued) input function to the system, and $\mathbf{Y}(t)$ is the output equation, a linear combination of the variables $x(t)$, where $\mathbf{B}$ is an appropriately dimensioned matrix. The (rate) constants in equation (6) satisfy:

$$k_{ij} \geq 0, \quad i \neq j$$
$$k_{ii} \leq 0 \tag{7}$$
$$|k_{ii}| \geq \sum_{\substack{j=1 \\ j\neq i}}^{m} k_{ji}$$

where $k_{ii} = -\sum_{\substack{j=0 \\ j \neq i}}^{m} k_{ji}$, which guarantee that all states are non-negative (for non-negative inputs $\mathbf{f}(t)$).

3. Fractional integrals and derivatives

Mathematical modelers dealing with dynamical systems are very familiar with derivatives of integer order, $\dfrac{d^m y}{dx^m}$, and their inverse operation, integrations, but they are generally much less so with fractional-order derivatives, for example $\dfrac{d^{\frac{1}{3}} y}{dx^{\frac{1}{3}}}$. One way to formally introduce fractional derivatives proceeds from the repeated differentiation of an integral power:

$$\frac{d^m}{dx^m} x^p = \frac{p!}{(p-m)!} x^{p-m} \tag{8}$$

For an arbitrary power p, repeated differentiation gives

$$\frac{d^m}{dx^m} x^\delta = \frac{\Gamma(\delta+1)}{\Gamma(\delta-m+1)!} x^{\delta-m} \tag{9}$$

with gamma functions replacing the factorials. The gamma functions allow for a generalization to an arbitrary order of differentiation α,

$$\frac{d^\alpha}{dx^\alpha} x^\delta = \frac{\Gamma(\delta)}{\Gamma(\delta-\alpha+1)!} x^{\delta-\alpha} \tag{10}$$

The extension defined by equation (10) corresponds to the Riemann–Liouville derivative. (Oldham and Spanier ; Miller and Ross).
A more elegant and general way to introduce fractional derivatives uses the fact that the m-th derivative is an operation inverse to m-fold repeated integration. Basic to the definition is the integral identity

$$\int_a^x \int_a^{y1} \cdots \int_a^{ym-1} f(y_m)\, dy_m \cdots dy_1 = \frac{1}{(m-1)!} \int_a^x (x-y)^{m-1} f(y)dy \tag{11}$$

Clearly, the equality is satisfied at $x=a$, and it is not difficult to see iteratively that the derivatives of both sides of the equality are equal. A generalization of the expression allows the definition of a fractional integral (FI) of arbitrary order via

$$J_a^\alpha f(x) = \frac{1}{\Gamma(\alpha)} \int_a^x (x-y)^{\alpha-1} f(y)dy \cdot \tag{12}$$

where again the gamma function is replacing the factorial. In this paper we are concerned with fractional time derivatives, and we take the lower limit in equation (12) to be zero. For this reason in the following we will drop the subscript a in the definition of the operators we consider, and use t, instead of x, to indicate the independent variable time. Starting from equation (12), one can construct several definitions for fractional differentiation. The fractional differential operator D_a^α is defined by

$$D_a^\alpha f(t) \stackrel{def}{=} J^{m-\alpha} D^m f(t) \tag{13}$$

where m is the smallest integer greater than α, $D^m = \dfrac{d^m}{dx^m}$ (m integer) is the classical differential operator, and $f(t)$ is required to be continuous and α-times differentiable in t. The operator D_a^α is named after Caputo (Caputo), who was among the first to use it in applications and to study some of its properties. It can be shown that the Caputo differential operator is a linear operator, i.e. that for arbitrary constants a and b,

$$D_a^\alpha \big(af(t) + bg(t)\big) = aD_a^\alpha f(t) + bD_a^\alpha g(t) \tag{14}$$

that it commutes:

$$D_a^\alpha D_a^\beta f(t) - D_a^\beta D_a^\alpha f(t) \tag{15}$$

and that it possesses the desirable property that:

$$D_a^\alpha c = 0 \tag{16}$$

for any constant c.

Having defined D_a^α, we can now turn to fractional differential equations (FDE), and systems of FDE. A FDE of the Caputo type has the form

$$D_a^\alpha \mathbf{y}(t) = \mathbf{f}(t, \mathbf{y}(t)), \tag{17}$$

where $\mathbf{y}(t)$ is a vector of dependent state variables, and $\mathbf{f}(t,\mathbf{y}(t))$ a, dimensionally conforming, vector valued function, satisfying a set of (possibly inhomogeneous) initial conditions

$$D^k \mathbf{y}(0) = \mathbf{y}_0^{(k)}, \quad k=0,1,\ldots,m\text{-}1 \tag{18}$$

It turns out that under some very weak conditions placed on the function $\mathbf{f}$ of the right-hand side of Eq. (17), a unique solution to Eqs. (17) and (18) does exist (Diethelm and Ford).

A typical feature of differential equations (both classical and fractional) is the need to specify additional conditions in order to produce a unique solution. For the case of Caputo fractional differential equations, these additional conditions are just the static initial conditions listed in (18) which are similar required by classical ordinary differential equations, and are therefore familiar. In contrast, for Riemann–Liouville fractional differential equations, these additional conditions constitute certain fractional derivatives (and/or integrals) of the unknown solution at the initial point $t=0$ (Kilbas and Trujillo), which are functions of t. These initial conditions are not physical; furthermore, it is not clear how such quantities are to be measured from experiment, say, so that they can be appropriately assigned in an analysis (Miller and Ross). If for no other reason, the need to solve fractional differential equations is justification enough for choosing Caputo's definition for fractional differentiation over the more commonly used (at least in mathematical analysis) definition of Liouville and Riemann, and this is the operator that we choose to use in the following.

3.1 Mittag-Leffler functions

Mittag-Leffler functions are generalizations of the exponential function (Erdélyi, Magnus et al.). The solutions of fractional order linear differential equations are often expressed in terms of Mittag-Leffler functions in similar way that the solutions of integer order linear differential equations are expressed in terms of the exponential function. The single parameter Mittag-Leffler function takes the form:

$$E_\alpha(z) = \sum_{i=0}^{\infty} \frac{z^i}{\Gamma(\alpha i + 1)} \tag{19}$$

while the two-parameters Mittag-Leffler function is:

$$E_{\alpha,\beta}(z) = \sum_{i=0}^{\infty} \frac{z^i}{\Gamma(\alpha i + \beta)} \tag{20}$$

The relationship with the exponential function is made clear by the relationships:

$$e^z = \sum_{i=0}^{\infty} \frac{z^i}{i!} = \sum_{i=0}^{\infty} \frac{z^i}{\Gamma(i+1)} = E_1(z) \tag{21}$$

The Laplace transform of the Mittag-Leffler functions are given by:

$$L\left\{ t^{\alpha k + \beta - 1} E_{\alpha,\beta}^{(k)}\left(-\lambda t^\alpha\right) \right\} = \frac{k! s^{\alpha-\beta}}{(s+\lambda)^{k+1}} \tag{22}$$

where $E_{\alpha,\beta}^{(k)}(z) = \dfrac{d^k}{dz^k} E_{\alpha,\beta}(z)$.

The solutions of fractional order linear differential equations are often expressed in terms of Mittag-Leffler functions in similar way that the solutions of integer order linear differential equations are expressed in terms of the exponential function. As shown in, e.g., (Bonilla, Rivero et al. ; Odibat) sums of Mittag-Leffler acquire a prominent role in the solutions of systems of fractional order differential equations, and, as we will see, compartmental models.

In the following to evaluate the single and two-parameters Mittag-Leffler function we implemented a FORTRAN 90 version the algorithm reported in (Gorenflo, Loutchko et al.).

Contrary to α, which has a strong influence on the overall shape of the curve for the case of the single parameter Mittag-Leffler function, the parameter β for has its most pronounced influence on the value of the function at $t = 0$.

The Mittag-Leffler function of the form $E_\alpha(-\lambda t)$ is non-negative and strictly non-increasing for $\lambda > 0$, $0 < \alpha < 1$, t > 0 (Podlubny), while for the function of the form $E_{\alpha,\beta}(-\lambda t^\beta)$ this is not the case, as it can be seen in Figure 1 for $\lambda = 1$. However, a remarkable property, especially in view of the following applications to system of fractional order differential equations, is that the function:

$$t^{1-\beta} E_{\alpha,\beta}\left(-\lambda t^{\beta}\right) \tag{23}$$

is non-negative and strictly non-increasing when $\lambda > 0$, $0 < \alpha < 1$, $0 < \alpha \le \beta \le 1$ (Gorenflo and Mainardi).

Figure 1. shows the Mittag-Leffler function corresponding to choice of parameters α and β reported in (Diethelm, Ford et al.):

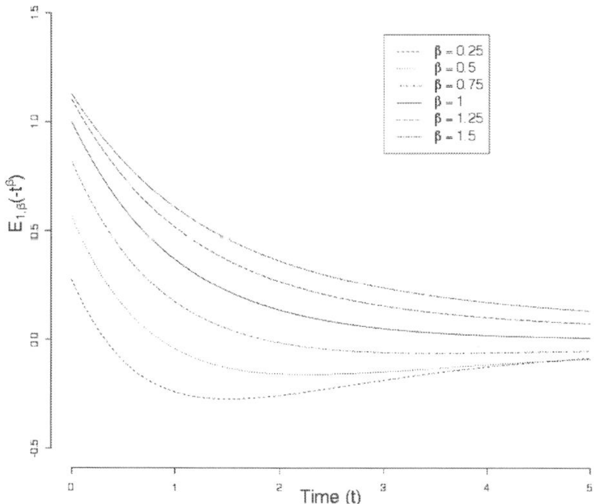

Fig. 1. The Mittag-Leffler function for $\alpha = 1$ and different values of β.

4. Commensurate fractional order linear compartmental models

Commensurate fractional order linear systems are described by a system of linear fractional differential equations (FDE) of the form (Bonilla, Rivero et al.):

$$D_{\blacktriangle}^{\alpha}\mathbf{x}(t) = \begin{pmatrix} D_{\blacktriangle}^{\alpha}x_1(t) \\ \ldots \\ D_{\blacktriangle}^{\alpha}x_m(t) \end{pmatrix} = \begin{pmatrix} k_{11} & \ldots & k_{1m} \\ \ldots & \ldots & \ldots \\ k_{m1} & \ldots & k_{mm} \end{pmatrix} \mathbf{x}(t) \tag{24}$$

$$\mathbf{x}(0) = \mathbf{x}_0$$

where now $D_{\blacktriangle}^{\alpha}$ indicates the Caputo fractional differential operator in respect to time ($D_{\blacktriangle}^{1}\mathbf{x}(t) = d\mathbf{x}(t)/dt$) (Caputo). These systems are called commensurate because all the differential equations are of the same fractional order, α, obtained, for $0 < \alpha \le 1$, exactly as for a standard (ODE) compartmental system.

To construct the solution of the system (24) (see e.g. (Bonilla, Rivero et al. ; Odibat)), we apply the Laplace transform to both sides of the system, to obtain

$$
\begin{pmatrix} s^\alpha x_1(s) - s^{\alpha-1} x_1(0) \\ \dots \\ s^\alpha x_m(s) - s^{\alpha-1} x_m(0) \end{pmatrix} = \begin{pmatrix} k_{11} & \dots & k_{1m} \\ \dots & \dots & \dots \\ k_{m1} & \dots & k_{mm} \end{pmatrix} \begin{pmatrix} x_1(s) \\ \dots \\ x_m(s) \end{pmatrix}
\tag{25}
$$

from which it follows that

$$
x_j(s) = \frac{\det\left(B_j(s)\right)}{\det\left(B(s)\right)}, j = 1,...,m
\tag{26}
$$

where

$$
\tag{27}
$$

$$
B(s) = \begin{pmatrix} k_{11} - s^\alpha & \dots & k_{1m} \\ \dots & \dots & \dots \\ k_{m1} & \dots & k_{mm} - s^\alpha \end{pmatrix}
$$

and $B_j(s)$ is the matrix formed by replacing the j-th column of $B(s)$ by the column $\left(s^{\alpha-1} x_1(0),...,s^{\alpha-1} x_m(0)\right)^T$ $\left(s^{\alpha-1} x_1(0),...,s^{\alpha-1} x_m(0)\right)^T$; $\det\left(B(s^{1/\alpha})\right)$ is a polynomial of degree m, that can be rewritten as $\det\left(B(s^{1/\alpha})\right) = (s - \lambda_1)^{q_1}...(s - \lambda_l)^{q_l}$; from equation (25) $\det\left(B_j(s^\alpha)\right)$ can be rewritten as $s^{\alpha-1}\left(P_1^j(s)x_1(0) + ...P_m^j(s)x_m(0)\right)$ where $P_1^j(s^{1/\alpha})$ is a polynomial of order m-1. Thus, we obtain

$$
x_j(s) = \frac{s^{\alpha-1}\left(P_1^j(s)x_1(0) + ...P_m^j(s)x_m(0)\right)}{(s^\alpha - \lambda_1)^{q_1}...(s^\alpha - \lambda_l)^{q_l}}
\tag{28}
$$

If we now apply a partial fraction decomposition to the j-th term of equation (28), we obtain:

$$
\frac{P_1^j(s)}{(s^\alpha - \lambda_1)^{q_1}...(s^\alpha - \lambda_l)^{q_l}} = \sum_{k=1}^{q_1} \frac{M_{i1}^{kj}}{(s^\alpha - \lambda_1)^{q_1}} + ... + \sum_{k=1}^{q_m} \frac{M_{il}^{kj}}{(s^\alpha - \lambda_l)^{q_l}}
\tag{29}
$$

Thus we can write:

$$
x_j(s) = \sum_{i=1}^{m} \left(\sum_{k=1}^{q_1} \frac{M_{i1}^{kj}}{(s^\alpha - \lambda_1)^{q_1}} + ... + \sum_{k=1}^{q_m} \frac{M_{il}^{kj}}{(s^\alpha - \lambda_l)^{q_l}} \right) s^{\alpha-1} x_i(0)
\tag{30}
$$

Applying the inverse Laplace transform to equation (30) and taking into account the Laplace trasfrom , we obtain the desired solution as a sum of single parameter Mittag-Leffner functions:

$$
x_j(t) = \sum_{i=1}^{m} \left(\sum_{k=1}^{q_1} M_{i1}^{kj} E_\alpha(-\lambda_1 t^\alpha) + ... + \sum_{k=1}^{q_m} M_{il}^{kj} E_\alpha(-\lambda_l t^\alpha) \right) x_i(0)
\tag{31}
$$

The solution to the initial value problem given by system of fractional order differential equations (24) represents the entire state of the system at any given time, is unique (as remarked by (Odibat) for the case of a linear system), and is continuous since it is a sum of continuous functions.

If the solution equation (34) is indicated by $\left(h_1(t),...,h_n(t)\right)^T$, then the initial value problem for the commensurate fractional order compartmental system,

$$D_*^\alpha \mathbf{x}(t) = \mathbf{A}\mathbf{x}(t) + \mathbf{f}(t) \tag{32}$$
$$\mathbf{x}(0) = \mathbf{x}_0$$

has the solution:

$$\mathbf{x}(t) = \mathbf{h}(t) + \int_0^t \mathbf{h}(t-\tau)\mathbf{f}(\tau)d\tau \tag{33}$$

Note that direct differentiation of terms of the form $\mathbf{x}(t) = \mathbf{u}E_\alpha\left(\lambda t^\alpha\right)$, substitution in equation (24), followed by removing the non-zero term $E_\alpha\left(\lambda t^\alpha\right)$ on both sides of the equation, and rearranging yields, $\mathbf{u}(\lambda \mathbf{I} - \mathbf{A}) = 0$, where $\mathbf{I}$ is the $m \times m$ identity matrix. Therefore, $\mathbf{x}(t) = \mathbf{u}E_\alpha\left(\lambda t^\alpha\right)$ is a solution of the system provided that λ is an eigenvalue and $\mathbf{u}$ an associated eigenvector of the characteristic equation associated with the matrix $\mathbf{A}$, that is

$$\mathbf{x}(t) = b_1\mathbf{u}_1^{(1)}E_\alpha(\lambda_1 t^\alpha) + b_2\mathbf{u}_2^{(2)}E_\alpha(\lambda_2 t^\alpha) + ... + b_m\mathbf{u}_m^{(m)}E_\alpha(\lambda_m t^\alpha) \tag{34}$$

where b_1, b_2, ... , b_m are arbitrary constants, λ_1, λ_2, ... , λ_m and $\mathbf{u}_1^{(1)}, \mathbf{u}_2^{(2)},...,\mathbf{u}_m^{(m)}$ are the eigenvalues and eigenvectors of the characteristic equation for (24).

It is interesting, because of its wide range of applications, to consider the case when the eigenvalues of the characteristic equation are real and distinct. When this property holds the solution to equation (32) for a unit impulse input of a substance given in the *j-th* compartment and observations taken in the same compartment, takes the form:

$$h_{jj}(t) = \theta_1 E_\alpha(\lambda_1 t^\alpha) + \theta_2 E_\alpha(\lambda_2 t^\alpha) + ... + \theta_m E_\alpha(\lambda_m t^\alpha) \tag{35}$$

where now $h_{jj}(t)$, with slight abuse of notation, is the unit-input response functions of compartment j for input in j. Equation (35) establishes a direct connection with the familiar multi-exponential response function corresponding to ordinary multi-compartment linear systems with distinct eigenvalues:

$$h_{jj}(t) = \theta_1 e^{\lambda_1 t} + \theta_2 e^{\lambda_2 t} + ... + \theta_m e^{\lambda_m t} \tag{36}$$

In both cases the parameters $\theta_1,..., \theta_m$, $\lambda_1,...,\lambda_m$ and α can be estimated from available input-output data, therefore effectively identifying the unit-impulse response corresponding to a *m*-order compartmental model that can be used to, e.g., predict the responses to arbitrary inputs making use of relationship (33) (Jacquez).

We now give an example of a possible use of fractional compartmental models to approximate data obtained from a system of unknown structure. To do so we generated error corrupted data using an eight compartments mammillary system based on the drug thiopental distribution in rats (Stanski, Hudson et al. ; Verotta, Sheiner et al.). The rate constants from the central compartment (blood) to the 7 peripheral compartments are: $k_{j1} =$ 1.80, 0.116, 0.126, 0.171, 2.43, 0.275, and 0.348 (min^{-1}), for $j=2,...,8$, respectively; the rate constants from the peripheral compartment to central are k_{j1} = 0.559, 0.172, 0.117, 0.0975, 4.84, 0.411, and 0.0499 (min^{-1}), for $j=2,...,8$, respectively; the exit rate from the elimination compartment (liver) is k_{02} = 0.0258 (min^{-1}), and the volume of the central compartment (for a 365 grams rat) is 9.89 (ml).

Figure 2 shows the fit of models equation (35) (solid line), and (36) (dashed line) with $m=2$, to the simulated data (open circles) obtained adding a proportional normally distribute error (according to a constant plus proportional error model). The parameters $\theta_1, \theta_2, \lambda_1, \lambda_2, \alpha$ are estimated from the data, with the constraints $\theta_1, \theta_2 > 0$, $\lambda_1, \lambda_2 < 0$, and $0 < \alpha \leq 1$, which guarantee that equation (35) (and (36)) is non-negative and non-increasing (strictly monotone) for $t \geq 0$.

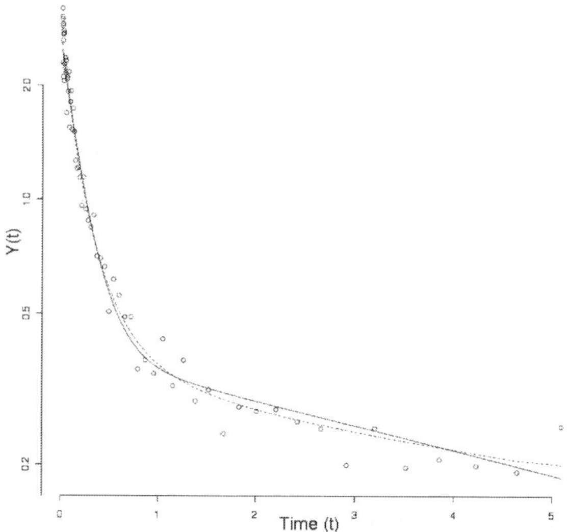

Fig. 2. The fit of the response function corresponding to integer (solid line) and commensurate fractional order (dashed line) two compartments system (dashed line) to simulated data (open circles). The data are generated using an eight compartments integer order mammillary system.

Note the added flexibility introduced by use of a sum Mittag-Leffler functions in respect to exponentials: the values of minus twice log-likelihood for the fit of the simulated data were -668.45, and -731.43, for models (35)-(36), respectively, a drop in the objective function that is highly significant according to, e.g. the Akaike criterion (Akaike). (We remark that this is an example provided to show the added flexibility introduced by the use of fractional differential equations: for this simulation, a sum of exponentials would fit the simulated data perfectly well when the number of exponential terms in the fitted response function is increased.)

5. Non-commensurate fractional order linear compartmental systems

In a non-commensurate fractional order linear system (Bonilla, Rivero et al.), the fractional order for each equation of the system are distinct (real positive) numbers $(\alpha_1, \cdots, \alpha_m)$. To obtain a non-commensurate compartmental system it would appear that all is required is to allow for a distinct fractional orders of differentiation in equations (24), to obtain the system of fractional differential equations:

$$\begin{pmatrix} D_{\blacktriangle}^{\alpha 1} x_1(t) \\ \dots \\ D_{\blacktriangle}^{\alpha m} x_m(t) \end{pmatrix} = \begin{pmatrix} k_{11} & \dots & k_{1m} \\ \dots & \dots & \dots \\ k_{m1} & \dots & k_{mm} \end{pmatrix} \mathbf{x}(t)$$

$$\mathbf{x}(0) = \mathbf{x}_0$$

(37)

where $0 < \alpha_i \le 1, i = 1, \dots, m$. Note however that now the flux of mass from the j-th to the i-th compartment, R_{ij}, appears inconsistently, since it is defined as am outgoing flux of fractional order α_j in the j-th differential equation, and appears as an incoming flux into compartment i as a rate of fractional order $\alpha_i \ne \alpha_j$. (As a consequence the equations in (37) do not necessarily satisfy mass-balance, even if the matrix $\mathbf{A}$ would guarantee mass balance in equation (24)). (The dimensions of the rate constants are also inconsistent in equation (37), since they change depending on the fractional differential equation they appear in.)
An example will clarify the problem. Consider the following second order model:

$$\begin{pmatrix} D_{\blacktriangle}^{\alpha 1} x_1(t) \\ D_{\blacktriangle}^{\alpha 2} x_2(t) \end{pmatrix} = \begin{pmatrix} -(k_{01} + k_{21}) & k_{12} \\ -k_{21} & -(k_{02} + k_{12}) \end{pmatrix} \begin{pmatrix} x_1(t) \\ x_2(t) \end{pmatrix}$$

(38)

in this representation the fluxes from the compartment to the outside of the system pose no problem, but the fluxes between compartments are not balanced: the outgoing flux from compartment 1 to 2 is at rate α_1, but it appears as incoming flux in compartment 2 at rate α_2, and vice-versa. In addition the rate constants are not expressed consistently in terms of their dimensions. In the first differential equation the units for the rate constant k_{01}, k_{12}, k_{21} are the fractional reciprocal of unit time (ut) of order α_1, while in the second the units for k_{02}, k_{12}, k_{21} are $(ut)^{-\alpha 2}$, which give inconsistent dimensions for the transfer rates between compartments k_{12}, k_{21}. $(ut)^{-1}$. Similarly to the suggestion reported in (Popovic, Atanackovic et al.), the problem of inconsistent units can be solved by normalizing the units of the rate constants in the system (that is left multiply equations (38), i.e. (37), by $diag\left(\tau^{\alpha 1}, \dots, \tau^{\alpha m}\right)$, where $\tau^{\alpha 1}, \dots, \tau^{\alpha m}$ are the characteristic time for each compartment, so that the elements in the matrix $\mathbf{A}$ have all dimensions $(ut)^{-1}$, see also (Dokoumetzidis, Magin et al.). However, the problem of balancing the fluxes is more fundamental and need to be addressed if one has to provide a general representation that allows a physical interpretation of the system.
We now describe three possible alternatives to represent non-commensurable fractional order compartmental systems.

5.1 Reducible systems

A possibility to solve the problem associated with equations (37) is to consider compartmental structures that include subsystems that do not transfer material to other parts of the system. For example if the matrix $\mathbf{A}$ in equation (37) can be put in the form:

$$\mathbf{A} = \begin{pmatrix} \mathbf{A}_{11} & 0 \\ \mathbf{A}_{21} & \mathbf{A}_{22} \end{pmatrix} \tag{39}$$

where $\mathbf{A}_{11}$ and $\mathbf{A}_{22}$ are square matrices of order m_1 and m_2, respectively ($m_1 + m_2 = m$), that is if the matrix $\mathbf{A}$ is reducible, the corresponding compartmental topology includes two subsystems, of which the first (of dimension m_1) does not transfer material to the second. We can then consider a representation in which each subsystem is characterized by one fractional rate, α_1 and α_2 respectively, obtaining the representation:

$$\left(D_{\blacktriangle}^{\alpha 1} x_1(t) \quad \ldots \quad D_{\blacktriangle}^{\alpha 1} x_{m1}(t) \quad D_{\blacktriangle}^{\alpha 2} x_{m1+1}(t) \quad \ldots \quad D_{\blacktriangle}^{\alpha 2} x_{m1+m2}(t) \right)^{T} = \begin{pmatrix} \mathbf{A}_{11} & 0 \\ \mathbf{A}_{21} & \mathbf{A}_{22} \end{pmatrix} \mathbf{x}(t) \tag{40}$$

where T indicates matrix transpose. The physical interpretation is that of two sub-systems that operate at distinct fractional rates, with the first receiving inputs from the second subsystem. A great number of situations can be modeled using reducible compartmental structures, for example cascades of chemical or metabolic reactions with one, or, multiple, irreversible steps; drug absorption, in which the intestine acts as a separated sub-system delivering substance/drug to the circulatory subsystem; administration o drugs using complex using external devices/formulation, e.g. nicotine patches or sustained release formulations (Pitsiu, Sathyan et al.) etc..

5.2 General representation

A second, and more general, alternative is to start from the commensurate system, equation (24), and introduce additional fractional kinetics in the form of departures from a reference fractional rate. Continuing with the second order model example reported above, we write:

$$D_{\blacktriangle}^{\alpha} \begin{pmatrix} x_1(t) \\ x_2(t) \end{pmatrix} = \begin{pmatrix} -(k_{01} + k_{21}) & k_{12} \\ -k_{21} & -(k_{02} + k_{12}) \end{pmatrix} \begin{pmatrix} x_1(t) \\ D_{\blacktriangle}^{\alpha-\alpha 2} x_2(t) \end{pmatrix} \tag{41}$$

where $\alpha = \alpha_1$. In this formulation the fluxes from compartment 1 to 2 ($k_{21} x_1(t)$) and from compartment 2 to 1 ($k_{12} D_{\blacktriangle}^{\alpha-\alpha 2} x_2(t)$) now appear as incoming/outgoing fluxes in fractional differential equations of the same order α, so that mass balance is satisfied. The general case is (a representation for non-commensurate fractional differential equation models that is, to the best of our knowledge, novel) takes the form:

$$D_{\blacktriangle}^{\alpha} \mathbf{x}(t) = \begin{pmatrix} k_{11} & \ldots & k_{1m} \\ \ldots & \ldots & \ldots \\ k_{m1} & \ldots & k_{mm} \end{pmatrix} \begin{pmatrix} D_{\blacktriangle}^{\alpha 1-\alpha} x_1(t) \\ D_{\blacktriangle}^{\alpha 2-\alpha} x_2(t) \\ \ldots \\ D_{\blacktriangle}^{\alpha m-\alpha} x_m(t) \end{pmatrix} \tag{42}$$

$$\mathbf{x}(0) = \mathbf{x}_0$$

where $\alpha = \alpha_1$, $0 < \alpha_i \leq 1$, $i = 1, \ldots, m$, and with no lack of generality, we order the indexes of the compartments so that $\alpha \geq \alpha_2, \ldots, \alpha_m$. The representation is now balanced in term of fluxes, and it is now consistent in terms of the units for the rate constants in $\mathbf{A}$, with the rates appearing in each column of the matrix now expressed with consistent dimensions

$\left[k_{ij} \right] = (ut)^{-(\alpha - \alpha j)}$. Equation (42) reduces to the commensurate fractional order differential equation compartmental case equation (24) when $\alpha_2, ..., \alpha_m = \alpha$.

The main problem with this representation is that it seems to require a numerical approximation to its solution. That is, no analytical solution for equation (42), of a form similar to, e.g., equation (34), could be found at the time of this writing.

The need to use numerical approximations has been present from the beginning of the modern investigation of fractional calculus, and analytical solutions to fractional differential and integral equations are known only for specific cases (see, e.g., the examples reported in (Magin ; Magin)). Algorithms dealing with fractional differential equations are reported in (Gorenflo ; Podlubny) but focus on solving Riemann–Liouville fractional differential equations and usually restrict the class of fractional differential equations to be linear with homogeneous initial conditions. The more general algorithms reported in (Diethelm, Ford et al. ; Diethelm, Ford et al.), could be adapted to solve the problem of integrating equation (42), and we have already adapted the algorithms to solve certain kinds fractional differential equations related to pharmacodynamics models (Verotta)).

The main modification of the fractional differential equation solver, which is of the Predict-Evaluate-Correct-Evaluate type, is to incorporate a fractional integrator to evaluate the terms of the form $D_{\blacktriangle}^{\alpha i - \alpha} x_i(t)$ $i = 2, ..., m$, on the right hand-side of equation (42) (note that, by construction, $-1 < \alpha_i - \alpha < 0$, so that all terms of the form $D_{\blacktriangle}^{\alpha i - \alpha} x_i(t)$ corresponds to fractional integrals); an algorithm for fractional integration can also be found in (Diethelm, Ford et al.)).

5.3 Response function representation

We return to the system of fractional differential equations (37) to show how certain analytical solutions can be used as response functions corresponding to compartmental systems.

In general an analytical solution to (37) does not exist, however solutions can be obtained if it is assumed that the fractional orders of the differential equations are rational numbers: $\alpha_i = r_i / q_i$ where p_i, q_i are integers, $i=1,..., m$. (Note that any real number can be approximated arbitrarily closely by a rational number and therefore one can approximate any system of differential equations with multiple fractional derivatives by a system fractional differential equations with orders that are as close as we choose to the original orders, a property that will apply in any case as soon as the orders are stored in a computer.)

The derivation of the solution follow the steps used for the commensurate case equations (25) – (31) (see (Diethelm and Ford ; Lakshmikantham and Vatsala ; Odibat), to arrive at the expression for $x_j(s)$ of the form:

$$x_j(s) = \sum_{i=1}^{m} \left(\sum_{k=1}^{q1} \frac{M_{i1}^{kj}}{(s^\gamma - \lambda_1)^{q1}} + ... + \sum_{k=1}^{qm} \frac{M_{il}^{kj}}{(s^\gamma - \lambda_1)^{ql}} \right) s^{\alpha i - 1} x_i(0) \tag{43}$$

Applying the inverse Laplace transform to equation (30) and taking into account the Laplace trasfrom (22), we obtain the solution for the non-commensurate fractional order system as a sum of two parameters Mittag-Leffner functions:

$$x_j(t) = \sum_{i=1}^{m} t^{\gamma - \alpha i} \left(\sum_{k=1}^{q1} M_{i1}^{kj} E_{\gamma, \gamma - \alpha i + 1}^{(k)}(-\lambda_1 t^\gamma) + ... + \sum_{k=1}^{qm} M_{il}^{kj} E_{\gamma, \gamma - \alpha m + 1}^{(k)}(-\lambda_1 t^\gamma) \right) x_i(0) \tag{44}$$

where $\gamma = 1/q$, $q = M.C.D(q_1,...,q_m)$ and $E_{\alpha,\beta}^{(k)}(z) = \dfrac{d^k}{dz^k} E_{\alpha,\beta}(z)$. If the eigenvalues of the characteristic equation for the system are real and distinct the solution simplifies to:

$$x_j(t) = \sum_{i=1}^{m} t^{\gamma-\alpha_i} \left(M_{i1}^j E_{\gamma,\gamma-\alpha j+1}(-\lambda_1 t^\gamma) + ... + M_{il}^j E_{\gamma,\gamma-\alpha m+1}(-\lambda_l t^\gamma) \right) x_i(0) \tag{45}$$

and in particular the unit-impulse response function for input/output in compartment j, takes the form:

$$h_{jj}(t) = t^{\gamma-\alpha j} \left[\theta_1 E_{\gamma,\gamma-\alpha j+1}(\lambda_1 t^\gamma) + \theta_2 E_{\gamma,\gamma-\alpha j+1}(\lambda_2 t^\gamma) + ... + \theta_m E_{\gamma,\gamma-\alpha j+1}(\lambda_m t^\gamma) \right] \tag{46}$$

which depends on γ, α_j but not on the fractional orders associated to the other compartments.

To show an application of this type of response functions we consider the same mammilary eight compartment model with input/output in compartment j. The input is now at a constant rate of 2 (um) for 0.5 (ut). As before we estimate $\theta_1, \theta_2, \lambda_1, \lambda_2, \alpha$ and in addition γ, with the constraints $\theta_1, \theta_2 > 0 =$, $\lambda_1, \lambda_2 < 0$, and $0 < \gamma, \alpha \leq 1$.

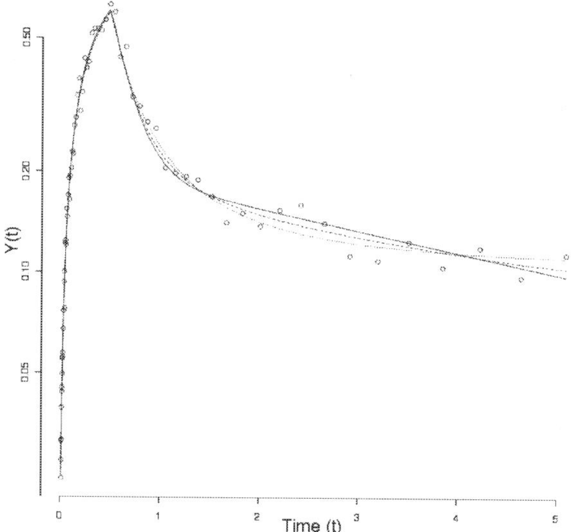

Fig. 3. The fit of the response function corresponding to ODE (solid line), commensurate FDE (dashed line), and non-commensurate FDE (widely dashed line) two compartments system to simulated data (open circles) generated using an eight compartments ODE mammillary model and step function input.

Note again the added flexibility introduced by use of a sum Mittag-Leffler functions in respect to a sum of exponentials: the decreases in objective function values (minus twice log-likelihood) for model (36) (commensurate system response function) and (46) (non-commensurate system response function) vs. model (35) (ordinary system response function), were -39.3 and -58.22, respectively, values that are again highly significant according to the Akaike criterion (Akaike), and that would select the response function corresponding to the non-commensurate system.

The response functions used above are of course not limited to input-output in the same compartment. The following figure shows the result of the fit to simulated data generated using the same mammillary model used in Figure 2 and 3 but with an added ninth compartment (gut) which receives the and delivers it to the central compartment (at a rate k_{19} =1 (min^{-1})), that is an example of a reducible system. The input consist now of two unit-impulses at time 0 and 2.3 (min) in the gut compartment. The competing models (for the response function in blood to a unit-input in the gut, $h_{19}(t)$) are the convolutions of the same models used in the previous examples with the mono-exponential $\theta_3 e^{\lambda_3 t}$ (that is, the open-loop unit-impulse response function for the gut). We estimate the same parameters as for the previous example, with the addition of λ_3 (constrained to be negative), and θ_3 fixed to one since it is not identifiable from the experiment.

Figure 4 shows the result of the fit of the models to the simulated data.

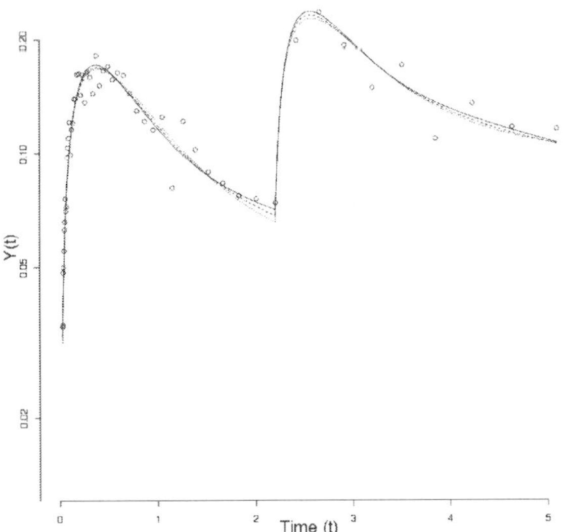

Fig. 4. See legend to Figure 3. Simulated data (open circles) correspond to two unit-impulse inputs in a peripheral compartment (gut).

The decreases in objective function for the commensurate and non-commensurate vs. the ordinary system response functions were -7.2 and -6.9, respectively. These values that are still significant according to the Akaike criterion (Akaike), and select the commensurate response function as "best" model. More importantly the narrowing distance between the likelihood of the models demonstrates how the choice of the input can influence model selection, and how a single unit-impulse input might result in the most informative model selection test input to discriminate between ordinary and fractional (linear) compartmental models.

6. Fractional order non-linear compartmental models

The generalization of commensurate fractional order linear compartmental systems to non-linear systems can be achieved simply by considering fluxes other than the linear case equation (4). For example a Michaelis-Menten compartmental model includes fluxes given by equations (5), and takes the form:

$$D_\blacktriangle^\alpha \mathbf{x}(t) = \begin{pmatrix} -\sum_{\substack{j=0 \\ j\neq 1}}^{m} \dfrac{a_{j1}}{b_{j1}+x_1} & \cdots & \dfrac{a_{1m}}{b_{1m}+x_m} \\ \cdots & \cdots & \cdots \\ \dfrac{a_{m1}}{b_{m1}+x_1} & \cdots & -\sum_{\substack{j=0 \\ j\neq m}}^{m} \dfrac{a_{jm}}{b_{jm}+x_m} \end{pmatrix} \mathbf{x}(t) \qquad (47)$$

$$\mathbf{x}(0) = \mathbf{x}_0$$

which is analogous to the ordinary differential compartmental model with Michelis-Menten elimination (Tong and Metzler). The non-commensurate case can be similarly be defined by the following representation:

$$D_\blacktriangle^\alpha \mathbf{x}(t) = \begin{pmatrix} -\sum_{\substack{j=0 \\ j\neq 1}}^{m} \dfrac{a_{j1}}{b_{j1}+x_1} & \cdots & \dfrac{a_{1m}}{b_{1m}+D_\blacktriangle^{\alpha m-\alpha}x_m(t)} \\ \cdots & \cdots & \cdots \\ \dfrac{a_{m1}}{b_{m1}+x_1} & \cdots & -\sum_{\substack{j=0 \\ j\neq m}}^{m} \dfrac{a_{jm}}{b_{jm}+D_\blacktriangle^{\alpha m-\alpha}x_m(t)} \end{pmatrix} \begin{pmatrix} x_1(t) \\ D_\blacktriangle^{\alpha 2-\alpha}x_2(t) \\ \cdots \\ D_\blacktriangle^{\alpha m-\alpha}x_m(t) \end{pmatrix} \qquad (48)$$

$$\mathbf{x}(0) = \mathbf{x}_0$$

where $\alpha_1 = \alpha$, $\alpha \geq \alpha_2,...,\alpha_m$. This model is well defined in terms of flux balances, and the dimensions of the constants a_{ij} and b_{ij}. The most general case of non-linear non-commensurate fractional order compartmental structure can be obtained from the general case equation

$$D_\blacktriangle^\alpha x_i(t) = R_{1o}(t) + \sum_{\substack{j=1 \\ j\neq i}}^{n} R_{ij}(D_\blacktriangle^{\alpha-\alpha j}x_j) - \sum_{\substack{j=1 \\ j\neq i}}^{n} R_{ji}(D_\blacktriangle^{\alpha-\alpha i}x_i) + R_{oi}(D_\blacktriangle^{\alpha-\alpha i}x_i) \qquad (49)$$

where again all the differential equations are of the same fractional order, allowing of mass balance and units consistency across equations.

7. Final remarks

The main purpose of this chapter is to discuss the use of systems of fractional differential equations to represent compartmental models. We first considered commensurate fractional differential equations compartmental systems, and show how they have a direct relationship with ordinary differential equation compartmental systems. We also showed how non-commensurate systems of fractional differential equations require special formulations to correspond to a compartmental system, and describe alternative way to represent such systems, including a novel general representation (equation (42)).

For both commensurate and non-commensurate compartmental systems we give expressions for response functions that can be used to describe input-output experiments, and satisfy physical constraints (non-negativity in particular). We show how sums of Mittag-Leffler functions with a single parameter, equation (35), are solutions for the response of compartmental system of commensurate fractional differential to impulse-inputs, while sums of two-parameters Mittag-Leffler functions, equation (46), are the corresponding solutions for a system of non-commensurate fractional differential equations. The corresponding unit-impulse response functions (sums of Mittag-Leffler functions, defined for each of the input/output possible combinations, $h_{ij}(t)$, $i, j = 1,...,m$) can be used to represent the response of a fractional order compartmental system to arbitrary inputs by means of the ordinary convolution operator, e.g. equation (33). This is in direct analogy with the use of unit-impulse response functions (consisting of sum of exponentials) used for ordinary compartmental models.

We also describe general formulations for fractional order non-linear kinetics compartmental models, and briefly discuss how such models could be implemented and solved using software algorithms.

In conclusion, while insight into the physiological interpretability of fractional compartmental system remains open to discussion, the technology is becoming available to investigate the application of these models to data sets that might show complex fractional kinetics. The bottleneck to initiate this kind of investigation is the development of appropriate software. In particular, while the evaluation of (sums of) Mittag-Leffler functions can be considered (to some extent) solved, the stable and reliable integration of system of fractional differential equations of the form (37), and even more so of the form of (42) or (49), is not non-trivial task, especially taking into consideration that the corresponding software needs to be interfaced with a non-linear regression program. The author is actively working on a set of routines that will interface with the open source program R and allow the use of multi-term Mittag-Leffler response functions as well as the integration of fractional compartmental models.

8. Acknowledgements

This work was supported in part by NIH grants R01 AI50587, GM26696

9. References

[1] (1998). Fractional differential systems: Models, methods, and applications. ESAIM Proceedings. Paris, SMAI.

[2] (2006). "The R project for Statistical Computing." http://www.r-project.org/.

[3] Akaike, H. (1974). "A new look at the statistical model identification problem." IEEE Trans Automat Contr 19: 716-723.

[4] Bagley, R. L. and P. J. Torvik (1983). "A theoretical basis for the application of fractional calculus to viscoelasticity." J. Rheol. 27: 201–210.

[5] Bonilla, B., M. Rivero, et al. (2007). "On systems of linear fractional differential equations with constant coefficients." Applied Mathematics and Computation 187: 68–78.

[6] Caputo, M. (1967). "Linear models of dissipation whose Q is almost frequency independent-II." Geophys. J. Roy. Astron. Soc. 13: 529–539.

[7] Caputo, M. and F. Mainardi (1971). "Linear models of dissipation in anelastic solids." Rivista del Nuovo Cimento 1: 161–198.

[8] Chern, J. T. (1993). "Finite element modeling of viscoelastic materials on the theory of fractional calculus." Ph.D. thesis, University Microfilms No. 9414260.

[9] Diethelm, K., J. M. Ford, et al. (2006). "Pitfalls in fast numerical solvers for fractional differential equations." J. Comput. Appl. Math. 186: 482–503.

[10] Diethelm, K. and N. J. Ford (2002). "Analysis of fractional differential equations." J. Math. Anal. Appl 265: 229–248.

[11] Diethelm, K., N. J. Ford, et al. (2005). "Algorithms for the fractional calculus: A selection of numerical methods." Comput. Methods Appl. Mech. Engrg. 194: 743–773

[12] Dokoumetzidis, A. and P. Macheras (2009). "Fractional kinetics in drug absorption and disposition processes." J Pharmacokinet Pharmacodyn 36: 165–178.

[13] Dokoumetzidis, A., R. Magin, et al. (2010). "A commentary on fractionalization of multi-compartmental models." J Pharmacokin Pharmacodyn.

[14] El-Sayed, A. M. A., S. Z. Rida, et al. (2009). "On the Solutions of Time-fractional Bacterial Chemotaxis in a Diffusion Gradient Chamber." Intern. J. Nonlin. Sci. 7: 485-492.

[15] Erdélyi, A., W. Magnus, et al. (1955). "§8.2." Higher transcendental functions, Bateman manuscript project 2.

[16] Forster, B., T. Blu, et al. (March 2006). "Complex B-Splines." Applied and Computational Harmonic Analysis 20(2): 261-282.

[17] Gaul, L., P. Klein, et al. (1991). "Damping description involving fractional operators." Mech. Syst. Signal Process.: 81–88.

[18] Glockle, W. G. and T. F. Nonnenmacher (1995). "A fractional calculus approach to self-similar protein dynamics." Biophys. J. 68 46-53.

[19] Gorenflo, R., Ed. (1997). Fractional calculus: Some numerical methods. Fractals and Fractional Calculus in Continuum Mechanics, CISM Courses and Lectures. Wien, Springer.

[20] Gorenflo, R., J. Loutchko, et al. (2002). "Computation of the Mittag-Leffler Function $E_{\alpha,\beta}(z)$ and its derivative." Fract. Calc. Appl. Anal. 5: 491–518.

[21] Gorenflo, R. and F. Mainardi (1997). Fractional calculus: Integral and differential equations of fractional order. Fractals and fractional calculus in continuum mechanics. CISM courses and lectures 378. G. Carpinteri and F. Mainardi. New York, Springer: 223-276.

[22] Gorenflo, R. and R. Rutman, Eds. (1995). On ultraslow and intermediate processes. Transform Methods and Special Functions. Sofia 1994, Science Culture Technology, Singapore.

[23] Jacquez, J. J. (1990). "Parameter estimation: local identifiability of parameters." Am J Physiol 258: E727-736.

[24] Kilbas, A. A. and J. J. Trujillo (2001). "Differential equations of fractional order: Methods, results and problems I." Appl. Anal. 78: 153–192.

[25] Lakshmikantham, V. and A. S. Vatsala (2008). "Basic theory of fractional differential equations, Nonlinear Anal. TMA." Nonlinear Anal. TMA 69(8): 2677–2682.

[26] Magin, R. L. (2004). "Fractional calculus in bioengineering Part 2." Crit Rev Biomed Eng. 32: 105–193.

[27] Magin, R. L. (2004). "Fractional calculus in bioengineering Part 3." Crit Rev Biomed Eng. 32: 195–377.

[28] Marks, R. J. and M. W. Hall (1981). "Differintegral interpolation from bandlimited signals samples." IEEE Trans. Acoust. Speech. Signal Process. 29: 872–877.

[29] Metzler, R., W. Schick, et al. (1995). "Relaxation in filled polymers: A fractional calculus approach." J. Chem. Phys. 103 7180–7186.

[30] Miller, K. S. and B. Ross (1993). An Introduction to the Fractional Calculus and Fractional Differential Equations. New York Wiley.

[31] Nonnenmacher, T. F. and R. Metzler (1995). "On the Riemann–Liouville fractional calculus and some recent applications." Fractals: 557–566.

[32] Odibat, Z. M. (2010). "Analytic study on linear systems of fractional differential equations." Computers and Mathematics with Applications 59: 1171–1183.

[33] Oldham, K. B. and J. Spanier (1974). The Fractional Calculus. San Diego, Calif., Academic Press.

[34] Olmstead, W. E. and R. A. Handelsman (1976). "Diffusion in a semi-infinite region with nonlinear surface dissipation." SIAM Rev. 18: 275–291.

[35] Pitsiu, M., G. Sathyan, et al. (2001). "A semiparametric deconvolution model to establish in-vivo in-vitro correlation applied to OROS Oxybutynin." J Pharm Sci 90: 1-11.

[36] Podlubny, I. (1994). "Fractional-order systems and fractional-order controllers." Tech. Report UEF-03-94, Institute for Experimental Physics, Slovak Academy of Sciences.

[37] Podlubny, I. (1999). Fractional differential equations. San Diego, Academic Press.

[38] Podlubny, I. (2002). "Geometric and physical interpretation of fractional integration and fractional differentiation." Fract. Calculus Appl. Anal. 5: 367–386

[39] Popovic, J. K., M. T. Atanackovic, et al. (2010). "A new approach to the compartmental analysis in pharmacokinetics: fractional time evolution of diclofenac." J Phamacokinet Pharmacodyn 37(2): 119-134.

[40] Sokolov, I. M., J. Klafter, et al. (November 2002). "Fracional Kinetics." Physics Today: 48-54.

[41] Stanski, D. R., R. J. Hudson, et al. (1984). "Pharmacodynamic modeling of Thiopental anesthesia." J Pharmacokin Biopharm 12: 223-240.

[42] Tong, D. D. M. and C. M. Metzler (1980). "Mathematical properties of compartment models with Michaelis-Menton type elimination." Math Biosci 48: 293-306.

[43] Unser, M. and T. Blu (March 2000). "Fractional Splines and Wavelets." SIAM Review 42(1): 43-67.

[44] Verotta, D. (2010). "Fractional Compartmental Models and Multi-Term Mittag-Leffler Response Functions." J Phamacokinet Pharmacodyn 37(2): 209-215.

[45] Verotta, D. (2011). "Fractional Dynamics Linear and Nonlinear Compartmental Models." In Preparation.

[46] Verotta, D., L. B. Sheiner, et al. (1989). "A semiparametric approach to physiological flow models." J Pharmacokinet Biopharm 17(4): 463-491.

[47] Xin, C. and L. Fawang (2007). "Numerical simulation of the fractional-order control system." Journal of Applied Mathematics and Computing 23(1-2)): 229-241.

2

Advances in Minimal Cell Models: a New Approach to Synthetic Biology and Origin of Life

Pasquale Stano
Biology Department, University of Roma Tre
Italy

1. Introduction

Minimal cells can be defined as those compartment-based systems containing the minimal and sufficient number of molecular components and still displaying the essential features (structure and functions) of natural living cells (Luisi et al., 2006). The research on minimal cells, in recent years, has attracted the attention of several groups, as witnessed by the increasing number of reports and reviews published on the subject, as well as a dedicated book (Stano & Luisi, 2011).

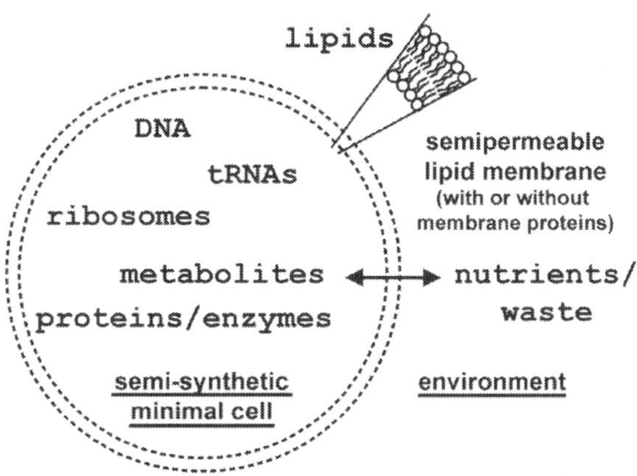

Fig. 1. Semi-synthetic minimal cells (SSMCs). The minimal number of genes, enzymes, RNAs, and low molecular-weight compounds are encapsulated into synthetic lipid-based compartments, such as in the case of lipid vesicles. The membrane acts as a boundary to confine the interacting internalized molecules, so that a "unit" is defined. Moreover, its semi-permeable character (possibly modulated thanks to the insertion of membrane proteins acting as selective pores, as in the case of hemolysin (Noireaux & Libchaber, 2004) or porins (Graff et al., 2001; Vamvakaki et al., 2005; Yoshimoto et al., 2005) allows the material exchange between the SSMC and its environment (nutrients uptake, waste release).

The concept of minimal cell in biology, however, is not new. Looking at the astonishing genetic and metabolic complexity of simplest living organisms, such as unicellular microbes, one might ask whether such complexity is really a necessary condition for life, or whether life is compatible with a simplest organization (Morowitz, 1992; Knoll et al., 1999). This question has its root in the field of origin and evolution of life, considering that modern sophisticated cells derive from million years of evolution, and that primitive cell could not be as complex as modern ones. Together with a theoretical approach, an experimental approach is needed to investigate the realm of minimal cells. Initially developed within the origin of life community, the construction of minimal cell models not only tries to answer the question of the minimal complexity for living organisms, but also focuses on the assembly steps (from separated molecules to organized compartments). At this aim, several studies have been carried out to build, like a chemist would do, minimal cell models from an appropriate compartment (such as micelles, or vesicles) and certain solutes. In addition to studies where very simple chemicals are used for building such structures, particularly promising seems the use of enzymes, RNAs, DNAs. In fact, it is easier to build typical cell functions from these "modern" components (Fig. 1). When evolved molecules, as those listed above, are used to construct a minimal cell in a synthetic compartment, such as a lipid vesicle, it is convenient to call these constructions as "semi-synthetic" minimal cells (SSMCs).

More recently, however, the "minimal cell project" became one of the well-recognized topics of synthetic biology (De Lorenzo & Danchin, 2008), where – however – it might assume a slightly different connotation. In fact, top-down approaches, typical of this new discipline, aim at reducing the complexity of extant cells by removing unessential parts, typically by genetic engineering and metabolic engineering. The progress in genome synthesis brought about the assembly of a synthetic genome, which has then transferred to a genome-deprived cell (Gibson et al., 2010). On the other hand, the semi-synthetic approach – which is discussed in this review - can be classified as bottom-up, because it starts from simple molecules and aims at constructing a cell. In the following, we will first introduce the theoretical framework for understanding what is the essential dynamics of a living cell, then quickly review some aspects of the minimal genome, and later discuss in details some recent experimental studies. Finally, we will see how SSMCs could be used in the future as biotechnological tool.

2. Minimal life from the autopoietic perspective

The starting point for the analysis of minimal living properties is the theory of autopoiesis (self-production). Developed by Humberto Maturana and Francisco Varela in the Seventies (Varela et al., 1974), autopoiesis is a theory that focuses on the essential dynamics of a single cell. It does not describe how a cell originates, but just how it functions. First of all it is recognized that a cell is a confined system composed by interacting molecules, and that the essential feature of a living cell is the maintenance of its own individuality. Several transformations take place inside its boundary. Thanks to the continuous construction and replacement of internal components (boundary molecules included), and thank to this internal activity only, the cell maintains its state within a range of parameters, which are compatible with the existence and the good functioning of the set of transformation occurring inside. So, despite the continuous regeneration of all its parts, an autopoietic cell maintains its individuality because the structural and functional organizations do not change. What is continuously changed is the material implementation of these organizations. It is easy to draw a minimal autopoietic unit that obeys to this mechanism (Fig. 2).

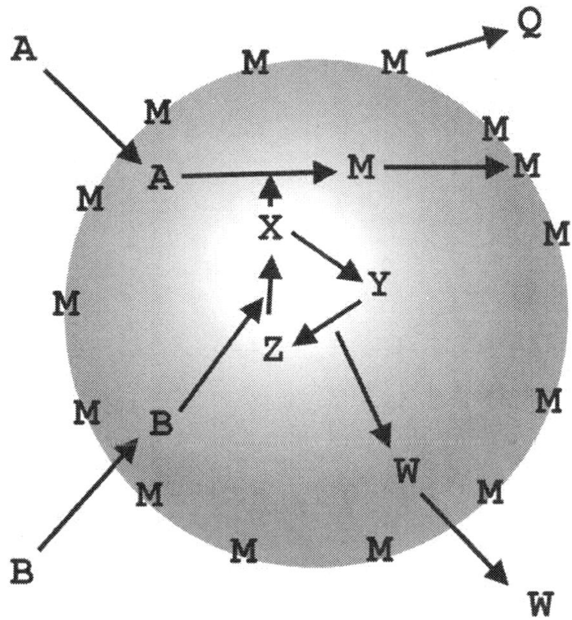

Fig. 2. Autopoietic dynamics as a guiding framework for constructing minimal cells. A set of compounds, here called M, X, Y, and Z are assembled in a self-bounded structure, i.e., a cell-like compartment. The functional and structural organization of these components foresees that X, Y, and Z act as mutual catalysts for their own production, and moreover X acts as catalyst for the formation of M (the boundary-forming component). Overall, this establishes a self-maintenance self-producing dynamics, based on chemical reactions and self-assembly, where all components of the autopoietic unit are generated from the autopoietic network and inside the autopoietic structure. In order to function, the autopoietic unit needs to uptake precursors from the environment (A, B), and release waste products (Q, W), therefore working as an open system. An autopoietic unit is in homeostatic regime when the rate of components building is equal to the rate of their decay.

Notice that in autopoiesis there is no need to specify the role of DNA, RNA, etc., these molecules act as elements of the internal networks aiming to keep the dynamic organization of internal transformations. The blueprint for the cell life consists of processes that produce all cell components that in turn produce the processes that produce such components, ... Clearly, all this occurs at the expense of energy and nutrients from the environment, so that the living cell is, thermodynamically speaking, an open system, yet characterized by its own operational closure (no external information is needed to organize and reproduce itself from inside).

The first attempts to create a minimal chemical autopoietic systems were done with simple supramolecular systems as micelles and reverse micelles, and later on with vesicles (for a recent review, see Stano & Luisi (2010a). The comment on such systems is outside the scope of this chapter, but it is useful to say that the current research of SSMCs construction still relies on the autopoiesis as a theoretical framework. The initial micelle/vesicle systems are taken as basic model to develop more complex design, based on the general scheme of Fig. 2, by substituting the abstract X, Y, Z, and M components with real (bio)chemical species that are able to function as a network of minimal complexity.

3. Minimal genome

	Mushegian & Koonin, 1999	Gil et al., 2004	Forster & Church, 2006
Amino acids and derivatives	3	0	0
Carbohydrate metabolism	21	14	0
Cell division and cell cycle	7	9	0
Cell wall and capsule	6	0	0
Cofactors, vitamins, prosthetic groups, pigments	8	15	0
DNA recombination and repair	5	3	0
DNA replication machinery	17	17	4
Electron transport chain	10	9	0
Fatty acids and lipids	7	7	0
Nucleoside and nucleotides	21	15	0
Protein biosynthesis	76	67	71
Protein folding	9	5	2
Protein turnover	1	2	0
RNA processing and modification	2	3	2
Transcription machinery	7	7	4
Transmembrane transport	19	4	0
tRNA synthesis and modification	26	27	25
Unclear/uncharacterized functions	6	4	1
TOTAL	*251*	*208*	*109*

Table 1. Comparison of three different versions of the "minimal genome". Adapted with minor modifications (additional columns have been removed) from Henry et al. (2010), with permission from Wiley.

When we ask how a SSMCs can be constructed, we look at what are the minimal number of functions that have to be implemented into a minimal cell. The concept of "minimal genome" is clearly linked to this aspect. If we would like to employ proteins as function-carrying molecules, the only way we know for generating them is decoding the DNA sequence. So, specifying what is the minimal number of functions in a minimal cell is equivalent to specify the minimal number of genes required to codify such functions. The concept of minimal genome is also a very well discussed topic in biology. Several studies and reviews have been published (Mushegian & Koonin, 1996; Gil et al., 2004; Islas et al., 2004; Forster & Church, 2006; Fehér et al., 2007; Moya et al., 2009; Henry et al., 2010). *Mycoplasma genitalium* genome (482 genes) is the smallest organism that can be grown in pure colture (Fraser et al., 1995). This means that its genome encodes for all functions required to sustain life at the cellular level. By comparing *M. genitalium* genome with those of other small microorganisms (endosymbionts such as *Buchnera*, or parasites as *Haemophilus influenzae* are taken into account) several studies have been published in order to find the common genes. The first minimal gene set (of about 250 genes) was generated by Mushegian & Koonin (1996), who compared *M. genitalium* and *H. influenzae* genomes, the only ones available at that time. In 2004, Moya and coworkers (Gil et al., 2004) performed a combined study including the comparison of reduced genomes from insect endosymbionts (*Buchnera*), reaching the conclusion that the minimal gene set is composed of about 200 genes. More recently, Forster & Church (2006) searched for the biochemical description of well-defined pathways that are needed to perform essential functions: in this way they describe a minimal gene set of about 100 genes. A comparison among these three minimal gene sets is shown in Table 1.

As it can be seen, the largest group of genes refers to protein biosynthesis, a pathway that involves the ribosomes, the essential and complex molecular machine that is the very cornerstone of cell function. tRNA-related genes are also numerous, as well as DNA replication. It is clear that the minimal genome essentially encodes the instruction for translating and duplicating the genetic information. Consider for example that in very permissive conditions a minimal cell could take energy and low molecular weight compounds from the environment, so that all genes required to accomplish these functions can be removed from the list of Table 1. It is evident that the essential part of the genome is that one dedicated to the production of internal components that reciprocally produce each other (from DNA to protein, from protein to DNA), plus those dedicated for boundary production, as requested by the autopoietic theory. In other words, under the constrains of current biochemical routes, the minimal genome should correspond to the minimal circuitry needed to establish a self-maintenance autopoietic dynamics. A further simplification would be possible only by removing these constraints, for example (hypothetically) by reducing the number of amino acids, or using a single enzyme for carrying out multiple functions, or by reducing the complexity of ribosomes.

Even considering the very small number of ca. 100 genes (Forster & Church, 2006), it is probably difficult to imagine the construction of a viable minimal cell. To date, the figure of 100 genes is still far beyond the current experimental approaches, where, as it will be shown in next paragraphs, a maximum of two genes have been simultaneously expressed inside lipid vesicles.

To conclude, it is useful to remark that the goal of defining (and synthesizing) a minimal genome can have – in addition to the theoretical conclusions for minimal cell bottom-up studies – some practical applications, especially in synthetic biology. For example, Henry et al. (2010) emphasize that the creation of a minimal organism (intended here as derived from the top-down approach) could be significant; for example the lack of mobile DNA elements might alleviate difficulties in controlling industrial bacteria strains; the lack of competing metabolic pathways might increase the net conversion yield of a desired product, or avoid the production of toxic side-products; the reduction of transcriptional regulation might favor the efforts to engineer minimal microorganisms.

4. A roadmap to the construction of minimal cells

The minimal genome analysis shown in the previous section has revealed that in order to build a SSMC – possibly living, although certainly "limping" (Luisi et al., 2006) – about 100 different genes must be encapsulated into lipid vesicles, and all of them must be transcripted and translated into functional proteins. At this aim, a new technology must be developed. To date, this technology comes from the convergence of *in vitro* cell-free systems and liposome technologies (Fig. 3A). Possibly, in the near future an key role could be played by microfluidics, which is experiencing a very strong growth with applications to several analytic and synthetic questions. In particular, the recent reports on the controlled formation of giant vesicles (GVs) inside (or with the help of) microfluidic devices is an interesting innovation (Walde et al., 2010; Sugiura et al., 2008; Ota et al., 2009; Matosevic & Paegel, 2011). From a more general viewpoint, one can imagine a sort of roadmap that leads to advanced minimal cell models, starting from very basic and already available technology (Fig. 3B).

The zero-th step is clearly the control of liposome technology, in terms of liposome formation and manipulation methods, solute encapsulation, and detection methods. This will be discussed more in details in next paragraph.

Then we can imagine a stepwise advancement, from simple enzyme-inside-vesicle systems to more elaborated ones, based on gene expression inside vesicles. Investigation on one-enzyme-containing vesicles is already well developed, and a variety of enzymes have been entrapped into lipid vesicles (alkaline phosphatase, amylase, asparaginase, chymotrypsin, elastase, galactosidase, lysozyme, pepsin, perossidase, glucose oxidase, glucose-6-phosphate-dehydrogenase, hexokinase, glucuronidase, phosphotriesterase, superoxide dismutase, tyrosinase, urease, carbonic anhydrase, luciferase, lipase, etc. For a recently published review, see Walde & Ichikawa, 2001). Much more limited is the progress in coentrapping several enzymes inside liposomes. After a pioneering study of Chang (1987) on cycling enzymatic routes inside non-lipid compartments (for example, the coupled system: urease, glutamate dehydrogenase, glucose dehydrogenase, glutamate-pyruvate transaminase), more recent approaches have been based on liposomes and polymersomes. Typical examples are glucose oxidase and horseradish peroxidase inside liposomes (Hill et al., 1997; Kaszuba & Jones, 1999) or polymersomes (Delaittre et al., 2009; van Dongen et al., 2009), possibly with the involvement of a third enzyme (a lipase). The other system that has received attention is the couple bacterhodopsin/ATP synthase (both membrane proteins) that have been reconstituted in liposomes at the aim of producing ATP after irradiation (Freisleben et al., 1995; Pitard et al., 1996; Choi and Montemagno, 2007). With respect to the control of vesicle membrane permeability, it is worth mentioning the increase of passive permeability of detergent-doped membranes (the concentration of detergent is below the lytic regime, Oberholzer et al., 1999; Treyer et al., 2002), and the use of α-hemolysis (Noireaux & Libchaber, 2004) or porines (Graff et al., 2001; Vamvakaki et al., 2005; Yoshimoto et al., 2005) as channel-forming agents.

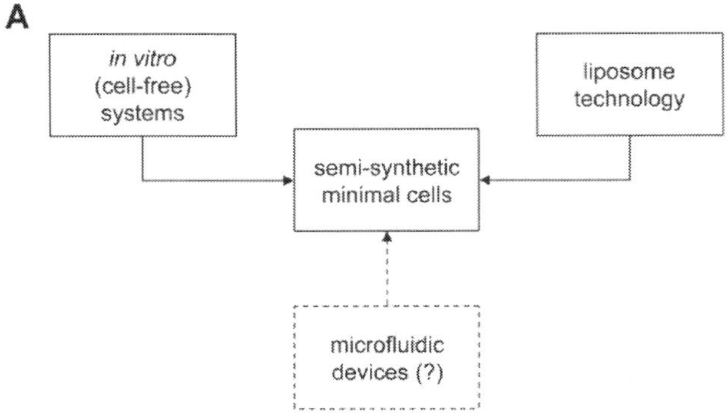

B

0. The basics
Liposome preparation and manipulation
Encapsulation of solutes
Detection methods

1. Enzyme reactions inside liposomes
One enzyme reactions
Multi-enyme reactions and cycles
Control of membrane permeability

2. Reactions involving nucleic acids
Transcription
Translation
Coupled transcription and translation
DNA replication
Ribosome production

3. "Core and shell" reproduction
Self-reproduction of "core" components
Lipid synthesis
Liposome growth and division

4. Toward complex functions
Omeostatic regime
Energy production
Control of gene expression
Genetic circuits
Sensing and communication

Fig. 3. Operative aspects of semi-synthetic minimal cell construction. (A) Technique currently involved for the production of minimal cells, mainly based on cell-free systems and liposome technology; the possible emerging role of microfluidic devices has been emphasized. (B) A possible roadmap for the stepwise construction of minimal cell models of increasing complexity.

The next step along the road map for SSMC construction is the realization of nucleic acid-based reactions inside liposomes. These have been already carried out, as in the case of DNA synthesis (PCR inside liposomes, Oberholzer et al., 1995), DNA transcription (RNA synthesis, Tsumoto et al., 2001), RNA replication (Kita et al., 2008), and the coupled transcription/translation reactions (protein synthesis). In particular, the protein synthesis is currently the most well studied reaction inside liposomes (for references, see paragraph Nr. 6). The current interest in protein synthesis fits well with the fact that this "module" is the most important for minimal cell construction, as also evident from the minimal genome analysis. The goal of ribosome production inside liposomes has not been reached, although some preliminary attempts have been anticipated (Jewett and Church, unpublished data presented at the 4th Synthetic Biology conference, Hong Kong 2008).

By further increasing the complexity of minimal cell models, one can think to move to cyclic reactions, where the focus is on the regeneration of cell components. These can be distinguished between "core" components (DNA, RNA, ribosomes) and "shell" components (lipids). Clearly, for a successful cell reproduction, both core and shell parts should be generated from inside, possibly in coordinate manner. Early enzyme-catalyzed attempts for synthesizing lipids inside liposomes have been reported (Schmidli et al., 1991). Kuruma et al. (2008) have recently updated this research by expressing the enzymes of interests (two acyl transferases) from the corresponding DNA sequences (see below). The ultimate goal of core-and-shell reproduction is the observation of SSMCs growth and division as it happens in natural cells, clearly with a reduced efficiency.

Finally, it is possible to conceive compartimentalized molecular systems that display really complex dynamics such as homeostasis (for a simple example, see Zepik et al., 2001), sensing and communication (Gardner et al., 2009), energy production, control of genetic transcription, for example thanks to genetic circuitry (Noireaux et al., 2003; Kim & Winfree, 2011).

5. The technology for building minimal cells

As stated before, the current SSMC technology is based on liposome technology and on cell-free systems. Essentially, the critical point is how to insert – possibly under control – the desired solutes inside lipid vesicles.

Let us start to consider the liposome technology. In principle there are three ways to operate (Fig. 4). The simplest approach consists into the spontaneous vesicle formation in a solution containing all solutes of interest (Fig. 4A). Currently, this is the most used procedure, and involves different preparation methods (lipid film hydration, freeze-dried liposome rehydration, ethanol injection (New, 1990). Current drug delivery technology, which is the most advanced branch of liposome technology, is also based on the spontaneous formation of solute-filled liposomes (only in the special case of weak bases an "active" loading method has been developed, Haran et al., 1993). In general terms, the apparently simple process described in Fig. 4A is still not well understood, due to the complexity of solute/membrane interactions and to not well-known subtle effects deriving from the interplay of chemical and physical factors during the vesicle formation. In fact, it has been recently discovered that a large heterogeneity exist in a population of vesicles prepared by these methods (Dominak & Keating, 2007; Lohse et al., 2008). A clear-cut evidence of superconcentration of proteins inside lipid vesicles has been reported in the case of ferritin (Luisi et al., 2010).

A second approach is described in Fig. 4B and consists into the physical insertion of the solutes of interest inside a large vesicle (giant vesicle), by means of microinjection methods. This technique was developed in the Nineties (Wick et al., 1996; Bucher et al., 1998; Fischer et al., 2002) and offer the unique possibility of a total control of vesicle composition, but it could be hindered by technical difficulties, and by the fact that only a limited number of vesicles can be prepared and observed.

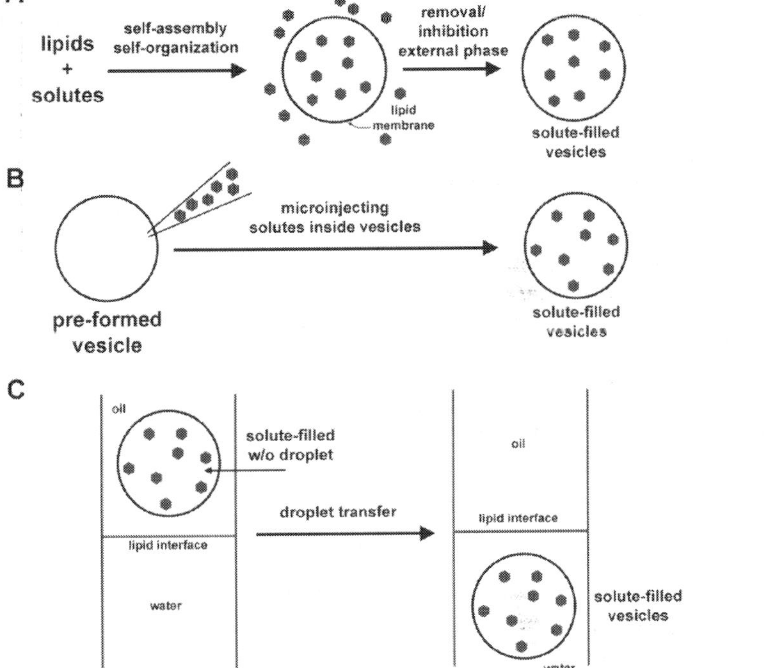

Fig. 4. Basic concepts for minimal cell constructions. (A) Liposomes are formed in a solution containing the solutes of interest; some liposomes will contain all required components. The external environment is quenched or changed in order to follow internalized reactions. (B) A microinjection procedure (for giant vesicles only) could allow the direct insertion of the mixture of interest into pre-formed vesicles. This procedure is clearly time-consuming and allows the study of a limited number of semi-synthetic minimal cells. (C) A very promising method for generating solute-filled vesicles starts from solute-filled water-in-oil droplets (which are easy to prepare) and transform them into giant vesicles according to a recently developed method (Pautot et al., 2003ab).

A third approach (Fig. 4C) is instead innovative and it is based on the recent success in preparing giant vesicles from water-in-oil droplets (Pautot et al., 2003ab). Some SSMC work has already be done by preparing vesicles with this method (Yamada et al., 2006; Pontani et al., 2009; Saito et al., 2009). Firstly, solutes are emulsified in a hydrocarbon solvent

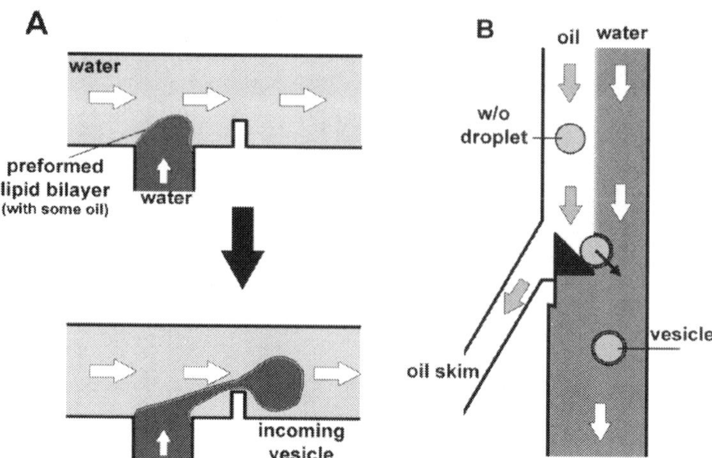

Fig. 5. Generation of giant vesicles in microfluidic devices. (A) The internal water phase (dark gray) is forced to flow into a microchannel containing the external water phase (light gray). Due to the presence of previously formed bilayer (possibly containing some oil), the two phases do not mix, and the lipid interphase is stretched to form a spherical vesicle that breaks after passage over an obstacle. Adapted with minor modifications (redrawn of the essential step of vesicle formation, removal of some details), from Ota et al., (2009), with permission from Wiley. (B) A microfluidic-generated water-in-oil droplet flow is tangentially poured near a water phase (dark gray) running in the same direction; lipids accumulate along the oil-water interface. An obstacle separate the oil flow from the aqueous flow forcing the water-in-oil droplet to cross the oil/water interface and form a vesicle, as shown in Fig. 4C. Adapted with minor modifications (redrawn of the essential step of vesicle formation, removal of some details), from Matosevic & Paegel (2011), with permission from the American Chemical Society.

containing lipids. This brings to the formation of lipid-stabilized water-in-oil droplets, where lipids are oriented with the polar head-groups toward the droplet core, and with the tails toward the apolar solvent. A second lipid-containing apolar solution is stratified over an aqueous phase, so that a lipid interface is spontaneously generated at the oil-water interface. When a water-in-oil droplet crosses the water-oil interface, it becomes covered by a second lipid layer, and is then transformed into a giant vesicle when reaches the aqueous phase (on the bottom). The process is enhanced by centrifugation, taking advantage of the higher water density when compared with hydrocarbons. In this way, solutes initially present in the water-in-oil droplets (where 100% compartmentalization is easy to achieve) are found inside giant vesicles. This methodology has been recently implemented in microfluidic devices.

The use of microfluidic devices seems very promising in SSMC technology, although not well developed when compared with the amount of work done with water-in-oil droplets. Here the advantage should be the control of vesicle production (uniform size, morphology, and lamellarity) and especially the control of vesicle content. Two interesting example of pioneering research are shown in Fig. 5. In the first case (Fig. 5A) giant vesicles have been

Component	Conc.	Component	Conc.
Translation factors		ProRS	1300 U/mL
IF1	2.7 µM	SerRS	1900 U/mL
IF2	0.4 µM	ThrRS	1300 U/mL
IF3	1.5 µM	TrpRS	630 U/mL
EF-G	0.26 µM	TyrRS	630 U/mL
EF-Tu	0.92 µM	ValRS	3100 U/mL
EF-Ts	0.66 µM		
RF1	0.25 µM	*Other enzymes*	
RF2	0.24 µM	MTF	4500 U/mL
RF3	0.17 µM	Ribosomes	1.2 µM
RRF	0.5 µM	Creatine kinase	4 µg/mL
		Myokinase	3 µg/mL
Amino acyl-tRNA synthetases (RSs)		Nucleoside diphosphate kinase	1.1 µg/mL
AlaRS	1900 U/mL	Pyrophosphatase	2 U/mL
ArgRS	2500 U/mL	T7 RNA polymerase	10 µg/mL
AsnRS	20 mg/mL		
AspRS	2500 U/mL	Energy sources	
CysRS	630 U/mL	ATP	2 mM
GlnRS	1300 U/mL	GTP	2 mM
GluRS	1900 U/mL	CTP	1 mM
GlyRS	5000 U/mL	UTP	1 mM
HisRS	630 U/mL	Creatine phosphate	20 mM
IleRS	2500 U/mL		
LeuRS	3800 U/mL	*Other components*	
LysRS	3800 U/mL	20 amino acids	0.3 mM
MetRS	6300 U/mL	10-formyl-5,6,7,8-tetrahydrofolic acid	10 mg/mL
PheRS	1300 U/mL	tRNAmix (Roche)	56 Abs260

*The components are solubilized in 50 mM HEPES-KOH pH 7.6; 100 mM potassium glutamate, 13 mM magnesium acetate, 2 mM spermidine, 1 mM DTT. One unit of activity was defined as the amount of enzyme that catalyzes the formation of 1 pmol of amino acyl-tRNA in 1min.

Table 2. Composition of the PURE system.* Adapted from Shimizu et al. (2005), with permission from Elsevier.

produced from two aqueous phases separated by a pre-formed lipid-rich micro-interface (Ota et al., 2009). The gentle flow of internal aqueous phase produces an elongated compartment that eventually breaks to give a giant vesicle. The proper design of an obstacle along the flow is essential to observe the desired pattern. Giant vesicles of average diameter 16.5 µm (± 3.7%) are produced. In the second case (Fig. 5B), the droplet transfer method has been implemented in a microfluidic channel (Matosevic & Paegel, 2011). Water-in-oil droplets, also generated within the microfluidic device tangentially flows against an aqueous phase which will constitute the external vesicle phase. A properly designed obstacle forces each water-in-oil droplet to cross the oil/water interface and become transformed into a vesicle, whereas the excess oil ends in a side channel. Vesicles with size from 20 to 70 µm can be generated, be vesicle size being controlled by the water-in-oil droplet size.

The second aspect to discuss is the *in vitro* cell-free technology. In the case of simple molecular systems, such as one enzyme-reaction or simple standardized sets (e.g., PCR reactants), it is clear that the experimenter perfectly controls the nature of the solute mix to be entrapped inside lipid vesicles. Different is the case of quite complicated mixtures as the whole transcription-translation kit. Early research on protein synthesis inside liposomes has been done by encapsulating home-made or commercial cell extracts (e.g., *Escherichia coli* extracts). For example, home-made S30 *E. coli* extracts, enriched of T7 RNA polymerase and in the presence of a DNA plasmid, were used to demonstrate for the first time the expression of functional green fluorescent protein (GFP) inside vesicles (Yu et al., 2001). Although successful, this approach – that was used by several researchers, has an intrinsic limitation. In fact, a cell extract can be compared to a "black box", in the sense that its chemical composition and the concentrations of the species are not known exactly. Ideally, since one would like to investigate the activity just as a function of the relative composition and concentration of the components, a more detailed knowledge of the protein synthesis kit would be advisable. In 2001, the group of Takuya Ueda at the Tokyo University has formulated a "reconstituted" cell-free transcription and translation system from purified *E. coli* components, named PURE system (Protein synthesis Using Recombinant Elements) (Shimizu et al., 2001). The composition of this kit is shown in Table 2. The PURE system contains 9 translation factors, 20 aa-tRNA synthetases (aaRSs), 6 additional enzymes (also needed for energy regeneration), ribosomes, tRNAs, and low molecular weight compounds. There are 36 different purified macromolecular compounds, and presumably 46 different tRNAs (Dong et al., 1996), for a total of 82 different macromolecules. If we sum the template DNA it results that 83 different compounds are involved in coupled transcription/translation reactions (prokaryote ribosomes consists in 3 rRNAs and 55 ribosomal proteins, and therefore the overall number of different macromolecular sequences in the PURE system actually sums up to 141).

Notice that for synthesizing a protein inside liposomes, these 83 different macromolecules (at least 1 molecule of each species) should be co-entrapped in the liposome. The probability of such occurrence has been calculated on the basis of Poisson distribution, as a function of vesicle size (Souza et al., 2009), and assumes an experimentally significant value (>5%) for vesicle diameters > 600 nm.

Starting from 2006, the use of PURE system for synthesizing a protein inside liposomes has become the benchmark for SSMCs construction (Sunami et al., 2006; Murtas et al., 2007), even if the yield of produced protein is typically one third when compared with cell extracts (Hillebrecht & Chong, 2008). Interestingly, we can look at the PURE system as a synthetic biology tool from the viewpoint of "standard biological parts" (http://partsregistry.org/Main_Page).

6. Four commented examples of current research

As stated above, the current state-of-the-art of SSMC research is the expression of functional proteins inside liposomes. This very specific research area has already a rich chronology of reports, starting from the pioneering study of Oberholzer et al. (1999), who synthesize poly(Phe) inside vesicles starting from the corresponding messenger poly(U). Several reviews have been compiled to discuss the advancement in the field (Luisi et al., 2006; Chiarabelli et al., 2009; Stano & Luisi, 2010b), so that these aspects will not be repeated here. Rather, here we would like to comment four very relevant studies done in the last years, namely: (i) genetic cascade reaction inside liposomes; (ii) inside-vesicle production of a pore forming compound; (iii) synthesis of membrane enzymes (lipid-producing) inside liposomes; and (iv) RNA replication by RNA-encoded Qβ-replicase.

The first case to discuss is a work from the Yomo group in Osaka, see Fig. 6A (Ishikawa et al., 2004). In this work the authors designed a two-stages cascade reaction inside liposomes. In particular the idea to have a DNA plasmid encoding for two proteins, namely T7 RNA polymerase and the green fluorescent protein (GFP). The corresponding genes, however, were under different promoters (SP6 and T7, respectively), so that the following cascade genetic reactions could happen. The plasmid was mixed with cell-free extracts enriched with SP6 RNA polymerase. This first polymerase acts on the first gene (SP6-*t7_rna_polymerase*) so that, after transcription, the T7RNA polymerase is effectively produced. In turn, this second polymerase acts on the second gene (T7-*gfp*), which ultimately gives the fluorescent protein, which was revealed via flow cytometry. Interestingly, this is the first case of controlled protein synthesis inside liposomes, not by transcription factors (this goal has not been reached yet), but by switching on the production of RNA polymerase.

The second interesting case comes from the Libchaber laboratory in New York, see Fig. 6B (Noireaux & Libchaber, 2004). The authors designed a lipid vesicle containing the gene of α-hemolysin (from *Staphylococcus aureus*). After the synthesis of α-hemolysin by a cell-free cell extract, encapsulated into vesicles, the hemolysin spontaneously self-assembles into a heptamer on the lipid membrane, creating a pore (cut-off ca. 3 kDa). Thanks to its peculiar size, the pore allows the exchange of low molecular-weight compounds of the transcription-translation kit (amino acids, nucleotides, etc.), without releasing large macromolecules. In this way, a long-lived "bioreactor" was created, capable of expressing proteins for ca. 100 hours, whereas in the absence of the pore, the reaction stops after ca. 4 hours (for this study, a fusion GFP-hemolysin protein was created). Interestingly, it was also shown that the co-expression of α-hemolysin and GFP as two separated proteins was possible.

The third case comes from our laboratory in Rome, see Fig. 7A (Kuruma et al. 2009). This work was inspired by a previous study done on lipid-synthesizing enzymes inside vesicles, at the aim of creating a lipid-producing liposome (Schmidli et al., 1991). The route consisted into four enzyme-catalyzed steps, namely, from glycerol-3-phosphate (G3P) to lysophosphatidic acid (LPA), to phosphatidic acid (PA), to diacylglycerol, to phosphatidylcholine (PC). The 1991 work consisted in inserting the four enzymes that catalyzes this route into lipid vesicles and observe the production of PC starting from G3P and acylcoenzymes A. However the efficient production of PC was observed only in the case of short chain acyls, and the yield was around 10%. In the work of Kuruma et al. (2009)

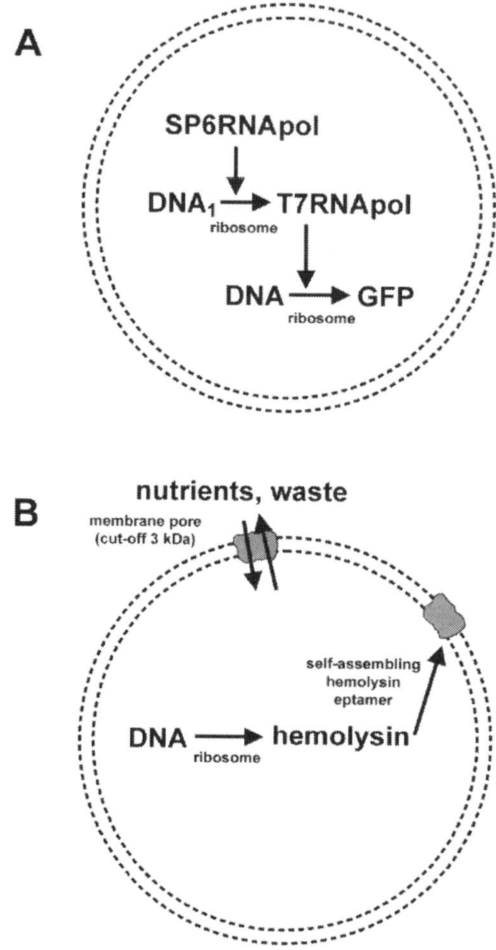

Fig. 6. Two remarkable examples of semi-synthetic minimal cell construction. (A) Genetic cascade reactions inside liposomes. SP6 RNA polymerase transcribes the mRNA of T7 RNA polymerase, which is then translated to give the functional protein. In turn, the newly formed T7 RNA polymerase transcribes the mRNA of GFP, which is then translated to give the fluorescent protein (Ishikawa et al., 2004). (B) α-Hemolysin, produced inside a giant vesicle from its DNA sequence, self-assembles into an heptamer which is a pore on the phospholipid membrane. Its low cutoff (3 kDa) allows the exchange of low molecular-weight solutes (entrance of energy-rich compounds, release of exhausted compounds) without releasing the internal macromolecular components (RNA polymerase, ribosomes, etc.). In this way the production of a marker protein (GFP) proceeded continuously for 4 days (Noireaux & Libchaber, 2004).

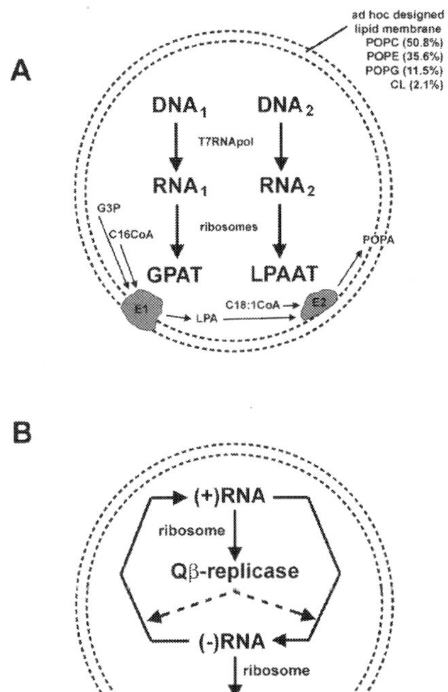

Fig. 7. Two remarkable examples of semi-synthetic minimal cell construction. (A) The two genes codifying for glycerol-3-phosphate acyltransferase (GPAT) and lysophosphatidic acid acyltransferase (LPAAT) have been co-entrapped into lipid vesicles with a cell-free protein synthesis kit. The two proteins, which – on the basis of their solubility – should be an integral membrane protein and a membrane associated one, respectively, catalyse the formation of 1-palmitoyl-2-oleoyl-*sn*-glycero-3-phosphatidic acid (POPA) from glycerol-3-phosphate (G3P) and palmitoyl- and oleoyl-coenzimeA (C16CoA, and C18:1CoA, respectively), through the intermediate lysophosphatidic acid (LPA). Notice that the membrane composition is a key factor for realizing a functional system (POPC: 1-palmitoyl-2-oleoyl-*sn*-glycero-3-phosphatidylcholine; POPE: 1-palmitoyl-2-oleoyl-*sn*-glycero-3-phosphatidylethanolamine; POPG: 1-palmitoyl-2-oleoyl-*sn*-glycero-3-phosphatidylglycerol; CL: cardiolipin) (Kuruma et al., 2009). (B) Starting from Qβ-replicase-codifying (+)RNA strand and cell-free protein expression system, the production of Qβ-replicase was carried out inside liposomes, so that the complementary (-)RNA strand is produced from nucleotides and (+)RNA template. In turn, (-)RNA acts as a template for the Qβ-replicase catalyzed (+)RNA strand synthesis. The correct production of (-)RNA is confirmed by the fact that it is designed to codify for β-galactosidase, whose ribosomal production was revealed by the increase of fluorescence due to the action of β-galactosidase on a fluorogenic substrate. This is an example of self-encoding genetic material, which codifies for its production inside lipid compartments (Kita et al., 2008).

a different approach was tested. In particular the genes for the first two enzymes, called glycerol-3-phosphate acyltransferase (GPAT) and lysophosphatidic acid acyltransferase (LPAAT), were co-entrapped with the PURE system inside lipid vesicles of proper membrane chemical composition. The latter was a critical parameter because the lipids used to build the liposome should: (1) form good compartments, (2) avoid chemical interference with the PURE system, (3) favor the folding, the membrane association and the activity of the two enzymes, which were, respectively, integral membrane enzyme and membrane associated enzyme (estimated on the basis of their solubility). It was shown that the two enzymes could be synthesized inside vesicles, although they required different redox environments (dithiothreitol vs oxidized glutathione) for acquiring the corresponding function. Despite this limitations and the low yield of functional protein synthesized inside liposomes, it was indeed possible to observe, in a discontinuous coupling assay, the production of a small amount 1-palmitoyl-2-oleoyl-*sn*-glycero-3-phosphatidic acid (POPA) from G3P and palmitoylCoA and oleoylCoA. Unfortunately, the low POPA yield did not allow a vesicle morphological change (for example, vesicle growth).

Finally, an important example of SSMC work comes again from the laboratory of Yomo and coworkers (Fig. 7B), and concerns the replication of RNA inside vesicles, operated by an enzyme (the Qβ-replicase) whose production was ultimately encoded in the RNA sequence itself (Kita et al., 2008). The aim of this study was the coupling between information-carrying molecules and the production of molecular machinery that replicate the information (RNA that codifies for an enzyme whose role is to replicate RNA). At this aim, a template (+)RNA strand was coencapsulated with the reconstituted transcription/translation kit. Its translation gives the functional enzyme Qβ-replicase, that in turn replicate the template (+)RNA into the complementary strand (-)RNA (and vice-versa). In a more complex design, the β-galactosidase, encoded on the (-)RNA strand was also successfully expressed, and its function verified by the production of a fluorescent product from a fluorogenic substrate.

7. Potential applications of minimal semi-synthetic cells and other microcompartimentalized systems

It is true that at the current stage the SSMC research appears still at its infancy, but it is also possible to devise some potential applications. Perhaps, one of the most interesting one has been proposed by LeDuc et al. (2007) (see also Zhang et al., 2008). The authors have proposed a potential application of a cell-like construction as "nanofactory" (Fig. 8). Shortly, inspired by the current research on liposomes as vehicle for drug delivery of drugs, and on their targeting based on antibody-functionalization, a kind of minimal cell could be constructed from a lipid shell and internalized metabolic/genetic circuits. Thanks to the surface recognition by specific intermolecular interactions, the "nanofactory", once injected in the body could find the target tissue and accumulate in the nearby. Once sit near the target tissue, thanks to membrane permeability an externally present chemical trigger might induce a cascade internal reaction that ultimately produce a toxic drug for the tissue, so that an *in situ* produced drug can act locally with minimal damage for other body parts.

A second interesting aspect is related to potential sensing capability of SSMCs. The possible applications of lipid vesicles as membrane models for toxicological assessment of xenobiotics has been described (Zepik et al., 2008). By combining surface-bound receptors with internalized triggered response, SSMCs biosensors could be constructed for sensing

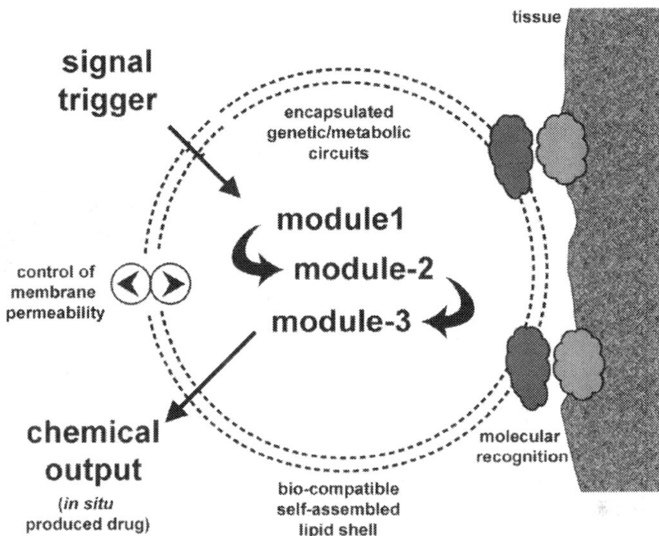

Fig. 8. A potential use of semi-synthetic minimal cells as drug nanofactories. The synthetic cell, injected in the body, might recognize specific tissues by means of surface-exposed antibodies, whereas its internal content consists into genetic/metabolic modules capable of detecting an externally present chemical trigger (a specific molecule that characterizes the tissue of interest), and work coordinately work to produce a chemical output, for example a drug. Entrance and release of compounds is achieved by controlling the membrane permeability by means of proteins or by changing lipid composition. Redrawn on the basis of a published figure (Le Duc et al., 2007).

different kind of molecules. An interesting study on the use of lipid vesicles for sensing odorants has been reported (Takeuchi et al., 2011). Although the core mechanism was based on chemical transformation, it is possible to think to similarly working biomolecular networks. From the technical viewpoint, the technological ability to generate *vesicle arrays* will certainly be important for exploiting SSMCs sensing (Christensen & Stamou, 2010).

In addition to long-term prospects for biotechnology, as the use of minimalized strains of bacteria-like objects for industrial applications, the construction of SSMCs also has a profound implication in basic science. First of all, it helps to demonstrated, if still needed, that a *minimal living organism* can be constructed from separated *non-living* parts, that is equivalent to say that life is an emerging property of a self-organized spatial-functional molecular ensemble, where the fact of being alive stems from the dynamical relations between components, rather than their chemical nature. All this is resonant with the autopoiesis introduced in the second paragraph. Secondly, and here we find the field where the SSMCs comes from, the construction of such structures has great relevance for origin of life studies. In this case the focus is shifted from the sophistication of biotechnology application to the assembly process itself, where the particular nature of used molecules merely represents a model of primitive solutes. What is important is the demonstration that sets of free molecules and lipids spontaneously assemble into a cell-like particle, and that functions – initially not present in the bulk phase, become characteristic of compartmentalized systems. For example, the recently reported discovery (Luisi et al., 2010) of self-concentration of proteins (and ribosomes, Souza et al., submitted manuscript) inside liposomes points to the emergence of functional cells from diluted compounds. In other words, the formation of liposomes implies the formation of concentrated/segregated phase that might exhibit enhanced functionality when compared with the analogous system in bulk (see also Souza et al., 2009). In this way, SSMCs studies might provide a rational explanation to the spontaneous emergence of living cells in primitive times.

In conclusion, the investigation on SSMCs is in a flourishing phase, and we are convinced that the development of this technology will be of great value for basic and applied science.

8. Acknowledgment

I greatly thank Pier Luigi Luisi for inspiring this research and for stimulating discussions. Funding/networking agencies: SYNTHCELLS project (Approaches to the Bioengineering of Synthetic Minimal Cells, EU-FP6 043359), HFSP (RGP0033/2007-C), ASI (I/015/07/0), PRIN2008 (2008FY7RJ4), SynBioNT, and COST Systems Chemistry action (CM0703).

9. References

Bucher, P., Fischer, A., Luisi, P. L., Oberholzer, T. & Walde, P. (1998). Giant Vesicles as Biochemical Compartments: The Use of Microinjection Techniques. *Langmuir* 14, 2712-2721.

Chang, M. T. S. (1987) Recycling of NAD(P) by multienzyme systems immobilized by microencapsulation in artificial cells. *Met. Enzymology* 136, 67-82.

Chiarabelli, C., Stano, P. & Luisi, P. L. (2009). Chemical approaches to synthetic biology. *Curr. Opin. Biotech.* 20, 492-497.

Choi, H. J. & Montemagno, C. D. (2007). Light-driven hybrid bioreactor based on protein-incorporated polymer vesicles. *IEEE Trans. Nanotech.* 6, 171-176.

Christensen, S. M. & Stamou, D. G. (2010). Sensing-Applications of Surface-Based Single Vesicle Arrays. *Sensors* 10, 11352-11368.

De Lorenzo, V. & Danchin, A. (2008). Synthetic biology: Discovering new worlds and new words. *EMBO Rep.* 9, 9.

Delaittre, G., Reynhout, I. C., Cornelissen, J. J. L. M. & Nolte, R. J. M. (2009). Cascade reactions in an all-enzyme nanoreactor. *Chem. Eur. J.* 15, 12600-12603.

Dominak, L. M. & Keating, C. D. (2007). Polymer Encapsulation within Giant Lipid Vesicles. *Langmuir* 23, 7148-7154.

Dong, H., Nilsson, L. & Kurland, C. G. (1996) Co-variation of tRNA abundance and codon usage in Escherichia coli at different growth rates. *J. Mol. Biol.* 260, 649-663.

Fehér, T., Papp, B., Pál, C. & Pósfai, G. (2007). Systematic Genome Reductions: Theoretical and Experimental Approaches. *Chem. Rev.* 107, 3498–3513.

Fischer, A., Franco, T. & Oberholzer, T. (2002). Giant Vesicles as Microreactors for Enzymatic mRNA Synthesis. *ChemBioChem* 3, 409-417.

Forster, A. C. & Church, G. M. (2006). Towards synthesis of a minimal cell. *Molecular Systems Biology* 2, 45.

Fraser, C. M., Gocayne, J. D., White, O. et al. (1995) The minimal gene complement of *Mycoplasma genitalium*. *Science* 270, 397–403.

Freisleben, H. J., Zwicker, K., Jezek, P., John, G., Bettin-Bogutzki, A., Ring, K. & Nawroth, T. (1995). Reconstitution of bacteriorhodopsin and ATP synthase from *Micrococcus luteus* into liposomes of the purified main tetraether lipid from *Thermoplasma acidophilum* : proton conductance and light-driven ATP synthesis. *Chem. Phys. Lip.* 78, 137-147.

Gardner, P. M., Winzer, K. & Davis, B. G. (2009). Sugar synthesis in a protocellular model leads to a cell signalling response in bacteria. *Nature Chem.* 1, 377-383.

Gibson, D. G., Glass, J. I., Lartigue, C., Noskov, V. N., et al. (2010). Creation of a bacterial cell controlled by a chemically synthesized genome. *Science* 329, 52-56.

Gil, R., Silva, F. J., Peretó, J. & Moya, A. (2004). Determination of the core of a minimal bacteria gene set. *Microbiol. Mol. Biol. Rev.* 68, 518–537

Graff, A., Winterhalter, M. & Meier, W. (2001). Nanoreactors from polymer-stabilized liposomes. *Langmuir* 17, 919-923.

Haran, G., Cohen, R., Bar, L. K. & Barenholz, Y. (1993). Transmembrane ammonium sulfate gradients in liposomes produce efficient and stable entrapment of amphipathic weak bases. *Biochim. Biophys. Acta* 1151, 201–215.

Henry, C. S., Overbeek, R. & Stevens, R. L. (2010). Building the blueprint of life. *Biotechnol. J.* 5, 695-704.

Hill, K. J., Kaszuba, M., Creeth, J. E. & Jones, M. N.(1997). Reactive liposomes encapsulating a glucose oxidase-peroxidase system with antibacterial activity. *Biochim. Biophys. Acta* 1326, 37-46.

Hillebrecht, J. R. & Chong, S. (2008). A comparative study of protein synthesis in in vitro systems: from the prokaryotic reconstituted to the eukaryotic extract-based. *BMC Biotechnology* 8, 58.

Ishikawa, K., Sato, K., Shima, Y., Urabe, I. & Yomo, T. (2004). Expression of a cascading genetic network within liposomes. *FEBS Lett.* 576, 387-390.

Islas, S., Becerra, A., Luisi, P. L. & Lazcano, A. (2004). Comparative genomics and the gene complement of a minimal cell. *Orig. Life Evol. Biosph.* 34, 243–256.

Kaszuba, M. & Jones, M. N. (1999). Hydrogen peroxide production from reactive liposomes encapsulating enzymes. *Biochim. Biophys. Acta* 1419, 221-228.

Kim, J. & Winfree, E. (2011). Synthetic in vitro transcriptional oscillators. *Molecular Systems Biology* 7, 465.

Kita, H., Matsuura, T., Sunami, T., Hosoda, K., Ichihashi, N., Tsukada, K., Urabe, I. & Yomo, T. (2008). Replication of genetic information with self-encoded replicase in liposomes. *ChemBiochem* 9, 2403-2410.

Knoll, A.; Osborn, M. J.; Baross, J.; Berg, H. C.; Pace, N. R. & Sogin, M. (eds.) (1999). *Size limits of very small microorganisms*. National Academic Press, Washington.

Kuruma, Y., Stano, P., Ueda, T. & Luisi, P. L. (2009). A synthetic biology approach to the construction of membrane proteins in semisynthetic minimal cells. *Biochim. Biophys. Acta* 1788, 567-574.

LeDuc, P. et al. (2007). Towards an in vivo biologically inspired nanofactory. *Nature Nanotech.* 2, 3-7.

Lohse, B., Bolinger, P.-Y. & Stamou, D. (2008). Encapsulation Efficiency Measured on Single Small Unilamellar Vesicles. *J. Am. Chem. Soc.* 130, 14372 –14373.

Luisi, P. L., Allegretti, M., Souza, T., Steineger, F., Fahr, A. & Stano, P. (2010). Spontaneous protein crowding in liposomes: A new vista for the origin of cellular metabolism. *ChemBiochem* 11, 1989-1992.

Luisi, P. L., Ferri, F. & Stano, P. (2006). Approaches to semi-synthetic minimal cells: a review. *Naturwissenschaften* 93, 1-13.

Luisi, P. L.; Stano, P. (Eds.) (2011). *The minimal cell. The Biophysics of Cell Compartment and the Origin of Cell Functionality.* Springer, Dordrecht.

Matosevic, S. & Paegel, B. M. (2011). Stepwise Synthesis of Giant Unilamellar Vesicles on a Microfluidic Assembly Line *J. Am. Chem. Soc.* 133, 2798–2800.

Morowitz, H. J. (1992). *Beginning of Cellular Life. Metabolism Recapitulates Biogenesis.* Yale University Press, New Haven.

Moya, A., Gil, R., Latorre, A., Peretó, J., Garcillán-Barcia, M. P. & de la Cruz, F. (2009). Toward minimal bacterial cells: evolution vs. design. *FEMS Microbiol. Rev.* 33, 225–235.

Murtas, G., Kuruma, Y., Bianchini, P., Diaspro, A. & Luisi, P. L. (2007). Protein synthesis in liposomes with a minimal set of enzymes. *Biochem. Biophys. Res. Commun.* 363, 12–17.

Mushegian, A. R. & Koonin, E. V. (1996). A minimal gene set for cellular life derived by comparison of complete bacterial genomes. *PNAS* 93, 10268-10273.

New, R. R. C. (1990). *Liposomes, a practical approach.* IRL Press at Oxford University Press, Oxford.

Noireaux, V. & Libchaber, A. (2004). A vesicle bioreactor as a step toward an artificial cell assembly. *PNAS* 101, 17669–17674.

Noireaux, V., Bar-Ziv, R. & Libchaber, A. (2003). Principles of cell-free genetic circuit assembly. *PNAS* 100, 12672–12677.

Oberholzer, T., Albrizio, M. & Luisi, P. L. (1995). Polymerase chain reaction in liposomes. *Chem. & Biol.* 2, 677-682.

Oberholzer, T., Meyer, E., Amato, I., Lustig, A. & Monnard, P.-A. (1999). Enzymatic reactions in liposomes using the detergent-induced liposome loading method. *Biochim. Biophys. Acta* 1416, 57-68.

Oberholzer, T., Nierhaus, K. H. & Luisi, P. L. (1999) Protein expression in liposomes. *Biochem. Biophys. Res. Comm.* 261, 238-241.

Ota, S., Yoshizawa, S. & Takeuchi, S. (2009). Microfluidic Formation of Monodisperse, Cell-Sized, and Unilamellar Vesicles. *Angew. Chem. Int. Ed.* 2009, 48, 6533-6537.

Pautot, S., Frisken, B. J. & Weitz, D.A. (2003a). Engineering asymmetric vesicles. *PNAS* 100, 10718–10721.

Pautot, S., Frisken, B. J. & Weitz, D.A. (2003b). Production of Unilamellar Vesicles Using an Inverted Emulsion. *Langmuir* 19, 2870 2879.

Pitard, B., Richard, P., Dunarach, M., Girault, G. & Rigaiud, J. L. (1996) ATP Synthesis by the F0F1 ATP synthase from Thermophilic bacillus PS3 reconstituted into liposomes with bacteriorhodopsin 1. Factors defining the optimal reconstitution of ATP synthases with bacteriorhodopsin. *Eur. J. Biochem.* 235, 769-778.

Pontani, L. L., van der Gucht, J., Salbreux, G., Heuvingh, J., Joanny J.-F. & Sykes, C. (2009). Reconstitution of an Actin Cortex Inside a Liposome. *Biophys. J.* 96, 192–198.

Saito, H., Kato, Y., La Berre, M., Yamada, A., Inoue, T., Yoshikawa, K. & Baigl, D. (2009). Time-Resolved Tracking of a Minimum Gene Expression System Reconstituted in Giant Liposomes. *ChemBioChem* 10, 1640-1643.

Schmidli, P. K., Schurtenberger, P. & Luisi, P. L. (1991). Liposome mediated enzymatic synthesis of phosphatidylcholine as an approach to self-replicating liposomes. *J. Am. Chem. Soc.* 113, 8127-8130.

Shimizu, Y., Inoue, A., Tomari, Y., Suzuki, T., Yokogawa, T., Nishikawa, K. & Ueda, T. (2001). Cell free translation reconstituted with purified components. *Nat. Biotechnol.* 19, 751-755.

Shimizu, Y., Kanamori, T. & Ueda, T. (2005). Protein synthesis by pure translation systems. *Methods* 36, 299–304.

Souza, T., Stano, P. & Luisi, P. L. (2009). The minimal size of liposome-based model cells brings about a remarkably enhanced biological activity. *ChemBiochem* 10, 1056-1063.

Stano, P. & Luisi, P. L. (2010a). Achievements and open questions in the self-reproduction of vesicles and synthetic minimal cells. *ChemComm.* 46, 3639-3653.

Stano, P. & Luisi, P. L. (2010b) Reactions in Liposomes. In: *"Molecular Encapsulation: Organic Reactions in Constrained Systems"* U. H. Brinker & J.-L. Mieusset (eds.) Wiley. Pp. 455-492.

Sugiura, S., Kuroiwa, T., Kagota, T., Nakajima, M., Sato, S., Mukataka, S., Walde, P. & Ichikawa, S. (2008). Novel Method for Obtaining Homogeneous Giant Vesicles from a Monodisperse Water-in-Oil Emulsion Prepared with a Microfluidic Device. *Langmuir* 2008, 24, 4581-4588.

Sunami, T., Sato, K., Matsuura, T., Tsukada, K., Urabe, I. & Yomo, T. (2006). Femtoliter compartment in liposomes for in vitro selection of proteins. Analytical Biochem. 357, 128–136.

Takeuchi, T., Montenegro, J., Hennig, A. & Matile, S. (2011). Pattern generation with synthetic sensing systems in lipid bilayer membranes. *Chem. Sci.* 2, 303–307.

Treyer, M., Walde, P. & Oberholzer, T. (2002). Permeability Enhancement of Lipid Vesicles to Nucleotides by Use of Sodium Cholate: Basic Studies and Application to an Enzyme-Catalyzed Reaction Occurring inside the Vesicles. *Langmuir* 18, 1043-1050.

Tsumoto, K., Nomura, S. M., Nakatani, Y. & Yoshikawa, K. (2001). Giant liposome as a biochemical reactor: transcription of DNA and transportation by laser tweezers. *Langmuir* 17, 7225–7228.

Vamvakaki, V., Fournier, D. & Chaniotakis, N. A. (2005). Fluorescence detection of enzymatic activity within a liposome based nano-biosensor. *Biosens. Bioelectr.* 21, 384-388.

Van Dongen, S. F. M., Nallani, M., Cornelissen, J. J. L. M., Nolte, R. J. M. & van Hest, J. C. M. (2009). A three-enzyme cascade reaction through positional assembly of enzymes in a polymersome nanoreactor. *Chem. Eur. J.* 15, 1107-1114.

Varela, F., Maturana, H. & Uribe, R. (1974). Autopoiesis: the organization of living systems, its characterization and a model. *BioSystems* 5, 187–195.

Walde, P. & Ichikawa, S. (2001). Enzymes inside Lipid Vesicles: preparation, reactivity and Applications. *Biomolecular Engineering* 18, 143-177.

Walde, P., Cosentino, K., Hengel, H. & Stano, P. (2010). Giant Vesicles: Preparations and Applications. *ChemBioChem* 11, 848-865.

Wick, R., Angelova, M., Walde, P. & Luisi, P. L. (1996). Microinjection into giant vesicles and light microscopy investigation of enzyme-mediated vesicle transformations. *Chem. & Biol.* 3, 105-111.

Yamada, A., Yamanaka, T., Hamada, T., Hase, M., Yoshikawa, K. & Baigl, D. (2006). Spontaneous Transfer of Phospholipid-Coated Oil-in-Oil and Water-in-Oil Micro-Droplets through an Oil/Water Interface. *Langmuir* 22, 9824 9828.

Yoshimoto, M., Wang, S., Fukunaga, K., Fournier, D., Walde, P., Kuboi, R. & Nakao, K. (2005). Novel immobilized liposomal glucose oxidase system using the channel protein OmpF and catalase. *Biotechnol. Bioeng.* 90, 231–238.

Yu, W., Sato, K., Wakabayashi, M., Nakatshi, T., Ko-Mitamura, E. P., Shima, Y., Urabe, I. & Yomo, T. (2001). Synthesis of functional protein in liposome. *J. Biosci. Bioeng.* 92, 590-593.

Zepik, H. H., Bloechliger, E. & Luisi, P. L. (2001). A Chemical Model of Homeostasis. *Angew. Chemie Int. Ed.* 40, 199-202.

Zepik, H. H., Walde, P., Kostoryz, E. L., Code, J. & Yourtee, D. M. (2008). Lipid vesicles as membrane models for toxicological assessment of xenobiotics. *Crit. Rev. Toxicol.* 38, 1-11.

Zhang, Y.; Ruder, W. C. & LeDuc. P. (2008). Artificial cells: building bioinspired systems using small-scale biology. *Trends Biotech.* 26, 14-20.

3

Wavelet Analysis for the Extraction of Morphological Features for Orthopaedic Bearing Surfaces

X. Jiang, W. Zeng and Paul J. Scott
University of Huddersfield,
United Kingdom

1. Introduction

Surface texture is one of the most critical factors and important functionality indicators in the performance of high precision and nanoscale devices and components. The functions that have been identified in various studies include wear, friction, lubrication, corrosion, fatigue, coating, paintability, etc. [1-3]. It is also reported that the wear rates of surfaces in operational service is determined by roughness, waviness and the multi-scalar topographic features of a surface, such as random peaks/pits and ridges/valleys. These functional topographical features will impact directly on wear mechanics and physical properties of a whole system, such as hip joint replacement system in bioengineering [4-9]. For example, during functional operation of interacting surfaces, peaks and ridges will act as sites of high contact stresses and abrasion; consequently wear particles and debris will be generated by such surface topographical features, whereas the pits and valleys will affect the lubrication and fluid retention properties. In this situation, a vitally important consideration for functional characterisation must be the appropriate separation of the different components of surfaces, which is not only to extract roughness, waviness and form error, but should also be extended to all multi-scalar topographical events over surfaces.

Effectively surface characterization can help to: (1) Judge whether the manufacturing process or manufacturing conditions are effective or out of control, i.e. whether specific events in the manufacturing processes have occurred; (2) Interpret functional properties of macro, micro and nano geometry, which reflect product properties, such as optical quality, tribological properties, service life, safety, reliability, etc.

Random process techniques have been applied to surface analysis in which the surface topography is assumed as a stationary random system. However, the disadvantage of random analysis is that significant events involved in a surface, such as large peaks/pits, or ridges/valleys will be smoothed over in the Fourier space of a signal, without indicating the location of the frequency events. The early studies of morphology features of surface topography only considered the roughness component of the surface texture. Comprehensive studies of surface characterisation have nearly always concentrated on roughness assessment and have mainly developed evaluation techniques based on parameters and filtration techniques. This is normally accomplished by using Fourier Transform based filtering techniques, especially Gaussian filtering. These techniques are employed to isolate the roughness/waviness frequency bands relevant to the surface, by breaking down a surface signal into a series of harmonics waves. In fact, Gaussian filtering, used to extract the roughness and waviness components, is based on two presuppositions. One presupposition is that before filtering, the irrelevant form and translation errors have been removed from the measured data set. The other is that the residual surface, obtained according to the first assumption, can be broken down into a series of harmonic components. If surfaces conform to these assumptions, the roughness and waviness can be identified well and extracted from the original surfaces with a modified amplitude resulting from the transmission characteristics of a Gaussian function. These types of filtering techniques only considered the frequency information and do not consider the spatial information of the features.

The Fourier transform has a very good localization property in the frequency domain, however, it lacks the ability to localize in the time/space domain. Figure 1 gives a typical example of the Fourier transform of two different signals with the same sine wave components, while one is stationary and the other is non-stationary. The two signals have the same Fourier transform spectrum. To distinguish between these two signals, the analysis tools need the time/space-frequency localization property.

There are two functions that link the time/space-frequency domain, the ambiguity function (in radar applications) and the Wigner distribution (in quantum mechanics). Both have been introduced to identify the topographic features of engineering surfaces by Whitehouse and Zheng [10, 11]. Among these tools, the Wigner distribution are more suitable as it preserves the absolute space and frequency parameters [10]. The energy distribution in the space-frequency plane, offered by the Wigner transform, can be used to identify the variation in surface topography. The result is that the information about the frequencies and their locations in a surface signal has been used to monitor and adjust the manufacturing processing. However, it is also well known that although the Wigner transform can offer an analysis of the energy distribution of a signal, it only yields an imperfect description of the energy distribution, due to its substantial interference terms.

During the last decade, novel filtering techniques based on robust Gaussian regression and multi-scale techniques have been developed. Robust filtering has been designed to eliminate problems in the evaluation of the roughness of multi-process surfaces such as plateau honed cylinder bores and is critical to the implementation of areal surface analysis techniques.

As previously stated, analysis of multi-scalar properties of a surface topography needs to not only provide both the frequencies of the signal and their location, but also to accurately recover and perfectly reconstruct these topographic features. Considering these needs of surface assessment, wavelet analysis has been applied to surface characterization [12-16].

Wavelet analysis employs time-frequency windows and offers the relevant time/space-frequency analysis. The Wavelet transform has been proved to be a powerful tool for various applications, for example, the Wavelet series expansions, developed for applied mathematics and signal processing. Multi-scalar features have been used in capturing, identifying and analyzing local, non-stationary processes; Sub-band coding used in encoding, compressing, reconstructing and modelling signals and images [17-20].

In this chapter three generations of biorthogonal wavelet transforms, for the extraction of morphological structures from micro/nano scalar surfaces in the field of bioengineering, will be introduced. The chapter's aim is to create a "tool box" of wavelet techniques capable of complex analysis and interpretation of surface topography data; leading to the extraction of functionally critical morphological features from micro/nano scalar surfaces of orthopaedic joints for in-vitro and clinically retrieved applications.

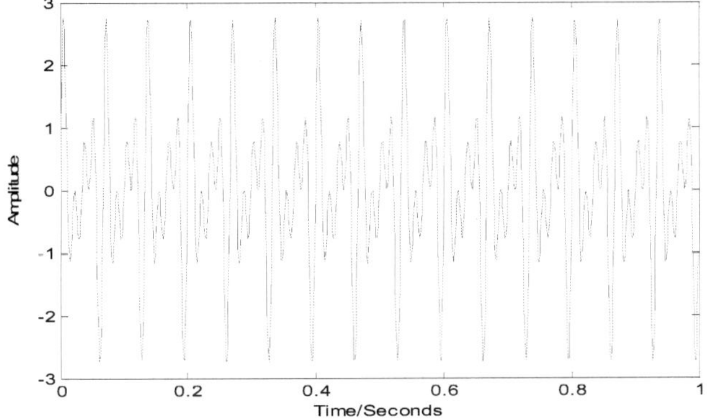

(a) a stationary signal composed of sine waves $\{30Hz, 45Hz, 60Hz\}$

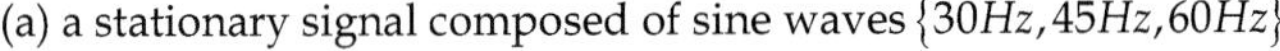

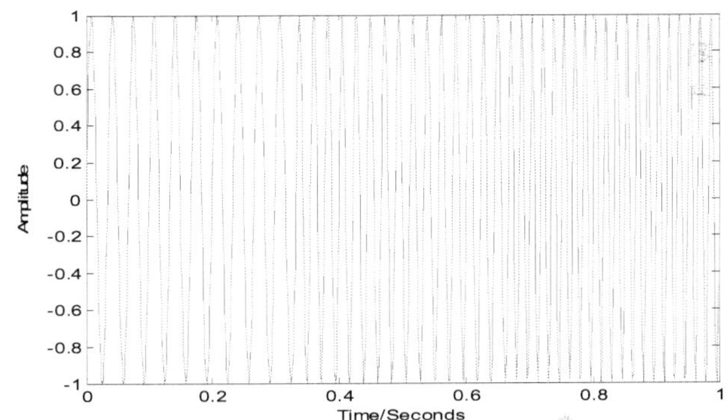

(b) a nonstationary signal composed of sinewaves $\{30Hz, 45Hz, 60Hz\}$

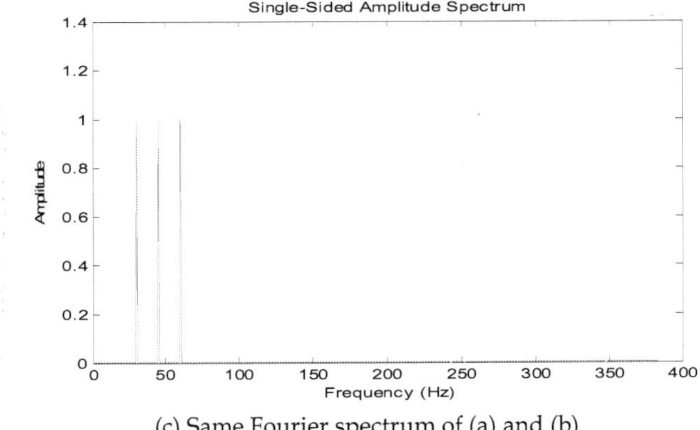

(c) Same Fourier spectrum of (a) and (b)

Fig. 1. Fourier transform of two different signals with the same sin components

2. Wavelet

A Wavelet $\psi(t)$ is a waveform that has compact support in both the space and frequency domains and whose integral is zero. Compare wavelets with sine waves, which are the basis of Fourier analysis. Sinusoids do not have limited duration — they extend from minus to plus infinity, where sinusoids are smooth and predictable, early wavelets tended to be irregular and asymmetric. The wavelet $\psi(t)$ function must satisfy the admissibility condition:

$$C_\psi = \int_{-\infty}^{\infty} \frac{\left|\hat{\psi}(\omega)\right|^2}{|\omega|} d\omega < \infty \tag{1}$$

which guarantees the existence of the inverse wavelet transform.
In wavelet analysis the signal is broken down into rescaled and shifted versions of the original waveform. This transfers space based information into scale based information, which represents the frequency and location properties of the original signal. The wavelet transform can be defined as [17-18]:

$$W_{a,b}(f) = \int_{-\infty}^{\infty} f(t) \frac{1}{\sqrt{a}} \psi^*(\frac{t-b}{a}) dt \tag{2}$$

where, $f(t)$ is the signal, a and b are the dilation and translation parameters for the function $\psi(t)$, which is a wavelet function or mother function. The wavelets, $\psi_{a,b}(t)$, (basis functions) can be obtained by dilation and translation of the mother wavelet $\psi(t)$. The parameter a can be used to illustrate dynamic transmission bandwidths or cover different frequency ranges, and b is the translation parameter that can be used to define the location of signal events. Typically one-dimensional continuous wavelets are expressed by [17-18]:

$$\psi_{a,b}(x) = \frac{1}{\sqrt{a}} \psi(\frac{x-b}{a}) \tag{3}$$

In the real world, most signals are discrete and the dilation and translation parameters both taking only discrete values. The rescaled wavelets, $\psi_{j,0}(t) = \psi(a^{-j}t)$ are dilated by the factor a^{-j}, this is illustrated in figure 2(a) where the parameter a takes the discrete value of two. j is the scalar parameter that can be used to illustrate dynamic transmission bandwidths or cover different frequency ranges. The shifted wavelets $\psi_{0,k}(t) = \psi(t-k)$ are translated by k (translation parameter) as shown in figure 2(b). By changing k, the time location centre has been moved. Typical one-dimensional wavelets $\psi_{j,k}(x)$ are dilated j times and shifted k times. The discrete wavelets are expressed by:

$$\psi_{j,k}(x) = a^{-j/2} \psi(a^{-j}x - k) \tag{4}$$

Thus a signal can be divided into different scales and different locations, with the signal data now being represented on a 'time-scalar plane'. Multi-resolution divides the frequencies into octave bands, from ω to 2ω. Frequencies shift upward by an octave when

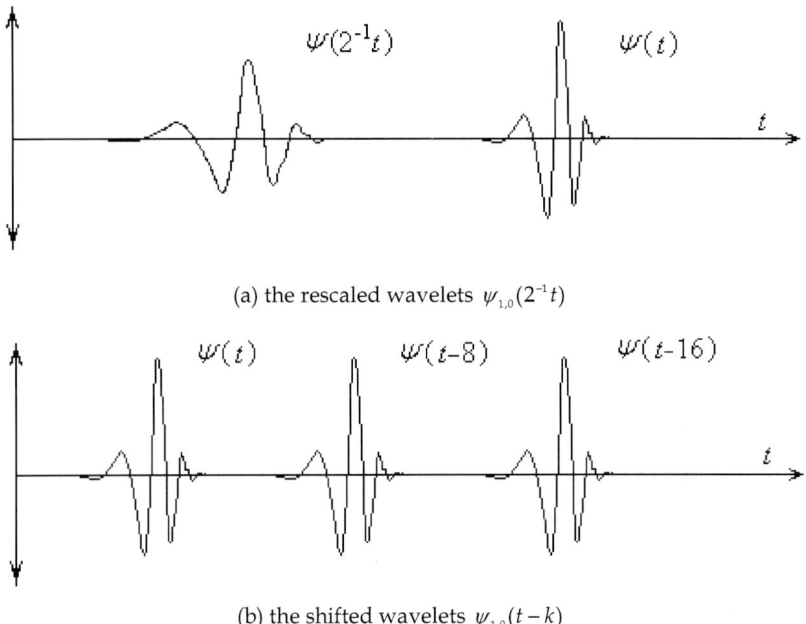

(a) the rescaled wavelets $\psi_{1,0}(2^{-1}t)$

(b) the shifted wavelets $\psi_{1,0}(t-k)$

Fig. 2. Example of a discrete wavelet

time is rescaled by two. Figure 3(c) shows how the time-frequency plane is partitioned naturally into rectangles of constant area.

For a discrete signal $z(x) \in L^2(Z)$, its discrete wavelet transform can be expressed by:

$$W_{j,k}(f) = \langle f(x), \psi_{j,k}(x) \rangle = 2^{-j/2} \sum_{j,k} f(x)\psi(2^{-j}x - k) \tag{5}$$

The signal $z(x)$ can be recovered with the following equation:

$$z(x) = \langle z(x), \psi_{j,k}(x) \rangle \Psi_{j,k}(x) \tag{6}$$

where, the $\Psi_{j,k} = (F * F)^{-1}\psi_{j,k}$.

A basic idea behind wavelet analysis is the transfer of a complex frequency analysis into a simple scalar analysis. Wavelet analysis employs time-frequency windows and offers the relevant time-frequency analysis, which uses long windows at low frequencies and short windows at high frequencies (as shown in figure 3c). Wavelet mathematics provides a mathematical microscope that "can divide functions into different frequency components, and then study each component with a resolution that is matched to its scale" [17-18]. The wavelet transform is an update of the classical Short-Time Fourier Transform (STFT) or Gabor Transform as well as the Wigner distribution [21-22]. In contrast, STFT uses a single size analysis window (shown in figure 3b), and offers the same accuracy of analysis in the whole time-frequency plane. The Wavelet transform is related to space-frequency analysis and is similar to the Wigner-Ville distribution, but is better able than the STFT to "zoom in" on very short-lived high frequency phenomena, such as transients in signals or singularities in functions. A point to be noted is that wavelet analysis takes the signal decomposition to a space-scalar (time-frequency) plane and separates then reconstructs these components in the space domain. In contrast, Wigner analysis also decomposes a signal into the space-frequency plane, then studies the energy distribution of these components but cannot reconstruct the modified signal perfectly.

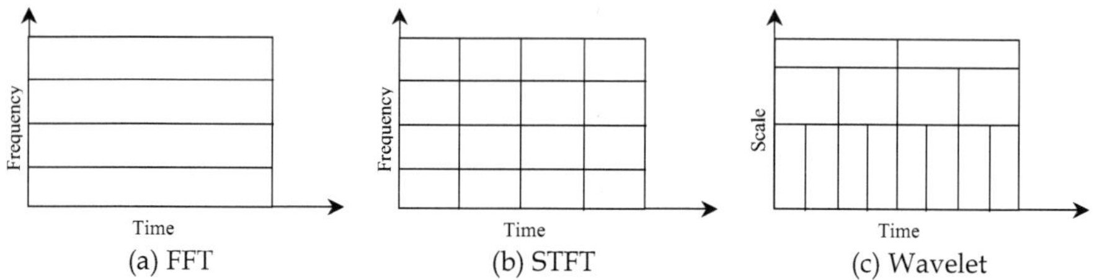

Fig. 3. The different windows of the FFT, STFT and WT

The Wavelet technique includes many different wavelet functions: each has its own properties and applications. Linear phase (symmetric wavelet) and finite pulse filtering are basic properties in surface analysis, only the Haar wavelet and the biorthogonal wavelet have both characteristics. Due to the fact that the Haar wavelet is binary and a discontinuous function makes it not suitable for surface analysis, only the biorthogonal wavelet are useful tools for surface decomposition.

3. The Bi-orthogonal wavelet for surface analysis

The main approach to the discrete wavelet transform is to use a two-channel filter bank. The dilation equations of the basis wavelets and the corresponding scaling functions have associated lowpass H_0 and highpass H_1 filter coefficients.

The basic property of the discrete wavelet function and the scale function for multi-scale analysis are their two-scale difference function [17-19]:

$$\begin{cases} \psi(x) = \sum_k 2h_0(k)\varphi(2x - k) \\ \varphi(x) = \sum_k 2h_1(k)\varphi(2x - k) \end{cases} \tag{7}$$

where, $\psi(.)$ is the wavelet and $\varphi(.)$ is the scale function. The two-scale difference functional relationships means that the wavelet function and the scale function at scale j can be represented as a linear combination of the scale functions at scale j+1. Here, the weighting coefficients $h_0(k)$, $h_1(k)$ are the impulse response of the lowpass filter H_0 and highpass H_1 respectively. For orthogonal wavelet transforms H_0 and H_1 are orthogonal. However, the frequency responses of the two-channel filters, the so-called analysing filters, are not ideal brick wall filters since normally there are overlaps that would lead to aliasing, amplitude modifications and phase distortions.

3.1 The first generation biorthogonal wavelets

The biorthogonal wavelet overcomes some of the defects of the orthogonal wavelet and other two-channel filters. In a biorthogonal wavelet the forward (H_0 and H_1) and inverse filters (G_0 and G_1) are different. The inverse filters G_0 and G_1 (the so-called synthesis filters), are specially designed to compensate for these errors of the analysing filters H_0 and H_1. When the frequency responses of the synthesis filters, G_0, G_1, are the inverses of the analysis filters, H_0, H_1 they satisfy [23]:

$$G_0(z)H_0(z) + G_1(z)H_1(z) = 2z^{-l}$$
$$G_0(z)H_0(-z) + G_1(z)H_1(-z) = 0$$

$$(8)$$

the errors in this analysis bank are cancelled. Where, l is time delay, the distortion term must be z^{-l}, and the aliasing term must be zero.

In the biorthogonal case, abandoning the orthogonal relationship between the lowpass analysis filters $H_0(z)$ and the highpass analysis filter $H_1(z)$, while keeping the biorthogonal conditions between $H_0(z), G_1(z)$, and $H_1(z), G_0(z)$. The biorthogonal condition can be expressed by the impulse responses, $h_0(k)$, $h_1(k)$, $g_0(k)$, $g_1(k)$, of $H_0(z), H_1(z), G_0(z), G_1(z)$, as:

$$\sum_k h_0(k)\tilde{g}_1(k-2n) = 0$$
$$\sum_k h_1(k)\tilde{g}_0(k-2n) = 0$$

$$(9)$$

Where, $\tilde{g}_0$ and $\tilde{g}_1$ are the time reverse of g_0 and g_1 respectively. Formulae (9) means $h_0(k)$ is orthogonal to $\tilde{g}_1(k-2n)$ and $h_1(k)$ is orthogonal to $\tilde{g}_0(k-2n)$. Here, $\tilde{g}_0(k-2n)$ and $\tilde{g}_1(k-2n)$ are the even shift $2n$ of $\tilde{g}_0(k)$ and $\tilde{g}_1(k)$ respectively.

According to Wavelet theory [17-18], the dilation equations of the analysis wavelet equation, $\tilde{\psi}(x)$, and synthesis wavelet function, $\psi(x)$ can be constructed from the impulse responses of the analysis and synthesis highpass filters respectively [23]:

$$\begin{cases} \tilde{\psi}(x) = \sum_0^N 2h_1(l)\tilde{\varphi}(2x-l) \\ \psi(x) = \sum_0^{\tilde{N}} 2g_1(k)\varphi(2x-k) \end{cases}$$

$$(10)$$

the scale equation are given by:

$$\begin{cases} \tilde{\varphi}(x) = \sum_0^{\tilde{N}} 2h_0(l)\tilde{\varphi}(2x-l) \\ \varphi(x) = \sum_0^N 2g_0(k)\varphi(2x-k) \end{cases}$$

$$(11)$$

The $\tilde{\varphi}(x)$, $\tilde{\psi}(x)$, $\varphi(x)$, $\psi(x)$ are constructed as a biorthogonal wavelet pair, their limit functions would inherit biorthogonality.

For areal surface, the analysis and synthesis scale functions, $\tilde{\varphi}(x,y)$ and $\varphi(x,y)$ can be expressed by :

$$\tilde{\varphi}(x,y) = \tilde{\varphi}(x)\tilde{\varphi}(y)$$
$$\varphi(x,y) = \varphi(x)\varphi(y) \tag{12}$$

Using the scale function pair, the analysis wavelets and synthesis wavelets are built by [27]:

$$\tilde{\psi}(x,y) = \begin{cases} \tilde{\psi}^h(x,y) = \tilde{\varphi}(x)\tilde{\psi}(y) \\ \tilde{\psi}^v(x,y) = \tilde{\psi}(x)\tilde{\varphi}(y) \\ \tilde{\psi}^d(x,y) = \tilde{\psi}(x)\tilde{\psi}(y) \end{cases} \quad \psi(x,y) = \begin{cases} \psi^h(x,y) = \varphi(x)\psi(y) \\ \psi^v(x,y) = \psi(x)\varphi(y) \\ \psi^d(x,y) = \psi(x)\psi(y) \end{cases} \tag{13}$$

where, $\tilde{\psi}^h(x,y)$, $\tilde{\psi}^v(x,y)$ and $\tilde{\psi}^d(x,y)$ give the analysis wavelets in the horizontal, vertical and diagonal directions, respectively. $\psi^h(x,y)$, $\psi^v(x,y)$ and $\psi^d(x,y)$ give the synthesis wavelets of the other three directions.

By employing a 2D biorthogonal wavelet pair, an $N \times N$ points areal discrete surface signal $z(x,y)$ can be transformed to a wavelet series by decomposing the signal data $z(x,y)$ on the 2D biorthogonal wavelets basis:

$$W_{j,k}(z) = \left\langle z(x,y), \tilde{\psi}_{j,k}(x,y) \right\rangle == (d_{1,k1},...d_{J,kJ}, a_{J,kJ}) \tag{14}$$

The $a_{J,kJ}$ are approximation coefficients of the signal $z(x,y)$ at the scale 2^{-J}, and the $d_{1,k1},...d_{J,kJ}$ are a group of detail coefficients of $z(x,y)$ at the scalar range of 2^{-j} ($j = 1 \sim J$, J is the maximal wavelet decomposition Level., and k_j refers to a subset $(1:\dfrac{N}{2^j})$). The $a_{j,k}$ are generated by the set of inner products [27]:

$$a_{j,k} = < A_{j-1}(x,y), \tilde{\varphi}_{j,k}(x,y) > \tag{15}$$

$A_{j-1}(x,y)$ is called a discrete scalar approximation of $z(x,y)$ at the scale $2^{-(j-1)}$, and represents the low frequency components of $z(x,y)$ at the scale $2^{-(j-1)}$. When the scale j is set to zero, the scalar approximation $A_0(x,y)$ equals $z(x,y)$. $A_j(x,y)$ at the scale 2^{-j} can be reconstructed by using the set $a_{j,k}$ ($j = 1 \sim J$, $k \in (1:\dfrac{N}{2^j})$), and applying the inverse wavelet transform:

$$A_j(x,y) = \sum_k a_{j,k}\varphi_{j,k}(x,y) = \sum_k < A_{j-1}(x,y), \tilde{\varphi}_{j,k}(x,y) > \varphi_{j,k}(x,y) \tag{16}$$

The detail coefficient, $d_{j,k}$, is a combination of three directional detail parts:

$$d_{j,k} = \begin{cases} d^h_{j,k} = < A_{j-1}(x,y), \tilde{\psi}^h_{j,k}(x,y) > \\ d^v_{j,k} = < A_{j-1}(x,y), \tilde{\psi}^v_{j,k}(x,y) > \\ d^d_{j,k} = < A_{j-1}(x,y), \tilde{\psi}^d_{j,k}(x,y) > \end{cases} \tag{17}$$

Similarly, the inverse wavelet transform of the detail coefficient $d_{j,k}$ ($j = 1 \sim J$, $k \in (1:\dfrac{N}{2^j})$) gives the high frequency components of $z(x,y)$ at the scale 2^{-j} and it is called a discrete detail $D_j(x,y)$ of $z(x,y)$ at the scale 2^{-j} :

$$D_j(x,y) = D_j^h(x,y) + D_j^v(x,y) + D_j^d(x,y)$$

$$= \sum_k d_{j,k}^h \psi_{j,k}^h(x,y) + \sum_k d_{j,k}^v \psi_{j,k}^v(x,y) + \sum_k d_{j,k}^d \psi_{j,k}^d(x,y) \tag{18}$$

The equations show that in the two-dimensional case, the approximation and detail coefficients are computed by separate filtering of the signal $z(x,y)$ along the abscissa and ordinate. The wavelet decomposition can thus be interpreted as a signal decomposition in a set of independent, spatially oriented frequency channels. The approximation coefficient $a_{j-1,k}$ is decomposed into $a_{j,k}$, $d_{j,k}^h$, $d_{j,k}^v$ and $d_{j,k}^d$. The coefficients $a_{j,k}$ corresponds to the lower frequencies, the coefficients $d_{j,k}^h$ correspond to the vertical high frequencies (horizontal edges), the coefficients $d_{j,k}^v$ the horizontal high frequencies (vertical edges) and the coefficients $d_{j,k}^d$ the high frequencies in both directions. Consequently, the areal surface $z(x,y)$ can be reconstructed using:

$$z(x,y) = A_1(x,y) + D_1(x,y) = A_2(x,y) + D_2(x,y) + D_1(x,y)$$

$$= \ldots\ldots = A_j(x,y) + D_j(x,y) + D_{j-1}(x,y) + \ldots\ldots + D_1(x,y) \tag{19}$$

3.2 The second generation biorthogonal wavelet

The fundamental idea behind the use of Biorthogonal wavelet filtering for surface analysis is to break down areal surface raw data set into a rescaled and shifted version of the original waveforms in a 'scalar space', extract the waveforms carrying different information and then reconstruct directly these components. The main advantage of using a biorthogonal wavelet is that it has a linear phase (leading to real output without aliasing and phase distortion) and a traceably located property so that the different component surfaces obtained can more naturally record the real surface. However due to the fact that wavelets (basic functions) are built by dilation and translation of the prototype wavelet which relied on the Fourier transform, and the fact that the Wavelet transform needs to be applied along three directions, horizontal, vertical, and diagonal, the theory and corresponding algorithm are very complex. Furthermore, there is still the boundary destruction inherent when using the Fourier transform.

For engineering application purposes the method used for surface characterisation must be simple and natural. In other words, the method of the separation and reconstruction of different components of the surface should have both the 'simplicity' of general surface filter and 'naturalness' of a biorthognal wavelet filter, as result of this, the second generation of biorthogonal wavelet, originally developed at Bell Laboratory [24-26], was introduced into surface analysis. The wavelet and scalar coefficients are only dependant on the measured raw data of a surface and the filtering and lifting factors calculated by a cubic spline interpolation in an interval (Dyn et al. 1987; Flowers 1995; and Stoer 1980). Compared to the first generation of Wavelet representation, the new model does not take the Fourier transform as a prerequisite. The Wavelet transform only embraces three stages, splitting, prediction and updating. The other advantage derived from this model is that there is no boundary destruction. Although the implementation of the lifting wavelet is completely different from the former model, it is much easier to understand and perform.

In the lifting scheme, the splitting procedure is to divide an original signal into even and odd subsets. The even subset is obtained by subsample the input signal at the even positions, and it can be seen as the initial approximation of the input signal. The odd subset is obtained by subsample the input signal at the odd positions. The lost information of the initial approximation is contained in the odd subsets. In the prediction step: the even subset is used to approximate the odd subset at the corresponding position by using interpolation on the neighbouring data. Then the difference between the odd subset and the approximated subset can be regarded as the output of the high pass filter, or the wavelet coefficients at this specified level. Finally, in the updating step: the final approximation of the input signal can be calculated by updating the even subset (initial approximation of the input signal) with the help of neighbouring wavelet coefficients obtained from last step. The output of this step is the approximation coefficients of low-pass filter. The benefit of using the lifting scheme is: (1) the lifting scheme allows a fully in-place calculation of the wavelet transform; (2) it is very simple as the inverse wavelet transform with the lifting scheme is only a reversal of the operations of the forward transform.

In this implementation, the analysis highpass filter, H_1, and synthesis lowpass filter, G_0, of the initial finite biorthogonal wavelet filter set, $\{ H_0, H_1, G_0, G_1 \}$ within the first generation are transferred to H_1', G_0' which can be found by the lifting scheme as[24-26]:

$$\begin{cases} H_1'(z) = H_1(z) + G_1(z)S(z^2) \\ G_0'(z) = G_0(z) - H_0(z)S(z^{-2}) \end{cases} \tag{20}$$

where the $S(z)$ is a Laurent polynomial. Substituting this new set to the equation (7), in order to perfectly reconstruct the signal, the frequency responses of analysis and synthesis filters of the second algorithm should also satisfy [24-26]:

$$G_1^2(z)S(z^2) - H_0^2(z)S(z^{-2}) = 0$$
$$G_1(z)G_1(-z)S(z^2) - H_0(z)H_0(-z)S(z^{-2}) = 0 \tag{21}$$

In this case, $S(z^2) = S(z^{-2})$. After lifting has been performed, the new biorthogonal wavelet pair can be found as:

$$\begin{cases} \tilde{\psi}'(x) = \tilde{\psi}(x) - \sum_0^{\tilde{N}} s(l)\tilde{\varphi}(x-l) \\ \psi'(x) = 2\sum_0^{\tilde{N}} g_1(k)\varphi'(2x-k) \end{cases} \tag{22}$$

and

$$\begin{cases} \tilde{\varphi}(x) = 2\sum_0^{\tilde{N}} h_0(l)\tilde{\varphi}(2x-l) \\ \varphi'(x) = 2\sum_0^{N} g_0(k)\varphi'(2x-k) + \sum_0^{\tilde{N}} s(k)\psi'(x-k) \end{cases} \tag{23}$$

where the coefficients, $s(\,\cdot\,)$, are from the Laurent polynomial $S(z)$, the order is based on how many neighbouring coefficients are chosen to interpolate in the prediction and updating steps. The power behind the lifting scheme is that $s(\,\cdot\,)$ can be used to fully control all wavelets and synthesis scaling functions.

The simplest wavelet transform using lifting scheme is when $N = \tilde{N} = 0$, which means in the predict and updating step, no neighbouring coefficients are used to interpolate and update. In this case, the wavelet transform does nothing but separates the input data into even and odd subset, which is called lazy wavelet.

In the lifting wavelet transform with $N = 2$ the coefficients, $s(l) = \left(\dfrac{1}{2}, \dfrac{1}{2}\right)$, the wavelet coefficient $d_{1,k}$ encodes the difference between the exact sample $A_{0,2k+1}$ and its linear prediction of two even neighbours $A_{0,2k}$, $A_{0,2k+2}$. The wavelet coefficients can be written as:

$$d_{j,k} := A_{j-1,2k+1} - 1/2(A_{j-1,2k} + A_{j-1,2k+1}) \tag{24}$$

Employing the dual lifting scheme, $\tilde{N} = 2$, that is, the coefficient $s'(k) = \left(\dfrac{1}{4}, \dfrac{1}{4}\right)$, the scalar coefficients $a_{1,k}$ of $A_{0,2k}$ would be updated by wavelet coefficients $d_{1,k}$ and $d_{1,k+1}$. The scalar coefficients can be expressed as [24-26]:

$$a_{j,k} = A_{j-1,2k} + 1/4(d_{j,k} + d_{j,k-1}) \tag{25}$$

3.3 The lifting wavelet algorithm

Cubic spline interpolation has been used and verified as an appropriate Laurent polynomial, $S(z)$ in the prediction stage since the resulting wavelet satisfies the requirements of surface analysis: excellent refinement accuracy, perfect reconstruction, minimum sampling condition (measured area) and a minimum computation.

The three stages of the Wavelet transform by the lifting scheme are depicted in figure 4. The fast lifting wavelet transform algorithm is simplified to [24-26]:

$$\begin{cases} A_{j,2k}, A_{j,2k+1} = Split(A_j) \\ d_{j+1,k} = A_{j,2k+1} - \rho(A_{j,2k}) \\ a_{j+1,k} = A_{j,2k} + \mu(d_{j+1,k}) \end{cases} \tag{26}$$

With this implementation of the forward Wavelet transform, $z(x)$ has been driven to subsets, d_{j+1} and a_{j+1} which records high and low frequency events at the scale $2^{-(j+1)}$ of $z(x)$. The whole decomposition of $z(x)$ is a simple repetitive scheme through rows and columns and all computations are carried out in-place. After the jth step of decomposition in the scalar domain, an original surface signal $z(x)$ is replaced with the wavelet series $d_1, \cdots, d_j, a_j$. It can be expressed by [27]:

$$\begin{aligned} W_j[z(x)] = (a_1, d_1) &= (a_2, d_2, d_1) \\ &= (a_j, d_j, d_{j-1}, \cdots, d_1) \end{aligned} \tag{27}$$

The above equation is Mallat's format arrangement of the wavelet transform coefficients. In practice, due to the lifting scheme's in-place calculation property, the coefficients are organised in-place. Let input data $z(i)$ with length $i = 0...N-1$, the output of the transform will also has length N, then $a_{j,}$ is the subset with index $n \cdot 2^j$, where $n = (0,...,\frac{N}{2^j}-1)$; d_j is the subset with index $n \cdot 2^j + 2^{j-1}$, where $n = (0,...,\frac{N}{2^j}-1)$.

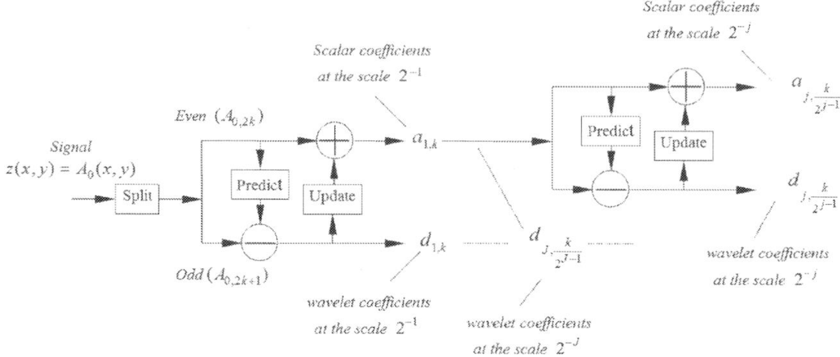

(a) the diagram of the lifting scheme

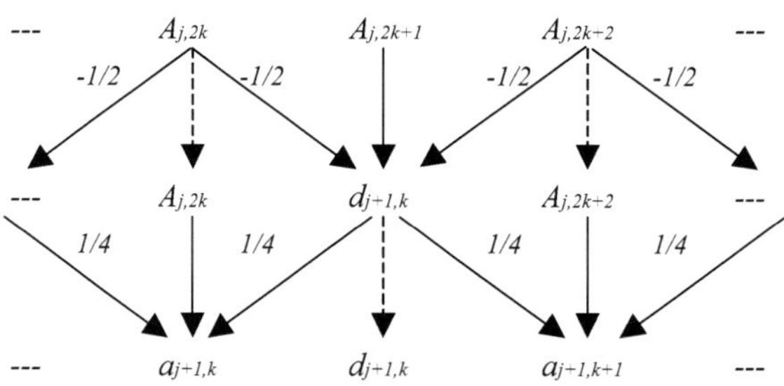

(b) the scheme of lifting Lazy wavelet

Fig. 4. Wavelet transform using the Lifting scheme

Figure 5 illustrates this schedule where j is the decomposed level of Wavelet transform in scalar domain [27].

3.4 Separation and extraction of frequency components of a surface

If a surface, $z(x,y)$ is assumed to consist of a series of superimposed frequency components:

$$z(x,y) = \eta(x,y) + \eta'(x,y) + \eta''(x,y) \tag{28}$$

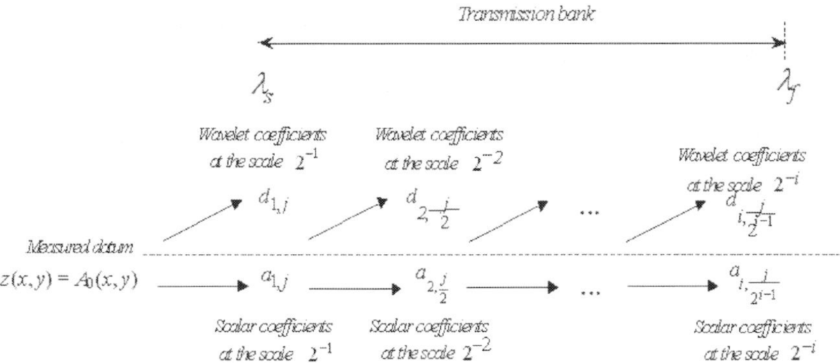

Fig. 5. Wavelet transform

These surface components can be separated in the scalar domain by band pass filters. The transmission bands are based on different cut-off wavelengths for different frequency components. The wavelet coefficients, $d_1, \cdots, d_{Jr}$ can be considered outputs of a high frequency band, $1/\lambda_s \sim 1/\lambda_c$, and referred to as the roughness component, $\eta(x,y)$. $1/\lambda_s$ indicates the high frequency limited by the sampling interval, and $1/\lambda_c$ is the roughness frequency limitation. The $d_{Jr+1}, \dots, d_{Jw}$ represents the output (waviness, $\eta'(x,y)$) of a sub-lowpass filter band $1/\lambda_c \sim 1/\lambda_{wc}$. $1/\lambda_{wc}$ is the waviness frequency limitation. a_{Jw} are scalar coefficients that represent the output (form error, $\eta''(x,y)$) of a lowpass filter band, $1/\lambda_{wc} \sim 1/l$; where l is the sample length when $l = l_x$ or l_y. Using an inverse Wavelet transform, these surfaces can be recovered flexibly and immediately in the different transmission bands in terms of functional analysis requirements. The inverse Wavelet lifting Transform is performed simply by reversing the order of the operations and toggling negative to positive for all operations [27]. The relationship between the cutoff frequency and the wavelet scale is shown in table 1.

$$\begin{cases} \text{Roughness: } \eta(x,y) = IW(\, d_1, \, \cdots, \, d_{Jr} \,) \\ \text{Waviness: } \eta'(x,y) = IW(\, d_{Jr+1}, \, \cdots, \, d_{Jw} \,) \\ \qquad \text{Form: } \eta''(x,y) = IW(\, a_{Jw} \,) \end{cases} \qquad (29)$$

Where, $Jr = \mathrm{int}\left(\log_2(\lambda_c / \Delta x) - 1\right)$ $Jw = \mathrm{int}\left(\log_2(\lambda_{wc} / \Delta x) - 1\right)$. Table 1 gives the relationship between the decomposition level and the cutoff wavelength.

In order to obtain $\eta(x,y)$ and $\eta'(x,y)$, the appropriate scalar coefficients, a_{Jw}, are set to zero and then the inverse transform is applied, and vice versa for $\eta''(x,y)$. In a similar way, a flexible reference surface can be obtained:

$$m(x,y) = z(x,y) - \eta(x,y) = z(x,y) - IW(d_1, \cdots, d_J) \qquad (30)$$

where J is a flexible selected decomposition level according to the cutoff wavelength of the reference surface. If a functional evaluation of surfaces is needed to cover all of the topographical information from roughness, through multi-scalar events, to waviness, a functional surface can be built by:

$$\eta(x,y) + \eta'(x,y) = IW(d_1, \cdots, d_{jw}) \tag{31}$$

Cut-off wavelength		The decomposition level of wavelet filter				
		3 $(2*2^3=16)$	4 $(2*2^4=32)$	5 $(2*2^5=64)$	6 $(2*2^6=128)$	7 $(2*2^7=256)$
Sampling interval (μm)	4	**0.064 (0.08)**	0.128	**0.256 (0.25)**	0.512	1.024
	8	0.128	**0.256 (0.25)**	0.512	1.024	2.048
	12	0.192	0.384	**0.768 (0.8)**	1.536	3.072
	16	**0.256 (0.25)**	0.512	1.024	2.048	4.096
	20	0.320	0.64	1.28	**2.56 (2.5)**	5.12
	24	0.384	**0.768 (0.8)**	1.536	3.072	6.144

Table 1. The relationship between the level of the wavelet decomposition and sampling interval

The identification of multi-scalar events in the roughness and waviness bands is important in order to study the functional performance of the areal surface topography. These multi-scalar topographical events, $\xi(x,y)$, such as peaks/pits and rings/valleys, hide in the roughness and waviness bands, and their wavelengths can cover a wide frequency range ($1/\lambda_s \sim 1/\lambda_{wc}$). Wavelet coefficient sets over the transmission bands "naturally" record the information concerning their amplitude and location. These events can be easily captured by using an amplitude threshold, T_j, to pick out the roughness and waviness. T_j is the value of an intersection of the probability curve of the cumulative amplitude distribution of each wavelet coefficient set. This process is based on an assessment that the amplitude distribution of each wavelet coefficient set, $d_{j,k}$, belonging to the roughness and waviness components, would obey a normal distribution. If the absolute value of the amplitude is equal to or larger than T_j, a thresholding estimator is applied.

$$d'_{j,k} = \begin{cases} 0 & |d_{j,k}| < T_j \\ d_{j,k} & |d_{j,k}| \geq T_j \end{cases} \tag{32}$$

Where, j is the decomposition level, $k = n \cdot 2^j + 2^{j-1}$ with $n = (0,...,\dfrac{N}{2^j} - 1)$ are the indices of the coefficients. The absolute value of the peak amplitude is smaller than T_j, the coefficient should be replaced by a zero. In the case of a detail coefficient being larger than or equal to T_j the coefficient should be retained. As a consequence, detail coefficients that represent only the information of the topographical events are obtained. From the experiments carried out, the threshold approaches the standard deviation of each wavelet coefficient set. The multi-scalar topographic features can then be built by using the 2D inverse discrete wavelet transform:

$$\xi(x,y) = IW(d'_1, \cdots, d'_j) \tag{33}$$

Where d' represent the thresholded wavelet coefficients.

3.5 Case studies on the bioengineering surfaces

Typical examples of much finer bioengineering surfaces from hip joint systems have been selected to show the behaviour of the spline wavelet filtering. During the Total Hip Replacement, a prosthetic head as shown in Figure 6a will replace the head of a bad or dead human femur. For the femoral heads of the hip joint system, it needs not only to support most of the body weight, but also to resist long term wear in service. The wear property of the femoral head surface is crucial for the lifetime of the whole hip joint system. The topographies of the femoral heads of the hip joint system were obtained by using a phase-shifting interferometer with a sampling area of $228 \times 203 \mu m$. The illustrations in Figure 7 show different topographies in polar and equatorial regions of worn heads.

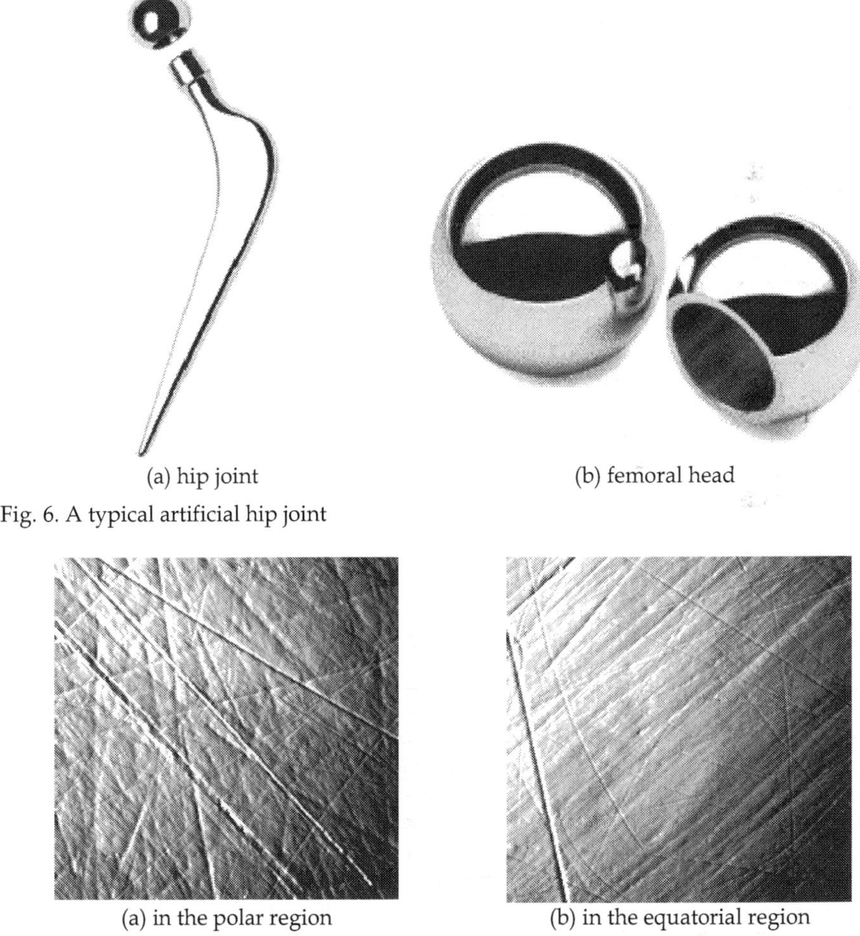

(a) hip joint   (b) femoral head

Fig. 6. A typical artificial hip joint

(a) in the polar region   (b) in the equatorial region

Fig. 7. The typical bearing surfaces of worn metallic femoral heads

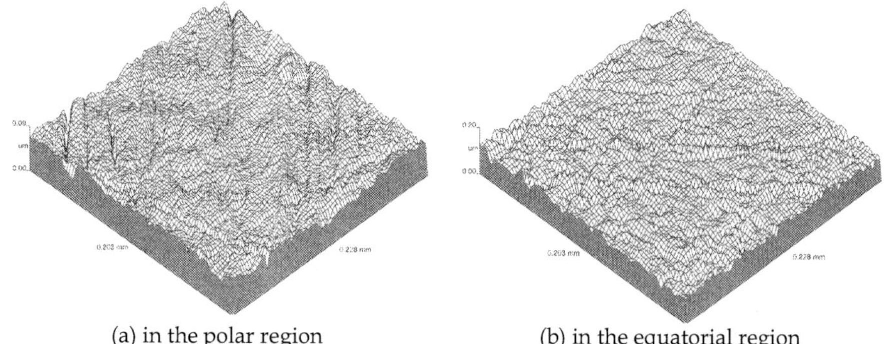

(a) in the polar region   (b) in the equatorial region

Fig. 8. Rough surfaces of worn metallic femoral heads with cut-off length $\lambda_c = 0.05mm$

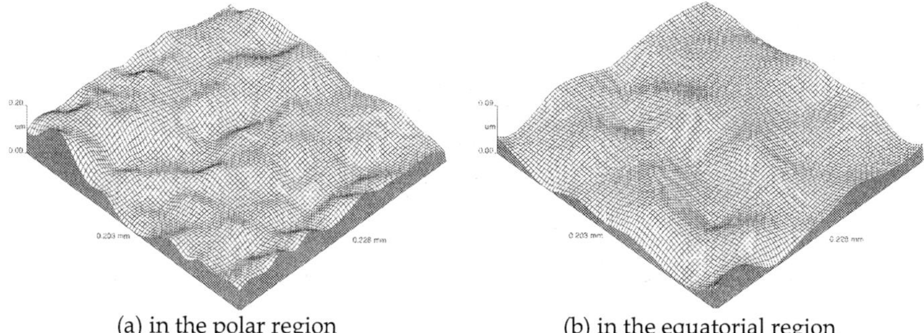

(a) in the polar region   (b) in the equatorial region

Fig. 9. Waviness surfaces of worn metallic femoral heads with cut-off length $\lambda_f = 0.10mm$

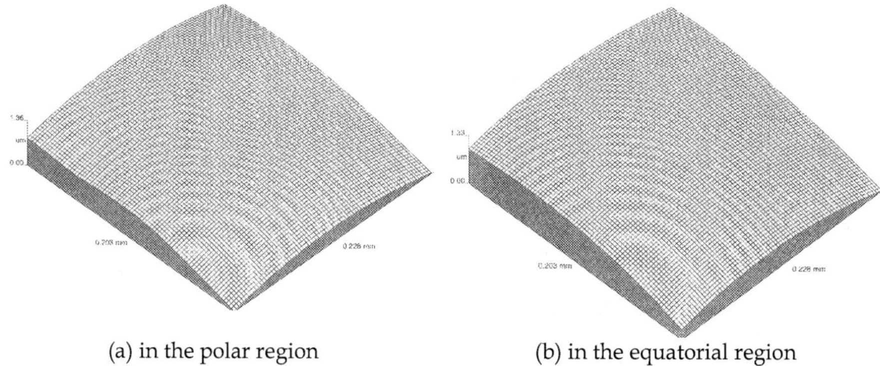

(a) in the polar region   (b) in the equatorial region

Fig. 10. Form error surfaces of worn metallic femoral heads

Whatever region the surface is measured, it is part of a ball, and so contains a very smooth spherical form. It also has two different kinds of scratches: regular and shallow scratches, possibly produced by the manufacturing processing; and random deeper scratches, resulting from functional performance in service. The latter scratches have a wider frequency band and higher amplitude, some with arc structures.

Using only one operation of wavelet filtering, the surface contents, roughness, waviness and form error can be detected and recognised. The outcomes of this are that rough, wavy and corresponding form error surfaces can be immediately separated and perfectly reconstructed within a flexible transmission bank. For instance, in order to indicate this transmission flexibility, the cut-off wavelengths of roughness may be selected as $\lambda_c = 0.05mm$. The cut-off length of waviness is limited by practical applications, in the above examples, $\lambda_f = 0.10mm$ is for the worn heads. From Figures 7-10 , it can be seen that the accuracy of the surfaces of these components cover the levels from micrometers to nanometers. There are no distortions caused by peaks/pits and scratches and boundaries, which can often be found in normal surface filtering.

The identification of the morphological features of surface topography of orthopaedic joint prostheses has been carried out in light of the Wavelet model outlined above. In the first set

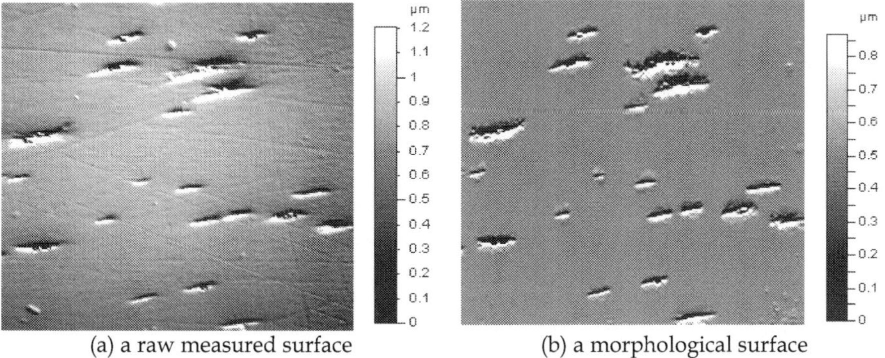

<table>
<tr><td>(a) a raw measured surface</td><td>(b) a morphological surface</td></tr>
</table>

Fig. 11. The measured and morphological surfaces of a worn ceramic head

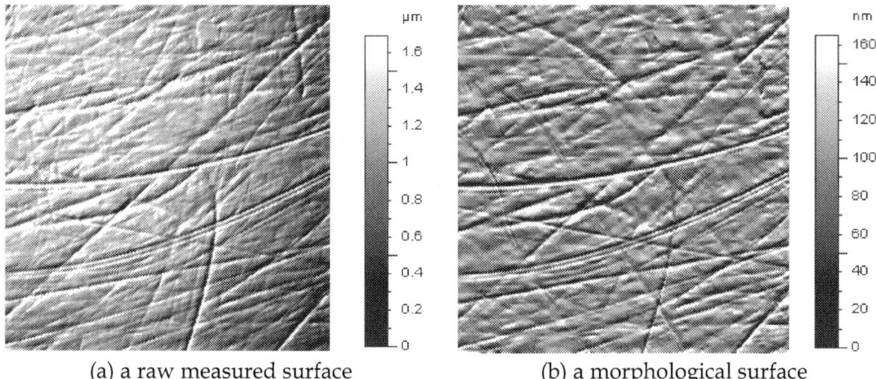

<table>
<tr><td>(a) a raw measured surface</td><td>(b) a morphological surface</td></tr>
</table>

Fig. 12. The measured and morphological surfaces of a worn metallic head

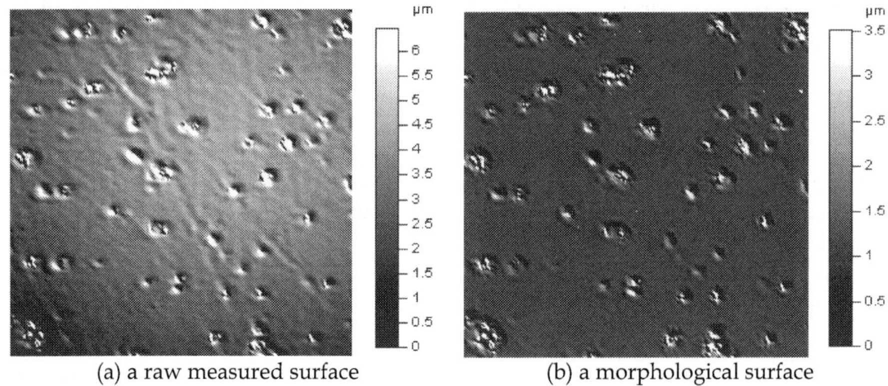

(a) a raw measured surface   (b) a morphological surface

Fig. 13. The measured and morphological surfaces of an unworn DLC head

of specimens, measurement datums of ceramic and metallic femoral heads were originally obtained by using a phase-shifting interferometer, with a 20X lens. The datum of the Diamond Like Carbon (DLC) specimen was obtained by employing a vertical-scanning interferometer using a 10X lens.

Figures 11-13(a) show lapped topographies in both the worn ceramic and metallic femoral heads with sampling area $300 \times 240 \mu m$. The worn ceramic surfaces look smooth and have some deep scratches. The worn metallic surfaces have two different kinds of scratches, regular shallow scratches, possibly produced by manufacturing processing and the random deeper scratches, resulting from functional performance in service. The latter scratches have a wider frequency band and higher amplitude, some with arc structures. The following Figure 13a is DLC surface with a morphological structure consisting of relatively large pits.

Figures 11-13(b) show how the Wavelet model has removed roughness, waviness and form deviation, revealing morphological surfaces. As illustrated, the morphological features are the dominant factors of the bearing surface structure of the new or wear heads, and roughness and waviness do not seem to heavily influence the functional performance of the head in service due to their relatively low levels. Examining these different surfaces (Figure.11-13), it can be seen that: (1) the components of waveforms of traces resemble each other with no relative phase shift in the sampling area; (2) the morphological information on the 3D bearing surfaces, conveyed by the wavelet model, is recorded completely and (3) with no running-in and running-out lengths. These morphological surfaces therefore have excellent refinement accuracy, which is suited to the need for assessment of a range of functional properties of the components and study of functional performance of bearing surfaces. The information and revealing the presence of these outliers, provided by Wavelet analysis can then be fed back to monitor manufacturing processes or study tribology and wear; or to study actual contact stress, loaded area and asperity volume, and additionally lubrication regimes occurring during the initial stages of wear.

4. The third generation complex wavelet model for surface feature extraction

The extraction of morphological features such as linear/curved scratches and plateaux with direction/objective properties require more consideration of how wavelets are used, for example: (1) *Automotive*. Plateau honed surfaces are produced by two machining processes. The final surface always includes the rough machining information (deep valleys). (2) *Biomedical engineering*. Surface topographies in different material heads are normally combined with morphological features from different wear stage in service, such as regular shallow scratches, random deeper scratches, arc scratches.

Morphological features in surface texture are not always of a large-amplitude and isolated. In many cases, morphological features have direction properties; they mix with the harmonic roughness and waviness and have similar amplitude scales and frequency bands. The first and second generation wavelet methods for extraction of such events in the surface texture is less effective at extracting these features. The reason is that the first and second generation wavelet models are used in there maximally decimated real discrete wavelet transform form, and lack the shift invariance property. This means that small shifts of surface signal (step height, linear and curve-like features) can cause large variations in the distribution of energy between real wavelet transform coefficients at different scales. The directional morphological features from these wavelet techniques are also destroyed by edge artifacts. Consequently, under the first two generation wavelet models, the accurate extraction of direction/objective morphological features is not efficient

A new method for extraction of direction/objective morphological features of surfaces has been proposed using a biorthogonal dual tree complex wavelet transform (DTCWT) [30,31]. It attempts to give affine invariance, with independence of the reference frame for the measurements, and also perfect reconstruction, limit redundancy and have efficient computation.

4.1 Theory of DT-CWT

For the extraction of direction/objective morphological features of surfaces, the DT-CWT attempts to give affined invariance, with independence from the reference frame for the measurements, and also perfect reconstruction, limit redundancy and efficient computation.

The complex wavelet model is based on the Z-transform theory of linear time invariant sampled systems. The main approach of DT-CWT operation is to use two-channel filter banks in the real and imaginary parts as shown in Figure 14. A Q-shift filter design technique is used to construct the lowpass FIR filter to satisfy a linear-phase and perfect reconstruction condition, as is demonstrated below [28-29]:

$$\begin{cases} H_{L2}(z) = H_L(z^2) + z^{-1}H_L(z^{-2}) \\ H_L(z)H_L(z^{-1}) + H_L(-z^{-1})H_L(-z) = 2 \end{cases} \tag{34}$$

The one-dimensional complex wavelet decomposes a signal $f(x)$ in terms of a complex shifted and dilated mother wavelet:

$$f(t) = \sum_{l \in Z} a_{J,l}\phi_{J,l}(x) + \sum_{j \leq J}\sum_{l \in Z} d_{j,l}\psi_{j,l}(x) \tag{35}$$

where $\phi_{J,l} = \phi_{J,l}^r + \sqrt{-1}\phi_{J,l}^i$, $\psi_{j,l} = \psi_{j,l}^r + \sqrt{-1}\psi_{j,l}^i$. J denotes the coarsest scale can be calculated by: $J = \text{int}\left(\log_2(\lambda_c / \Delta x) - 1\right)$, where λ_c is the cut-ff wavelength, and Δx is the sampling spacing. The $\psi_{j,l}^r$ and $\psi_{j,l}^i$ are real wavelets. $a_{j,l}$ and $d_{j,l}$ refer to the smooth coefficients at level J

and the detail components of coefficients at the level j, respectively. The complex wavelet transform is essentially a combination of two different real wavelet transforms, in one dimension, the $\{\phi_{j,l}^{r}, \phi_{j,l}^{i}, \psi_{j,l}^{r}, \psi_{j,l}^{i}\}$ can be interpreted as a wavelet tight frame with a redundancy factor of two. The real and imaginary parts of the complex wavelet transform are computed using separate filter bank structures in different trees.

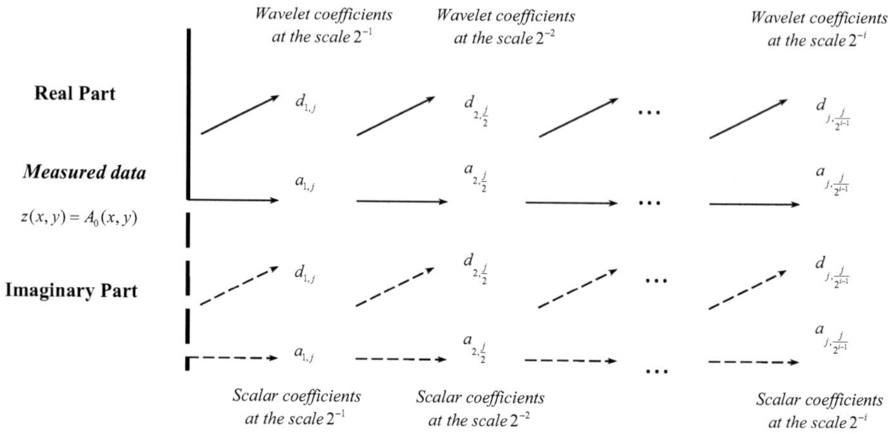

Fig. 14. Framework of the dual tree complex wavelet transform (DT-CWT)

For two dimensions, the decomposition of complex wavelet transform can be represented by the dilations and translations of a complex scaling function and six complex wavelet functions:

$$f\left(s_{1}, s_{2}\right) = \sum_{k \in Z^{2}} a_{J,k} \phi_{J0,k}\left(s_{1}, s_{2}\right) + \sum_{b \in B} \sum_{j \leq J} \sum_{k \in Z^{2}} d_{j,k}^{b} \psi_{j,k}^{b}\left(s_{1}, s_{2}\right) \tag{36}$$

where, $B = \{\pm 15°, \pm 45°, \pm 75°\}$ for six subbands' directions. For the separable two dimensional complex wavelet transform based on one dimensional complex ϕ, ψ, we have [28-30]:

$$\begin{cases} \psi^{+15^0} = \phi(s_1)\psi(s_2), & \psi^{-15^0} = \phi(s_1)\overline{\psi}(s_2) \\ \psi^{+45^0} = \psi(s_1)\psi(s_2), & \psi^{-45^0} = \psi(s_1)\overline{\psi}(s_2) \\ \psi^{-75^0} = \psi(s_1)\phi(s_2), & \psi^{-75^0} = \psi(s_1)\overline{\phi}(s_2) \end{cases} \tag{37}$$

where $\overline{\phi}, \overline{\psi}$ is the complex conjugate of ϕ, ψ. Similar to the multi-resolution analysis of real wavelets, the above decomposition can be summarized as a decomposition of L^2 space into a series of subspaces $\{W_1, W_2, \cdots, W_J, V_J\}$, here W_j and V_j respectively denote the wavelet bandpass subspace and the scale lowpass subspace at level j. They satisfy [28-30]:

$$V_{j-1} \subset V_j = V_{j+1} \oplus W_j \ , \ W_j \perp W_{j'}, j \neq j' \tag{38}$$

4.2 Transmission characteristics of DT-CWT

To extract different frequency components such as the roughness, waviness, form and form error accurately, the surface filter requires a near ideal transmission characteristics which include the zero or linear phase transmission and a 'steep transmission curve' amplitude transmission.

4.2.1 Phase transmission characteristics

A zero phase filter tends to preserve the shape of the surface components in the pass band region of the filter. This is very important for applications, as surface topographies are often composed of a wide range of frequency components. Zero or linear phase transmission ensures that the filtering results have no phase distortion and ensures the proper registration of different frequency components that make up these features. A digital filter has zero phase when its frequency response $H(\omega)$ is a real function so that $H(\omega) = H^*(\omega)$. From the symmetry property of the Fourier transform and considering that only real impulse response $h(n)$ is applied to real input to obtain real output, the necessary and sufficient condition for a zero phase filter is: $h(n) = h(-n)$.

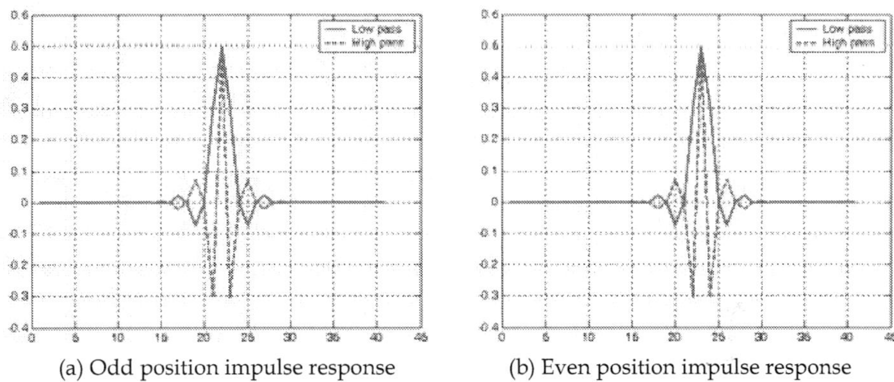

(a) Odd position impulse response   (b) Even position impulse response

Fig. 15. DT-CWT impulse response with 2 levels of decomposition

The difference between zero phase and linear phase property of filter lies only in the shift of the co-ordinate system origin which has no practical significance. From Eq.8 we know that to get zero or linear phase property, the filter impulse response or the filter weighting function should be symmetric. Figure 15 shows the impulse response of a DT-CWT low pass and high pass filter. From the figures it is clear that both the odd position and the even position impulse response have symmetrical shape, and the odd and even position impulse responses are very similar. For the real DWT filter, the odd and even position impulse responses are different. Therefore, the DT-CWT can produce the shift-invariance property with zero/linear phase, whereas the DWT can not produce the shift-invariance property [31].

4.2.2 Amplitude transmission characteristics

For surface analysis, filters need a 'steep transmission curve', which requires that the low-pass filter with a steep cut-off such that a negligible amount of higher frequency information passes and vice versa for high-pass filter. The "steep transmission curve" property is good for separation of different frequency components with a sharp attenuation at the cut-off wavelength, which leads to more precise separation of roughness, waviness and form components with no aliasing, and leads to the good feature location property.

The Gaussian filter is designed to have 50% transmission at the cutoff frequency. The high pass filter and low pass filter have complementary relations, which allow the calculation of waviness by simply subtracting the roughness from the form removed raw profile.

The DT-CWT also has very good amplitude transmission characteristics compared to the Gaussian filter. Figure 16 shows the highpass and lowpass amplitude transmission characteristics and the subband transmission characteristics of the first four decomposition levels. Figure 16(a) is the inverse transform of the level one DT-CWT transform of an impulse signal, and shows that the highpass and lowpass filter are intersected at the normalised frequency (frequency divided by the maximum frequency so it is normailised to the range (0,1)) 0.5 with 50% amplitude transmission. Accordingly, for an original input bandpass signal, similar to the Gaussian filter, 50% transmission at the cutoff frequency allows the calculation of waviness by simply subtracting the roughness from the raw profile. The DT-CWT has the "steep transmission curve" property so that it can separate out different frequency components effectively. Figure 16(b) shows the transmission characteristics of each of the four subbands levels from the DT-CWT decomposition. This shows that every adjacent subband has 50% transmission rate at the intersection frequency [31].

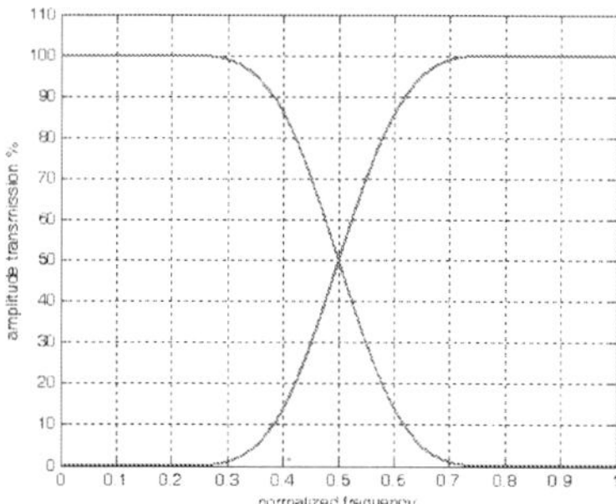

(a) Transmission characteristics

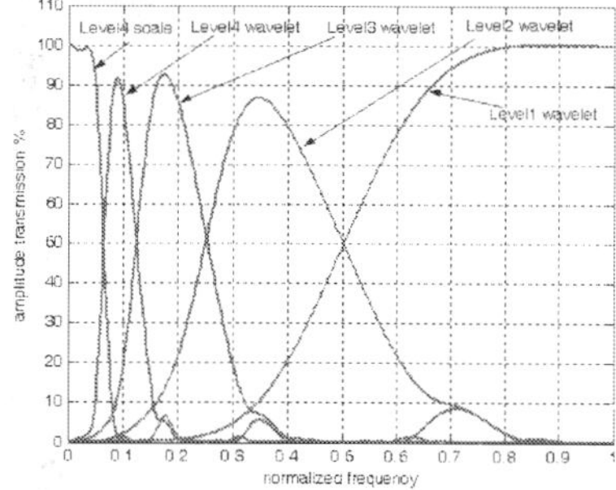

(b) Subband transmission characteristics

Fig. 16. Transmission characteristics of a DT-CWT (using normalized frequency)

Figure 17 compares the transmission characteristics of the DT-CWT and ISO Standard Gaussian filter using the same cutoff frequency. The continuous lines are the results from the DT-CWT and the dashed lines are the results from the Gaussian filter. Although there exists some very little and negligible ripples in the pass band, it is clear that the DT-CWT always have a steeper transmission curve than the Gaussian filter near the cutoff frequency. Therefore the DT-CWT has a better frequency resolution than the Gaussian filter.

Considering both the zero/linear phase property and the 'steep transmission curve' of the amplitude transmission characteristic, the DT-CWTs are suitable for the separation of different surface topographical features. Besides the separation of linear/curve-like features, the DT-CWT is also very good for the separation of roughness, waviness, form and form error such surface frequency components.

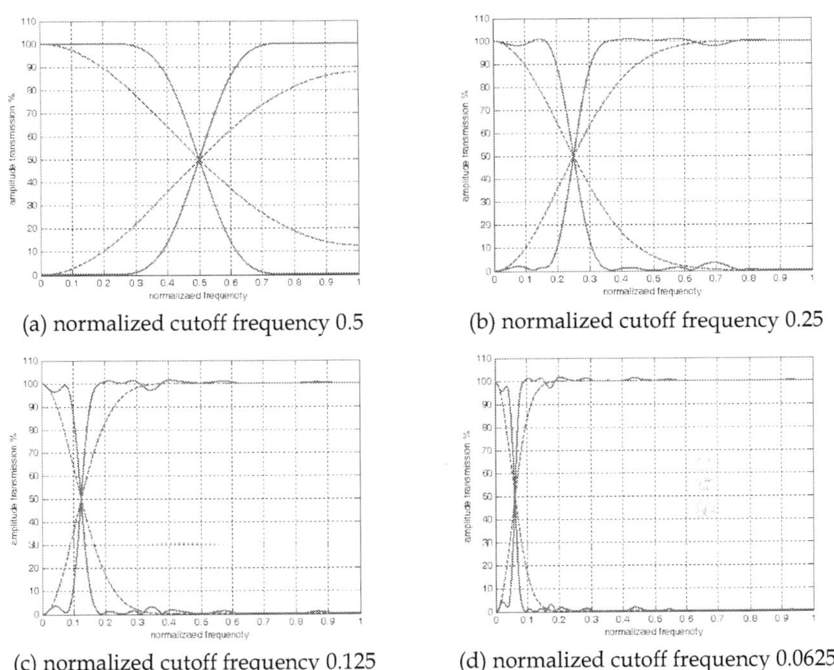

(a) normalized cutoff frequency 0.5   (b) normalized cutoff frequency 0.25

(c) normalized cutoff frequency 0.125   (d) normalized cutoff frequency 0.0625

Fig. 17. Comparison of the high pass and low pass transmission characteristics of the CWT and the Gaussian filter (continuous line: DT-CWT filter; dashed line: Gaussian filter)

4.2.3 Shift-invariant property

A linear feature and a circular feature have been simulated, as shown in Figure 18 and Figure 19. Figure 18(a) is a simulated 3D groove feature that is neither parallel nor vertical to the measurement direction. Figure 18(b) and (c) shows the results processed by a complex wavelet transform and a real wavelet transform respectively. From left to right in Figure 4.2, the decomposed components of the input signal are reconstructed by the complex wavelet coefficients at scalar level 1, 2 and 3 and the scaling function coefficients at scalar level 3 correspondingly. From the results, it can be seen that in each scale, the extracted feature from the input signal using the complex wavelet has a very smooth edge along the groove and the same amplitude energy distribution in every position. Comparison with Figure 18(b), the extracted features using the real wavelet as shown in Figure 18(c) have a significant shift aliasing, with irregular edges along the groove feature [28-31].

From new the wavelet model shown in the above section, it can be seen that the dual tree complex wavelet offers six bandpass components of complex coefficients at each level, which are strongly oriented at angles of $\pm 15^{0}$, $\pm 45^{0}$ and $\pm 45^{0}$. The special point of this kind of complex wavelet is that it provides a true directional selectivity for curve-like features. Figure 19(a) is a simulated 3D circular valley feature, and Figure 19(b) and (c) 6 shows the outcomes as reconstructed from complex wavelet and real wavelet coefficients respectively at scalar levels 1, 2 and 3.

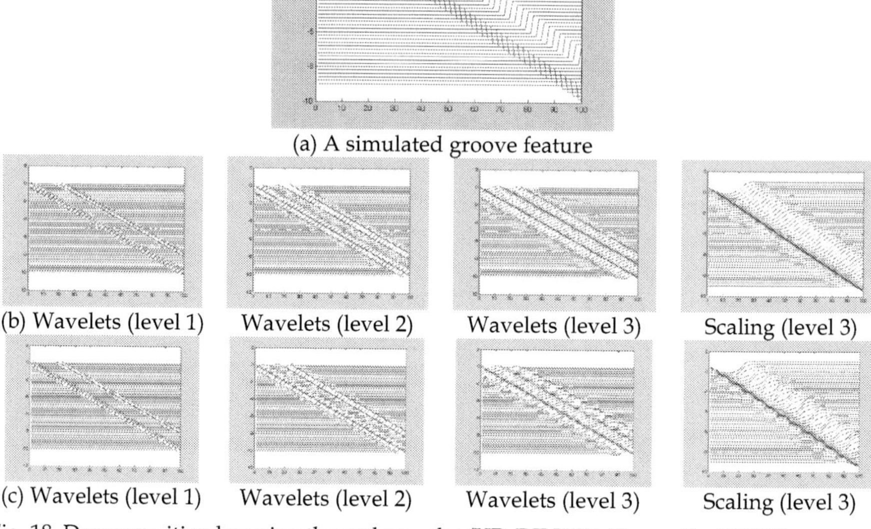

(a) A simulated groove feature

(b) Wavelets (level 1)   Wavelets (level 2)   Wavelets (level 3)   Scaling (level 3)

(c) Wavelets (level 1)   Wavelets (level 2)   Wavelets (level 3)   Scaling (level 3)

Fig. 18. Decomposition by using the real wavelet (UP: DTCWT, Down: Real DWT)

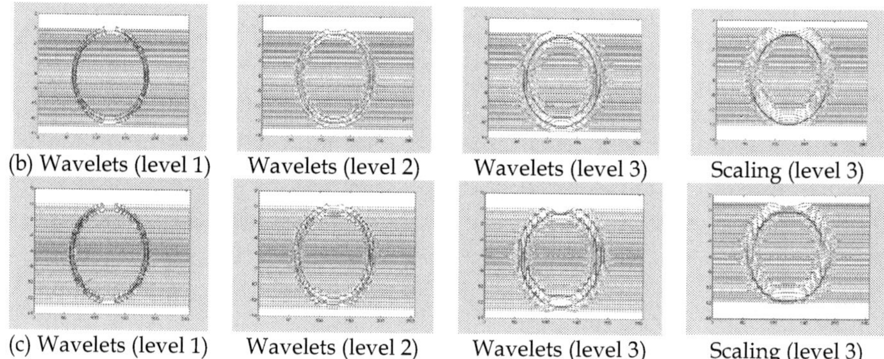

(a) A simulated 3D circular valley feature

(b) Wavelets (level 1)   Wavelets (level 2)   Wavelets (level 3)   Scaling (level 3)

(c) Wavelets (level 1)   Wavelets (level 2)   Wavelets (level 3)   Scaling (level 3)

Fig. 19. Decomposition by using the real wavelet (UP: DTCWT, Down: Real DWT)

4.3 Case studies

A group of typical examples of surfaces from engineering manufacture have been selected to show the metrological characteristics of the DT-CWT.

Figure 20 shows a typical peripheral milling surface profile filtered by the DT-CWT and the Gaussian filter with different cutoff wavelengths (0.64mm, 2.56mm). Both the DT-CWT and the Gaussian filter can extract the roughness mean line very well with no phase distortion. But the DT-CWT filtering results are smoother than the Gaussian filtering results, which shows that the DT-CWT filter has a steeper amplitude transmission characteristic than the Gaussian filter. Thus the DT-CWT has a better ability to separate different frequency components. Figure 21 and Figure 22 are grinding and turning surface profiles respectively. From the filtering results they also illustrate the good transmission characteristics including phase and amplitude properties of the DT-CWT.

The metrological characteristics of the DT-CWT investigated above can also be extended to areal surface analysis. Figure 23 is a worn hip joint prosthesis surface, the measured raw

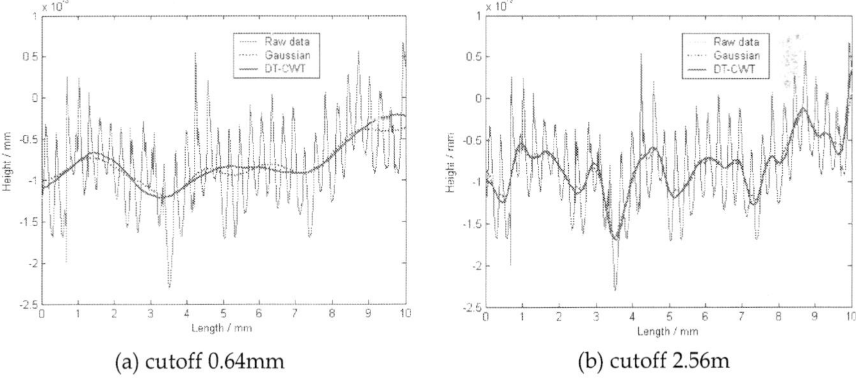

(a) cutoff 0.64mm   (b) cutoff 2.56m

Fig. 20. Milling surface profile filtered by DT-CWT and Gaussian filter

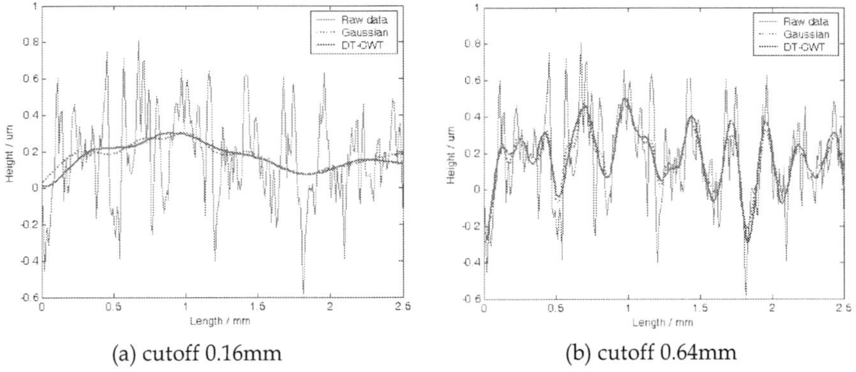

(a) cutoff 0.16mm   (b) cutoff 0.64mm

Fig. 21. Grinding surface profile filtered by DT-CWT and Gaussian

data includes waviness and form/form error that was introduced by designed shape and measurement error. By using DT-CWT the form, waviness and roughness have been successfully extracted and separated. It is also very clear that the roughness and waviness surfaces have no distortion.

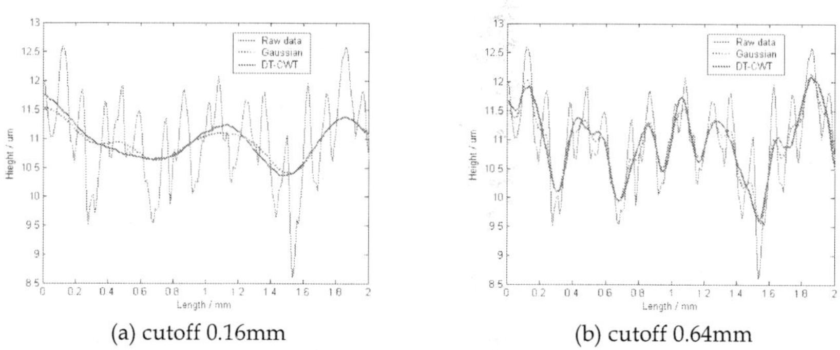

(a) cutoff 0.16mm

(b) cutoff 0.64mm

Fig. 22. Turning surface profile filtered by DT-CWT and Gaussian filter

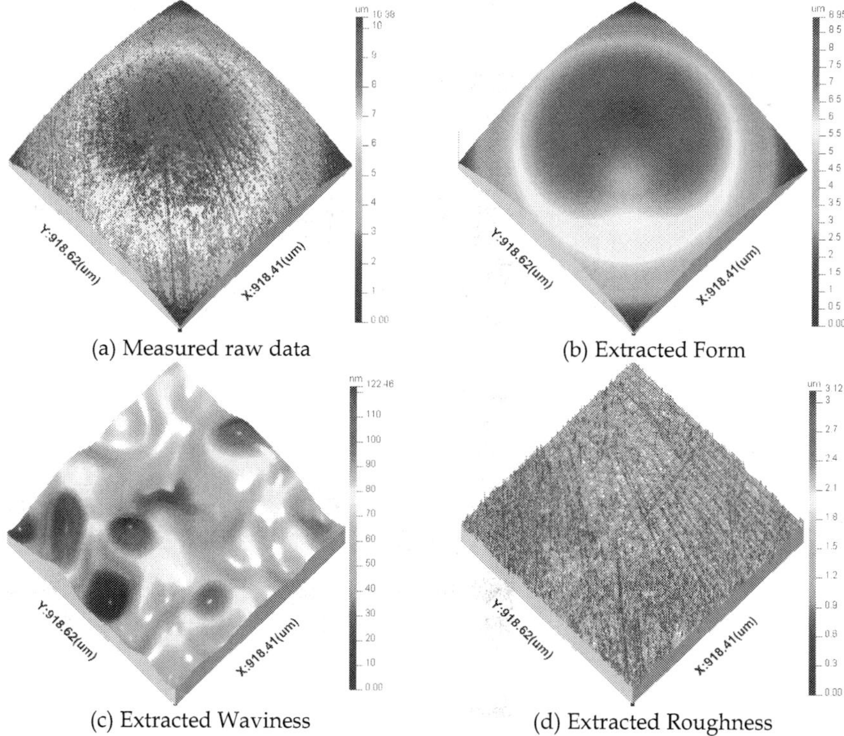

(a) Measured raw data

(b) Extracted Form

(c) Extracted Waviness

(d) Extracted Roughness

Fig. 23. Separation of the different surface components

The above research has shown that shift-invariance is a very important property. In general, the shift-invariance is independent of the reference frame for a surface measurement. In particular it is important for many approximation and filtering methods.

To verify the special properties of the third generation wavelet model, a group of typical examples of engineering and bioengineering surfaces are shown in Figures 4.7a ~ 4.10a, in which some significant features have been selected. These will be used to demonstrate the performance of the dual tree complex wavelet transform.

The first example shows two measured surface from different positions of the same worn hip-joint. Figure 24(a) (b) are raw measured surfaces, including the form, form deviation, waviness and roughness components. Figure 24(c) and (d) are the reconstroned feature surface from (a) and (b) respectively. the deep valleys are reconstructed using complex wavelets in which not only the form and some harmonic components are removed very effectively, but also the shape of the scratches are retained very well. There is no affine aliasing, no new artifacts and the edge of the deep vallys are preserved perfectly. The filtering results help to evaluate the worn rate of the hip joint at different position and further can help to improve the design.

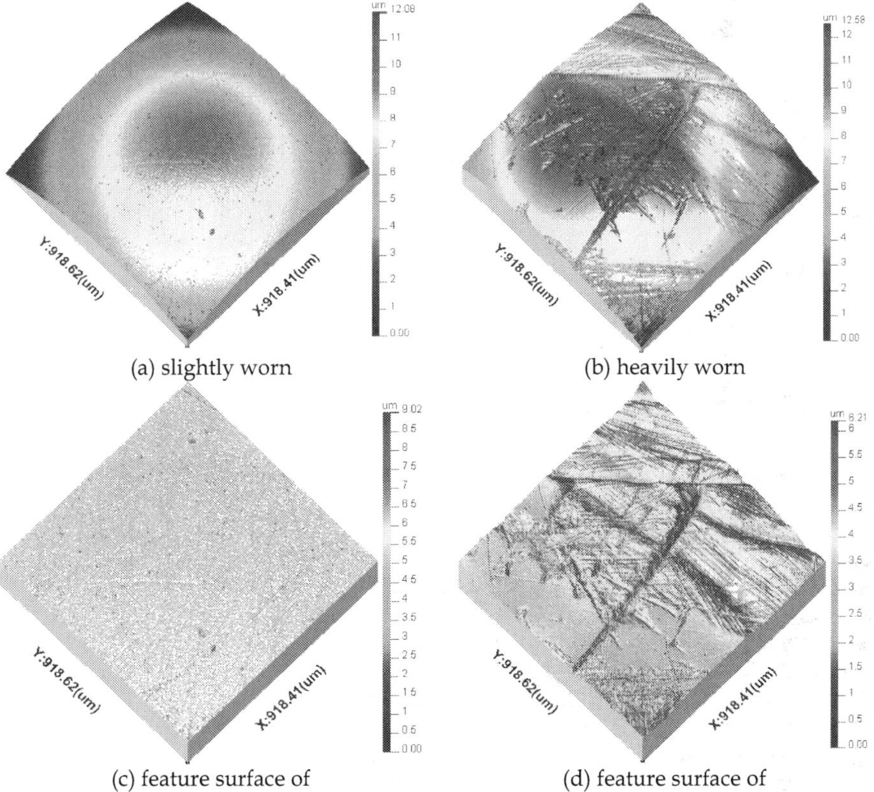

(a) slightly worn   (b) heavily worn

(c) feature surface of   (d) feature surface of

Fig. 24. Cylinder surface analysis using the complex wavelet

The following examples in Figure 25(a) to Figure 27(a) are a series of surfaces from different material femoral heads. Figure 25(a) shows a lapped topography surface in a worn metallic femoral heads. The worn metallic surfaces have two different kinds of scratches, regular shallow scratches, possibly produced by manufacturing processing and the random deeper scratches, resulting from functional performance in service. The latter scratches have a wider frequency band and higher amplitude, some with arc structures. Figure 26(a) is a surface topography from a diamond-like-carbon head with a morphological structure consisting of relatively large pits and a deep scratch. Figure 27(a) is a new ceramic surface which looks smooth and has some shorter and deeper scratches.

Looking at these surfaces, it must be noted that the most important factors that affect the life of the femoral head are the isolated peaks/pits and scratches rather than the nano-scalar roughness. From the function assessment point of view, the isolated peaks/pits and scratches will significantly interfere with the wear mechanics and tribological properties[6-9,27].

Using the complex wavelet, all the isolated features are extracted precisely as shown in Figure 25(b) to Figure 27(b), and the shape of the features are reconstructed very well and all the features' edges are preserve perfectly.

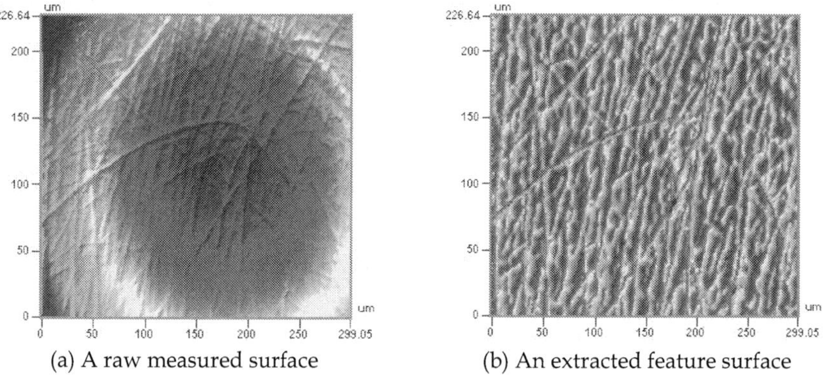

    (a) A raw measured surface   (b) An extracted feature surface

Fig. 25. Worn metallic surface analysis using the complex wavelet

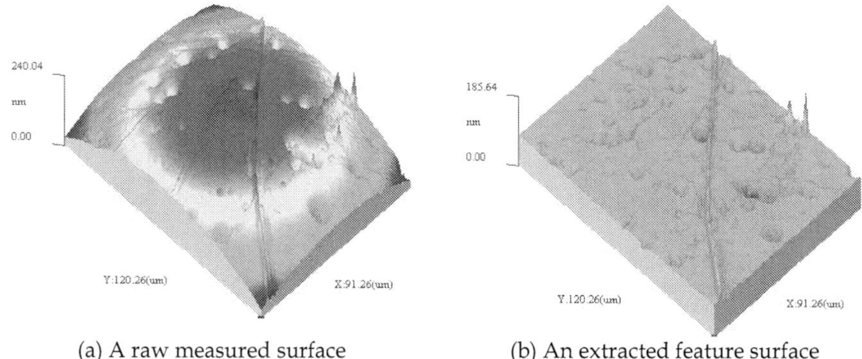

    (a) A raw measured surface   (b) An extracted feature surface

Fig. 26. Diamond like carbon surface analysis using the complex wavelet

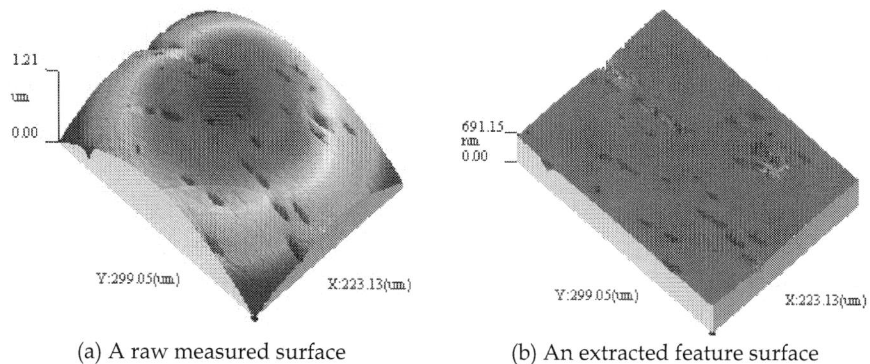

(a) A raw measured surface   (b) An extracted feature surface

Fig. 27. Ceramic surface analysis using the complex wavelet

5. Complex ridgelet transform

Wavelets have the good performance for piecewise smooth functions in one dimension, i.e., fully efficient at representing point-like isolated morphological features. Unfortunately, such is not the case for higher dimensional features because wavelets ignore the geometric properties of objects. So the function of wavelet techniques for extracting line/curve features is limited.

To extend the directional sensitivity for wavelet analysis, an approach based on a combination of the wavelet and Radon transform is used. This was initially developed for edge detection and compression in the field of signal processing.

Ridgelets were introduced by Candès and Donoho to deal effectively with line singularities by mapping a line singularity into a point singularity through the Radon transform, and then using the wavelet transform on each projection in the Radon transform domain [32-34]. The weakness of the traditional ridgelets transform is the lack of shift invariance due to the real DWT used.

5.1 Complex ridgelet transform

A bivariate complex ridgelet $\psi_{a,b,\theta}^c$ in R^2 can be defined as [32-37]:

$$\psi_{a,b,\theta}^c(x) = a^{-1/2}\psi^c\left((x_1\cos\theta + x_2\sin\theta - b)/a\right) \tag{39}$$

Here, $a > 0$ is a scale parameter, θ is an orientation parameter, and b is a location scalar parameter. This function is constant along lines $x_1\cos\theta + x_2\sin\theta = const$, while its transverse is a complex wavelet $\psi^c = \psi^r + \sqrt{-1}\psi^i$. Here, ψ^r and ψ^i are themselves real wavelets. If the real and imaginary part of complex wavelet can be viewed as two 'fat' points, then the complex ridgelet can be interpreted as two 'fat' lines so that it is specially adaptive to analysis the line ridge/valleys of surface topography. Figure 28 shows a real ridgelet function.

The continuous Complex Ridgelet Transform (CRIT) for an integrable bivariate function $f(x) \in L^2(R^2)$ is defined as:

$$\Re_f(a,b,\theta) = \int \psi^c_{a,b,\theta}(x) f(x) dx \tag{40}$$

The reconstruction formula is given as:

$$f(x) = \int_0^{2\pi} \int_{-\infty}^{\infty} \int_0^{\infty} \Re_f(a,b,\theta) \psi^c_{a,b,\theta}(x) \frac{da}{a^3} db \frac{d\theta}{4\pi} \tag{41}$$

Point and line singularities are related by the Radon transform. By applying the 1-D DT-CWT on the projections of the Radon transform the complex ridgelet transform can be rewritten as:

$$\Re_f(a,b,\theta) = \int R_f(\theta,t) a^{-1/2} \psi^c \left((t-b)/a \right) dt \tag{42}$$

The Radon transform is denoted as:

$$R_f(\theta,t) = \int f(x_1,x_2) \delta(x_1 \cos\theta + x_2 \sin\theta - t) dx_1 dx_2 \tag{43}$$

Where, δ is the Dirac delta function.

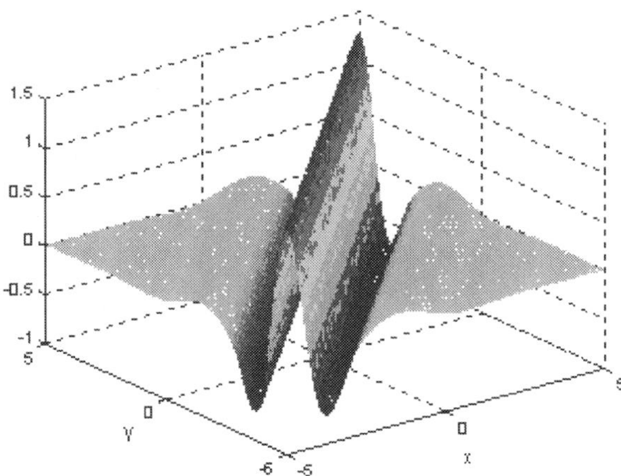

Fig. 28. A typical ridgelet function $\psi_{a,b,\theta}(x,y)$

5.2 Digital complex ridgelet transform

From above equations, one can see that the basic strategy for computing the CRIT is to first calculate the Radon transform $R_f(\theta,t)$, then to calculate the 1D-CWT of the projections $R_f(\theta, \cdot)$. For the calculation of the Radon transform, numerous digital methods have been devised. However, most of them were not designed to be invertible transforms for digital surfaces or images. Alternatively, the finite Radon transform (FRAT) theory provided an interesting solution for finite length signals. According to the practical requirements of surface characterisation, we use the digital form of the CRIT based on the FRAT and DT-CWT.

5.2.1 Finite Radon Transform [34,38-39]

The FRAT is defined as summations of surface pixels over a certain set of lines. These lines are defined by a finite geometry in a similar way as the lines for the continuous Radon transform in Euclidean geometry. Denote the group $G = Z_p^2$ to be the Cartesian product $Z_p \times Z_p$ of two exemplars from the cyclic group $Z_p = \{0,1,\cdots,p-1\}$ with addition modulo p, where p is a prime number, and let $N = \{0,1,\cdots,p\}$.
This group has $p+1$ nontrivial subgroups

$$
\begin{aligned}
H_i &= \{(k,l) \in G;\ li = k \ (\bmod p)\}, \quad 0 \leq i < p \\
H_p &= \{(k,0) \in G;\ k \in Z_p\},
\end{aligned}
\tag{44}
$$

The cosets of the factor group G/H_i are indexed by $j \in Z_p$ in the following way

$$
\begin{aligned}
H_i^j &= \{(k,l) \in G;\ li + j = k \ (\bmod p)\}, \quad 0 \leq i < p \\
H_p^j &= \{(k,j) \in G;\ k \in Z_p\}.
\end{aligned}
\tag{45}
$$

The Radon projection of a function f on G is given by

$$
\Lambda_i f\left(H_i^j\right) = \frac{1}{p}\sum_{k=0}^{p-1}\sum_{l=0}^{p-1} f(k,l)\delta_j\left(\pi_i(k,l)\right), \qquad i \in N, j \in Z_p
\tag{46}
$$

Here, the function π_i is a variance of the factor mapping of G on G/H_i. We have

$$
\begin{aligned}
\pi_i(k,l) &= k - li \ (\bmod p), \quad 0 \leq i < p \\
\pi_p(k,l) &= l
\end{aligned}
\tag{47}
$$

For surface topography or for an image, the coset H_i^j denotes the set of points that make up a line on the lattice G. Particularly H_0^j and H_p^j denote the horizontal and vertical lines respectively. π_i denotes the set of lines that go through a point $(k,l) \in G$. As in the Euclidean geometry, the line H_i^j on the affine plane G is uniquely represented by its slope $i \in N$ and its intercept $j \in Z_p$. H has p^2 points, $p^2 + p$ lines, every point $(k,l) \in G$ lies on $p+1$ lines, every line contains p points. Moreover, any two distinct points on G lie on just one line. For any given slope $i \in N$, there are p parallel lines to provide a complete cover of the plane G.
From the finite geometry property, for zero-average functions, the forward and inverse formula can be written as:

$$
F = \Lambda_i f = \frac{1}{\sqrt{p}} \sum_{(k,l) \in H_i^j} f(k,l) \quad = \frac{1}{\sqrt{p}} \sum_{(k,l) \in G} f(k,l)\delta_{H_i^j}(k,l) \quad = \left\langle f, \frac{1}{\sqrt{p}}\delta_{H_i^j} \right\rangle
$$

$$
f = V_i F = \frac{1}{\sqrt{p}} \sum_{(i,j) \in \pi_i(k,l)} F(i,j) \quad = \frac{1}{p} \sum_{(i,j) \in \pi_i(k,l)} \sum_{(k',l') \in H_i^j} f(k',l')
\tag{48}
$$

$$
= \frac{1}{p} \sum_{(k',l') \in G} f(k',l') + f(k,l) = f(k,l)
$$

Where $\delta_{H_i^j} : G \to R$ is the Dirac delta function, defined as:

$$\delta_{H_i^j}(k,l) = \begin{cases} 1 & (k,l) \in H_i^j \\ 0 & \text{elsewhere} \end{cases}$$

The inverse FRAT algorithm has the same structure and is symmetric with the algorithm of the forward FRAT.

5.2.2 Digital Ridgelet Transform

The complex wavelet basis is defined as:

$$\{w_m^{(i)}, \quad m \in Z_p\}, \quad i \in N \tag{49}$$

The digital FRIT can be integrated as:

$$CFRIT_f(i,m) = \langle \Lambda_i f, w_m^{(i)} \rangle = \sum_{j \in Z_p} w_m^{(i)}(j) \left\langle f, \frac{1}{\sqrt{p}} \delta_{H_i^j} \right\rangle = \left\langle f, \frac{1}{\sqrt{p}} \sum_{j \in Z_p} w_m^{(i)}(j) \delta_{H_i^j} \right\rangle \tag{50}$$

Therefore the basis functions of the discrete complex ridgelet transform can be written as

$$\rho_{i,m} = \frac{1}{\sqrt{p}} \sum_{j \in Z_p} w_m^{(i)}(j) \delta_{H_i^j} \tag{51}$$

Here, although the $\left\{ \frac{1}{\sqrt{p}} \delta_{H_i^j} \right\}$ is not an orthonormal system, if we take the $p+1$ orthonormal bases for $l^2(Z_p)$, $\{w_m^{(i)}, \ m \in Z_p\}$ with $w_0^{(i)} \equiv const$, the system $\{\rho_{i,m} : i = 0, \cdots, p; m = 1, \cdots, p-1\} \cup \{\rho_0\}$ is an orthonormal base for $l^2(Z_p)$ where $\rho_0(k,l) = 1/p, \forall (k,l) \in G$.

Figure 29 gives the basic procedure when using the proposed CFRIT to analysis surface.

5.3 Case studies

Figure 30 shows 16 shifted versions of the image (at the top) and their subspace reconstructed components in turn from the coefficients at levels $j \le j_0 = 4$ using the CFRIT (left) and real FRIT (right). In order to see the effects clearly, only the centre of the profiles of these images is shown. Each shift is displaced down a little to give a waterfall style display. The output of CFRIT is the modulus of the complex coefficients. Note that summing these components the input image can be reconstructed perfectly. Good shift invariance is seen from the fact that the shape and amplitude of each of the reconstructed components by CFRIT hardly varies as the input is shifted. In contrast, the reconstructed components using FRIT vary considerable with each shift.

Our next test shows the good performance of CFRIT for denoising an image with line singularities. We consider an artificial image with a deep scratch that is contaminated by an additive zero-mean Gaussian white noise of variance σ^2. The denoising includes the following steps: (1) Transform the noisy image using CFRIT; (2) Hard-thresholding of the

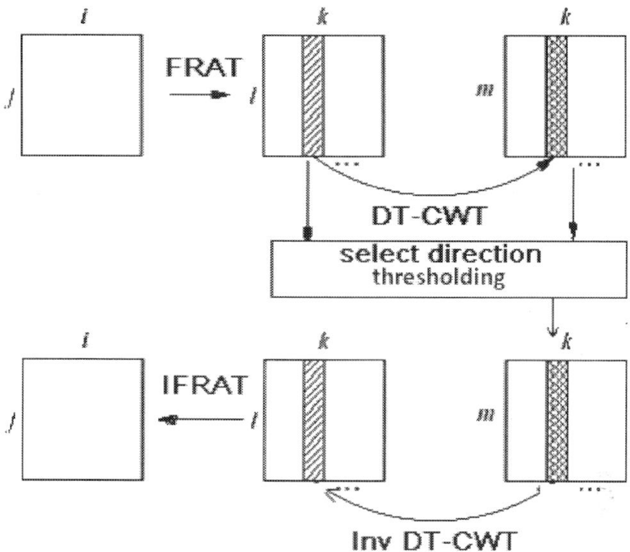

Fig. 29. Directional feature extraction

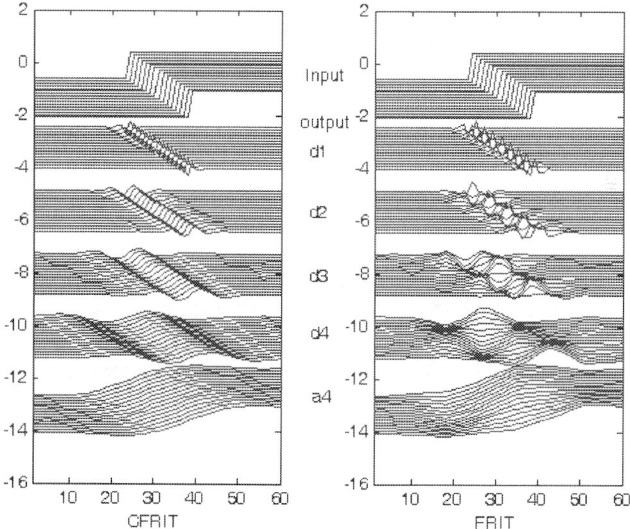

Fig. 30. The reconstructed components (centre profiles) at levels 1 to 4 of 16 shifted image with a stepped edge using the CFRIT (left) and real FRIT (right). Each shift is displaced down a little to give a waterfall style of display.

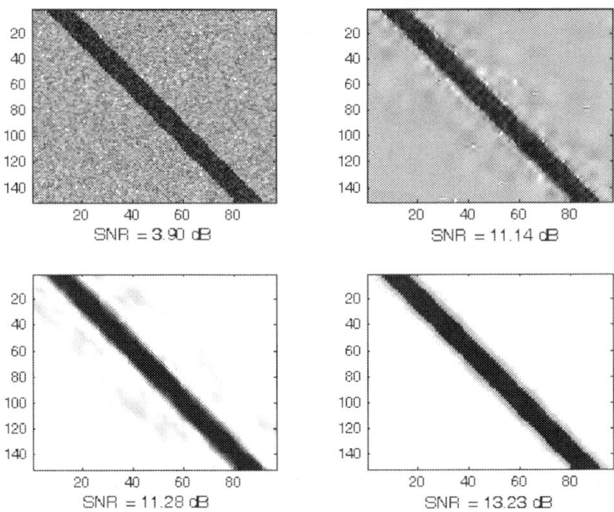

Fig. 31. Denoising an image with line singularities using the DWT (upper right), DT-CWT (lower left), and CFRIT (lower right). The upper left image is a noisy image.

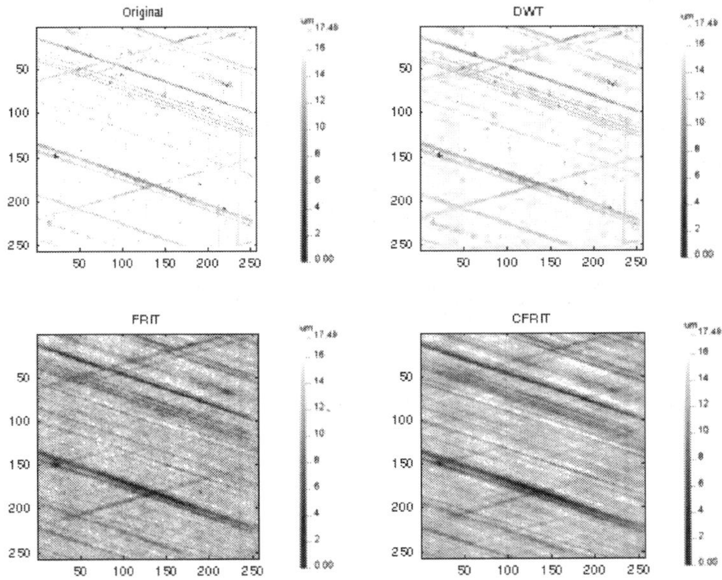

Fig. 32. Feature extraction of hone surface using the DWT (upper right), FRIT (lower left), and CFRIT (lower right). The upper left image is the original data (form removed).

coefficients using the universal threshold $T = \sigma\sqrt{2\log N}$ (where N is the number of pixels). (3) Reconstructing the thresholded coefficients. For comparison, the same uniform threshold value is also applied to the DWT and DT-CWT based algorithms.

It can be seen from Figure 31 that the CFRIT is effective in recovering straight edges, as well as in terms of the signal to noise ratio (SNR). The CFRIT reconstruction does not contain the undesirable artefacts along edge that one finds in the wavelet reconstruction. The simple thresholding scheme for CFRIT is effective in denoising the piecewise smooth image with line singularities. This is because the linear singularities are represented by a few significant coefficients in the CFRIT domain, whereas random noisy singularities are unlikely to produce the similar amplitude coefficients. The DT-CWT is relatively superior to the DWT in reducing the artefacts due to its shift invariance and good directional selectively. From the view of the hybrid approach, the CFRIT just combines the multiresolution analysis of DT-CWT with the anisotropy of the Radon transform.

Figure 32 shows a honed surface from an engine cylinder. As is well know, the most important features that influence the performance of cylinders are the deep scratches, the distribution and amplitude will considerably influence the flow of gas or air in the pressure balance of an engine. It can be seen that the DWT still exhibits numerical embedded blemishes (i.e., pits/peaks) in the extracted honed surface. Setting higher thresholds to remove these would cause even more of the intrinsic linear scratches to be destroyed or missed. In addition, the non-smooth aliasing along the scratches is clearly visible. In the FRIT result, although the pits/peaks have been removed efficiently, the edges of the linear scratches are not very smooth with aliasing due to the lack of the shift invariance property. In the CFRIT result, not only are the peaks/pits in the honed surface removed effectively, but also the shapes of the extracted scratches are well retained. There is also no affine aliasing, and the edges of the deep valleys are preserved perfectly.

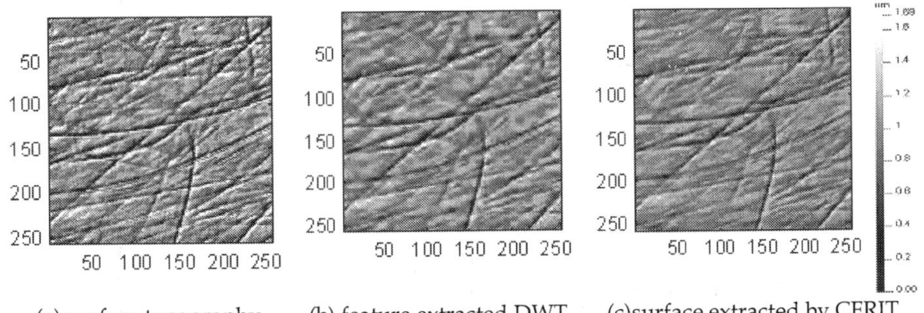

(a) surface topography   (b) feature extracted DWT   (c)surface extracted by CFRIT

Fig. 33. Comparison of the general wavelet, FRIT and block FRIT thresholding for the feature extraction on a metallic hip joint head surface with line and curve features

To compare the FRIT with the general wavelet thresholding for line and curve feature extraction, the above form removed surface was shown in Figure 33(a) as a grey topview image. The feature extraction results of the general wavelet thresholding, CFRIT are shown in Figures 33(b) and (c) respectively. Comparing the results of the general wavelet thresholding and the CFRIT, the features from the wavelet thresholding have blur margin and are discontinuous in some change points while those from CFRIT are smooth and continuous.

6. Conclusions

In this chapter, the early families of orthogonal wavelets and the three generations of biorthogonal wavelet transforms, developed later, for the extraction of morphological structures from micro/nano scalar surfaces in the field of bioengineering, have been introduced. The chapter's aim is to create a "tool box" of wavelet techniques capable of complex analysis and interpretation of surface topography data; leading to the extraction of functionally critical morphological features from micro/nano scalar surfaces of orthopaedic joints for in-vitro and clinically retrieved applications.

The orthogonal wavelets have been used for analysis of multi-scalar surfaces in engineering in which the phase distortions were neglected. The main advantages of biorthogonal wavelets are that they are projections (into an associated wavelet scale space) with linear phase (leading to real outputs without frequency aliasing and phase distortion) and a traceable location property).

The first generation biorthogonal wavelet transform overcame the phase distortion disadvantage of orthogonal wavelet transform, thus were suitable for surface filtering. The first generation model was successfully applied in de-noising areal surface stylus instruments.

The second generation model was built using the lifting scheme, which does not use the Fourier transform as a prerequisite, and can be implemented in three easy stages namely, splitting, prediction and updating. The models for second generation wavelets have helped the Bio-engineering industry in successful separation, extraction and reconstruction of isolated morphological features. For example, large peaks/pits in hip joint heads. The accuracy of the bearing surfaces of these medical components is from micrometers to nanometers. The reconstruction of morphological features has helped friction, contact stress and load area (which are evident during the initial stages of wear) to be more accurately assessed.

The first/second generation wavelet models were built by the real discrete wavelet transform that is less effective in the extraction of certain morphological features, such as linear/curved scratches and plateaux with direction/objective properties in micro-structured surfaces, for example, plateau honed surface, steel and stainless steel sheets, bio-engineered surfaces. Real discrete wavelet transforms have a lack of shift-invariance and have less sensitivity to direction and anisotropy, as a result, they have good performance at representing sparse zero-dimensional or point singularities, but not higher dimensional objects such as linear/curved scratches.

The third generation wavelet models are based on novel mathematics – the Complex Wavelet Transform (CWT). It is then proposed to explore in detail two types of tools: 1) The DT-CWT for shift-invariant extraction of point-like morphological features; and 2) The Complex Finite Ridglet transform (CFRIT) for linear/curve morphological features.

7. Acknowledgments

The authors would like to thank the EPSRC UK for supporting this research work under its programme EP/F032242/1. The corresponding author X. jiang gratefully acknowledges the Royal Society under a Wolfson-Royal Society Research Merit Award and the European Research Council for its 'Ideal Specific programme' ERC-2008-AdG 228117-Surfund.

8. References

[1] B. Bhushan. Tribology and mechanics of magnetic storage devices, 1996, 2nd edn, New York, Springer-Verlag

[2] T.R. Thomas, Rough surface, 1982, 1st edn, London, Longman

[3] D.J. Whitehouse, Surface Metrology, 1994,1st edn, Bristol and Philadephia, Institute of physics publishing

[4] M. Bachnick, M. Hasenpusch, H. Richter, U. Boenick, The effects of hardness and surface quality of metal tapers on the fracture load of ceramic ball heads in hip endoprostheses. Biomedizinische Technique, 1994, 39, 302-306

[5] T.W. Bauer, S. K. Taylor, M. Jiang, S.V. Medendorp, An indirect comparison of 3rd-body wear in retrieved hydroxyapatite-coated, porous, and cemented femoral components. Clinical Orthopaedics and Related Research, 1994, 298, 11-18

[6] J. Fisher, D. Dowson, H. Hamdzah, H. L. Lee, The effect of sliding velocity on the friction and wear of UHMWPE for use in total artificial joints. Wear, 1994, 175, 219-225

[7] Hall, R. M., Unsworth, A., Siney, P. and Wroblewski, B. M. 1996 The surface-topography of retrieved femoral heads. Journal of Materials Science-Materials in Medicine, 7, 739-744.

[8] T.E. McGovern, J. Black, R.M. Graham, M. Laberge, In-vivo wear of T16AL4V femoral heads - a retrieval study. Journal of Biomedical Materials Research, 1996, 32, 447-457

[9] A. Unswoth, Recent developments in the tribology of artificial joints. Tribology International, 1995, 28, 485-495

[10] D. J. Whitehouse, K. G. Zheng, The use of dual space-frequency functions in machine tool monitoring, Meas. Sci. Technol., 1992, 3, 796-80

[11] K. Zheng, D. J. Whitehouse, The application of the Wigner distribution function to machine tool monitoring, . Proc., Inst. Mech. Eng, 1992, 206, 249-264

[12] X. Chen, J. Raja, S. Simanapalli, Multi-scale analysis of engineering surfaces. Int. J. Mach. Tools Manufact., 1995, 35, 231-238

[13] T. Klimczak, Z. Hanzel-Powierza, Application of the filter with dynamically controlled transmission band in surface texture analysis. Ann. CIRP, 1995, 44, 505-508

[14] X. Q. Jiang, L. Blunt, K. J. Stout, Recent development in the characterisation technique for bioengineering surfaces, 1997 Proc. of Metrology and Properties of Engineering Surface, 2nd-4th Sweden.

[15] S. H. Lee, H. Zahouani, R. Caterini, T. G. Mathia, Multi-scale analysis of engineering surfaces. Int. J. Mach. Tools Manufact., 1998, 38, 581-589

[16] X. Liu, J. Raja, H. Sannareddy,1995 Assessment of plateau honed surface texture using Wavelet transform. Proc. of ASPE., 1995, 14, 672-675

[17] I. Daubechies, Ten lectures on Wavelets, 1992,1st edn, Philadelphia, SIAM

[18] C. K. Chui, An Introduction on Wavelets, 1992,Philadelphia, SIAM

[19] S. Mallat, A theory for multiresolution signal decomposition: The Wavelet representation, IEEE transaction on pattern analysis and machine intelligence, 1989, 11, 674-693.

[20] Y. Meyer, Wavelets: algorithms and applications, 1993, 1st edn, Philadelphia, SIAM

[21] J. B. Allen, L .R. Rabiner, A unified approach Short-Time Fourier analysis and synthesis, Proc. IEEE, 1977, 65, pp.1558-1564.

[22] D. Gabor, Theory of communication, J. of the IEE, 1946, 93, 429-457.

[23] G. Strang, T. Nguyen, Wavelets and filter banks, 1996, 1st edn, Wellesley-Cambridge Press.

[24] W. Swelden, Construction and application of Wavelets in numerical analysis, 1994, PhD Thesis.

[25] W. Swelden, The lifting scheme: a custom-design construction of biorthogonal Wavelets, 1995, Bell Laboratories.

[26] W. Swelden, The lifting scheme: a construction of second generation Wavelets, 1996, Bell Laboratories.

[27] X. Jiang, L. Blunt, K. J. Stout, Development of a lifting wavelet representation for surface characterization, Proc. R. Soc. Lond. A, 456(2000), 1-31

[28] N. Kingsbury, Image processing with complex wavelets, Phil. Tran. R. Soc. Lond. A, Vol.357, 1999, pp. 2543-2560.

[29] N. Kingsbury, Complex wavelets for shift invariant analysis and filtering of signals, Appli. Comput. Harmon. Anal., Vol.10, No.3, 2001, pp. 234-253.

[30] X. Jiang, L. Blunt, Third generation wavelet for the extraction of morphological features from micro and nano scalar surfaces, Wear, 2004, 257, 1235-1240

[31] W. Zeng, X. Jiang, P. Scott, Metrological characteristics of dual tree complex wavelet transform for surface analysis, Measurement Science and Technology, 16 (2005) 1410-1417

[32] E. J. Candès, D. L. Donoho, Ridgelets: a key to higher-dimensional intermittency, Phil. Trans. R. Soc. Lond. A., Vol.357, 1999, pp.2495-2509.

[33] D. L. Donoho, Orthonormal ridgelets and linear singularities, SIAM J. Math. Anal., Vol.31, No.5, 2000, pp. 1062-1099.

[34] M. N. Do, M. Vetterli, The finite ridgelet transform for image representation, IEEE Trans. Image Proc., Vol.12, No. 1, 2003, pp. 16-28

[35] X. Jiang, W. Zeng, P. Scott, J. Ma, L. Blunt, Linear feature extraction based on complex ridgelet transform, Wear, 2008, 264(5-6): 428-433

[36] W. Zeng, X. Jiang, P. Scott, Complex Ridgelets for the Extraction of Morphological Features on Engineering Surfaces, Journal of Physics: Conferences series, 2005, 13: 246-249

[37] J. Ma, X. Jiang, P. Scott, Complex ridgelets for shift invariant characterization of surface topography with line singularities, Physics Letters A (344) (2005) 423-431

[38] E. D. Bolker, The finite Radon transform, Contemp. Math., Vol. 63, 1987, pp.27-50.

[39] F. Matus, J. Flusser, Image representation via a finite Radon transform, IEEE Trans. Pattern Anal. Machine Intell., Vol.15, No.10, 1993, pp.996-1006.

4

Ten Years of External Quality Control for Cellular Therapy Products in France

Béatrice Panterne et al.[*]
[1]Direction des Laboratoires et des Contrôles,
Agence Française de Sécurité Sanitaire
des Produits de Santé (Afssaps)
France

1. Introduction

The French Health Products Agency (Afssaps) was created in 1998 within a national context of strengthened health monitoring and control. The Afssaps evaluates the safety, the efficacy and the quality of health products. The law 96-452 of May, 28th 1996 and the law 98-535 of July 1st 1998, published in the "Public Health Regulation" have established rules for the use of cell therapy products. The Laboratories and Controls Department (DLC) is in charge of external quality control of cell therapy products since 1999. In this French regular context and because haematopoietic cells are the main cellular products used in routine in Europe (Gratwohl et al., 2006) as they are in the US (Read & Sullivan, 2004), control was first developed for these. Haematopoietic stem cells are populations of primitive multipotent cells capable of self-renewal as well as differentiation and maturation into all haematopoietic lineages. They are found in small numbers in the bone marrow, in the mononuclear cell fraction of circulating blood (peripheral blood stem cells or PBSC) and in umbilical cord blood (umbilical or placental stem cells). For the twenty-five last years, important progresses have been obtained in the treatment of onco-hematology diseases, especially with the advent of therapeutic intensification following by a graft of haematopoietic stem cells (Copelan, 2006). The infused haematopoietic progenitor cell populations can originate from the recipient (autologous) or from another individual (allogeneic) but until now, most of the grafts are made in an autologous context. Also, 85% of the PBSC samples received for the control at Afssaps had an autologous status. Considering peripheral blood stem cells, the major criteria retained for clinical decision is the positive cell number for the CD34 antigen (Bender et al., 1993; Allan et al., 2002; Weaver et al., 1995). However, as there is no target value for CD34+ cells according to the nature of the product, there was a strong need for efficient methods to guarantee the CD34 value and thus make reliable the clinical decision. Numeration by flow cytometry method is currently

[*]Marie-Jeanne Richard[2], Christine Sabatini[1], Sophie Ardiot[1], Gérard Huyghe[1], Claude Lemarié[3], Fabienne Pouthier[4] and Laurence Mouillot[1]
[2]UMTCT EFS Rhône Alpes / CHU Grenoble, France
[3]Institut Paoli Calmettes, Centre de Thérapie Cellulaire, Marseille, France
[4]AICT, EFS Bourgogne Franche Comté, France

the most sensitive and rapid one, but variations due to the different techniques yet persist. For the clinical studies and the therapeutic use of these cell products, a standardization of the methods was undertaken with the assistance of learned societies. Recommendations for the CD34+ cell numeration were published in 1996 (Sutherland et al., 1996) followed by their adaptation to the single- platform (SP) methods (Keeney et al., 1998). Then, different teams using this methodology provided data showing the strong reliability of CD34+ evaluation (Barnett et al., 2000 ; Gratama et al., 2003 ; Chang.et al., 2004). However, double-platform method is still used to evaluate CD34+ cell quantity in haematopoitic graft products (Laroche et al., 2005; Timeus et al., 2006; Scaradavou et al., 2010).

Till now, one of the main criteria of survival after a cord blood allograft is the number of nucleated cells able to ensure the haematopoietic reconstitution (Gluckman et al., 1997). Also, the activity and the development of the "Réseau Français du Sang Placentaire" (RFSP) were done with strict criteria of quality where the number of total nucleated cells contained in a graft must be higher than 2.5 X 10^7/kg (Rubinstein et al., 1998) even $3.5.10^7$/kg at that time (Grewal et al., 2003). This criterion and its corollary (the initial volume of the graft must be higher than 80 ml) result in a therapeutic use of only 30 to 40% of collected umbilical cord blood. However, the amount of CD34+ seems increasingly important in the choice of the Cord Blood Unit (CBU). Since 2000, it is shown that the contents in progenitors (CFC) are a better indicator than the number of nucleated cells (Migliaccio et al., 2000). It was then established that an amount of CD34+ lower than $1.7.10^5$/kg was associated with a lower rate of engraftment and that the negative effect of a HLA mismatch (to the maximum 2) can, at least partially, be compensated by a CD34+ amount higher than this threshold (Wagner et al., 2002). Currently, if CD34 numeration of PBSC is validated by an international multicentric study (Gratama et al., 2003), and the number of viable CD34+ after thawing is associated with engraftment (Allan et al., 2002; Lee et al., 2008), recent inter-laboratory studies show that there exists still a great variability of results for placental bloods (Moroff et al., 2006; Brand et al., 2008). Other parameters like cell viability are able to inform about thawing yield and more to predict if this operation would be optimal. In so doing, we particularly focused on product composition as the granulated cell fraction containing polynuclear neutrophils. An analysis of this parameter measured in the product after leukapheresis showed an unexpected influence on cell viability. For several years, it has been shown that PBSC could be preserved for a 24H before freezing (Beaujean et al., 1996) as for CBU (Campos et al., 1995), and confirmed by Calmels et al (Calmels et al., 2007). However, Moroff and his collaborators have shown that this time preservation could be extended to 3 days before freezing (Moroff et al., 2004). Nowadays, a 24H storage before cryopreservation is a well established process. However, cell quality at thawing is not still homogeneous according to the product and, other parameters in addition to storage conditions should be studied to try to prevent loss of quality.

Another important tool is the standardization of bacteriological control under the impulse of the work carried out by Afssaps. The investigation on practices in French cell banks, which has taken place in 1998 to implement the external control of cell therapy products, contrary to the measurement of CD34, showed a very large heterogeneity of the practices for the microbiological control. There was sometimes the use of not adapted techniques probably leading to an under-detection of contaminated products by bacteria or fungi. We initially had a transposition of the methods used for infectious samples to the products of cellular therapy. The first ones do not have a limitation of volume, of germ concentration and a positive result is generally obtained within 24 to 48H. This is completely different for a cell therapy product where volume available is very small, and the concentration of germs, if there are, very low. At the time of implementation, based growth methods with rich medium as those for blood culture were used for the control of labile blood products (Brecher et al., 2001; McDonald et al., 2001; Dunne et al., 2005) as for bone marrow control (Schwella et al., 1998). Taking together, it was decided to evaluate an automated blood culture system for the microbiological control of cell therapy products. The work led at Afssaps with the active participation of French cell banks allowed us to develop specific recommendations for this control which has given the basis for the European Pharmacopoeia chapter 2.6.27 applied since 2007.

As a consequence of the French regulation for cell therapy products, an external quality control of haematopoietic stem cells (HSC) was implemented since 1999, based on an investigation for practices, a pilot and a feasibility study with all the French cellular therapy (CT) facilities and on collaborative studies for the bacteriological control. All the French CT facilities have participated to this control and it is now a robust tool used in order to check product specifications. Nowadays, external quality control takes place 2 to 3 times a year as a survey market where each French CT facility sends samples from graft product towards Afssaps. This chapter describes the main results obtained during the past 10 years since the implementation of the external quality control in 2000. These results are going to focus on the assessment site by site leading to a personalised follow-up, on the standardization of the CD34+ cell numeration, on the identification of parameters which could influence cell therapy product quality and on the standardization of the microbiological control for haematopoietic products.

2. Materials and methods

2.1 Materials

To organize external quality control for cell therapy products by the competent authority in France, facilities and technical assays have been implemented in 1998 in the Blood Products and Cell Therapy Unit at the Laboratories and Control Department (DLC), Afssaps in quality system management using the ISO17025 norma.

2.1.1 Cell samples for external quality control

Currently, 31 sites in France prepare haematopoietic products for grafts indicated in haematological diseases. After a pilot study to validate transportation conditions and technical procedures, these sites send samples of haematopoietic products according to the scheme designed by the Blood and Cell Therapy Products Control Unit at DLC, Afssaps. A procedure for the logistic and transportation has been defined with a transportation temperature from 4 to 12°C and a transfer from sites inferior to 24H (mean time is 16 hours). Each sample received is immediately analyzed for nucleated and CD34+ cell count, viability, CFU-GM progenitor and a microbiological control is performed using an automated blood culture system. In parallel, producers make the usual controls and send their results to Afssaps for comparison. In this context, 92% of the products sent to Afssaps were peripheral blood stem cells (PBSC); others were from bone marrow or from umbilical cord blood.

2.1.2 Cord blood units (CBU) for a multicentric study

Forty-two displaced units obtained from the AICT (Activity of Cellular and Tissue Engineering) at the French Blood Establishment in Besancon were sent by transport in nitrogen vapours to the 14 cellular therapy facilities participating to this study (gathered according to their localization). Each site received a non reduced in volume CBU and 2 miniaturized CBU. They have then to defrost these CBU the day chosen in dialogue with Afssaps. The thawing was carried out according to the usual processes on the sites. After taking away for control, the remainder of the defrosted unit was sent towards Afssaps in the transfer bag at 4-12°C transportation according to the procedure used within the framework of external control. At the receipt, the temperature, the conditioning and the aspect of the CBU were controlled and usual analyses as described below were performed.

2.1.3 Collaborative studies for microbiological control

Contaminated cell samples were prepared at the DLC. Cellular materials were obtained from the French Blood Establishment according to a contract, they generally consisted in residual buffy-coats or thawed mononuclear cells. Bacterial strains were prepared by standard practices and sowed in cell suspension at a determined concentration. A panel of samples was sent to each participant who performed the analysis according to its own procedure. Each kind of sample is also controlled by Afssaps after different times of preservation at 5±3°C to cover the delay of transportation and the beginning of the analysis for the different participants. Results are collected and analyzed by Afssaps. Participants are the usual laboratories who perform the microbiological control of cellular products for the French cell therapy producers. One study takes place per year with at least 32 participants.

2.2 Techniques used at Afssaps
2.2.1 Nucleated cell numeration

Nucleated cell counts were measured with an automated cell counter (MaxM, Beckman Coulter, Miami, FL) validated for leukapheresis products. In particular, it is necessary to check the linearity of counts because of the necessary dilution of leukapheresis products for most cases. Here, the nucleated cell count was used to define the working dilution for the StemKit procedure by flow cytometry.

2.2.2 Viability

Viability was determined by flow cytometry using a nucleic acid intercalating agent, 7-AAD (7-aminoactinomycine D). The percentage of 7-AAD positive cells was based on the use of the single-platform ISHAGE gating strategy developed by Keeney and Sutherland (Keeney et al., 1998) and determined during the CD34/CD45 labeling using StemKit methodology (Beckman Coulter, Miami, FL). The use of 7-AAD permits to check viable CD34+ cells especially for thawed products (Brocklebank & Sparrow, 2001).

2.2.3 CD34+ cell numeration

CD34/CD45 labelling using StemKit in the presence of StemCount fluorospheres to determine directly the CD34+ absolute count by flow cytometry and according to the ISHAGE guidelines (Sutherland et al., 1996) .Quickly, 20µl of a mix of CD45-FITC and CD34-PE antibodies and 20µl of 7-AAD (final concentration: 1µg/mL) were added to 100 µl of a cell suspension containing 15,000 to 30,000 cells/µl and incubated for 20 minutes at room temperature, in the dark. Lysis of red blood cells was then performed using chloride ammonium 1x (10x solution provided with StemKit) during an incubation of ten minutes at room temperature and in the dark. Then, 100 µl of FlowCount calibrated fluorospheres (Beckman Coulter) were added to allow the determination of the absolute CD34+ cell value.

Acquisition of sample tubes was performed according to the protocol designed by Keeney and Sutherland (Keeney et al., 1998).

2.2.4 CFU-GM evaluation

A clonogenic assay was performed using ready-to-use semi-solid media from StemCell Technologies, Vancouver, Canada and from Stem alpha, St-Genis l'Argentière, France. Several references of methylcellulose medium were used due to the external quality control context. Each sample of PBSC received at Afssaps (in mean 16h after the collection) was analyzed with the reference of medium used by the producer as much as possible to allow a better comparison of results. Samples were plated in duplicate in 35-mm culture dishes at a concentration targeted to 200 to 300 viable CD34+ cells/ml. The dishes were incubated at 37°C in a humidified 5% CO_2 incubator for 14 days. CFU-GM colonies were scored at the day 14 using an inverted microscope. The CFU-GM clonogenicity (number of CFU-GM/100 CD34+ cells) was determined for each sample.

2.2.5 Estimation of the granulated fraction

We have analyzed 271 fresh PBSC samples to study the influence of granulated cell fraction on the cell viability. The percentage of granulated cells was defined in a SSC/CD45 region during the CD45/CD34 procedure. This determination can be supplemented by a CD45/CD14/CD15 labelling (antibodies from Beckman Coulter) when this population is dense and in continuity with the monocytic population for more precision.The linear regression analysis between this CD15+ fraction and the granulated cell region obtained on a SSC/CD45 graph gave a correlation coefficient R^2 equal to 0.97 and this correlation was significant ($p<0.01$) between these 2 groups.

2.2.6 Bacteriological and fungal control

Automated growth-based method during 10 days on aerobic and anaerobic media (BactAlert system, BioMérieux). In case of positive controls, germs are identified on a Vitek 2 Compact automated system (BioMérieux) and an antibiogram is realized by the same apparatus with dedicated cards.

2.3 Techniques used by French producers

Techniques	Producers (at the time of the last survey in 2009)
Nucleated cell numeration	Automatic counting by an haematology automate
Viability	7-AAD for 25 sites, Trypan Blue for 3 sites and no viability determination before preservation for 2 sites
CD34+ cell numeration	Single platform method for 22 sites (StemKit : 16 sites, Procount or SCE : 5 sites, CD34 Count Kit : 1 site) and double-platform for 8 sites.
CFU-GM clonogenic assay	23 sites use a medium from StemCell Technologies and 6 from Stem alpha. One site does not perform this assay at this step For 12 of them (41%), they use a medium without Epo.

Bacteriological and fungal control	90% of the French sites implied in the cell therapy product preparation use an automated growth-based system (Bactec or BactAlert) with an incubation ranging from 7 to 10 days.

2.4 Statistical analysis

Means and standard deviations have been determined for each sample group with Excel 2003 software. The sample groups were compared using the comparison of the 2 observed means with $\alpha=0.05$ with Statgraphics software. Concerning the analysis of the results of numerations, being paired series, the existence of a significant correlation between the participating sites and Afssaps was analyzed using a variance analysis of linear regression using the Statgraphics software with a 5% risk. The differences (expressed in %) between the results of the site and those of Afssaps for each pair were determined and gathered according to various criteria (technique, thawing, cell nature...).

3. Results

3.1 External quality control of haematopoietic products
3.1.1 Results site by site from 10 years of control

The French Health Products Agency, Afssaps, is an evaluation and assessment agency for the whole of these products. For ten years, the Blood Products and Cell Therapy Unit is in charge of external quality control of cell therapy products, essentially peripheral blood stem cells after mobilization. This control allowed us to define quality points adapted to these products, following the measure variation between the producers and Afssaps. Upon those ten years, this work supplies the thirty French preparation sites of cell therapy products with a global statement of the control operation and a follow-up site by site. This allowed us to identify the strong points of the control with operations to carry on and/or to develop and to define controls to check the quality of cell products

To perform these controls, validation for viable CD34+ cell enumeration using a single-platform method was made. It was available for leukapheresis, bone marrow, cord blood cells and thawed products. After a ten year activity and for this site by site study, we have analysed 775 fresh peripheral blood stem cells (PBSC) samples from 30 French preparation facilities. The mean number of sent products per site was equal to 26 fresh PBSC with a range from 11 to 73 directly linked to the site activity. On these 775 received products, 31 (4%) were degraded at reception in Afssaps, generally because of not observed transport conditions and were excluded from this assessment. For each site, we have followed the variation between the laboratory and Afssaps, established control cards for cell numerations and CFU-GM assay. Correlations between CD34 and CFU-GM doses have been studied. For the nucleated cell numeration, 67% of the laboratories obtained a deviation lower than 10% with those made in Afssaps. For CD34 cell numeration, a mean deviation lower than 20% with the Afssaps value is obtained for each site and for 60% of them, it is lower than 15%. Finally, the mean deviation in 2009 was only 11% ± 7 between producers and Afssaps. These results show the good evolution for this parameter and constitute a very satisfactory result on the national plan. CFU-GM results show higher differences between the producer and Afssaps, however 79% of facilities give a mean deviation with Afssaps lower than 35%. About the correlation between the numerations of CD34+ cells and CFU-GM, 31% of the sites obtain a R^2 higher than 0.7. The results of comparisons between Afssaps and each site obtained for the ten years of external control activity are detailed in the following tables and graphics.

3.1.2 Mean deviation between French sites and Afssaps

After ten years of external quality control, deviations for pairs of results between the production site and Afssaps have been studied site by site for the different analysis and for all the fresh PBSC sent by each of them. Main results are presented on table 1 showing the range of the mean deviations between each site and Afssaps in comparison with the mean deviation for all the products received in 2009 between the producers and Afssaps for the corresponding analysis. Thus, each site can check if its result approaches the up-date mean observed in 2009.

Fresh PBSC analysis	Mean deviation for the all the 30 sites (%) (black bar on figure 1) and range [min-max]	Mean deviation in 2009 between Afssaps and sites for 62 fresh PBSC (%)
Nucleated cell numeration	8.9 ± 2.5 [3.9 – 13.8]	10.4 ± 9
CD34+ cell numeration	13.7 ± 2.8 [7.9 – 18.4]	11.1 ± 6.8
CFU-GM assay	30.3 ± 6.8 [17.7 – 46.9]	32.4 ± 21.9

Table 1. Mean deviations ± SD between French cell therapy sites and Afssaps. After ten years of external quality control, deviations for pairs of results between the production site and Afssaps have been studied site by site for the different analysis and for all the fresh PBSC sent by each of them.

The mean deviation between Afssaps and the site for each of the 30 ones are represented on the graphic 1 below. Each of them has received a personalised map with its own result appeared in a marked point.

The results show quite small variations (ranging between 4 and 14%) for the numeration of the total nucleated cells (TNC) because of the method used which is a simple and robust technique with a mean deviation equal to 9% ± 2.5. Good results were also obtained for CD34+ cell numeration with variations ranging between 8 and 18.4% and with a mean deviation for all sites equal to 13.7% ± 2.8. The most important variations are observed for the numeration of the CFU-GM whose experimental conditions make this test more difficult to standardise, the average deviation of 30.5% is however acceptable (regarding results obtained in Proficiency Testing Studies organised by StemCell Technologies with 150 to 200 participants, coefficient variation for the CFU-GM number are ranged from 30 to 50%). As the variations observed are close to 10% for the CD34 and TNC numerations and close to 25% for the CFU-GM assay and for some sites, the objective is to maintain or reach these values for all the producers.

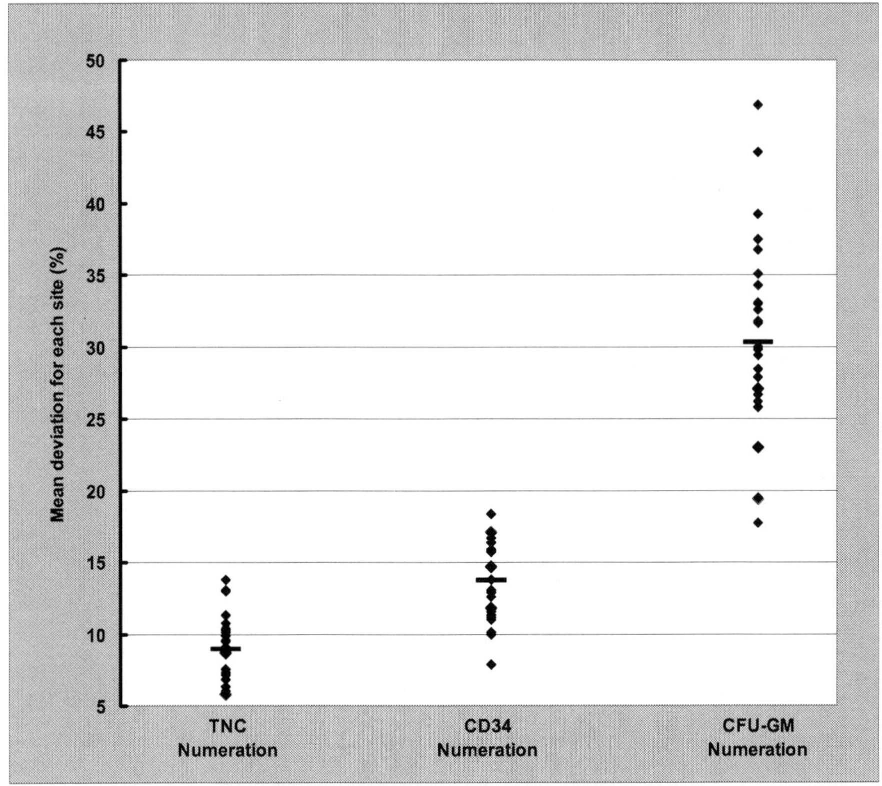

Fig. 1. Distribution of mean deviation between each site and Afssaps for 3 main analyses. Mean deviations for pairs of results between each site and Afssaps (%) and for each analysis is represented by a point (♦) and the mean deviation for all sites is represented by the bar (—).

3.1.3 Site follow-up

As an example, the following graphic shows the CD34 deviations between Afssaps and one French producer of PBSC. As the mean deviation for PBSC samples received in 2009 is equal to 11% ± 6.8, we used this result as a target value and looked at the CD34 delta repartition for each site so that the results are expected to be between -20% and +20%. A deviation equal to 0 means that Producer and Afssaps results are equal, a positive delta means that the Afssaps CD34+ result is higher than that of the producer whereas a negative one is lower. In this example (Figure 2), the mean deviation between the results from Afssaps and the producer for the 22 PBSC samples addressed for control is equal to 13% ± 9. The mean deviation when a double-platform method was used by the producer was 19.8% ± 10 (n=8) and was reduced to 9.3 ± 6.6 (n=14) when the producer changed for a single-platform method. These 2 results are significantly different (p=0.02). Indeed, since the use of a single-platform by the producer, no deviation higher than 20% was measured.

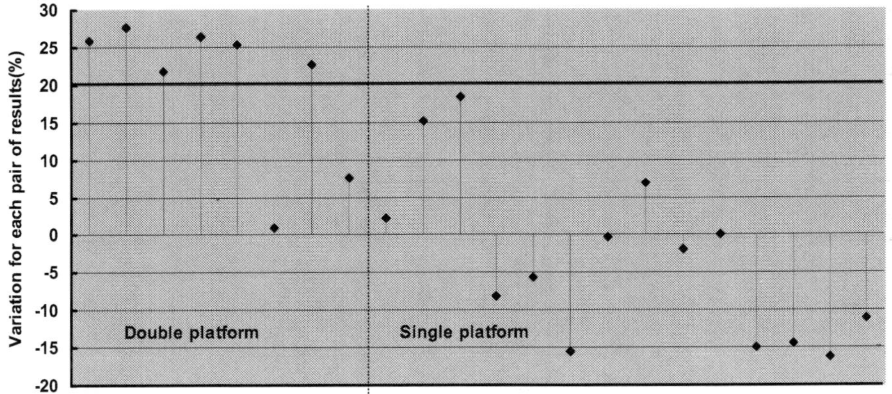

Fig. 2. Deviations between Afssaps and a French Producer site for the CD34+ cell determination in the 22 PBSC samples addressed. Each point (♦) represents the deviation (%) obtained for one pair of CD34+ results and allows to follow the evolution of results for that site along the time (from 2000 to 2009).

We also followed the results for CFU-GM assay site by site and sent personalised control map for that to each of them. For this previous site with the same fresh PBSC, even we observed a coefficient R^2 equal to 0.79, the mean deviation for the number of CFU-GM/kg between this French site and Afssaps remained quite high and was 43.6% ± 21.3 (n=19). However it was not the case for some other ones showing that different practices (sample preparation, targeted number of CD34+ cells per dish, medium choice and/or colony estimate) could influence the CFU-GM number result.

3.2 Standardization of CD34 evaluation through the French external quality control – Results from 2000 to 2010

On the basis of international recommendations (Sutherland et al., 1996), the first kits from manufacturers appeared whose objective was to bring a reliable and reproducible measurement, even for products with a very small percentage of CD34+ cells. Using this single-platform method, the DLC evaluated reproducibility of the CD34+ cell numerations by determining the difference expressed as a percentage between measurement of the producer and the measurement carried out in Afssaps for a fixed sample. Without a target value, this variation analysis made possible the establishment of a "normative" reference mark insofar as for approximately 80% of the controlled cellular samples, the difference between producers and Afssaps is lower than 20%. Also, the elements able to explain a higher deviation than 25% were explored, it can be elements related to the technique and/or the nature of the product, such as for example, problems of cellular stability. Thus, progressively with these controls, parameters are identified like being responsible of a lower reproducibility. Groups of defined samples according to these parameters are thus compared. At that time, as shown in table 2, mean deviation for fresh PBSC (n=789) between Afssaps and French facilities is equal to 14.7% ± 11.8 with a good correlation (R^2=0.88) but higher and around 25% for thawed products. For bone marrow, the mean deviation is equal to 19% ± 16.4, this higher result is probably due to the different approaches for the CD34+ region according to the more heterogeneous CD34+ population taking in account all the CD34+ cells (dim to bright cells) or only the brightest ones and also due to the different techniques as shown below. For all techniques, the mean deviation for CD34+ cells in thawed products between Afssaps and laboratories was 27.2% ± 21. To analyse the observed differences between Afssaps and producers according to the different kinds of products, we examined these results regarding the use of a single or a double-platform (DP) by laboratories to numerate CD34+ cells (see Table 3). For thawed PBSC, when producers have used a single platform, it was equal to 22% ± 16.5 for viable cells and only 14.7% ± 10.9 for total CD34+ cells.

Products	Mean deviation between Afssaps and Producers for the determination of CD34+ cell number	R^2
PBSC (n=789)	14.7% ± 11.8	0.88
Bone marrow (n=105)	19% ± 16.4	0.85
Thawed PBSC (n=150)	27.2% ± 20.7 (viable CD34+ cells) 20.5% ± 15.4 (total CD34+ cells)	0.73 0.83

Table 2. Mean deviation ± SD between Afssaps and Producers for the CD34+ cell evaluation according to different types of controlled haematopoietic products for this ten year period. A significant correlation between the participating sites and Afssaps has been obtained for each category with $p < 0.05$.

Regarding the CD34+ cell recovery observed at Afssaps (after transportation), the mean was 55% ± 24 (n=59) whereas it was equal to 72.6% ± 29.3 (n=60) for the producers. Moreover, the double-platform method could lead to recoveries after thawing higher than 100%, because of a fluctuating relative value for CD34+ cells. For PBSC, it is noted, indeed, that the mean CD34+ cell recovery was equal to 90.8% ± 28.6 (n=24) when a DP method is used whereas it was equal to 60.5% ± 23 with the SP method, coherent and in agreement with the data of the literature when robust tools are used (Calmels et al., 2007 ; Dauber et al., 2010)
For bone marrow, even the difference between the 2 evaluations is lower when a single-platform is used (15.9 ± 12), there is no significant difference with that obtained when producers used a double-platform method. As said before, because of a more heterogeneous CD34+ bone marrow population, some variations may persist regarding the limit of CD34+ region where dim cells could be taken in account or not. However, with single-platform, the mean variation observed is a satisfactory result. In the same way, no significant difference exists between SP and DP methods for viable CD34+ cells in thawed PBSC because of a certain loss of viability due to the delay between the analyses, but also, the difference with single-platform is lower. Finally, the mean deviations according to the use of a SP or a DP method are significantly different for total CD34+ cells and reduced with the SP method.

Products	Mean CD34 deviation between Producers and Afssaps (%)		p
	Double-Platform	Single-Platform	
Fresh PBSC	17.0 ± 14,9 (n=396)	12.9 ± 9,7 (n=386)	<0.0001
Bone marrow	20 ± 18 (n=57)	15.9 ± 12 (n=30)	0.2
Thawed PBSC (Viable CD34 cells/Prod)	27.5 ± 21,7 (n=87)	22.3 ± 16,5 (n=55)	0.1
(Total CD34 cells/Prod)	21.2 ± 15 (n=69)	14.7 ± 10,9 (n=36)	0.015

Table 3. Mean deviation ± SD between Afssaps and Producers for the CD34+ cell evaluation according to different types of controlled haematopoietic products for this ten year period gathered according to the flow cytometry method (DP or SP) used by those producers.

The good evolution of these results was noted by the follow-up of CD34 variations for fresh PBSC year per year: whereas it was equal to 19% ± 15 in 2000, it is no more than 11% ± 7 today (with p=0.0002) as shown in figure 3. The part of PBSC showing a difference between the 2 evaluations bigger than 20% is now lower than 15% of the products received from all French sites.

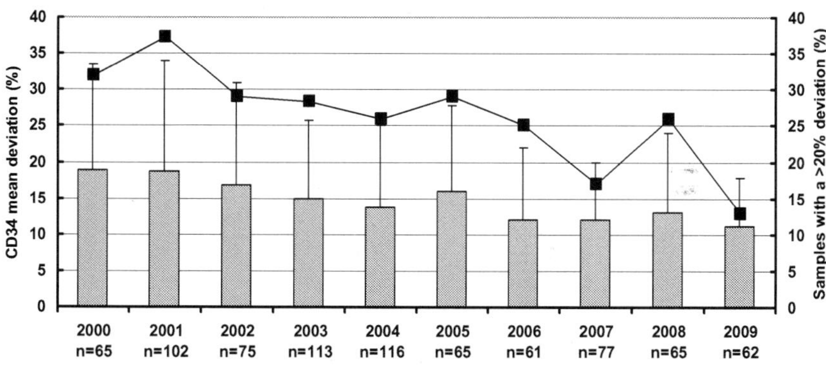

Fig. 3. Evolution of CD34 deviations between Afssaps and Producers for 10 years of external control. CD34 mean deviation (%) for each year since 2000 (grey bar) and part (%) of PBSC controlled samples with a difference Afssaps-Producer higher than 20% for CD34 evaluation (■).

3.3 Standardization of CD34 evaluation through a multi-centric study for cord blood products

In the same time, as the use of haematopoietic stem cell graft from cord blood increased and reached almost 20% of allogeneic grafts in 2009, the question of the thawed cord blood unit

(CBU) quality has been asked. Indeed a lot of cellular therapy facilities have started to thaw these products without being implied in their preparation which is done in cord blood bank. Thus, very little information of the resulting quality was known. In this context, we have organised a study with the French Society of Cell and Tissue Bioengineering (SFBCT) to ensure the inter-laboratory reproducibility of the quality controls practised by the banks during thawing. The cellular recoveries were analyzed according to the thawing techniques, according to the method used in flow cytometry: single-platform versus double-platform, or the product nature, i.e. in total blood or reduced volume. Concerning CD34+ cells numeration, the average deviation between the participating laboratories and Afssaps was 29% ± 23. When dividing laboratories depending on flow cytometry method used, the average deviation was 21% ± 16 for the laboratories using a SP method against 47% ± 25 for those using a DP method.

The CD34+ recoveries are equal to 82% ± 60 in J0 for the participating sites against 52% ± 20 for Afssaps. For the sites using a DP method, it is stressed that this output is particularly high with a rate of 126% ± 90 (n=15) whereas it is 62% ± 20 (n=32) for the sites using a SP method. There exists a significant difference between SP and DP with p<0.05 and, whereas dispersion is weak when a single-platform method is used, dispersion is increased with the DP method. Interestingly, it is noted that the average recovery on the sites using the simple-platform method, although superior, get closed to that found in Afssaps whereas with the double-platform the average recovery is higher than 100% with a very high dispersion as published (Laroche et al., 2005).

Methods	All techniques (n=48)	Double-platform method (n=15)	Single-platform method (n=33)
CD34 deviation between French sites and Afssaps (%)	29 ± 23	47 ± 25	21 ± 16
CD34 output after thawing at the participating sites (%)	82 ± 60	126 ± 90	62 ± 20

Table 4. Mean deviation ± SD between Afssaps and Producers for CD34+ cell evaluation in thawed cord blood and mean deviation ± SD for CD34+ cell recoveries after thawing obtained at the participating sites according to the technique used in flow cytometry.

As shown on figure 4, the strong difference for total cell viability (42.5%) and that of CD34+ cells (80%) because of a better resistance of these cells to cryopreservation, is one of the reason why double-platform method can generate over-estimate of CD34+ value and as a consequence to CD34+ cell recovery higher than 100%.

To illustrate the excellent quality of thawed cord blood transmitted to Afssaps, an example of CFU-GM is given on figure 5 showing very large colonies, the mean clonogenicity (CFU-GM number/ 100 CD34+ cells) in Afssaps was 14.9% ± 5 (n=44) and 15.6 ± 9.7 (n=36) for participating sites which are satisfactory results for thawed products. Finally, we analysed the different correlations obtained between the quantity of cells before cryopreservation

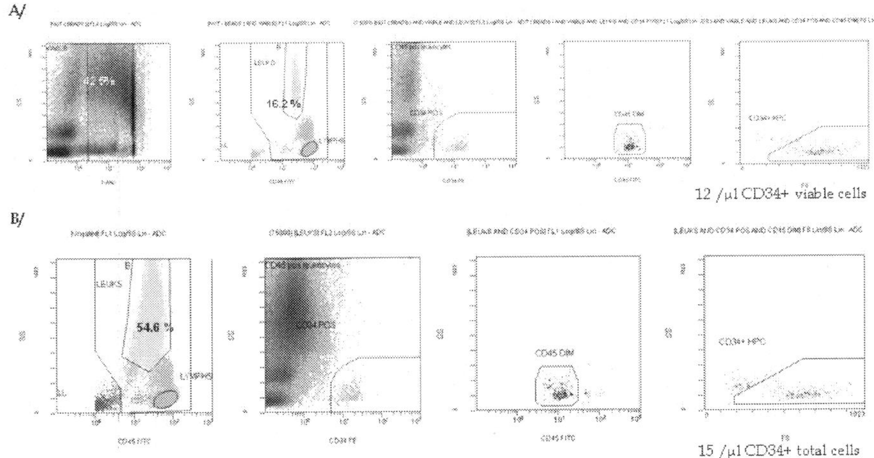

Fig. 4. An example of the analysis of a thawed CBU at Afssaps with StemKit labelling procedure and ISHAGE gating is done: A/ with 7-AAD. B/ without 7-AAD. On this CBU, the total viability of nucleated CD45+ cells was 42.5% whereas that of CD34+ cells was 80%.

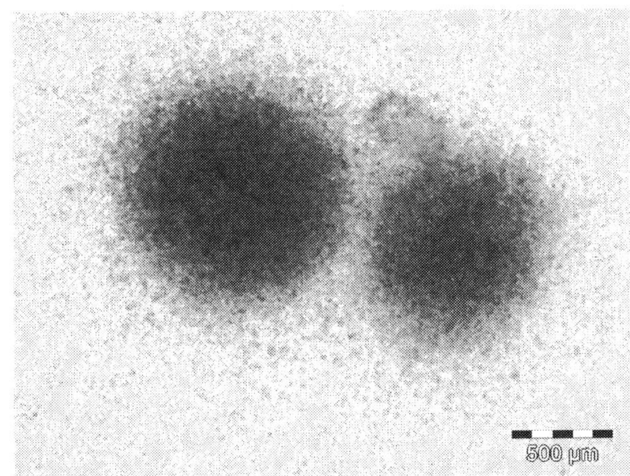

Fig. 5. CFU-GM grown on semi-solid media without Epo, H4535 from StemCell Technologies (Vancouver, Canada).

(determined by the cord blood bank AICT) and CD34+ cell number after thawing obtained by the 14 participating sites. It has been shown that content of total nucleated cells correlates with engraftment (Grewal et al., 2003 ; Rocha & Gluckman, 2009) but also that there exists discrepancy for CD34+ cell and CFU evaluation between the laboratory which have received the cord blood and the cord blood banks which have provided that material (Wagner et al., 2006). Here, in our multi-centric study as shown in figure 6, the linear regression obtained between TNC before cryopreservation (AICT) and CD34+ cell counts after thawing (participating sites) gave a quite low coefficient of determination with $R^2=0.42$. A similar linear regression was observed between CD34+ cell counts before cryopreservation (AICT) and CD34+ cell counts after thawing obtained by all the 14 participating sites with $R^2=0.46$. On the contrary, when the CD34+ counts obtained before cryopreservation were compared with the CD34+ counts obtained after thawing by the sites which used a SP method, a higher correlation was observed with $R^2=0.81$.

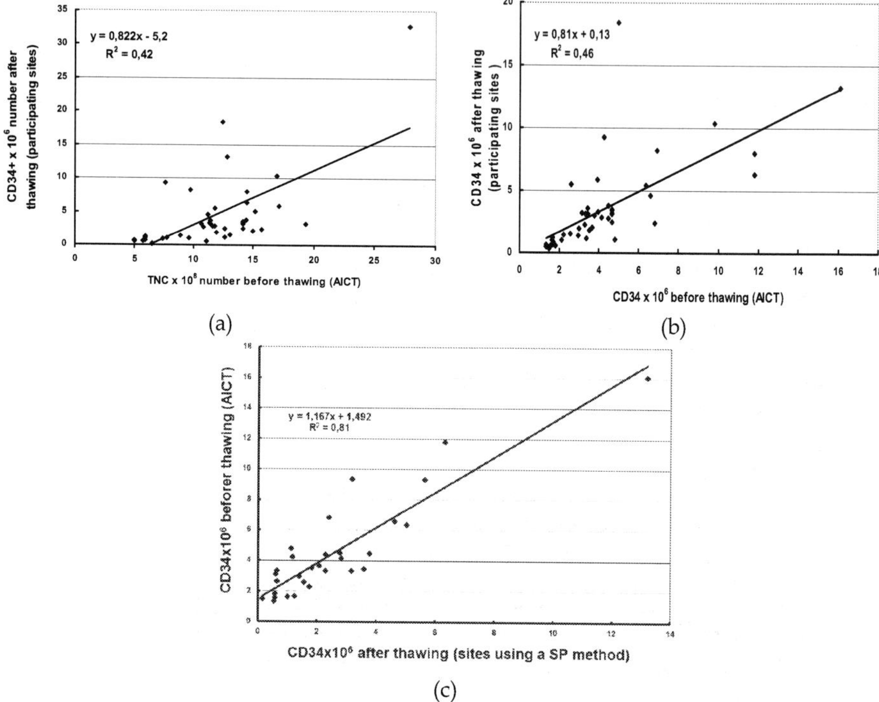

Fig. 6. Linear regression has been done for different contents of cells before and after cord blood cryopreservation. A: between total nucleated cell number before cryopreservation (AICT results) and CD34+ cell content after thawing (n= 46). B: between CD34+ cell number before (AICT results) and after cryopreservation with all the results from the participating sites (n=49). C: between CD34+ cell number before (AICT results) and after cryopreservation with the results from the participating sites who have used a single-platform method (n=30).

These good correlation between CD34+ cell numbers before and after cryopreservation is supported by the results of Lemarie et al. (Lemarie et al., 2007) as well as those of Yoo et al. (Yoo et al., 2007). All together, these results underline a good stability of viable CD34+ cells and a greater reliability of the single-platform methods for the CD34+ cell numeration for these thawed USP (Panterne et al., 2010) supported the CD34+ value as an important criteria of choice for CBU.

3.4 Standardization of CD34 evaluation through comparison of procedures (mono-centric study)

However, even when using a single-platform method to numerate CD34+ cells in cord blood cellular products, a relatively high variability persists (Moroff et al., 2006; Rivadeneyra-Espinoza et al., 2006). These variations could be related to technical variability between cytometers or to biological variability between different reagents. While many recommendations already exist, and have improved CD34+ cell enumeration standardisation (Barnett et al., 1999; Brocklebank & Sparrow, 2001), additional recommendation are necessary for cellular products with low viability such as thawed cord bloods units (CBU).

On thawed CBU, it is known for many years that inclusion of a viability dye is necessary (Brocklebank & Sparrow, 2001). However, Brand et al (Brand et al., 2008) recently showed that many different techniques are still in use (manual vs automated; 7AAD vs Trypan Blue), producing inconsistent results. Reproducibility and robustness of quality and potency controls on fresh and thawed CBU could be better standardized by using single-platform techniques and automated or semi-automated procedures to pilot cytometer setups and analyse results.

A recent study searching for a parameter to predict which CBU will engraft in double CBU transplant (Scaradavou et al., 2010) showed a correlation between CD34+ cell viability and engraftment when viability was measured with 7AAD but not with Trypan Blue. Of interest, this study also suggests the importance of using a threshold value for cell viability, which could be a better predictor of hematopoietic reconstitution than the viable CD34 + cell dose.

Currently, three complete diagnostic kits that include a viability dye are commercially available: these are the Stem-Kit™ enumeration kit (Beckman-Coulter) and SCE kit ™ (BD Biosciences), used with equipment from the same company, and the CD34 Count kit (Dako). We compared CD34+ cell counts in CBU, obtained 1) with 2 single-platforms diagnostic kits and 2) with manual vs automated procedures. Viable CD34+ cell enumeration in thawed CBUs was equivalent using SCE kit or Stem-Kit (R^2= 0.98; Mean deviations between techniques : 13% ± 10 ; n=5). Gating was manual (Cellquest software), based on the adapted ISHAGE guidelines for thawed samples (Brocklebank & Sparrow, 2001). These comparisons were done by the same technologist, on the same sample, on the same cytometer (FACS Calibur BD Biosciences). Comparisons previously performed on 44 fresh apheresis, fresh marrows, fresh CBUs and thawed apheresis samples were also equivalent (Lemarie et al., 2009). For both kits, samples were prepared following the manufacturer's recommendations, except for thawed products to which no erythrocyte lysis reagent was added (based on previous preclinical tests).

We then compared manual vs semi-automated procedures; for the latter, a dedicated software application proposes a grid for cell analysis. Viable CD34+ cell enumeration in thawed CBUs was equivalent using manual or semi-automated procedures to analyse results and set cytometers (R^2= 0.95 ; Mean deviations between techniques : 6 % ± 15 ; n=6). Again, these comparisons were done by the same technologist, on the same sample, with the same reagents (Stem-Kit™ Beckman-Coulter). We compared a manual technique using FACS Calibur cytometer with Cellquest software and manual settings (PMT and compensations) vs FC500 cytometer with semi automated Stem CXP software and auto standardization for parameters settings. Comparisons performed on 28 others fresh apheresis, fresh marrows and thawed apheresis were also equivalent (data not shown).

Similarly to what is observed for fresh bone marrows, thawed CBUs samples automated analysis often leads to overestimate CD34 cell counts, including in the count cells with unspecific CD34+ binding. For these kinds of samples, a negative sample helps to correctly position the CD34 gate. Software optimization in these specific samples could help to standardize CD34 enumeration.

3.5 Identification of parameters influencing cellular therapy product quality

Within the framework of this external quality control, we took particular interest to check parameters that could influence the cell viability and HSC recovery to evaluate the quality of the different products. Indeed, without apparent change in transportation, some samples can lose several points of viability. So, we studied different factors such as preservation duration and granulated cell fraction rate. For this study, we analyzed 271 fresh PBSC with a granulated fraction defined in a SSC/CD45 region. The mean viability at collection (producer evaluation) and the mean viability after transportation (Afssaps evaluation) were determined for those PBSC and analyzed according to product composition and storage time.

3.5.1 Viability and granulated cell quantity

As shown on the figure 7, as it is known, viability decrease between Producers (day of collection) and Afssaps (after transportation, mean delay equal to 16H) with a mean viability of 98.6% ± 1.5 before transportation and of 92.7% ± 6.4 after. A significant difference is observed between mean cell viability at day 1 post-collection (94.3% ± 5.3) and that of PBSC received at day 2 post-collection (89.9% ± 7.3) with p<0.0001. The cell viability at the time of collection (producer results) before transportation has also been analyzed to define the mean viability for these same groups of products. However for the quite same conditions of transportation, some PBSC have shown a viability loss higher than others. As the granulocytes are known as fragile cells, we looked at the viability according to the percentage of granulated leukocytes (PBSC containing mature and immature granulocytes) corresponding to about twice the median level of the granulated cell fraction ie 40% for those received at day 1 post-collection. Indeed, a significant difference is observed between mean cell viability for PBSC containing less than 40% of granulated cells (94.9% ± 3.4) and that of PBSC containing more than 40% of granulated cells (91.7% ± 9.3) with p=0.002. Moreover, when we looked at PBSC received at 48H post-collection (n=94), a stronger loss of viability was observed for the fraction with more than 40% of granulated cells with a mean viability equal to only 85.4% ± 9.1 whereas the mean viability for those with less than 40% of granulated cells was still equal to 91.4% ± 6 ; a significant difference was also observed with p=0.0005.

The determination of granulated cells in haematopoietic products appears now necessary since side effects have been associated with this cellular fraction during the re-injection according to its richness (Calmels et al., 2007 ; Milone et al., 2007 ; Fois et al., 2007 ; Cordoba et al., 2007). However, in order to follow the haematopoietic products with this criterion, it is required to have a robust determination of this cellular fraction. Indeed, it is necessary to establish a threshold allowing a better product management to limit the possible side effects. A study with inter-laboratories comparison and practices investigation have been done to evaluate this granulated cell numeration and a rather good correlation (R^2=0.8) has been obtained between Producers and Afssaps.

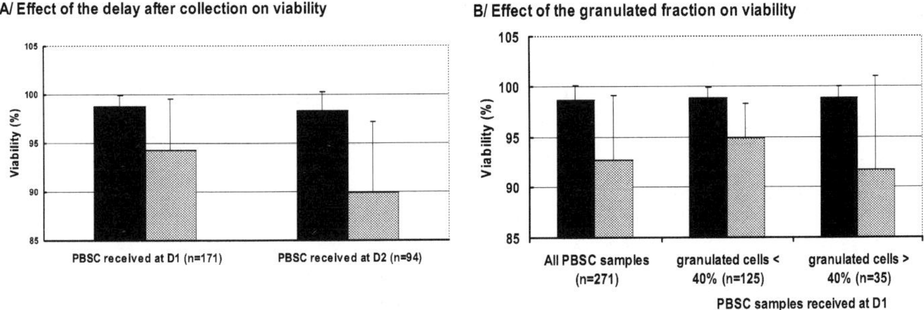

Fig. 7. Mean viability ± SD of PBSC is done before transportation by black bars corresponding to the producer viability at the day of collection and after transportation by grey bars corresponding to the viability observed at Afssaps for each group studied. A/ the mean viability at a day 1 delay post-collection is compared with that of a day 2 delay post-collection. B/ the mean viability for PBSC containing less than 40%of granulated cells (n=125) is compared with that of PBSC which contain more than 40% (n=35), those 2 groups are composed by PBSC received at day 1 post –collection.

3.5.2 Practice investigation

During leukapheresis performed with classical apparatus, leukocytes are separated from the erythrocytes, the platelets and the plasma of the donor or patient. The population of interest for the graft is contained in the mononuclear cell layer and leukapheresis permits to recover this fraction. Currently, it is still difficult to separate mononuclear cells from polynuclear cells without performing a density gradient. Even if new softwares allow a better recovery of the mononuclear cells compared to the semi-automated system, PBSC are still contaminated by granulocytes whatever the software used (Ravagnani et al., 1999). On the other hand, automated haematology analyzers still have difficulties to determine precisely the blood formula in a product obtained by leukapheresis after a mobilization by a growth factor. Indeed, these products can contain immature granulocytes like granulated metamyelocytes counted as non-MNCs because of their granularity. In this situation, there is a lack of robustness. Because of the need for a better evaluation of the granulated cell fraction, we have organized an investigation to evaluate the techniques used in 2009. Thirty producers were questioned and 26 questionnaires were returned leading to a participation rate of 86%. These cell producers observed a median rate for this cell fraction ranging from 15 to 45% for autologous cells corresponding to a median of 24.5% for all sites. For allogeneic cells, median rates were ranged from 12 to 30% corresponding to a median of 24.5% for all sites whereas the observed median rate for this cell fraction at Afssaps is equal to 22.6% (determined in SSC/CD45 region). Different techniques are still used by producers ie: manual cytology for 57.7% of them ; haematology analyser, 19% ; SSC/CD45 region, 38,5% ; and immunophenotyping for 8%. Finally, even the use of these different techniques, a significant correlation is observed between Afssaps and producers for this cell fraction evaluation with R^2=0.76 and p<0.05.

Considering the side effects observed with PBSC containing a high quantity of granulated cell and considering the loss of viability as soon as day 1 post-collection, taking in account the percentage of granulated cells as a quality marker is now a current practice in French cell facilities. This information leads to apply specific procedure for those products meeting a high quantity of granulated cell for example no cryopreservation after 24H and enhanced monitoring at administration for PBSC with a granulated fraction from 30 to 40%. Our results showed that increase in preservation duration, and granulated cell rate had a bad effect on viability. From this observation, we can think that samples with a minor granulated cell fraction can be stored for a longer time than the others.

3.6 Microbiological control of cellular products

To implement the external microbiological control, and to better know the techniques used, an investigation of practices has been done in 1998. This report led to the set up of a working group who prepare national recommendations for this control. In parallel, a validation of automated blood culture was led for these products. This technique enables us to use very low volumes for the inoculation of the bottles and can detect very weak contaminations HSC (1 to 10 germs/ml). The kind of germs able to contaminate, being essentially skin commensal germs (PBSC, bone marrow) or digestive flora (umbilical cord blood), automated blood culture systems allows a better detection than the classical European Pharmacopoeia 2.6.1 method for sterility testing. Our results are confirmed by the comparative study published by Khuu and his collaborators (Khuu et al., 2004) as well as that from Genzyme group for chondrocytes (Kielpinski et al., 2005) showing the superiority of blood culture system. Then, we organized collaborative studies by sending cellular products contaminated at various levels on one hand to know the performances of the methods used, and on the other hand to measure the impact of the recommendations. A panel of samples was sent since 2001 once or twice per year to each participant who performed the analysis according to its own procedure. Results are collected and analyzed by Afssaps. These studies allowed us to propose French recommendations for this control published in 2002. They deal with these different aspects: staff, environment, sample, analysis, results and methods validation. The size of the sample to control is probably the main problem due to the very small quantity of product available for the bacteriological control in most cases. They showed their usefulness by the improvement of performances when they are applied. This work was proceded and allowed us to propose recommendations at the European level through the European Pharmacopoeia (EP) and a new chapter, 2.6.27., "Microbiological control of cellular products" came into force in January 2007. In this chapter, a recommended list of germs has been established to validate the technique used, and it is particularly adapted for haematopoietic products because of their main therapeutic use. It may be necessary to modify this list depending on the cell origin and on the micro-organisms previously found or associated with the type of cells: Recommended micro-organisms for validation in 2.6.27 EP chapter

- *Aspergillus brasiliensis*
- *Bacillus subtilis*
- *Candida albicans*
- *Clostridium sporogenes*
- *Propionibacterium acnes*
- *Pseudomonas aeruginosa*
- *Staphylococcus aureus*
- *Streptococcus pyogenes*
- *Yersinia enterocolitica*

Other approaches to validation may also be used such as interlaboratory comparison. On this basis, we proposed to the French cell therapy banks to participate to collaborative studies organised by the Laboratories and Control Department at Afssaps.

3.6.1 Validation stage at Afssaps

Validation of automated blood culture system was first led for haematopoietic products. A first one has been done on a Vital apparatus and the following list of germs has been tested:

Staphylococcus aureus (ATCC 6538)	*Escherichia coli (ATCC 8739)*
Staphylococcus epidermidis (ATCC 49461)	*Pseudomonas aeruginosa (ATCC 9027)*
Streptococcus pneumoniae (ATCC 6303)	*Corynebacterium jeikeium (AGMED)*
Enterococcus faecium (CIP 5432)	*Bacillus cereus (ATCC 10876)*
Streptococcus pyogenes (ATCC 19615)	*Acinetobacter baumanii (ATCC 19606)*
Clostridium sporogenes (ATCC 19404)	*Propionibacterium acnes (ATCC 11827)*
Bacillus subtilis (ATCC 6633)	*Mycobacterium smegmatis (CIP 103599)*
Aeromonas hydrophila (AGMED)	*Aspergillus brasiliensis (ATCC 16404)*
Yersinia enterocolitica (ATCC 9610)	*Candida albicans (ATCC 2091)*

This technique enables us to use very low volumes for the inoculation of the bottles and can detect contaminations of about 1 to 10 germs/ml. Specificity, reproducibility and sensitivity were checked for 18 germs (bacteria and fungi) chosen because of their representativeness (usual germs found in these products when a contamination happens), for their different type of metabolism and according to the chapter 2.6.1 of the European Pharmacopoeia for the environment germs. Because of a change of the automated blood culture system, a second validation was performed with a comparison of results obtained by Vital apparatus which was the first automated growth-based system we used with the new apparatus BactAlert (BioMérieux). For aerobic bacteria, 100% of positive results were obtained as for the smallest contamination tested (46 UFC/ml), a correlation of 97% was obtained between the 2 systems and no false positive was observed. For anaerobic bacteria, all the contaminations were also detected in the condition that an anaerobic SN media is used to detect *Propionibacterium acnes* which is a very fastidious growing germ. A better detection of anaerobic bacteria with BactAlert apparatus was obtained for the low contaminations (<50 UFC/ml). The sensitivity threshold is around 10^3 micro-organisms per bag (ie at least 100ml) when 1% of the bag volume is tested.

3.6.2 Collaborative studies

The decision to organize collaborative studies for the French Microbiology laboratories that perform sterility testing of cell products was taken by the working group "Bacteriological control of cellular products" at Afssaps. The aim of these studies is to evaluate the performances of each laboratory to enhance the use of a standardized and validated control. At least, 30 laboratories participated in each proposed study on a voluntary basis and 12 studies have been organised since December, 2001. The first one was performed with cell samples contaminated by a rapid growing germ (*Staphylococcus epidermidis*) sent to each laboratory participant. This study was designed as a feasibility study where the detection of this germ should normally not be difficult. A negative control was also sent to validate the transportation and manipulation property. Ninety-six % of good answers were obtained and feasibility was demonstrated. The second one took place in 2002 and this time, a more fastidious growing germ *Propionibacterium acnes* was sent as well as dendritic cell samples contaminated by *Enterobacter cloacae*. The observed results were the following ones:

Propionibacterium acnes 100 UFC/ml	Blood culture media (n=23)	2.6.1 Pharmacopoeia media (n=7)	Agar medium (direct inoculation) (n=8)
Positive result (%)	82.6	71.4	25
Negative result (%)	17.4	28.6	75

Table 5. Positive and negative results obtained by the participants to the collaborative study in 2002 according to control method used.

For the false negative results, 83% were associated to a non respect of the recently diffused recommendations. For *Enterobacter cloacae*, 100% of good results were obtained. At Afssaps, we compared the time to detection obtained with blood culture system equal to 10 H ± 0.9 which was shorter than that obtained with the 2.6.1 Pharmacopoeia media equal to 24 H ± 0 (n=8). The same observation was made for *Propionibacterium acnes* with respectively 106 H ± 8 and 147.4 ± 20 (p<0.002).

Another example is done with the study performed in 2005 where the different blood culture methods used by the participants were compared (Table 6). Here, we showed that the mean time to detection, for all the 4 germs, was significantly shorter (p=0,037) with automated blood culture method (35.5 H ± 34.6) than with manual one (59 H ± 83). Important standard deviations of the detection time were observed principally with the manual blood culture method.

Mean detection time (H)	*Candida glabrata*	*Acinetobacter 15 UFC/ml*	*Acinetobacter 150 UFC/ml*	*Bacteroides fragilis*
BactAlert (n=13)	26 ± 12.8	30 ± 13.2	26 ± 11.8	49 ± 20
Bactec (n=8)	102 ±72.3	18 ± 6.4	14 ± 6.7	18 ± 4.8
Manual blood culture (n=7)	39 ± 36.7	101 ± 141	85 ± 127.4	43 ± 38.9

Table 6. Evaluation of the detection time for the 4 germs sent in 2005 towards the 31 participating laboratories according to the different blood culture method used by 28 of them.

Mean times to detection for *Acinetobacter* at 15 and 150 UFC/ml observed with automated blood culture were also significantly different and shorter of those obtained by manual culture in inter-laboratories conditions.

A summary of all the collaborative studies is shown on the following figure 8 from 2001 to 2009 and their interest taking into consideration the 2.6.27 chapter has been noted. In particular, these studies allow the bacteriology laboratories implied in the control of the cellular products to validate their method by inter-laboratory comparisons. From now on, all the germs recommended by the 2.6.27 E.P. monograph were addressed from 2001 till now through these studies. However, some laboratories could not take part in all the studies or failed to detect some germs, in particular, such as *Propionibacterium acnes* (60% of satisfactory results with the weak concentration) and *Aspergillus brasiliensis* (82% of satisfactory results).

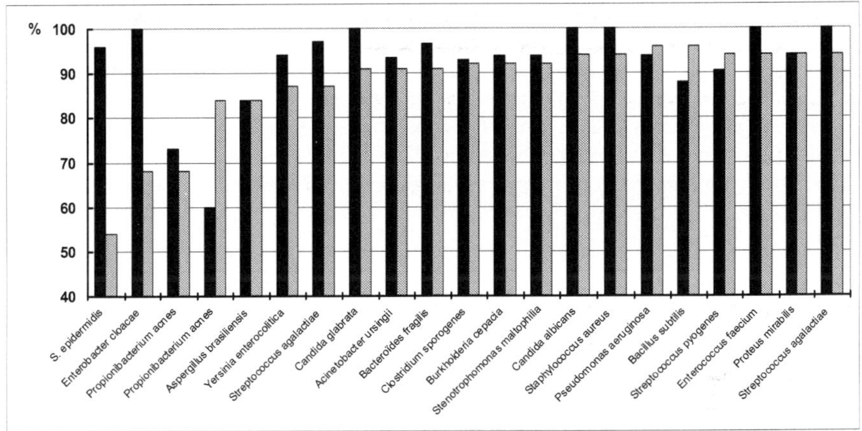

Fig. 8 Percent of satisfactory results obtained by the participant laboratories to collaborative studies performed from 2001 to 2009 (black bar) according to the germ in the contaminated samples sent by the DLC – Afssaps. Percent of blood culture media users for the microbiological control of cellular products (grey bar).

This figure shows that 17 germs chosen for the collaborative study (81% of the total), the percent of good results was superior or equal to 90% and that the users of blood culture media has increased from 54% to 95%. Nowadays, these studies continue in proposing germs which were not found yet by all the participants (either not detected or because the laboratory could not take part in a previous study) such as the germs which could be detected in an irregular way according to the technical requirements (temperature incubation for example for germs like *Pseudomonas fluorescens*). For the last one, we wish to have an inventory of technical conditions applied in the laboratories controlling the products of cellular therapy in France in order to have some information during the re-evaluation of the 2.6.27 monograph.

In order to make possible the use of these results by the laboratories at the time of audit or inspection for example, a laboratory by laboratory synthesis is in preparation.

3.6.3 Haematopoietic stem cell contamination collection

To better know the contaminations found in haematopoietic products and their frequency with an aim of defining new studies and also to analyze the different situations where contaminations happen, in 2004, we decided to collect the information about the contaminated products among the microbiological laboratories implied in the microbiological control of haematopoietic products (n=30). The following graphic (Figure 9) shows the rate of contamination for the different haematopoietic products in France from 2006 to 2009. With this follow-up, it is noticed that the contamination rate of the PBSC remains constant and from approximately 1% since 2006, that of bone marrows decreases since 2006, perhaps as a result of a better sensitizing of the teams for the harvesting whereas that of the Mononuclear Cells (MNC) exceeds 2% in 2009. The national contamination rate for bone marrow in 2007 and 2008 was the same as published by a mono-centre study (Vanneaux et al., 2007) and it was shown that contaminations came significantly from the harvesting and not from transformation in the cell facility. It is also observed a significant decrease (p<0.0001) of the contamination rate of cord blood units between 2008 to 2009, which could be explained by an improvement of the practices with the implication of new maternities in 2009 (in link with the openings of new cord blood banks) where enabling to the puncture of placental blood were recent (training, sharing experience between midwives, sensitizing to the procedures of disinfection...). For the cord blood units collected for banking, all units with a positive microbiological control at harvesting are discarded.

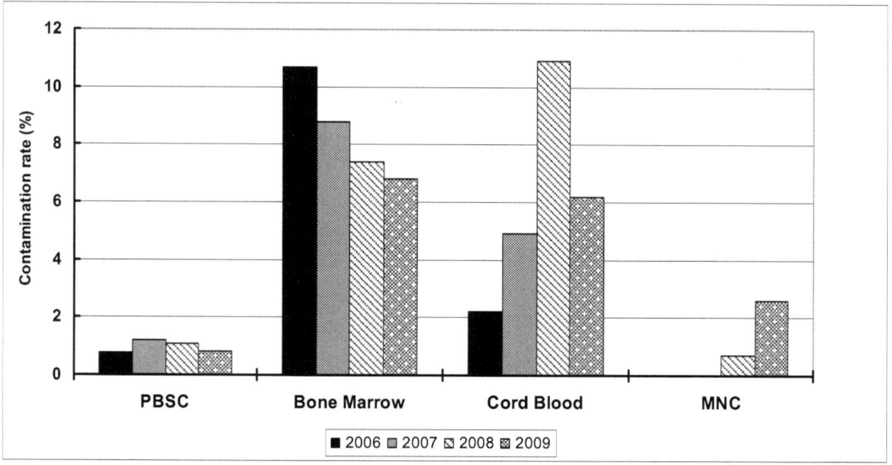

Fig. 9. Contamination rates are represented according to the year and to the haematopoietic product in 2006, 2007, 2008 and 2009.

It is also noted that the contamination rate of PBSC remains constant and of approximately 1% since 2006 and that of bone marrow decreases since 2006, perhaps the result of a better sensitizing of the collection teams whereas that of the MNC exceeds 2% in 2009
The germs found in the 1132 contaminated products for this 4 year period (for 54432 controlled haematopoietic products in the same time giving a global rate contamination equal to 2%) have been analysed regarding to the genus and as shown in table 7. The most frequent germs are, for PBSC and bone marrow, germs from skin as *Staphylococcus* and *Propionibacterium*; for umbilical cord blood, most frequent germs are from digestive tractus.
It is noted some homogeneity in the distribution of the germs by type of products with close rates from one year to another. This is probably the reflect of the exhaustiveness of the collection with a 90% feed-back for contamination declaim from sites preparing HSC as well as the sensitivity of the techniques implemented.
In this collection, we observed that 6.2% of the contaminations (70/1132) were detected after a 7 day incubation: where, for 70% of these cases, it was a contamination by *Propionibacterium acnes*, a very fastidious growing germ, the used technique was not an

automated blood culture system Moreover, we didn't have all the details concerning these cases, for example, the inoculum size being perhaps too small or the medium used wasn't the more adapted. Indeed, the performances of different blood culture media need to be checked because this germ doesn't grow with the same speed according to the medium. However, even if this type of micro-organism has a weak clinical relevance (Kamble et al., 2005) it is important to be able to detect all the contaminations to have a microbiological risk as low as possible, to check the process and to improve it, if necessary, in particular according to the contamination origin.

Year	PBSC	Bone Marrow	Umbilical cord Blood
2006	69% *Staphylococcus sp.* 8% *P.acnes*	57% *Staphylococcus sp.* 31% *P.acnes*	38% *Bacteroides sp.* 18% *Staphylococcus sp.* 13% *Corynebacterium* 10% *Streptococcus sp.*
2007	64% *Staphylococcus* 11% *Propionibacterium sp* 8% *Enterobacter*	50% *Staphylococcus* 34% *Propionibacterium sp*	39% *Bacteroides sp* 15% *Staphylococcus* 9% *Corynebacterium sp*
2008	62.6% *Staphylococcus sp.* 10.8% *Propionibacterium sp*	52.8% *Staphylococcus sp.* 35.8% *Propionibacterium sp*	28% *Bacteroides sp.* 26.3% *Staphylococcus sp.* 13% *Corynebacterium* 10.7% *Streptococcus sp.*
2009	61% *Staphylococcus sp.* 9.3% *Propionibacterium sp*	43.8% *Staphylococcus sp.* 32.8% *Propionibacterium sp*	28% *Bacteroides sp.* 21.9% *Staphylococcus sp.* 9.6% *Corynebacterium* 9.6% *Streptococcus sp.*

Table 7. Distribution according to main genus identified in the 1132 contaminated products for a 4 year period.

3.6.4 Recommendations for the microbiological control of cellular products

All these results allowed us to propose French guidelines[†] for this control nationally diffused to cell banks and bacteriological laboratories by Afssaps in 2002, and they showed their usefulness by the increase in performances throughout the various investigations. They deal with these different aspects: personnel, environment, sample, analysis, results and method validation. The size of the sample to control is probably the main problem due to the very small quantity of product available for the bacteriological control in most cases. The organization of collaborative studies by sending cellular products contaminated at various levels allowed to know the performances of the used methods and to measure the impact of

---

[†] Recommandations pour le contrôle bactériologique des produits de thérapie cellulaire- Produits hématopoïétiques. Afssaps, Direction des Laboratoires et des Contrôles – Afssaps /DLC. Référence GH/BP 2002-100 – Avril 2002.

the recommendations; we showed that improvement of the performances was in pair with the application of the recommendations. This work continued with new collaborative studies and a positioning of these recommendations at the European level (working group "Cell Therapy Products" near the European Pharmacopoeia) on the basis of 4 years of external control activity for cellular products. The chapter 2.6.27 "Microbiologic Control of cellular products" was published in the addendum 5.6 of the European Pharmacopoeia and is in application since 2007, January, the 1st . The objective is to promote standardized methods to guarantee the quality of the cellular products.

4. Conclusion and perspectives

Thirty dedicated synthesis have been prepared and sent to each site, so that they can look at their own results among all others obtained in the whole 30 sites. For the numeration and the percentage of CD34, a lower deviation than 20% with the Afssaps value is obtained for all the sites.

This assessment made it possible to appreciate some characteristics of this control and to evaluate the PBSC prepared in France by the 30 producing sites. This result follow-up can be integrated into the Quality Assurance System to attest the participation to the External Quality control. Moreover, these data can be provided at the time of a national request for authorization of a cellular therapy process/product in order to inform of the control validations obtained in inter-laboratory conditions. Finally, all these results show the good quality of the PBSC prepared in France and show the reliability of the used methods for the quality control of these haematopoietic products.

In conclusion, these works conducted by DLC at the French Health Product Agency, have left to increase technical standardization. External quality control allows comparing performances between various French facilities and constitutes an excellent indicator for CT laboratories. The single-platform methods are also relevant for the measurement of CD34+ cells in the thawed products where the sample brittleness makes the double-platform use more delicate (without beads) as well as for products from bone marrow because of the presence of erythroblasts (which are CD45 negative cells). Thus, external control results contribute to develop the use of standardized tools by showing their superiority as it confirms by several publications (Barnett et al., 2000 ; Gratama et al., 2003 ; Chang et al., 2004). Indeed, whereas single-platform methods were used only by 30% of French producers of HSC in 2000, there are now 75%. Finally, external control contributes to improve standardization of the used techniques, thanks to the comparative results established for cellular products received from all the production sites; and to guarantee the reliability of controls and decisions which result from this (particularly for the follow-up of the production systems). This control allowed the establishment of a follow-up for the variations site by site in the form of "control cards", though these cards not being of any lawful nature are used today by the producers in the files of requests for a process/ product authorization which they file in Afssaps in order to reinforce the data of quality.

The multi-centre laboratory study on thawed cord blood assesses the impact of the method used in flow cytometry to determine the number of viable CD34 cells. Finally, single-platform methods are especially useful for CD34+ cell enumeration in thawed CBUs. These methods are the only way to standardize sample preparation, viability measurement, gating and cytometers settings, and finally to reduce intra and inter laboratory variability. It is of particular interest in CBU transplantation, where CBUs are selected on cell counts produced by CB banks, and where clinical outpoint and thawed cellular products cell counts are correlated in multicentric studies.

The study about the granulated cell fraction effect on cell viability has taken a particular impact with the work done by the French biovigilance at Afssaps. Our findings allowed identifying product characteristics which could influence the product quality. The description of noxious effects linked to the granulated cell fraction on the PBSC quality (Calmels et al., 2007; Milone et al., 2007 ; Fois et al., 2007 ; Cordoba .et al., 2007 ; Martin-Henao et al., 2010) led to identify products which could present some risks at the administration. Following biovigilance declarations, highlighted the implication of this fraction in serious side effects of neurological type after the re-injection, an information and recommendation letter‡ from Afssaps was sent to all the French producers. The evaluation of the granulated cells in PBSC thus becomes a new quality indicator. This interaction biovigilance/quality control results in emitting "warning statements" in the certificates of analyses when the products are detected as fragile. As a consequence, it led to a particular management for these products on the sites (ie: cryopreservation at D0 only, premedication of the recipient…) and to sensitize to declare undesirable effects. Particular follow-up of that "risk" products could be done with additional control, precautions at the time of the administration, follow-up of the recipient. At last, as a good result, these measures led to a reduction in the declarations concerning this type of side effects on the national plan.

As a consequence, this work allowed us to propose recommendations at the European Pharmacopoeia to promote standardized methods to guarantee the quality of cellular therapy product. These recommendations are now validated and give rise to several monographs ie: 2323 "Human haematopoietic stem cells"; 2.7.23 "Numeration of CD34/CD45+ cells in haematopoietic products"; 2.7.24 "Flow cytometry"; 2.7.28 "Colony-forming cell assay for human haematopoietic progenitors cells"; 2.7.29 "Nucleated cell count and viability" and 2.6.27 "Microbiological control of cellular products"

Another important example is given with the implementation of a standardized microbiological control using blood culture system in more than 90% of the French cell banks leading to the monograph 2.6.27. The type of contaminated germs which can contaminate the HSC being primarily commensal germs, the blood culture media allow a better detection than media used with the method of the European Pharmacopoeia 2.6.1 and it is often more adapted because of the low volume available for this control. Our results are confirmed by the comparative study published by Khuu and collaborators (Khuu et al., 2004) showing the superiority of automated growth-based method. Those results are also confirmed by the complete validation data of this automated system done for chondrocytes (Kielpinski et al., 2005) leading to an approval of the use of this technique for the bacteriological control of this licensed product by FDA. At our level, a study for the bacteriological control of chondrocyte preparations initiated at the producer request made possible to test the use of the BactAlert apparatus in parallel of the use of TS/TR media. The results showed better results with the automated method (p<0.05).

All these results support the general use of automated growth-based methods for sterility testing of cell therapy products. They could be useful for the evaluation of cellular products at the European level. Indeed, several points should be considered:

---

‡ http://www.afssaps.fr/content/download/555/5823/version/3/file/csh.pdf

- Need for a rapid and sensitive control at the producer level even if the result is known after the release and the injection to the patient, it helps to better monitor performances of the production facilities if necessary. However, to improve this control, a revision of the 2.6.27 chapter is currently in progress with the development of rapid methods for sterility testing like PCR, flow cytometry or the detection of microcolonies. The Paul Erlich Institute (an Official Medicinal Laboratory Control for biological products) in Germany has already validated flow cytometry for the microbiological control of platelets (Karo et al., 2008) and it is in progress for the control of cell products.
- Considering the rare and single character of some products and the benefit/risk, the challenge is to have the best suitable test to be sure to take the decision even if a contamination has been found.
- Need for a method as growth-based one to allow identification
- Clinical sequelae following infusion of a microbial contaminated progenitors cells are rare (Padley et al., 2007 ; Kamble et al., 2005 ; Lowder & Whelton, 2003 ; Schwella et al., 1998 ; Klein et al., 2006)

All these considerations are in agreement with those of Padley and colleagues (Padley et al., 2007) who have reviewed the product culture results and clinical outcomes from 1998 to 2006 representing 7233 haematopoietic stem cell collections. Finally, we have done a study in 2009 with the French cord blood bank to evaluate performances of automated culture system after the thawing of the cord blood units. This system has already been validated for these products by an umbilical cord blood bank in Australia (Sparrow, 2004). All these results would allow us to complete these 2.6.27 recommendations.

In the future, we propose to study apoptosis, to establish CFU-GM recommendations for the assay validation which could complete the 2.7.28 European Pharmacopoeia chapter, to evaluate new methods for rapid germ detections and to initiate other analysis on human cells used in immunotherapy and regenerative therapy.

5. Acknowledgments

Sophie Ardiot was a student from ISIFC Genie Biomedical, 23 rue Alain Savary, Besançon , France. We thank the EFS Bourgogne - Franche Comté/ Besançon for the gracious provision of the displaced USP as well as the DLC of Afssaps for the logistical support. We also thank Vanessa Duval and Celia Maquin for their technical assistance for the functional assay at Afssaps and An Le and Brigitte Rogeau, Pharmacopoeia Unit at Afssaps, for their strong implication to bring French recommendations to the European Pharmacopoeia. We sincerely thank all the French producers of haematopoietic stem cells for their active participation. Thank to the 17 French Blood Establishment : Alpes Méditerranée / St-Laurent du Var - P. Philip, C.Tirtaine ; Aquitaine Limousin / Bordeaux - B.Dazey ; Bretagne / Rennes - C. Le Berre ; Bretagne / Brest - P. Delepine ; Bourgogne Franche-Comté / Besançon - C. Malugani ; Centre Atlantique / Tours - J. Domenech (Hôpital Bretonneau) ; Centre Atlantique / Poitiers - C. Giraud ; Ile de France / Créteil - H. Rouard ; Normandie / Caen - A. Batho ; Normandie / Bois-Guillaume - P. Pommier ; Nord de France / Lille - F. Boulanger ; Nord de France / Reims - E. Toulmonde ; Pays de Loire / Angers - N. Piard ; Pays de Loire / Nantes - A.-G. Léaute ; Pyrénées-Méditerranée / Toulouse - P. Bourin ; Rhône-Alpes / St-Ismier - V. Persoons ; Rhône-Alpes / Lyon - O. Hequet. Thank to the CTS des Armées / Clamart - J.-J. Lataillade and to the 12 Health Establishment : CH Mulhouse –

V. Rimelen, V. Reymond ; Centre Léon Bérard / Lyon – G. Clapisson ; CHRU Dupuytren / Limoges - M. Donnard ; CHU Clermont-Ferrand –N. Boiret ; CHU Montpellier Hôpital St-Eloi – J.L. Veyrune ; CHU Nancy – D. Bensoussan ; CHU Strasbourg / Hôpital de Haute Pierre – A. Bohbot ; Hôpital Necker Enfants malades / Paris – L. Dal Cortivo ; Hôpital de la Pitié-Salpétrière / Paris – M. Rosenzwajg ; Hôpital St-Louis / Paris – D. Rea ; Institut Gustave Roussy / Villejuif – V. Lapierre ; Institut Paoli-Calmettes / Marseille – B. Calmels. Lastly, we thank Sylvie Deguingand for helpful assistance for the English reviewing.

6. References

Gratwohl A, Baldomero H, Frauendorfer K, Urbano-Ispizua A, Joint, Accreditation, Committee, International, Society, for, Cellular, Therapy, European, Group, for, Blood, and, Marrow, Transplantation. (2006). EBMT activity survey 2004 and changes in disease indication over the past 15 years. *Bone Marrow Transplant*, Vol.37, N°12, pp.1069-1085,

Read E, Sullivan M. (2004). Cellular therapy services provided by blood centers and hospitals in the United States, 1999: an analysis from the Nationwide Blood Collection and Utilization Survey. *Transfusion*, Vol.44, N°4, pp.539-546,

Copelan E. (2006). Hematopoietic Stem-Cell Transplantation. *N Engl J Med*, Vol.354, pp.1813-1826,

Bender J, To L, Williams S, Schwartzberg L. (1993). Defining a therapeutic dose of peripheral blood stem cells. *J Hematother*, Vol.1, N°4, pp.329-341,

Allan DS, Keeney M, Howson-Jan K, Popma J, Weir K, Bhatia M, Sutherland DR, Chin-Yee IH. (2002). Number of viable CD34(+) cells reinfused predicts engraftment in autologous hematopoietic stem cell transplantation. *Bone Marrow Transplant*, Vol.29, N°12, (Jun), pp.967-972,

Weaver CH, Hazelton B, Birch R, Palmer P, Allen C, Schwartzberg L, West W. (1995). An analysis of engraftment kinetics as a function of the CD34 content of peripheral blood progenitor cell collections in 692 patients after the administration of myeloablative chemotherapy. *Blood*, Vol.86, N°10, (November 15), pp.3961-3969,

Sutherland DR, Anderson L, Keeney M, Nayar R, Chin-Yee I. (1996). The ISHAGE guidelines for CD34+ cell determination by flow cytometry. International Society of Hematotherapy and Graft Engineering. *J Hematother*, Vol.5, N°3, (Jun), pp.213-226, 1061-6128 (Print) 1061-6128 (Linking)

Keeney M, Chin-Yee I, Weir K, Popma J, Nayar R, Sutherland D. (1998). Single platform flow cytometric absolute CD34+ cell counts based on the ISHAGE guidelines. *Cytometry*, Vol.34, pp.61-70,

Barnett D, Granger V, Kraan J, Whitby L, Reilly JT, Papa S, Gratama JW. (2000). Reduction of intra- and interlaboratory variation in CD34+ stem cell enumeration using stable test material, standard protocols and targeted training. DK34 Task Force of the European Working Group of Clinical Cell Analysis (EWGCCA). *Br J Haematol*, Vol.108, N°4, (Mar), pp.784-792, 0007-1048 (Print) 0007-1048 (Linking)

Gratama J, Kraan J, Keeney M, Sutherland D, Granger V, Barnett D. (2003). Validation of the single-platform ISHAGE methods for CD34+ hematopoietic stem and progenitor cell enumeration in an international multicenter study. *Cytotherapy*, Vol.5, N°1, pp.55-65,

Chang A, Raik E, Marsden K, Ma DD. (2004). Australasian CD34+ quality assurance program and rationale for the clinical utility of the single-platform method for CD34+ cell enumeration. *Cytotherapy*, Vol.6, N°1, pp.50-61,

Laroche V, McKenna DH, Moroff G, Schierman T, Kadidlo D, McCullough J. (2005). Cell loss and recovery in umbilical cord blood processing: a comparison of postthaw and postwash samples. *Transfusion*, Vol.45, N°12, (Dec), pp.1909-1916, 0041-1132 (Print) 0041-1132 (Linking)

Timeus F, Crescenzio N, Sanavio F, Fazio L, Doria A, Foglia L, Pignochino Y, Berger M, Piacibello W, Madon E, Cordero di Montezemolo L, Fagioli F. (2006). Multilineage engraftment of refrozen cord blood hematopoietic progenitors in NOD/SCID mice. *Haematologica*, Vol.91, N°3, (Mar), pp.369-372, 1592-8721 (Electronic) 0390-6078 (Linking)

Scaradavou A, Smith KM, Hawke R, Schaible A, Abboud M, Kernan NA, Young JW, Barker JN. (2010). Cord blood units with low CD34+ cell viability have a low probability of engraftment after double unit transplantation. *Biol Blood Marrow Transplant*, Vol.16, N°4, (Apr), pp.500-508, 1523-6536 (Electronic) 1083-8791 (Linking)

Gluckman E, Rocha V, Boyer-Chammard A, Locatelli F, Arcese W, Pasquini R, Ortega J, Souillet G, Ferreira E, Laporte JP, Fernandez M, Chastang C. (1997). Outcome of cord-blood transplantation from related and unrelated donors. Eurocord Transplant Group and the European Blood and Marrow Transplantation Group. *N Engl J Med*, Vol.337, N°6, (Aug 7), pp.373-381, 0028-4793 (Print) 0028-4793 (Linking)

Rubinstein P, Carrier C, Scaradavou A, Kurtzberg J, Adamson J, Migliaccio AR, Berkowitz RL, Cabbad M, Dobrila NL, Taylor PE, Rosenfield RE, Stevens CE. (1998). Outcomes among 562 recipients of placental-blood transplants from unrelated donors. *N Engl J Med*, Vol.339, N°22, (Nov 26), pp.1565-1577, 0028-4793 (Print)

Grewal SS, Barker JN, Davies SM, Wagner JE. (2003). Unrelated donor hematopoietic cell transplantation: marrow or umbilical cord blood? *Blood*, Vol.101, N°11, (Jun 1), pp.4233-4244, 0006-4971 (Print)

Migliaccio AR, Adamson JW, Stevens CE, Dobrila NL, Carrier CM, Rubinstein P. (2000). Cell dose and speed of engraftment in placental/umbilical cord blood transplantation: graft progenitor cell content is a better predictor than nucleated cell quantity. *Blood*, Vol.96, N°8, (Oct 15), pp.2717-2722, 0006-4971 (Print)

Wagner JE, Barker JN, DeFor TE, Baker KS, Blazar BR, Eide C, Goldman A, Kersey J, Krivit W, MacMillan ML, Orchard PJ, Peters C, Weisdorf DJ, Ramsay NK, Davies SM. (2002). Transplantation of unrelated donor umbilical cord blood in 102 patients with malignant and nonmalignant diseases: influence of CD34 cell dose and HLA disparity on treatment-related mortality and survival. *Blood*, Vol.100, N°5, (Sep 1), pp.1611-1618, 0006-4971 (Print)

Lee S, Kim S, Kim H, Baek EJ, Jin H, Kim J, Kim HO. (2008). Post-thaw viable CD34(+) cell count is a valuable predictor of haematopoietic stem cell engraftment in autologous peripheral blood stem cell transplantation. *Vox sang*, Vol.94, N°2, (Feb), pp.146-152, 0042-9007 (Print) 0042-9007 (Linking)

Moroff G, Eichler H, Brand A, Kekomaki R, Kurtz J, Letowska M, Pamphilon D, Read EJ, Porretti L, Lecchi L, Reems JA, Sacher R, Seetharaman S, Takahashi TA. (2006).

Multiple-laboratory comparison of in vitro assays utilized to characterize hematopoietic cells in cord blood. *Transfusion*, Vol.46, N°4, (Apr), pp.507-515, 0041-1132 (Print)

Brand A, Eichler H, Szczepiorkowski ZM, Hess JR, Kekomaki R, McKenna DH, Pamphilon D, Reems J, Sacher RA, Takahashi TA, van de Watering LM. (2008). Viability does not necessarily reflect the hematopoietic progenitor cell potency of a cord blood unit: results of an interlaboratory exercise. *Transfusion*, Vol.48, N°3, (Mar), pp.546-549, 0041-1132 (Print)

Beaujean F, Pico J, Norol F, Divine M, Le Forestier C, Duedari N. (1996). Characteristics of peripheral blood progenitor cells frozen after 24 hours of liquid storage. *J Hematother*, Vol.5, pp.681-686,

Campos L, Roubi N, Guyotat D. (1995). Definition of optimal conditions for collection and cryopreservation of umbilical cord hematopoietic cells. *Cryobiology*, Vol.32, N°6, (Dec), pp.511-515, 0011-2240 (Print) 0011-2240 (Linking)

Calmels B, Lemarie C, Esterni B, Malugani C, Charbonnier A, Coso D, de Colella JM, Deconinck E, Caillot D, Viret F, Ladaique P, Lapierre V, Chabannon C. (2007). Occurrence and severity of adverse events after autologous hematopoietic progenitor cell infusion are related to the amount of granulocytes in the apheresis product. *Transfusion*, Vol.47, N°7, (Jul), pp.1268-1275, 0041-1132 (Print) 0041-1132 (Linking)

Moroff G, Seetharaman S, Kurtz J, Greco N, Mullen M, Lane T, Law P. (2004). Retention of cellular properties of PBPCs following liquid storage and cryopreservation. *Transfusion*, Vol.44, N°2, pp.245-252,

Brecher ME, Means N, Jere CS, Heath D, Rothenberg S, Stutzman LC. (2001). Evaluation of an automated culture system for detecting bacterial contamination of platelets: an analysis with 15 contaminating organisms. *Transfusion*, Vol.41, N°4, (Apr), pp.477-482, 0041-1132 (Print) 0041-1132 (Linking)

McDonald CP, Roy A, Lowe P, Robbins S, Hartley S, Barbara JA. (2001). Evaluation of the BacT/Alert automated blood culture system for detecting bacteria and measuring their growth kinetics in leucodepleted and non-leucodepleted platelet concentrates. *Vox sang*, Vol.81, N°3, (Oct), pp.154-160, 0042-9007 (Print) 0042-9007 (Linking)

Dunne WJ, Case L, Isgriggs L, Lublin D. (2005). In-house validation of the BACTEC 9240 blood culture system for detection of bacterial contamination in platelet concentrates. *Transfusion*, Vol.45, pp.1138-1142,

Schwella N, Rick O, Heuft HG, Miksits K, Zimmermann R, Zingsem J, Eckstein R, Huhn D. (1998). Bacterial contamination of autologous bone marrow: reinfusion of culture-positive grafts does not result in clinical sequelae during the posttransplantation course. *Vox sang*, Vol.74, N°2, pp.88-94, 0042-9007 (Print) 0042-9007 (Linking)

Brocklebank AM, Sparrow RL. (2001). Enumeration of CD34+ cells in cord blood: a variation on a single-platform flow cytometric method based on the ISHAGE gating strategy. *Cytometry*, Vol.46, N°4, (Aug 15), pp.254-261,

Dauber K, Becker D, Odendahl M, Seifried E, Bonig H, Tonn T. (2010) Enumeration of viable CD34(+) cells by flow cytometry in blood, bone marrow and cord blood:

results of a study of the novel BD stem cell enumeration kit. *Cytotherapy*, Vol.13, N°4, (Apr), pp.449-458, 1477-2566 (Electronic) 1465-3249 (Linking)

Rocha V, Gluckman E. (2009). Improving outcomes of cord blood transplantation: HLA matching, cell dose and other graft- and transplantation-related factors. *Br J Haematol*, Vol.147, N°2, (Oct), pp.262-274, 1365-2141 (Electronic) 1365-2141 (Linking)

Wagner E, Duval M, Dalle JH, Morin H, Bizier S, Champagne J, Champagne MA. (2006). Assessment of cord blood unit characteristics on the day of transplant: comparison with data issued by cord blood banks. *Transfusion*, Vol.46, N°7, (Jul), pp.1190-1198, 0041-1132 (Print) 0041-1132 (Linking)

Lemarie C, Esterni B, Calmels B, Dazey B, Lapierre V, Lecchi L, Meyer A, Rea D, Thuret I, Chambost H, Curtillet C, Chabannon C, Michel G. (2007). CD34(+) progenitors are reproducibly recovered in thawed umbilical grafts, and positively influence haematopoietic reconstitution after transplantation. *Bone Marrow Transplant*, Vol.39, N°8, (Apr), pp.453-460, 0268-3369 (Print)

Yoo KH, Lee SH, Kim HJ, Sung KW, Jung HL, Cho EJ, Park HK, Kim HA, Koo HH. (2007). The impact of post-thaw colony-forming units-granulocyte/macrophage on engraftment following unrelated cord blood transplantation in pediatric recipients. *Bone Marrow Transplant*, Vol.39, N°9, (May), pp.515-521, 0268-3369 (Print) 0268-3369 (Linking)

Panterne B, Richard MJ, Sabatini C, Pouthier F, Mouillot L, Bardey D, Boulanger F, Crea S, Dal Cortivo L, Decot V, Fleury-Cappellesso S, Giraud C, Lapierre V, Leaute AG, Le Berre C, Lemarie C, Piard N, Rapatel C, Rosenzwajg M. (2010). Quality control of defrosted cord blood units: results from an inter-laboratory study. *Transfus Clin Biol*, Vol.17, N°2, (Apr), pp.41-46, 1953-8022 (Electronic) 1246-7820 (Linking)

Rivadeneyra-Espinoza L, Perez-Romano B, Gonzalez-Flores A, Guzman-Garcia MO, Carvajal-Armora F, Ruiz-Arguelles A. (2006). Instrument- and protocol-dependent variation in the enumeration of CD34+ cells by flow cytometry. *Transfusion*, Vol.46, N°4, (Apr), pp.530-536, 0041-1132 (Print) 0041-1132 (Linking)

Barnett D, Janossy G, Lubenko A, Matutes E, Newland A, Reilly JT. (1999). Guideline for the flow cytometric enumeration of CD34+ haematopoietic stem cells. Prepared by the CD34+ haematopoietic stem cell working party. General Haematology Task Force of the British Committee for Standards in Haematology. *Clin Lab Haematol*, Vol.21, N°5, (Oct), pp.301-308,

Lemarie C, Bouchet G, Sielleur I, Durand V, Navarro F, Calmels B, Chabannon C. (2009). A new single-platform method for the enumeration of CD34+ cells. *Cytotherapy*, Vol.11, N°6, pp.804-806, 806 e801, 1477-2566 (Electronic) 1465-3249 (Linking)

Milone G, Mercurio S, Strano A, Leotta S, Pinto V, Battiato K, Coppoletta S, Murgano P, Farsaci B, Privitera A, Giustolisi R. (2007). Adverse events after infusions of cryopreserved hematopoietic stem cells depend on non-mononuclear cells in the infused suspension and patient age. *Cytotherapy*, Vol.9, N°4, pp.348-355, 1465-3249 (Print) 1465-3249 (Linking)

Fois E, Desmartin M, Benhamida S, Xavier F, Vanneaux V, Rea D, Fermand JP, Arnulf B, Mounier N, Ertault M, Lotz JP, Galicier L, Raffoux E, Benbunan M, Marolleau JP, Larghero J. (2007). Recovery, viability and clinical toxicity of thawed and washed haematopoietic progenitor cells: analysis of 952 autologous peripheral blood stem cell transplantations. *Bone Marrow Transplant*, Vol.40, N°9, (Nov), pp.831-835, 0268-3369 (Print) 0268-3369 (Linking)

Cordoba R, Arrieta R, Kerguelen A, Hernandez-Navarro F. (2007). The occurrence of adverse events during the infusion of autologous peripheral blood stem cells is related to the number of granulocytes in the leukapheresis product. *Bone Marrow Transplant*, Vol.40, N°11, (Dec), pp.1063-1067, 0268-3369 (Print) 0268-3369 (Linking)

Ravagnani F, Siena S, De Reys S, Di Nicola M, Notti P, Giardini R, Bregni M, Matteucci P, Gianni A, Pellegris G. (1999). Improved collection of mobilized CD34 hematopoietic progenitor cells by a novel leukapheresis system. *Transfusion*, Vol.39, N°1, pp.48-55,

Khuu H, Stock F, McGann M, Carter C, Atkins J, Murray P, Read E. (2004). Comparison of automated culture systems with a CFR/USP-compliant method for sterility testing of cell therapy products. *Cytotherapy*, Vol.6, N°3, pp.183-195,

Kielpinski G, Prinzi S, Duguid J, du Moulin G. (2005). Roadmap to approval: use of an automated sterility test method as a lot release test for Carticel®, autologous cultured chondrocytes. *Cytotherapy*, Vol.7, N°6, pp.531-541,

European Pharmacopoiea. Monograph 2.6.27. Microbiological control of cellular products. 7th edition. 2011

Vanneaux V, Fois E, Robin M, Rea D, de Latour RP, Biscay N, Chantre E, Robert I, Wargnier A, Traineau R, Benbunan M, Marolleau JP, Socie G, Larghero J. (2007). Microbial contamination of BM products before and after processing: a report of incidence and immediate adverse events in 257 grafts. *Cytotherapy*, Vol.9, N°5, pp.508-513, 1465-3249 (Print) 1465-3249 (Linking)

Kamble R, Pant S, Selby G, Kharfan-Dabaja M, Sehti S, Kratochvil K, Kohrt N, Ozer H. (2005). Microbial contamination of hematopoietic progenitor cell grafts - incidence, clinical outcome, and cost-effectiveness: an analysis of 735 grafts. *Transfusion*, Vol.45, pp.874-878,

Martin-Henao GA, Resano PM, Villegas JM, Manero PP, Sanchez JM, Bosch MP, Codins AE, Bruguera MS, Infante LR, Oyarzabal AP, Soldevila RN, Caiz DC, Bosch LM, Barbeta EC, Ronda JR. (2010). Adverse reactions during transfusion of thawed haematopoietic progenitor cells from apheresis are closely related to the number of granulocyte cells in the leukapheresis product. *Vox sang*, Vol.99, N°3, (Oct), pp.267-273, 1423-0410 (Electronic) 0042-9007 (Linking)

Karo O, Wahl A, Nicol SB, Brachert J, Lambrecht B, Spengler HP, Nauwelaers F, Schmidt M, Schneider CK, Muller TH, Montag T. (2008). Bacteria detection by flow cytometry. *Clin Chem Lab Med*, Vol.46, N°7, pp.947-953, 1434-6621 (Print) 1434-6621 (Linking)

Padley DJ, Dietz AB, Gastineau DA. (2007). Sterility testing of hematopoietic progenitor cell products: a single-institution series of culture-positive rates and successful infusion of culture-positive products. *Transfusion*, Vol.47, N°4, (Apr), pp.636-643, 0041-1132 (Print) 0041-1132 (Linking)

Lowder J, Whelton P. (2003). Microbial contamination of cellular products for hematolymphoid transplantation therapy: assessment of the problem and strategies to minimize the clinical impact. *Cytotherapy*, Vol.5, N°5, pp.377-390,

Klein MA, Kadidlo D, McCullough J, McKenna DH, Burns LJ. (2006). Microbial contamination of hematopoietic stem cell products: incidence and clinical sequelae. *Biol Blood Marrow Transplant*, Vol.12, N°11, (Nov), pp.1142-1149, 1083-8791 (Print) 1083-8791 (Linking)

Sparrow R. (2004). Microbial Screening of Umbilical Cord Blood Units by an Automated Culture System: Effect of Delayed Testing on Bacterial Detection. *Cytotherapy*, Vol.6, N°1, pp.23-29,

Part 2

Molecular and Cellular Engineering: Biomedical Applications

5

Hydrogels: Methods of Preparation, Characterisation and Applications

Syed K. H. Gulrez, Saphwan Al-Assaf
and Glyn O Phillips
*Glyn O Phillips Hydrocolloids Research Centre
Glyndwr University, Wrexham
United Kingdom*

1. Introduction

The terms gels and hydrogels are used interchangeably by food and biomaterials scientists to describe polymeric cross-linked network structures. Gels are defined as a substantially dilute cross-linked system, and are categorised principally as weak or strong depending on their flow behaviour in steady-state (Ferry, 1980). Edible gels are used widely in the food industry and mainly refer to gelling polysaccharides (i.e. hydrocolloids) (Phillips & Williams, 2000). The term hydrogel describes three-dimensional network structures obtained from a class of synthetic and/or natural polymers which can absorb and retain significant amount of water (Rosiak & Yoshii, 1999). The hydrogel structure is created by the hydrophilic groups or domains present in a polymeric network upon the hydration in an aqueous environment.

This chapter reviews the preparation methods of hydrogels from hydrophilic polymers of synthetic and natural origin with emphasis on water soluble natural biopolymers (hydrocolloids). Recent advances in radiation cross-linking methods for the preparation of hydrogel are particularly addressed. Additionally, methods to characterise these hydrogels and their proposed applications are also reviewed.

1.1 Mechanism of network formation

Gelation refers to the linking of macromolecular chains together which initially leads to progressively larger branched yet soluble polymers depending on the structure and conformation of the starting material. The mixture of such polydisperse soluble branched polymer is called 'sol'. Continuation of the linking process results in increasing the size of the branched polymer with decreasing solubility. This 'infinite polymer' is called the 'gel' or 'network' and is permeated with finite branched polymers. The transition from a system with finite branched polymer to infinite molecules is called 'sol-gel transition' (or 'gelation') and the critical point where gel first appears is called the 'gel point' (Rubinstein & Colby, 2003). Different types of gelation mechanism are summarised in Figure 1. Gelation can take place either by physical linking (physical gelation) or by chemical linking (chemical gelation). Physical gels can be sub categorised as strong physical gels and weak gels. Strong physical gel has strong physical bonds between polymer chains and is effectively permanent at a given set of experimental conditions. Hence, strong physical gels are analogous to chemical gels. Examples of strong physical bonds are lamellar microcrystals, glassy nodules or double and triple helices. Weak physical gels have reversible links formed from temporary associations between chains. These associations have finite lifetimes, breaking and reforming continuously. Examples of weak physical bonds are hydrogen bond, block copolymer micelles, and ionic associations. On the other hand, chemical gelation involves formation of covalent bonds and always results in a strong gel. The three main chemical gelation processes include condensation, vulcanisation, and addition polymerisation.

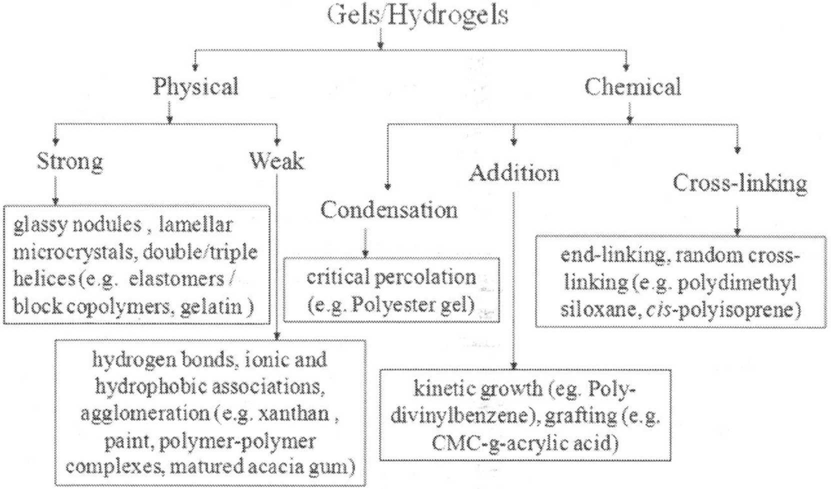

Fig. 1. Classification of gelation mechanism and relevant examples.

1.2 Classification of hydrogel

Hydrogels are broadly classified into two categories:

Permanent / chemical gel: they are called 'permanent' or 'chemical' gels when they are covalently cross-linked (replacing hydrogen bond by a stronger and stable covalent bonds) networks (Hennink & Nostrum, 2002). They attain an equilibrium swelling state which depends on the polymer-water interaction parameter and the crosslink density (Rosiak & Yoshii, 1999).

Reversible / physical gel: they are called 'reversible' or 'physical' gels when the networks are held together by molecular entanglements, and / or secondary forces including ionic, hydrogen bonding or hydrophobic interactions. In physically cross-linked gels, dissolution is prevented by physical interactions, which exist between different polymer chains (Hennink & Nostrum, 2002). All of these interactions are reversible, and can be disrupted by changes in physical conditions or application of stress (Rosiak & Yoshii, 1999).

1.3 Characteristic of hydrogel

The water holding capacity and permeability are the most important characteristic features of a hydrogel. The polar hydrophilic groups are the first to be hydrated upon contact with water which leads to the formation of primary bound water. As a result the network swells and exposes the hydrophobic groups which are also capable of interacting with the water molecules. This leads to the formation of hydrophobically-bound water, also called 'secondary bound water'. Primary and secondary bound water are often combined and called 'total bound water'. The network will absorb additional water, due to the osmotic driving force of the network chains towards infinite dilution. This additional swelling is opposed by the covalent or physical cross-links, leading to an elastic network retraction force. Thus, the hydrogel will reach an equilibrium swelling level. The additional absorbed water is called 'free water' or 'bulk water', and assumed to fill the space between the network chains, and/or the centre of larger pores, macropores, or voids. Depending on the nature and composition of the hydrogel the next step is the disintegration and/or dissolution if the network chain or cross-links are degradable. Biodegradable hydrogels, containing labile bonds, are therefore advantageous in applications such as tissue engineering, wound healing and drug delivery. These bonds can be present either in the polymer backbone or in the cross-links used to prepare the hydrogel. The labile bonds can be broken under physiological conditions either enzymatically or chemically, in most of the cases by hydrolysis (Hennink & Nostrum, 2002; Hoffman, 2002).

Biocompatibility is the third most important characteristic property required by the hydrogel. Biocompatibility calls for compatibility with the immune system of the hydrogel and its degradation products formed, which also should not be toxic. Ideally they should be metabolised into harmless products or can be excreted by the renal filtration process. Generally, hydrogels possess a good biocompatibility since their hydrophilic surface has a low interfacial free energy when in contact with body fluids, which results in a low tendency for proteins and cells to adhere to these surfaces. Moreover, the soft and rubbery nature of hydrogels minimises irritation to surrounding tissue (Anderson & Langone, 1999; Smetana, 1993).

The cross-links between the different polymer chains results in viscoelastic and sometimes pure elastic behaviour and give a gel its structure (hardness), elasticity and contribute to stickiness. Hydrogels, due to their significant water content possess a degree of flexibility similar to natural tissue. It is possible to change the chemistry of the hydrogel by controlling their polarity, surface properties, mechanical properties, and swelling behaviour.

1.4 Stimuli responsive hydrogels

Hydrogels can also be stimuli sensitive and respond to surrounding environment like temperature, pH and presence of electrolyte (Nho et al., 2005). These are similar to conventional hydrogels except these gels may exhibit significant volume changes in response to small changes in pH, temperature, electric field, and light. Temperature sensitive hydrogels are also called as thermogels (Jarry et al., 2002; Schuetz et al., 2008). These stimuli-sensitive hydrogels can display changes in their swelling behaviour of the network structure according to the external environments. They may exhibit positive thermo-sensitivity of swelling, in which polymers with upper critical solution temperature (UCST; temperature at which mixture of two liquids, immiscible at room temperature, ceases to separate into two phases) shrink by cooling below the UCST (Said et al., 2004). Some of the examples of stimuli sensitive hydrogels are poly (vinyl methyl ether) and poly (N-isopropyl acrylamide) gels, kappa-carrageenan-calcium based hydrogels, etc. (Bhardwaj et al., 2005; Sen, 2005). A summary of recent progress in biodegradable temperature sensitive polymers including polyesters, polyphosphazenes, polypeptides, and chitosan, and pH/temperature-sensitive polymers such as sulfamethazine-, poly(b-amino ester)-, poly(amino urethane)-, and poly(amidoamine)-based polymers is reviewed recently by Nguyen and Lee (2010). Recent progresses in the development and applications of smart polymeric gels have been reviewed extensively by Masteikova, Chalupova et al. (2003) and Chaterji, Kwon et al. (2007).

1.5 Xerogel & aerogel

A 'xerogel' is a solid formed from a gel by drying it slowly at about room temperature with unhindered shrinkage (Livage et al., 1988). Xerogels usually retain high porosity (25%) and enormous surface area (150–900 m²/g), along with very small pore size (1-10 nm). One such example of xerogel is boehmite AlO(OH)-monolithic gels with proposed application in space exploration and electronics (Yoldas, 1975). 'Aerogel' is derived from a gel (essentially by supercritical drying technique) in which the liquid component of the gel has been replaced with a gas. The result is an extremely low-density solid with several remarkable properties, most notably its effectiveness as a thermal insulator and its extremely low density. It is also called frozen smoke, solid smoke or blue smoke due to its translucent nature and the way light scatters in the material. Some of the examples are carbon and silicon aerogels which can be used in buildings double window glazing as transparent thermal super-insulators (Kistler, 1931).

2. Characterisation

An easy way to quantify the presence of hydrogel in a system is to disperse the polymer in water using a cylindrical vial and visually observe the formation of insoluble material. Visual monitoring of the solution viscosity by turning the universal up-side down can also provide quick measure of the bulk viscosity. reported in literature .

2.1 Solubility
2.1.1 Method A

Normally the hydrogel content of a given material is estimated by measuring its insoluble part in dried sample after immersion in deionised water for 16 h (Katayama, Nakauma 2006) or 48 h at room temperature (Nagasawa et al., 2004). The sample should be prepared at a dilute concentration (typically ~ 1%) to ensure that hydrogel material is fully dispersed in water. The gel fraction is then measured as follows:

$$Gel\ Fraction\ (hydrogel\%) = \left(\frac{W_d}{W_i}\right) * 100 \qquad (1)$$

Where, W_i is the initial weight of dried sample and W_d is the weight of the dried insoluble part of sample after extraction with water.

2.1.2 Method B

A more accurate measure of the insoluble fraction (also termed as hydrogel can be determined by measuring the weight retained after vacuum filtration. This is essentially the method prescribed by JECFA (Joint Expert Committee on Food Additives) for hydrocolloids which we have modified by changing the solvent from mild alkaline to water (Al-Assaf et al., 2009). The weight (W_1) of a 70 mm glass fibre paper (pore size 1.2 micron) is determined following drying in an oven at 105°C for 1 hour and subsequently cooled in a desiccator containing silica gel. Depending on the test material, 1-2 wt% (S) dispersion can be prepared in distilled water followed by overnight hydration at room temperature. The hydrated dispersion is then centrifuged for 2-5 minutes at 2500 rpm prior to filtration. Drying of the filter paper is carried out in an oven at 105°C followed by cooling to a constant weight (W_2). % Insoluble can then be calculated:

$$\% Hydrogel = \left(\frac{W_2 - W_1}{S} \right) * 100 \tag{2}$$

Depending on the test material different mesh size can be also used, e.g. the use of a 20-mesh steel screen (1041 µm) to determine the gel fraction (Yoshii & Kume, 2003).

2.2 Swelling measurement
2.2.1 Method A
The Japanese Industrial Standard K8150 method has been used to measure the swelling of hydrogels. According to this method the dry hydrogel is immersed in deionised water for 48 hours at room temperature on a roller mixer. After swelling, the hydrogel is filtered by a stainless steel net of 30 meshes (681 µm). The swelling is calculated as follows (Nagasawa et al., 2004):

$$Swelling = \frac{W_s - W_d}{W_d} \tag{3}$$

Where, W_s is the weight of hydrogel in swollen state and W_d is the weight of hydrogel in dry state. The terms 'swelling ratio' (Liu et al., 2005), 'equilibrium degree of swelling' (EDS) (Valles et al., 2000) or 'degree of swelling' (Liu et al., 2002a) has been used for more or less similar measurements.

2.2.2 Method B
Alternatively, to measure the swelling of hydrogel, in a volumetric vial (Universal) the dry hydrogel (0.05-0.1g) was dispersed into sufficiently high quantity of water (25-30 ml) for 48 hrs at room temperature. The mixture is then centrifuged to obtain the layers of water-bound material and free unabsorbed water. The free water is removed and the swelling can be measured according to Method A above.

2.2.3 Method C
The swelling can also be measured according to the Japanese Industrial Standard (JIS) K7223. The dry gel is immersed in deionized water for 16 h at room temperature. After swelling, the hydrogel was filtered using a stainless-steel net of 100-mesh (149 µm). Swelling is calculated as follows (Katayama et al., 2006):

$$Swelling = \frac{C}{B} * 100 \tag{4}$$

Where C is the weight of hydrogel obtained after drying and B is the weight of the insoluble portion after extraction with water.

2.3 FTIR

FTIR (Fourier Transform Infrared Spectroscopy) is a useful technique for identifying chemical structure of a substance. It is based on the principle that the basic components of a substance, i.e. chemical bonds, usually can be excited and absorb infrared light at frequencies that are typical of the types of the chemical bonds. The resulting IR absorption spectrum represents a fingerprint of measured sample. This technique is widely used to investigate the structural arrangement in hydrogel by comparison with the starting materials (2004; Mansur et al., 2004; Torres et al., 2003).

2.4 Scanning Electron Microscopy (SEM)

SEM can be used to provide information about the sample's surface topography, composition, and other properties such as electrical conductivity. Magnification in SEM can be controlled over a range of up to 6 orders of magnitude from about 10 to 500,000 times. This is a powerful technique widely used to capture the characteristic 'network' structure in hydrogels (Aikawa et al., 1998; Aouada et al., 2005; El Fray et al., 2007; 2004; Pourjavadi & Kurdtabar, 2007).

2.5 Light scattering

Gel permeation chromatography coupled on line to a multi angle laser light scattering (GPC-MALLS) is a widely used technique to determine the molecular distribution and parameters of a polymeric system. Hydrogel in a polymeric system can be quantified using this technique (Al-Assaf et al., 2007a). This technique is widely used in quantifying the hydrogels of several hydrocolloids such as gum arabic, gelatine and pullulan (Al-Assaf et al., 2006b; Al-Assaf et al., 2007b; 2006). It can be demonstrated how mass recovery data obtained from GPC-MALLS correlate with actual amount of hydrogel obtained for dextran radiation in solid state (Al-Assaf et al., 2006b) (Figure 2).

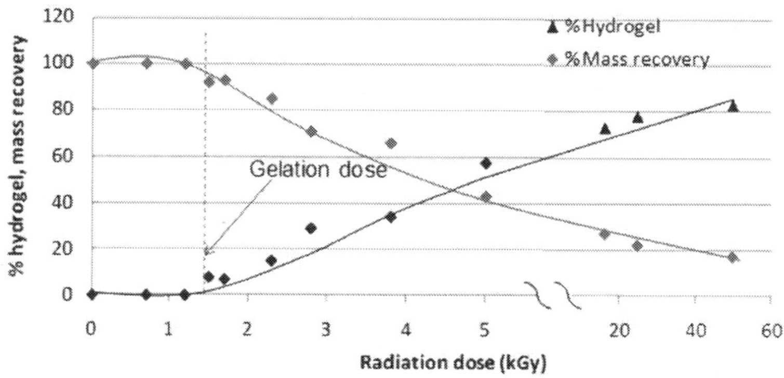

Fig. 2. Correlation between mass recovery data obtained from GPC-MALLS for dextran and amount of hydrogel formed as a function of radiation dose.

2.6 Sol – gel analysis

For radiation cross-linking, the sol-gel analysis is an important characterisation tool as it allows to estimate the parameters such as yield of cross-linking and degradation, gelation dose, etc. and to correlate these with some physico-chemical properties. The relation of sol fraction and absorbed dose according to the Charlesby–Pinner equation (Rosiak, 1998) is given in equation 5. This equation is widely reported for the linear polymers like carboxymethyl cellulose (Liu et al., 2002b).

$$s + \sqrt{s} = \frac{p_0}{q_0} + \frac{2}{q_0 \mu_{2,0} D} \tag{5}$$

Where, s is the sol fraction (s = 1-gel fraction). p_0 is the degradation density, average number of main chain scissions per monomer unit and per unit dose. q_0 is the cross-linking density, proportion of monomer units cross-linked per unit dose. $\mu_{2,0}$ is the initial weight average degree of polymerisation, and D is the radiation dose in Gy.

To avoid an inaccuracy resulting from unknown molecular weight distribution of used polymers, the Charlesby–Rosiak equation (Equation 6) is used. This equation allows for estimation of radiation parameters of linear polymers of any initial weight distribution as well as is applicable to systems when an initial material is monomer or branched polymer (Wach et al., 2003b).

$$s + \sqrt{s} = \frac{p_0}{q_0} + \left(2 - \frac{p_0}{q_0} \right) \frac{D_v + D_g}{D_n + D} \tag{6}$$

Where, D is the absorbed dose in Gy. D_g is gelation dose – a dose when the first insoluble gel appears. D_v is the virtual dose – a dose required to change the distribution of molecular weight of the certain polymer in such a way that the relation between weight-average and number-average molecular weight would be equal to 2. However, there is limitation to Charlesby–Pinner equation that it does not allow the chain reaction that occurs during the event of ionising radiation into its consideration and, so most of the experimental data of radiation polymerisation do not obey this equation. It is recently shown that chain reactions, rather than polydispersity and structure, explain most of the deviation from ideal Charlesby-Pinner behaviour of irradiated polymers (Jones et al., 1996).

To obtain the gelation dose, yield of cross-linking and scission, following equations are used:

$$G(x) = \frac{4.8 * 10^5}{M_{w,0} * D_g} \tag{7}$$

$$G(s) / G(x) = 2p_0 / q_0 \tag{8}$$

Where, G(x) and G(s) are radiation yield of cross-linking and of scission in mol J-1, respectively. $M_{w,0}$ is weight average molecular weights of initial polymer before irradiation. The above equations are valid for polymers with initial most probable molecular weight distribution and degree of polydispersity $M_{w,0} / M_{n,0} = 2$ (Rosiak et al., 2003; Wach et al., 2003b). For the degradation process occurring in a polymer solution when it is subjected to irradiation, the yield of scission (mol/J) can be calculated as:

$$G(s) = \frac{2c}{Dd} \left(\frac{1}{M_w} - \frac{1}{M_{w,0}} \right) \tag{9}$$

Where c is the concentration of polymer in solution (g/dm3); D is the absorbed dose (Gy); d is the solution density (kg/dm^3); $M_{w,0}$ and M_w are the weight-average molecular weight of polymer before and after irradiation, respectively. Degradation rate in irradiation is first-order reaction and the rate constants k can be evaluated from the following first order kinetic equation (Wasikiewicz et al., 2005):

$$\frac{1}{M_t} = \frac{1}{M_0} + \frac{kt}{m}$$
(10)

Where, M_0 and M_t are weight-average molecular weights before and after the treatment for t hours, respectively, m is the molecular weight of polymer monomer unit and k (h^{-1}) is the rate constant.

2.7 Rheology

The rheological properties are very much dependant on the types of structure (i.e. association, entanglement, cross-links) present in the system. Polymer solutions are essentially viscous at low frequencies, tending to fit the scaling laws: $G' \sim \omega^2$ and $G'' \sim \omega$. At high frequencies, elasticity dominates ($G' > G''$). This corresponds to Maxwell-type behaviour with a single relaxation time that may be determined from the crossover point and, this relaxation time increases with concentration. For cross-linked microgel dispersions, it exhibits G' and G'' being almost independent of oscillation frequency (Omari et al., 2006; Rubinstein & Colby, 2003). This technique has been used to characterize the network structure in seroglucan/borax hydrogel (Coviello et al., 2003), chitosan based cationic hydrogels (Kempe et al., 2008; Sahiner et al., 2006) and a range of other hydrocolloids (Al-Assaf et al., 2006b).

2.8 Other techniques

The main methods used to characterise and quantify the amount of free and bound water in hydrogels are differential scanning calorimetry (DSC) and nuclear magnetic resonance (NMR). The proton NMR gives information about the interchange of water molecules between the so-called free and bound states (Phillips et al., 2003) . The use of DSC is based on the assumption that only the free water may be frozen, so it is assumed that the endotherm measured when warming the frozen gel represents the melting of the free water, and that value will yield the amount of free water in the hydrogel sample being tested. The bound water is then obtained by difference of the measured total water content of the hydrogel test specimen, and the calculated free water content (Hoffman, 2002). Thermo-gravimetric analysis (Lazareva & Vashuk, 1995; Singh & Vashishth, 2008; Torres et al., 2003), X-ray diffraction (2008; Mansur et al., 2004), sol-gel analysis (Janik et al., 2008; Rosiak, 1998; Wach et al., 2003b; Xu et al., 2002) etc. are also used to confirm the formation of cross-linked network gel structures of hydrogel.

3. Application of hydrogel

Hydrogel of many synthetic and natural polymers have been produced with their end use mainly in tissue engineering, pharmaceutical, and biomedical fields (Hoare & Kohane,

2008). Due to their high water absorption capacity and biocompatibility they have been used in wound dressing, drug delivery, agriculture, sanitary pads as well as trans-dermal systems, dental materials, implants, injectable polymeric systems, ophthalmic applications, hybrid-type organs (encapsulated living cells) (Benamer et al., 2006; Nho et al., 2005; Rosiak et al., 1995; Rosiak & Yoshii, 1999). A list of hydrogels with their proposed corresponding applications is shown in Table 1.

Application	Polymers	References
Wound care	polyurethane, poly(ethylene glycol), poly(propylene glycol),	(Rosiak & Yoshii, 1999)
	poly(vinylpyrrolidone), polyethylene glycol and agar	(Benamer et al., 2006; Lugao & Malmonge, 2001; Rosiak et al., 1995)
	Xanthan, methyl cellulose	(2006)
	carboxymethyl cellulose, alginate, hyaluronan and other hydrocolloids	(Kim et al., 2005; Rosiak et al., 1995; Rosiak & Yoshii, 1999; Walker et al., 2003)
Drug delivery, pharmaceutical	poly(vinylpyrrolidone)	(Benamer et al., 2006; Rosiak et al., 1995)
	starch, poly(vinylpyrrolidone), poly(acrylic acid)	(Kumar et al., 2008; Spinelli et al., 2008)
	carboxymethyl cellulose, hydroxypropyl methyl cellulose	(Barbucci et al., 2004; Porsch & Wittgren, 2005)
	polyvinyl alcohol, acrylic acid, methacrylic acid	(Nho et al., 2005)
	chitosan, αβ-glycerophosphate	(Zhou et al., 2008)
	κ-carrageenan, acrylic acid, 2-acrylamido-2-methylpropanesulfonic acid	(Campo et al., 2009; Pourjavadi & Zohuriaan-Mehr, 2002)
	acrylic acid, carboxymethyl cellulose	(El-Naggar et al., 2006; Said et al., 2004)
Dental Materials	Hydrocolloids (Ghatti, Karaya, Kerensis gum)	(Al-Assaf et al., 2009)
Tissue engineering, implants	poly(vinylalcohol), poly(acrylic acid)	(Rosiak et al., 1995)
	hyaluronan	(Kim et al., 2005; Shu et al., 2004)
	collagen	(Drury & Mooney, 2003)
Injectable polymeric system	polyesters, polyphosphazenes, polypeptides, chitosan	(2010)
	β-hairpin peptide	(Yan et al., 2010)

Technical products (cosmetic, pharmaceutical)	Starch	(Trksak & Ford, 2008)
	gum arabic	(Al-Assaf et al., 2006b; Al-Assaf et al., 2007b; 2006; Katayama et al., 2008)
	xanthan, pectin, carrageenan, gellan, welan, guar gum, locust bean gum, alginate, starch, heparin, chitin and chitosan	(Phillips et al., 2003; Phillips et al., 2005)
Others (agriculture, waste treatment, separation, etc.)	Starch	(Jeremic et al., 1999; Trksak & Ford, 2008; Yoshii & Kume, 2003; Zhao et al., 2003b)
	xanthan, polyvinyl alcohol	(2002)
	poly (vinyl methyl ether), poly (N-isopropyl acrylamide)	(Bhardwaj et al., 2005; Sen, 2005)

Table 1. Applications of hydrogel, types of polymers and relevant references.

4. Methods to produce hydrogel

Cross-linked networks of synthetic polymers such as polyethylene oxide (PEO) (Khoylou & Naimian, 2009), polyvinyl pyrollidone (PVP) (Razzak et al., 2001), polylactic acid (PLA) (Palumbo et al., 2006), , polyacrylic acid (PAA) (Onuki et al., 2008), polymethacrylate (PMA) (Yang et al.), polyethylene glycol (PEG) (Singh et al.), or natural biopolymers (Coviello et al., 2007) such as alginate, chitosan, carrageenan, hyaluronan, and carboxymethyl cellulose (CMC) have been reported. The various preparation techniques adopted are physical cross-linking (Hennink & Nostrum, 2002), chemical cross-linking (Barbucci et al., 2004), grafting polymerisation (Said et al., 2004), and radiation cross-linking (Fei et al., 2000; Liu et al., 2002b). Such modifications can improve the mechanical properties and viscoelasticity for applications in biomedical and pharmaceutical fields (Barbucci et al., 2004; Nho & Lee, 2005; Rosiak et al., 1995; Rosiak & Yoshii, 1999). The general methods to produce physical and chemical gels are described below.

4.1 Physical cross-linking

There has been an increased interest in physical or reversible gels due to relative ease of production and the advantage of not using cross-linking agents. These agents affect the integrity of substances to be entrapped (e.g. cell, proteins, etc.) as well as the need for their removal before application. Careful selection of hydrocolloid type, concentration and pH can lead to the formation of a broad range of gel textures and is currently an area receiving considerable attention, particularly in the food industry. The various methods reported in literature to obtain physically cross-linked hydrogels are:

4.1.1 Heating/cooling a polymer solution

Physically cross-linked gels are formed when cooling hot solutions of gelatine or carrageenan. The gel formation is due to helix-formation, association of the helices, and forming junction zones (Funami et al., 2007). Carrageenan in hot solution above the melting transition temperature is present as random coil conformation. Upon cooling it transforms

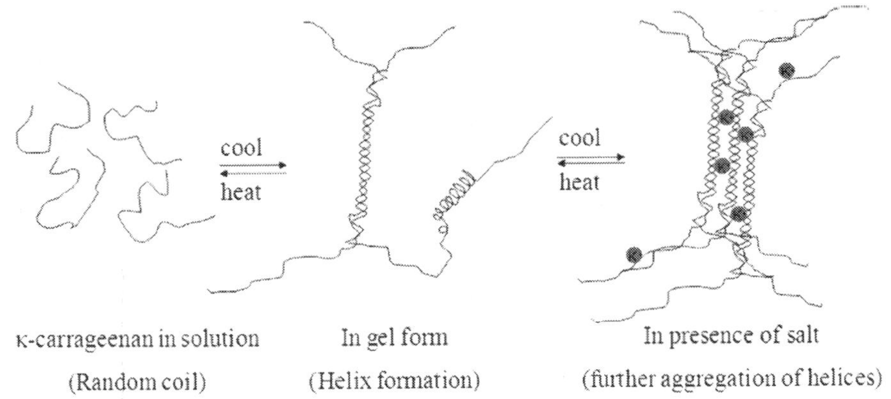

Fig. 3. Gel formation due to aggregation of helix upon cooling a hot solution of carrageenan.

to rigid helical rods. In presence of salt (K^+, Na^+, etc.), due to screening of repulsion of sulphonic group (SO_3^-), double helices further aggregate to form stable gels (Figure 3). In some cases, hydrogel can also be obtained by simply warming the polymer solutions that causes the block copolymerisation. Some of the examples are polyethylene oxide-polypropylene oxide (Hoffman, 2002), polyethylene glycol-polylactic acid hydrogel (Hennink & Nostrum, 2002).

4.1.2 Ionic interaction

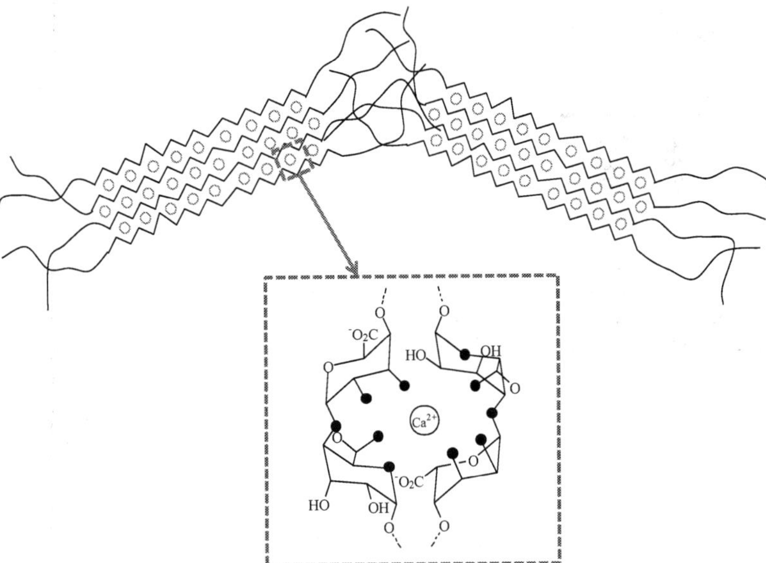

Fig. 4. Ionotropic gelation by interaction between anionic groups on alginate (COO^-) with divalent metal ions (Ca^{2+}).

Ionic polymers can be cross-linked by the addition of di- or tri-valent counterions. This method underlies the principle of gelling a polyelectrolyte solution (e.g. Na^+ alginate$^-$) with a multivalent ion of opposite charges (e.g. Ca^{2+} + $2Cl^-$) (Figure 4). Some other examples are chitosan-polylysine (Bajpai et al., 2008), chitosan-glycerol phosphate salt (Zhao et al., 2009), chitosan-dextran hydrogels (Hennink & Nostrum, 2002).

4.1.3 Complex coacervation

Complex coacervate gels can be formed by mixing of a polyanion with a polycation. The underlying principle of this method is that polymers with opposite charges stick together and form soluble and insoluble complexes depending on the concentration and pH of the respective solutions (Figure 5). One such example is coacervating polyanionic xanthan with polycationic chitosan (Esteban & Severian, 2000; 2001; 1999). Proteins below its isoelectric point are positively charged and likely to associate with anionic hydrocolloids and form polyion complex hydrogel (complex coacervate) (Magnin et al., 2004).

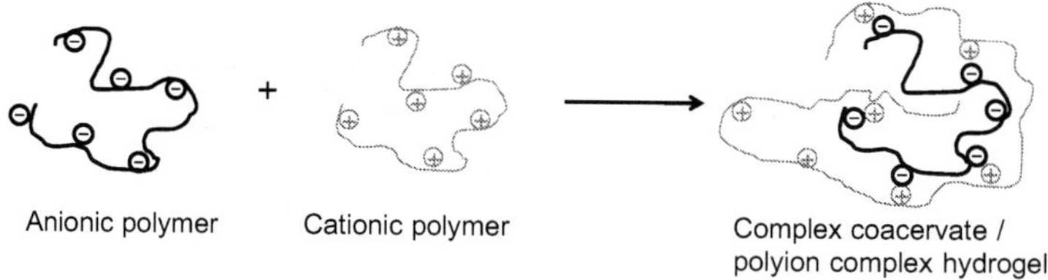

Fig. 5. Complex coacervation between a polyanion and a polycation.

4.1.4 H-bonding

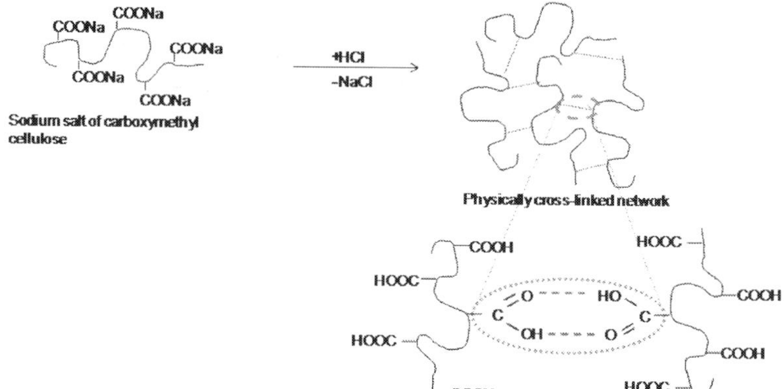

Fig. 6. Hydrogel network formation due to intermolecular H-bonding in CMC at low pH.

H-bonded hydrogel can be obtained by lowering the pH of aqueous solution of polymers carrying carboxyl groups. Examples of such hydrogel is a hydrogen-bound CMC (carboxymethyl cellulose) network formed by dispersing CMC into 0.1M HCl (Takigami et al., 2007). The mechanism involves replacing the sodium in CMC with hydrogen in the acid solution to promote hydrogen bonding (Figure 6). The hydrogen bonds induce a decrease of CMC solubility in water and result in the formation of an elastic hydrogel. Carboxymethylated chitosan (CM-chitosan) hydrogels can also prepared by cross-linking in the presence of acids or polyfunctional monomers (2008). Another example is polyacrylic acid and polyethylene oxide (PEO-PAAc) based hydrogel prepared by lowering the pH to form H-bonded gel in their aqueous solution (Hoffman, 2002). In case of xanthan-alginate mixed system molecular interaction of xanthan and alginate causes the change in matrix structure due to intermolecular hydrogen bonding between them resulting in formation of insoluble hydrogel network (2007).

4.1.5 Maturation (heat induced aggregation)

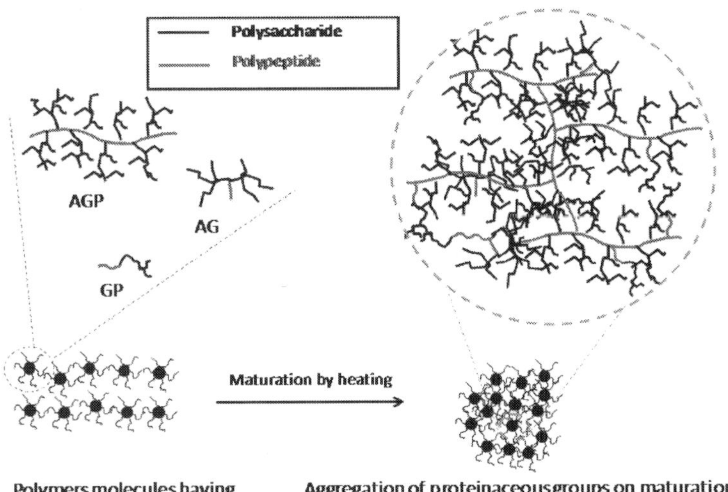

Fig. 7. Maturation of gum arabic causing the aggregation of proteinaceous part of molecules leading to cross-linked hydrogel network.

Gum arabic (Acacia gums) is predominately carbohydrate but contain 2-3% protein as an integral part of its structure (Williams & Phillips, 2006). Three major fractions with different molecular weights and protein content have been identified following fractionation by hydrophobic interaction chromatography with different molecular weights and protein content (Islam et al., 1997). These are arabinogalactan protein (AGP), arabinogalactan (AG) and glycoprotein (GP). Aggregation of the proteinaceous components, induced by heat treatment, increases the molecular weight and subsequently produces a hydrogel form with enhanced mechanical properties and water binding capability (Aoki et al., 2007a; Aoki et al., 2007b). The molecular changes which accompany the maturation process demonstrate that a hydrogel can be produced with precisely structured molecular dimensions. The controlling feature is the agglomeration of the proteinaceous components within the molecularly disperse system that is present in of the naturally occurring gum. Maturing of the gum leads to transfer of the protein associated with the lower molecular weight components to give larger concentrations of high molecular weight fraction (AGP) (Figure 7). The method has also been applied on to other gums such as gum ghatti and Acacia kerensis for application in denture care (Al-Assaf et al., 2009).

4.1.6 Freeze-thawing

Physical cross-linking of a polymer to form its hydrogel can also be achieved by using freeze-thaw cycles. The mechanism involves the formation of microcrystals in the structure due to freeze-thawing. Examples of this type of gelation are freeze-thawed gels of polyvinyl alcohol and xanthan (Giannouli & Morris, 2003; Hoffman, 2002; 2004).

4.2 Chemical cross-linking

Chemical cross-linking covered here involves grafting of monomers on the backbone of the polymers or the use of a cross-linking agent to link two polymer chains. The cross-linking of natural and synthetic polymers can be achieved through the reaction of their functional groups (such as OH, COOH, and NH_2) with cross-linkers such as aldehyde (e.g. glutaraldehyde, adipic acid dihydrazide). There are a number of methods reported in literature to obtain chemically cross-linked permanent hydrogels. Among other chemical cross-linking methods, IPN (polymerise a monomer within another solid polymer to form interpenetrating network structure) (2003) and hydrophobic interactions (Hennink & Nostrum, 2002) (incorporating a polar hydrophilic group by hydrolysis or oxidation followed by covalent cross-linking) are also used to obtain chemically cross-linked permanent hydrogels. The following section reviews the major chemical methods (i.e. cross-linker, grafting, and radiation in solid and/or aqueous state) used to produce hydrogels from a range of natural polymers.

4.2.1 Chemical cross-linkers

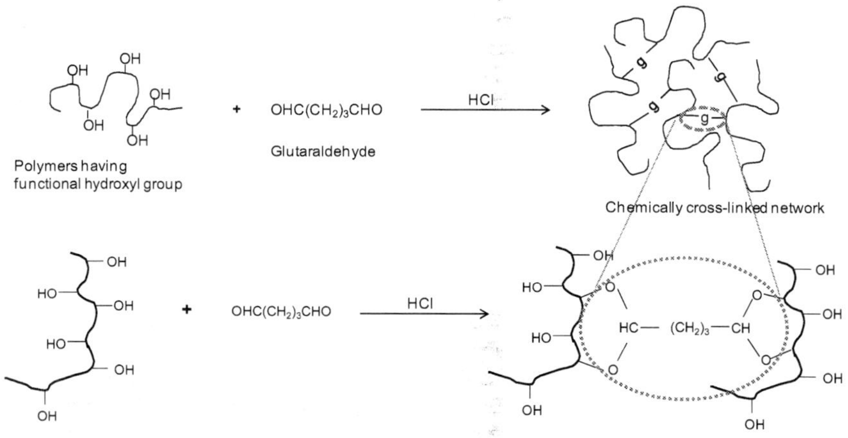

Fig. 8. Schematic illustration of using chemical cross-linker to obtain cross-linked hydrogel network.

Cross-linkers such as glutaraldehyde (2008), epichlorohydrin (2002), etc have been widely used to obtain the cross-linked hydrogel network of various synthetic and natural polymers.

The technique mainly involves the introduction of new molecules between the polymeric chains to produce cross-linked chains (Figure 8). One such example is hydrogel prepared by cross-linking of corn starch and polyvinyl alcohol using glutaraldehyde as a cross-linker (2008). The prepared hydrogel membrane could be used as artificial skin and at the same time various nutrients/healing factors and medicaments can be delivered to the site of action.CMC chains can also be cross-linked by incorporating 1, 3-diaminopropane to produce CMC-hydrogel suitable for drug delivery through the pores (2004). Hydrogel composites based on xanthan and polyvinyl alcohol cross-linked with epichlorohydrin in another example (2002). κ-carrageenan and acrylic acid can be cross-linked using 2-acrylamido-2-methylpropanesulfonic acid leading to the development of biodegradable hydrogels with proposed use for novel drug delivery systems (Pourjavadi & Zohuriaan-Mehr, 2002). Carrageenan hydrogels are also promising for industrial immobilisation of enzymes (Campo et al., 2009). Hydrogels can also be synthesized from cellulose in NaOH/urea aqueous solutions by using epichlorohydrin as cross-linker and by heating and freezing methods (Chang et al., 2010; Chang & Zhang, 2011).

4.2.2 Grafting

Grafting involves the polymerisation of a monomer on the backbone of a preformed polymer. The polymer chains are activated by the action of chemical reagents, or high energy radiation treatment. The growth of functional monomers on activated macroradicals leads to branching and further to cross-linking (Figure 9).

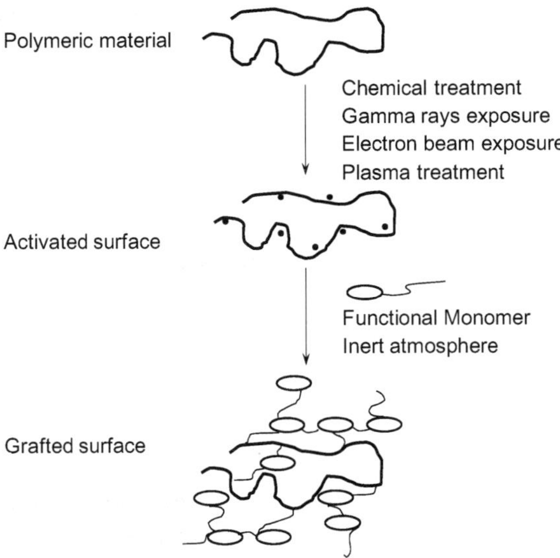

Fig. 9. Grafting of a monomer on preformed polymeric backbone leading to infinite branching and cross-linking.

4.2.2.1 Chemical grafting

In this type of grafting, macromolecular backbones are activated by the action of a chemical reagent. Starch grafted with acrylic acid by using N-vinyl-2-pyrrolidone is an example of this kind of process (Spinelli et al., 2008). Such hydrogels show an excellent pH-dependent swelling behaviour and possess ideal characteristic to be used as drug and vitamin delivery device in the small intestine.

4.2.2.2 Radiation grafting

Grafting can also be initiated by the use of high energy radiation such as gamma and electron beam. Said, Alla et al. (2004) reported the preparation of hydrogel of CMC by grafting CMC with acrylic acid in presence of electron beam irradiation, in aqueous solution. Electron beam was used to initiate the free radical polymerisation of acrylic acid on the backbone of CMC. Water radiolysis product will also be helpful to abstract proton form macromolecular backbones. Irradiation of both (CMC and monomer) will produce free radicals that can combine to produce hydrogel. They proposed the application of such acrylic acid based hydrogel for the recovery of metal ions like copper, nickel, cobalt, and lead. Also, they reported the application of hydrogels in dressings for temporary skin covers.

Zhai, Yoshii et al. (2002) also reported the preparation of starch based hydrogel by grafting polyvinyl alcohol PVA. Starch was first dissolved into water to form gel-like solution and then added to PVA solution, continuously stirred to form homogeneous mixture after heating at 90ºC for 30 mins. The result showed there was a grafting reaction between PVA and starch molecule besides the cross-linking of PVA molecule under irradiation. Amylose of starch was found to be a key reactive component. The properties of starch/PVA blend hydrogel too were governed by amylose component of starch.

Cai, Zhang et al. (2005) have reported the preparation of thermo- and pH-sensitive hydrogels by graft copolymerisation of chitosan (CS) and N-isopropylacrylamide (NIPA). The results showed that the grafting percentage and grafting efficiency increased with the increase of monomer concentration and total irradiation dose. The CS-g-NIPA hydrogels showed good thermo- and pH-sensitivity and swelling property.

4.3 Radiation cross-linking

Radiation cross-linking is widely used technique since it does not involve the use of chemical additives and therefore retaining the biocompatibility of the biopolymer. Also, the modification and sterilisation can be achieved in single step and hence it is a cost effective process to modify biopolymers having their end-use specifically in biomedical application (Lugao & Malmonge, 2001). The technique mainly relies on producing free radicals in the polymer following the exposure to the high energy source such as gamma ray, x-ray or electron beam. The action of radiation (direct or indirect) will depend on the polymer environment (i.e. dilute solution, concentrated solution, solid state).

4.3.1 Aqueous state radiation

Irradiation of polymers in diluted solution will lead to chemical changes as a result of 'indirect action' of radiation. Equation 11 shows that the radiation is mainly absorbed by water. The water radiolysis generates reactive free radicals which can interact with the polymer solute:

$$H_2O \xrightarrow{\text{ionising radiation}} {}^{\bullet}OH, H^{\bullet}, H_3O^+, H_2O_2, H_2, e^-_{aq} \tag{11}$$

Radiation chemical yield (G value) is defined as the number of a particular species produced per 100 eV of energy absorbed by the system from ionising radiation (Clark, 1963). This unit has been redefined in SI mode units by multiplying the old values by 1.036×10^{-7} in order to convert the yield to mol J^{-1}. The radiation chemical yield of these species are now well established as being 2.8, 0.6, 2.7, 0.7, 0.5 and 2.7×10^{-7} mol J^{-1} for $^{\bullet}OH$, $^{\bullet}H$, e^{-}_{aq}, H_2O_2, H_2 and H^{+} respectively (Sonntag, 1987).

A frequently used technique is to irradiate in nitrous oxide saturated solutions when the hydrated electrons (e^{-}_{aq}) are converted into $^{\bullet}OH$ radicals:

$$eaq - + N_2O \longrightarrow {}^{\bullet}OH + N_2 + {}^{-}OH \tag{12}$$

Under the above conditions the $^{\bullet}OH$ radical yield is 5.6×10^{-7} mol J^{-1} whereas the H atoms are formed with yield of $\sim 0.6 \times 10^{-7}$ mol J^{-1}.

Therefore radiation chemical techniques can be used for the quantitative generation of free radicals in aqueous solution. Table 2 gives details of natural polymers and monomers which have been irradiated in diluted solutions and solid state. Changes in molecular weight, rheology, viscometry, UV spectroscopy, and FT-IR have been used to follow the radiolysis reactions.

All the materials given in Table 2, irrespective of their structure and conformation degrade when irradiated in diluted aqueous solution. This is because at a low polymer concentration (i.e. below critical overlap concentration) the chain density of the polymer is not sufficient enough for the chain to recombine and form cross-link network. The two main radicals present in saturated aqueous system react with carbohydrates (RH) by abstracting carbon-bound H-atoms (Equation 13). The hydroxyl radical is not specific in its action and so there are radical sites formed at many position in a carbohydrate solute (Figure 10). In such systems it is the hydroxyl radical which is the main H-abstracting entity. The hydroxyl radicals react with hyaluronan with a rate constant $k_2 = 0.9 \times 10^9$ mol^{-1} dm^3 s^{-1}, whereas H atoms rate is a lower order of magnitude $k_2 = 7 \times 10^7$ mol^{-1} dm^3 s^{-1} (Myint et al., 1987). Figure 11 shows the various hydrolysis, rearrangement, and fragmentation reactions during aqueous radiolysis of cellobiose to gives possible chain break (Sonntag, 1987).

$$RH + {}^{\bullet}OH \left(H^{\bullet}\right) \longrightarrow R^{\bullet} + H_2O \left(H_2\right) \tag{13}$$

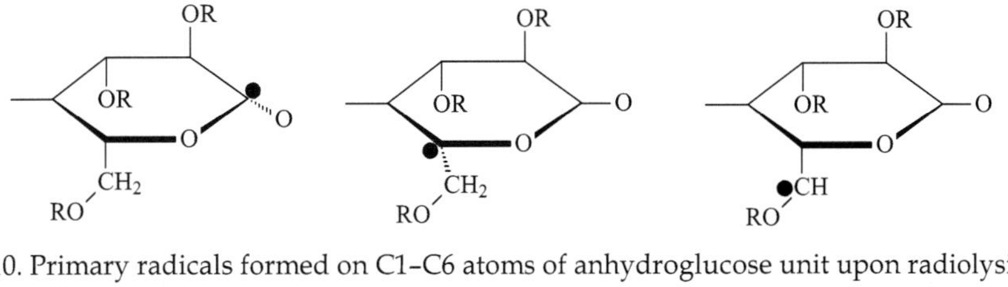

Fig. 10. Primary radicals formed on C1–C6 atoms of anhydroglucose unit upon radiolysis in absence of oxygen.

Material	References for degradation	
	In aqueous state	In solid state
Carboxymethyl cellulose (CMC)	(Choi et al., 2008; Fei et al., 2000; Liu et al., 2002b; Wach et al., 2003a; Yoshii et al., 2003)	(Fei et al., 2000; Liu et al., 2002b; Wach et al., 2001; Wach et al., 2003a; Yoshii et al., 2003)
Hydroxy ethyl cellulose	(Fei et al., 2000; Wach et al., 2003a)	(Fei et al., 2000; Wach et al., 2001)
chitin, chitosan & derivatives	(Ershov et al., 1993; Ershov, 1998; Jarry et al., 2001; Jarry et al., 2002; Yoshii et al., 2003)	(Wasikiewicz et al., 2005)
Cellulose & derivatives	(Ershov, 1998; Nakamura et al., 1985; Phillips, 1961; Phillips, 1963; Wach et al., 2002)	(Phillips & Moody, 1959; Wach et al., 2002)
Starch and derivatives	(Ershov, 1998; Nagasawa et al., 2004; Phillips, 1961; Yoshii & Kume, 2003; Yoshii et al., 2003; Zhai et al., 2003)	(Yoshii & Kume, 2003)
D-glucose	(Phillips, 1963; Schiller et al., 1998)	(Sharpatyi, 2003)
Hyaluronan & hyaluronic acid	(Al-Assaf et al., 1995; Al-Assaf et al., 2006a; Ershov, 1998; Phillips, 1961; Reháková et al., 1994; Stern et al.)	(Choi et al.; Reháková et al., 1994; Stern et al.)
Glucomannan, galactomannan	(Jumel et al., 1996)	(Sen et al., 2007)
Alginate	(Phillips, 1961)	(Wasikiewicz et al., 2005)
Carrageenan	(Abad et al., 2008; Abad et al., 2009)	(Abad et al., 2009; Relleve et al., 2005)
Dextran	(Phillips, 1961)	(Phillips & Moody, 1959)
Pectin	(Phillips, 1961; Zegota, 1999)	(Phillips & Moody, 1959)
Agar	(Abad et al., 2008; Phillips, 1961)	
Gum arabic	(Al-Assaf et al., 2006b; Katayama et al., 2006)	(Blake et al., 1988)
Xanthan, β-glucan	(Byun et al., 2008; Parsons et al., 1985)	

Table 2. List of references showing degradation of polysaccharide upon irradiation in dilute aqueous solution and in solid state.

Hydrated electrons (e^-_{aq}) formed upon water radiolysis react with the hydrocolloids only if the system contains no oxygen. They do not have the ability to abstract electrons from carbohydrate polymers, as for example carrageennan (Abad et al., 2007), hyaluronan (Myint et al., 1987) and CMC where the rate constant for the disappearance of the hydrated electron was measured as $4\text{-}5.2 \times 10^6$ $mol^{-1}dm^3s^{-1}$ (Wach et al., 2005). This rate constant approaches the normal disappearance rate of hydrated electrons in water alone in the absence of CMC, demonstrating that its reactivity with CMC is negligible.

In oxygenated solution the hydrated electron react with oxygen to produce superoxide radical (O·$_2$), (Equation 14).

$$e^-_{aq} + O_2 \longrightarrow O_2^- \quad \left(k_2 = 1.9 \times 10^{10}\, mol^{-1} dm^3 s^{-1}\right) \tag{14}$$

Additionally, in oxygenated solutions the hydrogen atoms form peroxyl radicals (Equation 15) which is unreactive with most organic compounds unless they contain weekly bonded hydrogen (Bielski & Gebicki, 1970).

$$^\bullet H + O_2 \longrightarrow HO_2^- --\left(k_2 = 1.9 \times 10^{10}\, mol^{-1} dm^3 s^{-1}\right) \tag{15}$$

The role of superoxide radicals have been considered to be important in arthritis diseased conditions due to their interaction with the body biopolymers. Two possible mechanisms for the generation of hydroxyl radicals through the reaction of superoxide radicals via metal catalaysed processes and its dismutation and subsequent reaction with hydrogen peroxide were reviewed (Al-Assaf et al., 1995).

In case of radiolysis of oxygenated solution of D-glucose, six primary peroxyl radicals are formed which rapidly undergo HO$_2$·. elimination and subsequently lead to chain break (Sonntag, 1987).

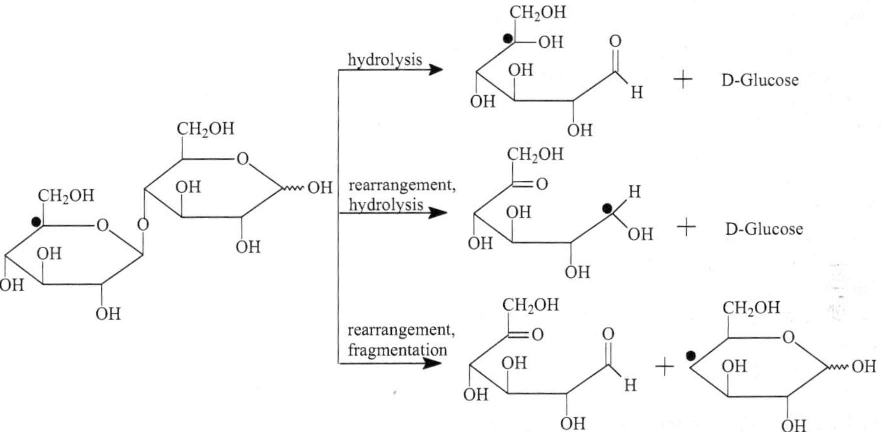

Fig. 11. Various hydrolysis, rearrangement, and fragmentation reactions during aqueous radiolysis of cellobiose.

4.3.2 Radiation in paste

The cross-linking of hydrocolloids in aqueous paste-like conditions state has received considerable attention recently. Under these conditions the concentration of the polymer is high such that both direct action of the radiation can form free radicals and also there is also sufficient water present to be radiolysed to form •OH and related radicals. There is thus a high concentration of radicals in close association with the original polymer and other secondary formed polymer radicals. Thus cross-linking to form new polymers can form by way of radical-radical reaction and polymer - polymer radical reactions. If the original polymer concentration is not sufficient to promote radical-radical reactions then degradation will result. The presence of water promotes the diffusion of macroradicals to combine and form cross-linked hydrogel network. Also, the radiolysis of water generate free radicals (hydrogen atoms and hydroxyl radicals), which increase the yield of macroradicals by abstracting H-atoms from the polymer chain. The concentration at which the modification can be achieved varies according to the structure, degree of substitution, distribution of substitution group and initial molecular weight. For example, a higher DS is effective for cross-linking of CMC due to the fact that intermolecular linkages are result of ether function (Shen et al., 2006; Wach et al., 2003a). Similar results have been reported on aqueous state irradiation of methylcellulose and hydroxypropyl cellulose (Horikawa et al., 2004; Wach et al., 2003b), carboxymethyl starch (Yoshii & Kume, 2003; Yoshii et al., 2003), gum arabic (Katayama et al., 2006), carboxymethylated chitin and chitosan (Wasikiewicz et al., 2006; Zhao et al., 2003a). The % hydrogel produced together with the proposed application from various investigations are summarised in Table 3.

Polymer	Maximum hydrogel (%)	Proposed application	Reference
Carboxymethyl cellulose	55% at 30 kGy	Wound care	(Fei et al., 2000; 2006).
	50% at 80 kGy		(Wach et al., 2001)
	40% at 100 kGy		(Xu et al., 2002)
	60% at 80 kGy		(Yoshii et al., 2003)
Carboxymethyl starch	70% at 10 kGy	Food and cosmetics	(Yoshii & Kume, 2003)
	40% at 2 kGy		(Nagasawa et al., 2004)
Carboxymethl chitosan	70% at 80kGy	Biomedical field	(Zhao et al., 2003a)
Gum Arabic	50-60% at 49.8 kGy	Food, cosmetic, agricultural, and hygienic materials	(Katayama et al., 2006)

Table 3. Radiation of different polymers in paste like condition with maximum amount of hydrogel obtained and their proposed applications.

4.3.3 Solid state radiation

Irradiation of hydrocolloids in solid state induces the radical formation in molecular chains as a result of the direct action of radiation. Here mainly two events take place (i) direct energy transfers to the macromolecule to produce macroradicals and (ii) generation of primary radicals due to the presence of water (moisture). During the solid state radiolysis of hydrocolloids, scission of glycosidic bond is the dominant reaction which eventually leads to decrease the molecular weight of macromolecules (Wach et al., 2003a). Generally, the degradation rates depend on the concentrations of reactants and temperature, like other chemical reactions. In addition, the rates depend on the purity, presence of substituted group and molecular weight of hydrocolloid (Makuuchi, 2010). The course of the degradation of carbohydrates in the solid state is illustrated in Figure 12. The main effects are fragmentation, hydrolysis (due to presence of moisture) or and rearrangement leading to low molecular weight products.

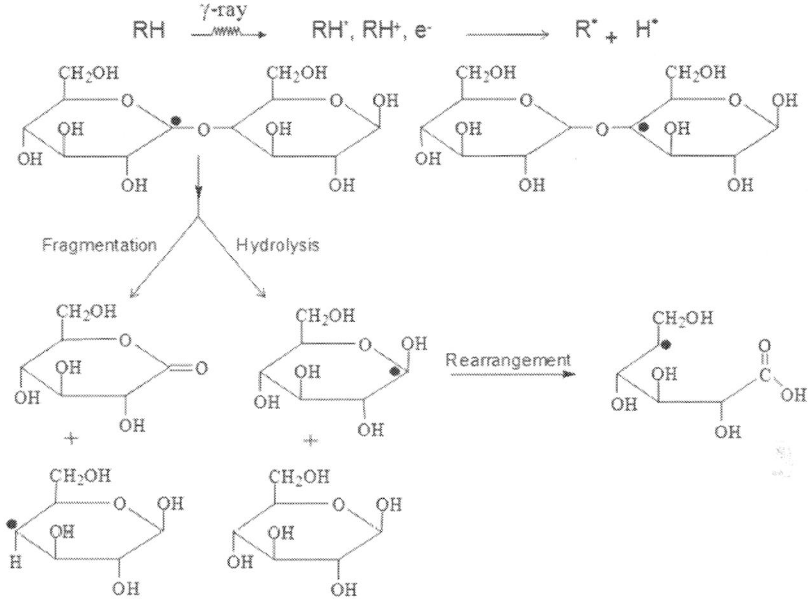

Fig. 12. Events in solid state radiation of hydrocolloids; the glycosidic bond cleavage and chain scission of cellobiose upon solid state radiation of hydrocolloids.

The reported radiation degradation yield (G_d) of κ-, ι-, and λ-carrageenans irradiated in solid and at 1% aqueous solution at atmospheric conditions were almost the same for all types of carrageenan. G_d was in the range of 2.3–2.7 x 10^{-7} mol J^{-1} and 1.0–1.2 x 10^{-7} mol J^{-1} for solid and aqueous state irradiation, respectively which shows the solid state radiation of carrageenan more susceptible to degradation. However, G_d was relatively low (0.3 x 10^{-7} mol J^{-1}) for paste-like state (4% concentration) probably due to simultaneous cross-linking place in such system (Abad et al., 2009). Similarly, the G_d in aqueous form was also affected by the conformational state of κ-carrageenan. The helical conformation gave a lower G_d (0.7 x 10^{-7} mol J^{-1}) than the coiled conformation (G_d = 1.2 x 10^{-7} mol J^{-1}). A helical structure has some interchain stabilisation effects which increases the possibility of free radical interchain cross-linking (Abad et al., 2010). For galactomannans the values are found relatively lower (0.85–1.07 x 10^{-7} mol J^{-1}) suggesting these hydrocolloids are less susceptible to degradation. Several hydrocolloids such as α-D-glucose (Moore & Phillips, 1971; Phillips, 1963; Phillips et al., 1966; Phillips, 1968), cellulose and derivatives (Fei et al., 2000; Horikawa et al., 2004), amylose and starch (Phillips & Young, 1966; Phillips, 1968; Yoshii & Kume, 2003), chitin and chitosan (Kuang et al., 2008; Wasikiewicz et al., 2005) have reportedly undergone degradation when subjected to solid state radiation. The results for a range of polysaccharides are shown in Table .

4.3.3.1 Cross-linking in solid state

The application of radiation processing of synthetic polymers to introduce structural changes by cross-linking and special performance characteristics is now a thriving industry.

In contrast treatment of polysaccharides and other natural polymers with ionizing radiation either in the solid state or in aqueous solution leads to degradation as described above. Therefore, a method to modify structure, without introducing new chemical groupings, could prove of advantage, particularly if the process could be achieved in the solid state. This has been possible in synthetic polymers by exposure to high energy ionizing radiation, arising mainly through the pioneering work of Charlesby (Rosiak & Yoshii, 1999). The method is now routinely used for the cross-linking of polymers. Polymer chains can be joined and a network formed. The method is used for crystal lattice modification for semiconductors and gemstones, etc., by which the crystalline structure of a material is modified. The sheathing on wire and cable is routinely cross-linked with radiation to improve a number of important properties and radiation cross-linked polymers are commonly used to make heat-shrinkable tubing, connectors, and films.

4.3.3.1.1 Natural polymers

Recently a process has been reported to modify natural polymers (e.g. hydrocolloids such as CMC, gum arabic, dextran, gelatine, etc) in solid state by high energy radiation (Al-Assaf et al., 2006b; Al-Assaf et al., 2007b) to obtain their hydrogel (Figure 13).

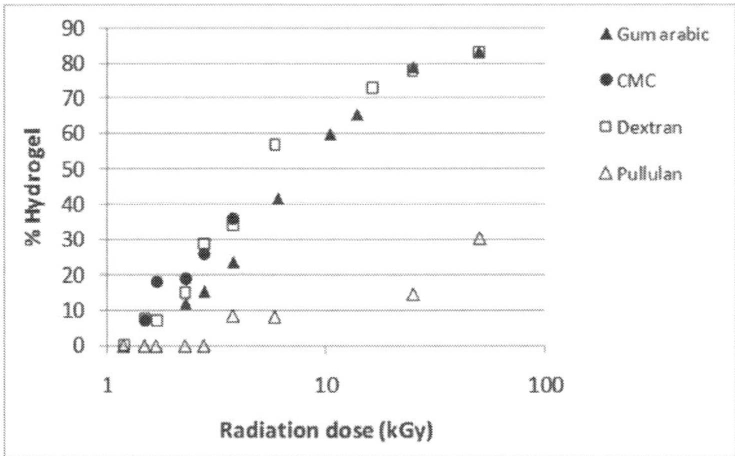

Fig. 13. Formation of hydrogel as a function of radiation dose for hydrocolloids irradiated in solid state in the presence of alkyne gas.

The new method allows the controlled modification of the structure of polysaccharide and other related materials in the solid state using ionizing radiation in the presence of a mediating alkyne gas. The method has been applied to a range of polysaccharides of differing origin and structure, to proteins either directly derived from animal connective tissue sources such as collagen, gelatin, and from human and animal products, such as casein, combinations of one or more such polysaccharides with proteins of plant origin. These polymers when irradiated in presence of acetylene gas, it leads to the cross-linking and hence formation of macromolecules with increased molecular weight and functionalities. Highly branched polysaccharide structures could produce a 4-fold increase in molecular weight with doses up to 10 kGy and hydrogels with doses up to 50 kGy, whereas straight chain structures can yield a similar change with doses as low as 1–3 kGy. Proteins require doses up to 25 kGy to achieve a similar result. The proposed cross-linking mechanism for solid state radiation is illustrated in Figure 14. For ease of presentation the two macromolecular chains are represented as R_1H and R_2H. The direct radiation action forms a free radical (•R_1) which then adds to the acetylene to give a radical with a double bond. This addition to the acetylene is slow and the reactive radical with a double bond abstracts hydrogen atom form a nearby polysaccharide chain to give two radicals, one on the original acetylene adduct and one on a nearby polysaccharide chain (•R_2). These

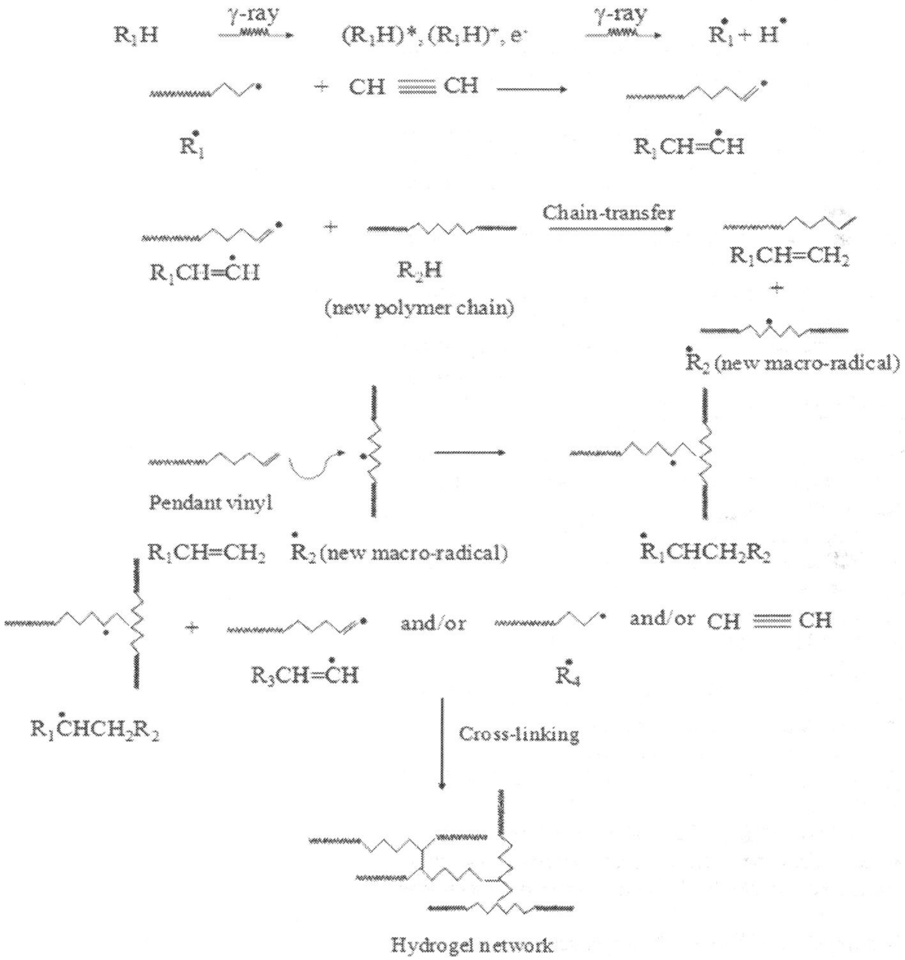

Fig. 14. Schematic representation of radiation cross-linking in solid state of polymers when irradiated in the atmosphere of acetylene.

recombine to give a cross-linked stable radical. This radical has fair degree of mobility and either recombines with acetylene, radical generated as a result of the action of ionizing radiation or another similar radical to form a cross-linked network (Al-Assaf et al., 2007b). Irradiation of carboxymethyl cellulose in solid state showed that the structural changes can again be achieved using the radiation processing. Result showed that initial mean M_w of 1.55 x 10^5, is increased three-fold to 4.44 x 10^5 Da. Moreover the polydispersity is increased from 2 to 2.8 with an increase in R_g from 36 to 52 nm. Hydrogel is formed at the higher doses and is visible in solution. Gelation of CMC solution can be controlled to give stable gels ranging in consistency from soft pourable to very firm. At a frequency of 0.1 Hz there is a 10-fold increase in G′ and G″. The method allows controlled increase in molecular weight and gel formation which are increased linearly with the radiation dose. Result on solid state radiation of dextran showed 83% of hydrogel formation at a dose of around 50 kGy. An increase in M_w from initial value of 2.34 x 10^6 Da to a maximum of 4.58 x 10^6 Da was observed. The modified dextran showed a marked increase in viscoelastic properties compared to it control. Radiation of another slightly blanched hydrocolloid, pullulan showed that on radiation processing the average Mw doubles from 3.17 x 10^5 to 6.81 x 10^5 Da and moreover, there is conversion of the original material to form hydrogel to an extent of 30% of the original material. Measurements of G′ shows the enhancement of the rheological properties in manner expected for the higher molecular weight polysaccharide. Result on a protein (gelatine) showed that using the solid state process, the molecular weight of gelatine can be increased in a controlled manner to produce a range of products with varying molecular weights and solution/gelling properties. The same behaviour has been achieved with casein in the form of its sodium salt. The modifications already demonstrated can be applied also to the widest range of commercial polysaccharides, including xanthan, pectin, carrageenan, gellan, welan, guar gum, locust bean gum, alginate, starch, heparin, chitin and chitosan (Phillips et al., 2003; Phillips et al., 2005).

A recent study on carrageenan modification in the solid state demonstrated that the hydrogel formation and the increase in viscoelasticity upon irradiation of κ-carrageenan are achieved without using a gelling agent (Gulrez et al., 2010). The optimum dose range to achieve modification is 5-10 kGy since at high dose degradation results in reduction of gel fraction. Irradiation of carrageenan led to production of nearly 78% hydrogel with an improvement in viscosity nearly four-fold to that of control material. The results showed improvement in viscoelasticity at moderate doses which can be defined as a result of increase in hydrodynamic radius of carrageenan gel solution. The results showed that radiation modified κ-carrageenan hydrogels are stronger than control sample. The strength of κ-carrageenan gels increased with increased radiation dose and reached to maximum at 5 kGy. The superior mechanical properties of the irradiated sample compared with the control can be explained as the aggregation of relatively longer superhelical rods in case of modified sample (Figure 15).

4.3.3.1.2 Synthetic/natural polymer blends

The same technique was applied on various mix systems of water soluble polymers of synthetic and natural origin and the result showed the synergistic effect on the functionalities of these mix systems. One such example is the radiation of mixture (1:1) of polyvinyl pyrrolidone (PVP) and gum arabic (GA) in solid state. The reheology measurement carried out for 10% aqueous solution of this system showed significant improvement in viscoelasticity of mixed polymers (synergy) compared to either of its constituents (Figure 16).

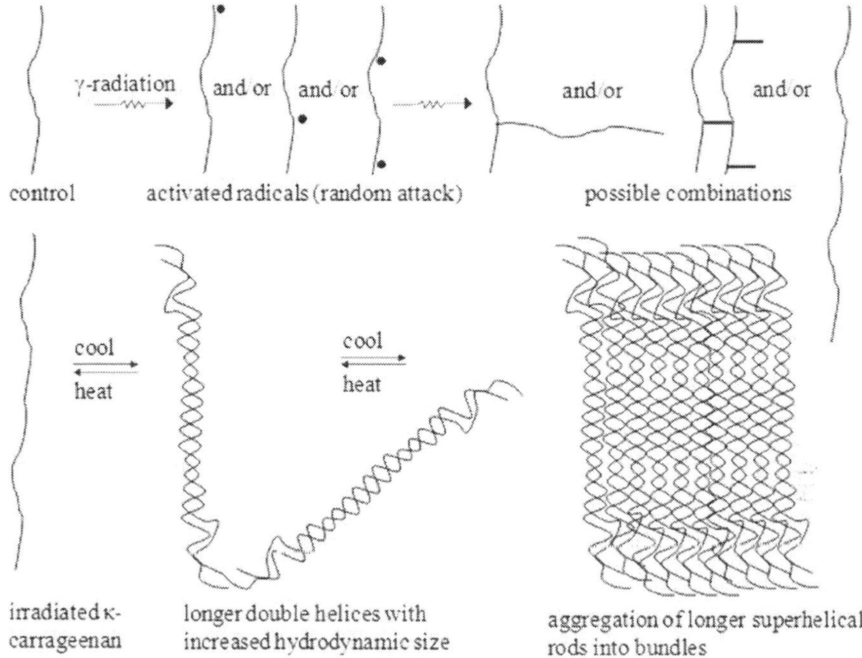

Fig. 15. Proposed mechanism for aggregation of superhelical rods into bundles on cooling the hot solution of modified κ-carrageenan hydrogels.

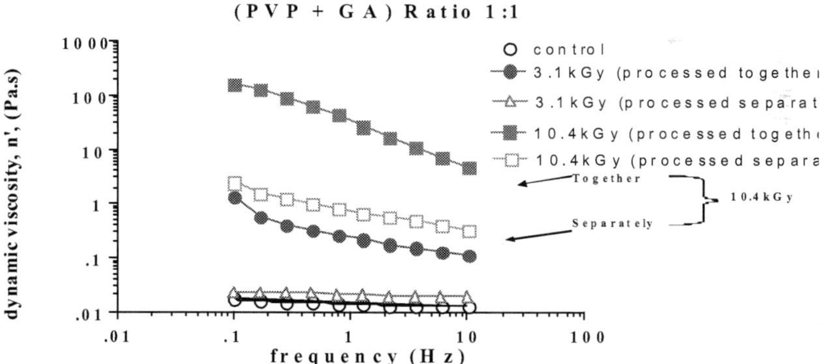

Fig. 16. Dynamic viscosity plotted as a function of oscillation frequency for 10% aqueous solution of PVP-GA blend system modified in solid state (Phillips et al., 2003).

5. Acknowledgment

The authors acknowledge the financial support in the form of PhD studentship given to SKG by Phillips Hydrocolloids Research Ltd.

6. References

Abad, L., Okabe, S., Shibayama, N., Kudo, H., Saiki, S., Aranilla, C., Relleve, L. & de la Rosa, A. (2008) Comparative studies on the conformational change and aggregation behavior of irradiated carrageenans and agar by dynamic light scattering. *International Journal of Biological Macromolecules* 42, 55-61.

Abad, L. V., Saiki, S., Kudo, H., Muroya, Y., Katsumura, Y. & de la Rosa, A. M. (2007) Rate constants of reactions of kappa-carrageenan with hydrated electron and hydroxyl radical. *Nuclear Instruments & Methods in Physics Research Section B-Beam Interactions with Materials and Atoms* 265, 410-413.

Abad, L. V., Kudo, H., Saiki, S., Nagasawa, N., Tamada, M., Katsumura, Y., Aranilla, C. T., Relleve, L. S. & Rosa, A. M. D. L. (2009) Radiation degradation studies of carrageenans. *Carbohydrate Polymers* 78, 100-106.

Abad, L. V., Kudo, H., Saiki, S., Nagasawa, N., Tamada, M., Fu, H., Muroya, Y., Lin, M., Katsumura, Y., Relleve, L. S., Aranilla, C. T. & DeLaRosa, A. M. (2010) Radiolysis studies of aqueous [kappa]-carrageenan. *Nuclear Instruments and Methods in Physics Research Section B: Beam Interactions with Materials and Atoms* 268, 1607-1612.

Aikawa, K., Matsumoto, K., Uda, H., Tanaka, S., Shimamura, H., Aramaki, Y. & Tsuchiya, S. (1998) Hydrogel formation of the pH response polymer polyvinylacetal diethylaminoacetate (AEA). *International Journal of Pharmaceutics* 167, 97-104.

Al-Assaf, S., Phillips, G. O., Deeble, D. J., Parsons, B., Starnes, H. & Von Sonntag, C. (1995) The enhanced stability of the cross-linked hylan structure to hydroxyl (OH) radicals compared with the uncross-linked hyaluronan. *Radiation Physics and Chemistry* 46, 207-217.

Al-Assaf, S., Navaratnam, S., Parsons, B. J. & Phillips, G. O. (2006a) Chain scission of hyaluronan by carbonate and dichloride radical anions: Potential reactive oxidative species in inflammation? *Free Radical Biology and Medicine* 40, 2018-2027.

Al-Assaf, S., Phillips, G. O. & Williams, P. A. (2006b) Controlling the molecular structure of food hydrocolloids. *Food Hydrocolloids* 20, 369-377.

Al-Assaf, S., Phillips, G. O., Aoki, H. & Sasaki, Y. (2007a) Characterization and properties of Acacia senegal (L.) Willd. var. senegal with enhanced properties (Acacia (sen) SUPER GUM(TM)): Part 1--Controlled maturation of Acacia senegal var. senegal to increase viscoelasticity, produce a hydrogel form and convert a poor into a good emulsifier. *Food Hydrocolloids* 21, 319-328.

Al-Assaf, S., Phillips, G. O., Williams, P. A. & Plessis, T. A. d. (2007b) Application of ionizing radiations to produce new polysaccharides and proteins with enhanced functionality. *Nuclear Instruments and Methods in Physics Research B* 265, 37-43.

Al-Assaf, S., Dickson, P., Phillips, G. O., Thompson, C. & Torres, J. C. (2009) Compositions comprising polysaccharide gums. In *World Intellectual property Organization, Vol. WO2009/016362 A2*, (ed. PCT), Phillips Hydrocolloid Research Limited (UK), Reckitt Benckiser (UK), United Kingdom.

Alupei, I. C., Popa, M., Hamcerencu, M. & Abadie, M. J. M. (2002) Superabsorbant hydrogels based on xanthan and poly(vinyl alcohol) - 1. The study of the swelling properties. *European Polymer Journal* 38, PII S0014-3057(02)00106-4.

Anderson, J. M. & Langone, J. J. (1999) Issues and perspectives on the biocompatibility and immunotoxicity evaluation of implanted controlled release systems. *Journal of Controlled Release* 57, 107-113.

Aoki, H., Al-Assaf, S., Katayama, T. & Phillips, G. O. (2007a) Characterization and properties of Acacia senegal (L.) Willd. var. senegal with enhanced properties (Acacia (sen) SUPER GUM(TM)): Part 2--Mechanism of the maturation process. *Food Hydrocolloids* 21, 329-337.

Aoki, H., Katayama, T., Ogasawara, T., Sasaki, Y., Al-Assaf, S. & Phillips, G. O. (2007b) Characterization and properties of Acacia senegal (L.) Willd. var. Senegal with enhanced properties (Acacia (sen) SUPER GUM(TM)): Part 5. Factors affecting the emulsification of Acacia senegal and Acacia (sen) SUPER GUM(TM). *Food Hydrocolloids* 21, 353-358.

Aouada, F. A., de Moura, M. R., Fernandes, P. R. G., Rubira, A. F. & Muniz, E. C. (2005) Optical and morphological characterization of polyacrylamide hydrogel and liquid crystal systems. *European Polymer Journal* 41, 2134-2141.

Bajpai, A. K., Shukla, S. K., Bhanu, S. & Kankane, S. (2008) Responsive polymers in controlled drug delivery. *Progress in Polymer Science* 33, 1088-1118.

Barbucci, R., Leone, G. & Vecchiullo, A. (2004) Novel carboxymethylcellulose-based microporous hydrogels suitable for drug delivery. *J. Biomater. Sci. Polymer Edn* 15, 607-619.

Benamer, S., Mahlous, M., Boukrif, A., Mansouri, B. & Youcef, S. L. (2006) Synthesis and characterisation of hydrogels based on poly(vinyl pyrrolidone). *Nuclear Instruments and Methods in Physics Research* B 248 284-290.

Bhardwaj, Y. K., Kumar, V., Acharya, A. & Sabharwal, S. (2005) Synthesis, characterization and utilization of radiation synthesized stimuli-responsive hydrogels and membranes. In *Radiation synthesis of stimuli-responsive membranes, hydrogels and adsorbents for separation purposes*, (ed. I. A. E. Agency), International Atomic Energy Agency, Vienna.

Bielski, B. H. & Gebicki, J. M. (1970) species in irradiated oxygenated water In *Advances in Radiation Chemistry Vol. 2*, (eds. M. Burton & J. L. Magee), pp. 177-280, Wiley Interscience.

Blake, S. M., Deeble, D. J., Phillips, G. O. & Plessy, A. D. (1988) The effect of sterlizing doses of g-irradiation on the molecular weight and emulsification properties of gum arabic. *Food Hydrocolloids* 2, 407-415.

Byun, E. H., Kim, J. H., Sung, N. Y., Choi, J. I., Lim, S. T., Kim, K. H., Yook, H. S., Byun, M. W. & Lee, J. W. (2008) Effects of gamma irradiation on the physical and structural properties of beta-glucan. *Radiation Physics and Chemistry* 77, 781-786.

Cai, L. B., Zuo, J. & Tang, S. (2005) A study on the nonergodic behavior of kappa-carrageenan thermoreversible gel by static and dynamic light scattering. *Acta Physico-Chimica Sinica* 21, 1108-1112.

Campo, V. L., Kawano, D. F., da Silva, D. B. & Carvalho, I. (2009) Carrageenans: Biological properties, chemical modifications and structural analysis - A review. *Carbohydrate Polymers* 77, 167-180.

Chang, C., Zhang, L., Zhou, J., Zhang, L. & Kennedy, J. F. (2010) Structure and properties of hydrogels prepared from cellulose in NaOH/urea aqueous solutions. *Carbohydrate Polymers* 82, 122-127.

Chang, C. & Zhang, L. (2011) Cellulose-based hydrogels: Present status and application prospects. *Carbohydrate Polymers* 84, 40-53.

Chaterji, S., Kwon, K. & Park, K. (2007) Smart polymeric gels: Redefining the limits of biomedical devices. *Prog. Polym. Sci.* 32, 1083-112.

Choi, J.-i., Kim, J.-K., Kim, J.-H., Kweon, D.-K. & Lee, J.-W. Degradation of hyaluronic acid powder by electron beam irradiation, gamma ray irradiation, microwave irradiation and thermal treatment: A comparative study. *Carbohydrate Polymers* 79, 1080-1085.

Choi, J. I., Lee, H. S., Kim, J. H., Lee, K. W., Lee, J. W., Seo, S. J., Kang, K. W. & Byun, M. W. (2008) Controlling the radiation degradation of carboxymethylcellulose solution. *Polymer Degradation and Stability* 93, 310-315.

Clark, G. L. (1963) Encyclopedia of X-rays and Gamma Rays, Chapman & Hall Ltd., London, UK.

Coviello, T., Coluzzi, G., Palleschi, A., Grassi, M., Santucci, E. & Alhaique, F. (2003) Structural and rheological characterization of Scleroglucan/borax hydrogel for drug delivery. *International Journal of Biological Macromolecules* 32, 83-92.

Coviello, T., Matricardi, P., Marianecci, C. & Alhaique, F. (2007) Polysaccharide hydrogels for modified release formulations. *Journal of Controlled Release* 119, 5-24.

Drury, J. L. & Mooney, D. J. (2003) Hydrogels for tissue engineering: scaffold design variables and applications. *Biomaterials* 24, 4337-4351.

El-Naggar, A. W. M., Alla, S. G. A. & Said, H. M. (2006) Temperature and pH responsive behaviours of CMC/AAc hydrogels prepared by electron beam irradiation. *Materials Chemistry and Physics* 95, 158-163.

El Fray, M., Pilaszkiewicz, A., Swieszkowski, W. & Kurzydlowski, K. J. (2007) Morphology assessment of chemically modified cryostructured poly(vinyl alcohol) hydrogel. *European Polymer Journal* 43, 2035-2040.

Ershov, B. G., Sukhov, N. L., Nudga, L. A., Baklagina, Y. G., Kozhevnikova, L. G. & Petropavlovskii, G. A. (1993) Radiation Destruction of Chitin. *Russian Journal of Applied Chemistry* 66, 540-545.

Ershov, B. G. (1998) Radiation-chemical destruction of cellulose and other polysaccharides. *Uspekhi Khimii* 67, 353-375.

Esteban, C. & Severian, D. (2000) Polyionic hydrogels based on xanthan and chitosan for stabilising and controlled release of vitamins, *Vol. WO0004086 (A1)* (ed. U. United States Patent), Kemestrie Inc [CA], USA.

Fei, B., Wach, R. A., Mitomo, H., Yoshii, F. & Kume, T. (2000) Hydrogel of biodegradable cellulose derivatives. I. Radiation-induced crosslinking of CMC. *Journal of Applied Polymer Science* 78, 278-283.

Ferry, J. D. (1980) Viscoelstic properties of polymers, pp. 486-544, John Wilet & sons, New York.

Funami, T., Hiroe, M., Noda, S., Asai, I., Ikeda, S. & Nishimari, K. (2007) Influence of molecular structure imaged with atomic force microscopy on the rheological behavior of carrageenan aqueous systems in the presence or absence of cations. *Food Hydrocolloids* 21, 617-629.

Giannouli, P. & Morris, E. R. (2003) Cryogelation of xanthan. *Food Hydrocolloids* 17, 495-501.

Gulrez, S., Al-Assaf, S. & Phillips, G. O. (2010) Cahracterisation and radiation modifcation of carrageenan in solid state. In *Radaiation Processing Technology Applications, Vol. 2*, (ed. R. K. Khandal), pp. 519-531, SRI, New Delhi.

Hennink, W. E. & Nostrum, C. F. v. (2002) Novel crosslinking methods to design hydrogels. *Advanced Drug Delivery Reviews* 54 13–36.

Hoare, T. R. & Kohane, D. S. (2008) Hydrogels in drug delivery: Progress and challenges. *Polymer* 49, 1993-2007.

Hoffman, A. S. (2002) Hydrogels for biomedical applications. *Advanced Drug Delivery Reviews* 43 3–12.

Horikawa, Y., Kawachi, S. & Honna, T. (2004) Sedimentation behavior of dispersed particles of clay and silt in acidic and alkaline suspensions of inorganic materials from various volcanic ash soils. *Soil Science and Plant Nutrition* 50, 19-25.

Hyeon, C. h. G., Seong, K. D. & Gyun, K. S. (2001) Process for producing insoluble hydrogel by using chitosan *Vol. KR20010061109 (A)*, (ed. K. Korean Patent Office), Hyosung T & C Co. Ltd [KR].

Islam, A. M., Phillips, G. O., Sljivo, A., Snowden, M. J. & Williams, P. A. (1997) A review of recent developments on the regulatory, structural and functional aspects of gum arabic. *Food Hydrocolloids* 11, 493-505.

Janik, I., Kasprzak, E., Al-Zier, A. & Rosiak, J. M. (2008) Radiation crosslinking and scission parameters for poly(vinyl methyl ether) in aqueous solution. *Nuclear Instruments & Methods in Physics Research Section B-Beam Interactions with Materials and Atoms* 208, 374-379.

Jarry, C., Chaput, C., Chenite, A., Renaud, M. A., Buschmann, M. & Leroux, J. C. (2001) Effects of steam sterilization on thermogelling chitosan-based gels. *Journal of Biomedical Materials Research* 58, 127-135.

Jarry, C., Leroux, J. C., Haeck, J. & Chaput, C. (2002) Irradiating or autoclaving chitosan/polyol solutions: Effect on thermogelling chitosan-beta-glycerophosphate systems. *Chemical & Pharmaceutical Bulletin* 50, 1335-1340.

Jeremic, K., Markov, S., Pekic, B. & Jovanovic, S. (1999) The influence of temperature and inorganic salts on the rheological properties of xanthan aqueous solutions. *Journal of the Serbian Chemical Society* 64, 109-116.

Jones, R. A., Ward, I. M., Taylor, D. J. R. & Stepto, R. F. T. (1996) Reactions of amorphous PE radical-pairs in vacuo and in acetylene: a comparison of gel fraction data with Flory-Stockmayer and atomistic modelling analyses. *Polymer* 37, 3643-3657.

Jumel, K., Harding, S. E. & Mitchell, J. R. (1996) Effect of gamma irradiation on the macromolecular integrity of guar gum. *Carbohydrate Research* 282, 223-236.

Katayama, T., Nakauma, M., Todoriki, S., Phillips, G. O. & Tada, M. (2006) Radiation-induced polymerization of gum arabic (Acacia sengal) in aqueous solution. *Food Hydrocolloids* 20, 983-989.

Katayama, T., Ogasawara, T., Sasaki, Y., Al-Assaf, S. & Phillips, G. O. (2008) Composition Containing Hydrogel Component Derived from Gum Arabic. In *US PTO, Vol. US2008038436 (A1)*, (ed. U. United States Patent), San Ei Gen, Japan, Phillips Hydrocolloid Research limited, UK.

Kempe, S., Metz, H., Bastrop, M., Hvilsom, A., Contri, R. V. & Mäder, K. (2008) Characterization of thermosensitive chitosan-based hydrogels by rheology and electron paramagnetic resonance spectroscopy. *European Journal of Pharmaceutics and Biopharmaceutics* 68, 26-33.

Khoylou, F. & Naimian, F. (2009) Radiation synthesis of superabsorbent polyethylene oxide/tragacanth hydrogel. *Radiation Physics and Chemistry* 78, 195-198.

Kim, S. J., Hahn, S. K., T, M. J. K., Kim, D. H. & Lee, Y. P. (2005) Development of a novel sustained release formulation of recombinant human growth hormone using sodium hyaluronate microparticles. *Journal of Controlled Release* 104 323–335.

Kistler, S. S. (1931) Coherent expanded aerogels and jellies. *Nature* 127, 741.

Kuang, Q. L., Zhao, J. C., Niu, Y. H., Zhang, J. & Wang, Z. G. (2008) Celluloses in an ionic liquid: the rheological properties of the solutions spanning the dilute and semidilute regimes. *Journal of Physical Chemistry B* 112, 10234-10240.

Kumar, S. V., Sasmal, D. & Pal, S. C. (2008) Rheological Characterization and Drug Release Studies of Gum Exudates of Terminalia catappa Linn. *Aaps Pharmscitech* 9, 885-890.

Lazareva, T. G. & Vashuk, E. V. (1995) Features of rheological and electrophysical properties of compositions based on polyvinyl alcohol and carboxymethylcellulose. *Mechanics of Composite Materials* 31, 524-532.

Liu, P., Zhai, M., Li, J., Peng, J. & Wu, J. (2002a) Radiation preparation and swelling behavior of sodium carboxymethyl cellulose hydrogels. *Radiation Physics and Chemistry* 63, 525-528.

Liu, P., Zhai, M., Li, J., Peng, J. & Wu, J. (2002b) Radiation preparation and swelling behavior ofsodium carboxymethyl cellulose hydrogels. *Radiation Physics and Chemistry* 63 525-528.

Liu, P., Peng, J., Li, J. & Wu, J. (2005) Radiation crosslinking of CMC-Na at low dose and its application as substitute for hydrogel. *Radiation Physics and Chemistry* 72, 635-638.

Livage, J., Henry, M. & Sanchez, C. (1988) Sol-Gel Chemistry of Transition Metal Oxides. *Prog. Solid State Chem* 18, 259.

Lugao, A. B. & Malmonge, S. M. (2001) Use of radiation in the production of hydrogels. *N uclear Instruments and Methods in Physics Research* B 185, 37-42.

Magnin, D., Lefebvre, J., Chornet, E. & Dumitriu, S. (2004) Physicochemical and structural characterization of a polyionic matrix of interest in biotechnology, in the pharmaceutical and biomedical fields. *Carbohydrate Polymers* 55, 437-453.

Makuuchi, K. (2010) Critical reviewofradiationprocessingofhydrogelandpolysaccharide. *Radiation PhysicsandChemistry* 79, 267-271.

Mansur, H. S., Orefice, R. L. & Mansur, A. A. P. (2004) Characterization of poly(vinyl alcohol)/poly(ethylene glycol) hydrogels and PVA-derived hybrids by small-angle X-ray scattering and FTIR spectroscopy. *Polymer* 45, 7193-7202.

Masteikova, R., Chalupova, Z. & Sklubalova, Z. (2003) Stimuli-sensitive hydrogels in controlled and sustained drug delivery. *Medicina* 39, 19-24.

Matthews, K. H., Stevens, H. N. E., Auffret, A. D., Humphrey, M. J. & Eccleston, G. M. (2006) Gamma-irradiation of lyophilised wound healing wafers. *International Journal of Pharmaceutics* 313, 78-86.

Moore, J. S. & Phillips, G. O. (1971) Radiation studies of aryl glucosides. *Carbohydrate Research* 16, 79-87.

Myint, P., Deeble, D. J., Beaumont, P. C., Blake, S. M. & Phillips, G. O. (1987) The reactivity of free radicals with hyaluronic acid: steady-state and pulse radiolysis studies. *Biochimica Et Biophysica Acta* 925, 194-202.

Nagasawa, N., Yagi, T., Kume, T. & Yoshii, F. (2004) Radiation crosslinking of carboxymethyl starch. *Carbohydrate Polymers* 58, 109-113.

Nakamura, Y., Ogiwara, Y. & Phillips, G. O. (1985) Free radical formation and degradation of cellulose by ionizing radiations. *Polymer Photochemistry* 6, 135-159.

Nguyen, M. K. & Lee, D. S. (2010) Injectable Biodegradable Hydrogels. *Macromol. Biosci.* 10, 563-579.

Nho, Y.-C. & Lee, J.-H. (2005) Reduction of postsurgical adhesion formation with hydrogels synthesized by radiation. *Nuclear Instruments and Methods in Physics Research* B 236 277-282.

Nho, Y.-C., Park, S.-E., Kim, H.-I. & Hwang, T.-S. (2005) Oral delivery of insulin using pH-sensitive hydrogels based on polyvinyl alcohol grafted with acrylic acid/methacrylic acid by radiation. *Nuclear Instruments and Methods in Physics Research* B 236 283–288.

Omari, A., Tabary, R., Rousseau, D., Calderon, F. L., Monteil, J. & Chauveteau, G. (2006) Soft water-soluble microgel dispersions: Structure and rheology. *Journal of Colloid and Interface Science* 302, 537-546.

Onuki, Y., Nishikawa, M., Morishita, M. & Takayama, K. (2008) Development of photocrosslinked polyacrylic acid hydrogel as an adhesive for dermatological patches: Involvement of formulation factors in physical properties and pharmacological effects. *International Journal of Pharmaceutics* 349, 47-52.

Palumbo, F. S., Pitarresi, G., Mandracchia, D., Tripodo, G. & Giammona, G. (2006) New graft copolymers of hyaluronic acid and polylactic acid: Synthesis and characterization. *Carbohydrate Polymers* 66, 379-385.

Parsons, B. J., Phillips, G. O., Thomas, B., Wedlock, D. J. & Clarkesturman, A. J. (1985) Depolymerization of Xanthan by Iron-Catalyzed Free-Radical Reactions. *International Journal of Biological Macromolecules* 7, 187-192.

Phillips, G. O. & Moody, G. J. (1959) The chemical action of gamma radiation on aqueous solutions of carbohydrates. *The International Journal of Applied Radiation and Isotopes* 6, 78-85.

Phillips, G. O. (1961) Advances in carbohydrates, Academic Press, UK.

Phillips, G. O. (1963) Molecular environment effects in the radiation decomposition of a-D-glucose. *Nature* 198, 282-283.

Phillips, G. O., Baugh, P. J. & Lofroth, G. (1966) Energy transport in carbohydrates. Part II. Radiation decomposition of D-glucose. *Journal of Chemical Society Section A*, 377-382.

Phillips, G. O. & Young, M. (1966) Energy transport in carbohydrates. Part III. Chemical effects of gamma radiation on cycloamyloses. *Journal of Chemical Society Section A*, 383-387.

Phillips, G. O. (1968) Energetics and Mechanisms in Radiation Biology. In *Radiation Effects on Carbohydrates*, (ed. G. O. Phillips), Academic Press, London.

Phillips, G. O. & Williams, P. A. (2000) Handbook of hydrocolloids. In *Starch*, (ed. P. Murphy), Woodhead Publishing limited, Cambridge, England.

Phillips, G. O., Plessis, T. A. D., SaphwanAl-Assaf & Williams, P. A. (2003) Biopolymers obtained by solid state irradiation in an unsaturated gaseous atmosphere, *Vol. 6,610,810*, (ed. U. S. Patent), Phillips Hydrocolloid Research limited, UK.

Phillips, G. O., Plessis, T. A. D., SaphwanAl-Assaf & Williams, P. A. (2005) Biopolymers obtained by solid state irradiation in an unsaturated gaseous atmosphere, *Vol. 6,841644*, (ed. U. S. Patent), Phillips hydrocolloid research Limited, UK, USA.

Pongjanyakul, T. & Puttipipatkhachorn, S. (2007) Xanthan–alginate composite gel beads: Molecular interaction and in vitro characterization. *International Journal of Pharmaceutics* 331, 61-71.

Porsch, B. & Wittgren, B. (2005) Analysis of calcium salt of carboxymethyl cellulose: size distributions of parent carboxymethyl cellulose by size-exclusion chromatography with dual light-scattering and refractometric detection. *Carbohydrate Polymers* 59, 27-35.

Pourjavadi, A. & Zohuriaan-Mehr, M. J. (2002) Modification of carbohydrate polymers via grafting in air. 2. Ceric-initiated graft copolymerization of acrylonitrile onto natural and modified polysaccharides. *Starch-Starke* 54, 482-488.

Pourjavadi, A. & Kurdtabar, M. (2007) Collagen-based highly porous hydrogel without any porogen: Synthesis and characteristics. *European Polymer Journal* 43, 877-889.

Razzak, M. T., Darwis, D., Zainuddin & Sukirno. (2001) Irradiation of polyvinyl alcohol and polyvinyl pyrrolidone blended hydrogel for wound dressing. *Radiation Physics and Chemistry* 62, 107-113.

Reháková, M., Bakos, D., Soldán, M. & Vizárová, K. (1994) Depolymerization reactions of hyaluronic acid in solution. *International Journal of Biological Macromolecules* 16, 121-124.

Relleve, L., Nagasawa, N., Luan, L. Q., Yagi, T., Aranilla, C., Abad, L., Kume, T., Yoshii, F. & dela Rosa, A. (2005) Degradation of carrageenan by radiation. *Polymer Degradation and Stability* 87, 403-410.

Rosiak, J. M., Ulanski, P. & Rzeinicki, A. (1995) Hydrogels for biomedical purposes. *N uclear Instruments and Methods in Physics Research* B 105, 335-339.

Rosiak, J. M. (1998) Gel / sol analysis ofirradiated polymers. *Radiation Physics and Chemistry* 51, 13-17.

Rosiak, J. M. & Yoshii, F. (1999) Hydrogels and their medical applications. *Nuclear Instruments and Methods in Physics Research* B 151, 56-64.

Rosiak, J. M., Janik, I., Kadlubowski, S., Kozicki, M., Kujawa, P., Stasica, P. & Ulanski, P. (2003) Nano-, micro- and macroscopic hydrogels synthesized by radiation technique. *Nuclear Instruments and Methods in Physics Research* B 208 325–330.

Rubinstein, M. & Colby, R. H. (2003) Polymer Physics, Oxford University Press, Oxford.

Sahiner, N., Singh, M., De Kee, D., John, V. T. & McPherson, G. L. (2006) Rheological characterization of a charged cationic hydrogel network across the gelation boundary. *Polymer* 47, 1124-1131.

Said, H. M., Alla, S. G. A. & El-Naggar, A. W. M. (2004) Synthesis and characterization of novel gels based on carboxymethyl cellulose/acrylic acid prepared by electron beam irradiation. *Reactive & Functional Polymers* 61, 397–404.

Schiller, J., Arnhold, J., Schwinn, J., Sprinz, H., Brede, O. & Arnold, K. (1998) Reactivity of cartilage and selected carbohydrates with hydroxyl radicals - An NMR study to detect degradation products. *Free Radical Research* 28, 215-228.

Schuetz, Y. B., Gurny, R. & Jordan, O. (2008) A novel thermoresponsive hydrogel based on chitosan. *European Journal of Pharmaceutics and Biopharmaceutics* 68, 19-25.

Sen, M. (2005) Radiation synthesis of stimuli responsive hydrogels and their use for the separation and enrichment of water pollutants. In *Radiation synthesis of stimuli-responsive membranes, hydrogels and adsorbents for separation purposes*, (ed. I. A. E. Agency), International Atomic Energy Agency, Vienna.

Sen, M., Yolacan, B. & Guven, G. (2007) Radiation-induced degradation of galactomannan polysaccharides. *Nuclear Instruments & Methods in Physics Research Section B-Beam Interactions with Materials and Atoms* 265, 429-433.

Severian, D., Hilda, G. & Itzhak, K. (1999) Supported polyionic hydrogels, *Vol. US 5858392 (A)* (ed. U. United States Patent), Yissum Res Dev Co., Israel Fiber Inst., Israel.

Sharpatyi, V. A. (2003) Radiation chemistry of polysaccharides: 1. Mechanisms of carbon monoxide and formic acid formation. *High Energy Chemistry* 37, 369-372.

Shen, X., Kitajyo, Y., Duan, Q., Narumi, A., Kaga, H., Kaneko, N., Satoh, T. & Kakuchi, T. (2006) Synthesis and Photocrosslinking Reaction of N-Allylcarbamoylmethyl Cellulose Leading to Hydrogel. *Polymer Bulletin* 56, 137-143.

Shu, X. Z., Liu, Y., Palumbo, F. S., Luo, Y. & Prestwich, G. D. (2004) In situ crosslinkable hyaluronan hydrogels for tissue engineering. *Biomaterials* 25 1339–1348.

Singh, A., Hosseini, M. & Hariprasad, S. M. Polyethylene Glycol Hydrogel Polymer Sealant for Closure of Sutureless Sclerotomies: A Histologic Study. *American Journal of Ophthalmology* 150, 346-351.e2.

Singh, B. & Vashishth, M. (2008) Development of novel hydrogels by modification of sterculia gum through radiation cross-linking polymerization for use in drug delivery. *Nuclear Instruments and Methods in Physics Research* B 266 2009-2020.

Smetana, K. (1993) Cell biology of hydrogels. *Biomaterials* 14, 1046-1050.

Sonntag, C. V. (1987) The chemical basis of radiation biology, Taylor & Francis, London.

Spinelli, L. S., Aquino, A. S., Lucas, E., d'Almeida, A. R., Leal, R. & Martins, A. L. (2008) Adsorption of polymers used in drilling fluids on the inner surfaces of carbon steel pipes. *Polymer Engineering and Science* 48, 1885-1891.

Stern, R., Kogan, G., Jedrzejas, M. J. & Soltés, L. The many ways to cleave hyaluronan. *Biotechnology Advances* 25, 537-557.

Takigami, M., Amada, H., Nagasawa, N., Yagi, T., Kasahara, T., Takigami, S. & Tamada, M. (2007) Preparation and properties of CMC gel. *Transactions of the Materials Research Society of Japan, Vol 32, No 3* 32, 713-716.

Torres, R., Usall, J., Teixido, N., Abadias, M. & Vinas, I. (2003) Liquid formulation of the biocontrol agent Candida sake by modifying water activity or adding protectants. *Journal of Applied Microbiology* 94, 330-339.

Trksak, R. M. & Ford, P. J. (2008) Sago-based gelling starches, *Vol. 7,422,638,* (ed. U. S. Patent), National Starch and Chemical Investment Holding Corporation (New Castle, DE), USA.

Valles, E., Durando, D., Katime, I., Mendizabal, E. & Puig, J. E. (2000) Equilibrium swelling and mechanical properties of hydrogels of acrylamide and itaconic acid or its esters. *Polymer Bulletin* 44, 109-114.

Wach, R. A., Mitomo, H., Yoshii, F. & Kume, T. (2001) Hydrogel of Biodegradeable Cellulose Derivatives. II. Effect of Some Factors on Radiation-Induced Crosslinking of CMC. *Journal of Applied Polymer Science* 81, 3030-3037.

Wach, R. A., Mitomo, H., Yoshii, F. & Kume, T. (2002) Hydrogel of Radiation-Induced Cross-Linked Hydroxypropylcellulose. *Macromol. Mater. Eng.* 287, 285-295.

Wach, R. A., Mitomo, H., Nagasawa, N. & Yoshii, F. (2003a) Radiation crosslinking of carboxymethylcellulose of various degree of substitution at high concentration in aqueous solutions of natural pH. *Radiation Physics and Chemistry* 68, 771-779.

Wach, R. A., Mitomo, H., Nagasawa, N. & Yoshii, F. (2003b) Radiation crosslinking of methylcellulose and hydroxyethylcellulose in concentrated aqueous solutions. *Nuclear Instruments and Methods in Physics Research* B 211, 533-544.

Wach, R. A., Kudoh, H., Zhai, M. L., Muroya, Y. & Katsumura, Y. (2005) Laser flash photolysis of carboxymethylcellulose in an aqueous solution. *Journal of Polymer Science Part a-Polymer Chemistry* 43, 505-518.

Walker, M., Hobot, J. A., Newman, G. R. & Bowler, P. G. (2003) Scanning electron microscopic examination of bacterial immobilisation in a carboxymethyl cellulose (AQUACEL) and alginate dressings. *Biomaterials* 24, 883-890.

Wang, M., Xu, L., Ju, X., Peng, J., Zhai, M., Li, J. & Wei, G. (2008) Enhanced radiation crosslinking of carboxymethylated chitosan in the presence of acids or polyfunctional monomers. *Polymer Degradation and Stability* 93, 1807-1813.

Wasikiewicz, J. M., Yoshii, F., Nagasawa, N., Wach, R. A. & Mitomo, H. (2005) Degradation of chitosan and sodium alginate by gamma radiation, sonochemical and ultraviolet methods. *Radiation Physics and Chemistry* 73, 287-295.

Wasikiewicz, J. M., Mitomo, H., Nagasawa, N., Yagi, T., Tamada, M. & Yoshii, F. (2006) Radiation crosslinking of biodegradable carboxymethylchitin and carboxymethylchitosan. *Journal of Applied Polymer Science* 102, 758-767.

Williams, P. A. & Phillips, G. O. (2006) Physicochemical characterisation of gum arabic arabinogalactan protein complex. *Food and Food Ingredients Journal of Japan* 211, 181-188.

Xu, G. Y., Chen, A. M., Liu, S. Y., Yuan, S. L. & Wei, X. L. (2002) Effect of C12NBr on the viscoelasticity of gel containing xanthan gum/Cr(III). *Acta Physico-Chimica Sinica* 18, 1043-1047.

Yan, C., Altunbas, A., Yucel, T., Nagarkar, R. P., Schneider, J. P. & Pochan, D. J. (2010) Injectable solid hydrogel: mechanism of shear-thinning and immediate recovery of injectable β-hairpin peptide hydrogels. *Soft Matter* 6, 5143-5156.

Yang, D., Zhang, J. Z., Fu, S., Xue, Y. & Hu, J. Evolution process of polymethacrylate hydrogels investigated by rheological and dynamic light scattering techniques. *Colloids and Surfaces A: Physicochemical and Engineering Aspects* 353, 197-203.

Yoldas, B. E. (1975) Alumina gels that form porous transparent Al_2O_3. *J. Mater. Sci.* 10, 1856-1860.

Yoshii, F. & Kume, T. (2003) Process for producing crosslinked starch derivatives and crosslinked starch derivatives produced by the same, *Vol. 6,617,448*, (ed. U. S. Patent), Japan Atomic Energy Research Institute (Tokyo, JP), USA.

Yoshii, F., Zhao, L., Wach, R. A., Nagasawa, N., Mitomo, H. & Kume, T. (2003) Hydrogels of polysaccharide derivatives crosslinked with irradiation at paste-like condition. *Nuclear Instruments and Methods in Physics Research B* 208 320–324.

Zegota, H. (1999) The effect of gamma-irradiation on citrus pectin in N2O and N2O/O-2 saturated aqueous solutions. *Food Hydrocolloids* 13, 51-58.

Zhai, M., Yoshii, F. & Kume, T. (2003) Radiation modification of starch-based plastic sheets. *Carbohydrate Polymers* 52, 311-317.

Zhai, M. L., Yoshii, F., Kume, T. & Hashim, K. (2002) Syntheses of PVA/starch grafted hydrogels by irradiation. *Carbohydrate Polymers* 50, 295-303.

Zhao, L., Mitomo, H., Nagasawa, N., Yoshii, F. & Kume, T. (2003a) Radiation synthesis and characteristic of the hydrogels based on carboxymethylated chitin derivatives. *Carbohydrate Polymers* 51, 169-175.

Zhao, L., Mitomo, H., Zhai, M. L., Yoshii, F., Nagasawa, N. & Kume, T. (2003b) Synthesis of antibacterial PVA/CM-chitosan blend hydrogels with electron beam irradiation. *Carbohydrate Polymers* 53, 439-446.

Zhao, Q. S., Ji, Q. X., Xing, K., Li, X. Y., Liu, C. S. & Chen, X. G. (2009) Preparation and characteristics of novel porous hydrogel films based on chitosan and glycerophosphate. *Carbohydrate Polymers* 76, 410-416.

Zhou, H. Y., Chen, X. G., Kong, M., Liu, C. S., Cha, D. S. & Kennedy, J. F. (2008) Effect of molecular weight and degree of chitosan deacetylation on the preparation and characteristics of chitosan thermosensitive hydrogel as a delivery system. *Carbohydrate polymers* 73, 265-273.

6

Chemical Mediated Synthesis of Silver Nanoparticles and its Potential Antibacterial Application

P.Prema
Post Graduate and Research Department of Zoology,
V.H.N.S.N. College
India

1. Introduction

1.1 Nanometer

A nanometer is a *unit of measure* just like inches, feet and miles. By definition a nanometer is one-billionth of a meter. A meter is about 39 inches long. A billion is a thousand times bigger than a million, as a number you write it out as 1,000,000,000. That is a big number and when you divide a meter into one billion parts, well that is very small.

A nanometer is used to measure things that are very small. Atoms and molecules, the smallest pieces of everything around us, are measured in nanometers.

1.2 Nanotechnology

Nanoscience is the study, and nanotechnology is the exploitation, of the strange properties of materials smaller than 100 nanometers (nm) to create new useful objects. This work is made possible by being able to manipulate structures at the size-scale of the atoms.

Nanotechnology, or, as it is sometimes called, Engineering at the Molecular Level, is multi-disciplinary area of applied science and engineering that deals with the design and manufacture of extremely small components and systems. They are built at the molecular level of matter, are characterized by large surface areas in comparison with their volumes, and have behaviors that are governed by the laws of quantum mechanics.

Here are the few definitions of Nanotechnology.

* **Nanotechnology** is the engineering of functional systems at the molecular scale.
* **Nanotechnology** is an emerging, interdisciplinary area of research with important commercial applications, and will, most assuredly, be a dominant technology in new-world economics.
* **Nanotechnology** refers to the projected ability to construct items from the bottom up, using techniques and tools being developed today to make complete, high performance products.
* **Nanotechnology** is a field of applied science focused on the design, synthesis, characterization and application of materials and devices on the nanoscale.
* **Nanotechnology** is a subclassification of technology in colloidal science, biology, physics, chemistry and other scientific fields and involves the study of phenomena and

manipulation of material at the nanoscale, in essence an extension of existing sciences into the nanoscale.

- **Nanotechnology** addresses our ability to understand and manipulate the physical and technological characteristics that govern the behavior of a class of systems that possess at least one physical dimension that is (typically) on the order of 100 nm or less.

Two main approaches are used in nanotechnology:

- **"bottom-up" approach** where materials and devices are built up atom by atom.
- **"top-down" approach** where they are synthesized or constructed by removing existing material from larger entities.

A unique aspect of nanotechnology is the vastly increased ratio of surface area to volume present in many nanoscale materials, which opens new possibilities in surface-based science, such as catalysis.

Lithography is a top-down fabrication technique where a bulk material is reduced in size to nanoscale pattern.

In contrast, bottom-up techniques build or grow larger structures atom by atom or molecule by molecule. These techniques include Chemical synthesis, self-assembly and positional assembly.

There are three distinct nanotechnologies:

"Wet" nanotechnology is the study of biological systems that exist primarily in a water environment. The functional nanometer-scale structures of interest here are genetic material, membranes, enzymes and other cellular components. The success of this nanotechnology is amply demonstrated by the existence of living organisms whose form, function, and evolution are governed by the interactions of nanometer-scale structures.

"Dry" nanotechnology derived from surface science and physical chemistry, focuses on fabrication of structures in carbon (for example, fullerenes and nanotubes), silicon, and other inorganic materials. Unlike the "wet" technology, "dry" techniques admit use of metals and semiconductors. The active conduction electrons of these materials make them too reactive to operate in a "wet" environment, but these same electrons provide the physical properties that make "dry" nanostructures promising as electronic, magnetic, and optical devices. Another objective is to develop "dry" structures that possess some of the same attributes of the self-assembly that the wet ones exhibit.

Computational nanotechnology permits the modeling and simulation of complex nanometer-scale structures. The predictive and analytical power of computation is critical to success in nanotechnology: nature required several hundred million years to evolve a functional "wet" nanotechnology; the insight provided by computation should allow us to reduce the development time of a working "dry" nanotechnology to a few decades, and it will have a major impact on the "wet" side as well. These three nanotechnologies are highly interdependent. The major advances in each have often come from application of techniques or adaptation of information from one or both of the others.

Nanomaterials display unique, superior and indispensable properties and have attracted much attention for their distinct characteristics that are unavailable in conventional macroscopic material. Their uniqueness arises specifically from higher surface to volume ratio and increased percentage of atoms at the grain boundaries. They represent an important class of materials in the development of novel devices that can be used in various physical, biological, biomedical and pharmaceutical applications.

1.3 Silver nanoparticle

Silver is a nontoxic, safe inorganic antibacterial agent used for centuries and is capable of killing about 650 type of diseases causing microorganisms. Silver has been described as being oligodynamic because of its ability to exert a bactericidal effect at minute concentrations. It has a significant potential for a wide range of biological applications such as antifungal agent, antibacterial agents for antibiotic resistant bacteria, preventing infections, healing wounds and anti-inflammatory. Silver ions (Ag^+) and its compounds are highly toxic to microorganisms exhibiting strong biocidal effects on many species of bacteria but have a low toxicity towards animal cells. Therefore, silver ions, being antibacterial component, are employed in formulation of dental resin composites, bone cement, ion exchange fibers and coatings for medical devices.

1.4 Applications of silver nanoparticles
1.4.1 Silver as a biocide

Silver (Ag) is a transition metal element having atomic number 47 and atomic mass 107.87. Its action as an antibiotic comes from the fact that it is a non-selective toxic "biocide." Silver based antimicrobial biocides are used as wood preservatives. In water usage, silver and copper based disinfectants are used in hospital and hotel distribution systems to control infectious agents (for example, *Legionella*). Silver together with copper, is commonly used to inhibit bacterial and fungal growth in chicken farms and in post harvested cleaning of oysters.

Silver based topical dressing has been widely used as a treatment for infections in burns, open wounds and chronic ulcers. The Silver nanoparticles and Ag^+ carriers can be beneficial in delayed diabetic wound healing as diabetic wounds are affected by many secondary infections. These nanoparticles can help the diabetic patients in early wound healing with minimal scars (Mishra *et al.*, 2008). Silver nitrate is still a common antimicrobial used in the treatment of chronic wounds (Wright *et al.*, 1999).

1.4.2 Colloidal silver

Scientists have discovered that the body's most important fluids are colloidal in nature, suspended ultra-fine particles. For example blood carries nutrition and oxygen to the body cells. An electro-colloidal method (electrical silver atoms) is used for the manufacture of colloidal silver. Colloidal silver appears to be a powerful, natural antibiotic and protective against infections. Acting as a catalyst, it reportedly disables the enzyme that one-celled bacteria, viruses and fungi need for their oxygen metabolism. They suffocate without corresponding harm occurring to human enzymes or parts of the human body chemistry. The result is the destruction of disease-causing organisms in the body and in the food.

1.4.3 Silver nanoparticles as a catalyst

A possible application of silver nanoparticles is the use as a catalyst Silver was prepared by the deposition−precipitation method and was found to be a novel visible light driven photocatalyst. The catalyst showed high efficiency for the degradation of nonbiodegradable azodyes and the killing of *Escherichia coli* under visible light irradiation ($\lambda > 420$ nm). The catalyst activity was maintained effectively after successive cyclic experiments under UV or visible light irradiation without the destruction of AgBr. On the basis of the characterization of X-ray diffraction, X-ray photoelectron spectroscopy, and Auger electron spectroscopy, the surface Ag species mainly exist as Ag^0 in the structure of all samples before and after reaction, and Ag^0 species scavenged h_{VB}^+ and then trapped e_{CB}^- in the process of photocatalytic reaction, inhibiting the decomposition of AgBr. The studies of ESR and H_2O_2 formation revealed that $\cdot OH$ and $O_2^{\cdot-}$ were formed in visible light irradiated aqueous $Ag/AgBr/TiO_2$ suspension, while there was no reactive oxygen species in the visible light irradiated Ag^0/TiO_2 system. The results indicate that AgBr is the main photoactive species for the destruction of azodyes and bacteria under visible light (Hu *et al.*, 2006). Silver nanoparticles immobilized on silica spheres have been tested for their ability to catalyze the reduction of dyes by sodium borohydride ($NaBH_4$). Catalysis of dyes was chosen because it is easy to detect a change in color when the dyes are reduced. In the absence of silver nanoparticles the sample was almost stationary showing very little or no reduction of the dyes. The reaction time showed to be strongly dependent on the concentration of silver nanoparticles. When the concentration was doubled, the reaction time was reduced to less than one third.

Silver nanoparticles act as an electron relay, aiding in the transfer of electrons from the BH^{-4} ions to the dyes, and thereby causing a reduction of the dyes. BH^{-4} ions are nucleophilic while dyes are electrophilic. It has been proven that nucleophilic ions can denote electron to metal particles, while an electrophilic can capture electrons from metal particles. It has been shown that BH^{-4} ions and dyes are simultaneously adsorbed on the surface of silver particles, when they were present together.

Silver nanoparticles have a strong tendency to agglomerate. This reduces the surface to volume ratio and thereby the catalytic effect. Therefore a stabilizing agent is often used to prevent agglomeration. However, the agent is adsorbed on the surface of the nanoparticles, shielding them from the oxidant and reductant and thereby inhibiting the catalysis. Instead a new method for stabilizing the nanoparticles is used. The intermolecular forces which keep the nanoparticles immobilized of the silica spheres, has proven to be strong enough to prevent the particles from forming aggregates.

1.4.4 Silver nanoparticles as a bactericidal agent

Another area where silver nanoparticles have proven to be effective is in controlling and suppressing bacterial growth. There have already been developed several applications which use the bactericidal effect of silver nanoparticles. Antibacterial properties of silver are documented since 1000 B.C., when silver vessels were used to preserve water. The first scientific papers describing the medical use of silver report the prevention of eye infection in neonates in 1881 and internal antisepsis in 1901. After this, silver nitrate and silver sulfadiazine have been widely used for the treatment of superficial and deep dermal burns of wounds and for the removal of warts (Rai *et al.*, 2009). Silver's mode of action is presumed to be dependent on Ag^+ ions, which strongly inhibit bacterial growth through suppression of respiratory enzymes and electron transport components and through interference with DNA functions (Li *et al.* 2006).

Silver in a nanometric scale (less than 100 nm) has different catalytical properties compared with those attributed to the bulk form of the noble metal, like surface Plasmon resonance, large effective scattering cross section of individual silver nanoparticles, and strong toxicity to a wide range of microorganisms (Elechiguerra *et al.*, 2005). Morones *et al.* (2005) defined the antibacterial activity of silver nanoparticles in four types of Gram negative bacteria:

Escherichia coli, Vibrio cholerae, Pseudomonas aeruginosa, and *Salmonella typhi* and suggested that silver nanoparticles attach to the surface of the cell membrane and disturb its function, penetrate bacteria, and release silver ions. Other groups determined a similar antibacterial activity in Gram positive bacteria, such as *Bacillus subtilis* (Yoon *et al.,* 2008), *Staphylococcus aureus* (Shrivastava *et al.,* 2007), and *Enterococcus faecalis* (Panacek *et al.,* 2006). Silver nanoparticles have also been found to exert antibacterial activity against some drug resistant bacteria (Birla *et al.,* 2009; Inoue *et al.,* 2010).

1.5 Synthesis of silver nanoparticles

One key aspect of nanotechnology concerns the development of rapid and reliable experimental protocols for the synthesis of nanomaterials over a range of chemical compositions, sizes, high monodispersity and large scale production. A variety of techniques have been developed to synthesize metal nanoparticles, including chemical reduction using number of chemical reductants such as $NaBH_4$, N_2H_4, NH_2OH, ethanol, ethylene glycol and N, N-dimethyl farmamide (DMF), aerosol technique, electrochemical or sonochemical deposition, photochemical reduction and laser irradiation technique.

Many methods, a chemical reduction method (Chou *et al.,* 2005), a polyol method (Lin & Yang, 2005) and radiolytic process (Shin *et al.,* 2004) have been developed for the synthesis of silver nanoparticles. Among the methods, chemical reduction was widely studied, due to its advantages of yielding nanoparticles without aggregation, high yield and low preparation cost (Kim *et al.,* 2004).

The chemical reduction method involves the reduction of $AgNO_3$ in aqueous solution by a reducing agent in the presence of a suitable stabilizer, which is necessary in protecting the growth of silver particles through aggregation . In the formation of silver nanoparticles by the chemical reduction method, the particle size and aggregation state of silver nanoparticles are affected by various parameters such as initial $AgNO_3$ concentration, reducing agent, $AgNO_3$ molar ratios and stabilizer concentration (Song *et al.,* 2009).

The silver nanoparticles were prepared by using chemical reduction method (Fang *et al.,* 2005). All solutions of reacting materials were prepared in double distilled water. In typical experiment 50ml of $1 \times 10^{-3}M$ $AgNO_3$ was heated to boiling. To this solution 5ml of 1% tri-sodium citrate was added drop by drop. During this process solution was mixed vigorously and heated until colour change was evident (pale brown). Then it was removed from the heating element and stirred until cooled to room temperature.

Mechanism of reaction could be expressed as follows:

$$2Ag^+ + C_6H_5O_7Na_3 + 2H_2O \rightarrow 4Ag^0 + C_6H_5O_7H_3 + 3Na^+ + H^+ + O_2 \uparrow$$

The aqueous solution air dried up to 3 days and produced the dry powdered particles that were taken for further analysis.

1.6 Detection of silver nanoparticles

The first step is to determine whether silver nanoparticles are actually synthesized or not. Characterization of the nanoparticles, which examines with includes size, shape, and quantity. A number of different measurement techniques can be used for this purpose, including Atomic Force Microscopy (AFM), Scanning Electron Microscopy (SEM), Absorbance Spectroscopy and Dynamic Light Scattering (DLS).

1.7 Bactericidal effect of silver nanoparticles

The growth of unwanted bacteria has for a long time been and is still a problem for the food industry and in the medical field. Therefore, there is a need for methods to kill or slow down the growth of unwanted bacteria. An interesting alternative methods is the use of metallic nanoparticles is silver nanoparticles.

In order to achieve an understanding of this effect, knowledge about the structure of bacteria is needed. In particular, the bacterial membrane and the contained proteins are of special interest, because the silver has to react with it in order to penetrate the bacteria.

Silver has for a long time been known to be toxic to a wide range of bacteria, and this has been utilized in various applications. Silver compounds are used as preservative in a variety of products and in the medical field to treat burns and infections. The bactericidal effect of silver can be divided into two groups; the reactive component being either silver ions or silver nanoparticles. Here a clear distinction between ions and particles has to be made. To clarify, silver ions are charged atoms (Ag^+) whereas silver nanoparticles are single crystals of nanosize dimensions.

In spite of the fact that the bactericidal effect of silver ions is well known and used it is still not fully understood. Experiments have shown that silver ions are able to make structural changes in the cell membrane. The membrane of bacteria contains lot of sulfate-containing enzymes. This inactivation makes the membrane vulnerable and easier to penetrate for silver ions. Inside the cell, silver ions continue to destroy different parts of the cell by interacting with sulfate-groups, which are often located in the active site of enzymes. This interaction with the active site causes an inactivation of the enzymes.

Silver ions are also able to interact with phosphorus-groups of molecules and experiments have shown that this can have severe effects. One example is the interaction between silver ions and the backbone of DNA, which makes the bacteria unable to replicate itself or transcribes mRNA for new proteins. All these changes slow down the growth of the bacteria and finally kill it.

However, the bactericidal mechanism of silver nanoparticles on bacteria is almost unknown. It has been proposed that the effect is caused by some of the same mechanisms that caused the bactericidal effect of silver ions (Alcamo, 1997).

1.8 Mechanism of the bactericidal effect of silver nanoparticles

The silver nanoparticles which show interaction with the bacteria were all between 1 and 10 nm. The reason for this size dependency is probably a combination of the particles ability to react with and penetrate the cell membrane and the higher surface to volume ratio of smaller particles. It is known that small (~ 5nm) metallic particles present electronic effects, which are defined as changes in the local electronic structure of the surface. These effects enhance the reactivity of the nanoparticle surfaces. When the size of the silver nanoparticles decreases the percentage of the interacting atoms increases, and this could be the explanation to why small (1-100 nm) silver nanoparticles are able to interact with the bacteria. With regard to the ability of silver nanoparticles to penetrate the cell membrane, is it reasonable to believe that small nanoparticles are more capable of penetrating the cell membrane than larger nanoparticles (Morones *et al.*, 2005).

The morphology of the interacting silver nanoparticles have also been studied, and the majority of the silver nanoparticles were either octahedral, multiple-twinned icosahedral or decahedral in shape. Previous experiments have demonstrated that the {111} facets exhibit high reactivity, which may explain why these types of particles are capable of interacting with bacteria. When silver nanoparticles are present in a solution they secrete a small amount of silver ions, which will have an additional contribution to the bactericidal effect of silver nanoparticles.

2. Characterization of nanoparticles

2.1 Visual inspection
The reduction of metal ions was roughly monitored by visual inspection of the solution by the method described by Fang *et al.* (2005). The conversion of the colourless reaction mixture to a brown colour clearly indicates the formation of silver nanoparticles.

2.2 UV-Vis spectroscopy
The reduction of metal ions was monitored by measuring the UV-Vis spectroscopy of the solution according to the method of Mie (1908), by the sampling of aliquots (3ml) of the aqueous component. The silver nanoparticles were measured in a wavelength ranging from 200-1100 nm. The UV-Vis spectroscopy measurements of silver nanoparticle was recorded on Shimadzu dual beam spectrophotometer (model UV-1650 PC) operated at a resolution of 1nm.

2.3 X-ray diffraction
The X-ray Diffraction patterns of silver nanoparticle were recorded according to the description of Wang (2000). Samples were air dried, powdered and used for XRD analysis. X-ray diffraction patterns were recorded in the scanning mode on an X`pert PRO PAN analytical instrument operated at 40 KV and a current of 30 mA with Cu K$\propto$ radiation (λ=1.5406 Å). The diffraction intensities were recorded from 35° to 79.93°, in 2θ angles. The diffraction intensities were compared with the standard JCPDS files. The software gave the information about the crystal structure of the particle.

The average size of the nanoparticles can be estimated using the Debye–Scherrer equation (Rau, 1962):

$$D = k\lambda \, / \, \beta\cos\theta \, ,$$

Where D = Thickness of the nanocrystal,
$\qquad k$ = Constant,
$\qquad \lambda$ = Wavelength of X-rays,
$\qquad \beta$ = Width at half maxima of (111) reflection at Bragg's angle 2θ,
$\qquad \theta$ = Bragg angle.

The size of the silver nanoparticle was made from the line broadening of the (111) reflection using the Debye-Scherrer formula. According to the formula,
$\qquad$ Constant (K) = 0.94
$\qquad$ Wave length (λ) = 1.5406 x 10^{-10}
$\qquad$ Full width at half maximum in radius (β) = 0.3553 x $\pi/180$
$\qquad$ Diffraction bragg angle 2θ = 38.1759
$\qquad \theta$ = 19.088
$\qquad D$ = 0.94 x 1.5406 x 10^{-10} x 180 / 0.3553 x cos 19.38 x 3.14
$\qquad$ = 24.778 x 10^{-9}

2.4 Scanning electron microscopy

Morphology and size of the silver nanoparticle was investigated with the Scanning Electron Microscope (JSM 35 CF JEOL) in a resolution of 60Å at 15 KV, magnification of 5.0 K. The scale was about 32 mm to 3.6 μm. The size of the particle can be calculated by using the scale provided in the micrograph.

3. Antibacterial activity of silver nanoparticles

3.1 Well diffusion technique

The following human bacterial pathogens *Pseudomonas* sp., *Escherichia coli*, *Staphylococcus aureus*, *Klebsiella pneumoniae*, *Shigella sp.*, and *Salmonella* sp., were grown on nutrient agar plate and maintained in the nutrient agar slants at 4ₒC. Overnight culture in the nutrient broth was used for the present experimental study.

The antibacterial activity of silver nanoparticle was determined by agar well diffusion method. In this method, sterile Mueller - Hinton Agar plate was prepared. Bacterial pathogens used in the present experiment were spread over the agar plate using sterile cotton swab. The plates were allowed to dry and a sterile well - cutter of diameter 6.0mm was used to bore wells in the agar plates. Subsequently, a 50μl of the nanoparticle suspension was introduced into wells of the inoculated Mueller – Hinton Agar plates. The plates were allowed to stand for 1h or more for diffusion to take place and then incubated at 37ₒC for 24 h and measured the diameter of inhibitory zones in mm (Lee *et al.*, 2010).

3.2 Bacterial killing kinetics using nanosilver

To examine the bacterial killing kinetics in the presence of silver nanoparticles, a modified method described by Pal *et al.* (2007) was followed. Since the following pathogens *Escherichia coli*, *Staphylococcus aureus,* and *Klebsiella pneumoniae* showed a better activity. The human bacterial pathogens were grown in 10ml of nutrient broth supplemented with different doses of nanosilver (silver content 1, 2, 3, 4, 5 and 6 μl) at 37°C without agitation.

Killing kinetic rates and bacterial concentrations were determined by measuring the colony forming unit in the nutrient agar plates. Percentage of bacterial growth inhibition was calculated as per the equation of Shahi *et al.* (2003).

$$BGI\,\% \;=\; (BC - BT)\; x\; 100\; /\; BC$$

Where,
BGI = Bacterial Growth Inhibition
BC = Number of Bacterial Colonies in the Control
BT = Number of Bacterial Colonies in the Treatment

4. Statistical analysis

The results obtained in the present experiments were subjected to statistical analysis such as Mean, Standard Deviation and Correlation Coefficient was performed by MS Excel.

5. Observation

5.1 Visual inspection of silver nanoparticles

The appearance of a pale brown colour in solution and mirror like illumination on the walls of the Erlenmeyer flask clearly indicated the formation of silver nanoparticles in the reaction mixture (Fig. 1). The colour of the solution was due to the excitation of surface plasmon vibrations in the silver nanoparticles. The powder form of silver nanoparticles prepared by the chemical reduction method is predicted in Fig. 2.

Fig. 1. The Silver Nanoparticles formed in the chemical reduction method

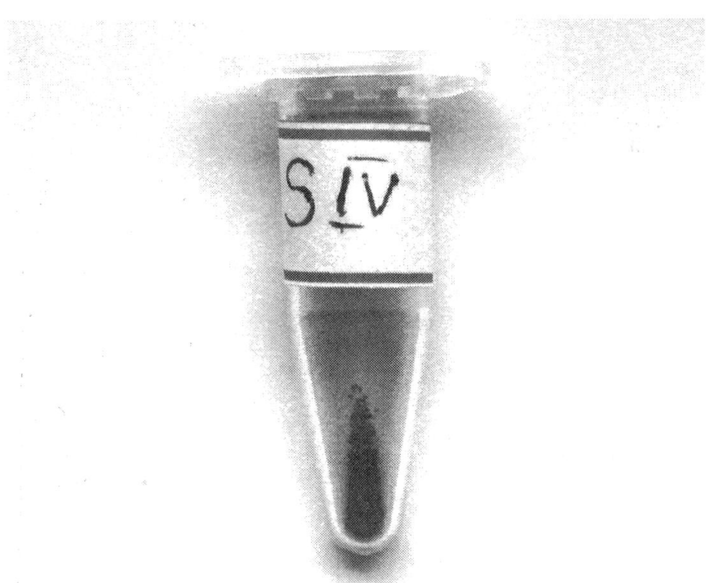

Fig. 2. Silver Nanoparticles in the powder form

5.2 UV-Vis spectroscopy of silver nanoparticles

The UV-Vis spectra of the silver nanoparticles showed a well defined surface plasmon band centered at around 420 nm (Fig. 3), which is the characteristic of silver nanoparticles and clearly indicate the formation of nanoparticles in solution. A minimum at ~320nm correspond to the wavelength at which the real and imaginary parts of the dielectric function of silver almost vanish. The plasmon bands are broad with an absorption tail in the longer wavelengths, which could be in practice due to the size distribution of the particle. The exact position of absorbance depends on a number of factors such as the dielectric constant of the medium, size of the particle etc.

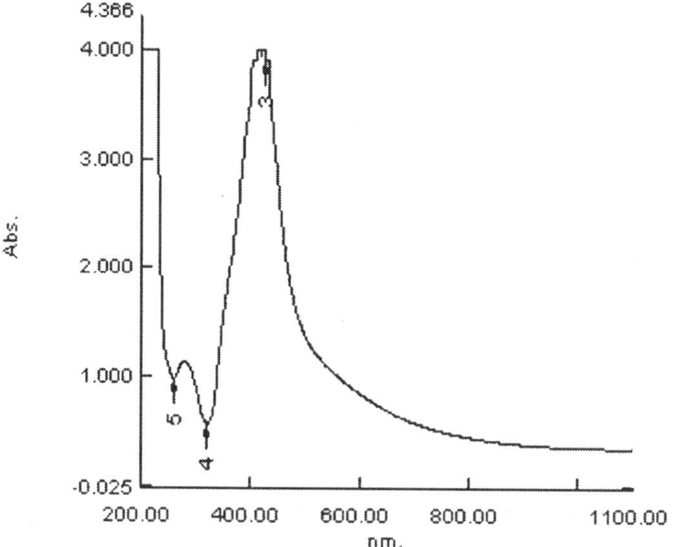

Fig. 3. UV-Vis Spectrum of Silver Nanoparticles

5.3 X-ray diffraction of silver nanoparticles

The intensive diffraction peak at a 2θ value of 38.18° from the (111) lattice plane of face centered cubic (fcc) silver unequivocally indicates that the particles are made of pure silver. Three additional broad bands are observed at 44.32° (2θ), 64.50° (2θ), and 77.05° (2θ) they correspond to the (200), (220) and (311) planes of silver respectively (Fig. 4). Other spurious diffractions are due to crystallographic impurities. Table 1 explains the X-ray diffraction peak list of silver nanoparticles. In the obtained spectrum, the Bragg peak position and their intensities were compared with the standard JCPDS files. The result shows that the particles have a cubic structure. The size of the silver nanoparticles was found to be 25 nm.

5.4 Scanning electron microscopy of silver nanoparticles

The scanning electron micrograph of silver nanoparticles is depicted in Fig. 5. The micrograph shows that the particles have a spherical nature and the average size (mean ± SD) of the particles can be calculated as 21.22 ± 5.17nm (Table 2).

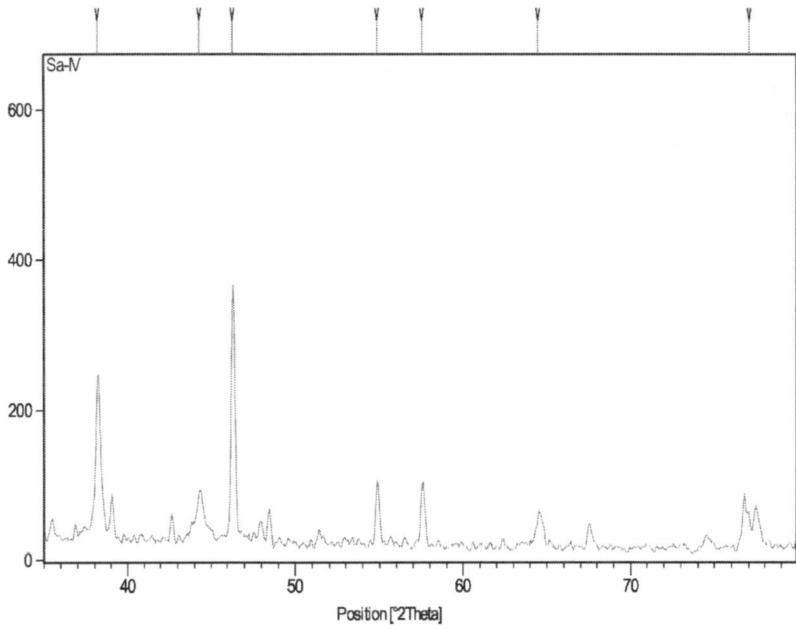

Fig. 4. X-ray Diffraction Spectrum of Silver Nanoparticles Prepared by Chemical Reduction
Method

Pos. [°2Th.]	Height [cts]	FWHM [°2Th.]	d-spacing [Å]	Rel. Int. [%]
38.1759	200.85	0.3553	2.35551	31.98
44.3197	56.29	0.6764	2.04219	8.96
46.2811	338.76	0.1716	1.96011	53.95
54.8576	83.39	0.1698	1.67221	13.28
57.5224	89.84	0.1913	1.60092	14.31
64.5007	45.22	0.4181	1.44354	7.20
77.0516	55.22	0.9851	1.23670	8.79

Table 1. X-ray Diffraction Peak List of Silver Nanoparticles

5.5 Antibacterial activity of silver nanoparticles

Silver has long been considered as a powerful and natural antibiotic and antibacterial agent.
Silver nanoparticles exhibit antimicrobial properties against bacterial pathogens with close
attachment of the nanoparticles themselves with the microbial cells.

The antimicrobial activity of silver nanoparticle have been investigated against *Escherichia
coli, Staphylococcus aureus, Klebsiella pneumoniae ,Pseudomonas* sp., *Shigella* sp., and *Salmonella*
sp., Silver nanoparticles obtain very strong inhibitory (+++) action against *Staphylococcus
aureus* followed by *Escherichia coli, Klebsiella pneumonia* and *Pseudomonas* sp (Table 3). After
sufficient incubation the nanoparticles showed an inhibition zone near to 10mm, in,
Escherichia coli, Staphylococcus aureus sp., *Klebsiella pneumoniae* and*Pseudomonas* sp (Fig.6).

Fig. 5. Scanning Electron Micrograph of Silver Nanoparticles

S.No.	Particle size of silver (nm)
1.	28.80
2.	16.46
3.	23.04
4.	23.04
5.	11.52
6.	23.04
7.	14.40
8.	16.46
9.	23.04
10.	23.04
11.	28.80
12.	23.04
Total	254.68
Mean (Average)	21.22
Standard Deviation	± 5.17

Table 2. Size of Silver Nanoparticles Fabricated by Chemical Reduction Method

Human pathogens	Silver
Escherichia coli	++
Staphylococcus aureus	+++
Klebsiella pneumoniae	++
Pseudomonas sp	++
Salmonella sp.,	-
Shigella sp.,	-

+++ Very strong suppression
++ Strong suppression
-No suppression

Table 3. Antibacterial Activity of Silver Nanoparticles

5.6 Bacterial killing kinetics using nanosilver

The antibacterial properties of the colloidal silver were tested against gram positive and gram negative bacteria. The viable bacteria were monitored by counting the number of colony forming units from the appropriate dilution on nutrient agar plates. The survival fraction was determined calculating the colony forming units per milliliters of the culture. The activity was examined after 24 hrs. It was observed that the silver nanoparticles exhibit the killing rate of *Escherichia coli* to 77.86 %, *Staphylococcus aureus* to 81.8 % and *Klebsiella pneumoniae* to 70.17 % of viability. The silver nanoparticles had a better activity against *Staphylococcus aureus* (Table 4). The decrease in number of viable cells with increasing amounts of silver nanoparticles in solution can be fitted with correlation coefficient. The correlation coefficient between silver nanoparticles and selected bacterial pathogens is provided in Table 5. It revealed that there is a strong negative correlation of silver nanoparticles against selected pathogens such as *E.coli*, *Staphylococcus aureus* and *Klebsiella pneumoniae* (-0.975, -0.993 and -0.998 respectively).

Silver nanoparticle concentration (µl)	*Escherichia coli*	*Staphylococcus aureus*	*Klebsiella pneumoniae*
0	280	264	295
1	220 (21.4%)	243 (7.95%)	268 (9.15%)
2	180 (35.7%)	182 (31.06%)	225 (23.7%)
3	164 (41.43%)	159 (39.77%)	186 (36.95%)
4	148 (47.14%)	126 (52.27%)	147 (50.17%)
5	124 (55.71%)	98 (62.88%)	118 (60.0%)
6	62 (77.86)	48 (81.82%)	88 (70.17%)

Table 4. Bacterial Growth and Killing Kinetics (%) in the Presence of Nanosilver

Correlation	'r' value
Silver nanoparticle Vs *E.coli*	-0.975
Silver nanoparticle Vs *Staphylococcus aureus*	-0.993
Silver nanoparticle Vs *Klebsiella pneumoniae*	-0.998
Silver nanoparticle Vs *Pseudomonas* Sp.	-0.997

Table 5. Correlation Coefficient between Silver Nanoparticles and Selected bacterial pathogens

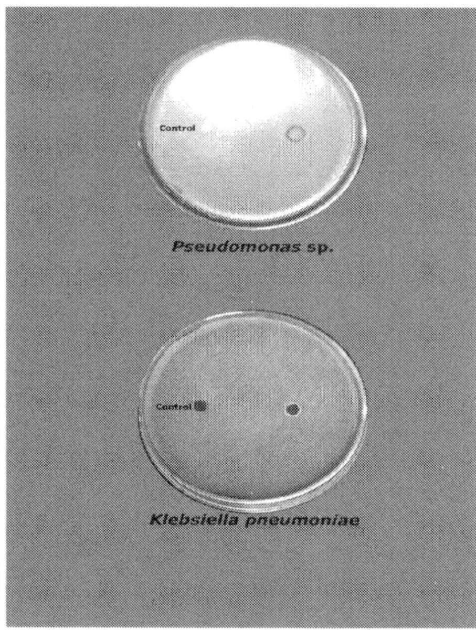

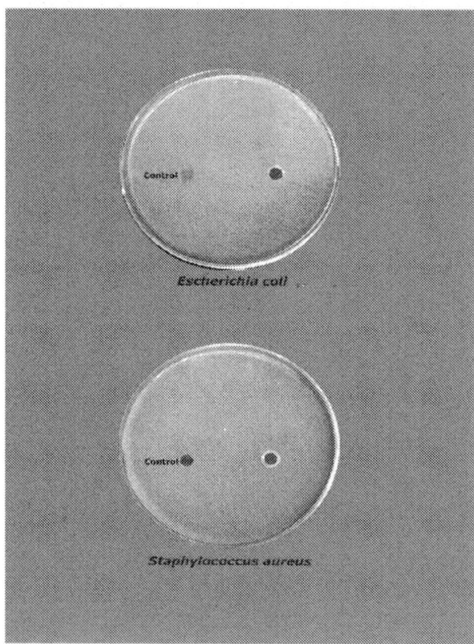

Fig. 6. Antimicrobial activity of silver nanoparticles against bacterial pathogens used in the experiment

6. Conclusion

Silver nanoparticles exhibit a broad size distribution and morphologies with highly reactive facets. The major mechanism through which silver nanoparticles manifested antibacterial properties is by anchoring to and penetrating the bacterial cell wall, and modulating cellular signalling by dephosphorylating putative key peptide substrates on tyrosine residues. Silver nanoparticles act primarily in three ways against Gram-negative bacteria:

1. nanoparticles mainly in the range of 1–10 nm attach to the surface of the cell membrane and drastically disturb its proper function, like permeability and respiration;
2. they are able to penetrate inside the bacteria and cause further damage by possibly interacting with sulfur- and phosphorus-containing compounds such as DNA;
3. nanoparticles release silver ions, which have an additional contribution to the bactericidal effect of the silver nanoparticles (Feng *et al.*, 2000). Although bacterial cell lysis could be one of the reasons for the observed antibacterial property, nanoparticles also modulate the phosphotyrosine profile of putative bacterial peptides, which could thus affect bacterial signal transduction and inhibit the growth of the organisms. The effect is dose dependent and is more pronounced against gram negative organisms than gram-positive ones. The antibacterial effect of nanoparticles is independent of acquisition of resistance by the bacteria against antibiotics. However, further studies must be conducted to verify if the bacteria develop resistance towards the nanoparticles and to examine cytotoxicity (Braydich-Stolle *et al.*, 2005) of nanoparticles towards human cells before proposing their therapeutic use. Finally, this is an important area of research that deserves all our attention owing to its potential application in the fight against multi-drug resistant microorganisms.

7. References

Alcamo, I. E. (1997). Fundamentals of Microbiology 5th edition.Benjamin Cummings.

Birla, S. S., Tiwari, V. V., Gade, A. K., Ingle, A. P., Yadav, A. P., & Rai, M. K. (2009). Fabrication of silver nanoparticles by Phoma glomerata and its combined effect against *Escherichia coli, Pseudomonas aeruginosa* and *Staphylococcus aureus. Lett Appl Microbiol*, 48, (173-179).

Braydich-Stolle, L., Hussain, S., Schlager, J. J., & Hofmann, M. C. (2005). *In vitro* cytotoxicity of Nanoparticles in Mammalian Germline stem cells. *Toxicol. Sci*, 88 (2), (412-419).

Chou, K. S., Lu, Y. C., & Lee, H. H. (2005). Effect of alkaline ion on the mechanism and kinetics of chemical reduction of silver. *Mater. Chem. Phys*, 94, (429-433).

Elechiguerra, J. L., Burt, J. L., Morones, J. R., Camacho-Bragado, A., Gao, X., Lara, H. H. (2005). Interaction of silver nanoparticles with HIV-1. *J.Nanobiotechnology*, 3, 6.

Fang, J., Zhang, C., &Mu, R. (2005). The study of deposited silver particulate films by simple method for efficient SERS. *Chemical Physics Letters*, 401, (271-275).

Feng, Q. L., Wu, J., Chen, G. Q., Cui, F. Z., Kim, T. M., & Kim, J. O. (2000). A mechanistic study of the antibacterial effect of silver ions on *Escherichia coli* and *Staphylococcus aureus. J. Biomed. Mater. Res*, 52, (662-668).

Hu, C., Lan, Y., Qu, J., Hu, X., & Wang, A. (2006). Ag/AgBr/TiO2 Visible light photocatalyst for destruction of Azodyes and Bacteria. *J. Phys. Chem*, 110(9), (4066-4072).

Inoue, Y., Uota, M., Torikai, T., Watari, T., Noda, I., Hotokebuchi, T. (2010). Antibacterial properties of nanostructured silver titanate thin films formed on a titanium plate. *J Biomed Mater Res*, 92A (3), 1171-1180.

Kim, K. D., Han, D. N., & Kim, H. T. (2004). Optimization of experimental conditions based on the Taguchi robust design for the formation of nanosized silver nanoparticles by chemical reduction method. *Chem. Eng. J,* 104 (1-3), (55-61).

Lee, S. M., Song, K. C., & Lee, B. S. (2010). Antibacterial activity of silver nanoparticles prepared by a chemical reduction method. *Korean J. Chem. Eng,* 27(2), (688-692).

Li, Y., Leung, P., Yao, L., Song, Q. W., & Newton, E. (2006). Antimicrobial effect of surgical masks coated with nanoparticles. *J. Hosp Infect,* 62, (58-63)

Lin, W. C., & Yang, M. C. (2005). Novel Silver/Poly (vinyl alcohol) Nanocomposites for surface-enhanced Raman Scattering-active substrates. *Macromol. Rapid Commun,* 26 (24), (1942-1947).

Mie, G. (1908). Contributions to the optics of turbid media, especially colloidal metal solutions. *Ann. Phys,* 25, 377-445.

Mishra, M., Kumar, H., Singh, R. K., & Tripathi, K. (2008). Diabetes and nanomaterials. *Digest Journal of Nanomaterials and Biostructures,* 3 (3), (109-113).

Morones, J. R., Elechiguerra, J. L., Camacho, A., Holt, K., Kouri, J. B., Ram'ırez, J. T., & Yacaman, M. J. (2005). The bactericidal effect of silver nanoparticles. *Nanotechnology,* 16, (2346-2353).

Pal, S., Tak, Y. K., & Song, J. M. (2007). Does the antimicrobial activity of silver nanoparticles depend on the shape of the nanoparticle? A study of the Gram negative bacterium *Escherichia coli. Appl. Environ. Microbiol,* 73 (6), (1712-1720).

Panacek, A., Kvitek, L., Prucek, R., Kolar, M., Vecerova, R., & Pizurova, N., Sharma, V. K., Naveena, T., & Zboril, R. (2006). Silver colloid nanoparticles: synthesis, characterization, and their antibacterial activity. *J. Phys. Chem.,* 110 (33), (16248-16253).

Rai, M., Yadav, A., & Gade, A. (2009). Silver nanoparticles as a new generation of antimicrobials. *Biotechnol Adv,* 27, (76-83).

Shahi, S. K., Patra, M. (2003). Microbially synthesized bioactive nanoparticles and their formulation active against human pathogenic fungi. *Rev. Adv. Mater. Sci,* 5, (501-509).

Shin, H. S., Yang, H. J., Kim, S. B., & Lee, M. S. (2004). Mechanism of growth of colloid silver nanoparticles stabilized by polyvinyl pyrrolidone in γ-irradiated silver nitrate solution, *J. Colloid Interf. Sci,* 274, (89-94).

Shrivastava, S., Bera, T., Roy, A., Singh, G., Ramachandrarao, P., & Dash, D. (2007). Characterization of enhanced antibacterial effects of novel silver nanoparticles. *Nanotechnology,* 18, (103-112).

Song, K. C., Lee, S. M, Park, T. S., & Lee, B. S. (2009). Preparation of colloidal silver nanoparticles by chemical reduction method. *Korean J. Chem. Eng,* 26(1), (153-155).

Wang, Z. L. (2000). In: Characterization of nanophase materials, Z. L. Wang (ed.), Wiley-VCH, Weinheim, Germany. pp. (37-80).

Wright, I. B., Lame, K., Hansen, D., & Burrell, R. E. (1999). Efficacy of topical silver against fungal burn wound pathogens. *Am J Inf Cont,* NA 7, 27, (344-350),

Yoon, K. Y., Byeon, J. H., Park, J. H., Ji, J. H., Bae, G. N., & Hwang, J. (2008). Antimicrobial characteristics of Silver Aerosol Nanoparticles against *Bacillus subtilis* Bioaerosols. *Environ Eng Sci,* 25, (289-293).

7

Polymer-Mediated Broad Spectrum Antiviral Prophylaxis: Utility in High Risk Environments

Dana L. Kyluik, Troy C. Sutton, Yevgeniya Le and Mark D. Scott
*Canadian Blood Services and the Department of Pathology and Laboratory Medicine
and Centre for Blood Research at the University of British Columbia
Canada*

1. Introduction

Viral infections are a significant cause of morbidity and mortality in humans throughout the world. However, modern medicine has a very limited ability to prevent viral diseases. While traditional vaccination strategies have been highly successful against a subset of viruses, the antigenic variation of viruses as well as the shear number of viral pathogens has limited the efficacy of this approach.

This observation is exemplified by the finding that while most common respiratory infections are caused by Rhinoviruses, Coronaviruses, Adenoviruses and Orthomyxoviruses, a number of other viral families are also frequently implicated. Indeed, over 300 serologically distinct viruses are known to cause the pathology associated with the *'common cold'* and *'flu'*. [Spector, 1995] Furthermore, vaccinations have yet to prove effective against the single viral family (Rhinoviruses) commonly implicated in >60% of common colds; again due to the extreme antigenic variability found within even this single viral family. As a result, there are currently no broad-spectrum anti-viral prophylactics (either prescription or over-the-counter) capable of preventing or interrupting the progression of viral infections.

However, the safe, low cost, low technology, and non-toxic bioengineering of the terminally differentiated nasal pharyngeal epithelial host cells may provide a radically new antiviral prophylactic approach that gives rise to a transient, broad-spectrum, prophylaxis against virally transmitted respiratory infections (Figure 1). [McCoy & Scott, 2005, Sutton & Scott, 2010] This polymer-based technology is derivative of the polymer-based "immunocamouflage" technology of blood cells being actively developed within the Canadian Blood Services to reduce the risk of transfusion reactions and alloimmunization to donor red blood cells. [Scott *et al.*, 1997, Scott & Murad, 1998, Murad *et al.*, 1999a, Murad *et al.*, 1999b, Bradley *et al.*, 2001, Bradley *et al.*, 2002, Bradley & Scott, 2007, Rossi *et al.*, 2010b]

As schematically shown in Figure 1, the non-toxic bioengineering of the nasal cavity attenuates or prevents viral respiratory infections at the primary site of infection - the nasopharyngeal cell surface of the upper respiratory tract. Surprisingly to some, the primary mode of viral entry in respiratory diseases is via accidental inoculation of the nasal passage via contaminated hands. As demonstrated in Figure 1A, the initial inoculum (1) is typically

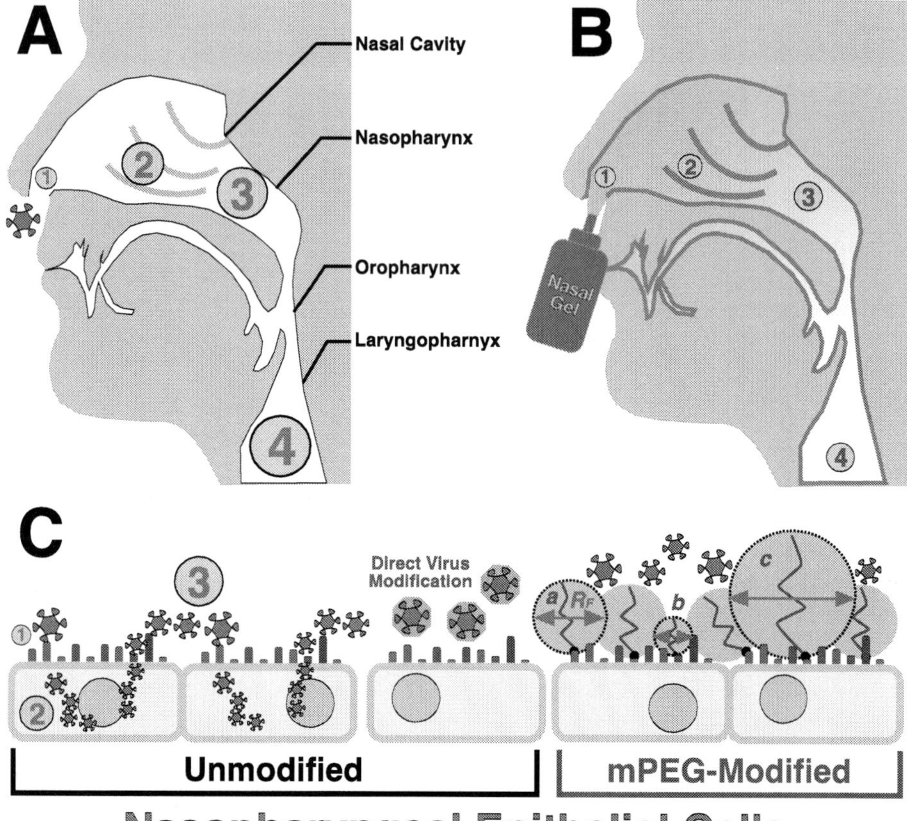

Fig. 1. Antiviral effects of nasopharyngeal immunocamouflage by activated polymer gel on disease pathogenesis. Panel A: Normal viral pathogenesis. Panel B: Effect of mPEG-Nasal Gel on viral pathogeniesis. The relative efficacy of the antiviral barrier is denoted by the intensity of the blue shading. In Panels A and B, the size of the number relects the relative viral number and disease progression: (1) Initial Inoculum; (2) invasion of adjacent cells; (3) production of progeny virus; and (4) disease progression to lower respiratory tract. Panel C: The antiviral effects of grafted polymer are shown at the epithelial cell membrane-environment interface. The efficacy of the grafted polymer is shown by the zone of protection induced by the small (*b*), medium (*a*) and large (*c*) polymers. R_F defines the radius of gyration of the grafted polymer.

small but, upon successful invasion, replicates and produces progeny virus (2) that invade adjacent cells (3) producing further progeny that may remain in the upper respiratory tract or may progress to the lower respiratory tract (4). [Winther, 2011] However, as shown in Figure 1B, the application of the activated mPEG-gel within the nasal cavity covalently modifies the terminally differentiated epithelial cells producing a physical and charge neutralization barrier preventing viral recognition of known and unknown viral receptors. Consequent to this camouflage, the successful tissue invasion by the initial viral inoculum (1) is significantly reduced and few progeny virus (2) are produced. Subsequent replication cycles (3) are also reduced decreasing the severity or onset of disease in both the upper and lower (4) respiratory tracts. The relative efficacy of the mPEG-antiviral barrier is denoted by the intensity of the blue shading and decreases with distance from the nostril opening.

The antiviral effects of grafted polymer occurs at the epithelial cell membrane-environment interface (Figure 1C). The efficacy of the grafted polymer is dependent upon the size/topography of the viral receptor and the size (molecular weight; m.w.) of the polymer. The size of the polymer governs the Flory radii (R_F; root mean square of end to end length of the polymer chain; radius of gyration) of the covalently bound polymers and the immunocamouflage of viral receptors. This is shown by the zone of protection induced by the small (b), medium (a) and large (c) polymers. Note that cell free virus can also be modified by the polymer gel resulting in an inability of the PEGylated virus to bind to its receptor. The direct PEGylation of the progeny virus further reduces the risk of clinical disease.

In contrast to traditional vaccine approaches, this novel intranasal antiviral prophylactic provides immediate, albeit transient, protection against a broad spectrum of respiratory viral pathogens. It is these pathogens that can, and do, create massive healthcare emergencies in '*at risk*' populations (displaced individuals, first responders, aid workers and healthcare providers) in over-crowded refugee centers and health care facilities throughout the world. Indeed, as will be demonstrated, polymer grafting to cells results in a global, multivalent, and non-specific inhibition of viral invasion that is practical and highly suitable for rapid distribution and easily used by the target populations in a variety of environments.

2. PEGylation: Inhibition of virus-receptor recognition and binding

Our earlier studies (*e.g.*, erythrocytes, leukocytes and pancreatic islets) demonstrated that cells and tissues are readily amenable to the covalent grafting of low immunogenicity polymers to the cell membrane or tissue capsule. [Chen & Scott, 2001, Chen & Scott, 2003, Chen & Scott, 2006, Murad *et al.*, 1999a, Murad *et al.*, 1999b, Scott & Chen, 2004, Scott *et al.*, 1997] Successful immunocamouflage of cells, tissues and viruses can be accomplished by a number of polymer species such as methoxypoly(ethylene glycol) [mPEG] and hyperbranched polyglycerols [HPG]. [Scott *et al.*, 1997, Bradley *et al.*, 2002, Le & Scott, 2010, Rossi *et al.*, 2010a, Rossi *et al.*, 2010b]

Of these polymers, mPEG is the best characterized and is synthesized from poly(ethylene glycol) [$HO-(CH_2CH_2O)_n-CH_2CH_2OH$]. [Roberts *et al.*, 2002] The first –OH group is used to covalently attach the PEG-moiety to a linker compound that in turn is used to covalently modify cell membrane proteins. Because the second terminal -OH group of PEG confers some residual chemical reactivity, this is replaced by a –CH_3 moiety, to form activated mPEG: $CH_3-(CH_2CH_2O)_n-CH_2CH_2-Linker$. Multiple chemical linker groups are currently available for the grafting of mPEG to proteins. In general, the majority of these linker agents covalently react with lysine residues on membrane surface proteins (Figure 2). Consequent to this polymer grafting, our studies have demonstrated that cell charge is camouflaged and cell:cell (*e.g.*, Rouleaux formation) and receptor-ligand interactions (*e.g.*, allorecognition and antibody binding) are inhibited resulting in immunological camouflage (*i.e.*, immunocamouflage). [Scott *et al.*, 1997, Scott & Murad, 1998, Murad *et al.*, 1999a, Murad *et al.*, 1999b, Chen & Scott, 2001, Chen & Scott, 2003, Chen & Scott, 2006, Le & Scott, 2010]

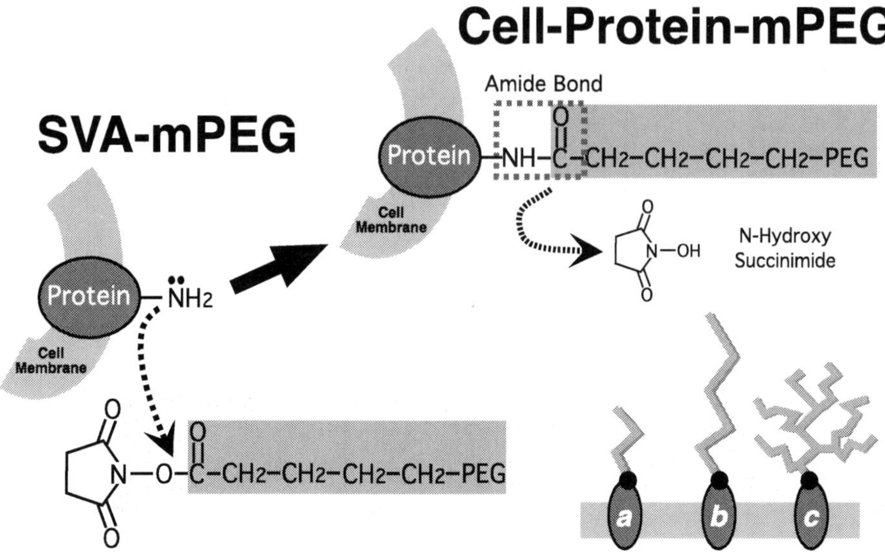

Fig. 2. Mechanism of grafting of activated mPEG to cell membrane proteins. Shown is the reaction for succinimidyl valerate methoxypoly(ethylene glycol) [SVA-mPEG]. The polymer component can be linear (*a, b*) of variable length (m.w.; *a, b* denote short and long polymers, respectively) or can be highly complex as denoted by the branched (*c*) structure. The linker chemistry is also highly variable but typically targets lysine residues on membrane proteins. Lipid anchored PEGs also exist but can negatively impact membrane fluid mechanics adversely affecting cell viability. Some work has been done with hyperbranched polyglycerols (HPG) which resemble (*c*). [Le & Scott, 2010, Rossi *et al.*, 2010b]

Importantly, viral binding to its host cell is highly analogous to receptor-ligand interactions. For viral infections to occur, viruses must bind to receptor(s) located on the cell surface (Figure 1C). Hence the immunocamouflage of either the virus or host cell would theoretically inhibit viral invasion and subsequent disease due to both charge camouflage and the steric hindrance resulting from the molecular intra-chain flexibility and rapid mobility of the heavily hydrated PEG chains (Figure 3). Moreover, the global camouflage of the cell surface effectively masks both known and unknown viral receptors resulting in a nonspecific, non-immunological, broad-spectrum antiviral effect. [McCoy & Scott, 2005, Sutton & Scott, 2010]

The biophysical basis of this antiviral effect is schematically shown in Figures 3-4. The PEG layer obscures the inherent electrical charge associated with surface proteins since the charged molecules become buried beneath the viscous, hydrated, neutral PEG layer. [Szleifer, 1997, Satulovsky *et al.*, 2000, Bradley *et al.*, 2002, Bradley & Scott, 2004, Le & Scott, 2010] As schematically shown in Figure 3, the surface of a generic cell may have a net negative charge due to the proteins and carbohydrates present on the cell surface. The

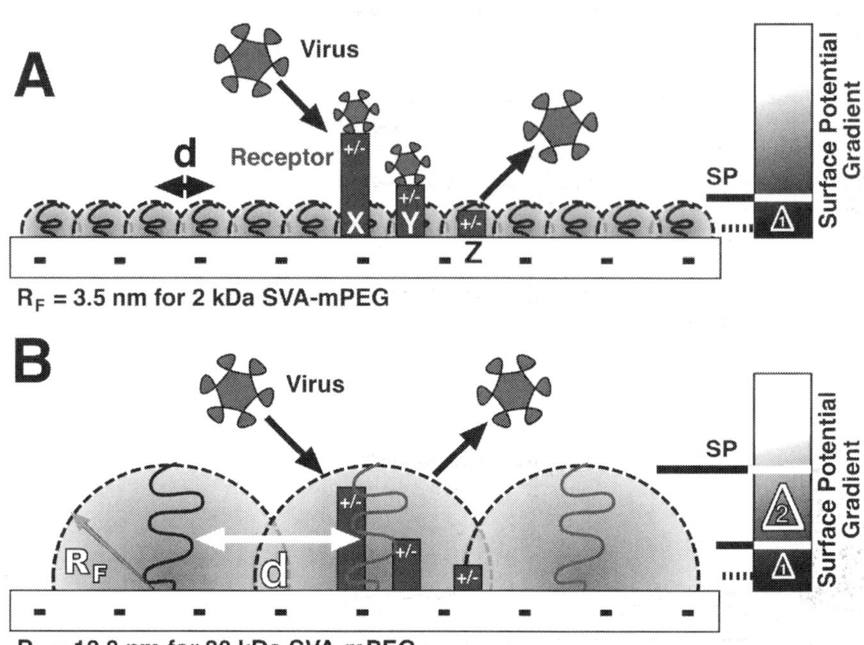

Fig. 3. Immunocamouflage of the cell surface is driven by both charge camouflage and steric interference induced by the grafted polymer. Surface charge camouflage is primarily driven by polymer-mediated extension of the shear plane (SP) of the surface towards a region of decreased surface potential (Surface Potential Gradient). In the absence of polymer, the inherent shear plane (SP) of a cell is typically located 1-3 nm above the surface. The extension of SP is proportional to the hydrodynamic thickness of the polymer layer, which in turn is governed by the Flory radii (R_F; root mean square of end to end length of the polymer chain; radius of gyration) of the covalently bound polymers. Thus, 20 kDa polymers (large R_F; B) provide improved charge camouflage over 2 kDa polymers (small R_F; A). Delta (Δ) is the difference in the surface potential at the shear plane of a particle modified with the short ($\Delta 1$) *vs.* the long polymer ($\Delta 2$). The receptors X, Y and Z denote putative viral receptors extending different distances from the cell surface. Modified from [Le & Scott, 2010].

positively charged counter-ions from a bulk aqueous solution migrate and interact with the surface to neutralize the surface charge. This creates a Surface Potential Gradient, with the electric potential being the highest at the surface and decreasing with the distance away from the surface. The Shear Plane (SP) is defined as the region around the surface, where counter-ions behave as if they are physically attached to the cell and roughly approximates a neutral net charge. Polymer grafting alters the location of the shear plane relative to the membrane surface and this change is directly influenced by the size and density of the grafted polymer. As demonstrated in Figure 3A, a low molecular weight polymer (*e.g.*, 2

kDa) characterized by a small Flory radius (R_F approximating the radius of gyration of the polymer; 2 kDa mPEG has an R_F ~3.5 nm) will move the shear plane away from the surface of the cell ($\Delta 1$). [Heuberger *et al.*, 2005, Damodaran *et al.*, 2010,] However, the grafted polymer may or may not camouflage a viral receptor (X, Y, Z) depending on its topographical location relative to the SP. In contrast, a high molecular weight polymer (*e.g.*, 20 kDa; R_F = 13.8 nm; Figure 3B) will significantly move the SP away from the membrane ($\Delta 2$) and is more likely to result in the charge camouflage of known and unknown viral receptors (X, Y and Z). Moreover, the steric (physical) interference of the grafted polymer can also physically prevent the stable interaction of a virus with its surface receptor (Figure 3). The steric effect is maximized when chains are grafted at higher density; *i.e.*, with a small separation between the chains (d). High density grafting is difficult to achieve with polymers possessing a large R_F (*e.g.*, 20 kDa mPEG; B) as the initially bound chains sterically inhibit the approach and binding of additional polymers.

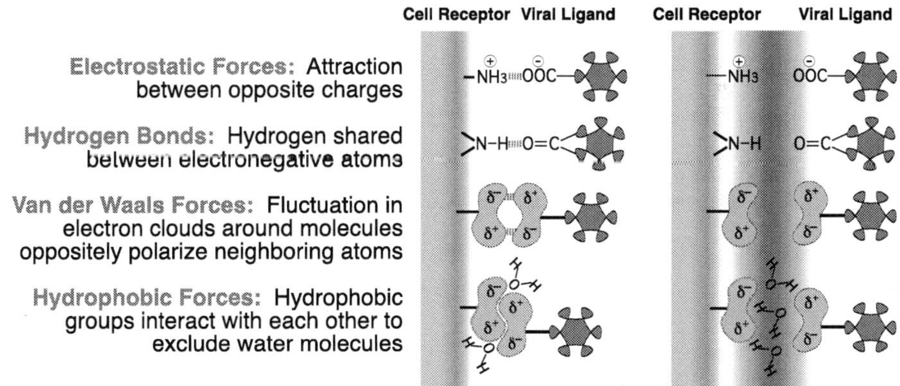

Fig. 4. Non-covalent forces mediate receptor-ligand interactions. Partial charges of electric dipoles are designated by δ. Electrostatic forces decrease as $1/r^2$, where r is the distance between the two atoms. Van der Waals forces are much weaker than electrostatic forces, because the strength of the interaction diminishes as $1/r^6$. Covalent bonds do not occur between a receptor and its ligand.

The highly malleable nature of mPEG polymers results in a broad range of possibilities of enhancing its antiviral effects via the use of both linear or branched molecules over an extraordinarily wide range of molecular weights and grafting densities (Figure 2). Of biologic importance, the absolute effects arising from both the migration of the SP within the Surface Potential Gradient and/or the steric hindrance of viral attachment need only be minor as the non-covalent forces that mediate receptor-ligand interactions are relatively weak and easily disrupted by the biophysical changes mediated by the grafted polymer (Figure 4). Thus, the bioengineering of the nasal pharyngeal epithelial cells with an mPEG nasal gel may provide significant opportunities to attenuate or block viral invasion of the initial viral inoculum, as well as any progeny, thereby reducing the both disease progression and severity.

A critical concern of this approach is the safety (acute and chronic) of the polymer. PEG is generally viewed by the US FDA as a safe compound and is widely used in food, cosmetic

and pharmaceutical formulations. Previous studies have demonstrated a lack of toxicity of PEG with polymer lengths greater than 400 Da. Human experiments dating from the late 1940's demonstrated both orally and intravenously administer PEG (*Carbowax®*) had no acute toxicity. [Shaffer & Critchfield, 1947, Smyth *et al.*, 1947] Indeed, intravenously administered PEGs (1 and 6 kDa) were readily excreted by the human kidney. Within 12 hours, approximately $87.2 \pm 2.3\%$ and $96.3 \pm 2.4\%$ of the 1 and 6 kDa PEG, respectively, were recovered from the urine. More recent (and ethical) studies with PEG-derivatized proteins (*e.g.*, PEG-deaminase and PEG-hemoglobin) have also demonstrated that these modified compounds (even when of xenobiotic origin) and the PEG itself have few systemic consequences and the PEG moiety is similarly excreted via the urine as found in the 1947 studies. [Veronese & Mero, 2008] PEGylated intact cells have also proven to exert no overt toxicity in a murine model. Repeated transfusions of PEGylated erythrocytes in mice have demonstrated no adverse events (*e.g.*, no hemolysis, splenomegaly, weight loss) even when >80% of their red cell mass was mPEG-RBC. [Murad *et al.*, 1999b] Moreover, there was no evidence of anti-PEG antibodies despite the massive infusion of mPEG. Indeed, after two or more decades of clinical use of PEGylated proteins, the PEG moiety has proven to be both effective in prolonging vascular retention and safe to the recipients of this therapy. [Veronese & Harris, 2002, Veronese & Mero, 2008, Veronese & Pasut, 2005]

3. Broad spectrum antiviral prophylaxis

Virally mediated respiratory diseases remain a critical problem for humans despite all our advances in pharmacology and vaccine development over the last 150 years. [Spector, 1995] This is in large part due to the shear number of pathogens (>300 serologically distinct viruses) capable of causing the respiratory pathology associated with the *'common cold'* and *'flu'*. Many of these viral respiratory diseases are characterized by rapid onset and high communicability, especially when introduced into situations characterized by high population densities and poor sanitation. Indeed, in refugee centers created following natural (*e.g.*, earthquake or flooding) or man-made (*e.g.*, war, bioterrorism or incarceration) disasters, epidemic respiratory disease invariably arises within a very short time span and, along with diarrhea, is a major cause of morbidity and mortality – especially amongst the young and old. Under such emergency conditions, vaccinations (even if immunologically plausible) would be of limited utility as there would be insufficient time to adequately vaccinate the *'herd'*. Rather, what is needed under these circumstances is an effective prophylactic therapy that confers significant individual, thus *'herd'*, protection immediately upon application. [Katriel & Stone, 2010, Van Effelterre *et al.*, 2010, Paulke-Korinek *et al.*, 2011]

Importantly, for the intervention to be effective, it does not have to inactivate 100% of a viral threat. Rather, such prophylactic intervention must reduce the viral exposure to sub-infective levels and/or inhibit person-to-person transmission. As shown in Figure 5, all viruses have a threshold infective level. This may range from 1-2 virions per person for an extremely contagious (but not necessarily lethal) virus to several hundred or thousand viral particles in order to cause disease. The biologic objective of an activated mPEG-Gel would be to reduce the successful invasions of a respiratory virus to a level below the infectious dose necessary to cause disease. Moreover, when viewed in the context of *'herd immunity'* it is also not necessary to successfully protect 100% of the at risk population to achieve significant protection of the at risk population. As shown in Figure 6, an efficacy of 50% or

75% against a theoretical respiratory virus would yield a dramatic reduction in disease progression within an at risk population.

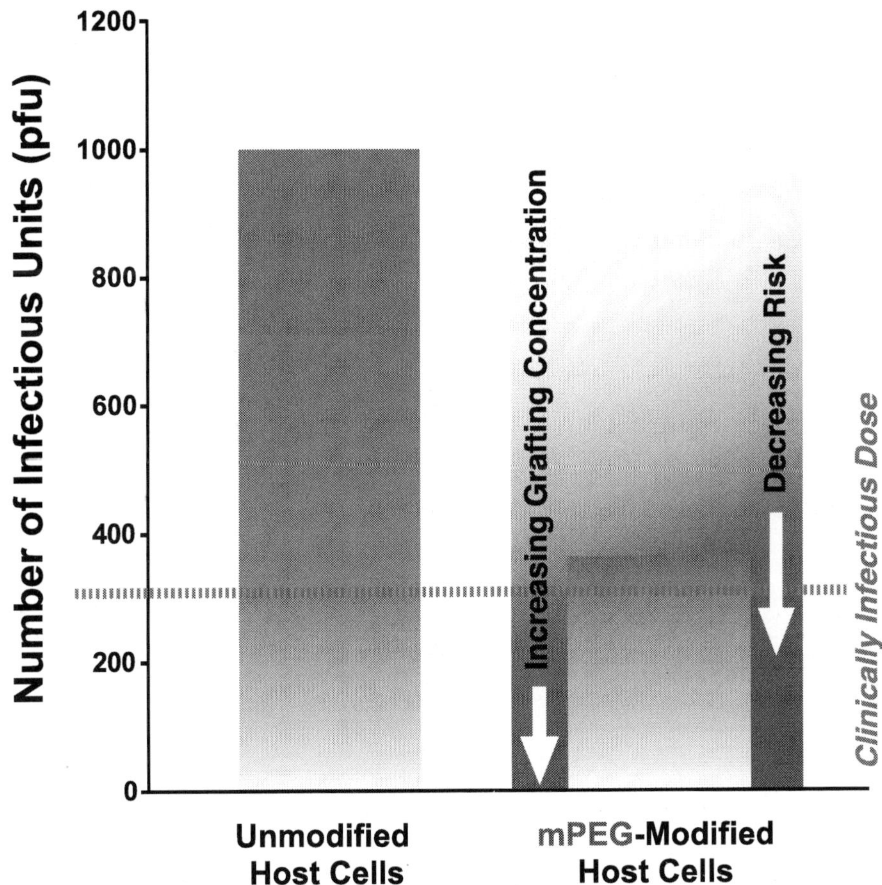

Fig. 5. The goal of mPEG-modification of either the host cells or virus is not 100% protection against viral invasion, but rather reducing the inoculation dose to below the clinically infectious dose. For example, in studies of healthy adult humans, a viral dose of 112-448 pfu (plaque forming units; a measure of viral load) of Respiratory Syncytial Virus (RSV) Long Strain A induced clinical disease in 20 of 41 volunteers. Reduction to less than this critical viral load would reduce risk of disease. [Kravetz *et al.*, 1961, Murata & Falsey, 2007]

With these key concepts in mind, we have been pursuing the functional utility and formulation of an anti-viral prophylactic polymer gel against a broad range of viruses *in vitro* (Table 1). Such protection could target either the virus or the potential target as shown in Figure 1C. Direct modification of the virus, while highly effective via this technology, is not practical for respiratory viruses (in contrast to viruses contained within blood products).

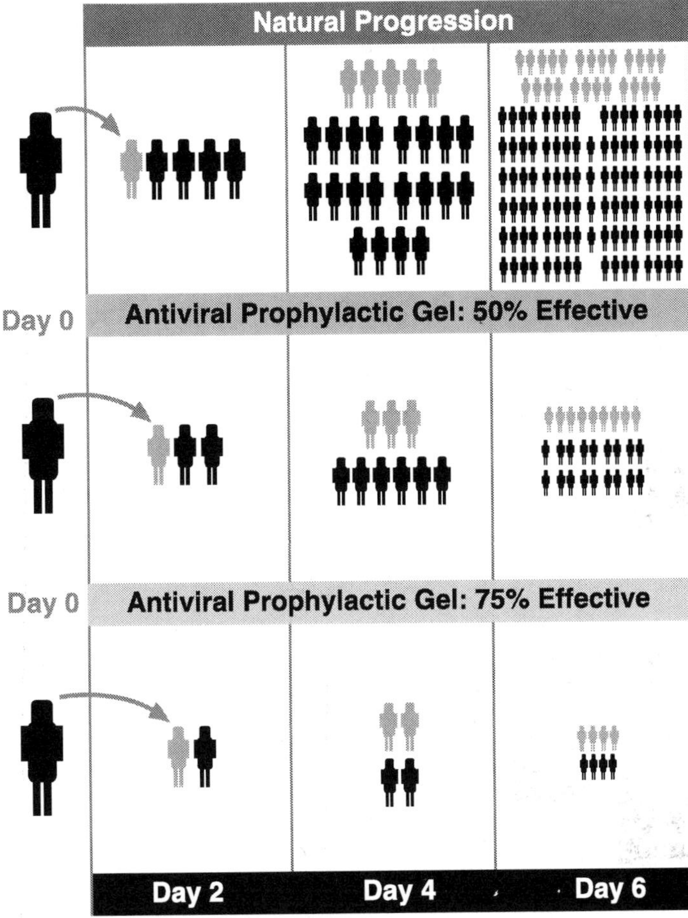

Fig. 6. Disease transmission can be dramatically reduced even if the prophylactic nasal gel is not 100% effective at the level of the individual (*Herd Immunity*). As shown, the natural progression of a hypothetical respiratory disease is such that 1 infected person transmits disease to 4 healthy individuals every 2 days. Diagrammed is the effect that a 50% or 75% effective nasal gel would have on an at risk population over 6 days. Previously infected individuals are denoted in grey. This model assumes that an individual does not die and remains contagious for a minimum of 4 days.

However, our studies suggest that the application of the activated polymer gel within the nasal cavity could prove to be a highly effective antiviral prophylactic (Figure 1B). To further explore this prophylactic approach, experimental viruses were chosen to include both enveloped and non-enveloped viruses, receptor and fusion mediated viral invasion, large and small viruses and to be representative of known human respiratory viruses (*e.g.*, Rhinoviruses, Coronaviruses, Adenoviruses).

Virus and Family	Virus Structure	Virion Size	Mode of Entry	Receptor	Receptor (nm)
Mouse Adenovirus (MAV) Adenoviridae	Naked Icosahedral Capsid	70-90 nm	Receptor Mediated Endocytosis	mCAR	4.6 nm
Rat Coronavirus (RCV) Coronaviridae	Enveloped Helical Capsid	80-160 nm	Fusion	Not identified - other family members use APN	13.5nm (APN)
Theiler's Murine Encephalo-myelitis Virus (TMEV) Picornaviridae	Naked Icosahedral Capsid	28-30 nm	Receptor Mediated Endocytosis	Not identified - other family members use ICAM-1	18.7 nm (ICAM-1)
Respiratory Syncytial Virus (RSV) Paramyxoviridae	Enveloped Helical Capsid	150-300 nm	Fusion	Not Identified	Unknown
Simian Virus 40 (SV40) Papovavirida	Naked Icosahedral Capsid	45-55 nm	Receptor Mediated Endocytosis	MHC-1	7 nm

Murine Homologue of Coxsackie and Adenovirus Receptor (mCAR); Aminopeptidase N (APN); Intracellular Adhesion Molecule-1 (ICAM-1); and Major Histocompatibility Molecule-1 (MHC-1). SV40 is not a respiratory virus but is a well characterized experimental model suitable for experimental study. Receptor size references as follows: MHC-1 [Bjorkman *et al.*, 1987]; mCAR [He *et al.*, 2001]; APN [Hussain *et al.*, 1981]; EGFR [Mi *et al.*, 2008]; ICAM-1 [Jun *et al.*, 2001]

Table 1. Partial Listing of Experimental Viral Models Utilized in the Evaluation of the Proposed Antiviral Prophylactic Nasal Gel.

3.1 Antiviral nasal gel

To determine if an activated mPEG-gel could inhibit viral invasion and proliferation, *in vitro* studies were conducted using the viruses described in Table 1. These studies examined both direct polymer modification of the viruses themselves as well as their host (target) cells.

To determine if viruses were amenable to PEGylation by the chemistry noted in Figure 2, an analysis of envelope and capsid proteins was done. The amino acid composition and sequence of the envelope and capsid proteins suggested that all viruses had suitable targets for lysine based grafting. For example, analysis of a number of human RSV isolates demonstrate that both the F and G capsid glycoproteins are lysine rich and provide an excellent substrate for direct viral modification with mPEG. Infection assays confirmed this analysis. These experiments demonstrated that direct covalent modification of the virion by mPEG resulted in an almost complete abrogation of viral infection and proliferation [Sutton & Scott, 2010]. However some significant variability in the efficacy of small and large polymers was noted. As shown in Figure 7A, at low grafting concentrations (≤ 2 mM) small molecular weight mPEG demonstrated superiority over large polymers when modifying the virus (RSV) directly. At these low virus:mPEG ratios the improved efficacy of the 2 kDa polymer may be partially due to the distribution of virus specific proteins within the viral envelope. On the RSV virion the G (attachment) and F (fusion) proteins form glycoprotein spikes that are 6-10 nm apart and extent 11-20 nm from the virion surface. Thus, given the distance between glycoprotein spikes on the RSV virion, grafting of 2 kDa mPEG will likely result in the direct modification of a greater number of glycoprotein spikes. Grafting of 20 kDa mPEG will likely result in fewer spikes being modified as the initially bound mPEG will exclude other polymer chains from grafting to spikes in close proximity. At higher grafting concentrations, the self-exclusion effect of grafted mPEG is partially overcome resulting in high levels of surface modification and prevention of viral binding to the host cell. Importantly, soluble mPEG (*i.e.*, unable to covalently modify the virus) had no antiviral effect. In support of these observations, our previous studies clearly demonstrated that binding of mPEG-modified RSV or SV40 to the target cells is dramatically decreased as determined by immunofluorescence staining 90 minutes post challenge at 4°C (to prevent internalization).[McCoy and Scott, 2005; Sutton and Scott, 2010]

While the utility of direct viral modification might be beneficial in blood banking environments bereft of modern viral detection methodology, this application would have little or no practical value in the prevention of respiratory disease. To this end, the *in vitro* PEGylation of the host cell monolayers were conducted. These studies have demonstrated that the selection of the polymer molecular weight and grafting concentration are important in preventing viral infection.

In contrast to Figure 7A, small chain polymers were completely ineffective when used to directly modify the RSV host cell – even at very high grafting concentrations (Figure 7B). This suggests that the unknown membrane receptor for RSV extends well away from the membrane surface and the 2 kDa polymer is unable to effectively camouflage it. However, cellular grafting of small polymers may provide partial protection against some viruses. As shown in Figure 8, low molecular weight mPEG (3.4 kDa) when grafted to host cells does provide anti-viral protection over multiple logs of viral challenge when using SV40. Even at very high levels SV40 challenge (*e.g.*, 10^{11} pfu/ml) grafted polymer (20 > 3.4 kDa) offered very significant protection against infection and viral proliferation as evidenced by the significant decrease in infected cells. For example, unmodified host cells infected with 10^7 pfu/mL were 27% infected, while those modified by 1.2 and 15 mM 3.4 kDa mPEG were 6% and 0.6% infected. In contrast, the 20 kDa polymer at the 1.2 and 15 mM grafting concentrations when challenged with 10^7 pfu/mL resulted in infection rates of only 2% and 0.05%, respectively. In contrast to RSV, the viral receptor for SV40 is known to be MHC-1, a molecule that extends approximately 7 nm from the cell membrane. The protection offered by the small polymer is inferior to that offered by the 20 kDa mPEG but is still significant relative to unmodified cells. As with RSV, and in all models tested, soluble mPEG (*i.e.*, unable to covalently modify the host cell surface) had no antiviral effect.

The observed relationship between polymer molecular weight and the physical size of the grafting target was consistently observed across multiple viral models. Further analysis of the relationship was examined using a latex bead – plasma adsorption model to which mPEGs of various molecular weights were grafted (Figure 9). Using small latex beads (1.2 µm), short chain polymers were more effective as noted by the decreased amount of adsorbed fluorescently labeled plasma protein. In contrast, with large 8 µm latex beads (or 8µm RBC) large chain polymers are most effective at preventing the adsorption of the fluorescently labeled plasma. In agreement with this finding, PEGylation of viruses (0.02-200 µm) also demonstrate that short chain polymers (2 kDa) are more effective at preventing viral invasion and infection while large chain polymers provide superior protection when grafted to host cells (*e.g.,* RSV; Figure 7).

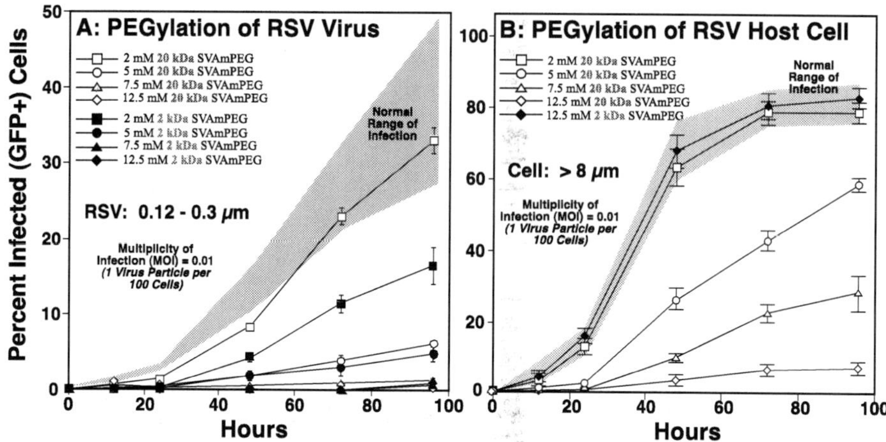

Fig. 7. PEGylation of virus (A) or host cell (B) can prevent viral invasion and proliferation. Shaded areas denote normal range of infection for control studies. An MOI of 0.01 was used for both panels. Shown are the mean ± SD of a minimum of 3 independent experiments. Data modified from: [Sutton & Scott, 2010]

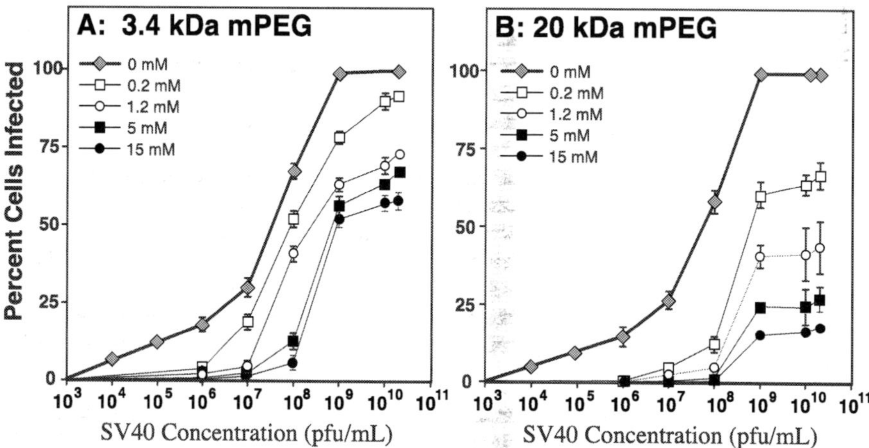

Fig. 8. In contrast to RSV, infection by other viruses is partially blocked by the grafting of short chain polymers to the host cells. CV-1 cells were modified with the indicated concentrations of activated mPEGs of 3.4 (A) or 20 (B) kDa and challenged over a broad range of viral titers. Percent infected cells was determined at 72 hours via T antigen staining. The mean ± standard deviation of 3 independent experiments are shown. Data modified from: [McCoy & Scott, 2005]

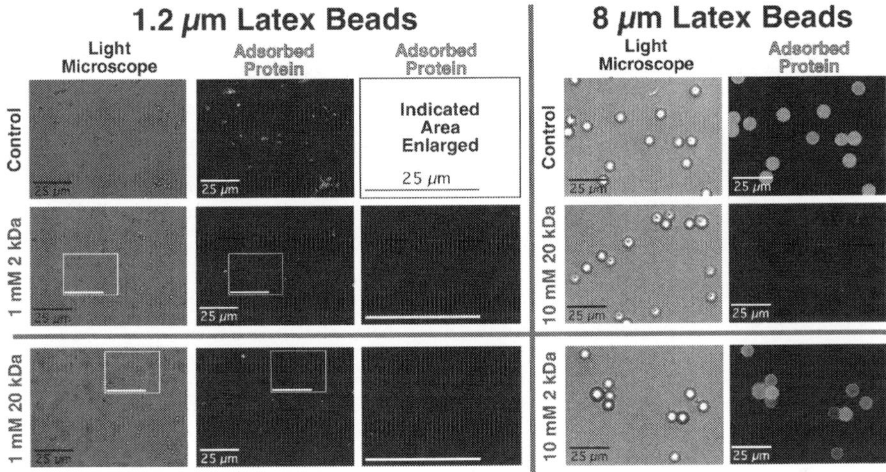

Fig. 9. Paradoxically, the size of the PEGylation target influences which polymer size most effectively imparts immunocamouflage. Modified from Le and Scott (2010). [Le & Scott, 2010]

To determine if a nasal gel would adequately cover the tissue of interest, *in vivo* studies were conducted. To assess the area of coverage, a polymer gel formulation was labeled with India Ink and administered intranasally (150μl per nostril) in a Guinea Pig model of Respiratory Syncytial Virus (RSV) infection. These studies demonstrated full coverage of the nasal cavity and even significant protection to the level of the lung (Figure 10).

Indeed, studies using a pigmented nasal gel demonstrated that a gel can provide protection to both the upper and lower respiratory tract in a Guinea Pig model of respiratory Syncytial Virus (RSV). Dissection of the Guinea Pig nasal cavity demonstrated uniform staining of the cavity surface. Moreover, as shown in Figure 10, application of a PEG-based gel containing India Ink resulted in significant protection even within the lung tissue.

To date, we have demonstrated that the covalent grafting of mPEG to virus itself or to the host/target cell results in a highly effective broad-spectrum antiviral prophylaxis. Consequent to the immunocamouflage of the host cell, viral entry and propagation by members the *Picornaviridae, Adenoviridae, Cornaviradae, Pneumovirus* (all representative of common human respiratory viruses) families as well as members of the *Papovaviridae* and *Hepesviridae* families are virtually (≥90%) blocked even at high viral titers. Furthermore, this antiviral prophylaxis is effective against: 1) enveloped and non-enveloped viruses; 2) receptor-mediated and fusiogenic viruses; 3) small and large viruses; 4) DNA and RNA viruses; and 5) viruses with known and unknown viral receptors. The mechanism of this protection is biophysical (not immunological) in nature and dependent upon charge camouflage and steric hindrance induced by the grafted polymer. Importantly, due to the basic biophysical nature of this antiviral effect, it is not prone to biological adaptation (*i.e.,* sequence mutations obviating the efficacy of antibodies) by the virus.

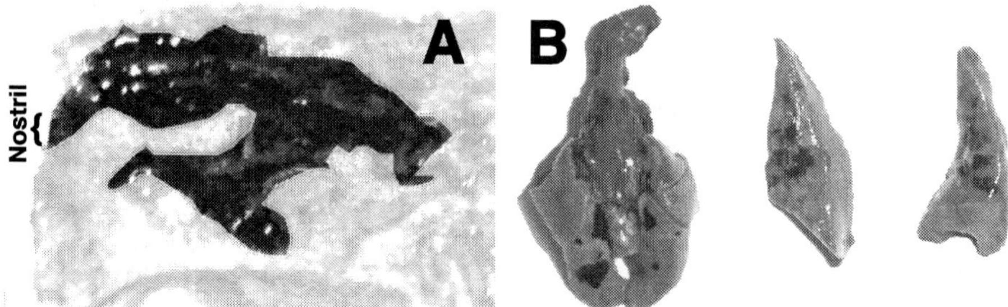

Fig. 10. Penetration of the nasal gel within the nasal cavity (A) and to the lower respiratory tract (B) was examined within a Guinea Pig model of Respiratory Syncytial Virus (RSV) infection. Shown are the nasal cavity and whole lung and sagittal sections of the lung lobes one hour after intranasal administration of 300 µl (150 per nostril) of India ink labelled polymer gel. The non-sinus tissues is partially masked. Note the distribution of the ink, indicative of the successful delivery of the nasal polymer to the nasal cavity and regions of the lung.

3.2 Alternative preventative or therapeutics approaches

Current methods to prevent respiratory viral disease range from homeopathic drug treatments (*e.g.*, mega doses of vitamin C, oral zinc and Echinacea) to pharmacological interventions such as anti-viral drugs and vaccination. Many of these methods have been proven ineffective (vitamin C) or are not always practical or effective (*i.e.*, mass vaccination) in preventing disease. [Block & Mead, 2003, Yale & Liu, 2004]

Potential pharmacological means of preventing viral infections are to block either viral entry, or proliferation, within the host cell. However again there are no United States FDA-approved *broad spectrum* prophylactic therapies or anti-viral treatments capable of preventing either of these events. Most approved pharmacological antiviral agents exhibit high specificity. Among the best described prescription antiviral agent is Amantadine (FDA approved in 1966), which is used to prevent the uncoating of influenza A. Within cells, Amantadine specifically inhibits viral uncoating by blocking the activity of the proton channel of the influenza A M2 protein preventing the endosomal pH change required for uncoating to occur. [Kandel & Hartshorn, 2001] However, Amantadine has only a brief window of efficacy and is highly specific to influenza A (*i.e.*, completely ineffective against other influenza strains much less different viral families). Most recently the U.S. Center for Disease Control (CDC) recommends against the use of Amantadine due to viral resistance. In its place, the CDC has recommended the use of Tamiflu® (Oseltamivir; FDA approved 1999) to prevent or treat influenza A and B infections. Oseltanmivir is a neuraminidase inhibitor that is specific to influenza and blocks the exit of the progeny influenza viruses from the host cell, thus limiting viral replication and disease progression. However, the efficacy of Tamiflu® has recently been reassessed ([Jefferson *et al.*, 2009]) and found to offer only modest prophylaxis when administered within 24 hours of viral infection. Other prescription antiviral drugs demonstrate similar limited or absent efficacy against other respiratory viral families (Rhinoviruses, Coronaviruses, and Adenoviruses) associated with cold or flu causation.

Thus, patients typically turn to over-the-counter (OTC) cough and cold remedies which focus on the relief of symptoms associated with the common cold. Currently in North America alone an estimated \$3-6 billion (US) is spent on symptom relief. A few OTC compounds do attempt to prevent or, after infection, attenuate viral infection. These commercial products include compounds such as ZICAM Nasal Gel (active ingredient: zinc gluconate) and ColdFX (active ingredient: proprietary natural extract containing poly-furanosyl-pyranosyl-saccharides in a concentration of greater than 80%). [Hirt et al., 2000, McElhaney et al., 2006] For drug efficacy, ZICAM must be applied intranasally every 2-4 hours. However, Zicam's proven mode of action is highly specific and, in laboratory studies, only prevents infection by those rhinoviruses that use ICAM1 as the viral receptor. [Bella & Rossmann, 1999] Hence, it is completely ineffective against the majority of cold viruses that do not use ICAM1 (e.g., Coronaviruses and Adenoviruses). More recently, in 2009 ZICAM's original intranasal formulation was removed from the market due to adverse side effects. [Lim et al., 2009] ColdFX's mode of action and viral range of efficacy are not as well documented, though clinical studies do suggest variable degrees of efficacy in small population studies. Studies cited by ColdFX suggest that the natural ingredients stimulate the immune system (e.g., activate macrophages, increase natural killer (NK) cell numbers) to both prevent initial viral infection and to more efficiently kill off virally infected cells to prevent viral proliferation. [Wang et al., 2004, Miller et al., 2009] However, direct experimental evidence for these purported actions is limited.

Other experimental approaches have also been investigated but are also are highly specific to single viral strains and/or families. For example, two other experimental methods of preventing rhinoviral infection have been studied which, like ZICAM, targets the rhinovirus:ICAM-1 interaction. These are the intranasal application of soluble Intracellular Adhesion Molecule (sICAM) or anti-ICAM-1 antibodies (rhinovirus receptor murine monoclonal antibody; RRMA). Like ZICAM, these experimental drugs require intranasal application every 2-4 hours. [Marlin et al., 1990, Huguenel et al., 1997, Turner et al., 1999] Both of these drugs do demonstrate some in vitro and in vivo (murine) efficacy against rhinoviruses. Randomized human trials with sICAM demonstrated that intranasal administration of sICAM 7 hours prior to or 12 hour post rhinovirus challenge resulted in a relative reduction in cold symptoms but very weak prevention (92% of the placebo-treated subjects and 85% of the sICAM-treated subjects became infected). [Turner et al., 1999] Similarly, human trials with RRMA demonstrated no reduction in rhinovirus infection or disease severity. [Hayden et al., 1988, Sperber & Hayden, 1989] Furthermore, repeated administration of the mouse derived RRMA would also lead to rapid immunization against the drug. Importantly, unlike the described mPEG nasal gel, both sICAM and RRMA are highly specific to only those Rhinoviruses that utilize ICAM-1 as the viral receptor. For example, when sICAM was tested against herpes simplex virus type 1 (HSV-1), coxsackie virus B1 or poliovirus no observable protective effect was noted. [Marlin et al., 1990]

Thus, the antiviral prophylactic polymer nasal gel illustrated here represents the only broad spectrum antiviral prophylactic agent described to date. Application of the activated nasal gel is uncomplicated and requires a minimal (~3-5 minutes) application time to effectively camouflages known and unknown viral receptors. After application, the individual then simply "blows their nose" into a tissue to remove any residual carrier fluid. Current data suggests that a single daily application provides maximal protection and retains significant efficacy for up to 60 hours. A brief comparison of a cross sample of the above approaches with our proposed antiviral nasal gel is shown in Table 2A/B.

A. Anti-Viral Activity Against Respiratory Viruses

Viral Family	mPEG Nasal Gel	RRMA[a]	Soluble ICAM	Zicam	Traditional Vaccination
Adenoviridae	+	-	-	-	+/-
Coronaviridae	+	-	-	-	+/-
Herpesviridae	+	-	-	-	+/-
Papovaviridae	+	-	-	-	+/-
Picornaviridae	+	+*/-	+*/-	+*/-	-
Retoviridae	+	-	-	-	+/-
Paramyxoviridae	+	-	-	-	+/-

B. Phamacological Traits

	Gel	RRMA[a]	Soluble ICAM	Zicam	Traditional Vaccination
Non-Specific Protection	Yes	No	No	No	No
Non-Immunogenic	Yes	?	?	+	+/-
Lack of Inherent Resistance	Yes	?	?	+/-	+/-
Longevity of Protection by a Single Application	24-48 Hr	2-4 Hr*	2-4 Hr*	2-4 Hr*	**

[a] Anti-ICAM Antibody; +/- Highly Strain Specific; * Only against Rhinoviruses utilizing ICAM-1 as viral receptor; ** Protection if generated, can last weeks to years.

Table 2. Comparison of the antiviral (A) and pharmacological (B) traits of the described mPEG-Nasal Gel to other experimental and commercial antiviral agents.

3.3 Non-respiratory antiviral applications

Are there other applications for an antiviral prophylactic gel? Humans are beset by a host of viral pathogens and common to all of these pathogens is the need to gain entry into its target cell in order to proliferate. Not all viruses or modes of entry (*e.g.*, *i.v.* drug use) would be amendable to an antiviral gel. However, the use of an antiviral gel in preventing or reducing the risk of an initial infection by the Human Immunodeficiency Virus (HIV) could be envisioned. [Piguet & Steinman, 2007, Cutler & Justman, 2008, Boily *et al.*, 2009, Garg *et al.*, 2010]

In both male-female and male-male transmission, the HIV virus typically enters the host via Langerhans cells, a resident epidermal dendritic cell, dermal dendritic cells, and/or resident CD4+ lymphocytes all of which are present within the mucosal and sub-epithelial tissues of the anogenital region (Figure 11A). Viral entry is gained primarily via two receptors located on the surface of the cell: CD195 (CCR5) and CD184 (CXCR4). [Piguet & Steinman, 2007] If dendritic (antigen presenting) cells capture the virus, it is either endocytosed or can be transferred directly to CD4+ CD195+ T lymphocyte via an infection synapse. Proliferation of the virus within T cells is followed by infection of additional lymphocytes via either CD184 or CD195. Importantly, all of these events are analogous to cellular invasion by respiratory viruses and amendable to disruption by grafted polymer. In a model analogous to spermicide or lubricant, an activated polymer gel could be applied either vaginally or anally prior to at-risk behavior creating an antiviral barrier on the epithelial cell surfaces of the mucosal tissue (Figure 11B) and preventing viral binding via the biophysical mechanisms already described. As shown in Figure 12, proof-of-concept studies clearly demonstrate that grafting of mPEG to CD184 and CD195 positive cells inhibits recognition of these markers by high affinity antibodies.

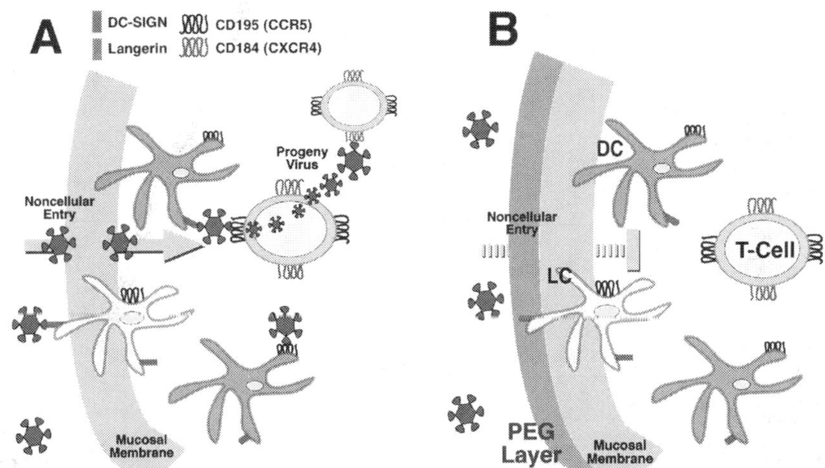

Fig. 11. Immunocamouflage of HIV receptors by activated mPEG antiviral gel. Panel A) Normal mode of HIV invasion; Panel B) Blockade of HIV binding by activated mPEG gel. Langerhan Dendritic Cells (LC); Dermal Dendritic Cells (DC); CD4+ T Lymphocyte (T-Cell). Langerin, a lectin present on LCs is a primary attachment point for HIV. A different lectin (DC-SIGN) is present on other DC and serves as a receptor for HIV. Noncellular entry may occur via epithelial cell injury (*e.g.*, microabrasions).

Because the risk of HIV transmission per single interaction is relatively low (estimated to be 0.04% for female-to-male; 0.08% for male-to-female; and 1.7% for receptive anal intercourse; Boily *et al.*, 2009), the application of an effective activated mPEG-based mucosally applied antiviral gel would further reduce the risk of transmission. Moreover, PEG itself is a lubricant and would reduce the risk of traumatic tissue injury and microabrasions – known risk factors in HIV transmission. The lubricant effect of the mPEG-gel would be of particular importance with regards to anal transmission as the anal epithelium ranges from a multi- to single-layer epithelial lining prone to tearing (in contrast to the vagina which is a multicellular stratified squamous epithelium and resistant to traumatic injury). This simple anatomical difference in large part underlies the differential transmission risk associated with anal intercourse. Finally, the immunocamouflage of the infected dendritic cells and macrophages would also make the cells less prone to activation by other pathogens due to the camouflage of key surface receptors. In the absence of activation, HIV proliferation within infected cells is greatly diminished decreasing viral shedding.

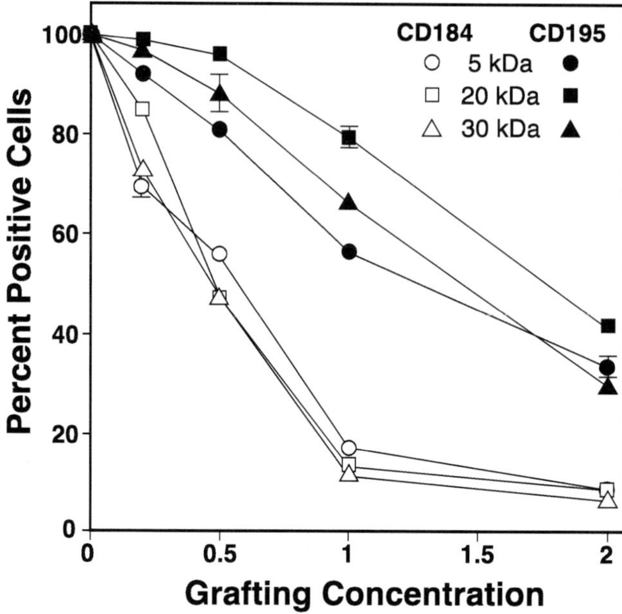

Fig. 12. Immunocamouflage of HIV receptors by activated mPEG antiviral gel. As shown, PEGylation of dendritic cells and lymphocytes results in the immunocamoufalge of CD184 and CD195 as detected by high affinity monoclonal antibodies to the CD determinants. The efficacy of the immunocamouflage is concentration dependent.

Indeed, there are several topical microbiocidal gels in clinical trials that aim to prevent against HIV infection either non-specifically (surfactants and acidifying agents) or specifically (viral entry and reverse transcriptase inhibitors). Some surfactants, such as nonoxynol-9, experimentally were shown to result in viral disruption but also exerted toxicity to host cells. Further clinical trials of nonoxynol-9 in HIV endemic areas actually demonstrated increased risk of HIV infection with its use. [Rustomjee *et al.*, 1999, Van Damme *et al.*, 2002, Herold *et al.*, 2011] These studies suggested that the surfactant-induced mucosal injury appeared to mimic the presence of microabrasions already known to increase HIV transmission. Other surfactants such as sodium lauryl sulfate (invisible condom) disrupt enveloped and non-enveloped viruses and are associated with decreased (relative to nonoxynol-9) mucosal toxicity. [Howett & Kuhl, 2005, Mbopi-Keou *et al.*, 2010] In contrast to surfactants, the mPEG antiviral gel is of low toxicity to cells while simultaneously providing significant camouflage of the mucosal surface thereby preventing or decreasing the initial viral interaction with resident dendritic cells and T lymphocytes.

4. Conclusion

The safe, low cost, low technology, non-toxic, and transient bioengineering of the terminally differentiated nasal pharyngeal epithelial host cells may provide a radically new antiviral prophylactic approach. Moreover, this approach may be applicable to other non-respiratory viruses in which a polymer gel can be applied. This antiviral protection arises from the ability of the grafted polymer to directly impede cellular invasion and subsequent proliferation thus abrogating the disease process at its initial stage. Moreover, this technology is effective against: 1) enveloped and non-enveloped viruses; 2) receptor-mediated and fusiogenic viruses; 3) small and large viruses; 4) DNA and RNA viruses; and 5) viruses with known and unknown viral receptors.

The envisioned use of this prophylactic technology is via a simple intranasal gel application in at risk individuals in acute, high transmissibility, environments (*e.g.,* nosocomial spread, refugee centers, bioterrorism, airplanes, or with visiting grandchildren). For example, within refugee centers, epidemiological analyses demonstrate that, alongside malnutrition and diarrhea, acute respiratory infections are a major cause of morbidity and mortality (*e.g.,* Figure 6). Under identical circumstances, traditional vaccines, even if they were available, would not be able to prevent widespread dissemination of disease while this novel intranasal anti-viral prophylactic provides immediate, albeit transient, protection against a broad spectrum of respiratory viral pathogens.

Application of the activated nasal gel is surprisingly uncomplicated and requires a minimal (~3-5 minutes) application time to the host (*e.g.,* nasopharyngeal) cells but effectively camouflages known and unknown viral receptors thus protecting against a broad spectrum of viral pathogens. Of note, nasal epithelial cells are terminally differentiated and slough off after about 4 days and are biologically *outside* the body resulting in minimal systemic exposure to the mPEG polymer. The polymer itself has an extremely safe toxicity profile and is commonly used in current pharmaceuticals (most often as a stabilizer or binding agent), cosmetics and food preparations. After application, the individual then simply "blows their nose" into a tissue to remove any residual carrier fluid. Furthermore, preliminary data suggests that a single daily application provides maximal protection and even retains significant efficacy for up to 60 hours.

Respiratory pathogens can, and do, create massive healthcare emergencies in at risk populations. The development of this inexpensive, easy to use intranasal antiviral gel would be of significant health and economic benefit to both the at risk individual and to governmental and health agencies addressing respiratory disease crises. Importantly, the term '*at risk*' is relative and may range from environmental or political upheavals to more mundane environments such as air travel and visiting grandchildren.

5. Acknowledgment

This work was supported by grants from the Canadian Blood Services, Health Canada and the Canadian Blood Services-Canadian Institutes of Health Research (CBS-CIHR) Partnership Fund. The views expressed herein do not necessarily represent the view of the federal government of Canada. We thank the Canada Foundation for Innovation and the Michael Smith Foundation for Health Research for infrastructure funding at the University of British Columbia Centre for Blood Research. Dr. Sutton is now a Postdoctoral Fellow at the University of Maryland, USA.

6. References

Bella, J., Rossmann, M. G. (1999) Review: rhinoviruses and their ICAM receptors. *J Struct Biol,* 128(1), 69–74.

Bjorkman, P. J., Saper, M. A., Samraoui, B., Bennett, W. S., Strominger, J. L., Wiley, D. C. (1987) Structure of the human class I histocompatibility antigen, HLA-A2. *Nature*, 329(6139), 506–512.

Block, K. I., Mead, M. N. (2003) Immune system effects of echinacea, ginseng, and astragalus: a review. *Integr Cancer Ther*, 2(3), 247–267.

Boily, M. C., Baggaley, R. F., Wang, L., Masse, B., White, R. G., Hayes, R. J., Alary, M. (2009) Heterosexual risk of HIV-1 infection per sexual act: systematic review and meta-analysis of observational studies. *Lancet Infect Dis*, 9(2), 118–129.

Bradley, A. J., Murad, K. L., Regan, K. L., Scott, M. D. (2002) Biophysical consequences of linker chemistry and polymer size on stealth erythrocytes: size does matter. *Biochim Biophys Acta*, 1561(2), 147–158.

Bradley, A. J., Scott, M. D. (2004) Separation and purification of methoxypoly(ethylene glycol) grafted red blood cells via two-phase partitioning. *J Chromatogr B Analyt Technol Biomed Life Sci*, 807(1), 163–168.

Bradley, A. J., Scott, M. D. (2007) Immune complex binding by immunocamouflaged [poly(ethylene glycol)-grafted] erythrocytes. *Am J Hematol*, 82(11), 970–975.

Bradley, A. J., Test, S. T., Murad, K. L., Mitsuyoshi, J., Scott, M. D. (2001) Interactions of IgM ABO antibodies and complement with methoxy-PEG-modified human RBCs. *Transfusion*, 41(10), 1225–1233.

Chen, A. M., Scott, M. D. (2001) Current and future applications of immunological attenuation via pegylation of cells and tissue. *BioDrugs*, 15(12), 833–847.

Chen, A. M., Scott, M. D. (2003) Immunocamouflage: prevention of transfusion-induced graft-versus-host disease via polymer grafting of donor cells. *J Biomed Mater Res A*, 67(2), 626–636.

Chen, A. M., Scott, M. D. (2006) Comparative analysis of polymer and linker chemistries on the efficacy of immunocamouflage of murine leukocytes. *Artif Cells Blood Substit Immobil Biotechnol*, 34(3), 305–322.

Cutler, B., Justman, J. (2008) Vaginal microbicides and the prevention of HIV transmission. *Lancet Infect Dis*, 8(11), 685–697.

Damodaran, V. B., Fee, C. J., Ruckh, T., Popat, K. C. (2010) Conformational studies of covalently grafted poly(ethylene glycol) on modified solid matrices using X-ray photoelectron spectroscopy. *Langmuir*, 26(10), 7299–7306.

Garg, S., Goldman, D., Krumme, M., Rohan, L. C., Smoot, S., Friend, D. R. (2010) Advances in development, scale-up and manufacturing of microbicide gels, films, and tablets. *Antiviral Res*, 88 Suppl 1S19–29.

Hayden, F. G., Gwaltney, J. M. J., Colonno, R. J. (1988) Modification of experimental rhinovirus colds by receptor blockade. *Antiviral Res*, 9(4), 233–247.

He, Y., Chipman, P. R., Howitt, J., Bator, C. M., Whitt, M. A., Baker, T. S., Kuhn, R. J., Anderson, C. W., Freimuth, P., Rossmann, M. G. (2001) Interaction of coxsackievirus B3 with the full length coxsackievirus-adenovirus receptor. *Nat Struct Biol*, 8(10), 874–878.

Herold, B. C., Mesquita, P. M., Madan, R. P., Keller, M. J. (2011) Female Genital Tract Secretions and Semen Impact the Development of Microbicides for the Prevention of HIV and Other Sexually Transmitted Infections. *Am J Reprod Immunol*, 65(3), 325–333.

Heuberger, M., Drobek, T., Spencer, N. D. (2005) Interaction forces and morphology of a protein-resistant poly(ethylene glycol) layer. *Biophys J*, 88(1), 495–504.

Hirt, M., Nobel, S., Barron, E. (2000) Zinc nasal gel for the treatment of common cold symptoms: a double-blind, placebo-controlled trial. *Ear Nose Throat J*, 79(10), 778–80, 782.

Howett, M. K., Kuhl, J. P. (2005) Microbicides for prevention of transmission of sexually transmitted diseases. *Curr Pharm Des*, 11(29), 3731–3746.

Huguenel, E. D., Cohn, D., Dockum, D. P., Greve, J. M., Fournel, M. A., Hammond, L., Irwin, R., Mahoney, J., McClelland, A., Muchmore, E., Ohlin, A. C., Scuderi, P. (1997) Prevention of rhinovirus infection in chimpanzees by soluble intercellular adhesion molecule-1. *Am J Respir Crit Care Med*, 155(4), 1206–1210.

Hussain, M. M., Tranum-Jensen, J., Noren, O., Sjostrom, H., Christiansen, K. (1981) Reconstitution of purified amphiphilic pig intestinal microvillus aminopeptidase. Mode of membrane insertion and morphology. *Biochem J*, 199(1), 179–186.

Jefferson, T., Jones, M., Doshi, P., Del Mar, C. (2009) Neuraminidase inhibitors for preventing and treating influenza in healthy adults: systematic review and meta-analysis. *BMJ*, 339b5106.

Jun, C. D., Carman, C. V., Redick, S. D., Shimaoka, M., Erickson, H. P., Springer, T. A. (2001) Ultrastructure and function of dimeric, soluble intercellular adhesion molecule-1 (ICAM-1). *J Biol Chem*, 276(31), 29019–29027.

Kandel, R., Hartshorn, K. L. (2001) Prophylaxis and treatment of influenza virus infection. *BioDrugs*, 15(5), 303–323.

Katriel, G., Stone, L. (2010) Pandemic dynamics and the breakdown of herd immunity. *PLoS One*, 5(3), e9565.

Kravetz, H. M., Knight, V., Chanock, R. M., Morris, J. A., JOHNSON, K. M., RIFKIND, D., UTZ, J. P. (1961) Respiratory syncytial virus. III. Production of illness and clinical observations in adult volunteers. *JAMA*, 176657–663.

Le, Y., Scott, M. D. (2010) Immunocamouflage: the biophysical basis of immunoprotection by grafted methoxypoly(ethylene glycol) [mpeg]. *Acta Biomater*, 62631–2641.

Lim, J. H., Davis, G. E., Wang, Z., Li, V., Wu, Y., Rue, T. C., Storm, D. R. (2009) Zicam-induced damage to mouse and human nasal tissue. *PLoS One*, 4(10), e7647.

Marlin, S. D., Staunton, D. E., Springer, T. A., Stratowa, C., Sommergruber, W., Merluzzi, V. J. (1990) A soluble form of intercellular adhesion molecule-1 inhibits rhinovirus infection. *Nature*, 344(6261), 70–72.

Mbopi-Keou, F. X., Trottier, S., Omar, R. F., Nkele, N. N., Fokoua, S., Mbu, E. R., Domingo, M. C., Giguere, J. F., Piret, J., Mwatha, A., Masse, B., Bergeron, M. G. (2010) A randomized, double-blind, placebo-controlled Phase II extended safety study of two Invisible Condom formulations in Cameroonian women. *Contraception*, 81(1), 79–85.

McCoy, L. L., Scott, M. D. (2005) Broad spectrum antiviral prophylaxis: Inhibition of viral infection by polymer grafting with methoxypoly(ethylene glycol). In *Antiviral Drug*

Discovery for Emerging Diseases and Bioterrorism Threats (T. PF, ed.), Wiley & Sons, Hoboken, NJ, *379–395*.

McElhaney, J. E., Goel, V., Toane, B., Hooten, J., Shan, J. J. (2006) Efficacy of COLD-fX in the prevention of respiratory symptoms in community-dwelling adults: a randomized, double-blinded, placebo controlled trial. *J Altern Complement Med*, 12(2), 153–157.

Mi, L. Z., Grey, M. J., Nishida, N., Walz, T., Lu, C., Springer, T. A. (2008) Functional and structural stability of the epidermal growth factor receptor in detergent micelles and phospholipid nanodiscs. *Biochemistry*, 47(39), 10314–10323.

Miller, S. C., Delorme, D., Shan, J. J. (2009) CVT-E002 stimulates the immune system and extends the life span of mice bearing a tumor of viral origin. *J Soc Integr Oncol*, 7(4), 127–136.

Murad, K. L., Gosselin, E. J., Eaton, J. W., Scott, M. D. (1999a) Stealth cells: prevention of major histocompatibility complex class II-mediated T-cell activation by cell surface modification. *Blood*, 94(6), 2135–2141.

Murad, K. L., Mahany, K. L., Brugnara, C., Kuypers, F. A., Eaton, J. W., Scott, M. D. (1999b) Structural and functional consequences of antigenic modulation of red blood cells with methoxypoly(ethylene glycol). *Blood*, 93(6), 2121–2127.

Murata, Y., Falsey, A. R. (2007) Respiratory syncytial virus infection in adults. *Antivir Ther*, 12(4 Pt B), 659–670.

(2009) Over-the-counter medications: Zicam nasal products may cause loss of sense of smell. *Child Health Alert*, 273.

Paulke-Korinek, M., Kundi, M., Rendi-Wagner, P., de Martin, A., Eder, G., Schmidle-Loss, B., Vecsei, A., Kollaritsch, H. (2011) Herd immunity after two years of the universal mass vaccination program against rotavirus gastroenteritis in Austria. *Vaccine*,

Piguet, V., Steinman, R. M. (2007) The interaction of HIV with dendritic cells: outcomes and pathways. *Trends Immunol*, 28(11), 503–510.

Roberts, M. J., Bentley, M. D., Harris, J. M. (2002) Chemistry for peptide and protein PEGylation. *Adv Drug Deliv Rev*, 54(4), 459–476.

Rossi, N. A., Constantinescu, I., Brooks, D. E., Scott, M. D., Kizhakkedathu, J. N. (2010a) Enhanced cell surface polymer grafting in concentrated and nonreactive aqueous polymer solutions. *J Am Chem Soc*, 132(10), 3423–3430.

Rossi, N. A., Constantinescu, I., Kainthan, R. K., Brooks, D. E., Scott, M. D., Kizhakkedathu, J. N. (2010b) Red blood cell membrane grafting of multi-functional hyperbranched polyglycerols. *Biomaterials*, 31(14), 4167–4178.

Rustomjee, R., Abdool Karim, Q., Abdool Karim, S. S., Laga, M., Stein, Z. (1999) Phase 1 trial of nonoxynol-9 film among sex workers in South Africa. *AIDS*, 13(12), 1511–1515.

Satulovsky, J., Carignano, M. A., Szleifer, I. (2000) Kinetic and thermodynamic control of protein adsorption. *Proc Natl Acad Sci U S A*, 97(16), 9037–9041.

Scott, M. D., Chen, A. M. (2004) Beyond the red cell: pegylation of other blood cells and tissues. *Transfus Clin Biol*, 11(1), 40–46.

Scott, M. D., Murad, K. L. (1998) Cellular camouflage: fooling the immune system with polymers. *Curr Pharm Des*, 4(6), 423–438.

Scott, M. D., Murad, K. L., Koumpouras, F., Talbot, M., Eaton, J. W. (1997) Chemical camouflage of antigenic determinants: stealth erythrocytes. *Proc Natl Acad Sci U S A*, 94(14), 7566–7571.

Shaffer, C. B., Critchfield, F. H. (1947) The absorption and excretion of the solid polyethylene glycols ("Carbowax" compounds). *Journal of the American pharmaceutical Association*, 36152–157.

Smyth, H. E., Carpenter, C. P., Shaffer, C. B. (1947) The toxicity of high molecular weight polyethylene glycols; chronic oral and parental administration. *Journal of the american pharmaceutical association*, 36157–160.

Spector, S. L. (1995) The common cold: current therapy and natural history. *J Allergy Clin Immunol*, 95(5 Pt 2), 1133–1138.

Sperber, S. J., Hayden, F. G. (1989) Protective effect of rhinovirus receptor blocking antibody in human fibroblast cells. *Antiviral Res*, 12(5-6), 231–238.

Sutton, T. C., Scott, M. D. (2010) The effect of grafted methoxypoly(ethylene glycol) chain length on the inhibition of respiratory syncytial virus (RSV) infection and proliferation. *Biomaterials*, 31(14), 4223–4230.

Szleifer, I. (1997) Protein adsorption on surfaces with grafted polymers: a theoretical approach. *Biophys J*, 72(2 Pt 1), 595–612.

Turner, R. B., Wecker, M. T., Pohl, G., Witck, T. J., McNally, E., St George, R., Winther, B., Hayden, F. G. (1999) Efficacy of tremacamra, a soluble intercellular adhesion molecule 1, for experimental rhinovirus infection: a randomized clinical trial. *JAMA*, 281(19), 1797–1804.

Van Damme, L., Ramjee, G., Alary, M., Vuylsteke, B., Chandeying, V., Rees, H., Sirivongrangson, P., Mukenge-Tshibaka, L., Ettiegne-Traore, V., Uaheowitchai, C., Karim, S. S., Masse, B., Perriens, J., Laga, M. (2002) Effectiveness of COL-1492, a nonoxynol-9 vaginal gel, on HIV-1 transmission in female sex workers: a randomised controlled trial. *Lancet*, 360(9338), 971–977.

Van Effelterre, T., Soriano-Gabarro, M., Debrus, S., Claire Newbern, E., Gray, J. (2010) A mathematical model of the indirect effects of rotavirus vaccination. *Epidemiol Infect*, 138(6), 884–897.

Veronese, F. M., Harris, J. M. (2002) Introduction and overview of peptide and protein pegylation. *Adv Drug Deliv Rev*, 54(4), 453–456.

Veronese, F. M., Mero, A. (2008) The impact of PEGylation on biological therapies. *BioDrugs*, 22(5), 315–329.

Veronese, F. M., Pasut, G. (2005) PEGylation, successful approach to drug delivery. *Drug Discov Today*, 10(21), 1451–1458.

Wang, M., Guilbert, L. J., Li, J., Wu, Y., Pang, P., Basu, T. K., Shan, J. J. (2004) A proprietary extract from North American ginseng (Panax quinquefolium) enhances IL-2 and IFN-gamma productions in murine spleen cells induced by Con-A. *Int Immunopharmacol*, 4(2), 311–315.

Winther, B. (2011) Rhinovirus infections in the upper airway. *Proc Am Thorac Soc*, 8(1), 79–89.

Yale, S. H., Liu, K. (2004) Echinacea purpurea therapy for the treatment of the common cold: a randomized, double-blind, placebo-controlled clinical trial. *Arch Intern Med,* 164(11), 1237–1241.

8

Solid Lipid Nanoparticles: Technological Developments and in Vivo Techniques to Evaluate Their Interaction with the Skin

Mariella Bleve, Franca Pavanetto
and Paola Perugini
Department of Drug Sciences, University of Pavia
Italy

1. Introduction

Nanotechnology is an emerging science involving manipulation of matter at the nanometer scale [Stern et al, 2008]. Nanoscience research has shown remarkable growth over the past 10 years, which is expected to continue for the foreseeable future. Discovery, development and implementation of nanotechnologies are driven by the social desire for smaller products with enhanced capabilities. As such, nanotechnology research and development has been increasing steadily. The discovery and development of nanotechnology is evident in food products, pesticides, consumer products and medicine. The application of nanotechnology to medicine is termed nanomedicine, which spans from nanodiagnostics to nanorobotic treatments and novel nanoparticle drug-delivery systems. The most prominent use of nanoparticles in medicine is the development of novel drug-delivery systems. The use of nanotechnology in drug delivery could revolutionize current therapies and is set for rapid advancements. This is due to the unique properties of nanomaterials, including large surface:mass ratio (i.e., large functional surface), ease in engineering tissue-targeted nanoparticles, and higher loading capacity due to reduced drug expulsion during storage compared with micro-sized systems, which increases their ability to carry natural and synthetic chemical compounds [Buse et al, 2010].

Although opportunities to develop nanotechnology based efficient drug delivery systems extend into all therapeutic classes of pharmaceuticals, the development of effective treatment modalities for the respiratory, central nervous system and cardiovascular disorders remains a financially and therapeutically significant need. Many therapeutic agents have not been successful because of their limited ability to reach to the target tissue. In addition, the faster growth opportunities are expected in developing delivery systems for anti-cancer agents, hormones and vaccines because of safety and efficacy shortcomings in their conventional administration modalities. For example, in cancer chemotherapy, cytostatic drugs damage both malignant and normal cells alike. Thus, a drug delivery strategy that selectively targets the malignant tumor is very much needed. Additional problems include drug instability in the biological milieu and premature drug loss through rapid clearance and metabolism. Similarly, high protein binding of certain drugs such as protease inhibitors limits their diffusion to the brain and other organs. However, nanotechnology for drug delivery applications may not be suitable for all drugs, especially those drugs that are less potent because the higher dose of the drug would make the drug delivery system much larger, which would be difficult to administer [Sahoo et al, 2003].

The Royal Society and Royal Academy of Engineering define nanoparticles as particulate matter that has a size of 100 nm or less, while British Standards (British Department of Trade and Industry) define nanoparticles based upon the point at which the properties of nanoparticles differ from the bulk material – typically particulate with a scale less than 100 nm [[Buse et al, 2010]. However, not all particles used in drug delivery comply with this definition. Medical formulations of 100–500 nm are also considered nanoparticles since they share the same functionality in pharmaceutical applications and are governed by quantum effects instead of Newtonian physics [Bawa , 2005].

One nanoparticle class that has been widely used in drug delivery is lipid-based nanoparticles. Compared to liposomes and emulsions, solid particles possess some advantages, e.g. protection of non-incorporated active compounds against chemical degradation and more flexibility in modulating the release of the compound. Advantages of liposomes and emulsions are that they are composed of well tolerated excipients and they can easily be produced on a large scale, the pre-requisite for a carrier to be introduced to the market. At the beginning of the 1990s, the advantages of solid particles, emulsions and liposomes were combined by the development of the 'solid lipid nanoparticles' (SLN) [Muller et al,2002; Bunjes et al, 2006].

Solid lipid nanoparticles protect the drug from the environment and prevent its degradation whilst increasing its bioavailability. Therefore, two direct benefits can be achieved; first, due to the improved encapsulation efficiency, less of the active drug is required during the formulation process. Second, the desired effects of the drug are expedited due to an increase in the initial release, which is a result of the homogenization production process of SLNs. Both heating the lipid/water mixture and increasing the total surfactant concentration causes a partitioning of the active drug into the water phase due to an increase in its water solubility. During cooling, the solubility of the drug in water is decreased, allowing for the drug to partition back into the lipid phase [Üner, 2006]. The crystallization of the lipid core will occur while a high percentage of the drug is in the water phase. As a result, a considerable amount of the drug is accumulated on the surface of SLNs, hence the initial burst release of the drug after administration. The increase in the initial drug release can be offset by performing homogenization at lower temperatures (i.e., using cold homogenization) preventing drug expulsion and crystallization during cooling or using a greater amount of surfactant, which improves the lipid solubility of the drug entities [Buse, 2010]. This chapter describes the more innovative methods to produce and to characterize SLN and to assure their safety after cutaneous application.

In fact, after skin application of SLN, a lipidic film onto the skin is formed. Such a film can have an occlusive effect that increases drug penetration or increases water content in the upper epidermis layers.

Potential systemic effects after dermal application of SLN should be considered in order to obtain a safe topical product. Skin penetration and systemic absorption should be estimated with the intention to assess the risk of using nanoparticles in topical products. Visualization of colloidal systems after skin application is essential to evaluate their interaction with cutaneous structures. This chapter provides informations about the non-invasive bioengineering techniques to study the SLN distribution and interaction on the skin.

2. Solid lipid nanoparticles

2.1 Ingredients

General ingredients include solid lipid(s), emulsifier(s) and water. The term lipid is used here in a broader sense and includes triglycerides (e.g. tristearin, tricaprin, tripalmitin, partial glycerides (e.g. Imwitor, glycerol behenate), fatty acids (e.g. stearic acid), steroids (e.g. cholesterol) and waxes (e.g. cetyl palmitate) [Buse, 2010, Choi et al, 2008, Chen et al, 2006, Harivardhan et al, 2006].

A clear advantage of SLN with respect to polymeric nanoparticles is that the lipid matrix is made from physiological lipids which decrease the danger of acute and chronic toxicity. The vast majority of solid lipids are naturally occurring lipids (performing various physiological functions) and as a result they have lower cytotoxicity than synthetic polymers.

Critical parameters for nanoparticle formation will be different for different lipids. Examples include the velocity of lipid crystallization, the lipid hydrophilicity (influence on self-emulsifying properties) and the shape of the lipid crystals (and therefore the surface area). It is also noteworthy, that most of the lipids used represent a mixture of several chemical compounds. The composition might therefore vary from different suppliers and might even vary for different batches from the same supplier. However, small differences in the lipid composition (e.g. impurities) might have considerable impact on the quality of SLN dispersion (e.g. by changing the zeta potential, retarding crystallization processes etc.). For example, lipid nanodispersions made with cetyl palmitate from different suppliers had different particle sizes and storage stabilities.

All classes of emulsifiers (with respect to charge and molecular weight) have been used to stabilize the lipid dispersion (e.g., poloxamers, Tween 80, soya lecithin and sodium dodecyl sulphate). The choice of the emulsifiers and their concentration is of great impact on the quality of the SLN dispersion. It has been found that the combination of emulsifiers might prevent particle agglomeration more efficiently. The choice of the emulsifier depends on the administration route and is more limited for parenteral administrations. High concentrations of the emulsifier reduce the surface tension and facilitate the particle partition during homogenization. The decrease in particle size is connected with a tremendous increase in surface area. Therefore, kinetic aspects have to be considered [Uner, 2006].

Recently, the incorporation of a small amount of liquid lipids led to a new nanoparticle structures, the Nanostructured Lipid Carriers (NLC) in which the liquid lipid phase can be embedded into the solid matrix or to be localized at the surface of solid particles [Schafer-Korting et al, 2007].

The simplistic production methods and structural composition of SLNs has increased the appeal for their use in pharmaceutical formulations. For example, the inclusion of lipids that have 'generally recognized as safe' (GRAS) status minimizes cytotoxic effects whilst maintaining the overall stability of the SLN structure. However, structural stabilization of SLNs requires the incorporation of surfactants. While some surfactants may increase the cytotoxicity, nontoxic surfactants (e.g., lecithin) can be utilized to maintain the desired low cytoxicity of SLNs [Cevc, 2004].

2.2 Preparation methods

Many different techniques for the production of lipid nanoparticles have been described in the literature. These methods are high pressure homogenization [Liedtke, 2000; Jahnke, 1998; Wissing et al, 2004], microemulsion technique [Gasco, 1993; Gasco, 1997; Priano et al, 2007], emulsification-solvent evaporation [Sjöström & Bergenstahl, 1992], emulsification-solvent diffusion method [Hu et al, 2002; Trotta et al, 2003], solvent injection (or solvent displacement) method [Schubert, & Müller-Goymann, 2003], phase inversion [Heurtault et al, 2002], multiple emulsion technique [Garcy'-Fuentes et al, 2002], ultrasonication [Pietkiewicz, & Sznitowska, 2004; Puglia et al, 2008] and membrane contractor technique [Charcosset et al, 2005].

However, high pressure homogenization technique has many advantages compared to the other methods, e.g. easy scale up, avoidance of organic solvents and short production time. High pressure homogenizers are widely used in many industries including the pharmaceutical industry, e.g. for the production of emulsions for parenteral nutrition. Therefore, no regulatory problems exist for the production of topical pharmaceutical and cosmetic preparations using this production technique. It can be considered as being industrially the most feasible one.

Lipid nanoparticles can be produced by either the hot or cold high pressure homogenization technique. In the hot homogenization method the lipid melt containing the active compound is dispersed in a hot surfactant solution of the same temperature by high speed stirring. The obtained emulsion (generally called pre-emulsion) is then passed through a high pressure homogenizer adjusted to the same temperature. In the cold homogenization method, the active containing lipid melt is cooled down. After solidification the mass is crushed and ground to obtain lipid microparticles. The lipid microparticles are then dispersed in a cold surfactant solution yielding a cold pre-suspension of micronized lipid particles. For hot and cold homogenization is not required the use of organic solvents, in this way the cytoxicity is further reduced.

2.3 SLN characterization

After preparation, it has to be ensured that the particles obtained have the desired properties and are thus suitable for the intended type of administration. The most obvious parameters to be investigated are the (colloidal) particle size and the (solid) state of the particle matrix. Other important features include the surface characteristics, the particle shape, and in particular, the interaction with incorporated drugs. Due to the complexity of the systems, a combination of different characterization techniques is the most promising approach to obtain a realistic image of the sample properties [Pardeike J. et al, 2009].

Amongst the methods used to measure particle size, Photon Correlation Spettroscopy (PCS) is the most widely employed. Particle size of the SLNs is influenced by different factors such as lipid matrix, drug to lipid ratio, surfactant blend, viscosity of lipid and aqueous phase and production parameters.

The zeta potential is used as a measure of surface charge. This is valuable in preventing aggregation and imparts the physical stability to formulation. At higher zeta potential, particle aggregation is less likely to occur, due to electrical repulsion.

Shape and Morfology of lipid nanoparticles were usually investigated using Scanning Electron Microscopy (SEM), Transmission Electron Microscopy (TEM) and Atomic Force Microscopy (AFM).

The cristallinity and the polymorphic behavior of lipids strongly influence drug incorporation and release rate. Crystalline solids lipids have an olderly arrangement of units. The presence of emulsifiers, the preparation method, and the high dispersity as well as the small particle size of the colloidal system influences crystallinity of lipids in the SLNs. Basic techniques used to investigate of the lipid are Differential Scanning Calorimetry (DSC) and X-Ray Diffractometry (XRD) [Rizwan et al, 2009].

About the encapulation capability there are basically three different models for the incorporation of active ingredients into SLN: homogeneous matrix model, drug-enriched shell model and drug-enriched core model. The structure obtained is a function of the formulation composition (lipid, active compound, surfactant) and of the production conditions (hot vs. cold homogenisation)[Mehnert&Mader, 2001, Haskell, 2006].

Solid lipid nanoparticles (SLN) are able to incapsulate a great amount of lipophilic compounds such as steroids, retinol, sunscreens but for most drugs, especially hydrophilic ones, the payload is very low. This effect can be due either to the crystalline structure of the lipid matrix, or to the low solubility of hydrophilic substances into lipidic phase.

Two approaches were developed to improve the payload of hydrophilic compounds. The first approach was the development of oil loaded SLN, also described as nanostructured lipid carriers (NLC). The alternative strategy, applied recently from Perugini et al., is to modify the lipid matrix by incorporation of amphiphilic substances such as phosphatydilcholine, polyglyceryl-3-diisostearate and sorbitan [Perugini et al, 2010b].

In addition to usual particle characterization, to ensure the safety of topical preparations *in vitro* cytotoxicity of SLN should be evaluated. Cellular damage results in loss of the metabolic cell function. The tetrazolium salt MTT, using keratinocytes line, is widely used to quantitate by colorimetric assay the cytotoxicity of preparations. The tetrazolium salts are metabolically reduced to highly colored end products called formazans. The colorless MTT is cleaved to formazan by the succinate-tetrazolium reductase system which belongs to the mitochondrial respiratory chain and is active only in viable cells. [Weyenberg et al, 2007]. Perugini et al. calculated, by linear regression analysis of data, the EC50 values of SLN aqueous suspension in order to define non toxic concentration of nanoparticles suspension that can to be use in topical formulation [Perugini et al, 2010a].

Furthermore, the two most frequent manifestation of skin toxicity are irritant contact dermatitis and allergic contact dermatitis. Keratinocytes, which represent the 95% of epidermal cells in both human and mouse skin, are a rich source of cytokines and they actively participate to skin inflammatory and immunological reactions. Among the cytokines produced by keratinocytes, IL-1 is one of the most interesting, since it is produced constitutively by keratinocytes and retained in normal conditions into the cell. In general, cytokines have been identified as useful tools to discriminate between irritant and allergic contact dermatitis [Corsini et al, 1998]. Perugini et al. using a specific sandwich ELISA to evaluate IL-1 release from keratinocytes after SLN application [Perugini et al, 2010].

3. Skin

The skin (Latin: cutis) consists of 250 Am (<4000 Am) thick inner skin region (=dermis), and of 50 Am (<100 Am) thick outer skin region (=epidermis). The fact that the skin is one of best biological barriers is mainly due to the outermost part of the epidermis, the skin horny layer (Latin: the stratum corneum, SC) [Walters, 2002].

The SC serves as the principal barrier against the percutaneous penetration of chemicals and microbes and is capable of withstanding mechanical forces [Madison, 2003]. It is further involved in the regulation of water release from the organism and into the atmosphere, known as transepidermal water loss (TEWL).

The SC forms a continuous sheet of protein-enriched cells (corneocytes) connected by corneodesmosomes and embedded in an intercellular matrix enriched in non-polar lipids and organized as lamellar lipid layers. The final steps in keratinocyte differentiation are associated with profound changes in their structure, resulting in their transformation into the flat and anucleated corneocytes of the SC, which are loaded with keratin filaments and surrounded by a cell envelope composed of cross-linked proteins (cornified envelope proteins) as well as a covalently bound lipid envelope. Extracellular non-polar lipids surround the corneocytes to form a hydrophobic matrix.

During the final stages of normal differentiation, keratins are aligned into highly ordered and condensed arrays through interactions with filaggrin, a matrix protein. The role of filaggrin in skin barrier homeostasis is only partially known. Filaggrin aggregates the keratin filaments into tight bundles. This promotes the collapse of the cell into a flattened shape, which is characteristic of corneocytes in the cornified layer [Palmer et al, 2006, Proksch et al, 2008]. Together, keratins and filaggrin constitute 80–90% of the protein mass of mammalian epidermis [Roop, 1995, Nemes &Steinert, 1999].

The cornified cell envelope (CE) is a tough protein / lipid polymer structure formed just below the cytoplasmic membrane and residing on the exterior of the corneocytes. It consists of two parts: a protein envelope and a lipid envelope. The protein envelope contributes to the biomechanical properties of the CE as a result of cross-linking of specialized cornified envelope structural proteins, including involucrin, loricrin, trichohyalin and to the class of small proline-rich proteins by both disulphide bonds and N-epsilon-(gamma-glutamyl)lysine isopeptide bonds formed by transglutaminases [Candi et al, 2005, Roop, 1995].

Desmosomes on corneocytes are called (corneo)desmosomes. Desmosomes, which connect the keratinocytes of the granular layer with the corneocytes of the SC, have been called transition desmosomes and are exclusively found at that location [Koch &Franke, 1994, Al-Amoudi et al, 2005]. Adjacently interconnected corneocytes are important for SC cohesion and are shed during the desquamation process in the SC. The corneocytes provide mechanical and chemical protection and, together with their intercellular lipid surroundings, confer water impermeability [Madison 2003]. The cohesiveness of, and the communication between, the viable epidermal cells, the cell-cell interaction, is maintained in a fashion similar to the cell-matrix connection, except that desmosomes replace hemidesmosomes. Adherens junctions are also located at keratinocyte-keratinocyte borders. At the desmosomal junction there are two transmembrane glycoprotein types: desmogleins and desmocollins wich are associated with the cytoplasmatic plaque proteins and provide a linkage to keratin intermediate filaments. On the other hand, in the adherens junction, classic cadherins act as transmembrane glycoproteines and these are linked to the actin filament α-, β- and χ-catenins. Thus, in the epidermis, the desmosomes are responsible for interconnecting individual cell keratin cytoskeletal structure, thereby creating a tissue very resistant to shearing forces [Prow et al, 2011].

The major lipid classes in the SC are ceramides, free fatty acids and cholesterol [Downing et al, 1987]. Ceramides are amide-linked fatty acids containing a long-chain amino alcohol called sphingoid base and account for 30 to 40% of SC lipids. The SC contains at least nine different free ceramides [Huchida &Hamanaka, 2006], two of which are ceramide A and ceramide B, covalently bound to cornified envelope proteins, most importantly to involucrin [Bouwstra et al, 2006]. The epidermis contains free fatty acids as well as fatty acids bound in triglycerides, phospholipids, glycosylceramides and ceramides. The chain length of free fatty acids ranges from C12 to C24. Saturated and monounsaturated fatty acids are synthesized in the epidermis, while others must be obtained from food and blood flow. Cholesterol is the third major lipid class in the SC. Although basal cells are capable of resorbing cholesterol from circulation, most cholesterol in the epidermis is synthesized in situ from acetate.

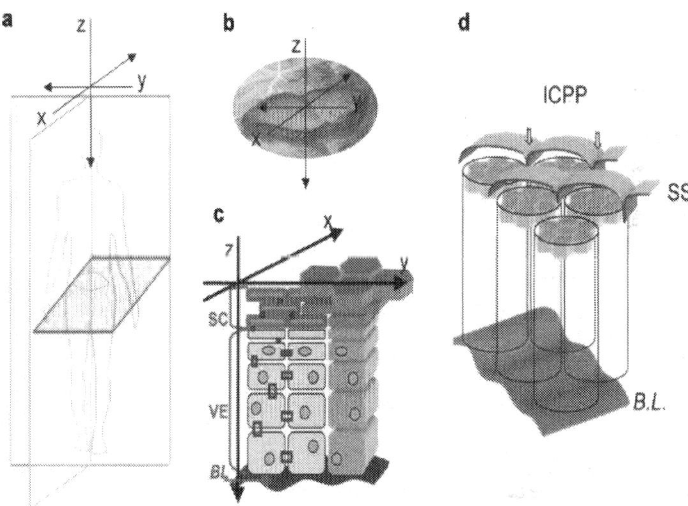

Fig. 1. Structure of epidermis. This figure shows (a) a human body representation in relation with the X–Y–Z coordinate system. The area of abdominal specimen origin is also shown (below the belly); (b) an abdominal skin specimen deposited flat on a table, where stratum corneum is upward and in the X–Y plane; (c) a schematic representation of epidermis in the Y–Z plane, where stratum corneum is SC, viable epidermis VE, and basal lamina B.L. Squares between keratinocytes and corneocytes respectively represent desmosomes and corneodesmosomes. Hexagonal shape of corneocytes is shown in the X–Y plane, where an attempt of a 3D perspective is provided. (d) Representation of corneocyte clusters and their relationships with basal lamina (B.L.) and skin surface (SS). Skin produces furrows along cluster perimeters, where the intercluster penetration pathway (ICPP) is supposed to exist[Baroli 2010]

There are still many unanswered questions about the exact way in which the SC lipids are organized at the molecular level and this is an active area of research [Bouwstra et al, 2003]. Understanding the physical structure of the membranes is critical to understanding their function as a barrier, both to water and to other substances, and ultimately to understanding the mechanisms of barrier disruption in a variety of skin diseases Numerous biophysical studies of SC structure suggest the presence of coexisting liquid crystalline and gel phase domains in the membranes of the SC. This concept was suggested by Forslind (1994) [Forslind, 1994] and presented as the "domain mosaic" model; a new model for the existence of fluid phases within the lamellae, the "sandwich model", was presented by Bouwstra et al (2000) [Bouwstra et al, 2000]. Norlen, however, has proposed a different "single gel phase" model feeling more consistent with the documented barrier properties of the SC [Norlen 2001, Baroli 2010].

Although the SC is recognized as the most important physical barrier, the nucleated epidermal layers are also significant in barrier function. Clinical observations confirm the importance of the nucleated epidermal layers in skin barrier function by preventing both excessive water loss and the entry of harmful substances into the skin [Honari 2004]. Tight junctions (TJ) are cell–cell junctions which connect neighbouring cells, control the paracellular pathway of molecules (barrier function) and separate the apical from the basolateral part of a cell membrane (fence function). In human epidermis, various TJ proteins have been identified, including occludin, claudins 1, 4, and 7, JAM-1 (junctional adhesion molecule-1), zonula occludens protein 1 and MUPP-1 (multi-PDZ protein-1) [Brandner et al 2006a, Brandner et al 2006b]. In human skin, TJ proteins and/or discrete TJ-like structures are found in the interfollicular epidermis as well as in the skin appendages [Pummi 2001, Wilke et al 2006, Madison 2003].

3.1 Skin as a site for particle delivery

Skin is a widely used route of delivery for local and systemic drugs and is potentially a route for their delivery as nanoparticles. Whilst nanoparticle drug delivery has been touted as an enabling technology, its potential in treating local skin and systemic diseases has yet to be realised. Most drug delivery particle technologies are based on lipid carriers, i.e. solid lipid nanoparticles and nanoemulsions of around 300 nm in diameter, which are now considered microparticles.

The skin has historically been used for the topical delivery of compounds but it is only since the 1970s with the advent of transdermal patches that it has widely been used as a route for systemic delivery. Nanoparticle delivery to the skin is being increasingly used to facilitate local therapies. The nanoparticle definition designated by the National Nanotechnology Initiative has been adopted by the American National Standards Institute as particles with all dimensions between 1 nm and 100 nm. Fig. 2 shows that the potential sites for targeting nanoparticles include the surface of the skin, furrows, and hair follicles [Souto & Muller 2008]. A recent review by Baroli discusses nanoparticle penetration largely from the skin structure perspective debate of nanoparticle penetration [Baroli 2010, Sawant & Dodiya 2008, Medina et al 2007].

The theory and practical aspects of percutaneous penetration of drugs, particulate material and contaminants have been covered in a number of excellent reference texts. Souto and Muller maintained that submicron-sized particles show adhesiveness when in contact with surfaces. This property has been demonstrated for polymeric nanoparticles and for liposomes.

Regarding lipid nanoparticles, it has been published that approximately 4% of lipid nanoparticles with a diameter of approximately 200 nm should form theoretically a monolayer film when 4 mg of formulation is applied per cm^2. Being hydrophobic in character, this mono-layered film has an occlusive action on the skin retarding the loss of

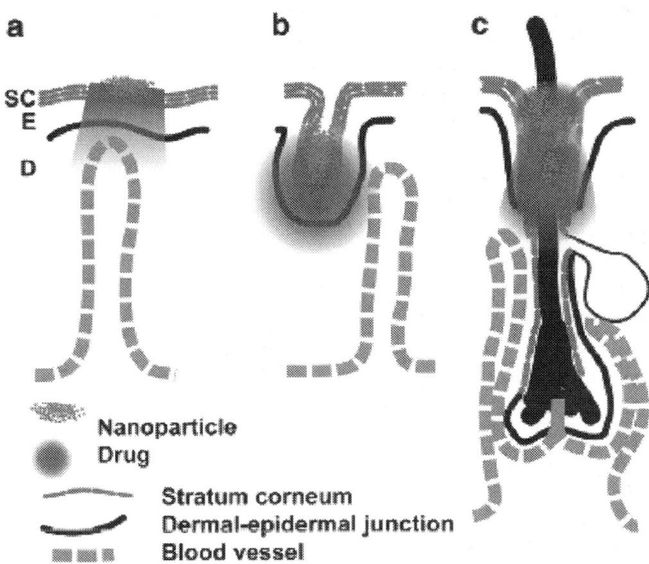

Fig. 2. Sites in skin for nanoparticle delivery. Topical nanoparticle drug delivery takes place in three major sites: stratum corneum (SC) surface (panel a), furrows (dermatoglyphs) (panel b), and openings of hair follicles (infundibulum) (c). The nanoparticles are shown in green and the drug in red. Other sites for delivery are the viable epidermis (E) and dermis (D)[Prow et al, 2011].

moisture caused by evaporation which can result in reduction of corneocyte packing and opening of inter-corneocyte gaps and thus facilitates the drug penetration into deeper layers of skin [Desai et al 2010, Wissing & Muller 2003]. The loss of water content from the SLN induces crystal modification of SLN matrix and this can induce drug expulsion and penetration [Lombardi Borgia et al 2005]. The occlusion effect of these nanoparticles is dependent on the applied sample volume, particle size, crystallinity and lipid concentration [Teeranachaideekul et al 2008]. The space filled with air in a layer of optimal packing density is independent on the particle size, which is considered to be 24% if assuming a three-dimensional hexagonal packing of ideal spherical-like particles [Desai et al 2010].

At the other hand the promise of nanoparticle-mediated drug delivery into the epidermis and dermis without barrier modification has met with little success. Where the barrier is compromised, however, such as in aged or diseased skin, there may be potential for enhanced particle penetration. Transport of substances across the SC occurs mainly by passive diffusion and based on the dual-compartment bricks and mortar structure of the SC, interrupted by appendages, is considered to occur via three possible routes. These are the transcellular, the intercellular and the appendageal routes. For most penetrants, the intercellular route is favoured. Small molecules are able to move freely within the inter-cellular spaces and diffusion rates are governed largely by their lipophilicity, but also physicochemical properties such as molecular weight or volume, solubility and hydrogen bonding ability [Potts & Guy 1995]. However, the free movement of macromolecules or particles may be physically restricted within the lipid channels, which have been estimated by van de Merwe et al. to be 19 nm [Van der Merwe et al 2006] and by Baroli et al. to be 75 nm [Baroli et al 2007]. This suggests that for such materials, the SC could present an additional barrier that is not present for small molecules.

Polar and non-polar solutes were originally thought to permeate through the SC via separate routes [Scheuplein 1965], with polar solutes taking a transcellular route and more lipophilic solutes going via the intercellular lipids. However a perception of the difficulty of repeated partitioning between lipophilic and hydrophilic compartments in the SC led to this pathway being regarded as unlikely in most cases. There is considerable interest in targeted follicular delivery with tailored drug formulations [Grice et al 2010] or nanoparticle-bound drugs [Lademann et al 2007,Souto &Muller 2008].

Hair follicles were regarded as insignificant as potential routes for drug delivery, covering only 0.1% of the human skin surface area, their complex vascularisation and deep invagination with a thinning SC has led to a reappraisal of this view. Work has been done on assessing the contribution of the follicular route to drug penetration, as well as targeted delivery [Essa et al 2002, Teichmann et al 2006]. The deposition of these systems in hair follicles was observed by some authors and follicular penetration of solid particles has been demonstrated [Rolland et al 1993, Lademann et al 2009a]. Lademann discussed the finding that 300–600 nm particles penetrated follicles best on massage as a consequence of the distance between the scales on the hairs, and suggested that the movement of the hair acted as a geared pump to push the particles into the follicle [Lademann et al 2009b]. They viewed follicles as an efficient reservoir for nanoparticle-based drug delivery [Lademann et al 2007]. It is possible also that some follicles may be blocked by a "plug" of sebum or closed leading to particle penetration being impeded. Significant penetration of 40 nm nanoparticles beyond the follicles into epidermal cells can occur when the hair sheath has been pulled out [Vogt et al 2006, Ryman-Rasmussen et al 2007, Schroeter 2010, Muller et al 2002].

3.2 Toxicity of nanocarrier systems

Humans have been exposed to nanoparticles throughout their evolutionary phases; however, this exposure has been increased to a great extent in the past century because of the industrial revolution. The growing use of nanotechnology in high-tech industries is likely to become another way for humans to be exposed to intentionally generate engineered nanoparticles. However, the same properties (small size, chemical composition, structure, large surface area and shape), which make nanoparticles so attractive in medicine, may contribute to the toxicological profile of nanoparticles in biological systems [Muller 2000, Koo et al 2005, Vega-Villa et al 2008].

Because of increased use of nanotechnology, the risk associated with exposure to nanoparticles, the routes of entry and the molecular mechanisms of any cytotoxicity need to be well understood.

In fact, these tiny particles (<1000 nm in size) are able to enter the body through the skin, lungs or intestinal tract, depositing in several organs and may cause adverse biological reactions by modifying the physiochemical properties of living matter at the nanolevel [Muller 2000, Monteiro-Riviere & Baroli 2010].

The likelihood of nanoparticle penetration across the skin has recently been reviewed by the Scientific Committee on Consumer Products (SCCP) who conclude that in relation to dermal exposure:

- There is evidence of some skin penetration into viable tissues (mainly into the stratum spinosum in the epidermal layer, but eventually also into the dermis) for very small particles (less than 10 nm), such as functionalised fullerenes and quantum dots.
- When using accepted skin penetration protocols (intact skin), there is no conclusive evidence for skin penetration into viable tissue for particles of about 20 nm and larger primary particle size as used in sunscreens with physical UV-filters.
- The above statements on skin penetration apply to healthy skin (human and porcine). There is an absence of appropriate information for skin with impaired barrier function, e.g. atopic skin or sunburned skin. A few data are available on psoriatic skin.
- There is evidence that some mechanical effects (e.g. flexing) on skin may have an effect on nanoparticle penetration.
- There is no information on the transadnexal structures penetration for particles under 20 nm. Nanoparticles of 20 nm and above penetrate deeply into hair follicles, but no penetration into viable tissue has been observed.

The statement that nanoparticles of 20 nm in diameter do not penetrate into viable tissue is controversial. There have been reports showing nanoparticles 20 nm penetrating through the stratum corneum(SC), to the viable epidermis. The reasons for this important disparity may be due to differences in nanoparticle constituents, models, and methodologies [SCCP opinion 2007].

4. Evaluation of nanoparticles interaction with skin

Among the different methods that exist for the evaluation of the efficacy of topical products, the non-invasive biophysical measurements have the advantages of being non-invasive, non traumatizing, causing minimal discomfort and not altering the skin function. In addition, they permit to detect defined parameters, which in most cases cannot be discriminated by visual scoring [Berardesca et al 1995].

In order to evaluate nanoparticles interaction with the skin and their efficacy after cutaneous application, a number of techniques can be involved, namely:

- Assessment of the cutaneous electrical properties and of the water evaporation through the epidermis (TEWL) in relation to the water content of the outer skin layers;
- Evaluation of the mechanical/visco-elastic properties of the skin (measurement of the skin reaction to friction, torsion and suction)
- Instrumental evaluation of skin topography: surface/texture and desquamation (digital image analysis, silicone replicas and profilometry)
- Analyzing some spectroscopic and optical properties of the skin (ultrasound, near infrared and Raman spectroscopy, confocal microscopy).

4.1 Electrical and mechanical properties of the outer skin layers
4.1.1 Stratum corneum hydration and transepidermal water loss

Skin capacitance represents one of the most popular method to assess the water content in the stratum corneum by an indirect way, rising with the increase of skin hydration/moisturization. However, the relationship between this parameter and the water content of the skin is not linear but rather more complex, as other factors such as ions, the dipolar structure of protein as well as the different strength of water binding to keratin or the presence of hair on the skin surface influence the measurement.

Until now, only a few works are performed on the "in vivo" evaluation of solid lipid nanoparticles efficacy after topical administration and all of them demonstrated that an increase of stratum corneum (SC) hydration was obtained after SLN application [Muller et al 2002, Muller et al 2007].

The mechanism with which nanoparticles can produced this effect could be by the formation of an occlusive film on the skin, by the restoration of the skin lipid barrier or by a humectant effect retaining the water of the formulation avoiding the water evaporation from the deeper layers of the skin. The simultaneous measurement of the transepidermal water loss (TEWL) is very important in order to understand the real mechanism of nanoparticles.

In general, in the healthy skin, there is proportionality in the relation between TEWL and SC hydration. A decrease in TEWL, parallel with an increase in the stratum corneum hydration, is observed after the application of occlusive substances (petrolatum) on the skin. Elevated TEWL is detected immediately after application of moisturizing agents as a consequence of the evaporation of the water incorporated in the cosmetic product and it is not due to the increase in the SC hydration [Darlenski & Fluhr 2011]. For these reasons, a combination of more non-invasive technique is advised as more reliable and accurate approach.

4.1.2 Assessment of the skin mechanical properties

Skin mechanical properties are changing with aging in relation to the alterations of the dermal tissue composition [Smalls et al 2006]. The in vivo mechanical properties of the skin were studied by different methods based mainly on torsion stress and suction.

One of the most common instrument used in dermatological and cosmetic fields is the Cutometer (Courage&Khazaka electronic, Cologne , Germany) equipped with a 2 or 6 mm measuring probe. The time/strain mode used consists with a 2-s application of a constant negative pressure of 500 mbar, followed by a 2-s relaxation period. Each measurement consisted of three consecutive cycles. A typical skin deformation curve is illustrated in Figure 3.

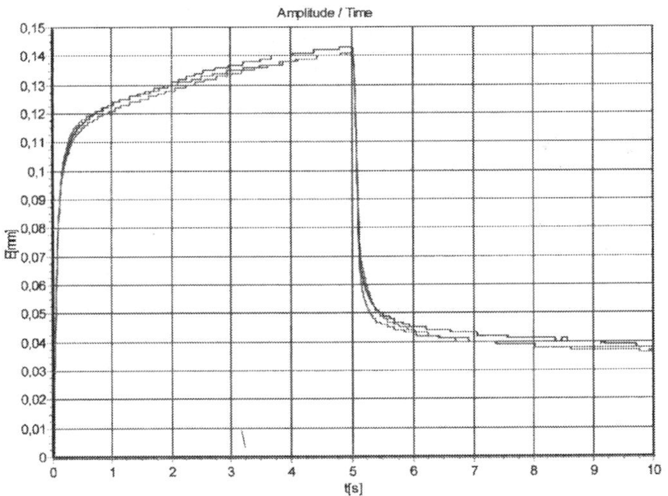

Fig. 3. Typical skin deformation curve

From these curves the following parameters were analyzed: R0 final distension (skin distensibility), R2 skin elasticity including viscous deformation and R5 skin elasticity without viscous deformation.

Until now, only a few work studied the influence of SLN application in changes of skin mechanical properties [Muller et al 2002, Muller et al 2007].

4.2 Instrumental evaluation of the skin topography

The examination of the skin surface topography with its scaling and roughness can be performed by several non-invasive techniques.

4.2.1 Replica

Many testing procedures used to examine the skin require biopsies or some other surgical manipulation. Skin replica is a technique used for the indirect topographic measurement of the skin and the procedure consists in placing a thin layer of silicone based resin on test skin surface in order to obtain an imprint [Ryan et al 1983, Hatzis 2004].

Replica technique is a non-invasive procedure developed for replicating human skin. Skin replica is simple, accurate and allows for records of skin condition to be kept [Hatzis 2004]. Nevertheless the use of this technique requires systematic methodology development and rigorous adherence to experimental protocol, as artefacts can be readily introduced [Ryan et al 1983]. Therefore, the quality of replicas is a limiting factor for the accuracy of the visualization. If the silicone is spilled over the borders of the ring or if the ring's thickness varies or if air bubbles are present in the gel the measurement results are heavily influenced. For these reasons is fundamental to chose a fluid material enough to penetrate the furrows, but not excessively fluid as to avoid overflowing outside the interested area. Furthermore a constant pressure of about the gel applied on the skin must be applied in order to obtain replicas with reproducible thickness [Bogner et al 2007].

Perugini et al demonstrated how the association of skin replica and Scanning Electron Microscopy (SEM) can be used for in vivo visualization of solid lipid nanoparticles on the skin surface [Perugini et al, 2011]. The results highlighted that particle with different matrix composition had a different behaviour. They studied two SLN suspension: the first contained phosphatidylcholine in the lipophilic phase of nanoparticles (BR1) while the second formulation contained cetearyl glucoside, a nonionic surfactant (TE1). The last formulation is, therefore, more apolar and could be more similar to the silicone material of the replica (Figure 4), for this reason nanoparticles seem to be on surface or incorporated into the replica material.

In the same study the behaviour of lipid systems applied on the skin of volunteers with different age was investigated. In the case of the child skin (8 years old) the particles seem to be distributed mainly on the microreliefs of the silicone resina: primary and secondary lines of the skin [Hashimoto 1974]. The behaviour changes when we observe with the same magnification the replica of the woman (38 years old). In this replica the particles distribution seems more homogeneous and particles are present both in relieves and in furrows (Figure 5).

To explain these results it's useful to remember that skin relieves reflect the organization of the deeper layers of the skin and changes at the level of the dermis will affect skin relieves. Skin of a child and skin of a woman are very different. The number of primary and secondary lines decreases progressively with age increasing and the polygonal surface delimited by these lines becomes less homogeneous and defined [De Paepe et al 2000]. These changes could be the reason of the different system distribution on the skin surface.

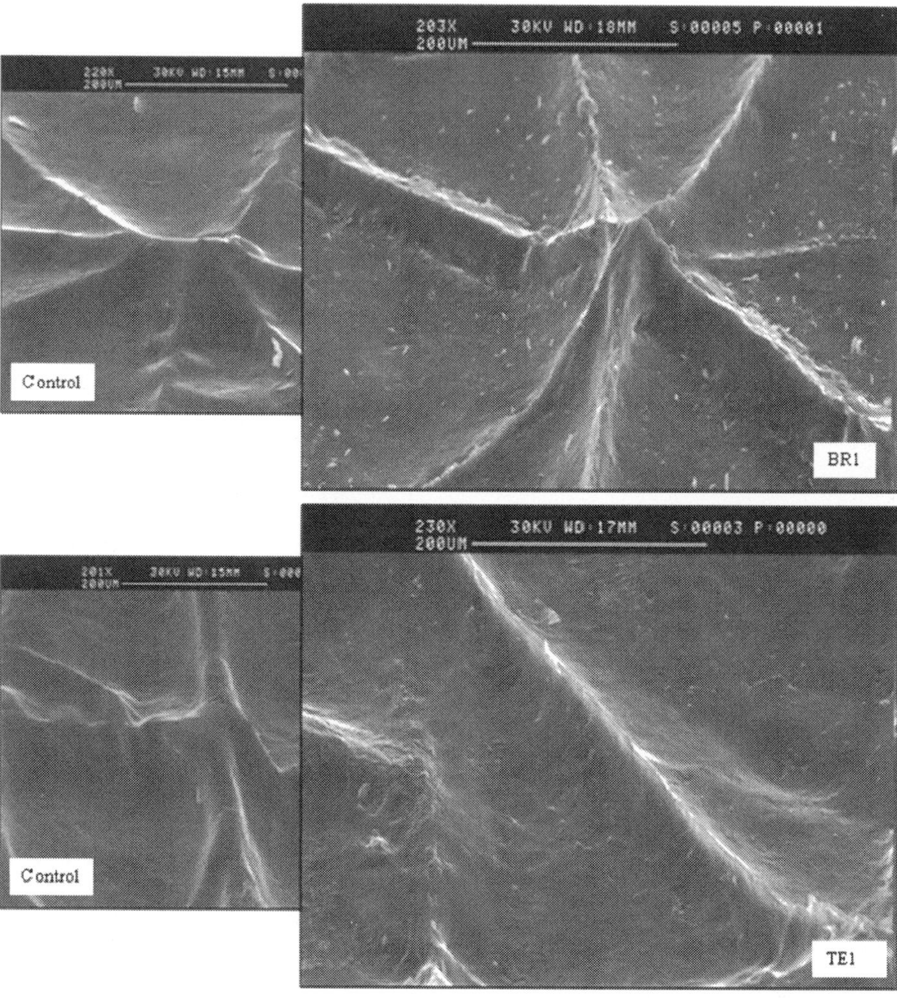

Fig. 4. Scanning electron micrographs of replicas obtained 1 h after solid lipid nanoparticle (batches BR1 and TE1) application (on the right) and the corresponding control replicas (on the left).

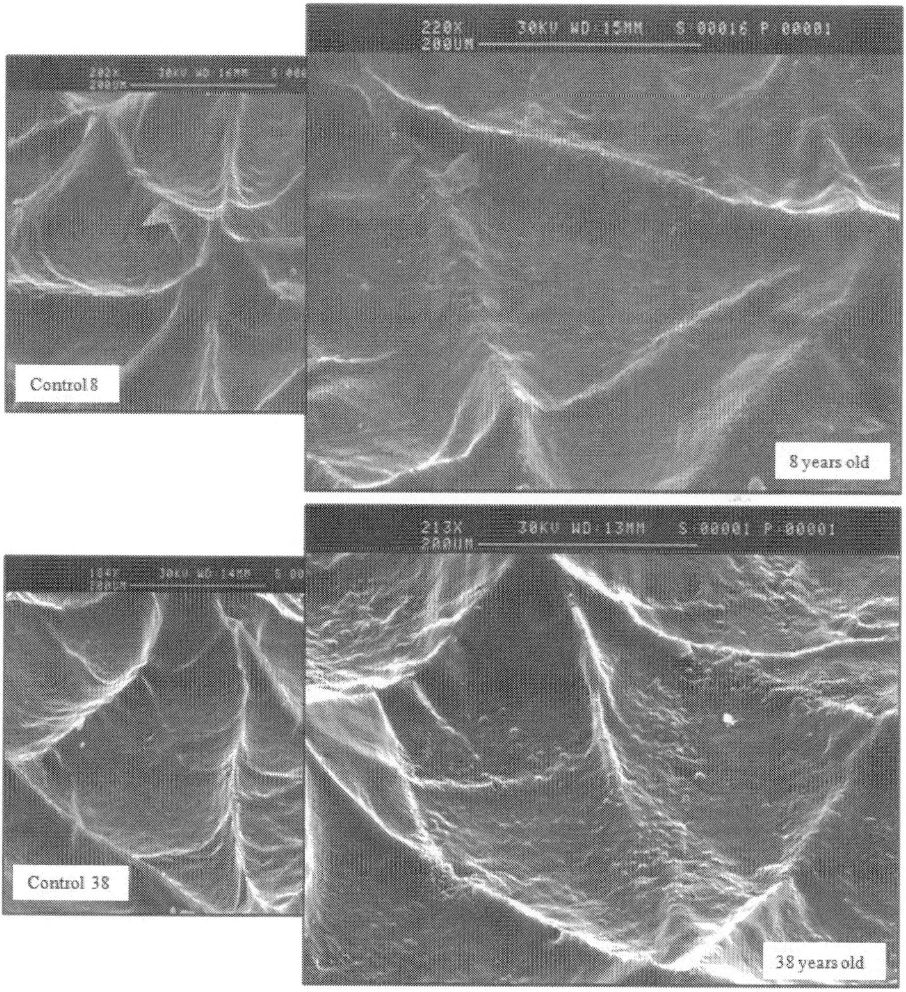

Fig. 5. Scanning electron micrographs of replicas obtained after the application of TE systems on the skin of an 8-year-old child and a 38-year-old woman.

4.2.2 Three-dimensional (3D) imaging of the skin

Three-dimensional (3D) imaging is a new measuring method in medicine. It uses optical projections, a high resolution video camera, and computer software to rapidly generate images and measurements of skin topography. The method has been used successfully in wound care, dermatological laser treatments and its fundamental physics have been validated.

The skin surface structure can to be analyze in the non-contact mode using the 3D optical system Primos (GFMesstechnik GmbH, Teltow, Germany) as described in detail by Jacobi et al. [Bloemen et al 2011]. This system is based on the digital stripe projection technique, which is used as an optical measurement process. A parallel stripe pattern is projected onto the skin surface and depicted on the CCD chip of a camera through an optical system. The measurement system consists of a freely movable optical measurement head (with an integrated micro-mirror projector, a projection lens system, and a CCD recording camera), together with an evaluation computer. The 3D effect is achieved by the minute elevation differences on the skin surface, which deflect the parallel projection stripes. The measurements of these deflections represent qualitative and quantitative measurements of the skin profile. The instrument determine the roughness, which is based on the depth and the density of the furrows and wrinkles of the skin.

The average roughness (Ra) is the mathematical average value of profile amounts within the total measuring length, and represents the roughness of the skin surface structure (Figure 6). High roughness values corresponded to deep furrows and wrinkles with a high density. This parameter is based on the sampling of all the points characterizing the profile, so it represents a true average because it is very significant in regard to the roughness of the skin.

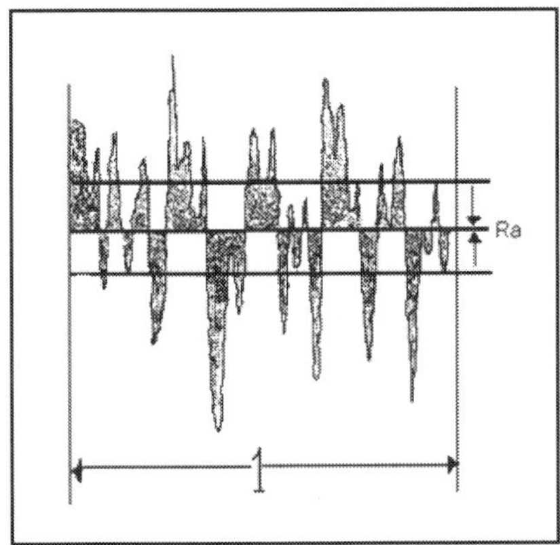

Fig. 6. Roughness mean value (Ra)

The Ra is two-dimensional parameter. For skin with very anisotropic characteristics, the value of this parameter can be very different depending on the direction along which the profile being analyzed (Figure 7). Ra has a higher value in perpendicular orientation to respect to the grooves (X direction in the figure 2); in fact, Rax> Ray. It must, therefore, to process simultaneously a large number of profiles that follow all possible directions and calculate the average of several measurements obtained.

The number of lines to be measured is an important factor; it is therefore very important to try as many lines as possible, to encompass the entire area of the captured image in the analysis; the higher the number of available profiles, the greater will be the precision of measurement.

Furthermore the Primos system evaluate, by matching, the same skin area in the different subsequent analysis times. 3D image of the skin, before and after SLN application, could be used to study SLN behavior. In order to evaluate the distribution of particle systems with different size on human skin, a 3D skin images before and after application of microparticles and nanoparticles formulations, were acquired.

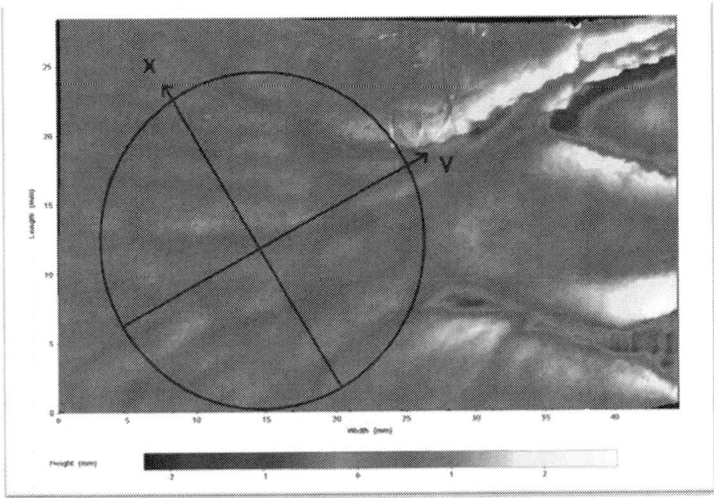

Fig. 7. Microrelief image. In particular directions along which the profile being analyzed

Microparticles seem to be mainly distributed in the furrows filling the microrelieves. In this case the value Ra after microparticles application (Ra= 36 µm) decrease considerably compared to skin no treated (Ra = 65 µm), as shown in the Figure 8.

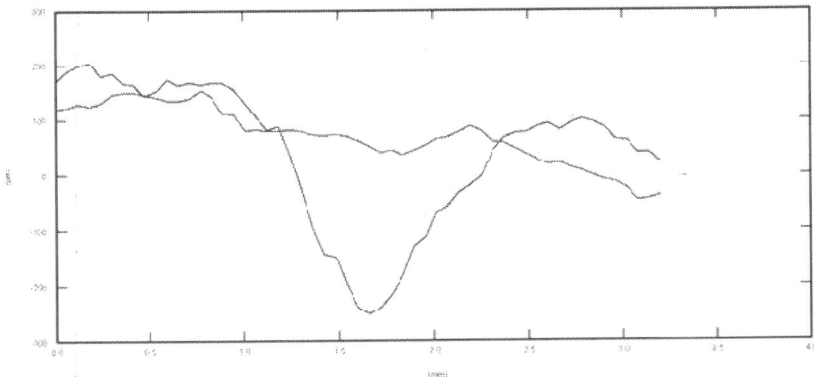

Fig. 8. Single profile of microrelief before (black curve) and after 1 minutes (blue curve) of microparticles application.

On the contrary, Ra value acquisition after nanoparticles application doesn't reveal a significant modification of the average roughness of microrelief (Ra before SLN application = 30 µm and Ra after SLN application = 38 µm) . This result could be probably due to the SLN nanosize that permits an homogeneous distribution both in relieves and in furrows supporting the film forming theory about the solid lipid nanoparticle interaction with the skin after cutaneous application (Figure 9).

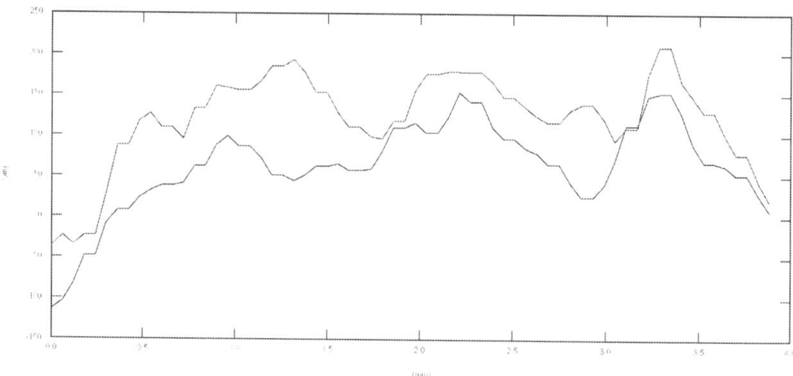

Fig. 9. Single profile of microrelief before (black curve) and after 1 minutes (blue curve) of solid lipid nanoparticle application.

The SLN form an adhesive film on the skin the furrows became less depth and the roughness parameters should reduce. Otherwise, penetration of the SLN in the stratum corneum through different routes should not lead to changes in roughness parameters. This hypothesis should be further investigated by combining this technique with other non-invasive methods of skin analysis.

4.3 High-frequency ultrasound and skin

The high-frequency scanners available today operate at frequencies between 20 MHz and 1-2 GHz. The optimal frequency range for dermatological questions is between 20 and 100 MHz [Gammal et al 1995].

Using a 20 MHz transducer it is possible to visualize structures up to 6-7 mm in depth. This means that the zones of diagnostic interest are covered i.e. epidermis, corium, and one portion of subcutaneous fatty tissue. Particularly, if the subcutis is not very well developed, evaluation of the muscle fasciae is also possible (Figure 10) .

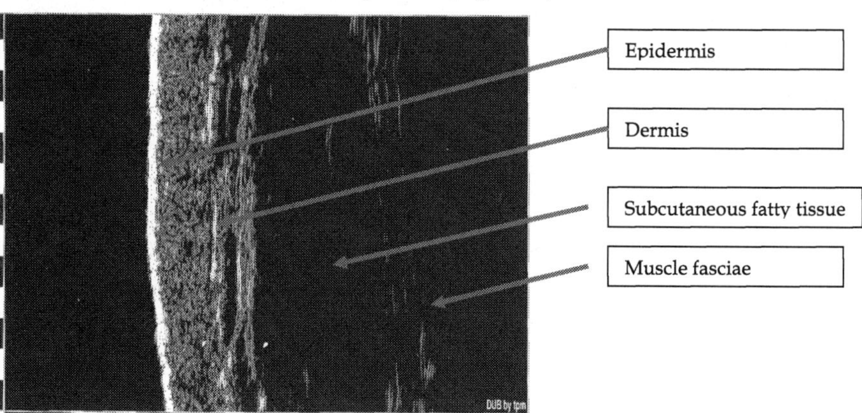

Fig. 10. Ultrasound B scan of a back skin of a 32 year old woman.

Until now, no one has studied the interaction of nanoparticle systems with the skin by ultrasound scanning. Instead, the use of ultrasound may be helpful in investigating the effects of SLN application on the different layers of the skin.

4.4 Skin microcirculation

Skin microcirculation is rather complex. The skin viability depends upon the nutritional circulation, which cannot be assessed by conventional macrocirculatory methods that evaluate total blood supply. The indisputable advantage of the microcirculatory methods is to provide information directly in diseased skin areas and assess the effectiveness of vasoactive drugs where they are supposed to act.

Several techniques are available today to evaluate the skin microcirculation. Among them, capillaroscopy and transcutaneous measurement of the partial oxygen pressure are of special interest because they provide information which is directly useful in clinical practice [Li et al 2006].

The blood supply to the skin is provided by a network of arterioles, capillaries and venules organized into a superficial and a deep plexus. The assessment of skin microcirculation is of valuable interest in cosmetology in the quantification of the sun protection factor, skin irritation and efficacy of antiredness treatments. Skin microcirculation can be measured by means of different techniques, based mainly on the quantification of optical and thermal properties of the skin which are modified by the amount of blood perfusion. Relevant and reproducible data can be obtained only through the understanding of the biophysical background of the technique(s) utilized. Standardization of measuring conditions and procedures is particularly required for blood flow assessment. [Berardesca et al., 2002]

5. Conclusion

Micro and nano sytems are gaining great attention in biomedical applications such as design of drug carrier devices. Among them SLN have lots of advantages over conventional systems since they enhance the delivery, extend the bioactivity of the drug by protecting them from environmental effects in biological media, show minimal side effects, demonstrate high performance characteristics, and are more economical since minimum amount of expensive drugs are used.

After skin application of SLN three pathways of SLN penetration across the deeper layers of skin have been identified. That can to be exploited in order to promote the penetration of active ingredients into the dermis; in this way the active can to carry out a direct therapeutic and cosmetic action.

Potential systemic effects after dermal application of SLN should be considered in order to obtain a safe topical product. Skin penetration and systemic absorption should be estimated with the intention to assess the risk of using nanoparticles in topical products. Visualization of colloidal systems after skin application is essential to evaluate their interaction with cutaneous structures.

Non-invasive bioengineering techniques have become indispensable tools both in the development of drugs and cosmetics and in clinical dermatology. These techniques enable researchers to study the structure and function of human skin objectively and quantitatively.

These methods should help scientists working on formulations containing SLN in order to better understand the fate of nanoparticles after topical application and the effectiveness of their application to skin disorders and diseases.

6. References

(2007). Sccp- Opinion on safety nanomaterials in cosmetic products, public consultation on the 14th plenary of 18. (December 2007), pp. (1-63), European Commission

Al-Amoudi, A., Dubochet, J., Norle'n, L. (2005). Nanostructure of the epidermal extracellular space as observed by cryo-electron microscopy of vitreous sections of human skin. *The journal of investigative dermatology,* Vol. 124, No. 4, (April 2005), pp. (764–777), ISSN 0022-202X

Baroli, B. (2010). Penetration of nanoparticles and nanomaterials in the skin: fiction or reality?. *Journal of pharmaceutical sciences,* Vol. 99, No. 1, (January 2010), pp. (21-50), 0022-3549

Baroli, B., Ennas, M.G., Loffredo, F., Isola, M., Pinna, R., Lopez-Quintela, M.A. (2007). Penetration of metallic nanoparticles in human full-thickness skin. *The Journal of Investigative Dermatology,* Vol. 127, No. 7, (July 2007), pp. (1701–1712), ISSN 0022-202X

Bawa, R. (2005). Patent and nanomedicine. *Nanomedicine,* Vol. 2, No. 3, (June 2007), pp. (351-374), ISSN 1549-9634

Berardesca E., Elsner, P., Wilhelm, K.P., Maibach, H.I. (1995), Bioengineering of the skin: methods and Instrumentation. ISBN 0849383749 CRC Boca Raton.

Berardesca, E., Lévêque, J., Masson , P. (2002) . EEMCO Guidance for the Measurement of Skin Microcirculation. *Skin Pharmacology and Applied Skin Physiology,* Vol. 15 , No. 6 , (2002) , pp. (442-456), ISSN 1422-2868

Bloemen, M.C., van Gerven, M.S., van der Wal , M.B., Verhaegen, P.D., Middelkoop, E. (2011). An objective device for measuring surface roughness of skin and scars. *Journal of the American Academy of Dermatology ,* Vol. 64 , No. 4 , (April 2011) , pp. (706-15) , ISSN 0190-9622

Bogner, A., Jouneau ,PH., Thollet, G., Basset, D., Gauthier ,C. (2007) . A history of scanning electron microscopy developments: towards "wet-STEM" imaging. *Micron, Vol.38, No.4,* (2007) , pp. (390-401) , ISSN 0968-4328

Bouwstra, J.A., Dubbelaar , F.E.R. , Gooris G.S. , Ponec M. (2000). The lipid organization in the skin barrier. *Acta Dermato -Venereologica ,* Vol. 208, (July 2000), pp. (23-30), ISSN 0001-5555

Bouwstra, J.A., Honeywell-Nguyen, L., Gooris, G.S., Ponec, M. (2003). Structure of the skin barrier and its modulation by vesicular formulations. *Progress in lipid research,* Vol. 42, No. 1, (January 2003), pp. (1-36), ISSN 0163-7827

Bouwstra, J.A., Pilgrim, K., Ponec, M. (2006). Structure of the skin barrier, In: *Skin Barrier,* Elias, P.M., Feingold, K.R., pp. (65-95), Ed.: Taylor and Francis, ISBN 9780824758158, New York

Brandner, J. M., Kief, S. , Wladykowski, E., Houdek, P., Moll, I. (2006a). Tight junction proteins in the skin. *Skin pharmacology and physiology ,* Vol. 19 , No. 2 , (May 2006), pp. (71–77), ISSN 1660-5527

Brandner, J. M., Proksch, E. (2006b). Epidermal barrier function: role of tight junctions, In: *Skin Barrier,* Elias P. M., Feingold, K. R., pp. (191–210), Taylor and Francis, ISBN 9780824758158, New York

Bunjes, H. (2006). Manufacture, characterization and Applications of Solid Lipid Nanoparticles as Drug Delivery Systems, In: *Drugs and the pharmaceutical Sciences*, Gupta, R.M., Kompella, U.B., Vol. 158, pp. (213-268), Ed. Taylor & Francis group, ISBN 1574448579.

Buse, J., El-Aneed, A. (2010). Properties, engineering and applications of lipid-based nanoparticle drug-delivery systems: current research and advances. *Nanomedicine*, Vol. 5, No. 8, (October 2010), pp. (1237-60), ISSN 1549-9634

Candi, E., Schmidt, R., Melino, G. (2005). The cornified envelope: a model of cell death in the skin. *Nature Reviews Molecolar cell biology*, Vol. 6, No. 4, (April 2005), pp. (328–340), ISSN 1471-0072

Cevc, G. (2004). Lipid vesicles and other colloids as drug carriers on the skin. *Advanced drug delivery reviews*, Vol. 56, No. 5, (March 2004), pp. (675-711), ISSN 0169-409X

Charcosset, C., El-Harati, A., Fessi, H. (2005). Preparation of solid lipid nanoparticles using a membrane contactor. *Journal of controlled release*, Vol. 108, No. 1, (September 2005), pp. (112–120), ISSN 0168-3659

Chen, H., Chang, X., Du, D., Liu, W., Liu, J., Weng, T., Yang, Y., Xu, H., Yang, X. (2006). Podophyllotoxin-loaded solid lipid nanoparticles for epidermal targeting. *Journal of Controlled Release*, Vol. 110, No. 2, (January 2006), pp. (296–306), ISSN 0168-3659

Choi, SH., Jin, SE., Lee, MK., Lim, SJ., Park, JS., Kim, BG., Ahn, WS., Kim, CK. (2008). Novel cationic solid lipid nanoparticles enhanced p53 gene transfer to lung cancer cells. *European Journal of Pharmaceutics and Biopharmaceutics*, Vol. 68, No. 3, (March 2008), pp. (545–554), ISSN 0939-6411

Corsini, E., Primavera, A., Marinovich, M., Galli, CL. (1998) Selective induction of cell-associated interleukin-1alpha in murine keratinocytes by chemical allergens, *Toxicology*, Vol. 129, No. 2-3, (Aug 1998), pp. (193-200), ISSN 0300-483X.

Darlenski, R., Fluhr, J.W. (2011), Moisturizers and emollients, In: Practical aspects of cosmetic testing, Joachim W. Fluhr Editor, pp (123-141), Springer, ISBN 978-3-642-05066-4, Berlin.

De Paepe, K., Lagarde, J.M., Gall, Y., Roseeuw, D., Rogiers , V. (2000) . Microrelief of the skin using a light transmission method. *Archives of Dermatological Research* , Vol. 292 , No. 10, (October 2000) , pp. (500-10) , ISSN 0340-3696

Desai, P., Patlolla, R.R., Singh, M. (2010). Interaction of nanoparticles and cell-penetrating peptides with skin for transdermal drug delivery. *Molecular Membrane Biology*, Vol. 27, No. 7 (October 2010), pp. (247-59), ISSN 0968-7688

Downing, D.T., Stewart, M.E., Wertz, P.W., Colton, S.W., Abraham, W., Strauss, J.S. (1987). Skin lipids: an update. *The journal of investigative dermatology*, Vol. 88, No. 3, (March 1987), pp. (2s–6s), ISSN 0022-202X

El-Harati, A.A., Charcosset, C., Fessi, H. (2006). Influence of the formulation for solid lipid nanoparticles prepared with a membrane contactor. *Pharmaceutical Development and Technolology*, Vol. 11, No. 2, (2006), pp. (153–157), ISSN 1083-7450

Essa, E.A., Bonner, M.C., Barry, B.W. (2002). Human skin sandwich for assessing shunt route penetration during passive and iontophoretic drug and liposome delivery. *The Journal of Pharmacy and Pharmacology*, Vol. 54, No. 11, (November 2002), pp. (1481–1490)

Forslind ,B. (1994). A domain mosaic model of the skin barrier. *Acta Dermato-Venereologica* , Vol. 74, No. 1, (January 1994), ISSN 0001-5555

Gammal, S., Auer, T., Hoffman, K., Altmeyer, P., (1995) High-resolution ultrasound of human epidermis, In Handbook of noninvasive methods and the skin, vol 1, J. Serup and G. Jemec Eds, pp (125-131), CRC Press, Boca Raton.

Garcy´-Fuentes, M., Torres, D., Alonso,M. (2002). Design of lipid nanoparticles for the oral delivery of hydrophilic macromolecules. *Colloid Surfaces B. Biointerfaces*, Vol. 27, No. 2-3, (February 2003), pp. (159–168), ISSN 0927-7765

Gasco, M.R. (1993). Method for producing solid lipid microspheres having a narrow size distribution. *USA Patent 5 250 236*, (May 1993), 188-837.

Gasco, M.R. (1997). Solid lipid nanospheres from warm microemulsion. *Pharmaceutical Technology Europe*, Vol. 9, (1997), pp. (32–42), ISSN 1753-7967

Grice, J.E., Ciotti,S., Weiner, N., Lockwood, P., Cross, S.E., Roberts, M.S. (2010). Relative uptake of minoxidil into appendages and stratum corneum and permeation through human skin in vitro. *Journal of Pharmaceutical Sciences*, Vol. 99, No. 2, (February 2010),pp. (712–718), ISSN 0022-3549

Hashimoto, K. (1974). New methods for surface ultrastructure: Comparative studies of scanning electron microscopy, transmission electron microscopy and replica method. *Internal Journal of Dermatolology*, Vol. 13, No.6 , (1974) , pp. (357-81), ISSN 0011-9059

Haskell, R.J., (2006). Physical Characterization of Nanoparticles, In: *Drugs and the Pharmaceutical Sciences*, Beuita, S., Vol. 159, pp. (103-138), Ed. Taylor & Francis group, ISBN 9780849374555, New York

Hatzis, J. (2004). The wrinkle and its measurement - A skin surface Profilometric method. *Micron*, Vol. 35, No. 3, (2004), pp.(201-219), ISSN 0968-4328

Heurtault, B., Saulnier, P., Pech, B., Proust, J.E., Benoit, J.P. (2002). A novel phase inversion-based process for the preparation of lipid nanocarriers. *Pharmaceutical Research*, Vol. 19, No. 6, (June 2002), pp. (875–880), ISSN 0724-8741

Honari, S. (2004). Topical therapies and antimicrobials in the management of burn wounds. *Critical Care Nursing Clinics of North America* , Vol. 16, No. 1, (March 2004) ,pp. (1-11) , ISSN 0899-5885

Hu, F.Q., Yuan, H., Zhang, H.H., Fang, M. (2002). Preparation of solid lipid nanoparticles with clobetasol propionate by a novel solvent diffusion method in aqueous system and physicochemical characterization. *International Journal of Pharmaceutics*, Vol. 239, No. 1-2, (June 2002), pp. (121–128), ISSN 0378-5173

Jahnke, S. (1998). The theory of high pressure homogenization, In: *Emulsions and nanosuspensions for the formulation of poorly soluble drugs*, Muller, R.H., Benita, S., Bohm, B., pp. (177– 200), Medpharm Scientific Publishers ISBN 3887630696, Stuttgart.

Koch, P.J., Franke, W.W. (1994). Desmosomal cadherins: another growing multigene family of adhesion molecules. *Current opinion in cell biology*, Vol. 6, No. 5, (October 1994), pp. (682–687), ISSN 0955-0674

Koo, O.M., Rubinstein, I., Onyuksel, H. (2005). Role of nanotechnology in targeted drug delivery and imaging: a concise review. *Nanomedicine*, Vol. 1, No. 3 (September 2005), pp. (193-212), ISSN 1549-9634

Lademann, J., Patzelt, A., Richter, H., Antoniou, C., Sterry, W., Knorr, F. (2009b) Determination of the cuticula thickness of human and porcine hairs and their potential influence on the penetration of nanoparticles into the hair follicles. *Journal of Biomedical Optics*, Vol. 14 ,No. 2, (April 2009), pp. (10-14)

Lademann, J., Richter, H., Teichmann, A., Otberg, N., Blume-Peytavi, U., Luengo, J., Weiss, B., Schaefer, U. F., Lehr, C.M., Wepf, R., Sterry, W. (2007). Nanoparticles—an efficient carrier for drug delivery into the hair follicles. *European Journal of Pharmaceutics and Biopharmaceutics*, Vol. 66 ,No. 2, (May 2007), pp. (159–164), ISSN 0939-6411

Ladermann, J., Meinke, M., Sterry, W., Patzelt, A. (2009a). How safe are nanoparticles?. *Hautarzt*, Vol. 60, No. 4, (April 2009), pp. (305-9), ISSN 0017-8740

Li,L., Mac-Mary, S., Marsaut, D., Sainthillier, J.M., Nouveau, S., Gharbi, T. (2006) Age related changes in skin topography and microcirculation, *Arch. Dermatol. Res.*, Vol297, No. 9, pp (412-416), ISSN 0340-3696

Liedtke, S., Wissing, S., Müller, RH., Mäder, K., (2000). Influence of high pressure homogenisation equipment on nanodispersions characteristics. *International Journal of Pharmaceutics*, Vol. 196, No. 2, (10 March 2000), pp. (183–185), ISSN 0378-5173

Lombardi Borgia, S., Regehly, M., Sivaramakrishnan, R., Mehnert, W., Korting, H.C., Danker, K., Röder, B., Kramer, K.D., Schäfer-Korting, M. (2005). Lipid nanoparticles for skin penetration enhancement-correlation to drug localization within the particle matrix as determined by fluorescence and parelectric spectroscopy. *Journal of Controlled Release*, Vol. 110, No. 1 (December 2005), pp. (151–163), ISSN 0168-3659

Madison, KC., (2003). Barrier function of the skin: "la raison d'être" of the epidermis. *The journal of investigative dermatology*, Vol. 121, No. 2, (August 2003), pp. (231-41), ISSN 0022-202X

Medina, C., Santos-Martinez, M.J., Radomski, A., Corrigan, O.I., Radomski, M.W. (2007). Nanoparticles: pharmacological and toxicological significance. *British Journal of Pharmacology*, Vol. 150, No. 5, (March 2007) ,pp. (552-8), ISSN 0007-1188

Mehnert, W., Mäder, K. (2001). Solid lipid nanoparticles: production, characterization and applications. *Advanced Drug Delivery Reviews*, Vol. 47, No. 2-3, (April 2001), pp. (165-96), ISSN 0169-409X

Miiller, R.H. (2000). Nanoparticles (SLN) as a carrier system for the controlled release of drugs, In: *Handbook of Pharmaceutical Controlled Release*, Wise, D.L., pp. (377), Marcel Dekker, New York

Monteiro-Riviere, N. A., Baroli, B. (2010). Nanomaterial penetration, In: *Toxicology of the Skin*, N.A. Monteiro- Riviere , pp. (333–346), Informa Healthcare USA, Inc., New York, NY

Muller, R.H., Peterson R.D., Hommoss, A., Pardeike, J. (2007) Nanostructured lipid carriers (NLC) in cosmeticc dermal products, Advances Drug Delivery Reviews, Vol 59, pp (522-530), ISSN 0169-409X.

Müller, R.H., Radtke, M., Wissing, S.A. (2002). Solid lipid nanoparticles (SLN) and nanostructured lipid carriers (NLC) in cosmetic and dermatological preparations. *Advanced Drug Delivery Reviews*, Vol. 54, No. 1, (November 2002), pp. (131-55), ISSN 0169-409X

Nemes, Z., Steinert, P.M. (1999). Bricks and mortar of the epidermal barrier. *Experimental and Molecular Medicine*, Vol. 31, No. 1, (March 1999), pp. (5–19), ISSN 1226-3613

Norlen , L. (2001). The lipid organization in the skin barrier. *Journal of Investigative Dermatology*, Vol. 117, pp. (830-836)

Palmer, C.N., Irvine, A.D., Terron-Kwiatkowski, A., Zhao, Y., Liao, H., Lee, S.P., Goudie, D.R., Sandilands, A., Campbell, L.E., Smith, F.J., O'Regan, G.M., Watson, R.M.,

Cecil, J.E., Bale, S.J., Compton, J.G., DiGiovanna, J.J., Fleckman P, Lewis-Jones S, Arseculeratne G, Sergeant A, Munro C.S., El Houate B, McElreavey K, Halkjaer L.B., Bisgaard, H., Mukhopadhyay, S., McLean, W.H. (2006). Common loss-of-function variants of the epidermal barrier protein filaggrin are a major predisposing factor for atopic dermatitis. *Nature Genetics*, Vol. 38, No. 4, (April 2006), pp. (441–446), ISSN 1061-4036

Pardeike J., Hommoss, A., Müller, RH. (2009). Lipid nanoparticles (SLN, NLC) in cosmetic and pharmaceutical dermal products. *International Journal of Pharmaceutics, Vol. 366*, No. 1-2, (January 2009), pp. (170-84), ISSN 0378-5173

Perugini, P., Bruni, G., Vettor, M., Bleve, M., Corsini, E., Pavanetto, F. (2010a). Preparation, Characterization and in Vitro Tolerability of SLN intended for the regeneration of hyaluronan into skin preliminary investigation, Proceedings of *7th World meeting on Pharmaceutics, Biopharmaceutics and Pharmaceutical Technology*, Malta, March 2010

Perugini, P., Tomasi, C., Vettor, M., Dazio, V., Conti, B., Genta, I., Pavanetto, F. (2010b). Influence of SLN matrix modification on "in vitro" and "in vivo" nanoparticle performances. *International Journal of Pharmacy and Pharmaceutical Sciences*, Vol. 2, No. 3 , (February 2010) , pp. (37-42) , ISSN 0975-1491

Perugini, P., Vettor, M., Bleve , M., Bruni , G., Mondelli, A., Secchi , G.F., Pavanetto, F. (2011). Preliminary evaluation of particle systems visualization on the skin surface by scanning electron microscopy and transparency profilometry. *Skin Research and Technology* , doi: 10.1111/j.1600-0846.2011.00529.x. , ISSN 0909-752x

Pietkiewicz, J., Sznitowska, M., (2004). The choice of lipids and surfactants for injectable extravenous microspheres. *Pharmazie*, Vol. 59, No. 4, (April 2004), pp. (325–326), ISSN 0031-7144

Potts, R.O., Guy, R.H. (1995). A predictive algorithm for skin permeability: the effects of molecular size and hydrogen bond activity. *Pharmaceutical Research*, Vol. 12, No. 11, (November 1995), pp. (1628–1633), ISSN 0724-8741

Priano, L., Esposti, D., Esposti, R., Castagna, G., De Medici, C., Fraschini, F., Gasco, M.R., Mauro, A. (2007). Solid lipid nanoparticles incorporating melatonin as new model for sustained oral and transdermal delivery systems. *Journal of nanoscience and nanotechnology*, Vol. 7, No. 10, (October 2007), pp. (3596–3601), ISSN 1533-4880

Proksch, E., Brandner, JM., Jensen, JM. (2008). The skin: an indispensable barrier. *Experimental dermatology*, Vol. 17, No. 12, (December 2008), pp. (1063-72), ISSN 0906-6705

Prow, TW., Grice, J.E., Lin, L.L., Faye, R., Butler, M., Becker, W., Wurm, E.M., Yoong, C., Robertson, T.A., Soyer , H.P., Roberts, M.S. (2011). Nanoparticles and microparticles for skin drug delivery. *Advanced Drug Delivery Reviews*, (February 2011), ISSN 0169-409X

Puglia, C., Blasi, P., Rizza, L., Schoubben, A., Bonina, F., Rossi, C., Ricci, M. (2008). Lipid nanoparticles for prolonged topical delivery: an in vitro and in vivo investigation. *International Journal of Pharmaceutics*, Vol. 357, No. 1-2, (February 2008), pp. (295–304), ISSN 0378-5173

Pummi, K., Malminen, M., Aho, H., Karvonen, S.L., Peltonen, J., Peltonen, S. (2001). Epidermal tight junctions: ZO-1 and occludin are expressed in mature, developing, and affected skin and in vitro differentiating keratinocytes. *The journal of investigative dermatology*, Vol. 117, No. 5, (November 2001), pp. (1050–1058), 0022-202X

Reddy, L., Vivek, K., Bakshi, N., Murthy, RS., (2006). Tamoxifen citrate loaded solid lipid nanoparticles (SLN™): preparation, characterization, *in vitro* drug release, and pharmacokinetic evaluation. *Pharmaceutical Development and Technology*, Vol. 11, No. 2, (2006), pp. (167–177), ISSN 1083-7450

Rizwan, M., Aqil, M., Talegaonkar, S., Azeem, A., Sultana, Y., Ali. A. (2009). Enhanced transdermal drug delivery techniques: an extensive review of patents. *Recent patents on drug delivery & formulation*, Vol. 3, No. 2, (2009), pp. (105-24), ISSN 1872-2113

Rolland, A., Wagner, N., Chatelus, A., Shroot, B., Schaefer, H. (1993). Site-specific drug delivery to pilosebaceous structures using polymeric microspheres. *Pharmaceutical Research*, Vol. 10, No. 12, (December 1993), pp. (1738-44), ISSN 0724-8741

Roop, D. (1995). Defects in the barrier. *Science*, Vol. 267, No. 5197, (27 January 1995), pp. (474–475), 0036-8075

Ryan, RL., Hing ,SAO., Theiler ,RF. (1983). A replica technique for the evaluation of human skin by scanning electron microscopy. *Journal of Cutaneous Pathology* , Vol. 10 , No. 4 , (August 1983) , pp. (262-276), ISSN 0303-6987

Ryman-Rasmussen, J.P., Riviere, J.E., Monteiro-Riviere, N.A. (2007). Variables influencing interactions of untargeted quantum dot nanoparticles with skin cells and identification of biochemical modulators. *Nano Letters*, Vol. 7, No. 5, (May 2007) , pp. (1344–1348), ISSN 1530-6984

Sahoo, S.K., Labhasetwar, V. (2003). Nanotech approaches to drug delivery and imaging. *Drug Discovery Today*, Vol. 8, No. 24, (December 2003), pp. (1112-20), ISSN 1359-6446

Sawant, K.K., Dodiya S.S. (2008). Recent advances and patents on solid lipid nanoparticles. *Recent patents on drug delivery and formulation*, Vol. 2, No. 2, (2008), pp. (120-35), ISSN 1872-2113

Scheuplein, R.J. (1965). Mechanism of percutaneous adsorption. I. Routes of penetration and the influence of solubility. *The journal of Investigative Dermatology*, Vol. 45, No. 5 (November 1965), pp. (334–346), ISSN 0022-202X

Schroeter, A. (2010). New nanosized technologies for dermal and transdermal drug delivery. A review. *Journal of biomedical nanotechnology*, Vol. 6, No. 5, (October 2010), pp. (511-28)

Schubert, M.A., Müller-Goymann, C.C. (2003). Solvent injection as a new approach for manufacturing lipid nanoparticles—evaluation of the method and process parameters. *European journal of Pharmaceutics and Biopharmaceutics*, Vol. 55, No. 1, (January 2003), pp. (125–131), ISSN 0939-6411

Sjöström, B., Bergenstahl, B. (1992). Preparation of submicron drug particles in lecithin-stabilized o/w emulsions. I. Model studies of the precipitation of cholesteryl acetate. *International Journal of Pharmaceutics*, Vol. 88, No. 1-3, (December 1992), pp. (53–62), ISSN 0378-5173

Smalls, L.K., Wickett, R.R., Visscher, M.O. (2006) Effect of dermal thickness, tissue composition, and body site on skin biomechanical properties, Skin research and Technology, Vol 12, pp (43-49), ISSN 0909-752x

Souto, E.B., Müller, R.H. (2008). Cosmetic features and applications of lipid nanoparticles (SLN, NLC). *International Journal of Cosmetic Science*, Vol. 30, No. 3, (June 2008), pp. (157-165), ISSN 0142-5463

Stern, S.T., McNeil, SE. (2008). Nanotechnology Safety Concerns Revisited. *Toxicological Sciences*, Vol. 101, No. 1, (June 2007), pp.(4–21), ISSN 1096-6080

Teeranachaideekul, V., Boonme, P., Souto, E.B., Müller, R.H., Junyaprasert, V.B. (2008). Influence of oil content on physicochemical properties and skin distribution of Nile red-loaded NLC. *Journal of Controlled Release*, Vol. 128, No. 2 , (June 2008), pp. (134–141), ISSN 0168-3659

Teichmann, A., Otberg, N., Jacobi, U., Sterry, W., Lademann, J. (2006). Follicular penetration: development of a method to block the follicles selectively against the penetration of topically applied substances. *Skin Pharmacology and Physiology*, Vol. 19, No.4, (2006), pp. (216–223),

Trotta, M., Debernardi, F., Caputo, O. (2003). Preparation of solid lipid nanoparticles by a solvent emulsification-diffusion technique. *International Journal of Pharmaceutics*, Vol. 257, No. 1-2, (May 2003), pp. (153–160), ISSN 0378-5173

Uchida, Y., Hamanaka, S. (2006). Stratum corneum ceramides: function, origins, and therapeutic implications, In: *Skin Barrier*, Elias, P.M., Feingold, K.R., pp. (43-64), Ed. Taylor and Francis, ISBN 9780824758158, New York

Üner, M. (2006). Preparation, characterization and physico-chemical properties of solid lipid nanoparticles (SLN) and nanostructured lipid carriers (NLC): their benefits as colloidal drug carrier systems. *Pharmazie*, Vol. 61, No. 5, (May 2006), pp. (375–386). ISSN 0031-7144.

Van der Merwe, D., Brooks, J.D., Gehring, R., Baynes, R.E., Monteiro-Riviere, N.A., Riviere J.E. (2006). A physiologically based pharmacokinetic model of organophosphate dermal absorption, *Toxicological Sciences*, Vol.89, No. 1 (January 2006), pp. (188–204), ISSN 1096-6080

Vega-Villa, K.R., Takemoto, J.K., Yáñez, J.A., Remsberg, C.M., Forrest, M.L., Davies, N.M. (2008) Clinical toxicities of nanocarrier systems. *Advanced Drug Delivery Reviews* (22 May 2008), Vol. 60, No. 8, pp. (929-38), ISSN 0169-409X

Vogt, A., Combadiere, B., Hadam, S., Stieler, K.M., Lademann, J., Schaefer, H., Autran, B., Sterry, W., Blume-Peytavi, U. (2006) 40 nm, but not 750 or 1,500 nm, nanoparticles enter epidermal CD1a+ cells after transcutaneous application on human skin. *Journal of Investigative Dermatology*, Vol. 126, No. 6 , (June 2006), pp. (1316–1322), ISSN 0022-202X

Walters, K.A. (2002). The structure and function of skin, In: *Drugs and the pharmaceutical Sciences*, Walters, K.A., Roberts, M.S., Vol. 119, pp. (1-40), Ed. Marcel Dekker, ISBN 0824798899, New York

Weyenberg, W., Filev,P., Van den Plas,D., Vandervoort,J., De Smet,K., Sollie, P., Ludwig A. Weyenberg, W., (2007). Cytotoxicity of submicron emulsions and solid lipid nanoparticles for dermal application, *Pharmaceutical Nanotechnology*, Vol. 337, (January 2007), pp. (291–298), ISSN 0975-7384

Wilke, K., Wepf, R., Keil, F.J., Wittern, K.P., Wenck, H., Biel, S.S. (2006). Are sweat glands an alternate penetration pathway? Understanding the morphological complexity of the axillary sweat gland apparatus. *Skin Pharmacology and Physiology*, Vol. 19, No. 1, (2006), pp. (38-49), ISSN 1660-5527

Wissing, S.A., Kayser, O., Müller, R.H. (2004). Solid lipid nanoparticles for parenteral drug delivery. *Advanced drug delivery reviews*, Vol. 56, No. 9, (May 2004), pp. (1257–1272), ISSN 0169-409X

Wissing, S.A., Müller, R.H. (2003) Cosmetic applications for solid lipid nanoparticles (SLN). *International Journal of Pharmaceutics*, Vol. 254, No. 1, (18 March 2003), pp. (65–68), ISSN 0378-5173

9

Bioprocess Design: Fermentation Strategies for Improving the Production of Alginate and Poly-β-Hydroxyalkanoates (PHAs) by *Azotobacter vinelandii*

Carlos Peña, Tania Castillo,
Cinthia Núñez and Daniel Segura
Instituto de Biotecnología,
Universidad Nacional Autónoma de México (UNAM)
México

1. Introduction

Industrial interest in microbial polymers has been stimulated by their unique properties and the opportunity to develop new materials, which can be used for specific applications in medical and pharmaceutical industries. *Azotobacter vinelandii* produces two polymers of biotechnological importance; alginate, an extracellular polysaccharide, and poly-β-hydroxybutyrate (PHB), an intracellular polyester of the polyhydroxyalkanoates (PHAs) family (Galindo et al., 2007). Alginates are linear polysaccharides composed of variable amounts of (1–4)-β-D-mannuronic acid and its epimer, α-L-guluronic acid. Alginates present a wide range of applications, acting as stabilizing, thickening, gel or film-forming agents, in various industrial fields. Currently, new applications are being discovered for these polymers, such as their use as a source of soluble fiber, or in medical products. One example is found in the use of alginate gel beads as entrapment devices for transplantation of e. g. insulin producing cells and tissue engineering (Hernández et al., 2010).

The intracellular polyester PHB and other PHAs have been drawing attention because they are biodegradable and biocompatible thermoplastics, which can be processed to create a wide variety of consumer products, including plastics, films, and fibers (Aldor & Keasling, 2003). Recently, and based on their properties of biocompatibility and biodegradability, new attractive applications for PHAs have been proposed in the medical field, where the chemical composition and product purity are critical (Williams & Martin, 2005).

The subjects covered in this chapter include research concerning the production of alginate and PHB by *A. vinelandii*, particularly the molecular regulation of the production of these polymers, the influence of fermentation parameters on the production and composition of alginate and PHB, some reports about the scaling-up of the process and downstream processing, and finally, novel fermentation strategies that could be applied for the production of alginate and PHAs by *A. vinelandii*.

2. Structure and physical properties of alginate and PHAs

Alginates are polysaccharides constituted by variable amounts of β-D-mannuronic acid and its C5-epimer α-L-guluronic acid linked by 1-4 glycosidic bonds (Figure 1). The monomers are distributed in blocks of continuous mannuronate residues (M), guluronate residues (G) or alternating residues (MG) (Smidsrod & Draget,1996). *Azotobacter* alginates are true block co-polymers, composed of homopolymeric regions of M and G residues, separated by regions of alternating structure (Clementi, 1997). Unlike alginates produced by algae, bacterial alginates are acetylated to a variable extent at positions O-2 and/or O-3 of the mannuronate residues (Skjak-Braek et al., 1986). The variability in molecular mass, monomer block structure and acetylation influence the physicochemical and rheological characteristics of the polymer.

The capability of alginate to confer viscosity in solution is dependent on its molecular mass (MM). The MM of algal alginates has been found to range from 48 to 186 kDa (Donnan & Rose, 1950). In contrast, some alginates isolated from *A. vinelandii* present MM in the range of 80 to 4,000 kDa (Galindo et al., 2007).

The gelling properties of alginate are based on the affinity of the molecule towards certain ions, especially Ca^{++}, and the ability to bind these ions selectively and cooperatively. Selective ion binding is related to the content of G residues, in particular the length of the G-blocks. Alginates rich in G residues show an increased ionic binding, resulting in enhanced mechanical rigidity (Grant et al., 1973). Alginates with a low M/G ratio form strong and brittle gels, while alginates with a high M/G ratio form weaker and softer, but more elastic gels (Skjak-Braek et al., 1986).

Bacterial alginates which are O-acetylated, have a polyelectrolyte behaviour different from that of the non-acetylated algal alginates. This is because acetyl groups strongly perturb stereoregular sequences and produce significant conformational effects. In addition, the presence of acetyl groups diminishes the binding capacity and the selectivity coefficient for cations, affecting the gelling properties of alginate. The presence of O-acetyl groups in bacterial alginates might represent an advantage for certain applications, as it has been demonstrated that they improve the viscosity and enhance the swelling ability of the polymer (Clementi, 1997; Peña et al., 2006).

On the other hand, polyhydroxyalkanoates (PHAs) are aliphatic polyesters generally composed of β-hydroxy fatty acid monomers in which the carboxyl group of one monomer forms an ester bond with the hydroxyl group of the neighboring monomer (Madison & Huisman, 1999; Figure 1). The MM of PHAs is dependent on the bacterial species used and culture conditions but is generally on the order of 50 to 1,000 kDa (Madison & Huisman, 1999). At present, more than 150 different hydroxyalkanoate constituents have been reported in PHAs, as homopolyesters or as copolyesters (Steinbuchel & Lutke-Eversloh, 2003). These highly diverse polymers can be classified according to the size of the comprising monomers. PHAs containing monomers with C4–C5 atoms are categorized as short-chain-length PHAs (scl-PHAs). In contrast, medium-chain-length PHAs (mcl-PHAs) are composed of C6–C14 β-hydroxy fatty acids (Lee, 1996). Most bacteria synthesize either scl-PHAs or mcl-PHAs (Madison & Huisman, 1999). scl-PHAs have properties close to conventional plastics, while the mcl-PHAs are regarded as elastomers and rubbers (Suriyamongkol et al., 2007).

Polyhydroxybutyrate (PHB) is the more abundant PHA and has been studied extensively. This polymer has some mechanical properties similar to conventional plastics like polypropylene or polyethylene, although it is highly crystalline and stiff, leading to brittleness and low elongation to break (Khanna & Srivastava, 2005). Initial biotechnological developments were aimed at making PHAs easier to process. Because the monomeric composition of a PHA is crucial for its mechanical properties, the incorporation in the PHB polymer of secondary monomer units(s) such as β-hydroxyvalerate (3HV) improves the characteristics of the material obtained. For example, a random copolymer of 3HB and 3HV is more ductile, easier to mold, and tougher than PHB homopolymer (Taguchi & Doi, 2004), and it can be used to prepare films with excellent water and gas barrier properties reminiscent of polypropylene, and can be processed at a lower temperature while retaining most of the other mechanical properties of PHB.

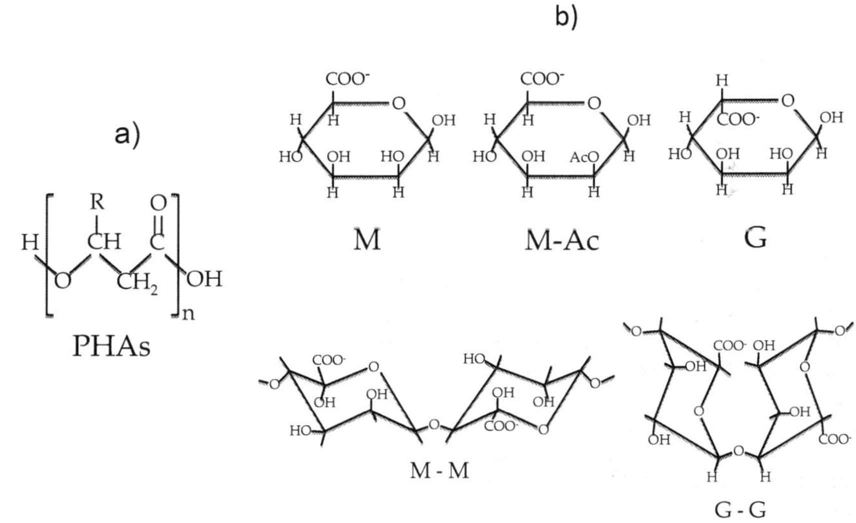

Fig. 1. Chemical structure of PHAs (a) and Alginate (b). M, mannuronic acid; G, guluronic acid; Ac, acetyl. R, alkyl group.

3. Novel applications of alginate and PHAs

3.1 Novel applications of alginates

Novel alginates applications have been focused on pharmaceutical and biomedical fields, because they are non-toxic, biocompatible, non-immunogenic, hydrophilic and biodegradable material (Augst et al., 2006; Hernández et al., 2010). Alginate hydrogels can be used as bulking materials for *in vivo* and *in vitro* cell immobilization, like drug controlled delivery system, for tissue engineering (Augst et al., 2006; Hernández et al., 2010), and alginate conjugates have also been tested as antigens to control cystic fibrosis (Kashef et al., 2006), and other bacterial infections (H. Sun et al., 2007). During the last three decades, microencapsulation using alginate has been investigated to deliver and protect from the host immune system, not only drugs, but also transplanted cells (Hoesli et al., 2011). The materials used for these molecular/cell immobilization require to be biocompatible and bio-inert, with certain size and shape. In addition, alginates have been used as scaffold for tissue engineering, and chemical modifications that increase alginate biocompatibility or cell adhesion have been developed. The polysaccharide can be modified by coupling proteins and peptides, allowing the control of cell attachment (Augst et al., 2006; Hernández et al., 2010). The biomimetic gel design provides *in vivo* long-term functionality and higher mechanical stability (Hernández et al., 2010).

Some therapeutic applications of alginate microencapsulation are related with drug delivery. For low-molecular weight drugs, regulating drug-alginate interactions in alginate gels allows the control of drug release. This is especially important for drugs that have severe side effects, like antineoplastic agents. Besides, some proteins with therapeutic activities can be alginate-microencapsulated to improve their efficacy and targeting, because alginate encapsulation facilitate a localized delivery without adverse side effects. Alginate microencapsulation has been proven with basic fibroblast growth factor (bFGF) and vascular endothelial growth factor (VEGF). The release of VEGF is controlled by the dissolution of the ionic binding complex between alginate and VEGF and subsequent diffusion, showing a constant release rate for several weeks (Augst et al., 2006).

Alginate has been successfully used for cell microencapsulation, which is of great importance for Diabetes Mellitus type 1 treatment, where several efforts have been made for regulated insulin supply for treating insulin-dependent patients (Hernández et al., 2010). Moreover, there are several new strategies developed to improve the cell-alginate immobilization process (Hoesli et al., 2011), as well as immune protection and oxygen supply to avoid hypoxia problems during transplants (Ludwig et al., 2010). In the tissue engineering field, alginate has been used for bone regeneration therapy using co-immobilization of human osteoprogenitors and endothelial cells in studies *in vivo* and *in vitro* (Hernández et al., 2010). Other important applications of alginate for tissue engineering are related to neurological and cardiologic tissue regeneration (Hernández et al., 2010).

Pseudomonas aeruginosa is the most common pathogen responsible for morbidity and mortality in cystic fibrosis patients. During infection, this bacterium produces alginate, which is an important virulence factor (Kashef et al., 2006). For this reason, alginate has been used for vaccine design, against *P. aeruginosa*. Vaccines based on purified alginate bring poor immunogenicity (Dörig & Pier, 2008); however, when alginate is conjugated with proteins, the immune response could be enhanced (Dörig & Pier, 2008; Kashef et al., 2006). Kashef et al., (2006) designed an alginate-tetanus toxoid conjugate non-toxic, non pyrogenic, which was able to protect mice against a lethal dose of mucoid *P. aeruginosa*. It is important to point out that alginate viscosity plays an important role for the *P. aeruginosa* protection during the infection process. Because of this, it has been proposed to induce changes in the rheology of the alginate by addition of alginate olygoelectrolytes conformed by G blocks only (Draget & Taylor, 2011). These oligoguluronates reduce the mechanical response of the polymer synthesized by *P. aeruginosa* in patients diagnosed with cystic fibrosis (Draget & Taylor, 2011).

Alginates have also been studied for development of novel immunotherapy strategies for cancer treatment using dendritic cells which are potent initiators of immune response. Calcium cross-linked alginate gels carrying dendritic cells initiated the immune response and allowed the migration of the immune cells through the alginate gel (Hori et al., 2008, 2009).

3.2 Novel applications for PHAs

PHAs have received much attention as candidates to produce biodegradable plastics compatible with the environment, due to their material properties (similar to those of well-known plastics such as polypropylene), their production from renewable sources, and their inherent biodegradability in various environments (Taguchi & Doi, 2004). These biopolyesters are attractive to replace non-biodegradable plastics, especially for those products that usually have single-use applications, such as food packaging.

Material	Application	Desirable properties	Reference
Alginate	Drug delivery	Gel formation/ Biocompatible/ Bioinert/ Biodegradable/ Drug alginate interaction	Augst et al., 2006
	Cell immobilization	Gel formation/ Biocompatible/ Bioinert/ Diffusion	Hernández et al., 2010 Hoesli et al., 2011
	Tissue Engineering	Gel formation/ Biocompatible/ Bioinert/Diffusivity	Hernández et al., 2010
	Vaccines against *Pseudomonas aeruginosa*	Immunogenicity	Kashef et al., 2006
	Mucus alteration	Low Viscosity	Draget & Taylor, 2011
	Cancer therapy	Gel formation/ Biocompatible/ Bioinert/ Diffusivity	Hori et al., 2008
PHA	Bioplastics	Thermoplasticity, physical and mechanical resistance/ Biodegradable	Chen, 2009
	Tissue Engineering	Thermoplasticity, physical and mechanical resistance/ Biocompatible	Wu et al., 2009 Grage et al., 2009
	Drug delivery	Biocompatible/ Biodegradable	Chen, 2009
	Cell microencapsulation and nanoencapsulation	Biocompatible/ Biodegradable	Grage et al., 2009
	Affinity support	Biocompatible	Lee et al., 2005 Wang et al., 2008
	Affinity chromatography	Biocompatible	
	Biomarkers/ Biosensors	Biocompatible	Grage et al., 2009
	Biofuels	Biodegradability/ Methyl esterification capability	Zhang et al., 2009

Table 1. Novel applications of Alginate and PHAs.

Some of the monomers present in PHAs are known to be present in human and animals. For example, the monomeric component of PHB (β-hydroxybutyrate) is a ketone body normally found in human blood (Williams & Martin, 2005). The biocompatibility, together with the adjustable mechanical properties, and controllable biodegradability of PHAs have raised interesting applications in the medical field. These polymers have been used in artificial organ construction, drug delivery, tissue repair, and nutritional/therapeutic uses (Chen & Wu, 2005; Freier, 2006; Grage et al., 2009; Valappil et al., 2006; Williams & Martin, 2005; Wu et al., 2009; Zinn et al., 2001). Because several PHAs are available now in sufficient quantity, some of them have been used in biocompatibility studies *in vivo* (Valappil et al., 2006). Some of the medical devices tested with different degrees of success include meniscus repair devices, staples, screws, bone plating systems, cardiovascular patches, stents and nerve guides (Wu et al., 2009).

Very interesting applications for PHAs are found in the fabrication of drug delivery devices. Their biocompatibility, combined with their biodegradation, make them good candidates for this purpose (Chen, 2009). The possibility to create PHAs of various monomeric compositions and molecular weights makes possible the fine control of their degradation rate (Wu et al., 2009). Several drugs have been entrapped or microencapsulated in PHA homopolymers or copolymers, such as, anticancer agents, antihypertensives, hormones, vaccines, etc. (Williams & Martin, 2005).

More recently, new applications have been reported for PHAs, not just as a material but for the PHA granules themselves as micro-nano beads, resulting in applications for protein purification and specific drug delivery (Grage et al., 2009). Affinity protein purification methods make use of an affinity tag fused to the protein of interest. The interaction of the tagged protein with an immobilizing matrix allows the separation of the protein. PHA granules have been used as an inexpensive affinity support, while the phasins, PHA synthase, or PHA depolymerase, proteins naturally associated with the granules, work as the affinity tag for the purification or immobilization of proteins (Lee et al., 2005; Wang et al., 2008). Combining the fusion of the protein of interest with the phasin protein (PhaP), intein mediated self-cleavage, and PHA synthesis in recombinant *Escherichia coli*, specific proteins can be produced together with their insoluble matrix, and after cell disruption, precipitation and self-cleavage, the purified protein is released (Banki et al., 2005; Mee at al., 2008). A similar method using *Cupriavidus necator* instead of *E. coli* has been developed (Barnard et al., 2005).

PHA nanoparticles have also been used in drug delivery, target specific therapy and as biomarkers or biosensors (Grage et al., 2009). Using the same principle of affinity binding of PHA synthase to PHA granules, Brockelbank et al., (2006) demonstrated the display of antigen fragments at the surface of PHA beads, showing their potential to be used in immunoglobulin G (IgG) purification from human serum. The functional display of antigen or antibodies fragments at the bead surface can be used for diagnostic or therapeutic applications (Grage et al., 2009). Fusion of PHA synthase with streptavidin has shown that these PHA beads can also be used for ELISA, DNA purification, enzyme immobilization and flow cytometry (Peters & Rhem, 2008). Engineered proteins for inorganics, like gold or silica, or IgG, have also been fused to the PHA synthase, and displayed at the surface of PHA granules, so these biobeads can be used for medical bioimaging procedures as inorganic contrast agents (Grage et al., 2009; Jahns et al., 2008).

With respect to targeted drug delivery, Yao et al., (2008) demonstrated that phasins can be fused to ligands recognized by tissue specific receptors, and that the ligand-PhaP-nanobeads are taken up by the correct tissue *in vivo*, delivering drugs loaded in the PHA beads.

Other applications for PHAs include their use as fine chemicals. A high diversity of carboxylic acids, all in the (R)- configuration, can be obtained by depolymerization or by chemical degradation of PHAs, and these can be bulk chemicals for various applications (Chen, 2009). Some of them have been used as starting material for the synthesis of antibiotics, vitamins, aromatics and pheromones (Ruth et al., 2007).

A new field of application for PHAs has been devised in the energy industry. These polymers can be used as biofuels (Chen, 2009). The conversion of PHB or mcl-PHAs to their methyl ester derivatives by acid catalyzed hydrolysis, allowed their use as fuels in blends with ethanol, gasoline, and diesel, with reasonable combustion heats (Zhang et al., 2009).

4. *Azotobacter vinelandii*

A. vinelandii is a gamma Proteobacteria having a strictly respiratory type of metabolism with oxygen as the terminal electron acceptor. Nitrogen is fixed at either microaerobic or fully aerobic conditions. Its growth is heterotrophic where sugars, alcohols and salts of organic acids are used as carbon source (Kennedy et al., 2005). Sugars are metabolized trough the Entner-Doudoroff pathway (Conway, 1992). The genus *Azotobacter* is distinguished by the ability to form metabolically dormant cyst in stationary phase or upon induction of vegetative cells with 0.2% of n-butanol. The cysts are significantly more resistant than vegetative cells to desiccation (Socolofsky & Wyss, 1962). Alginate is a component of the envelope that protects the cyst, and is essential for the resistance to desiccation (Campos et al., 1996). Upon induction of encystment intracellular accumulation of poly-β-hydroxybutyrate (PHB) occurs at an exponential rate; however recent data demonstrated that PHB was not essential for the formation of mature cysts (Segura et al., 2003a).

The majority of nitrogen fixing bacteria are capable of reducing N_2 only in anaerobic or microaerobic conditions. In contrast, *A. vinelandii* is an obligate aerobe capable of fixing N_2 even at high concentration of O_2. This is possible because this bacterium can adjust oxygen consumption rates to help maintain low levels of cytoplasmic oxygen, which is otherwise detrimental not only for nitrogenase, but also to other oxygen-sensitive enzymes, a process that has been called respiratory protection (Poole & Hill, 1997). In addition to the respiratory protection of the nitrogenase, another way to keep the cytoplasm anaerobic is to prevent the O_2 transfer into the cell. The polysaccharide alginate is believed to form a coating around the cell, acting as a physical O_2 barrier (Sabra et al., 2000). This barrier has also been reported to protect the cell from heavy metals toxicity, as an ion-exchange system with high affinity to Ca^{++}, or to provide a negatively charge coating which creates a barrier against attack and adverse environmental conditions (Clementi, 1997).

5. Genetics and biosynthesis of alginates and PHAs in *A. vinelandii*

5.1 Biosynthesis of alginates by *A. vinelandii*

The pathway for alginate synthesis has been well established and it is conserved among brown algae, *Pseudomonas* and *Azotobacter* spp. (Lynn & Hassid, 1966; Pindar & Bucke, 1975). Fructose 6-P, the precursor of this metabolic pathway, is converted by four enzymatic reactions to GDP-mannuronic acid (for a detailed review see Galindo et al., 2007; Remminghorst & Rehm 2006). Polymerization of GDP-mannuronic acid is conducted by an inner membrane mannuronate polymerase (Alg8) and its activity is regulated by another inner membrane protein (Alg44), essential for alginate biosynthesis (Galindo et al., 2007).

The resultant poly-mannuronic acid is then modified by a periplasmic O-acetylase complex (AlgI, AlgV, AlgF) and some of the non-acetylated mannuronate residues are epimerized to guluronate by a periplasmic mannuronate epimerase (AlgG). The polymer is then exported through the outer membrane via the pore-forming protein AlgJ where the activity of several extracellular C-5 epimerases (AlgE1-7), present only in *A. vinelandii* generate alginates with different amounts of alternating structures and/or G-block lengths (Ertesvag et al., 1999). The molecular factors determining the molecular mass of the alginate remain largely unknown, but it has been suggested that it is the result of the polymerase and/or lyase activities on the polymer. In *A. vinelandii*, the existence of five alginate lyases showing different sequence cleavage specificity and cellular locations have been reported (Gimmestad et al., 2009; Trujillo-Roldán et al., 2003). The periplasmic AlgL protein is involved in the biosynthesis of the polysaccharide, while the extracellular AlgE7 (a bi-functional alginate lyase and C-5 epimerase) and AlyA3 enzymes are involved in the release of the polymer from the cell surface and in the rupture of the cyst coat during germination, respectively. The function of the lyases AlyA1 and AlyA2 remains unknown. However, AlyA2 activity was shown to be essential for vegetative growth (Gimmestad et al., 2009).

The *algD* gene encodes the key enzyme catalyzing the generation of the alginate monomer GDP-mannuronic acid (for a recent review see Galindo et al., 2007). Expression of *algD* is highly controlled by several global regulators such as the stress response sigma factor AlgU and the signalling transduction cascade conformed by the two-component system GacA/GacS and the stationary growth phase sigma factor RpoS protein, which also positively control PHB synthesis (Castañeda et al., 2000; 2001). *A. vinelandii* mutants have been constructed with the aim of generating alginates with different physicochemical properties. An *algL* mutant was shown to produce an alginate with a molecular mass higher than that of the wild type strain (Trujillo-Roldan et al., 2003). Furthermore a mutation in the *algF* gene, encoding one of the subunits of the O-acetylase complex, resulted in the production of a non-acetylated alginate, similar to that of algal origin (Vazquez et al., 1999). As the polymers PHB and alginate compete for the supplied carbon source we generated mutants with a total blockade in PHB synthesis and in which the production of alginate was increased several fold. The contrary was also true for the production of PHB as a total blockade in the synthesis of alginate increased the accumulation of PHB (Segura et al., 2003b).

5.2 Biosynthesis of PHAs by *A. vinelandii*

PHB in *A. vinelandii*, as in most bacteria, is synthesized in three steps from acetyl-CoA (Figure 2; Manchak & Page 1994). β-Ketothiolase catalyzes the first reaction, i.e. the condensation of two molecules of acetyl-CoA to form acetoacetyl-CoA. This product is subsequently reduced by an NADPH dependent acetoacetyl-CoA reductase to stereoselectively produce (R)-β-hydroxybutyryl-CoA. Finally, PHB synthase catalyzes the polymerization of (R)-β-hydroxybutyryl-CoA releasing free CoA.

In *A. vinelandii*, the genes coding for these enzymes are contained in the PHB biosynthetic operon *phbBAC*, which codes for the acetoacetyl-CoA reductase, β-ketothiolase, and PHB synthase respectively (Peralta-Gil et al., 2002; Segura et al., 2000, 2003a). In the same DNA region, other genes related to PHB synthesis were also found: *phbR*, which codes for a transcriptional regulator; *phbP*, a putative granule-associated protein; and *phbF*, a putative regulator of *phbP* (Peralta-Gil et al., 2002; Segura et al., 2003a).

When *A. vinelandii* UWD is grown on medium supplemented with n-alkanoates, such as valerate, heptanoate or nonanoate, a copolymer of poly(Hydroxybutyrate-Co-Hydroxyvalerate) (PHBV) is synthesized (Page et al., 1992). The recent analysis of the *A. vinelandii* genome sequence (Setubal et al., 2009) demonstrated the presence of *phaJ*, a gene coding for a (*D*)-specific enoyl-Coenzyme A hydratase that is responsible for the channeling of enoyl-CoA derivatives from the fatty acid oxidation pathway to PHAs synthesis in several pseudomonads (Fukui et al., 1998). Thus, this enzyme would be producing the hydroxyvalerate precursors for PHBV synthesis in *A. vinelandii* (Figure 2).

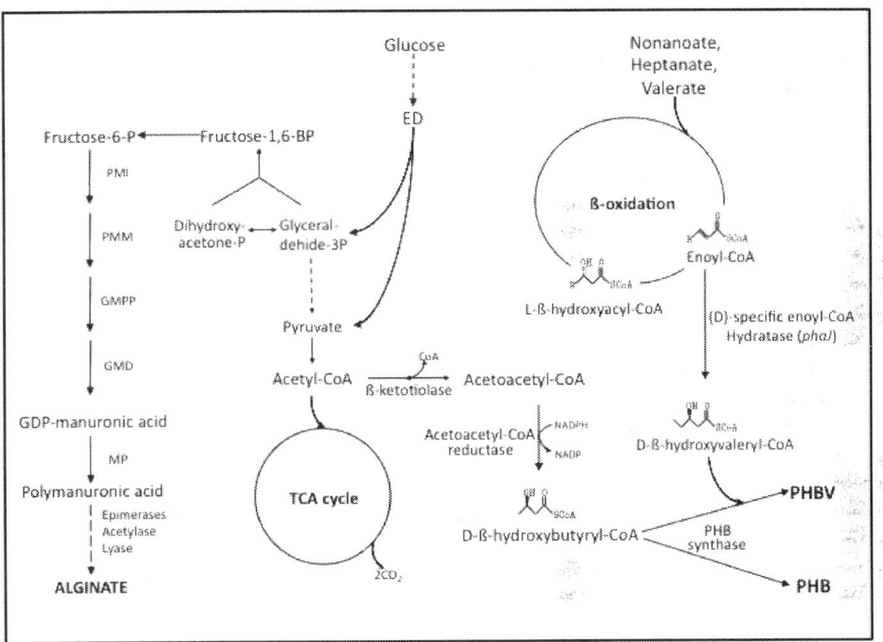

Fig. 2. Metabolic pathways involved in the synthesis of alginate and PHAs in *A. vinelandii*. PMI, phosphomannose isomerase; PMM, phosphomannose mutase; GMPP, guanosine diphosphomannose pyrophosphorylase; GMD, GDP-mannose dehydrogenase; MP, mannuronate polymerase; ED, Entner-Doudoroff pathway.

The control of PHB biosynthesis in *Azotobacter* was one of the first to be studied (Senior et al., 1972; Senior & Dawes, 1971, 1973). The main condition triggering PHB synthesis is oxygen limitation, which leads to high concentrations of NADH and NADPH, that in turn inhibit TCA cycle enzymes, increasing the concentration of acetyl-CoA available for PHB biosynthesis (Manchak & Page, 1994; Senior et al., 1972; Senior & Dawes, 1973).
The regulation of PHB synthesis in *A. vinelandii* is complex and additional regulatory systems are involved. PhbR, a regulator of the AraC family of transcriptional regulators, activates transcription of the *phbBAC* operon, and the stationary growth phase sigma factor RpoS is involved in the control of transcription of *phbR* (Peralta-Gil et al., 2002). The nitrogen-related phosphotransferase system (PTS[Ntr]), formed by proteins EI[Ntr], Npr, and

IIA[Ntr] regulates PHB synthesis through a phosphorelay from phosphoenolpyruvate, where the IIANtr protein acts as negative regulator of PHB synthesis in its non-phosphorylated state (Segura & Espin, 1998; Noguez et al., 2008). It has also been reported that the FNR-like regulatory protein called CydR (Wu et al., 2001) and the iron-regulatory small RNA named ArrF (Pyla et al., 2009) control PHB synthesis in response to the redox state of the cell (oxygen), and the availability of iron respectively.

6. Fermentation parameters affecting the production and the composition of alginate

For several decades the synthesis of alginate and PHB by *A. vinelandii* has been the subject of study, either in batch (Clementi et al., 1997; Page & Cornish 1993; Parente et al., 2000; Peña et al., 2000,2011; Sabra et al., 1999; Trujillo-Roldán et al., 2004), continuous (Díaz-Barrera et al., 2009, 2010; Sabra et al., 2000), fed batch cultures (Chen & Page, 1997; Mejía et al., 2010; Priego-Jiménez et al., 2005) and systems with immobilized cells in membrane reactors (Saude & Junter, 2002). In the following sections we will describe and discuss the most recent advances regarding the influence of fermentation parameters, which determine the production and composition of alginate and PHAs.

6.1 Influence of the dissolved oxygen tension (DOT) and the oxygen transfer rate (OTR) on the quantity and quality of alginate

Many studies have shown that aeration and mixing are critical parameters for optimizing the production of microbial polysaccharides (Galindo et al., 2007). It is known that under low dissolved oxygen tension (DOT), the organism accumulates the intracellular storage polymer, PHB; whereas at high DOT, *A. vinelandii* uses the carbon source mainly for biomass production. Efficient conversion of sucrose to alginate is achieved only if the oxygen is accurately controlled between 1 and 10% of oxygen saturation (Parente et al., 2000; Peña et al., 2000; Sabra et al., 2000; Trujillo-Roldán et al., 2003). The DOT also affects the composition and molecular mass of the alginate produced by *A. vinelandii*. Studies in bioreactor, under oxygen controlled conditions (Peña et al., 2000; Sabra et al., 2001; Trujillo-Roldán, 2004), indicate that the mean molecular mass (MMM) of the polymer, is strongly influenced by the DOT and the stirring speed of the culture. For example, in cultures conducted at low agitation speed (300 rpm) and DOT of 5 % the MMM of the polymer reached a maximum of 680 kDa. In contrast, at high agitation speed (700 rpm), the MMM increased to a plateau at low DOT (1– 3 %) and then decreased at higher DOT (5 %) (Peña et al., 2000). On the other hand, Sabra et al., (2000) reported that in phosphate-limited continuous culture, both the MM and the L-guluronic acid content increased with the DOT, reaching a maximal MM of 800 kDa and a guluronic acid content of 50 % in the cultures conducted at 10 % of air saturation. Those authors proposed that under nitrogen- fixing conditions, the bacterium builds a slimy layer or alginate capsule around the cells, to overcome the oxygen stress and to protect the nitrogenase system, causing a decrease in alginate biosynthesis.

A. vinelandii is known for its high respiratory activity (Post et al., 1983) and in cultures without DOT control, the oxygen transfer becomes the limiting factor for growth. Without DOT control, the cultures operate at DOT near zero. Under this condition, a parameter that has been used for studying alginate production is the oxygen transfer rate (OTR). In this line, Díaz-Barrera et al., (2007, 2009) reported that the alginate yield and the MMM of the polymer were linked to the OTR of the culture. They found that the MMM of the alginate increased as OTR_{max} decreased, observing that the MMM obtained at 3.0 mmol L^{-1} h^{-1} was 7.0 times higher (1560 kDa) than at 9.0 mmol L^{-1} h^{-1} (220 kDa; Figure 3). It is important to quote that in previous reports the cultures were oxygen limited and under such conditions the carbon source was only partially oxidized, which forced the cells to follow anaerobic pathways with the consequent production of PHB.

More recently, Lozano et al., (2011) reported a study about the evolution of the MMM of the alginate produced by *A. vinelandii* ATCC 9046 in terms of the maximum oxygen transfer rate (OTR_{max}) in cultures where the dissolved oxygen tension (DOT) was kept constant. An increase in the agitation rate (from 300 to 700 rpm) caused a significant increase in the OTR_{max} (from 17 to 100 mmol L^{-1} h^{-1} for DOT of 5 %, and from 6 to 70 mmol L^{-1} h^{-1} for DOT of 0.5 %). This increase in the OTR_{max} improved alginate production, as well as the specific alginate production rate. In contrast, the mean molecular mass (MMM) of the alginate isolated from cultures developed under non-oxygen limited conditions increased by decreasing the OTR_{max}, reaching a maximum of 550 kDa at an OTR_{max} of 17 mmol L^{-1} h^{-1} . However, in the cultures developed under oxygen limitation (0.5 % DOT), the MMM of the polymer was practically the same (around 200 kDa) at 300 and 700 rpm and it remained constant throughout the cultivation (Lozano et al., 2011).

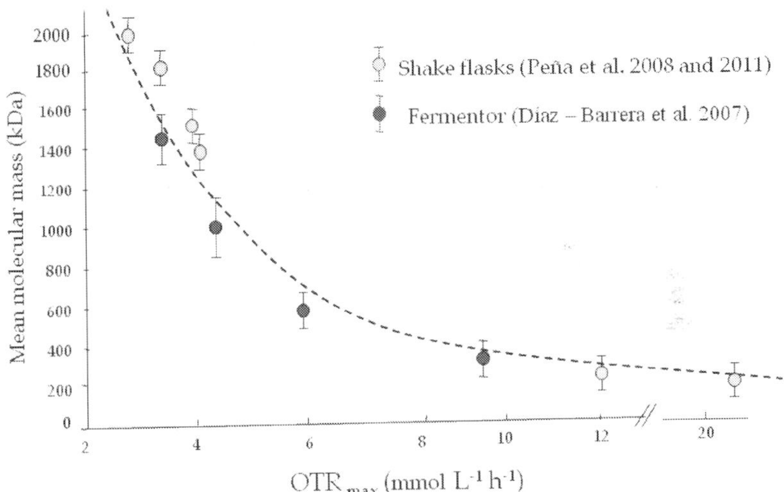

Fig. 3. Mean molecular mass of alginate as a function of the oxygen transfer rate in cultures of *A. vinelandii* in shake flasks and stirred fermentor.

6.2 Influence of the medium components

It is widely known that the components of the culture medium play an important role in the production of alginate by *A. vinelandii*. Most studies have focused on assessing the influence of calcium, phosphate and nitrogen on alginate yield and its molecular mass (MM) (Parente et al., 2000; Sabra et al. 1999). Sabra et al., (1999) found that an excess of phosphate in the culture medium (400 mg L^{-1}) caused a decrease in alginate yield. In the case of nitrogen the results reported are contradictory, probably because the strain and medium composition used were different (Clementi, 1997; Sabra et al., 2001). Recently, Zapata-Vélez & Trujillo-Roldán (2010) reported that in cultures of *A. vinelandii* in shake flasks, the best nitrogen sources for alginate production were peptone, tryptone, and yeast extract, yielding a maximal alginate concentration of 4.0 ± 0.4 g L^{-1}. Those authors found that the highest MMM was obtained in cultures grown with peptone ($1,520 \pm 110$ kDa), whereas cultures grown with yeast extract, tryptone, ammonium acetate, and ammonium sulphate, showed values between 1,400 and 1,100 kDa. On the other hand, a lower MMM was obtained under N$_2$-fixing conditions (625 ± 110 kDa).

Our group reported the influence of (3N-morpholino)-propane-sulfonic acid (MOPS), a component used in the medium to keep a constant pH, on the quality of the alginate in terms of the chemical composition and rheological behaviour of alginate-reconstituted solutions (Peña et al., 2006). This compound had an important effect on the acetyl content and physicochemical properties of this polymer. A two-fold higher acetylation degree of alginate was obtained when 13.6 mM MOPS was supplemented to the medium. The higher acetylation resulted in greater viscosity of the alginate solutions, but it exhibited less pronounced pseudoplastic behaviour. These changes in the functional properties of the polymer can have great value in terms of specific applications of alginate in food and pharmaceutical fields.

6.3 Effect of the specific growth rate

Another important culture parameter for the synthesis of alginate is the specific growth rate (Díaz-Barrera et al., 2009, 2010; Priego-Jiménez et al., 2005). Priego-Jiménez et al., (2005) using exponentially fed-batch cultures, found that the specific growth rate of *A. vinelandii* negatively affects the MM of the alginate and to some extent, the alginate/biomass and alginate/sucrose ratio. This effect was particularly pronounced at very low specific growth rates (0.03 h^{-1}), where the $Y_{P/X}$, $Y_{P/S}$ and the MMM were up to 2.3, 10 and 14 times higher, respectively, than those obtained at a specific growth rate of 0.21 h^{-1} (the value found in conventional batch cultures). More recently, Díaz-Barrera et al., (2010) reported, in chemostat cultures, that the alginate MM increased from 800 to 1800 kDa when the dilution rate increased from 0.05 to 0.1 h^{-1} at a low inlet sucrose concentration (5 g L^{-1}). In contrast, at high sucrose concentration, the MM increased from 1230 to 2500 kDa when the dilution rate, and therefore, the specific growth rate were decreased in the same range. According to the authors, this behaviour is linked to changes in the specific sucrose uptake rate (Díaz-Barrera et al., 2010).

7. Parameters that affect PHAs production in *A. vinelandii*

Commercial production of PHAs requires not only high yields and productivities, but also a well defined chemical composition. Fermentation parameters affect the amount of PHAs produced by *Azotobacter*, and their chemical characteristics, such as the kind of polymer produced (PHB homopolymer or PHBV heteropolymer); the MM; and finally, the monomer ratio and distribution along the PHA heteropolymer chain.

PHAs production in different organisms is induced under nutrient limitation (Verlinden et al., 2007). For *Azotobacter* species, oxygen limitation is the most efficient way to induce PHB production (Galindo et al., 2007; Senior & Dawes, 1971, 1973; Verlinden et al., 2007). Besides oxygen limitation, changes in the carbon and nitrogen sources can affect PHB biosynthesis by *Azotobacter* (Myshkina et al., 2008; Page, 1992). Also, the addition of alkanoates to the medium allows the synthesis of PHAs with different monomer composition (Myshkina et al., 2010; Page et al., 1992).

7.1 Oxygen limitation

When *Azotobacter* grows under oxygen limitation there is a reduction in the activity of the tricarboxylic acid cycle (TCA), and the molecules of acetyl-CoA are channeled to PHB production, and the synthesis of PHB acts like an electron sink (Page & Knosp, 1989; Senior et al., 1972). The positive effect of oxygen limitation on PHB production (based on yield and PHB content), has been reported for *A. vinelandii* in batch cultures, using wild type strains UW (Page & Knosp, 1989) and ATCC9046 (Peña et al., 2011), and the PHB overproducer mutant strain UWD (Page & Knosp, 1989). Changes in oxygen concentration have been successfully used in fed batch cultures of *A. vinelandii* UWD (Chen & Page, 1997; Page et al., 2001). Chen & Page (1997), enhanced biomass production of this strain using high aeration during the first stage of the culture and then, at the second stage, aeration was lowered, promoting PHB formation. At the end of the culture, PHB concentration reached 36 g L^{-1} of PHB, in contrast to 25 g L^{-1} of PHB reported by Page & Cornish (1993) in fed-batch cultures without aeration changes.

On the other hand, there are few reports related to the effect of oxygen on the composition of the PHB produced by *Azotobacter*. Quagliano & Miyazaki (1997) observed in fermentations of *A. chroococcum* 6B that changes in the aeration rate from 0.5 to 2.5 vvm could negatively impact the molecular weight of the PHB with a 10 fold decrease from 1100 to 111 kDa. Myshkina et al., (2008) observed, in cultures of *A. chroococcum* 7B in shake flasks, that the MM increased from 1480 to 1670 kDa when the agitation rate decreased from 250 to 190 rpm. In contrast, the yields of PHB ($Y_{PHB/Biomass}$) at both agitation rates were the same (0.75 g $_{PHB}$ g$^{-1}_{biomass}$). These authors also evaluated the effect of strict microaerobic and anaerobic conditions during stationary phase, on PHB yield and on its MM. Under these conditions, the PHB content decreased to 2.6 and 1.7 g L^{-1}, respectively, but the MM increased, reaching 2215 kDa at the strict anaerobic condition (Myshkina et al., 2008).

7.2 Medium composition: carbon and nitrogen sources

The high production cost is the main limiting factor for the use of PHAs for commercial purposes. An alternative to reduce costs is the use of cheaper feedstock (Page, 1992). Several attempts have been made to improve the culture media composition which depends on the microorganism (Table 2). Although *Azotobacter* is a nitrogen fixing bacteria, addition of a fixed nitrogen source can improve PHAs production. Quagliano & Miyasaki (1997), found that increasing the C:N ratio improved PHB yields, although the MM of the polymer dropped eight fold. Page & Cornish (1993) observed that organic nitrogen sources like fish peptone improved PHB production. In addition to the use of low cost nitrogen sources, the use of low price carbon sources is a good alternative to reduce production costs (Page, 1992; Verlinden et al., 2007). Page (1992) found that in shake flasks cultures the addition of 0.5 % (P/V) of beet molasses, increased the PHAs content in *A. vinelandii* UWD to 7 g L^{-1}, in contrast to the 1 g L^{-1} obtained when using sucrose at 2%(P/V) as a sole carbon source. Although beet molasses were the best carbon source, this strain was also able to grow and produce PHAs using cane molasses, malt extract or corn syrup. However, Myshkina et al., (2008) observed that for *A. chroococcum* 7B, molasses did not improve growth or PHB production, while the best PHB yields were obtained with commercial sugar and vinasses. These authors also reported that the MM of the PHB could be affected by the carbon source, reaching the highest MM when using glucose, food sugar or starch (1660, 1490 and 1310 kDa).

Strain	Carbon source	Nitrogen source	PHB (gL⁻¹)	$Y_{P/S}$	Mw (kDa)	Type of culture	Reference
A. vinelandii UWD	Glucose (1%) and acetate (15mM)	NH_4^+	2.37	0.25	N.D	Flasks	Page & Knosp, 1989
A. vinelandii UWD	Beet Molasses (5%) and sucrose (2%)	NH_4^+	7	0.35	N.D	Flasks	Page, 1992
A. vinelandii UWD	Glucose (5%) acetate (15mM)	NH_4^+ Fish peptone (1%)	25	0.65	1700	Fed batch culture	Page & Cornish, 1993
A. chroococcum 7B	Glucose (4%)	-----	4	0.1	1660	Flasks	Myshkina et al., 2008
A. chroococcum 7B	Glucose (2%) Acetate (20mM)	-----	4	0.2	1100	Flasks	Myshkina et al., 2008
A. chroococcum 7B	Molasses (4%) Sucrose (2%)	-----	1	0.05	N.D.	Flasks	Myshkina et al., 2008

Table 2. PHB production by *Azotobacter* grown with different sources of carbon and nitrogen.

7.3 Addition of alkanoates

For several *Azotobacter* spp., the addition of alkanoates to the growth media for PHAs production allows the synthesis of polymers with specific composition (Durner et al., 2000; Gónzalez-López et al., 1996; Myshkina et al., 2010; Page et al., 1992; Sun et al., 2007; Zinn et al., 2003). The effect of alkanoates addition is dependent on the strain and its metabolism. In cultures of *A. vinelandii* UWD, the addition of odd alkanoates (C_5-C_9) allowed the biosynthesis of the copolymer PHBV. Although there was copolymer production using the three different substrates (valerate, heptanoate and nonaoate), the highest proportion of HV was achieved with valerate (Page et al., 1992).

For *A. chroococcum* the biosynthesis of PHBV by addition of not only five carbon valeric acid but of other organic acids (propanoic and hexanoic), was reported. However the best yields and the highest HV content were obtained with valerate (Myshkina et al., 2010). The addition of alkanoates with more than five carbons to *Azotobacter* cultures did not allow biosynthesis of hydroxyalkanoates of a higher monomer chain length (Myshkina et., al 2010; Page et al., 1992). Although *Azotobacter* and *Pseudomonas* are genetically related (Setubal et al., 2009), the PHAs metabolism in *Azotobacter* differs completely from that observed in most *Pseudomonas* species which are efficient producers of mcl-PHAs (Durner et al., 2000; Hartmann et al., 2005; Sun et al., 2007). However, the close genetic relationship between *Pseudomonas* and *A. vinelandii* could be useful for genetic improvement of *A. vinelandii* strains for the biosynthesis of mcl-PHAs, as has been successfully reported for *E. coli* (Sun et al., 2007; Verlinden et al., 2007).

8. Scaling up of alginate and PHAs production

Trying to reproduce in agitated tank the results obtained in plates or in shake flasks, is troublesome and the variables involved are poorly understood. This is particularly important, because the MM of the polymer drops dramatically when the alginate process is scaled up from shake flasks to fermentors (Peña et al., 1997, 2000). Both, the power input (P/V) and the oxygen transfer rate (OTR), have been used as scaling up parameters (Peña et al., 2008; Reyes et al., 2003). Recently, our group has studied both the evolution of the specific power consumption and oxygen transfer rate, occurring in shake flasks cultures of *A. vinelandii* (Peña et al., 2007).

These studies have revealed that the power consumption increased exponentially during the course of the fermentation (up to 1.4 kW m^{-3}) due to an increase in the viscosity of the culture broth. Taking these data as a starting point, a scale-up strategy based on the evolution of the power input observed in shake flasks has been evaluated, trying to reproduce in a stirred fermentor culture the MMM of the alginates obtained in shake flasks (Peña et al., 2008). Simulating the evolution of the power input in 14 L fermentors, allowed us to reproduce the MMM and molecular mass distributions of the alginate obtained in shake flasks (Figure 4), a situation that had not been possible to achieve before using other criteria (i.e., initial power input (Reyes et al., 2003)).

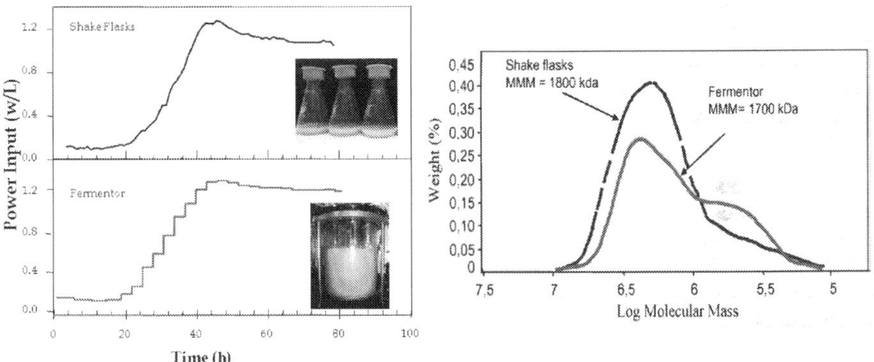

Fig. 4. Scaling-up of alginate production using the evolution of power input as a criterion.

Currently PHAs bacterial production, at industrial scale, is mainly conducted using the strain *C. necator*, until now the most cost-effective fermentative process for the copolymer PHBV (Verlinden, 2007; Wang et al., 2011). In addition, several species of *Pseudomonas* have been described as potential producers of mcl-PHAs (Durner et al., 2000; Hartmann et al., 2005; Sun et al., 2007; Wang et al., 2011). For both fermentation processes the scaling up strategies have focused not only on increasing biomass and polymer content using low cost feedstock, but also on improving chemical properties of the biopolymer. The main strategies proposed in this line are related to the use of fed batch and continuous cultures (Sun et al., 2007; Verlinden et al., 2007;). The PHAs production is usually operated as fed batch cultures, with an initial growth phase in rich medium in order to obtain a high cell density culture, followed by a product accumulation phase usually under substrate limiting conditions (Verlinden et al. 2007; Sun et al., 2007). It is worthy to point out that for fed batch cultivations it is important to define the composition of the media used for starting the culture, the feeding composition, the type of limitation used for inducing PHA synthesis, and the time at which the feeding should be started. The adequate selection of these parameters can improve the yields and composition of the PHA produced.

Fed batch cultures of *A. vinelandii* UWD using beet molasses have successfully improved PHAs content from 7 g L^{-1} obtained in flasks, to up to 25-36 g L^{-1} in fed batch cultures, reaching a maximum yield of 0.65 g $_{PHA}$ g^{-1}$_{Carbon source}$ (Chen & Page, 1997; Page et al., 2001). Continuous fermentation is an alternative to increase PHAs productivity (Sun et al., 2007). These fermentations have been successfully used to improve mcl-PHAs production parameters with *Pseudomonas* (Sun et al., 2007) and scl-PHAs with *C. necator* (Zinn et al., 2003). Using *C. necator* DSM 428, Zinn et al., (2003) evaluated the PHBV copolymer production under dual (C,N) limitation in chemostat cultures, obtaining controlled composition of the polymer in the range of 0-62 mol % HV. The cultivations in chemostat allowed a constant production of PHB/HV and an accurate control of the polymer composition.

In the case of mcl-PHAs production, Hartmann et al., (2005) evaluated chemostat cultures of *Pseudomonas putida* GPo1 under dual (C, N) limitation. They found, that the monomeric composition of mcl-PHAs was independent of the C:N ratio in the feed media, but it was dependent on the dilution rate. They also found that at low dilution rates, the fraction of aliphatic monomers in the mcl-PHAs was slightly higher than at high dilution rates.

9. Design of novel processes to improve alginate and PHA production

Alginate production by fermentation using the *A. vinelandii* bacterium could be a feasible strategy; however, polymer concentrations in batch cultures reported so far are very low, with maximum concentrations of alginate in the range of 3–5 g L^{-1} (Parente et al., 2000; Peña et al., 2000; Sabra et al., 2000). It is important to point out that the final alginate concentration is the most important parameter in determining the economics of the process, as it is related to the recovery cost, especially during the precipitation step (Peña et al., 2008). Several fermentation strategies have been reported in the literature with the aim of improving the quantity and/or quality of the alginate (Asami et al., 2004; Cheze-Lange et al., 2002; Mejia et al., 2010; Saude & Junter, 2002). For example, Cheze-Lange et al., (2002) reported the advantages of continuous production of bacterial alginate by *A. vinelandii*, coupled to a system of membranes of varying nominal pore sizes. According to these authors, the yields of alginate with respect to sucrose were significantly higher compared to the batch process. However, the MM of the polymer and the polydispersity were very similar to those of the alginate obtained from batch experiments. Asami et al., (2004) found that the productivity and the fraction of GG-blocks of the alginate produced by *A. vinelandii* in a bubble column were higher than those obtained in shake flasks. They observed that the production of GG-blocks in the late exponential growth phase was higher than that obtained in the stationary phase. However, the authors did not explain the reasons for the difference in the fraction of GG blocks under varying conditions.

Our research group has carried out studies in fed-batch and multistage fermentation processes (Mejía et al., 2010; Priego-Jiménez et al., 2005) that are able to achieve high biomass concentration, in order to take advantage of the higher specific-alginate production capacities of mutant such as the AT6 strain. Employing a high oxygen concentration (10%) allowed obtaining a maximum biomass concentration of 7.5 g L^{-1} in the first stage of the cultivation. In the second stage, the cultures were limited by oxygen (oxygen close to 0%) and fed with a sucrose solution at high concentration. Under those conditions, the growth rate decreased considerably and the cells used the carbon source mainly for alginate biosynthesis, obtaining a maximum concentration of 9.5 g L^{-1}, after 50 h of cultivation. Alginate concentration obtained from the AT6 strain was two fold higher than that obtained using the wild-type strain (ATCC 9046) and was the highest reported in the literature (Mejía et al., 2010).

Most of the studies using *Azotobacter* spp. for PHA production have been implemented using batch cultures. Because PHAs are intracellular products and their synthesis occurs under growth limiting conditions, like oxygen limitation, fed-batch fermentation or multistage cultures have been the methods used to achieve high cell densities containing the highest possible amount of PHA with several PHA producers (Akaraonye et al., 2010). On the other hand, continuous cultivation is an interesting alternative strategy because a high productivity can be reached, especially for strains with a high specific growth rate (Akaraonye et al., 2010). However, it is difficult to balance biomass concentration, PHAs content and productivity, because changes in the dilution rate can have opposite effects on cell growth and PHAs synthesis. This is due to the requirement for a nutrient limitation to induce polymer synthesis, at growth rates below the maximum specific growth rate. The ability to produce PHAs under non limiting growth conditions of some *A. vinelandii* strain could represent an advantage to establish continuous culture processes (Page & Knosp, 1989).

Two-stage continuous cultures can help establishing a good growth/PHAs synthesis compromise. Jung et al., (2001) used a two-stage continuous cultivation system with two fermentors connected in series, producing cells at a specific growth rate in the first compartment, and establishing conditions to accumulate PHA at higher rates in a second compartment, with a relatively long residence time. Dilution rates of 0.21 h^{-1} in the first fermentor and 0.16 h^{-1} in the second fermentor yielded a volumetric PHAs productivity of 0.06 g_{PHA} L^{-1} h^{-1}, a high productivity for cultures grown on alkanes.

Processes using cheaper substrates have the potential to lower the production costs of PHAs production, but for the use of some of these substrates additional processing is needed. Cerrone et al., (2010) reported an interesting strategy to simultaneously produce PHB and treat olive mil wastewater using *A. vinelandii* UWD, *A. vinelandii* ATCC 12387, or *A. chroococcum*. It consisted on a phase of anaerobic digestion of the olive mil wastewater during 27 h, in order to produce short-chain fatty acids, to later inoculate the *Azotobacter* strains for the aerobic PHA producing stage. Similar strategies could be implemented for the utilization of other agro-industrial residues allowing the use of cheap substrates and additionally reducing costs by linking PHA production with waste disposal processes.

10. Down stream processes

Alginate and PHAs have been proposed for novel applications in pharmaceutical and biomedical fields. However, for these applications it is necessary to ensure products with a high purity, and in most cases with a defined chemical composition. Chemical composition of these polymers can be controlled by the fermentation stage, but their purity will be determined by the down stream processing. Moreover, costs and efficiency of purification procedures could affect the whole process feasibility.

The extraction process of alginate from *Azotobacter*, starts with the supernatant recovery from the fermentation broth by centrifugation. Afterwards, the supernatant is treated with $NaCl_2$, followed by an acidification-hydration step, and finally it is precipitated with isopropanol. This final product could be dried and milled (Sabra & Zeng, 2009). However, alginates for applications in the biomedical and pharmaceutical fields need to be non-immunogenic, and this extraction process does not ensure a high purity of the product. The immunogenic response could be due to the presence of polyphenols, endotoxins or proteins (Ménard et al., 2010). With the purpose of removing these impurities Ménard et al., (2010) proposed the introduction of size exclusion chromatography (SEC). Using this method, the authors observed a reduction of up to 90 % of residual protein contaminants in commercial alginate and therefore a decrease in the immunogenicity of the alginate beads prepared. Once that alginate is purified, it could be modified to enhance or change its physicochemical properties, by either chemical (Yang et al., 2011) or enzymatic methods (Morch et al., 2008). Alginate downstream modifications could include, acetylation, addition of aminoacids and/or proteins, deacetylation, epimerization, oxidation, sulfation and copolymerization.

The PHAs extraction processes require the separation of the cells containing the polymer by centrifugation. The recovery of intracellular PHAs could be carried out by solvent extraction using acetone, chloroform, methylene chloride or dichloroethane. Although this method is the most used, it is also expensive and environmentally unfriendly (Verlinden et al., 2007; Yasotha et al., 2006). Besides, several alternative methods have been developed to improve PHAs purification. Enzymatic digestion does not need hazardous solvents and it shows high selectivity. Yasotha et al., (2006) proposed an enzymatic method coupled to an ultrafiltration system and achieved a final PHAs purity of 92.6 % with a recovery of almost 90 %. However, this method could be very expensive. Another interesting alternative for PHAs recovery was proposed by Page & Cornish (1993), using fish peptone like nitrogen source for the growth of *A. vinelandii*. Fish peptone enhanced PHB production and led to the production of pleomorphic and osmotically sensitive cells. They took advantage of this cell fragility to simplify PHAs extraction method using NH_4OH at 45°C. With this method PHB was recovered with a 94% of purity. Finally, Hejazi et al., (2003), developed a method based on supercritical fluid disruption of the cells using supercritical CO_2 at 200 atm, with a PHB recovery of 89%. Although this method also uses organic solvents, it requires less than the traditional extraction method.

11. Conclusions

Based on a better understanding of the biosynthesis and regulation of alginate and PHAs in *A. vinelandii*, as well as on the development of new cell culture systems for biopolymers production, it is possible to propose new fermentation strategies to obtain alginate and PHAs with specific chemical characteristics and more defined material properties. These materials could be used in specific applications in pharmaceutical and biomedical fields. In summary, this chapter has shown that the use of a multidisciplinary approach, integrating molecular and bioengineering aspects, would allow the optimization of both alginate and PHAs production using *A. vinelandii*.

12. Acknowledgment

Financial support of DGAPA-UNAM (grants IN218201, IN216700, and IN221809) and CONACyT (grant 57220, 101643 and 127979) is gratefully acknowledged. The authors acknowledge R. Rodríguez Bahena for his assistance on computer support.

13. References

Akaraonye, E.; Keshavarz, T. & Roy, I. (2010). Production of polyhydroxyalkanoates: the future Green materials of choice. *Journal of Chemical Technology and Biotechnology, 85,* 732-743.

Aldor, I. & Keasling, J. (2003). Process design for microbial plastic factories: metabolic engineering of polyhydroxyalkanoates. *Current Opinion on Biotechnology, 14,* 475-483.

Asami, K.; Aritomi, T.; Tan, Y. & Ohtaguchi, K. (2004). Biosynthesis of polysaccharide alginate by *Azotobacter vinelandii* in a bubble column. *Journal of Chemical Engineering Japan, 37,* 1050-1055.

Augst, A.; Joon-Kong, H. & Mooney, D. (2006). Alginate hydrogels as biomaterials. *Macromolecular Bioscience, 6,* 623-633.

Banki, M.; Gerngross, T. & Wood, D. (2005). Novel and economical purification of recombinant proteins: Intein-mediated protein purification using in vivo polyhydroxybutyrate (PHB) matrix association. *Protein Sciences, 14(6),* 1387–1395.

Barnard, G.; McCool, J.; Wood, D. & Gerngross, T. (2005). Integrated recombinant protein expression and purification platform based on *Ralstonia eutropha*. *Applied and Environmental Microbiolology, 71,* 5735–5742.

Brockelbank, J.; Peters, V. & Rehm, B. (2006). Recombinant *Escherichia coli* strain produces a ZZ domain displaying biopolyester granules suitable for Immunoglobulin G purification. *Applied and Environmental Microbiology, 72(11),* 7394–7397.

Campos, M.; Martinez-Salazar, J.; Lloret, L.; Moreno, S.; Núñez, C.; Espín, G. & Soberon-Chávez, G. (1996). Characterization of the gene coding for GDP-mannose dehydrogenase (*algD*) from *Azotobacter vinelandii*. *Journal of Bacteriology, 178,* 1793-1799.

Castañeda, M.; Guzmán, J.; Moreno, S. & Espin, G. (2000). The GacS sensor kinase regulates alginate and poly-beta-hydroxybutyrate production in *Azotobacter vinelandii*. *Journal of Bacteriology, 182,* 2624-2628.

Castañeda, M.; Sanchez, J.; Moreno, S.; Nunez, C. & Espin, G. (2001). The global regulators GacA and sigma(S) form part of a cascade that controls alginate production in *Azotobacter vinelandii*. *Journal of Bacteriology, 183,* 6787-6793.

Cerrone, F.; Sánchez-Peinado, M.; Juárez-Jiménez, B.; González-López, J. & Pozo, C. (2010). Biological teatment of two-phase olive mil wastewater (TPOMW, alpeorujo): Polyhydroxyalkanoates (PHAs) production by *Azotobacter* strains. *Journal of Microbiology and Biotechnology, 20,* 594-601.

Chen, G. & Page, W. (1997). Production of poly-β-hydroxybutyrate by *Azotobacter vinelandii* in a two-stage fermentation process. *Biotechnology Technology, 11,* 347-350.

Chen, G. & Wu, Q. (2005). The application of polyhydroxyalkanoates as tissue engineering materials. *Biomaterials, 26,* 6565–6578.

Chen, G. (2009). A microbial polyhydroxyalkanoates (PHA) based bio-and materials industry. *Chemistry Society Reviews, 38*, 2434–2446.

Cheze-Lange, H.; Beunard, D.; Dhulster, P.; Guillochon, D.; Caze, A.; Morcellet, M.; Saude, N. & Junter.; G. (2002). Production of microbial alginate in a membrane bioreactor. *Enzyme Microbiology Technology, 30*, 656-661.

Clementi, F. (1997). Alginate production by *Azotobacter vinelandii*. *Critical Reviews in Biotechnology, 17*, 327-361.

Conway, T. (1992). The Entner-Doudoroff pathway: History, physiology and molecular biology. *FEMS Microbiology Reviews, 9*, 1-27.

Díaz-Barrera, A.; Peña, C. & Galindo, E. (2007). The oxygen transfer rate influences the molecular mass of the alginate produced by *Azotobacter vinelandii*. *Applied Microbiology Biotechnology, 76*, 903-910.

Díaz-Barrera, A.; Silva, P.; Ávalos, R. & Acevedo, F. (2009). Alginate molecular mass produced by *Azotobacter vinelandii* in response to changes of the O_2 transfer rate in chemostat cultures. *Biotechnology Letters, 31*, 825-829.

Díaz Barrera, A.; Silva, P.; Berrios, J. & Acevedo, F. (2010). Manipulating the molecular weight of the alginate produced by *Azotobacter vinelandii* in continuous cultures. *Bioresource Technology, 101*, 9405-08.

Donnan, F. & Rose, R. (1950). Osmotic pressure, molecular weight, and viscosity of sodium alginate. *Canadian Journal Research, 28 (B)*, 105-113.

Dörig, G. & Pier G. (2008). Vaccines and immunotherapy against *Pseudomonas aeruginosa*, *Vaccine, 26*, 1011-1024.

Draget, I. & Taylor, C. (2011). Chemical, physical and biological properties of alginates and their biomedical implications. *Food Hydrocolloids, 25*, 251-256.

Durner, R. ; Witholt, B. & Egli, T. (2000). Accumulation of Poly R-β-Hydroxyalkanoates in *Pseudomonas oleovorans* during growth with octanoate in continuous culture at different dilution rates, *Applied and Environmental Microbiology, 66(8)*, 3408-3414.

Ertesvag, H.; Hoidal, H.; Schjerven, H.; Svanem, B. & Valla, S. (1999). Mannuronan C-5 epimerases and their application for *in vitro* and *in vivo* design of new alginates useful in biotechnology. *Metabolic Engineering, 1*, 262-269.

Freier, T. (2006). Biopolyesters in Tissue Engineering Applications. *Advances in Polymer Science. 203*, 1-61.

Fukui, T.; N. Shiomi, & Y. Doi. (1998). Expression and characterization of (*R*)-specific enoyl coenzyme A hydratase involved in polyhydroxyalkanoate biosynthesis by *Aeromonas caviae*. *Journal of Bacteriology, 180*, 667–673.

Galindo, E.; Peña, C. ; Núñez, C. ; Segura, D. & Espin, G. (2007). Molecular and bioengineering strategies to improve alginate and polyhydroxyalkanoate production by *Azotobacter vinelandii*. *Microbial Cell Factories, 6*, 1-16.

Gimmestad, M.; Ertesvag, H.; Heggeset, T.;Aarstad, O.; Svanem, B. & Valla, S. (2009). Characterization of three new *Azotobacter vinelandii* alginate lyases, one of which is involved in cyst germination. *Journal of Bacteriology, 191*, 4845-4853.

González-López, J. ; Pozo, C. ; Martínez-Toledo, M. ; Rodelas, B. & Salmeron, V. (1996). Production of polyhydroxyalkanoates by *Azotobacter chroococcum* H23 in wasteater from olive oil mills (Alpechin), *International Biodeterioration and Biodegradation, 38 (3-4)*, 271-276.

Grage, K.; Jahns, A.; Parlane, N.; Palanisamy, R.; Rasiah, I.; Atwood, J. & Rehm B. (2009). Bacterial polyhydroxyalkanoate granules: biogenesis, structure, and potential use as nano-/micro-Beads in biotechnological and biomedical applications. *Biomacromolecules, 10,* 660–669.

Grant, G.; Morris, E.; Rees, D.; Smith, P.; & Thom, D. (1973). Biological interactions between polysaccharide and divalent cations: the egg box model. *FEBS Letters, 32,* 195-198.

Hartmann, R. ; Hany, R. ; Pletscher, E. ; Ritter, A. ; Witholt, B. & Zinn M. (2005). Tailor-made olefinic medium chain lenght poly R-hydroxyalkanoates by *Pseudomonas putida* Gpo1 : Batch versus chemostat production, *Biotechnology and Bioengineering, 93 (4),* 737-746.

Hejazi, P. ; Vasheghani-Farahani, E. & Yamini, Y. (2003). Supercritical fluid disruption of *Ralstonia eutrofa* for Poly-β-Hydroxybutyrate recovery, *Biotechnology Progress, 19,* 1519-1523.

Hernández R. ; Orive G. ; Murua A. ; & Pedraz J. (2010). Microcapsules and microcarriers for in situ cell delivery, *Advanced Drug Delivery Reviews, 62 (7-8),* 711-730.

Hoesli, C. ; Raghuram, K. ; Kiang, R. ; Mocinecová, D. ; Hu, X.L ; Johnson, J. ; Lacík, I. ; Kieffer, T. & Piret, J. (2011). Pancreatic cell immobilization in alginate beads produced by emulsion and internal gelation, *Biotechnology and Bioengineering, 108 (2),* 424-434.

Hori, Y.; Winans, A.; Huang, C.; Horrigan, E. & Irvine, D. (2008), Injectable dendritic cell carrying alginate gels for immunization and immunotherapy. *Biomaterials, 29,* 3671-3682.

Hori, Y.; Stern, P. J.; Hynes, R. & Darrell, J. (2009). Engulfing tumors with synthetic extracellular matrices for cancer immunotherapy. *Biomaterials, 30 (35),* 6757-6767.

Jahns, A.; Haverkamp, R. & Rehm, B. (2008). Multifunctional Inorganic-Binding Beads Self-Assembled Inside Engineered Bacteria. *Bioconjugate Chemistry, 19 (10),* 2072–2080.

Jung, K.; Hazenberg, W.; Prieto, M. & Witholt, B. (2001). Two-stage continuous process development for the production of medium-chain-length poly(3-hydroxyalkanoates). *Biotechnology and Bioengineering, 72,* 19-24.

Kashef, N.; Behzadian-Nejad, Q.; Najar-Peerayeh, S.; Mousavi-Hosseini, K.; Moazzeni, M. & Gholamreza E. (2006). Synthesis and characterization of *Pseudomonas aeruginosa* alginate-tetanus toxoid conjugate. *Journal of Medical Microbiology, 55,* 1441-1446.

Kennedy, C.; Rudnick, P.; MacDonald, T.; & Melton, T. (2005). Genus *Azotobacter*. In: *Bergey´s Manual of Systematic Bacteriology,* G. M. Garrita, 384-401, vol. 2 part B, Springer-Verlag, New York, NY.

Khanna, S. & Srivastava, A. (2005). Recent advances in microbial polyhydroxyalkanoates. *Process Biochemistry, 40,* 607–619.

Lee S. (1996). Bacterial Polyhydroxyalkanoates. *Biotechnology and Bioengineering 49,* 1-14.

Lee, S.; Park, J.; Park, T.; Lee, S.; Lee, S. & Park, J. K. (2005). Selective Immobilization of Fusion Proteins on Poly(hydroxyalkanoate) Microbeads. *Analytical Chemistry, 77,* 5755–5759.

Lozano, E.; Galindo, E. & Peña, C. (2011). The quantity and molecular mass of the alginate produced by *Azotobacter vinelandii* under oxygen-limited and non oxygen-limited conditions are determined by the maximal oxygen transfer rate (OTR$_{max}$). *Microbial Cell Factories, 10,* 1-13.

Ludwig, B.; Zimmerman, B.; Steffen, A.; Yavriants, K.; Azarov, D.; Reichel, A.; Vardi, P.; Grman, T.; Shabtay, N.; Rote, A.; Evron, Y., Neufeld, T.; Mimon, S.; Ludwig, S.; Brendel, M.; Bornstein, S. & Barkai, U. (2010). A novel device for islet transplantation providing immune protection and oxygen supply. *Hormone Metabolism Research, 42,* 918-922.

Lynn, A. & Hassid, W. (1966). Pathway of alginic acid synthesis in the marine brown alga, *Fucus gardneri* Silva. *Journal of Biological Chemistry, 241,* 5284-5297.

Madison, L. & Huisman, G. (1999). Metabolic Engineering of Poly(3-Hydroxyalkanoates): From DNA to Plastic. *Microbiology and Molecular Biology Reviews, 63(1),* 21-53.

Manchak, J. & Page, W. (1994). Control of polyhydroxyalkanoates synthesis *in Azotobacter vinelandii* strain UWD. *Microbiology, 140,* 953-963.

Mee, C.; Banki, M. & Wood, D. (2008). Towards the elimination of chromatography in protein purification: Expressing proteins engineered to purify themselves. *Chemical Engineering Journal, 135 (1-2),* 56-62.

Mejía, M.; Segura, D.; Espín, G.; Galindo, E. & Peña, C. (2010). Two stage fermentation process for alginate production by *Azotobacter vinelandii* mutant altered in poly-β-hydroxybutyrate (PHB) synthesis. *Journal of Applied Microbio*logy, *108,* 55-61.

Ménard, M.; Dusseault, J.; Langlois, G.; Baille, W.; Tam, S.; Yahia, L.; Zhu, X.; & Hallé, J-P. (2010). Role of protein contaminants in the inmmunogenicity of alginates. *Journal of Biomedical Materials Research part B: Applied Biomaterials,* DOI:10.1002/jbm.b.31570.

Morch, Y.; Holtan, S.; Donatti, I.; Strand, B. & Skjak-Braek. (2008). Mechanichal properties of C-5 epimerized alginates. *Biomacromolecules, 9,* 2360-2368.

Myshkina, V.; Nikolaeva, D.; Makhina, T.; Bonartsev, A.; & Bonartseva, G. (2008). Effect of growth conditions on the molecular weight of polyhydroxybutyrate produced by *Azotobacter chroococcum* 7B, *Applied Biochemistry and Microbiology, 44 (5),* 482-486.

Myshkina, V.; Ivanov, E.; Nikolaeva, D.; Makhina T.; Bonartsev A.; Filatova E.; Ruzhitsky A. & Bonartseva G. (2010). Biosynthesis of poly-β-hydroxybutyrate-co-hydroxyvalerate copolymer by *Azotobacter chroococcum* strain 7B. *Applied biochemistry and microbiology, 46 (3),* 289.296.

Noguez, R.; Segura, D.; Moreno, S.; Hernández, A.; Juárez, K. & Espín, G. (2008). Enzyme I[Ntr], NPr and IIA[Ntr] are involved in regulation of the poly-β-hydroxybutyrate biosynthetic genes in *Azotobacter vinelandii. Journal of Molecular Microbiology and Biotechno*logy *15,* 244-254.

Page, W. & Knosp, O. (1989). Hyperproduction of polyhydroxybutyrate during exponential growth of *Azotobacter vinelandii* UWD, *Applied and Environmental Microbiology, 55(6),* 1334-1339.

Page, W.; Manchak, J. & Rudy, B. (1992). Formation of poly(hydroxybutyrate-co-hydroxyvalerate) by *Azotobacter vinelandii* UWD. *Applied and Environmental Microbiology, 58,* 2866-2873.

Page, W. (1992). Production of polyhydroxyalkanoates by *Azotobacter vinelandii* UWD in beet molasses culture, *FEMS Microbiology Reviews, 103,*149-158.

Page, W. & Cornish, A. (1993). Growth of *Azotobacter vinelandii* UWD in Fish Peptone Medium and Simplified Extraction of Poly-β-hydroxybutyrate. *Applied and Environmental Microbiology, 59,* 4236-4244.

Page, W.; Tindale, A.; Chandra, M. & Kwon E. (2001). Alginate formation in *Azotobacter vinelandii* UWD during stationary phase and the turnover polyhydroxybutyrate. *Microbiology*, 147, 483-490.

Parente, E.; Crudele, M.; Ricciardi, A.; Mancini, M.; Clementi F. (2000). Effect of ammonium sulphate concentration and agitation speed on the kinetics of alginate production by *Azotobacter vinelandii* DSM576 in batch fermentation. *Journal of Industrial Microbiology and Biotechnology*, 25, 242-248

Peña, C.; Campos, N.; Galindo, E. (1997) Changes in alginate molecular mass distributions, broth viscosity and morphology of *Azotobacter vinelandii* cultured in shake flasks. *Applied Microbiology and Biotechnology*, 48, 510-515.

Peña, C.; Trujillo-Roldan, M. & Galindo, E. (2000). Influence of dissolved oxygen tension and agitation speed on alginate production and its molecular weight in cultures of *Azotobacter vinelandii*. *Enzyme and Microbiology Technology*, 27, 390-398.

Peña, C.; Hernandez, L. & Galindo, E. (2006) Manipulation of the acetylation degree of *Azotobacter vinelandii* alginate by supplementing the culture medium with 3-(N-morpholino)-propane-sulfonic acid. *Letters of Applied Microbiology*, 43, 200-204.

Peña, C.; Peter, C.; Büchs, J. & Galindo, E. (2007) Evolution of the specific power consumption and oxygen transfer rate in alginate-producing cultures of *Azotobacter vinelandii* conducted in shake flasks. *Biochemical Engineering Journal*, 36: 73-80.

Peña, C.; Millán, M. & Galindo, E. (2008). Production of alginate by *Azotobacter vinelandii* in a stirred fermentor simulating the evolution power input observed in shake flasks. *Processes of Biochemistry*, 43, 775-778.

Peña, C.; Galindo, E.; Büchs, J. (2011). The viscosifying power, degree acetylation and molecular mass of the alginate produced by *Azotobacter vinelandii* in shake flasks are determined by the oxygen transfer rate. *Process Biochemistry*, 46, 290-297.

Peralta-Gil, M.; Segura, D.; Guzmán, J.; Servin-Gonzalez, L. & Espin, G. (2002). Expression of the *Azotobacter vinelandii* poly-beta-hydroxybutyrate biosynthetic *phbBAC* operon is driven by two overlapping promoters and is dependent on the transcriptional activator PhbR. *Journal of Bacteriology*, 184, 5672-5677.

Peters, V. & Rehm, B. (2008). Protein engineering of streptavidin for *in vivo* assembly of streptavidin beads. *Journal of Biotechnology*, 134, 266–274.

Pindar, D. & Bucke, C. (1975). The biosynthesis of alginic acid by *Azotobacter vinelandii*. *Biochemistry Journal*, 152, 617-622.

Poole, R. & Hill, S. (1997). Respiratory protection of nitrogenase activity in *Azotobacter vinelandii*, roles of the terminal oxidases. *Biosciences Reports*, 17, 303-317.

Post, E.; Kleiner, D. & Oelze, J. (1983). Whole cell respiration and nitrogenase activities in *Azotobacter vinelandii* growing in oxygen controlled continuous culture. *Archives of Microbiology*, 134, 68-72.

Priego-Jiménez, R.; Peña, C.; Ramírez, O.T.; Galindo, E. (2005). Specific growth rate determines the molecular weight of the alginate produced by *Azotobacter vinelandii*. *Biochemistry Engineering Journal*, 25, 187-193.

Pyla, R.; Kim, T.; Silva, J. & Jung, Y. (2009). Overproduction of poly-beta-hydroxybutyrate in the Azotobacter vinelandii mutant that does not express small RNA ArrF. *Applied Microbiology and Biotechnology*, 84 (49), 717-724.

Quagliano, J. & Miyazaki, S. (1997). Effect of aeration and carbon/nitrogen ratio on the molecular mass of the biodegradable polymer poly-β-hydroxybutyrate obtained from *Azotobacter chroococcum* 6B, *Applied microbiology and biotechnology, 48,* 662-664.

Remminghorst, U. & Rehm, B. (2006). Bacterial alginates: from biosynthesis to applications. *Biotechnology Letters, 28,* 1701-1712.

Reyes, C.; Peña, C. & Galindo, E. (2003). Reproducing shake flasks performance in stirred fermentors: production of alginates by *Azotobacter vinelandii. Journal of Biotechnology, 105,* 189-198.

Ruth, K.; Grubelnik, A.; Hartman, R.; Egli, T.; Zinn, M.; & Ren Q. (2007). Efficient production of (R)-β-hydroxycarboxylic acids by biotechnological conversion of Polyhydroxyalkanoates and their purification. *Biomacromolecules. 8(1),* 279-286.

Sabra, W.; Zeng, A.; Sabry, S.; Omar, S. & Deckwer, W-D. (1999). Effect of phosphate and oxygen concentrations on alginate production and stoichiometry of metabolism of *Azotobacter vinelandii* under microaerobic conditions. *Applied Microbiology and Biotechnology, 52,* 773-780.

Sabra, W.; Zeng, A.; Lunsdorf, H. & Deckwer, W-D. (2000). Effect of oxygen on formation and structure of *Azotobacter vinelandii* alginate and its role in protecting nitrogenase *Applied and Environmental Microbiology, 66,* 4037-4044.

Sabra, W.; Zeng, A. & Deckwer, W.D. (2001). Bacterial alginate: physiology, product quality and process aspects. *Applied Microbiology and Biotechnology, 56,* 315-325.

Sabra, W. & Zeng, A. (2009). Microbial production of alginates: Physiology and Process Aspects, In: *Alginates: Biology and Applications,* Bernd H. A. Rehm, 153-173, Microbiology Monographs Vol. 13, Springer Verlag, DOI: 10.1007/978-3-540-92679-5, Berlín-Heidelberg.

Saude, N. & Junter, G. (2002). Production and molecular weight characteristics of alginate from free and immobilized-cell cultures of *Azotobacter vinelandii. Process Biochemistry 2 (37),* 895-900.

Segura, D. & Espin, G. (1998). Mutational inactivation of a gene homologous to *Escherichia coli* ptsP affects poly-beta-hydroxybutyrate accumulation and nitrogen fixation in *Azotobacter vinelandii. Journal of Bacteriology, 180,* 4790-4798.

Segura, D.; Vargas, E. & Espín, G. (2000). Beta-ketothiolase genes in *Azotobacter vinelandii. Gene, 260,*113-120.

Segura, D.; Cruz, T. & Espin, G. (2003a). Encystment and alkylresorcinol production by *Azotobacter vinelandii* strains impaired in polybeta- hydroxybutyrate synthesis. *Archives of Microbiology, 179,* 437-443.

Segura, D.; Guzmán, J.; & Espín, G. (2003b). *Azotobacter vinelandii* mutants that overproduce poly-beta-hydroxybutyrate or alginate. *Applied Microbiology and Biotechnology, 63,* 159-163.

Senior, P. & Dawes, E. (1971). Poly-β-hydroxybutyrate biosynthesis and the regulation of glucose metabolism in *Azotobacter beijerinckii. Biochemistry Journal, 125,* 55–66.

Senior, P.; Beech, G.; Ritchie, G. & Dawes, E. (1972). The role of oxygen limitation in the formation of poly-β-hydroxybutyrate during batch and continuous culture of *Azotobacter beijerinckii. Biochemistry Journal, 128,* 1193–1201.

Senior, P. & Dawes, E. (1973). The regulation of poly-β-hydroxybutyrate metabolism in *Azotobacter beijerinckii. Biochemistry Journal, 134,* 225–238.

Setubal, J.; dos Santos, P.; Goldman, B.; Ertesvåg, H.; Espin, G.; Rubio, L.; Valla, S.; Almeida, N.; Balasubramanian, D.; Cromes, L.; Curatti, L.; Du, Z.; Godsy, E.; Goodner, B.; Hellner-Burris, K.; Hernandez, J.; Houmiel, K.; Imperial, J.; Kennedy, C.; Larson, T.; Latreille, P.; Ligon, L.S.; Lu, J.; Mærk, M.; Miller, N.; Norton, S.; O'Carroll, I.; Paulsen, I.; Raulfs, E.; Roemer, R.; Rosser, J.; Segura, D.; Slater, S.; Stricklin, S.; Studholme, D.; Sun, J.; Viana, C.; Wallin, E.; Wang, B.; Wheeler, C.; Zhu, H.; Dean, D.; Dixon, R. & Wood, D. (2009). The genome sequence of *Azotobacter vinelandii*, an obligate aerobe specialized to support diverse anaerobic metabolic processes. *Journal of Bacteriology. 191(14)*, 4534-4545.

Skjak-Braek, G.; Grasdalen, H. & Larsen, B. (1986). Monomer sequence and acetylation pattern in some bacterial alginates. *Carbohydrates Research, 154*, 239-250.

Smidsrod, O. & Draget, K. (1996). Chemistry and physical properties of alginates. *Carbohydrates European, 14*, 6-12.

Socolofsky, M. & Wyss, O. (1962). Resistance of the *Azotobacter* cyst. *Journal of Bacteriology, 84*, 119-124.

Steinbüchel, A. & Lütke-Eversloh, T. (2003). Metabolic engineering and pathway construction for biotechnological production of relevant polyhydroxyalkanoates in microorganisms. *Biochemical Engineering Journal, 16*, 81–96.

Sun. H.; Pan,.; Zhigang, Y. & Shi, M. (2007). The immune response and protective efficacy of vaccination with oral microparticle *Aeromonas sobria* vaccine in mice. *International Immunopharmacology, 7*, 1259-1264.

Sun, Z.; Ramsay J.; Guay M. & Ramsay B. (2007). Fermentation process development for the production of medium chain length poly-β-hydroxyalkanoates. *Applied Microbiology Biotechnology, 75*, 475-485.

Suriyamongkol, P.; Weselake, R.; Narine, S.; Moloney, M. & Shah, S. (2007). Biotechnological approaches for the production of polyhydroxyalkanoates in microorganisms and plants — A review. *Biotechnology Advances. 25*, 148–175.

Taguchi, S. & Doi, Y. (2004) Evolution of polyhydroxyalkanoate (PHA) production system by "Enzyme Evolution": successful case studies of directed evolution. *Macromolecules Bioscience, 4*, 145-156.

Trujillo-Roldán, M.; Moreno, S.; Segura, D.; Galindo, E. & Espín, G. (2003). Alginate production by an *Azotobacter vinelandii* mutant unable to produce alginate lyase. *Applied Microbiology Biotechnology, 60*, 733-737.

Trujillo-Roldán, M.; Moreno, S.; Espín, G. & Galindo, E. (2004). The roles of oxygen and alginate-lyase in determining the molecular weight of alginate produced by *Azotobacter vinelandii*. *Applied Microbiology Biotechnology, 63*, 742-747.

Valappil, S.; Misra, K.; Boccaccini, A. & Roy, I. (2006). Biomedical applications of polyhydroxyalkanoates, an overview of animal testing and in vivo responses. *Expert Review of Medical Devices, 3 (6)*, 853-868.

Vázquez, A.; Moreno, S.; Guzmán, J.; Alvarado, A.; & Espín, G. (1999). Transcriptional organization of the *Azotobacter vinelandii algGXLVIFA* genes: characterization of *algF* mutants. *Gene, 232*, 217-222.

Verlinden, R.; Hill, D.; Kenward, M.; Wiliams, C. & Radecka, I. (2007). Bacterial synthesis of biodegradable polyhydroxyalkanoates, *Journal of Applied Microbiology, 102*, 1437-1449.

Wang, Z.; Wu, H.; Chen, J.; Zhang, J.; Yao, Y. & Chen, G.-Q. (2008). A novel self-cleaving phasin tag for purification of recombinant proteins based on hydrophobic polyhydroxyalkanoate nanoparticles. *Lab Chip*, 8, 1957–1962.

Wang, H-H.; Zhou, X.; Liu, Q. & Chen G-Q. (2011), Biosynthesis of polyhydroxyalkanoate homopolymers by *Pseudomonas putida, Applied Microbiology and Biotechnology, 89,* 1497-1507.

Williams, S. & Martin, D. (2005). Applications of Polyhydroxyalkanoates (PHA) in Medicine and Pharmacy. *Biopolymers Online*. DOI: 10.1002/3527600035.bpol4004.

Wu, G.; Moir, A.; Sawers, G.; Hill, S. & Poole R. (2001). Biosynthesis of polybeta-hydroxybutyrate (PHB) is controlled by CydR (Fnr) in the obligate aerobe *Azotobacter vinelandii. FEMS Microbiology Letters, 194,* 215-220.

Wu, Q.; Wang, Y. & Chen G. (2009). Medical application of microbial biopolyesters Polyhydroxyalkanoates. *Artificial Cells, Blood Substitutes, and Biotechnology, 37,* 1-12.

Yang, J.-S.; Xie, Y.-J. & He, W. (2011). Research progress on chemical modification of alginate: A review. *Carbohydrate polymers, 84,* 33-39.

Yao, Y., Zhan, X.; Zhang, J.; Zou, X.; Wang, Z.; Xiong, Y.; Chen, J. & Chen, G. (2008). A specific drug targeting system based on polyhydroxyalkanoate granule binding protein PhaP fused with targeted cell ligands. *Biomaterials, 29(36),* 4823-4830.

Yasotha, K.; Aroua, M.; Ramachandran, K. & Tan, I. (2006). Recovery of medium-chain-length polyhydroxyalkanoates (PHAs) through enzymatic digestion treatments and ultrafiltration, *Biochemichal Engineering Journal, 30,* 260-268.

Zapata-Vélez, A. & Trujillo-Roldán, M. (2010). The lack of a nitrogen source and/or the C/N ratio affects the molecular weight of alginate and its productivity in submerged cultures of *Azotobacter vinelandii. Annals of Microbiology, 60,* 661-668.

Zhang, X.; Luo, R.; Wang, Z.; Deng, Y. & Chen, G. (2009). Application of (R)-3-hydroxyalkanoate methyl esters derived from microbial polyhydroxyalkanoates as novel biofuels. *Biomacromolecules, 10,* 707-711.

Zinn, M.; Witholt, B. & Egli, T. (2001). Occurrence, synthesis and medical application of bacterial Polyhydroxyalkanoate. *Advanced Drug Delivery Reviews, 53,* 5–21.

Zinn, M.; Weilenmann, H.; Hany, R.; Schmid, M. & Egli T. (2003). Tailored synthesis of poly R-β-hydroxybutyrate-co-hydroxyvalerate (PHB/HV) in *Ralstonia euthropha* DSM 428, *Acta Biotechnologica, 23,* 309-316.

10

Research and Development of Biotechnologies Using Zebrafish and Its Application on Drug Discovery

Yutaka Tamaru[1,2,3], Hisayoshi Ishikawa[1],
Eriko Avşar-Ban[1],Hajime Nakatani[1,2],
Hideo Miyake[1,2,3] and Shin'ichi Akiyama[4]
[1]Department of Life Science, Graduate School of Bioresources,
[2]Laboratory of Applied Biotechnology, Venture Business Laboratory,
[3]Department of Bioinformatics, Life Science
Research Center, Mie University
[4]Department of Nephrology, Graduate School of
Medicine, Nagoya University,
Japan

1. Introduction

The zebrafish, *Danio rerio*, a small minnow from the Indian subcontinent, was first purchased from pet stores in the 1970s and propagated in the laboratory for its attractive attributes such as year-round breeding, large clutch sizes and transparent embryos[1]. It grew in popularity as an experimental system, and in the 1980s and 1990s, a critical mass of researchers began to develop the tools necessary to perform large-scale genetic screens and genomic analyses. Since then, the zebrafish research community has grown to include thousands of researchers, trained largely in the fields of developmental genetics and, more recently, functional genomics. The primary goal of the work carried out by these researchers is to use zebrafish to define the genetic mechanisms underlying vertebrate development, in many cases with direct application to human health. Now, zebrafish has several features that make them an ideal vertebrate model, for example their small size, the ease of breeding, short generation intervals, the embryos are transparent and their early development is well-characterized[2-6]. Moreover, zebrafish has recently been successfully incorporated into large-scale genetic screens due to the optical clarity of the embryos and their accessibility to various experimental techniques throughout development. The attractiveness of the zebrafish as a model organism is enhanced by the biological availability of continuously improving genomic tools and methodologies for functional characterization of the genes. In addition, transparent zebrafish embryos are well suited to manipulations involving DNA or mRNA injection, cell labeling, and transplantation. Once the scheduled zebrafish genome project is complete, targeted genetic manipulations in zebrafish would be able to become even more desirable. Since adult zebrafish only grow up to 30-50 mm in length, they can be kept a lot of population in relatively small spaces. Moreover, zebrafish are easy to maintain and breed under laboratory conditions, they have short generation times (about 3 months) and can reproduce for about 1.5 years. A number of embryos can be obtained at one time, because female fish easily lay 100–200 eggs in each spawning. After the eggs are fertilized among a pair of zebrafish, the embryos develop rapidly and the formation of somatic structures is achieved within 2-3 days of post-fertilization (**Figure 1**).

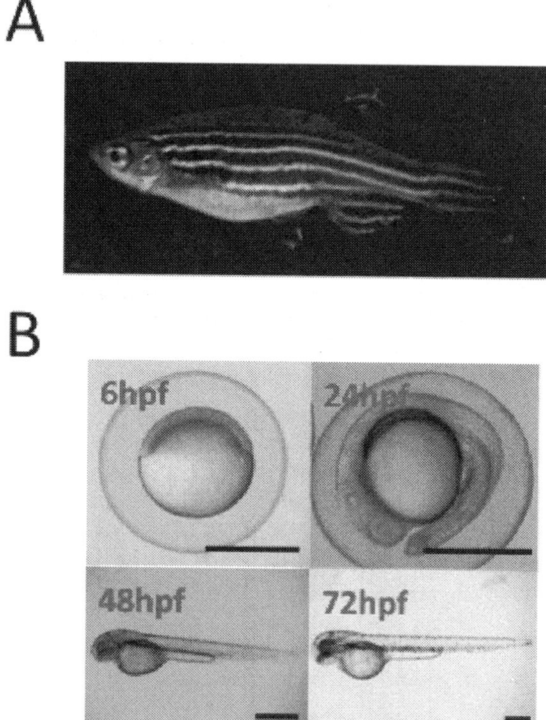

Photo images: female adult zebrafish (A) and zebrafish embryos (B) at 6, 24, 48 and 72 hpf, respectively. Scale bar, 500 μm.

Fig. 1. Zebrafish and its embryogenesis

Forward genetics has been applied, successfully, using methods for large-scale mutagenesis and screening for altered phenotypes, resulting in the discovery of more than 2000 mutations that perturb the normal development of zebrafish[7-9]. In addition to these advantages, their embryonic developmental processes are easily observed in live because of transparent embryos. Methods for standard (non-targeting) germline transgenesis of zebrafish are established[10,11], with several modifications for increasing their efficiency also reported[12-16]. One advantage of zebrafish transgenics having compared with the mammalian counterpart technology is that reproduction involves external fertilization and embryo development, eliminating the need for surgical intervention. Nowadays, the zebrafish are becoming a useful genetic model and starting to be employed in various researches such as infection desiease[17], cancer research[18], chemical genetic screening[19], toxicology[20], and proteome[21]. Some researchers noted on zebrafish as an *in vivo* protein expression system, which can be applied for useful protein production[22], while they used for genetic model are spreading.

2. Omics research in zebrafish

Modern biomedical research greatly benefits from large-scale genome-sequencing projects ranging from studies of viruses, bacteria, and yeast to multicellular organisms. There are currently many organisms whose genomes are undergoing systematic sequencing by the next-generation sequencer. The zebrafish genome-sequencing project has been started in 2001 at the Sanger Institute, and all the genome sequence will become available near the future. Zebrafish microarrays have been produced that contain either DNA fragments derived from expressed sequence tag (EST) and cDNA libraries[23], or from oligonucleotide libraries based on all the genes or transcriptional units predicted from bioinformatic analysis of the entire zebrafish genome. At present, 14,000-22,000 zebrafish genes are included on commercially available arrays (Agilent, Affymetrix, Compugen/Sigma-Aldrich, MWGBiotech and Qiagen/Operon) offering a standardized toolset for zebrafish transcriptional profiling. Recently, microRNA expression profiles have been characterized[24] adding this new family of control factors for gene expression to the zebrafish toolbox repertoire.

An important challenge facing life sciences is to quantitatively describe the bewildering complexity of living organisms[25], both to appreciate the elegance of nature and to make medically relevant predictions. Indeed, the scope of this complexity is vast. Even the function of a single mammalian cell typically involves coordinated activities among over 20,000 genes, 100,000 proteins[26], and thousands of small-molecule lipids, carbohydrates and metabolites, each of which may be expressed at differing levels over time. These components interact in physical complexes and functional modules that operate at many levels of organization[25]. On the other hand, the classic method for reverse engineering a system is to poke a component with a stick and then to characterize the effect of the perturbation[26]. An alternative is to poke many components simultaneously and at random, repeating the experiment over many random sets of components[27]. Conveniently, the genetic variation that occurs naturally within a population is a source of multifactorial perturbation[28,29]. The use of natural genetic variation to probe the causal network that links genotype and phenotype has grown recently as large data sets have been generated for many experimental model species, crops and humans[30-32].

Activity-based profiling (ABP) of proteomes is a powerful strategy for identifying the functional participants in complex biological processes[33]. The recent development of ABP, in which a chemical probe can be used to label and isolate an enzyme from a complex mixture, provides associated with a particular biological activity, thereby taking a step toward their functional identification[34,35]. Moreover, although transcriptional profiling assesses changes in the amount of RNA transcripts in response to a perturbation in environment of an organism, organ, or cell[36], the abundance of the encoded protein cannot be predicted from the abundance of the transcript. Chromatographic, electrophoretic, and mass spectroscopic methods have also been developed to separate and quantify the amount of individual proteins in proteomes[37]. However, the absolute amount of a protein is also, at best, an indirect indicator of its function. The biological potency and activity of a protein cannot be predicted from its abundance; posttranslational modification (phosphorylation, acetylation, or glycosylation) often is the switch for turning the biological activity of a protein on or off. Therefore, protein microarray provides a new strategy for assessing the in vitro interactions of selected members if a proteome with selected ligands[38]. Yet this approach is limited by the availability of relevant proteins and ligands. The zebrafish is also suitable for chemical genomics, in part as a result of the permeability of its embryos to small molecules and consequent avoidance of external confounding maternal effects[39]. The use of zebrafish in high-throughput (HTP) screens of small molecules may allow time-series analyses that could be particularly useful for studying variable gene expression in early development and for toxicogenomic studies. On the other hand, genetic suppressor screens may identify second-site mutations that modify the effect of an existing genetic mutation[40]. In this case, zebrafish larvae are most commonly used for whole-organism screens. Adult zebrafish are popular, too, but their mobility and larger size make them less convenient to use. Embryos develop quickly: within three days of fertilization a zebrafish has a vascular system, a beating heart, the fish equivalent of a pancreas and kidneys. Even better, the larvae, as well as some mutant adult strains, are transparent, facilitating imaging[41].

Metabolomics is an emerging tool that can be used to gain insights into cellular and physiological responses. In principle, the metabolome, particularly the unbiased metabolome, would be more diverse and dynamic in terms of chemical and physical properties of metabolites than the transcriptome and proteome. Therefore, the analysis of the metabolome would be suitable for describing the dynamic changes that occur during embryogenesis. However, there have been no reports on the practical application of metabolomics for determining the mechanisms underlying specific biological processes in higher organisms. Therefore, early embryogenesis was a suitable period for determining whether metabolomics can be used to understand complex biological processes. We first identified and profiled 63 types of metabolites from 24 developmental stages, i.e., from 1-cell stage to 48 h postfertilization (hpf), of zebrafish embryos by using gas chromatography/mass spectrometry (GC/MS) method[42]. Analysis of the GC/MS data with partial least square (PLS) regression clearly indicated a good correlation between metabolomes and developmental stages. Next, we developed a model for predicting embryonic stages on the basis of the metabolome. Thus, zebrafish model is a practical tool to analyze the biological processes in early development.

3. Studies on activity-based profiling with disease-associated proteins using zebrafish

Proteomic technology can be very useful in development of production processes for therapeutic proteins by use of genetically engineered animal cells[43,44] or human stem cells[45]. However, the analysis of proteomes is significantly more challenging that of genomes. In particular, there is greater diversity in proteins at the amino acid composition level; the proteome is dynamic, both spatially and temporally; and a wide range of variation of protein concentrations exists within cells[46]. Moreover, proteomic analysis is substrate limited, because methods for protein amplification are not available. Therefore, two main areas of this field are 'profiling' and 'functional' proteomics. Profiling proteomics encompasses the description of the whole proteome of an organism (by analogy with the genome) and includes organelle mapping and differential measurement of expression levels between cells or conditions. Functional proteomics characterizes protein activity, interactions and the presence of posttranslational modifications.

We are focusing on posttranslational modifications in our laboratory and have recently reported protein *O*-mannosyltransferases (POMTs) in zebrafish[47]. POMTs (POMT1 and POMT2) catalyze the first step in *O*-mannosyl glycan synthesis[48], and defects in human POMT1 (hPOMT1) or hPOMT2 result in Walker–Warburg syndrome (WWS), an autosomal recessive disorder associated with severe congenital muscular dystrophy, abnormal neuronal migration and eye anomalies[49,50]. Although zebrafish are superior for vertebrates or human *in vivo* model, the mice are the most commonly employed vertebrate's model. However, with their advantages of easy manipulation under laboratory conditions, availability of genome information, and the easy establishment of transgenic fish, the zebrafish is gradually spreading into a wide variety of studies as a handier model animal than mouse. In this study, injection of antisense morpholino oligonucleotides of zebrafish POMT1 (zPOMT1) and zPOMT2 resulted in several severe phenotypes including bended body, edematous pericaridium and abnormal eye pigmentation. Immunohistochemistry using anti-glycosylated α-dystroglycan antibody (IIH6) and morphological analysis revealed that the phenotypes of zPOMT2 knockdown were more severe than those of zPOMT1 knockdown, even though the IIH6 reactivity was lost in both zPOMT1 and zPOMT2 morphants. On the other hand, only when both zPOMT1 and zPOMT2 were expressed in human embryonic kidney 293T cells, high levels of protein *O*-mannosyltransferase activity were detected, indicating that both zPOMT1 and zPOMT2 were required for full enzymatic activity. Moreover, either heterologous combination, zPOMT1 and hPOMT2 or hPOMT1 and zPOMT2, resulted in enzymatic activity in cultured cells. These results indicate that the protein *O*-mannosyltransferase machinery in zebrafish and humans is conserved and suggest that zebrafish may be useful for functional studies of protein *O*-mannosylation. More recently, Dr. Kunkel's group has reported that two known zebrafish dystrophin mutants, *sapje* and *sapje*-like (*sapc*/100), represent excellent small-animal models of human muscular dystrophy[51]. Using these dystrophin-null zebrafish, they have screened the Prestwick chemical library for small molecules that modulate the muscle phenotype in these fish. With a quick and easy birefringence assay, they have identified seven small molecules that influence muscle pathology in dystrophin-null zebrafish without restoration of dystrophin expression. Finally, three of seven candidate chemicals restored normal birefringence and increased survival of dystrophin-null fish.

4. Recent genetic engineering in zebrafish

The transgenic fish technology is employed in diverse areas of biological researches including analysis of regulatory elements, gene over-expression, tracing of cellular lineages, mutagenesis and protein analysis. The method of gene transfer into vertebrate embryos is commonly performed by microinjection into embryo at the one cell stage. However, in the most of the mammalian's cases, it is generally difficult to obtain the embryos at quite early stage, and more difficult to maintain externally those isolated embryos. In the case of zebrafish, a huge number of embryos at one cell stage are easily available at one time because eggs are external-fertilized and spawned hundreds of eggs weekly. In general, microinjection into zebrafish embryos is relatively easier than that of other fish because of their soft chorion. Therefore, it is easy to imagine that a large numbers of injections will be needed for developing protein expression in zebrafish. To improve performance of injection by hand, we are developing auto-injection machine for zebrafish eggs (**Figure 2**). This injection system can currently operate 100 pL per embryo level injection, and the injectioin speed is 20 eggs per minute.

Fig. 2. Fully automated injection system for zebrafish

Techniques for reverse genetic approaches in zebrafish are limited to mRNA knockdown strategies using modified antisense oligomers (morpholinos) [52] and TILLING for point mutations by detection of heterozygosity in a locus of interest, and subsequent sequencing, among a library of chemically mutagenized gametes. On the other hand, conventional gene targeting, a powerful technique for gene disruption in mouse embryonic stem cells[53], often requires positive-negative selection with cytotoxic drugs[54], which is inapplicable in the context of a vertebrate embryo. In 2008, the use of zinc-finger nucleases (ZFNs) for somatic and germline disruption of genes in zebrafish, in which targeted mutagenesis was previously intractable, have been repoted[55,56]. ZFNs induce a targeted double-strand break in the genome that is repaired to generate small insertions and deletions. Therefore, only co-injection of mRNAs encoding these ZFNs into one-cell-stage zebrafish embryos led to mutagenic lesions at the target site that were transmitted through the germ line with high frequency. In near future, the use of engineered ZFNs to introduce heritable mutations into a genome obviates the need for embryonic stem cell lines and should be applicable to most animal species for which early stage embryos are easily accessible.

5. Development of protein expression vectors in zebrafish

The plasmid DNA has been used for expression of exogenous gene in wide variety of animals. For the zebrafish, the pXeX vector might be first used for protein expression in zebrafish, which is originally used for protein expression in *Xenopus* embryo[57], containing the transcription regulatory regions of the *Xenopus laevis* elongation factor-1 alpha gene (EF-1 alpha) and SV40 polyadenilation signaling. Amsterdam *et al.* cloned green fluorescent protein (GFP) into pXeX vector (pXeX-GFP) and expressed GFP in zebrafish embryos by plasmid injection into fertilized eggs[58]. Moreover, they constructed pXIG vector which is originally constructed for expression in zebrafish embryos, based on the backbone of pXeX vector. They inserted rabbit beta-globin IVS2 into the promoter region of pXeX vector, and then followed by GFP's open reading frame. Using the pXIG vector, they expressed GFP in the whole body of transgenic zebrafish and observed more frequent generation of transgenic fish than that of pXeX-GFP injectant.

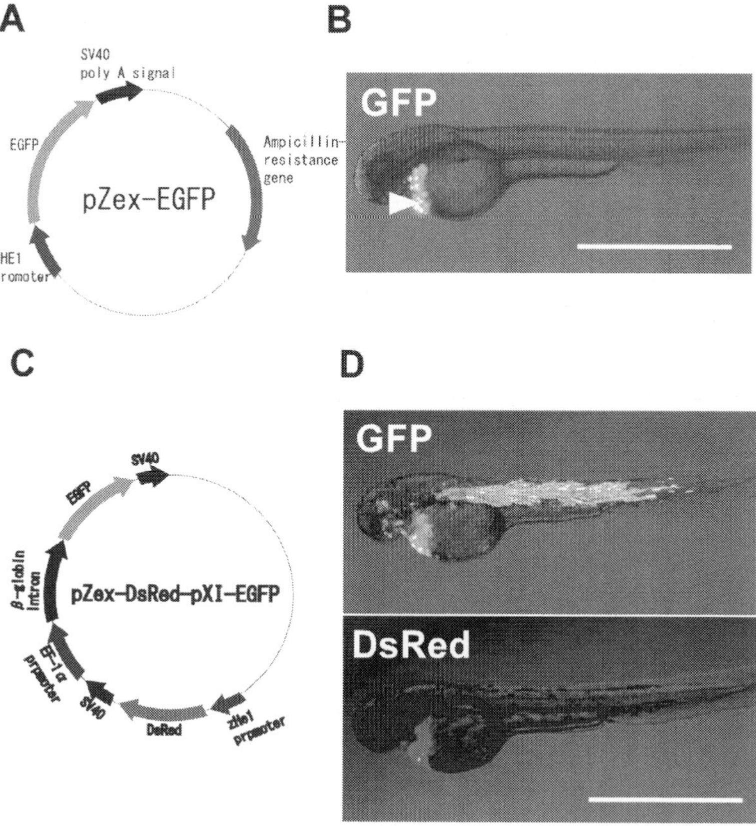

The pZex-EGFP vector (A) or pZex-DsRed-pXI-EGFP tandem vector (C) was injected into zebrafish embryos and pZex-EGFP expression in hatching gland at 48 hpf (arrow head in B) or pZex-DsRed-pXI-EGFP expression in zebrafish Embryos at 48 hpf (D) was observed. Note that expression of GFP in panel D is ubiquitous, while the expression of DsRed was limited in hatching gland cells. Scale bars, 500 μm.

Fig. 3. Protein expression vectors and their expression in zebrafish embryos

We constructed the pZex vector derived from the pXI vector in our laboratory (**Figure 3A**). This vector included the promoter region of zebrafish he1 (hatching enzyme 1) gene and GFP is expressed in hatching gland cells during only early developmental stages up to 72 hrs post-fertilization (hpf) (**Figure 3B**). Furthermore, since tissue-specific and stage-specific protein expression by pZex can be possible in zebrafish embryos, even some apoptosis-related protein is able to express. Although one of the critical problems for protein expression in zebrafish embryos is expression efficiency, most target proteins were easily expressed by pZex in more than 30% of injected embryos. Furthermore, we constructed a pXI-EGFP-pZex-DsRed vector tandemly connected with both pXI-EGFP and pZex-DsRed, (**Figure 3C**). EGFP and DsRed can be successfully expressed in each promoter-dependent manner (**Figure 3D**). These constructs can be applied for the identification of embryos expressing target proteins. Thus, we can choose efficiently the embryos expressing the target protein only observed by monitoring fluorescence.

6. Zebrafish as a model for combinatorial bioengineering

In recent years, the importance of the target proteins with therapeutic potential and drug discovery is getting more and more increasing. For example, several monoclonal antibodies have already applied to human cancer therapy because of their minimum side effects and specificity to the target disease. For the purpose of developing the novel molecular target drugs, the spatiotemporal protein-protein interactions in normal or abnormal tissue has been attempted to analyze extensively. In addition, the effective production of such a functional mammalian protein in large scale and at low cost will be also demanded as spreading the use of these proteins in human therapy or researches like protein structure analysis for novel drug discovery.

Although expression and preparation of target proteins in large scale has been tried in bacterial cells, bacterial recombinant proteins often lost their native properties. It is due to the differences of protein synthesis system between eukaryotic cells and prokaryotic cells. That is, protein synthesis on endoplasmic reticulum (ER) follows by various posttranslational modifications such as glycosylation, phosphorylation, and N-terminus conjugation of several lipids in eukaryotic cells. Accordingly, such posttranslational modifications never occur in prokaryotic cells. On the other hand, the posttranslational modifications are often critical for the correct folding or functions of mammalian proteins. For this reason, the mammalian proteins for pharmaceutical agent or protein structure analysis has been produced by eukaryotic cells or extracted from mammalian tissues. However, these methods are not efficient and often less expensive. Therefore, several alternative ways to produce mammalian proteins more efficient than using cell cultures has been studied and one successful example are to secrete the protein in the milk of transgenic mammals, like a pig[59,60]. However, maintenance of such a large mammal needs large spaces and high cost. In addition, it is originally unable to produce and keep various kinds of transgenic mammals.

The zebrafish are easy to maintain large population in a small space, lay thousands of eggs weekly, and can generate and reproduce transgenic fishes easily. Therefore, we introduced and described the advantage of zebrafish researches. In order to apply this tool to combinatorial bioengineering in the post-genomic era, we attempt to use the ability and

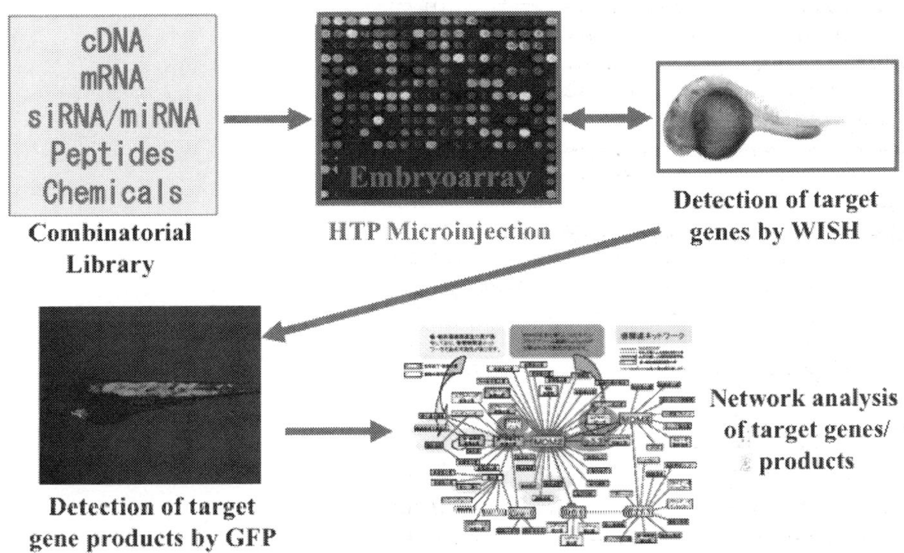

Fig. 4. Scheme of combinatorial bioengineering using zebrafish embryogenesis

potentiality of zebrafish "embryoarray" as protein sources (**Figure 4**). In fact, there are many and various kinds of libraries for not only genes but also natural or artificial compounds. For instance, if complete cDNAs encoding a total of human genes were able to transfer into the zebrafish, human protein library would be obtained and could be stably expressed in all generations of transgenic zebrafish with their native properties. Thus, we believe that transgenic zebrafish have brought us remarkable advances in many areas of biological researches. Therefore, we would like to emphasize the additional advantages that the target proteins expressed in zebrafish would have a proper conformation, activity and posttranslational modifications. The effective production of such functional mammalian proteins will become gradually important as increasing attention to developing pharmaceutical proteins.

7. Zebrafish and its potential application on drug discovery

The low-cost and high clutch-size zebrafish is, at the embryonal and larval stages, optically transparent, permitting visualization of pathogens and lesions in real time[61], as well as offering exciting possibilities for high-throughput imaging[62]. Zebrafish are also amenable to forward genetic screening, or reverse genetics techniques such as injection of morpholinos (inhibitory of mRNA translation)[63,64]. More recently, it is clear that much can be learned about Tuberculosis (TB) from the study of *Mycobacterium marinum* infections in zebrafish, and the use of this pathogen offers practical advantages when compared to *M. tuberculosis*, such as lower biosafety restrictions and faster growth rate[65]. That notwithstanding, it was of interest to study the human pathogen, *M. tuberculosis*, directly in zebrafish via robotic injection system. Importantly, they use reference compounds to validate their system in the testing of molecules that prevent tuberculosis progression, making it highly suited for investigating novel anti-tuberculosis compounds *in vivo*. Thus, by introducing advanced biotechnologies into zebrafish, we are confident that our approach will contribute to the novel knowledge of drug discovery and could be helpful for the development of new medicines.

8. Acknowledgements

This work was supported by grants from the Wakayama Prefecture Collaboration of Regional Entities for the Advancement of Technological Excellence (Y.T) and SENTAN (Y.T), Japan Science and the Technology Agency and the New Energy and Industrial Technology Development Organization (02A09003d) (Y.T).

9. References

[1] Schilling TF, Webb J. Considering the zebrafish in a comparative context. J Exp Zool 2007; 308B: 515–522.

[2] Alestrom P, Holter JL, Nourizadeh-Lillabadi R. Zebrafish in functional genomics and aquatic biomedicine. Trends Biotechnol 2006; 24: 15-21.

[3] Streisinger G, Walker C, Dower N, Knauber D, Singer F. Production of clones of homozygous diploid zebra fish (Brachydanio rerio). Nature 1981; 291: 293-296.

[4] Kimmel CB. Genetics and early development of zebrafish. Trends Genet 1989; 5: 283-288.

[5] Nusslein-Volhard C. Of flies and fishes. Science 1994; 266: 572-574.

[6] Sprague J, Clements D, Conlin T, Edwards P, Frazer K, Schaper K, Segerdell E, Song P, Sprunger B, Westerfield M. The zebrafish information network (ZFIN): the zebrafish model organism database. Nucleic Acids Res 2003; 31(1): 241-243.

[7] Currie PD. Zebrafish genetics: mutant cornucopia. Curr Biol 1996; 6: 1548-1552.

[8] Holder N, McMahon A. Genes from zebrafish. Nature 1996; 384: 515-516.

[9] van Eeden FJ, Granato M, Odenthal J, Haffter P. Developmental mutant screens in the zebrafish. Methods Cell Biol 1999; 60: 21-41.

[10] Stuart GW, McMurray JV, Westerfield M. Replication, integration and stable germline transmission of foreign sequences injected into early zebrafish embryos. Development 1988; 103 (2): 403-412.

[11] Collas P, Alestrӧm P. Nuclear localization signals: a driving force for nuclear transport of plasmid DNA in zebrafish. Biochem Cell Biol 1997; 75 (5): 633-640.

[12] Collas P, Alestrom P. Nuclear localization signals enhance germline transmission of a transgene in zebrafish. Transgenic Res 1998; 7: 303-309.

[13] Liang MR, Alestrӧm P, Collas P. Glowing zebrafish: single luciferase transgene integration, transmission and expression promoted by nuclear localization signals. Mol Reprod Dev 2000; 55 (1): 8-13.

[14] Thermes V, Grabher C, Ristoratore F, Bourrat F, Choulika A, Wittbrodt J, Joly JS. I-SceI meganuclease mediates highly efficient transgenesis in fish. Mech Dev 2002; 118 (1-2): 91-98.

[15] Davidson AE, Balciunas D, Mohn D, Shaffer J, Hermanson S, Sivasubbu S, Cliff MP, Hackett PB, Ekker SC. Efficient gene delivery and gene expression in zebrafish using the Sleeping Beauty transposon. Dev Biol 2003;263(2):191-202.

[16] Shin J, Park HC, Topczewska JM, Mawdsley DJ, Appel B. Netural cell fate analysis in zebrafish using olig2 BAC transgenics. Methods Cell Sci 2003; 25 (1-2): 7-14.

[17] Carvalho R, de Sonneville J, Stockhammer OW, Savage ND, Veneman WJ, Ottenhoff TH, Dirks RP, Meijer AH, Spaink HP. A high-throughput screen for tuberculosis progression. PLoS One. 2011; 6 (2): e16779.

[18] White RM, Cech J, Ratanasirintrawoot S, Lin CY, Rahl PB, Burke CJ, Langdon E, Tomlinson ML, Mosher J, Kaufman C, Chen F, Long HK, Kramer M, Datta S, Neuberg D, Granter S, Young RA, Morrison S, Wheeler GN, Zon LI. DHODH modulates transcriptional elongation in the neural crest and melanoma. Nature 2011; 471 (7339): 518-522.

[19] Kaufman CK, White RM, Zon L. Chemical genetic screening in the zebrafish embryo. Nat Protoc. 2009; 4 (10): 1422-1432.

[20] Weigta S, Hueblera N, Streckerb R, Braunbeckb T, Broscharda TH. Zebrafish (Danio rerio) embryos as a model for testing proteratogens. Toxicology 2011;281:25-36.

[21] Link V, Shevchenko A, Heisenberg CP. Proteomics of early zebrafish embryos. BMC Develop Biol 2006;6:1.

[22] Hwang G., Müller M, Rahman, Darren W. Williams, Paul J, Murdock K, Pasi J, Goldspink G, Farahmand H, Maclean N. Fish as Bioreactors: Transgene Expression of Human Coagulation Factor VII in Fish Embryos. Mar. Biotechnol. 2004; 6: 485-492.

[23] Handley-Goldstone HM, Grow MW, Stegeman JJ. Cardiovascular gene expression profiles of dioxin exposure in zebrafish embryos. Toxicol Sci 2005; 85: 683-693.

[24] Wienholds E, Kloosterman WP, Miska E, Alvarez-Saavedra E, Berezikov E, de Bruijn E, Horvitz HR, Kauppinen S, Plasterk RH. MicroRNA expression in zebrafish embryonic development. Science 2005; 309: 310-311.

[25] Stelling J, Sauer U, Szallasi Z, Doyle FJ III, Doyle J. Robustness of cellular functions. Cell 2004;118:675-685.

[26] Koonin EV, Wolf YI, Karev G.P. The structure of the protein universe and genome evolution. Nature 2002; 420: 218-223.

[27] Rockman MV. Reverse engineering the genotype–phenotype map with natural genetic variation. Nature 2008; 456: 738-744.

[28] Jansen RC, Nap JP. Genetical genomics: the added value from segregation. Trends Genet 2001; 17: 388-391.

[29] Jansen RC. Studying complex biological systems using multifactorial perturbation. Nature Rev Genet 2003;4:145-151.

[30] Brem RB, Yvert G, Clinton R, Kruglyak L. Genetic dissection of transcriptional regulation in budding yeast. Science 2002;296:752–755.

[31] Schadt EE, Monks SA, Drake TA, Lusis AJ, Che N, Colinayo V, Ruff TG, Milligan SB, Lamb JR, Cavet G, Linsley PS, Mao M, Stoughton RB, Friend SH. Genetics of gene expression surveyed in maize, mouse and man. Nature 2003; 422: 297–302.

[32] Rockman MV, Kruglyak L. Genetics of global gene expression. Nature Rev Genet 2006; 7: 862–872 (2006).

[33] Gerlt JA. "Fishing" for the functional proteome. Nat Biotechnol 2002; 20: 786–787.

[34] Cravatt BF, Sorensen EJ. Chemical strategies for the global analysis of protein function. Curr Opin Chem Biol 2000; 4: 663-668.

[35] Greenbaum D, Medzihradszky KF, Burlingame A, Bogyo M. Epoxide electrophiles as activity-dependent cysteine protease profiling and discovery tools. Chem Biol 2000; 7: 569-581.

[36] Schena M, Shalon D, Davis RW, Brown PO. Quantitative monitoring of gene expression patterns with a complementary DNA microarray. Science 1995; 270: 467-470.

[37] Aebersold R, Goodlett DR. Mass spectrometry in proteomics. Chem Rev 2001; 101: 269-295.

[38] MacBeath G, Schreiber SL. Printing proteins as microarrays for high-throughput function determination. Science 2000; 289: 1760-1763.

[39] Pichler FB, Laurenson S, Williams LC, Dodd A, Copp BR, Love D. Chemical discovery and global gene expression analysis in zebrafish. Nat Biotechnol 2003; 21: 879-883.

[40] Peterson RT, Shaw SY, Peterson TA, Milan DJ, Zhong TP, Schreiber SL, MacRae CA, Fishman MC. Chemical suppression of a genetic mutation in a zebrafish model of aortic coarctation. Nat Biotechnol 2004; 22: 595-599.

[41] Baker M. Screening: the age of fishes. Nat Meth 2011; 8 (1): 47-51.

[42] Hayashi S, Akiyama S, Tamaru Y, Takeda Y, Fujiwara T, Inoue K, Kobayashi A, Maegawa S, Fukusaki E. A novel application of metabolomics in vertebrate development. Biochem Biophys Res Commun. 2009; 386 (1):268-272.

[43] Gupta P, Lee KH. Genomics and proteomics in process development: Opportunities and challenges. Trends Biotechnol 2007; 25: 324-330.

[44] Al-Fageeh MB, Marchant RJ, Carden M., Smales CM. The cold-shock response in cultured mammalian cells: Harnessing the response for the improvement of recombinant protein production. Biotech Bioeng 2005; 93:829-835.

[45] Li Y, Powell S, Brunette E, Lebkowski J, Mandalam R. Expansion of human embryonic stem cells in defined serum-free medium devoid of animal-derived products. Biotechnol Bioeng 2005; 91: 688-698.

[46] Choudhary J, Grant SGN. Proteomics in postgenomic neuroscince: the end of the beginning. Nat Neurosci 2004; 7: 440-445.

[47] Avsar-Ban E, Ishikawa H, Manya H, Watanabe M, Akiyama S, Miyake H, Endo T, Tamaru Y. Protein O-mannosylation is necessary for normal embryonic development in zebrafish. Glycobiology. 2010; 20 (9): 1089-1102.

[48] Manya H, Chiba A, Yoshida A, Wang X, Chiba Y, Jigami Y, Margolis RU, Endo T. Demonstration of mammalian protein O-mannosyltransferase activity: coexpression of POMT1 and POMT2 required for enzymatic activity. Proc Natl Acad Sci USA 2004; 101 (2): 500-505.

[49] Beltran-Valero de Bernabe D, Currier S, Steinbrecher A, Celli J, van Beusekom E, van der Zwaag B, Kayserili H, Merlini L, Chitayat D, Dobyns WB, Cormand B, Lehesjoki AE, Cruces J, Voit T, Walsh CA, van Bokhoven H, Brunner HG.

Mutations in the O-mannosyltransferase gene POMT1 give rise to the severe neuronal migration disorder Walker–Warburg syndrome. Am J Hum Genet. 2002; 71 (5): 1033-1043.

[50] van Reeuwijk J, Janssen M, van den Elzen C, Beltran-Valero de Bernabe D, Sabatelli P, Merlini L, Boon M, Sche"er H, Brockington M, Muntoni F, Huynen MA, Verrips A, Walsh CA, Barth PG, Brunner HG, van Bokhoven H. POMT2 mutations cause α-dystroglycan hypoglycosylation and Walker-Warburg syndrome. J Med Genet. 2005; 42 (12): 907-912.

[51] Kawahara G, Karpf JA, Myers JA, Alexander MS, Guyon JR, Kunkel LM. Drug screening in a zebrafish model of Duchenne muscular dystrophy. Proc Natl Acad Sci USA. 2011; 108 (13): 5331-5336.

[52] Nasevicius A, Ekker SC. Effective targeted gene 'knockdown' in zebrafish. Nat Genet 2000; 26: 216-220.

[53] Thomas KR, Folger KR, Capecchi MR. High frequency targeting of genes to specific sites in the mammalian genome. Cell 1986; 44: 419-428.

[54] Sedivy JM, Joyner AL. Gene Targeting. (Oxford University Press, Oxford, 1992).

[55] Meng X, Noyes MB, Zhu LJ, Lawson ND, Wolfe SA. Targeted gene inactivation in zebrafish using engineered zinc-finger nucleases. Nat Biotechnol 2008; 26:695-701.

[56] Doyon Y, McCammon JM, Miller JC, Faraji F, Ngo C, Katibah GE, Amora R, Hocking TD, Zhang L, Rebar EJ, Gregory PD, Urnov FD, Amacher SL. Heritable targeted gene disruption in zebrafish using designed zinc-finger nucleases. Nat Biotechnol 2008; 26: 702-708.

[57] Johnson AD, Krieg PA. pXeX, a vector for efficient expression of cloned sequences in Xenopus embryos. Gene1994; 147: 223-226.

[58] Amsterdam A, Lin S, Hopkins N. The Aequorea victoria green fluorescent protein can be used as a reporter in live zebrafish embryos. Dev Biol 1995; 171: 123-129.

[59] Paleyanda RK, Velander WH, Lee TK, Scandella DH, Gwazdauskas FC, Knight JW, Hoyer LW, Drohan WN, Lubon H. Transgenic pigs produce functional human factor VIII in milk. Nat Biotechnol 1997; 15: 971-975.

[60] Van Cott KE, Lubon H, Gwazdauskas FC, Knight J, Drohan WN, Velander WH. Recombinant human protein C expression in the milk of transgenic pigs and the effect on endogenous milk immunoglobulin and transferring levels. Transgenic Res 2001; 10: 43-51.

[61] Lesley R, Ramakrishnan L. Insights into early mycobacterial pathogenesis from the zebrafish. Curr Opin Microbiol 2008; 11 (3): 277-283.

[62] Pardo-Martin C, Chang TY, Koo BK, Gilleland CL, Wasserman SC, Yanik MF. High-throughput in vivo vertebrate screening. Nat Methods 2010; 7 (8): 634-636.

[63] Amsterdam A, Hopkins N. Mutagenesis strategies in zebrafish for identifying genes involved in development and disease. Trends Genet 2006; 22 (9): 473–478.

[64] Nasevicius A, Ekker SC. Effective targeted gene 'knockdown' in zebrafish. Nat Genet 2000; 26 (2): 216-220.

[65] Carvalho R, de Sonneville J, Stockhammer OW, Savage ND, Veneman WJ, Ottenhoff
     TH, Dirks RP, Meijer AH, Spaink HP. A high-throughput screen for tuberculosis
     progression. PLoS One. 2011; 6 (2): e16779.

11

Liver Regeneration: the Role of Bioengineering

Pedro M. Baptista, Dipen Vyas and Shay Soker
Wake Forest Baptist Health,
Wake Forest Institute for Regenerative Medicine,
Winston-Salem, NC
USA

1. Introduction

An estimated two million people die of terminal liver disease every year. The World Health Organization calculates that over six hundred and fifty million people worldwide suffer from some form of liver disease, including thirty million Americans. On a worldwide base, approximately one to two million deaths are accounted to liver related diseases annually. Around the globe, China has the world's largest population of Hepatitis B patients (approximately 120 million) with five hundred thousand people dying of liver illnesses every year(1, 2). In the US alone, five hundred thousand critical liver problem episodes are reported every year requiring hospitalization with great burden to the patients and a huge cost to the health care system. In the European Union and United States of America alone, over eighty one thousand and twenty six thousand people died of chronic liver disease in 2006, respectively(1, 3). For these patients, liver transplantation is presently the only proven therapy able to extend survival for end-stage liver disease. It is also the only treatment for severe acute liver failure and to some forms of inborn errors of metabolism. Nevertheless, the waiting list for liver transplantation is long and many patients will not survive long enough to receive an organ due to the dramatic shortage of donors or lack of eligibility(1).

A good example of this is that in 2007 there were almost seventeen thousand candidates on the US waiting list for liver transplantation. From those, only 30% were actually transplanted by the end of the year, with an average waiting time of more than 400 days. In the same year, nearly one thousand and three hundred people died while waiting for a suitable donor, with no real therapeutic alternative available to save their lives. Moreover, for those patients with fulminant hepatic failure, a severe liver disease with 60-90% mortality, depending on the etiology, only 10% received a transplant. Altogether, liver transplantation still has a relatively high mortality of 30-40% at 5-8 years with 65% of the deaths occurring in the first 6 months. Patients who have undergone transplantation have to also use lifelong immunosuppressive therapy, with sometimes severe side effects(4).

There are innumerous etiologies of end-stage chronic liver disease that lead to transplantation and approximately 80% of the candidates in the liver transplantation waiting list have a primary diagnosis of liver cirrhosis. Fortunately, some of the causes of these diseases are currently preventable. An excellent example is the successful vaccination programs in many countries around the world against Hepatitis B virus, which have considerably reduced the incidence of chronic carriers and viral induced cirrhosis(5). However, close to 20% of the livers transplanted in the USA and 30% in Europe have a preventable underlying cause, alcoholic liver disease. Furthermore, approximately 45% of deaths due to liver cirrhosis in the USA are associated with alcohol abuse(1, 3, 4). Patients with pathologies like hepatic cancer, congenital malformations and metabolic diseases, and acute hepatic necrosis make up the remaining percentage of the list.

The success of liver transplantation has resulted in a progressively increasing demand for such treatment. Nevertheless and as mentioned above, the availability of donor organs has remained stable, resulting in the number of potential recipients far exceeding organ supply. Due to this, several strategies have been explored in the past decade or so with the aim to increase access to liver transplantation. These consist of obtaining organs from non-heart-beating and live donors; and/or using split liver technique and livers from expanded donor criteria. In addition, the introduction of the Model for End-Stage Liver Disease (MELD) score system implemented on February 27 of 2002 in the United States tremendously helped Organ Procurement Organizations to prioritize patients waiting for a liver transplant. The MELD score is a numerical scale used for adult liver transplant candidates that ranges between 6 (less ill) and 40 (gravely ill). The number is calculated using the most recent laboratory tests for bilirubin, INR and creatinine(6) and the individual score determines how urgently a patient needs a liver transplant within the next 3 months.

The implementation of more adequate and efficient allocation systems, development of better immunosupressive regimens, and the increase of living donors have all helped to raise overall patient survival and graft survival in the past decade in the United States. The number of transplanted livers also increased to an all time high in 2006, with a marked decrease on the waiting time for liver transplantation after MELD score system implementation, especially for the sickest patients.

The best example of the impact of these measures is the increase of 6% (86% in 2007) and 16% (87% in 2007) of the unadjusted 1-year graft survival for deceased donor and living donor liver recipients between 1998 and 2007, respectively. This explains also the improvement of 3% (89% in 2007) and 11% (91% in 2007) of the unadjusted 1-year patient survival for deceased donor and living donor liver recipients for the same period, respectively(7). However, these numbers decrease significantly when we consider 5-year patient survival, showing how much remains to be done. In 2007 it was 74% and 79% for deceased donor and living donor liver recipients, respectively. These numbers decrease even further for the 10-year patient survival, where in 2007 there were 61% and 71% patient survival for deceased donor and living donor liver recipients, respectively. One important note is that patient survival was higher than graft survival ~5%, due to the opportunity for repeated liver transplantation in the event of graft failure(8).

These numbers highlight the need for innovative therapies that can increase patient survival, as well as lower costs to the health care systems. Naturally acquired tolerance research and its experimental clinical induction is a good example of this. The identification of molecular signatures in naturally tolerant patients to whom immunosuppression could be stopped, and of tolerance induction, through lymphocyte depletion or T lymphocyte co-stimulation blockade, are some of the most advanced approaches to decrease the complications of immunosuppression(9).

Additionally, artificial liver devices which can efficiently remove accumulated lethal toxins from blood or plasma by using membrane filtration and/or adsorbents have been developed as support devices. Liver Dialysis Device, Molecular Adsorbent Recirculating

System (MARS) and Prometheus are the most widely used artificial support systems. These devices have been widely used in clinical trials across Europe and Asia, and have showed some benefits to the patients, but unfortunately have little or no significant improvement in patient survival.

2. Bioartificial Liver Devices (BALs)

Extracorporeal liver support devices have been developed in the past few decades to support the failing liver resulting from different complications. These devices were created initially for the management of patients waiting for a suitable donor for orthotopic liver transplantation. Recent advances in the design of these devices have made it possible to utilize them for recovering the liver from an acute injury. Thus, these devices can either bridge the patients to liver transplantation or can fully avoid the need for it (10).

Although artificial liver devices have been able to provide temporary support to the patients with acute liver failure by detoxifying the blood or plasma, they have major limitations in replacing synthetic and metabolic functions of liver (11). Thus, attempts have been made to develop bioartificial liver systems, which can provide both metabolic and synthetic hepatic functions along with detoxification. BALs generally utilize primary hepatocytes or hepatoma cell lines as a biological component and a hollow fiber or porous matrix membranes on which the functional hepatocytes are coated (11, 12). Hepatocytes from various sources have been investigated for use in BALs. Primary human hepatocytes have been widely studied as an ideal cell source due to their biocompatibility but they are scarcely available and their proliferative capacity *in vitro* is limited (12, 13). Animal cell sources such as porcine primary hepatocytes are being investigated due to ease of availability and their ability to maintain metabolic functions similar to human hepatocytes. However, concerns regarding immunological reaction to the animal proteins and transmission of disease exist (14). Nonetheless, porcine hepatocytes remain a popular choice as a hepatocyte source in various BAL systems.

A bioreactor is a critical component of BALs and thus has a major impact on the efficacy of these systems. The bioreactor should be capable of providing a suitable environment for hepatic cells to survive and remain functional along with an adequate interface between blood and hepatocytes for mass transport (15). The bioreactors should also be flexible enough for scale up and customization according to the patient's needs. Currently available bioreactor systems need structural optimization and modifications even though there have been recent advances in this technology. It should be highlighted that no bioreactor system is currently approved for patient use, although some have been used in clinical trials (16). Table 1 lists all BAL devices currently under clinical investigation.

So far over 200 patients have been treated with HepatAssist and over 40 patients treated with Extracorporeal Liver Assist Device (ELAD), making them the most common BALs used in clinical trials so far (22, 23). In all these cases, most patients were bridged to liver transplantation while some patients fully recovered thus avoiding the need of transplantation. ELAD is the only BAL system which utilizes a human hepatocyte cell line; most of the other BAL systems use porcine hepatocytes as a cell source. ELAD uses the immortalized C3A cell line derived from human hepatoma cell line HepG2 (24). The cells are located in the extracapillary space of hollow fiber cartridges (200 gram total cells in four cartridges). The membrane is impermeable to immunoglobulins, blood cells and C3A cells. The blood flows through the lumen of cartridges as the ultrafiltrated plasma from the membrane comes in direct contact with hepatocytes (11). HepatAssist incorporates approximately 5-7 billion cryopreserved porcine hepatocytes attached to microcarriers and loaded onto a hollow fiber. The separated plasma passes through a charcoal column and oxygenator prior to entering the hollow fibers in the bioreactor. An upgraded version of HepatAssist known as Hepamate contains 14×10^9 porcine hepatocytes. The membrane pores are 0.15μm in size which prevents a physical contact between human cells and porcine hepatocytes (10, 23). Most of the BALs listed above are undergoing clinical trials in the USA, Europe and Asia. The goal of these clinical trials is to assess the safety and efficacy of these devices in treatment of various terminal liver diseases. Currently, none of the BALs have been approved by the FDA for clinical use.

Device	Reference
Extracorporeal Liver Assist Device (ELAD)	(17)
HepatAssist	(18)
Bioartificial Liver Support System (BLSS)	(19)
The Academic Medical Center – Bioartificial Liver (AMC-BAL)	(20)
Modular Extracorporeal Liver Support device (MELS)	(21)

Table 1. Summary of developed and published bioartificial devices.

Recent developments in artificial and bioartificial devices have shown a potential for use of these devices in the management of patients with acute liver failure. However, considerable technical challenges and regulatory issues remain to be addressed in order to efficiently utilize these devices in the clinic. Artificial liver devices have demonstrated the ease of use and cost effectiveness along with showing improvement in biochemical parameters and clinical symptoms by detoxifying the blood or plasma, but it has a major limitation of not replacing critical metabolic and synthetic functions of liver. BALs developed over the past decade have been designed to provide these functions of the liver along with detoxification. BALs hold a promising future as they have shown potential by efficiently treating several patients across different clinical trials. Many challenges exist in BAL technology including the debate on ideal cell source, the requirement of a large number of cells, maintainance of the functional hepatocytes for a longer period of time in a bioreactor, complexity of the design and high cost. The aforementioned challenges have delayed the entry of BAL systems in the clinic. Nonetheless, there are plenty of optimized designs of liver support devices that are undergoing development and clinical trials. This is an unmistakable sign of optimism in this area of critical care management.

3. Cell therapies

Hepatocyte transplantation is certainly in the vanguard of new therapeutic strategies. The first successful hepatocyte transplantation was performed in June 1992 to a French Canadian woman with familial hypercholesterolemia. After *ex vivo* transduction with a retrovirus encoding for the human LDL receptor, the patient's hepatocytes were infused through the inferior mesenteric vein into the liver. LDL and HDL levels improved throughout the next 18 months and transgene expression was detected in a liver biopsy(25). Following this first accomplishment, other patients followed through. However, not all the patients treated had a clear benefit from the procedure(26).

Since then,several other metabolic diseases have been treated with hepatocyte transplantation with various degrees of success(27-31). It has also been used as a support treatment to acute(32-34) and chronic liver diseases(33-36) in bridging severely ill patients to orthotopic liver transplantation (OLT). Low efficacy and lack of long-term therapeutic effect have been common in all these procedures. These failures could be explained by the relatively small number of hepatocytes that engraft in the recipient liver due to quality, quantity and possibly immunosuppression protocols(37). However, transplantation of a number of hepatocytes corresponding to 1-5% of the total liver mass has been able to show a positive impact in transplanted patients, even if for a limited period of time(37).

Due to the shortage of available human hepatocytes for transplantation, other cell sources have been used. Specifically, bone marrow derived mesenchymal stem cells(38), hematopoietic stem cells(39, 40) and fetal liver progenitor or stem cells (41) have shown to improve, to a certain extent, the condition of cirrhotic patients. The latter cell type holds an enormous potential for cell or regenerative medicine therapies due to their high expansion capabilities and differentiation into hepatocytes and biliary epithelium(42).

Cell Type	Disease	References
Hepatocytes	Familial Hypercholesterolaemia	(25, 26)
Hepatocytes	Crigler-Najjar syndrome Type I	(27)
Hepatocytes	Severe Ornithine Transcarbamylase Deficiency	(28)
Hepatocytes	Crigler-Najjar Syndrome Type 1	(29)
Hepatocytes	Glycogen Storage Disease Type 1a	(30)
Hepatocytes	Peroxisomal Biogenesis Disease	(31)
Hepatocytes	Acute Liver Disease	(32-34)
Hepatocytes	Chronic Liver Disease	(33-36)
BM-Mesenchymal SC	Chronic Liver Disease	(37)
Hematopoietic SC	Chronic Liver Disease	(39, 40)
Fetal Liver Progenitor/ Stem Cells	Chronic Liver Disease	(41)

Table 2. List of cell therapy procedures that were performed in clinical trials

Recent data suggests that human embrionic (hES) and induced pluripotent (iPS) stem cells hold great promise to regenerative applications in every medical field. Specifically for the liver, several studies have established the required pathways to differentiate a hES or iPS into a hepatic fate by using defined soluble growth factor signals that mimic embryonic development(43, 44). These cells, once transplanted into rodent livers were able to engraft and express several normal hepatic functions(45, 46). Still, more extensive characterization, as well as further safety evaluation, are needed to determine whether these cells can fully function as primary adult hepatocytes.

4. Tissue and organ bioengineering

Tissue engineering is one of the most promising fields in regenerative medicine. As described in 1993 by Robert Langer and Joseph Vacanti, it is the conjugation of biomaterials (synthetic or naturally derived) with cells, in order to generate tissue constructs that can be implanted into patients to substitute a lost function, and maintain or gain new functions(47). The current paradigm is suitable for the engineering of thin constructs like the bladder, skin or blood vessels. Although, in the specific case of the liver, the 3D architecture and dense cellular mass requires novel tissue engineering approaches and the development of vascularized biomaterials, in order to support thick tissue masses and be readily transplantable. Additionally to the vascular support for large tissue mass, hepatocyte function maintenance represents the ultimate aim in any organ engineering or regenerative medicine strategy for liver disease.

Hepatocytes are known to be attachment-dependent cells and lose rather quickly their specific functions without optimal media, ECM composition, and cell-cell adhesion. Also, function and differentiation of liver cells are influenced by the 3D organ architecture(48).

In the last two decades innumerous strategies for the culture of adult hepatocytes in combination with several types of 3D, highly porous polymeric matrices have been attempted(49-53). Nevertheless, lack of vasculature, restriction in cell growth and function are common due to the limitations in nutrient and oxygen diffusion. Finally, some of these problems have been partially resolved with the development of bioreactors that provide continuous perfusion of culture media and gases allowing a 3D culture configuration and hepatocyte function maintenance(54-56).

The tissue engineering concept has several advantages over the injection of cell suspensions into solid organs. The matrices provide sufficient volume for the transplantation of an adequate cell mass up to whole-organ equivalents. Transplantation efficiency could readily be improved by optimizing the microarchitecture and composition of the matrices as well as by attaching growth factors and extracellular matrix molecules to the polymeric scaffold, helping to recreate the hepatic microenvironment(48). The use of naturally derived matrices has also proved to be very helpful in hepatocyte culture(51). These matrices, besides preserving some of the microarchitecture features of the tissues that they are derived from, also retain bioactive signals (e.g., cell-adhesion peptides and matrix bound growth factors) required for the retention of tissue-specific gene expression(57, 58). Additionally, cell transplantation into polymeric matrices is, in contrast to cell injection into tissues and organs, a reversible procedure since the cell-matrix-constructs may be removed if necessary. Finally, heterotopic hepatocyte transplantation in matrices has already been demonstrated in long-term studies(59, 60), even so initial engraftment rates are suboptimal. One of the reasons for this is the absolute requirement of the transplanted hepatocytes for hepatotrophic factors that the liver constantly receives through its portal circulation(61). Thus, the development of a tissue engineered liver construct capable of being orthotopically transplanted is essential.

Apart from cellular therapies, other early developments of experimental approaches are not showing results that will indicate clinical translation in the next few years. However, two experimental approaches are worth mentioning. They already display a higher functional level and may have the potential for succesful clinical translation. The first experimental approach is the "cell sheet" technology developed by Okano *et al.* in Japan(62). Its simple configuration and fabrication allows for the stacking of up to four hepatocyte cell sheets that can readily engraft and provide a specific metabolic relief to the recipient(63). This technology has already been applied successfully to one patient with heart failure. Other technology that shows great promise is tissue and organ decellularization. Our lab and others have been able to generate several decellularized scaffolds for tissue engineering applications like tissue engineering of urethra(64), heart valves(65), blood vessel(66). More recently, Ott *et al.* reported a novel method of perfusion decellularization that is able to generate whole organ scaffolds. The use of this method allowed the decellularization of a whole heart that was later repopulated with neonatal rat cardiomyocytes. This bioengineered heart was able to contract up to 2% of the normal contractile function(67). This approach may have a tremendous potential for the field of organ bioengineering.

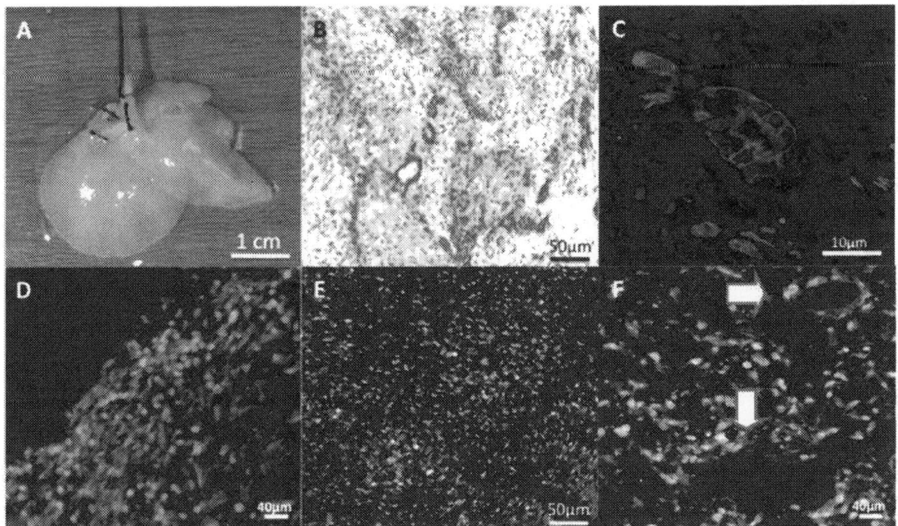

Fig. 1. Human bioengineered livers are highly cellular and display some of the functions observed in native hepatic tissue. (A) Macroscopic appearance of a seeded liver bioscaffold 7 days after seeding with primary human fetal liver and endothelial cells. (B) H&E showing broad recellularization of the bioscaffold with the formation of biliary ductal structures. (C) Immunofluorescence staining for cytokeratin 19 (red) showing a biliary duct formed within the bioscaffold with a visible lumen. (D) Immunofluorescence staining of cytochrome P450 3A (orange) and (E) albumin (purple) showing groups of hepatocytic lineage cells expressing these more mature hepatic markers. (F) Immunofluorescence staining of eNOS (purple) showing hUVECs coating and spreading from vascular structures (arrows). All nuclei stained with YO-PRO1 (green) or DAPI (blue).

We have also developed a similar perfusion decellularization method for the liver, reported for the first time in June 2005 (68), which culminated with two papers published recently(69, 70). We applied this technique to liver, pancreas, intestine and kidney generating decellularized organ scaffolds for organ bioengineering(69, 71). These bioscaffolds preserve their tissue microarchitecture and an intact vascular network that can be readily used as a route for recellularization by perfusion of culture medium with different cell populations. In the particular case of the bioengineered liver, the generated hepatic tissue is clearly visible by the naked eye after 7 days in the bioreactor (Fig. 1A). This tissue is highly cellular (Fig. 1B) and displays biliary duct structures positive for cytokeratin 19 (Fig. 1C), as well as clusters of hepatocytes expressing cytochrome P450 3A and albumin (Fig. 1D, E, respectively). The seeded human endothelial umbilical vein cells (hUVECs) are also observed coating and spreading from vascular structures and express endothelial cell nitric oxyde synthase (eNOS) (Fig. 1F).

In a similar approach, Uygun *et al.* decellularized rat livers and repopulated them with rat primary hepatocytes, showing promising hepatic function and the ability of heterotopically transplanting these bioengineered livers into animals for up to eight hours(72). Nevertheless, we were able to take this a step further in our lab by using human primary fetal liver progenitor/stem and endothelial cells to bioengineer a vascularized human liver. Additionally to the markers showed above, these bioengineered livers displayed some of the functions of a native human liver (albumin and urea secretion, CYP450 enzyme expression, etc) and also exhibiting an endothelialized vascular network that prevented platelet adhesion and aggregation, critical for blood vessel patency after transplantation(71). Hence, this technology has the potential to translate in the future into the bioengineering of human size livers, which may offer readily available organs for drug discovery applications and for transplantation, overcoming organ shortage.

5. Conclusion

OLT is the only cure for end-stage liver disease and inborn errors of metabolism. Due to lack of appropriate organ donors the new experimental procedures presented here have the potential to replace liver transplantation and become the standard of care. Regenerative medicine has indeed the potential to simplify and reduce the morbidity or impact of the "life changing" procedure of OLT.

6. References

[1] CDC. Centers for Disease Control and Prevention Database. In; 2007.

[2] WHO. World Health Organization - Global Burden of Disease: 2004 update (2008). In: WHO publications; 2008.

[3] Eurostat. Eurostat's Harmonised Regional Statistical Database. In; 2007.

[4] OPTN. Transplant Database. In; 2011.

[5] Kao JH, Chen DS. Global control of hepatitis B virus infection. Lancet Infect Dis 2002;2:395-403.

[6] OPTN. MELD score. In; 2011.

[7] Wolfe RA, Merion RM, Roys EC, Port FK. Trends in organ donation and transplantation in the United States, 1998-2007. Am J Transplant 2009;9:869-878.

[8] Thuluvath PJ, Guidinger MK, Fung JJ, Johnson LB, Rayhill SC, Pelletier SJ. Liver transplantation in the United States, 1999-2008. Am J Transplant;10:1003-1019.

[9] Turka LA, Wood K, Bluestone JA. Bringing transplantation tolerance into the clinic: lessons from the ITN and RISET for the Establishment of Tolerance consortia. Curr Opin Organ Transplant;15:441-448.

[10] Carpentier B, Gautier A, Legallais C. Artificial and bioartificial liver devices: present and future. Gut 2009;58:1690-1702.

[11] Park JK, Lee DH. Bioartificial liver systems: current status and future perspective. J Biosci Bioeng 2005;99:311-319.

[12] Cao S, Esquivel CO, Keeffe EB. New approaches to supporting the failing liver. Annu Rev Med 1998;49:85-94.

[13] Pless G. Artificial and bioartificial liver support. Organogenesis 2007;3:20-24.

[14] Stange J, Mitzner S. Cell sources for bioartificial liver support. Int J Artif Organs 1996;19:14-17.

[15] Tilles AW, Berthiaume F, Yarmush ML, Tompkins RG, Toner M. Bioengineering of liver assist devices. J Hepatobiliary Pancreat Surg 2002;9:686-696.

[16] Yu CB, Pan XP, Li LJ. Progress in bioreactors of bioartificial livers. Hepatobiliary Pancreat Dis Int 2009;8:134-140.

[17] Ellis AJ, Hughes RD, Wendon JA, Dunne J, Langley PG, Kelly JH, Gislason GT, et al. Pilot-controlled trial of the extracorporeal liver assist device in acute liver failure. Hepatology 1996;24:1446-1451.

[18] Demetriou AA, Brown RS, Jr., Busuttil RW, Fair J, McGuire BM, Rosenthal P, Am Esch JS, 2nd, et al. Prospective, randomized, multicenter, controlled trial of a bioartificial liver in treating acute liver failure. Ann Surg 2004;239:660-667; discussion 667-670.

[19] Patzer JF, 2nd, Mazariegos GV, Lopez R, Molmenti E, Gerber D, Riddervold F, Khanna A, et al. Novel bioartificial liver support system: preclinical evaluation. Ann N Y Acad Sci 1999;875:340-352.

[20] Flendrig LM, la Soe JW, Jorning GG, Steenbeek A, Karlsen OT, Bovee WM, Ladiges NC, et al. In vitro evaluation of a novel bioreactor based on an integral oxygenator and a spirally wound nonwoven polyester matrix for hepatocyte culture as small aggregates. J Hepatol 1997;26:1379-1392.

[21] Sauer IM, Kardassis D, Zeillinger K, Pascher A, Gruenwald A, Pless G, Irgang M, et al. Clinical extracorporeal hybrid liver support--phase I study with primary porcine liver cells. Xenotransplantation 2003;10:460-469.

[22] Brophy CM, Nyberg SL. Extracorporeal treatment of acute liver failure. Hepatol Res 2008;38:S34-S40.

[23] McKenzie TJ, Lillegard JB, Nyberg SL. Artificial and bioartificial liver support. Semin Liver Dis 2008;28:210-217.

[24] Adham M. Extracorporeal liver support: waiting for the deciding vote. ASAIO J 2003;49:621-632.

[25] Grossman M, Raper SE, Kozarsky K, Stein EA, Engelhardt JF, Muller D, Lupien PJ, et al. Successful ex vivo gene therapy directed to liver in a patient with familial hypercholesterolaemia. Nat Genet 1994;6:335-341.

[26] Grossman M, Rader DJ, Muller DW, Kolansky DM, Kozarsky K, Clark BJ, 3rd, Stein EA, et al. A pilot study of ex vivo gene therapy for homozygous familial hypercholesterolaemia. Nat Med 1995;1:1148-1154.

[27] Fox IJ, Chowdhury JR, Kaufman SS, Goertzen TC, Chowdhury NR, Warkentin PI, Dorko K, et al. Treatment of the Crigler-Najjar syndrome type I with hepatocyte transplantation. N Engl J Med 1998;338:1422-1426.

[28] Horslen SP, McCowan TC, Goertzen TC, Warkentin PI, Cai HB, Strom SC, Fox IJ. Isolated hepatocyte transplantation in an infant with a severe urea cycle disorder. Pediatrics 2003;111:1262-1267.

[29] Ambrosino G, Varotto S, Strom SC, Guariso G, Franchin E, Miotto D, Caenazzo L, et al. Isolated hepatocyte transplantation for Crigler-Najjar syndrome type 1. Cell Transplant 2005;14:151-157.

[30] Muraca M, Gerunda G, Neri D, Vilei MT, Granato A, Feltracco P, Meroni M, et al. Hepatocyte transplantation as a treatment for glycogen storage disease type 1a. Lancet 2002;359:317-318.

[31] Sokal EM, Smets F, Bourgois A, Van Maldergem L, Buts JP, Reding R, Bernard Otte J, et al. Hepatocyte transplantation in a 4-year-old girl with peroxisomal biogenesis disease: technique, safety, and metabolic follow-up. Transplantation 2003;76:735-738.

[32] Strom SC, Fisher RA, Thompson MT, Sanyal AJ, Cole PE, Ham JM, Posner MP. Hepatocyte transplantation as a bridge to orthotopic liver transplantation in terminal liver failure. Transplantation 1997;63:559-569.

[33] Strom SC, Chowdhury JR, Fox IJ. Hepatocyte transplantation for the treatment of human disease. Semin Liver Dis 1999;19:39-48.

[34] Strom SC, Fisher RA, Rubinstein WS, Barranger JA, Towbin RB, Charron M, Mieles L, et al. Transplantation of human hepatocytes. Transplant Proc 1997;29:2103-2106.

[35] Combs C, Brunt EM, Solomon H, Bacon BR, Brantly M, Di Bisceglie AM. Rapid development of hepatic alpha1-antitrypsin globules after liver transplantation for chronic hepatitis C. Gastroenterology 1997;112:1372-1375.

[36] Mito M, Kusano M, Kawaura Y. Hepatocyte transplantation in man. Transplant Proc 1992;24:3052-3053.

[37] Fisher RA, Strom SC. Human hepatocyte transplantation: worldwide results. Transplantation 2006;82:441-449.

[38] Kharaziha P, Hellstrom PM, Noorinayer B, Farzaneh F, Aghajani K, Jafari F, Telkabadi M, et al. Improvement of liver function in liver cirrhosis patients after autologous mesenchymal stem cell injection: a phase I-II clinical trial. Eur J Gastroenterol Hepatol 2009;21:1199-1205.

[39] Salama H, Zekri AR, Zern M, Bahnassy A, Loutfy S, Shalaby S, Vigen C, et al. Autologous hematopoietic stem cell transplantation in 48 patients with end-stage chronic liver diseases. Cell Transplant 2010.

[40] Zacharoulis D, Milicevic MN, Helmy S, Jiao LR, Levicar N, Tait P, Scott M, et al. Autologous infusion of expanded mobilized adult bone marrow-derived CD34+ cells into patients with alcoholic liver cirrhosis. Am J Gastroenterol 2008;103:1952-1958.

[41] Khan AA, Shaik MV, Parveen N, Rajendraprasad A, Aleem MA, Habeeb MA, Srinivas G, et al. Human fetal liver derived stem cell transplantation as supportive modality in the\ management of end stage decompensated liver cirrhosis. Cell Transplantation 2010.

[42] Schmelzer E, Zhang L, Bruce A, Wauthier E, Ludlow J, Yao HL, Moss N, et al. Human hepatic stem cells from fetal and postnatal donors. J Exp Med 2007;204:1973-1987.

[43] Gouon-Evans V, Boussemart L, Gadue P, Nierhoff D, Koehler CI, Kubo A, Shafritz DA, et al. BMP-4 is required for hepatic specification of mouse embryonic stem cell-derived definitive endoderm. Nat Biotechnol 2006;24:1402-1411.

[44] Gadue P, Huber TL, Paddison PJ, Keller GM. Wnt and TGF-beta signaling are required for the induction of an in vitro model of primitive streak formation using embryonic stem cells. Proc Natl Acad Sci U S A 2006;103:16806-16811.

[45] Basma H, Soto-Gutierrez A, Yannam GR, Liu L, Ito R, Yamamoto T, Ellis E, et al. Differentiation and transplantation of human embryonic stem cell-derived hepatocytes. Gastroenterology 2009;136:990-999.

[46] Liu H, Kim Y, Sharkis S, Marchionni L, Jang YY. In vivo liver regeneration potential of human induced pluripotent stem cells from diverse origins. Sci Transl Med 2011;3:82ra39.

[47] Langer R, Vacanti JP. Tissue engineering. Science 1993;260:920-926.

[48] Mooney D, Hansen L, Vacanti J, Langer R, Farmer S, Ingber D. Switching from differentiation to growth in hepatocytes: control by extracellular matrix. J Cell Physiol 1992;151:497-505.

[49] Fiegel HC, Kaufmann PM, Bruns H, Kluth D, Horch RE, Vacanti JP, Kneser U. Hepatic tissue engineering: from transplantation to customized cell based liver directed therapies from the laboratory. J Cell Mol Med 2008;12:56-66.

[50] Kim SS, Sundback CA, Kaihara S, Benvenuto MS, Kim BS, Mooney DJ, Vacanti JP. Dynamic seeding and in vitro culture of hepatocytes in a flow perfusion system. Tissue Eng 2000;6:39-44.

[51] Lin P, Chan WC, Badylak SF, Bhatia SN. Assessing porcine liver-derived biomatrix for hepatic tissue engineering. Tissue Eng 2004;10:1046-1053.

[52] Linke K, Schanz J, Hansmann J, Walles T, Brunner H, Mertsching H. Engineered liver-like tissue on a capillarized matrix for applied research. Tissue Eng 2007;13:2699-2707.

[53] Tong JZ, Bernard O, Alvarez F. Long-term culture of rat liver cell spheroids in hormonally defined media. Exp Cell Res 1990;189:87-92.

[54] Gerlach J, Unger J, Hole O, Encke J, Muller C, Neuhaus P. [Bioreactor for long-term maintenance of differentiated hepatic cell functions]. ALTEX 1994;11:207-215.

[55] Torok E, Pollok JM, Ma PX, Kaufmann PM, Dandri M, Petersen J, Burda MR, et al. Optimization of hepatocyte spheroid formation for hepatic tissue engineering on three-dimensional biodegradable polymer within a flow bioreactor prior to implantation. Cells Tissues Organs 2001;169:34-41.

[56] Torok E, Vogel C, Lutgehetmann M, Ma PX, Dandri M, Petersen J, Burda MR, et al. Morphological and functional analysis of rat hepatocyte spheroids generated on poly(L-lactic acid) polymer in a pulsatile flow bioreactor. Tissue Eng 2006;12:1881-1890.

[57] Kim BS, Baez CE, Atala A. Biomaterials for tissue engineering. World J Urol 2000;18:2-9.

[58] Voytik-Harbin SL, Brightman AO, Kraine MR, Waisner B, Badylak SF. Identification of extractable growth factors from small intestinal submucosa. J Cell Biochem 1997;67:478-491.

[59] Kaufmann PM, Kneser U, Fiegel HC, Kluth D, Herbst H, Rogiers X. Long-term hepatocyte transplantation using three-dimensional matrices. Transplant Proc 1999;31:1928-1929.

[60] Johnson LB, Aiken J, Mooney D, Schloo BL, Griffith-Cima L, Langer R, Vacanti JP. The mesentery as a laminated vascular bed for hepatocyte transplantation. Cell Transplant 1994;3:273-281.

[61] Starzl TE, Francavilla A, Halgrimson CG, Francavilla FR, Porter KA, Brown TH, Putnam CW. The origin, hormonal nature, and action of hepatotrophic substances in portal venous blood. Surg Gynecol Obstet 1973;137:179-199.

[62] Yang J, Yamato M, Shimizu T, Sekine H, Ohashi K, Kanzaki M, Ohki T, et al. Reconstruction of functional tissues with cell sheet engineering. Biomaterials 2007;28:5033-5043.

[63] Ohashi K, Yokoyama T, Yamato M, Kuge H, Kanehiro H, Tsutsumi M, Amanuma T, et al. Engineering functional two- and three-dimensional liver systems in vivo using hepatic tissue sheets. Nat Med 2007;13:880-885.

[64] El-Kassaby AW, Retik AB, Yoo JJ, Atala A. Urethral stricture repair with an off-the-shelf collagen matrix. J Urol 2003;169:170-173; discussion 173.

[65] Lee DJ, Steen J, Jordan JE, Kincaid EH, Kon ND, Atala A, Berry J, et al. Endothelialization of heart valve matrix using a computer-assisted pulsatile bioreactor. Tissue Eng Part A 2009;15:807-814.

[66] Amiel GE, Komura M, Shapira O, Yoo JJ, Yazdani S, Berry J, Kaushal S, et al. Engineering of blood vessels from acellular collagen matrices coated with human endothelial cells. Tissue Eng 2006;12:2355-2365.

[67] Ott HC, Matthiesen TS, Goh SK, Black LD, Kren SM, Netoff TI, Taylor DA. Perfusion-decellularized matrix: using nature's platform to engineer a bioartificial heart. Nat Med 2008;14:213-221.

[68] Baptista PM SM, Atala A, Soker S. A Novel Whole Organ Bioscaffold System for Tissue Engineering and Regenerative Medicine Applications. In: 3rd International Society for Stem Cell Research International Meeting; 2005 June 15-18, 2005; San Francisco, CA, USA; 2005.

[69] Baptista PM, Orlando G, Mirmalek-Sani SH, Siddiqui M, Atala A, Soker S. Whole organ decellularization - a tool for bioscaffold fabrication and organ bioengineering. Conf Proc IEEE Eng Med Biol Soc 2009;2009:6526-6529.

[70] Baptista PM, Siddiqui MM, Lozier G, Rodriguez SR, Atala A, Soker S. The use of whole organ decellularization for the generation of a vascularized liver organoid. Hepatology 2011;53:604-617.

[71] Uygun BE, Soto-Gutierrez A, Yagi H, Izamis ML, Guzzardi MA, Shulman C, Milwid J, et al. Organ reengineering through development of a transplantable recellularized liver graft using decellularized liver matrix. Nat Med 2010.

12

Platelet Rich Plasma in Reconstructive Periodontal Therapy

Selcuk Yılmaz, Gokser Cakar
and Sebnem Dirikan Ipci
*Yeditepe University, Faculty of Dentistry, Istanbul,
Turkiye*

1. Introduction

1.1 Regenerative periodontal therapy

The goal of periodontal therapy is to improve periodontal health and thereby to satisfy the patient's esthetic and functional needs or demands. To achieve this goal, most periodontal treatments aim to reduce probing depths and maintain or improve attachment levels and these parameters are used as surrogates of improved tooth retention. Conventional periodontal therapy includes non-surgical treatment as well as a variety of surgical approaches. In such treatments, histologic analysis revealed that periodontal healing occurs with repair rather than regeneration (Listgarten & Rosenberg, 1979). In repair, long junctional epithelium exists between the treated root surface and alveolar bone (Caton & Greenstein, 1993). However, over the last three decades, the major goal of periodontal therapy has been shifted from repair to reconstruction of periodontal tissues thereby reversing the damage to the periodontium caused by the disease process.

"True periodontal regeneration" is the reformation of a functionally oriented periodontal ligament with collagen fibers inserting in both regrown alveolar bone and reformed cementum over a previously diseased root surface. The first evolutionary stage of periodontal regeneration focused on using a variety of bone graft materials. A number of techniques and autogenic, allogenic, xenogenic and alloplastic bone graft materials have been used for regeneration purpose (Brunswold & Mellonig, 1993). Although significant clinical improvements in terms of probing depth reduction, attachment and bone gains were obtained, the results of the histological studies reported that new attachment achieved by bone grafts was usually a result of the formation of long junctional epithelium with slight or no new connective tissue attachment and negligible new cementum formation. Since these techniques have had limited success, more effective regenerative approaches have been suggested that utilize tissue-engineering techniques.

The concept of tissue engineering in periodontics began with guided tissue regeneration (GTR), a mechanical approach utilizing nonresorbable or bioabsorbable membranes to regenerate periodontal defects. GTR is a technique in which the placement of an occlusive membrane guides progenitor cells, residing in the periodontal ligament to repopulate the osseous defects in order to form new tooth supporting tissues (Nyman et al., 1982). The evidence, in fact, demonstrated that treatment of two- and three- wall intrabony defects with GTR has yielded successful clinical results in numerous studies and could promote periodontal regeneration in terms of true new attachment with nonresorbable as well as bioabsorbable barrier membranes (Cortellini & Tonetti, 2000). Since the improvement of the fibrous attachment level seems to be easier to obtain than a corresponding improvement in bone level with the utilization of GTR technique, it is utmost important that the space underneath the barrier must be preserved for an adequate period of time during healing for complete periodontal regeneration to occur. On the other hand, in cases where the membrane collapsed into the defects, reduced amounts of bone were formed due to the lack of space for progenitor cell population. In the light of the above, bone grafting technique has been an option in creating a space for the regenerating tissues underneath the membranes and also suggests the use of additional osteoconductive and/or partially osteoinductive properties of the graft materials. Combined usage of bone graft materials with GTR technique generally resulted in similar or more bone gain with the reported GTR studies alone (Paolantonio, 2002; Nygaard-Ostby et al., 2008). Although GTR and combined techniques successfully promote regrowth of the destroyed periodontium, application of the method is often difficult while there is substantial variation in clinical predictability, degree of efficacy, and histologic outcomes.

Shortcomings associated with GTR and advances in molecular/developmental biology set the ground for a new concept in periodontal regeneration by emphasizing the importance of biologic mediators.

Enamel matrix protein is one of the biologic mediators used for regeneration purpose. The discovery of the presence of the enamel matrix layer between the peripheral dentin and the developing cementum, periodontal ligament and alveolar bone formation, has provided the fundamental concept for enamel matrix protein derivative (EMD)-supported tissue engineering in regenerative periodontal therapy. General conclusions about the clinical relevance of EMD are limited by the high level of heterogeneity across the studies. Some studies have reported superior effects of EMD, other studies failed to present any additional effects (Trombelli, 2005). Histological data indicate that the application of EMD on the diseased root surfaces enhances the formation of a new connective tissue attachment (i.e. new cementum with inserting collagen fibers) and of new alveolar bone (Bosshardt, 2008). However, it has been suggested that there exists a possible limitation to the regenerative capability of EMD, related to its semi-fluid consistency and lack of space making effect (Mellonig, 1999). Therefore, combining EMD with a graft material, will overcome the problem of flap collapse and space maintenance when using it alone. Thus, more recently prominence has been given to the use of EMD in combination with graft materials (Scheyer et al., 2002; Bosshardt, 2008). While bone grafts intended to promote bone formation, their combination with EMD would designate a biological effect on the cascade of events leading to periodontal regeneration. Some studies indicate that the clinical outcomes of EMD may be improved when used in combination with bone grafts with respect to EMD alone (Kuru et al., 2006; Yılmaz et al., 2010b). In contrast, limited evidence seems to demonstrate no additional effect of EMD over the regenerative potential of the bone grafts (Scheyer et al., 2002).

2. Polypeptide growth factors

Recently, there has been a tremendous interest in polypeptide growth factors (PGFs) as another biologic mediator in periodontal regeneration (Giannobile, 1996). They are an enchanting group of agents in regeneration because of their regulatory effects on proliferation and differentiation of cells from bone and connective tissues. PGFs have the ability to regulate biological events including cell adhesion, migration, proliferation and differentiation. Among all PGFs, platelet derived growth factor (PDGF) and transforming growth factor-β (TGF-β) have been studied most extensively. PDGF and TGF-β have been shown to promote cell growth and differentiation *in vitro* and periodontal regeneration in animals (Lynch et al., 1991; Rutherford et al., 1992, Sporn & Roberts, 1992). However, human studies concerning the effects of PGFs on periodontal regeneration *in vivo* are limited. In these studies, PDGF has been shown to exert a substantial effect on periodontal regeneration as measured by attachment gain and bone fill in human intrabony and furcation defects (Howell et al., 1997; Camelo et al., 2003). Although promising results have been obtained, the routine use of PDGF and TGF-β as therapeutic agents for periodontal regeneration is not reality yet. Recently, a convenient approach to obtain not only these PGFs but also epithelial growth factor, vascular endothelial growth factor, insulin-like growth factor-1, basic fibroblast growth factor, hepatocyte growth factor is the use of autologous platelets.

3. What is platelet rich plasma?

Platelet rich plasma (PRP) is a preparation, serving as an autologous source of highly concentrated doses of platelets. The term PRP includes a high concentration of platelets obtained by a single or double step centrifugation of autologous blood (Tamimi et al., 2007). PRP preparations have also been designated as 'platelet pellet', 'autologous platelet concentrate' or 'platelet gel' (Marx, 2001).

Although it is not well clarified, PRP presents its effects through enhancement of soft and hard tissue healing processes. Since the actual amount of regenerated tissue and the course of soft tissue healing are dependent on the individual healing potential, which is significantly influenced by the presence and amount of the PGFs naturally available in the wound, the increased local concentrations of PGFs with the application of PRP in periodontal wound site enhance the healing outcome (Christgau et al., 2006a). In the early stages of the healing process, PGFs within PRP attract undifferentiated mesenchymal cells within the fibrin matrix and trigger cell division. Proliferation of connective tissue progenitors, stimulation of fibroblasts and osteoblast activity and angiogenesis are crucial steps in healing process and regeneration and it is assumed that local delivery of PRP seems to be responsible for all these cascade of events (Marx et al., 1998; Cheung & Griffin, 2004).

3.1 Preparation of platelet rich plasma

PRP can be easily prepared from patient's own blood by centrifugation and separated into three fractions: platelet-poor plasma (PPP) (fibrin glue or adhesive); PRP; and red blood cells. Platelets are enriched by 338% in the PRP preparation. PPP is the upper layer obtained after blood centrifugation in which platelet counts are negligible and is composed of acellular plasma containing fibrinogen and plasmatic growth factors. PRP is the bottom layer and is a volume of autologous plasma that has a platelet concentration approximately 3-4 times higher than baseline levels (150000-450000/µl).

Different procedures have been established for the preparation of PRP including Curasan PRP kit (Curasan, Kleinostheim, Germany), platelet concentration collection system (3i/Implant Innovations, Palm Beach Gardens, FL) and Smart PreP (Harvest Technologies Corp., Plymouth, MA, USA). A variety of factors influence the reliability of these systems including cell separator used, centrifugation steps, amount of blood collected preoperatively, baseline platelet concentration, amount of platelet concentrate obtained, final platelet concentration, type of blood anticoagulant and platelet activator used (Del Fabbro et al., 2010). Any of these factors may play a major role in PRP activation and affect the expected outcome in terms of biologic properties. Smart PreP is an FDA approved system in which PRP, PPP and autologous thrombin can be obtained. The unique property of the system is its autologous thrombin that is not present in any of the other systems which is an important step in the activation of release of PGFs from platelets. Since the activator of the system is autologous, there is no risk of disease transmission.

When PRP is taken into account for use in clinical practice, clinicians may encounter with several issues, including the clinician's proficiency in drawing the necessary blood, the cost of the procedure, the efficacy of centrifugation machine in obtaining appropriate concentrations of platelets, extra time and steps to prepare the coagulated PRP for its actual use. The centrifugation process is considered to be critical since different platelet counts correspond to the differences in centrifugation machines and techniques (Hanna et al., 2004; Weibrich et al., 2002, 2003). The PRP processed by means of Smart PreP system, ensures high percentages of platelet concentration with less blood drawn from the patient and with less complex procedures. In this system, for the PRP preparation, one hour before surgery, 20 ml of blood was drawn from the patient through a venipuncture in the antecubital vein. The drawn blood was mixed with 2ml of anticoagulant solution and was then processed through a centrifuge to obtain 3ml of the PRP according to the manufacturer's instructions. Immediately before application, this PRP was mixed with autologous thrombin. For the preparation of autologous thrombin, 10 ml of blood was also drawn from the antecubital vein. The drawn blood was mixed with 1ml of anticoagulant solution and following 45 minutes of incubation period, was centrifuged to obtain 1ml of autologous thrombin. A delivery syringe was used to mix the PRP with autologous thrombin (Yılmaz et al., 2007, 2009, 2010a, 2011).

3.2 The use of platelet rich plasma in the treatment of periodontal intrabony defects

During the last decade, there has been an increasing interest on the use of PRP in intraoral therapy including periodontal defects. The effectiveness of PRP in combination with different types of grafting materials, with or without EMD/GTR membranes, has been evaluated in regenerative periodontal therapy. Review of the literature reveals that, mostly, PRP is combined with bone grafts in the treatment of intrabony defects. Since the space maintenance of the defect is a crucial factor in periodontal regeneration, PRP's gel-like consistency may complicate the healing leading to flap collapse. Yılmaz et al. (unpublished data), evaluated the clinical and radiographic results of PRP alone in the treatment of intrabony defects. The authors concluded that successful results were obtained only in narrow and deep defects (Figures 1a-f).

Studies comparing the treatment of intrabony defects with PRP combined with different types of bone grafts with or without GTR to open flap debridement, GTR or grafts alone demonstrated significantly greater clinical attachment gain and defect fill following the combination approach (De Obarrio et al., 2000; Camargo et al., 2002; Lekovic et al., 2002; Hanna et al., 2004; Okuda et al., 2005; Quyang & Qiao, 2006). On the other hand, very recent controlled clinical studies demonstrated similar results when PRP, bone grafts and EMD/GTR were compared to bone grafts and GTR (Camargo et al., 2009; Christgau et al., 2006b; Papli&Chen, 2007; Yassıbağ-Berkman et al., 2007; Piemontese et al., 2008; Döri et al., 2007a, 2007b, 2008a, 2008b, 2009; Harnack et al., 2009). Although it is difficult to draw

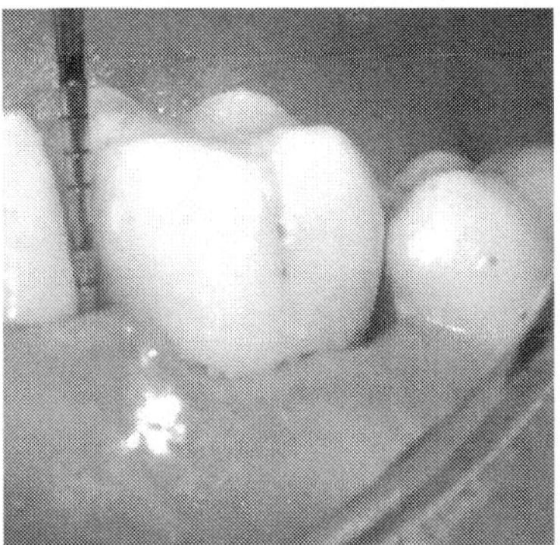

Fig. 1. (a) Initial clinical view of the intrabony defect and probing depth

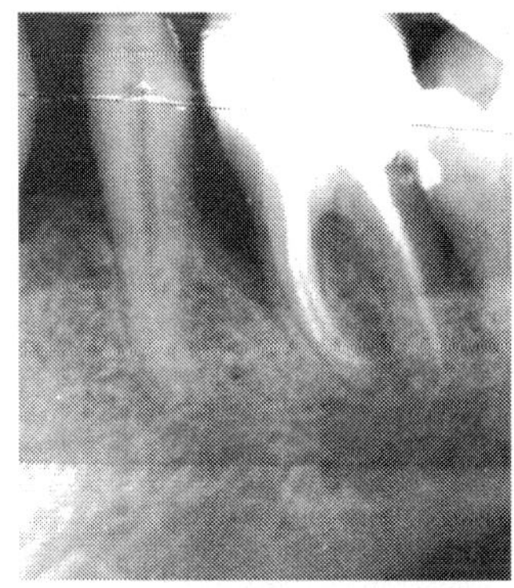

Fig. 1. (b) Initial radiographic view of the intrabony defect

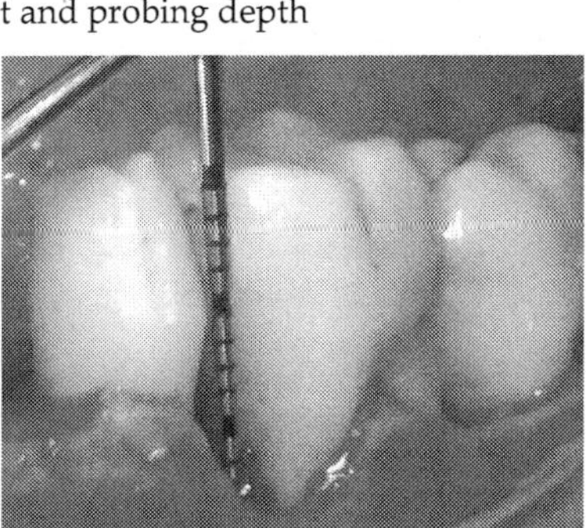

Fig. 1. (c) Intrasurgical measurement of the intrabony defect

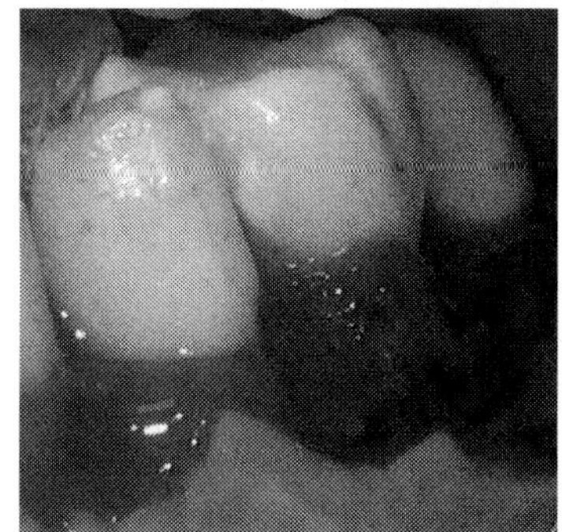

Fig. 1. (d) Application of PRP alone

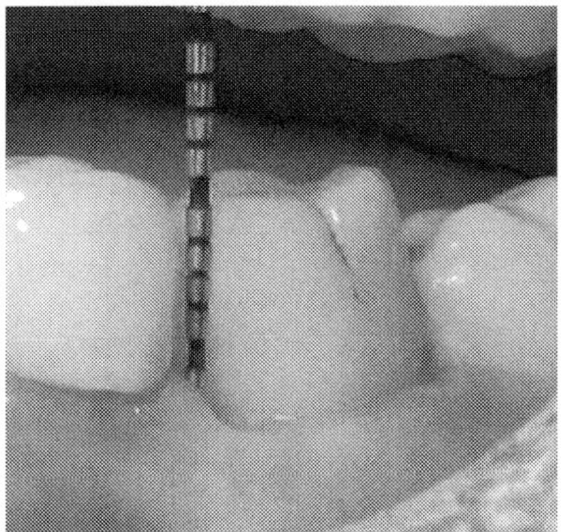

Fig. 1. (e) 12-months clinical view of the intrabony defect and probing depth

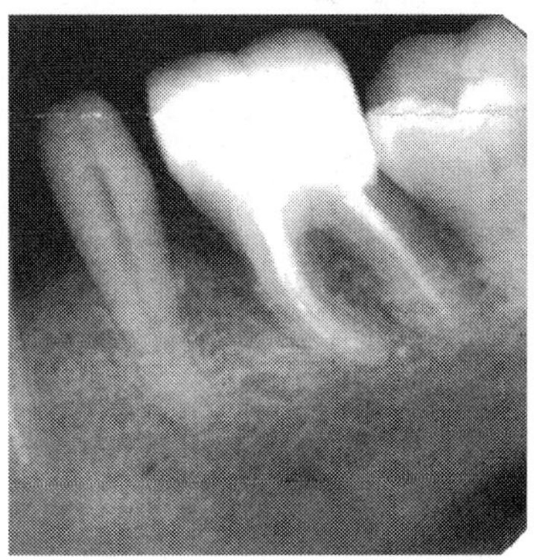

Fig. 1. (f) 12-months radiographic view of the intrabony defect

general conclusions and conflicting results exist, the pre-clinical and clinical data of PRP seems to be promising.

The clinical researches on PRP and their contradictional statements warrant further investigations to contribute to the PRP-supported regenerative therapy. In this process, trials evaluating the efficacy of PRP in combination with different regenerative materials can still add valuable information for the clinician in decision making regarding effective and predictable treatment alternatives for periodontal regeneration. Therefore a series of studies have been performed on the effect of PRP in the treatment of periodontal intrabony defects by Yılmaz et al. (2007, 2009, 2010a, 2011).

In these studies, outcome variables were soft and hard tissue measurements. For all patients, the following clinical parameters were recorded preoperatively and at 12 months postoperatively by the same calibrated examiner. A calibration exercise was carried out to obtain acceptable intra-examiner reproducibility (Sculean et al., 2005). Intraexaminer calibration was performed as follows: five patients, not included in the study, with at least four teeth of probing depth ≥ 5 mm on at least one aspect of each tooth, were evaluated by the examiner on two seperate sessions with 48 h interval. The examiner was accepted as calibrated if measurements at baseline and at 48 h were similar to the millimeter at $\geq 90\%$. Plaque index was measured according to Silness & Löe (1964) and sulcus bleeding index according to Mühlemann & Son (1971). Probing depth, relative attachment level, marginal recession and probing bone level were measured to the nearest millimeter with a calibrated periodontal probe (PCP 15 UNC, Hu-Friedy, Chicago, IL, USA) using an individual occlusal stent as a reference point for probe placement. Occlusal stents for positioning measuring probes were fabricated with cold-cured acrylic resin on a cast model obtained from an alginate impression. It was produced so that it covered the occlusal surfaces of the tooth being treated and the occlusal surfaces of at least one tooth in the mesial and distal directions. It was also extended apically on the buccal and lingual surfaces to cover the coronal third of the teeth. Six grooves were placed so that the post-surgical measurements could be at the same position and angulation as those made prior to surgery. Probing depth was the distance between the free gingival margin and the probeable bottom of the pocket, relative attachment level was the distance between the probeable bottom of the pocket and the edge of the stent, recession was the distance between the free gingival margin and the edge of the stent and probing bone level was the distance between the probeable bone crest and the edge of the stent (Figure 2).

Probing bone level was measured under local anesthesia by transgingival probing (sounding). The probe was forced through the soft tissue toward the bone until definite resistance was met (Kersten et al., 1992). Plaque index was evaluated at 4 periodontal sites (mesio-buccal, mid-buccal, disto-buccal, mid-lingual) whereas other measurements were made at 6 points (mesio-buccal, mid-buccal, disto-buccal, mesio-lingual, mid-lingual, and disto-lingual). Measurements where the edge of the stents was taken as the reference point were relative values (relative attachment level, recession, probing bone level) to evaluate the attachment loss/gain, marginal soft tissue level change, and clinical bone loss/gain.

Pre-operative and 12-month post-operative intra-oral radiographs were taken by the paralleling technique using a film holder device (RWT® Standard Film Holder System; bite blocks, indicator arms, aiming rings; Kentzler-Kaschner Dental GmbH, Ellwagen/Jagst, Germany) for the evaluation of radiographic bone level connected to an acrylic dental splint (individual) to achieve identical film placement at each evaluation with the aim of standardization. The film holder was coupled to the X-ray tube via an adapter (RWT®

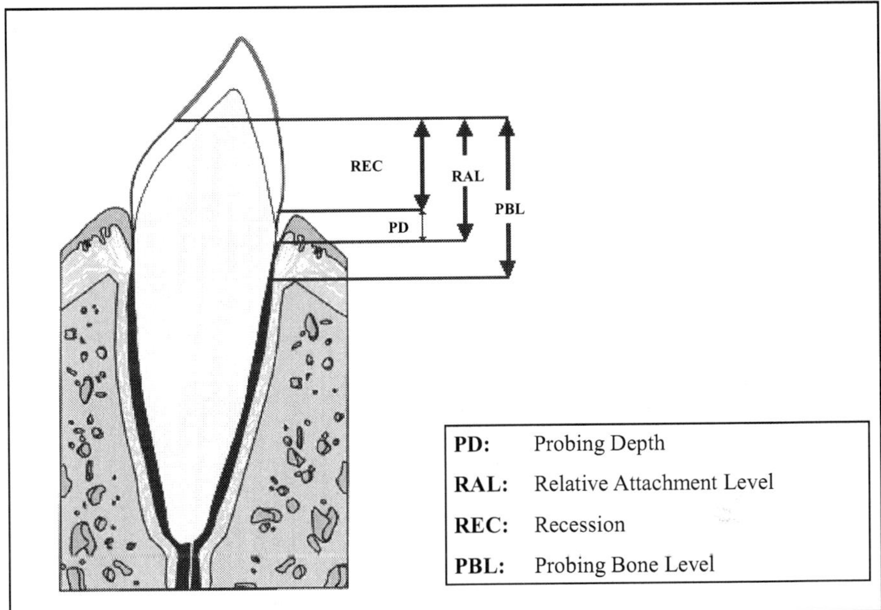

Fig. 2. Clinical measurements

Standard Film Holder System; aiming rings; Kentzler-Kaschner Dental GmbH, Ellwagen/Jagst, Germany). Pre- and post-operative radiograph pairs were independently assessed on a light box by 3 experienced clinicians who were not told which radiograph was which. The mode (most frequent) count was accepted. When measuring radiographic bone level, the 3 investigators were blinded with respect to the clinical measurements and had to reach agreement in terms of the location of both anatomical and bone loss landmarks. Radiographic measurements were obtained utilizing an adhesive millimeter grid (X-ray Grid, 3-4 cm, Meyer Haake GmbH, Oberursel, Germany). The differences between pre- and post-operative radiographic bone level measurements were considered as the radiographic bone loss/gain.

Besides these clinical and radiographic measurements, during operations, distance from the edge of the occlusal stent to the bottom of the defect (A) and distance from the edge of the stent to the most coronal extension of the alveolar bone crest (B) were measured. The intrabony component of the defects was defined as A-B (Figure 3).

The pioneering study was a case report evaluating the clinical, radiographic and re-entry results of a generalized aggressive periodontitis patient with wide intrabony periodontal defects treated with combined PRP and bovine derived xenograft (BDX) (Yılmaz et al., 2007). At 12 months postoperatively, clinical and radiographic measurements together with re-entry results showed marked improvements from baseline with increased stabilization of whole dentition including the hopeless teeth. In the light of this report, the authors emphasized that the surgical technique together with the materials used may be a possible solution for extensive bone loss in patients with generalized aggressive periodontitis (Figures 4a-g).

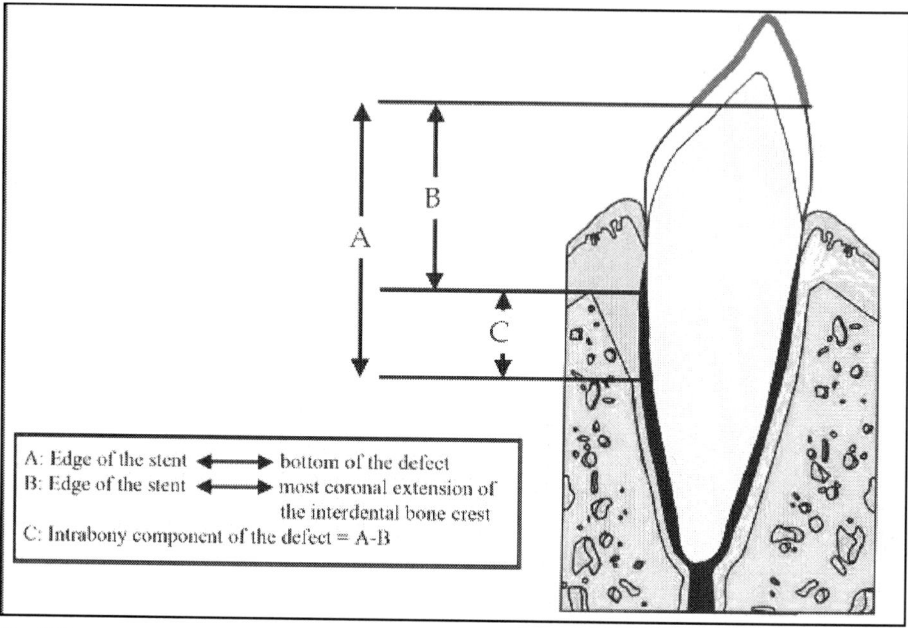

Fig. 3. Intrasurgical measurements

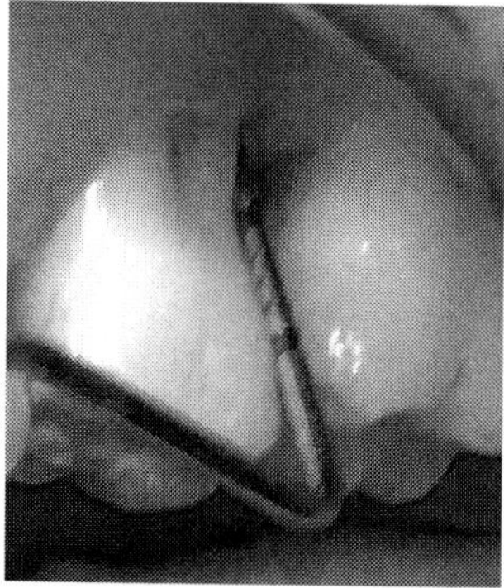

Fig. 4. (a) Initial clinical view of the intrabony
defect and probing depth

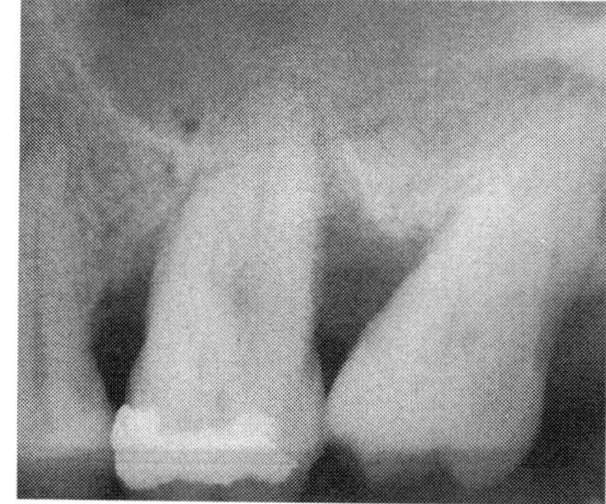

Fig. 4. (b) Initial radiographic view of the
intrabony defect

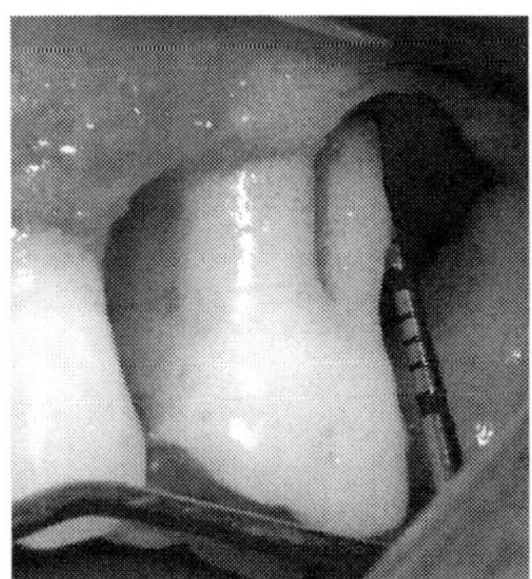

Fig. 4. (c) Intrasurgical measurement of the intrabony defect

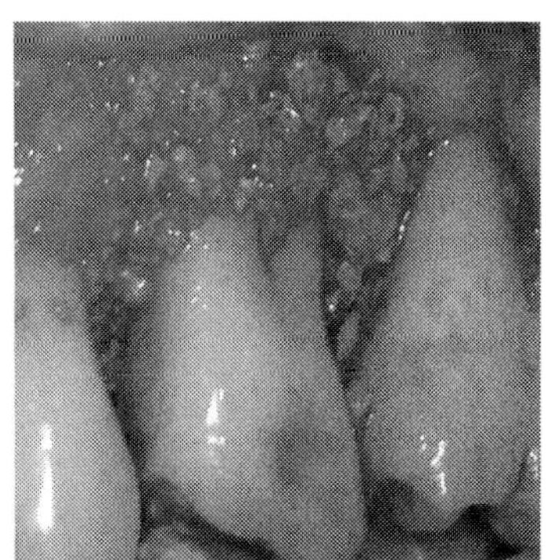

Fig. 4. (d) Application of PRP and BDX combination

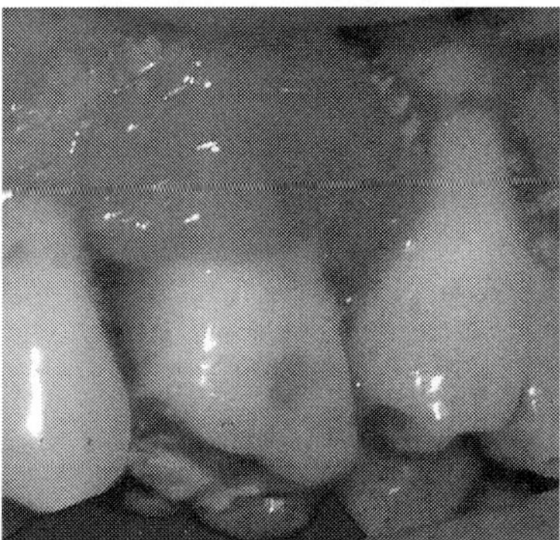

Fig. 4. (e) Application of PRP

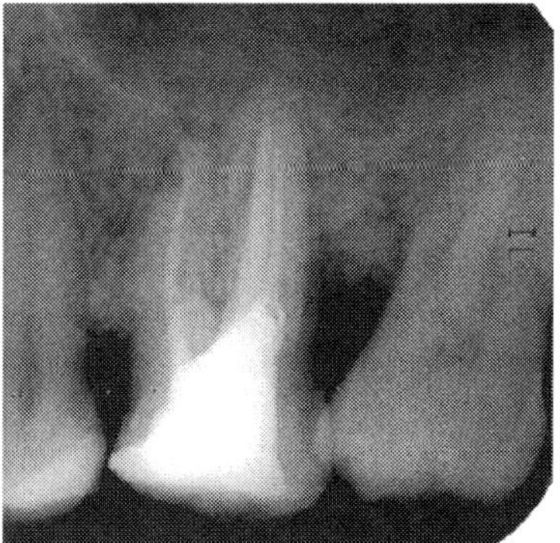

Fig. 4. (f) 12-months radiographic view of the intrabony defect

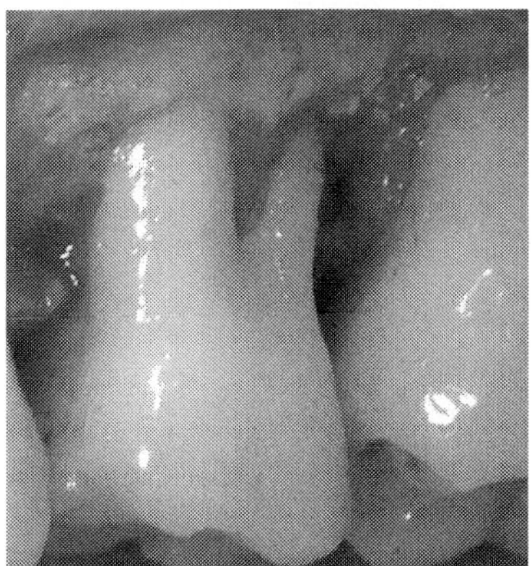

Fig. 4. (g) Re-entry at 12-months

The consecutive study investigated the effectiveness of PRP and BDX combination in the treatment of deep intrabony defects with an emphasis on the evaluation of early wound healing (Yılmaz et al., 2009). A total of 85 intrabony defects with an intrabony component of ≥3mm were selected in 20 advanced chronic periodontitis patients. Defects were surgically treated with PRP/BDX. At baseline and 12 months after surgery, the following parameters were recorded: plaque and sulcus bleeding indices, probing depth, relative attachment level, marginal recession, probing bone and radiographic bone levels. Postoperative healing was evaluated by an early healing index at 1 and 2 weeks after surgery. At 12 months, all clinical and radiographic parameters were improved ($p<0.001$). The mean changes at 12 months were: probing depth reduction of 4.78±1.20 mm, attachment gain of 4.24±1.03 mm, recession of 0.54±0.34 mm, clinical bone gain of 3.75±0.97 mm, and radiographic bone gain of 3.79±1.02 mm, respectively. Two weeks after surgery, primary closure was maintained in 95% of the defect sites. Treatment with a combination of PRP and BDX leads to a significantly favorable clinical and radiographic improvement in deep intrabony periodontal defects (Figures 5a-h).

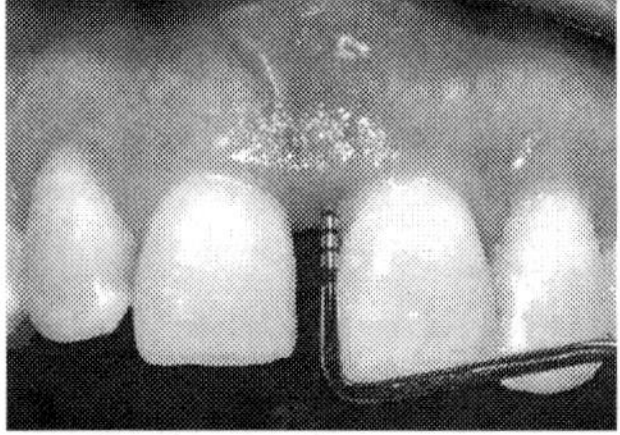

Fig. 5. (a) Initial clinical view of the intrabony defect and probing depth

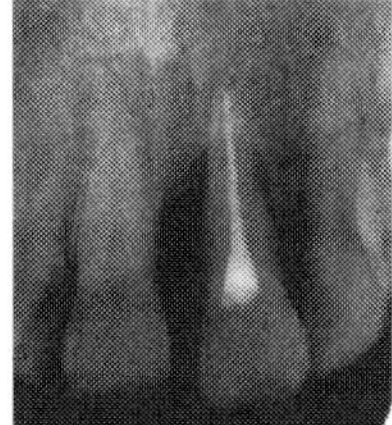

Fig. 5. (b) Initial radiographic view of the intrabony defect

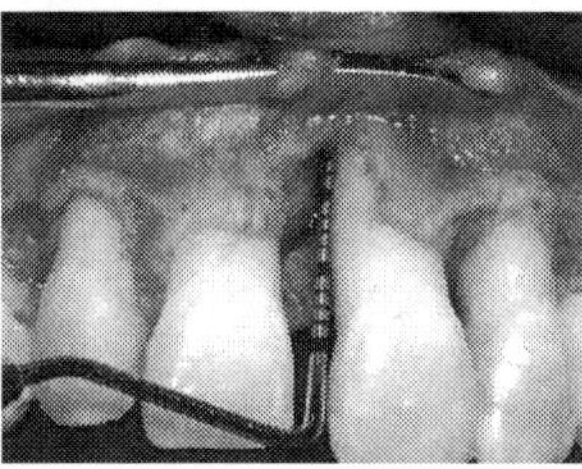

Fig. 5. (c) Intrasurgical measurement of the intrabony defect

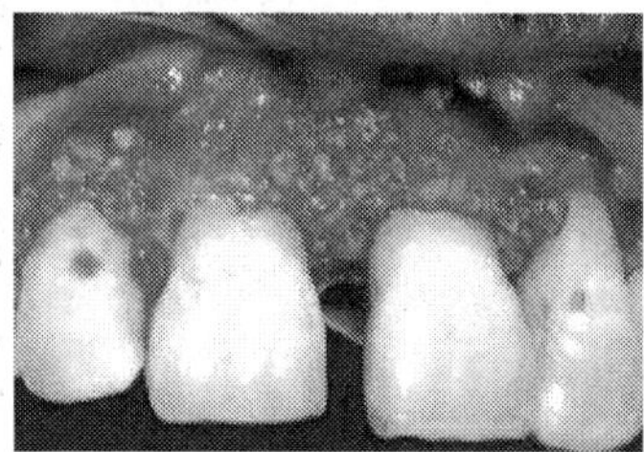

Fig. 5. (d) Application of PRP and BDX combination

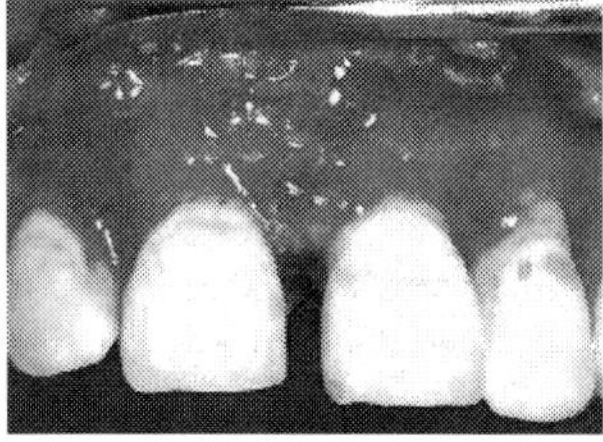

Fig. 5. (e) Application of PRP

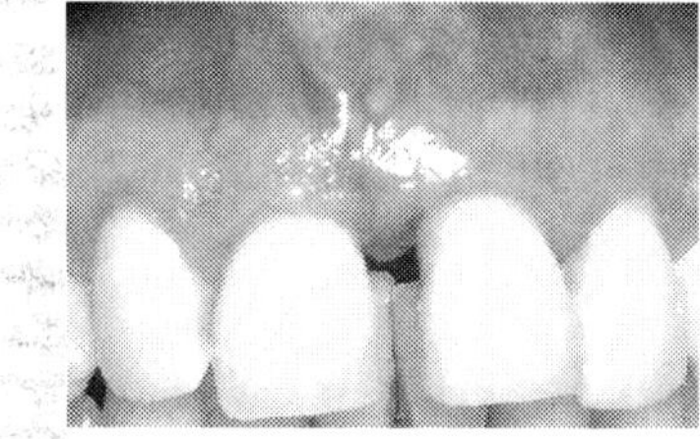

Fig. 5. (f) 12-months clinical view of the intrabony defect

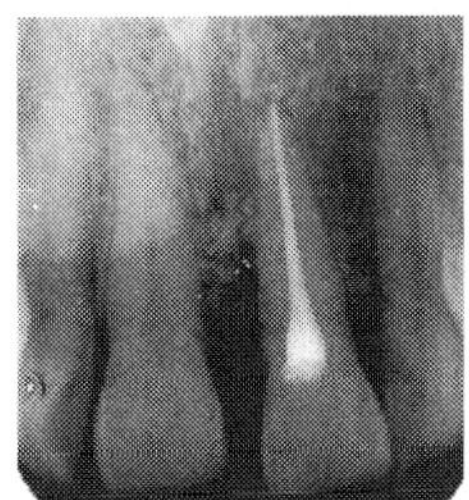

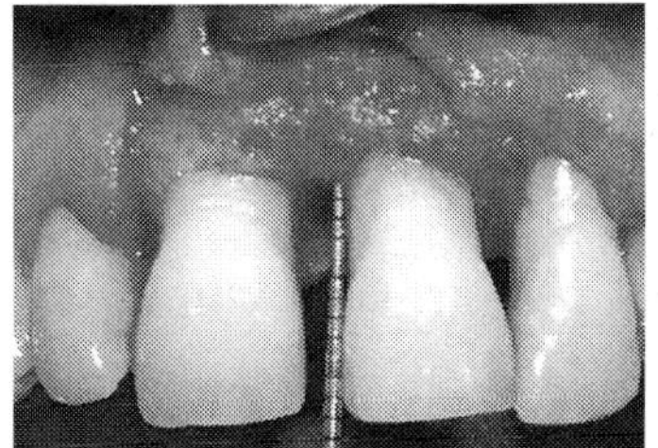

Fig. 5. (g) 12-months radiographic view of the intrabony defect

Fig. 5. (h) Re-entry at 12-months

The most common environmental risk factor jeopardizing the outcomes of periodontal regenerative therapy is smoking. Smokers showed a significantly less favourable response compared with non-smokers after regenerative procedures (Tonetti et al., 1995; Trombelli & Scabbia, 1997; Trombelli et al., 1997; Zucchelli et al., 2002; Stavropoulos et al., 2004). The precise mechanisms by which smoking interferes with periodontal regenerative healing are not completely understood. It can be hypothesized that any substance that might jeopardize the function of cells capable of periodontal regeneration could also impair tissue repair and regeneration (Balaji, 2008). Smoking byproducts such as nicotine and cotinine may inhibit the attachment, proliferation and chemotaxis of human periodontal ligament fibroblasts (Giannopoulou et al., 1999; James et al., 1999). In the literature, there is a body of clinical evidence supporting the negative influence of smoking on the outcome of regenerative procedures, mostly GTR (Tonetti et al., 1995; Trombelli & Scabbia, 1997; Trombelli et al., 1997; Zucchelli et al., 2002; Stavropoulos et al., 2004). However, there are no data on the effect of smoking status on the clinical and radiographic outcomes of a procedure based on the usage of PRP in intrabony defects. In order to clarify this issue, the healing response of intrabony defects following regenerative treatment with PRP/BDX was evaluated in smokers and non-smokers (Yılmaz et al., 2010a). A total of 24 advanced chronic periodontitis patients, 12 smokers and 12 non-smokers, with 113 intrabony defects with an intrabony component of $\geq$3mm were included in this study. Defects were surgically treated with PRP/BDX. At baseline and 12 months after surgery, plaque and sulcus bleeding indices, probing depth, relative attachment level, marginal recession, probing and radiographic bone levels were recorded. Considering the soft tissue measurements, smokers and non-smokers presented a mean probing depth reduction of 3.97±0.76 mm and 4.63±0.52 mm, recession of 0.76±0.44 mm and 0.50±0.12 mm and attachment gain of 3.26±0.42 mm and 4.06±0.40 mm, respectively. Evaluation of the hard tissue findings revealed that the mean clinical and radiographic bone gains in smokers and non-smokers were 2.83±0.47 mm and 3.63±0.38 mm, 2.98±0.38 mm and 3.67±0.48 mm, respectively. Inter-group differences for probing depth reduction (p<0.05), attachment (p<0.001), clinical (p<0.001) and radiographic bone gains (p<0.001) were found to be significant between smokers and non-smokers. These results emphasized that treatment outcome following PRP/BDX application in intrabony defects is impaired with smoking.

All patients participating in these studies tolerated the surgical procedures well. No complications such as infection, abscess formation and tissue necrosis were observed at any treated site. Additionally, an early healing index representing the early wound healing was evaluated (Wachtel et al., 2003). This index not only differentiates different degrees of exposure, but also records the amount of fibrin formation when complete closure is present. Clinical experience has shown that the most rapid and uneventful healing is associated with no or minimal fibrin formation. In all patients, at 1 and 2 weeks, almost all sites were completely closed. It has been reported that biological, physical and chemical properties of PRP may effect the wound healing (Okuda et al., 2003). During the early stages of wound healing, PGFs lead to a cellular and molecular events that result in wound healing in an orchestrated manner (Wikesjö et al., 1992). The effects of PGFs on cells and high content of fibrinogen (fibrin glue) that promotes a favorable scaffold for cellular migration are essential steps in the regeneration of periodontal defects. The PRP preparation presents a sticky characteristic, which works as a hemostatic and stabilizing agent and may aid the immobilization of the blood clot and bone graft in the defect area (Kawase et al., 2003). Blood clot immobilization has been suggested to be an important event in the early phases of wound healing in periodontal regenerative procedures (Polson & Proye, 1983). Moreover, establishing nontension primary wound closure of various soft tissue flaps is paramount for optimal postoperative wound healing (Kuru et al., 2006). Regenerative surgical procedures that require clinical flap manipulation also require excellence in suturing. When the proper suture technique is used, primary intention healing occurs. Accurate apposition of surgical flaps is significant to patient comfort, blood clot stabilization and prevention of unnecessary bone destruction. Interrupted suture techniques achieve excellent clinical results when used for wound closure with tension-free flaps. In these studies, interrupted sutures were used and flaps were placed back to their original places. The sutures were free of tension, obtaining a complete coverage of the intrabony defects. Collectively, the enhanced wound healing potential of PRP and primary wound closure may explain the improved early healing index results found in these studies. A shortcoming encountered with the currently available modalities of periodontal regenerative therapy is limited predictability. Even though the various modalities of osseous grafting, GTR, and, in particular, the combination of both, have been shown to be effective in promoting clinical and histologic periodontal regeneration, complete restoration of the attachment apparatus in every treated defect is still not a reality (Cortellini & Tonetti, 2000). Furthermore, it is very difficult and expensive to use such materials when full mouth bone defects are present. The ability to incorporate PGFs into the periodontal wound healing site with the application of PRP provides a promising approach to reach the established regenerative goals (Christgau et al., 2006a, 2006b). It is also possible to treat multiple intrabony defects in the same mouth with PRP prepared with the blood drawn from the same patient. Bone grafts as autografts, allografts, xenografts and synthetic materials have been shown to improve attachment levels and promote defect fill in humans (Garret, 1996). However, they usually result in the development of a long junctional epithelium between the root surface and gingival connective tissue (Caton & Zander, 1976). Therefore, while materials intended to promote bone formation play an important role in periodontal therapy, their combination with agents like PRP, capable of enhancing cell-mediated phenomena in periodontal wound healing has the potential to optimize the outcome of periodontal regeneration.

Smoking, as a risk factor, affects the results of regenerative periodontal therapy with PRP/BDX (Yılmaz et al., 2010a). Smokers presented less favourable soft and hard tissue healing when compared to non-smokers. This result may be attributed to the hypothesis that there may be a differential susceptibility of PRP to the negative effects of smoking (due to nicotine and its cytotoxic and vasoactive effects). In the literature, PRP was accepted as a practical source for PGFs required in regenerative procedures (Marx et al., 1998). Therefore, any factor that might jeopardize the regenerative potential of PRP by adversely affecting PGF production could also influence the regenerative outcomes (Balaji, 2008). It has been reported that smoking down-regulates hydroxyproline and collagen production (Jorgensen et al., 1998). Hydroxyproline and collagen are essential for the production and maintenance of connective tissue. The presence of nicotine on root surfaces in smokers has also been documented (Cuff et al., 1989). This nicotine can be stored in fibroblasts, which alters fibroblast function and proliferation (Peacock et al., 1993; Tipton & Dabbous, 1995). When fibroblasts are exposed to nicotine, cellular changes can occur (Hanes et al., 1991). This altered function of fibroblasts due to nicotine exposure could also be the cause of the poor periodontal wound healing. These combined effects of smoking may lead to less favourable outcomes following regenerative periodontal procedures.

3.3 Platelet rich plasma versus platelet poor plasma

Few studies have attempted to evaluate the effects of PPP which is poor in platelets. An animal experimental study provides an evidence of the positive role of PPP as a potential osteoinductive biologic tissue adhesive (Abiraman et al., 2002). In a recent animal study, it has been shown that PPP enhances bone formation based on radiographic and histologic findings (Findikcioglu et al., 2009). It has also been demonstrated that PPP is capable of stimulating osteoblastic proliferation, increasing DNA synthesis and performing a mitogenic effect of osteoblasts/periodontal ligament cells (Hamdan et al., 2009; Findikcioglu et al., 2009). Despite the fact that pre-clinical data appear promising, only one clinical study is available in the literature for the use of PPP in the treatment of intrabony defects. In this study, the healing outcomes of intrabony defects following treatment with PRP versus PPP combined with BDX were assessed (Yılmaz et al., 2011). Using a split mouth design, a total of 79 intrabony defects with an intrabony component of ≥3 mm in 20 patients were treated either with PRP/BDX or PPP/BDX. At baseline and 12 months after surgery, plaque and sulcus bleeding indices, probing depth, relative attachment level, recession, probing and radiographic bone levels were recorded. After 12 months, PRP/BDX and PPP/BDX groups, presented a mean probing depth reduction of 3.87±0.86 mm and 3.76±0.80 mm, recession of 1.35±0.68 mm and 1.58±0.54 mm, attachment gain of 2.51±0.97 mm and 2.18±0.87 mm, clinical bone gain of 2.18±0.86 mm and 2.09±0.89 mm, and radiographic bone gain of 2.11±0.87 mm and 2.19±0.96 mm, respectively. Inter-group differences were found to be insignificant. These results suggest that the outcomes of the treatment following PRP/BDX and PPP/BDX applications in intrabony defects are similar. When the platelet counts are taken into consideration, plasma poor in platelets, seems to demonstrate similar clinical efficacy as the plasma rich in platelets (Figures 6a-h).

PRP is an application of tissue engineering and can be considered as a storage vehicle for growth factors. However, previously demonstrated platelet count related actions may not solely reflect the mechanism of PRP and additional components of PRP and PPP also may have important biological activities during healing. It was suggested that PRP contains high concentrations of several growth factors such as PDGF and TGF-β, which may strongly modulate the regeneration process (Kawase et al., 2005; Christgau et al., 2006a). On the other hand, there may have been sufficient amounts of PGFs naturally occuring in the periodontal wound. Platelet activation in response to tissue damage during surgical periodontal procedures forms a platelet plug and blood clot to provide hemostasis and secretion of biologically active proteins. Thus, the therapeutical local application of PRP might have a little effect on the periodontal wound healing in terms of growth factor content (Christgau et al., 2006a). Other than growth factors, it was also suggested that because of its fibrinogen content PRP reacts with thrombin to induce fibrin clot formation (Camargo et al., 2005). This

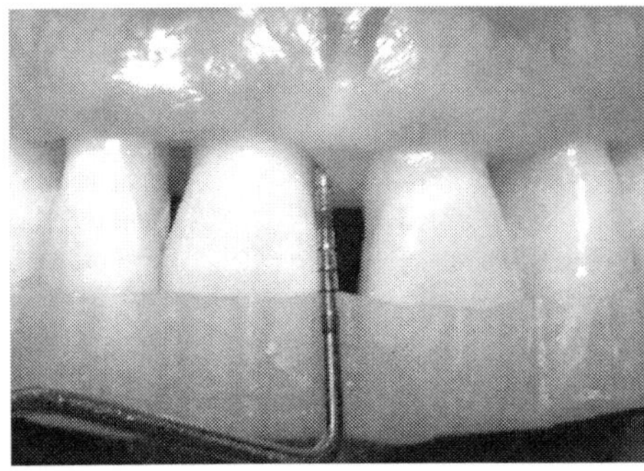

Fig. 6. (a) Initial clinical view of the intrabony defect and probing depth

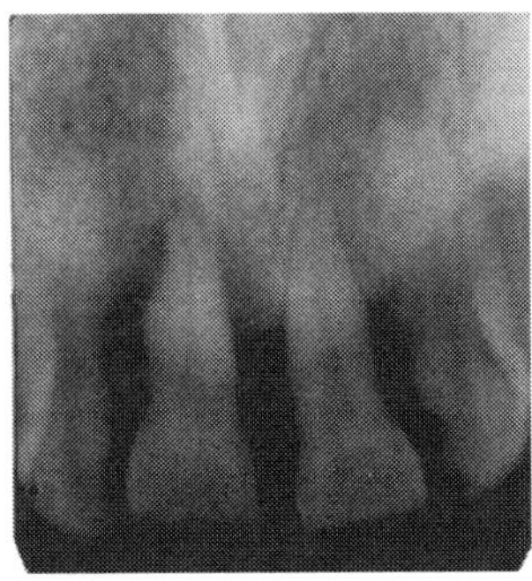

Fig. 6. (b) Initial radiographic view of the intrabony defect

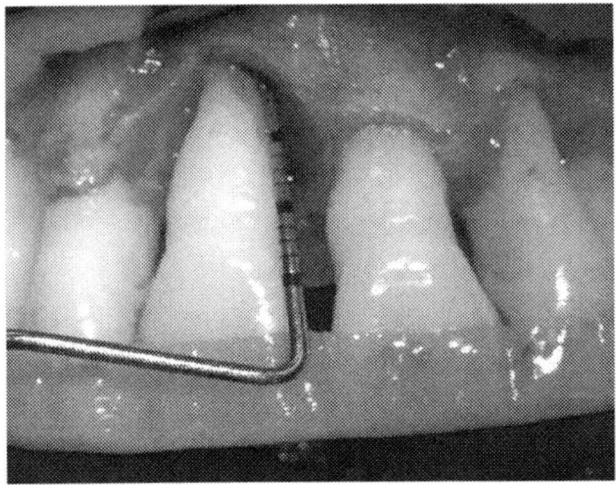

Fig. 6. (c) Intrasurgical measurement of the intrabony defect

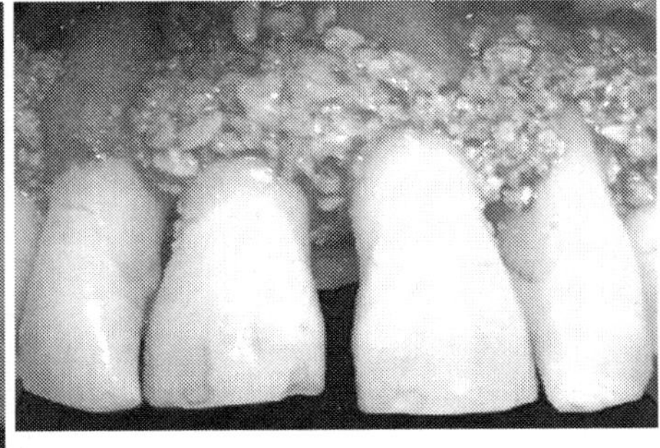

Fig. 6. (d) Application of PPP and BDX combination

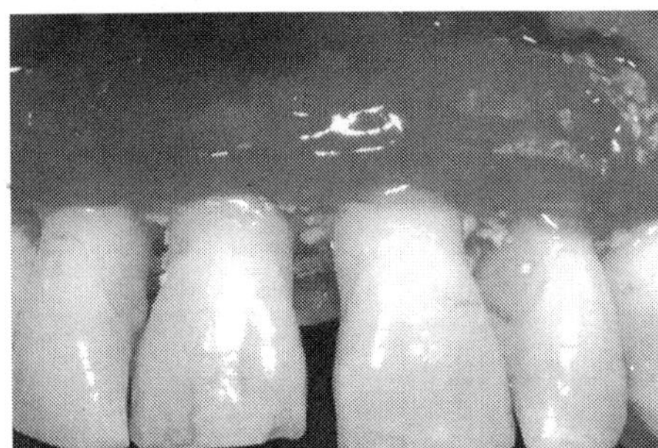

Fig. 6. (e) Application of PPP

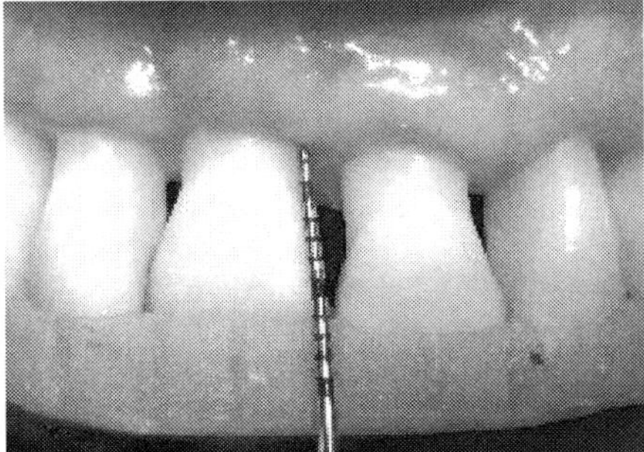

Fig. 6. (f) 12-months clinical view of the intrabony defect

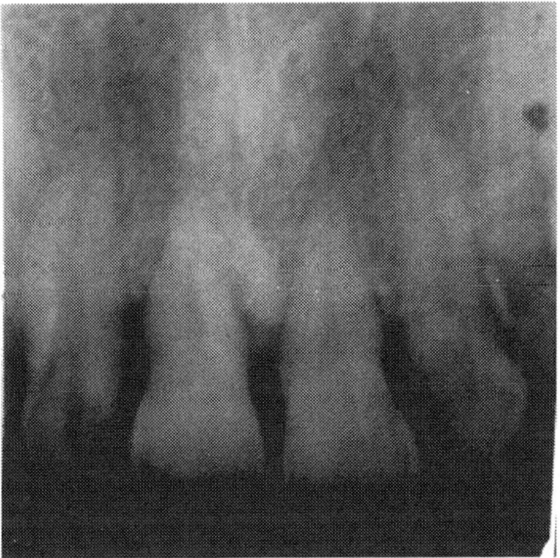

Fig. 6. (g) 12-months radiographic view of the intrabony defect

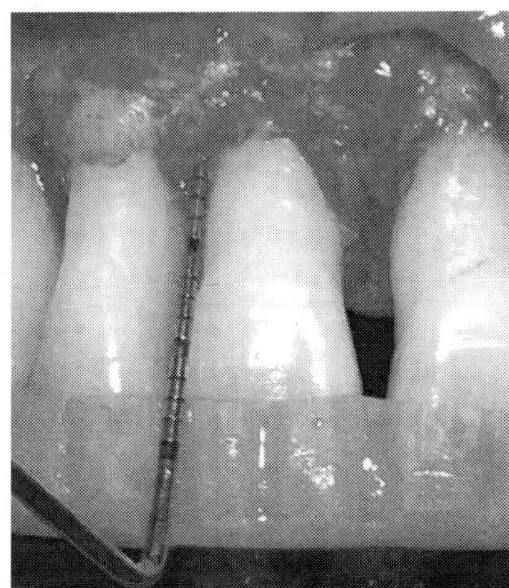

Fig. 6. (h) Re-entry at 12-months

reaction is capable of upregulating collagen synthesis in the extracellular matrix and provides a scaffold for cellular migration and adhesion (Camargo et al., 2005). PPP contains much more fibrinogen than PRP (Gosain & Lyon, 2002). It is of interest to note that this fibrinogen has firmer consistency that leads to its biologic sealant activity and adhesive potential (Gosain & Lyon, 2002). These features may have contributed to the similar clinical healing results obtained in both groups. Apart from the aforementioned knowledge, Creeper et al. (2009) stated that both PRP and PPP are capable of inducing differentiation of the periodontal ligament cells and osteoblasts that are critical for periodontal and bone regeneration and PPP also contains certain growth factors comparable to those found in PRP.

4. Conclusion

Although complete periodontal regeneration is unpredictable with any regenerative therapy currently used, the use of spesific biomaterials/biologicals show strong potential in improving clinical results in periodontal defects. Part of the problem is that it is still unclear how periodontal disease effects the supporting bone's regenerative potential and what specific biologic factors are involved. In recent years, however, clinicians have begun to learn much more about how periodontal regeneration works on a cellular and molecular level. This is a key step to developing strategies and materials that allow clinicians to promote periodontal regeneration predictably. As more is learned about the biologic process of periodontal wound healing and regeneration, new materials and techniques, are expected to make the task of periodontal regeneration even more predictable. It is likely that some combination of techniques may eventually prove to yield the best results. Overall, reconstruction of lost periodontal tissues with these new approaches results with functionally and esthetically acceptable teeth not only for the dentist but also for the patient.

5. References

[1] Abiraman, S.; Varna, H.K.; Umashankar, P.R. & John A. (2002). Fibrin glue as an osteoinductive protein in a mouse model. *Biomaterials*, Vol.23, pp.3023-3031, ISSN 0142-9612.

[2] Balaji, S.M. (2008). Tobacco smoking and surgical healing of oral tissues: a review. *Indian Journal of Dental Research*, Vol.19, pp.344-348, ISSN 0970-9290.

[3] Bosshardt, D.D. (2008). Biological mediators and periodontal regeneration: A review of enamel matrix proteins at the cellular and molecular levels. *Journal of Clinical Periodontology*, Vol.35, pp.87-105, ISSN 0303-6979.

[4] Brunswald, M.A. & Mellonig, J.T. (1993). Bone grafts and periodontal regeneration. *Periodontology 2000*, Vol.1, pp.80-91, ISSN 0906-6713.

[5] Camargo, P.M.; Lekovic, V.; Weinlaender, M.; Vasilic, N.; Madzarevic, M. & Kenney, E.B. (2002). Platelet-rich plasma and bovine porous bone mineral combined with guided tissue regeneration in the treatment of intrabony defects in humans. *Journal of Periodontal Research*, Vol.37, pp.300-306, ISSN 0022-3484.

[6] Camargo, P.M.; Lekovic, V.; Weinlaender, M.; Vasilic, N.; Madzarevic, M. & Kenney, E.B. (2005). A reentry study on the use of bovine porous bone mineral, GTR, and platelet-rich plasma in the regenerative treatment of intrabony defects in humans. *International Journal of Periodontics and Restorative Dentistry*, Vol.25, pp. 49–59, ISSN 0198-7569.

[7] Camargo, P.M.; Lekovic, V.; Weinlaender, M.; Divnic-Resnik, T.; Pavlovic, M. & Kenney, E.B. (2009). A surgical reentry study on the influence of platelet-rich plasma in enhancing the regenerative effects of bovine porous bone mineral and guided tissue regeneration in the treatment of intrabony defects in humans. *Journal of Periodontology*, Vol.80, pp.915-923, ISSN 0022-3492.

[8] Camelo, M.; Nevins, M.L.; Schenk, R.K.; Lynch, S.E. & Nevins, M. (2003). Periodontal regeneration in human class II furcations using purified recombinant human platelet-derived growth factor- BB (rhPDGF-BB) with bone allograft. *International Journal of Periodontics and Restorative Dentistry*, Vol.23, pp.213–225, ISSN 0198-7569.

[9] Caton, J.G. & Zander, H.A. (1976). Osseous repair of an infrabony pocket without new attachment of connective tissue. *Journal of Clinical Periodontology*, Vol.3, pp.54-58, ISSN 0303-6979.

[10] Caton, J.G. & Greenstein, G.G. (1993). Factors related to periodontal regeneration. *Periodontology 2000*, Vol.1, pp.9-15, ISSN 0906-6713.

[11] Cheung, W.S. & Griffin, T.J. (2004). A comparative study of root coverage with connective tissue and platelet concentrate grafts: 8 month results. *Journal of Periodontology*, Vol.75, pp.1678-1687, ISSN 0022-3492.

[12] Christgau, M.; Moder, D.; Hiller, K.A.; Dada, K.A.; Schmitz, G. & Schmalz, G. (2006a). Growth factors and cytokines in autologous platelet concentrate and their correlation to periodontal regeneration outcomes. *Journal of Clinical Periodontology*, Vol.33, pp.837-845, ISSN 0303-6979.

[13] Christgau, M.; Moder, D.; Wagner, J.; Glassl, M.; Hiller, K.A.; Wenzel, A. & Schmalz, G. (2006b). Influence of autologous platelet concentrate on healing in intra-bony defects following guided tissue regeneration therapy: A randomized prospective clinical split-mouth study. *Journal of Clinical Periodontology*, Vol.33, pp.908–921, ISSN 0303-6979.

[14] Cortellini , P. & Tonetti, M. (2000). Focus on intrabony defects: guided tissue regeneration. *Periodontology 2000*, Vol.22, pp.104-132, ISSN 0906-6713.

[15] Creeper, F.; Lichanska, A.M.; Marshall, R.I.; Seymour, G.J. & Ivanovski, S. (2009). The effect of platelet-rich plasma on osteoblast and periodontal ligament cell migration, proliferation and differentiation. *Journal of Periodontal Research*, Vol.44, pp.258-265, ISSN 0022-3484.

[16] Cuff, M.J.; McQuade, M.J.; Scheidt, M.J.; Sutherland, D.E. & Van Dyke, T.E. (1989). The presence of nicotine on root surfaces of periodontally diseased teeth in smokers. *Journal of Periodontology*, Vol.60, pp.564-569, ISSN 0022-3492.

[17] De Obarrio, J.J.; Araúz-Dutari, J.I.; Chamberlain, T.M. & Croston, A. (2000). The use of autologous growth factors in periodontal surgical therapy: Platelet gel biotechnology – Case reports. *International Journal of Periodontics and Restorative Dentistry*, Vol.20, pp.486-497, ISSN 0198-7569.

[18] Del Fabbro, M.; Bortolin, M.; Taschieri, S. & Weinstein, R. (2010). Is platelet concentrate advantageous for the surgical treatment of periodontal diseases? A systematic review and meta-analysis. *Journal of Periodontology*, Epub ahead of print, ISSN 0022-3492.

[19] Döri, F.; Huszár, T.; Nikolidakis, D.; Arweiler, N.B.; Gera, I. & Sculean, A. (2007a). Effect of platelet rich plasma on the healing of intrabony defects treated with an anorganic bovine bone mineral and expanded polytetrafluoroethylene membranes. *Journal of Periodontology*, Vol.78, pp.983-990, ISSN 0022-3492.

[20] Döri, F.; Húszár, T.; Nikolidakis, D.; Arweiler, N.B.; Gera, I. & Sculean, A. (2007b). Effect of platelet-rich plasma on the healing of intra-bony defects with a natural bone mineral and a collagen membrane. *Journal of Clinical Periodontology*, Vol.34, pp.254-261, ISSN 0303-6979.

[21] Döri, F.; Huszár, T.; Nikolidakis, D.; Arweiler, N.B.; Gera, I. & Sculean, A. (2008a). Effect of platelet-rich plasma on the healing of intrabony defects treated with beta tricalcium phosphate and expanded polytetrafluoroethylene membranes. *Journal of Periodontology*, Vol.79, pp.660-669, ISSN 0022-3492.

[22] Döri, F.; Nikolidakis, D.; Huszár, T.; Arweiler, N.B.; Gera, I. & Sculean, A. (2008b). Effect of platelet-rich plasma on the healing of intrabony defects treated with an enamel matrix protein derivative and a natural bone mineral. *Journal of Clinical Periodontology*, Vol.35, pp.44-50, ISSN 0303-6979.

[23] Döri, F.; Kovacs, V.; Arweiler, N.B.; Huszár, T.; Gera, I.; Nikolidakis, D. & Sculean, A. (2009). Effect of platelet-rich plasma on the healing of intrabony defects treated with an anorganic bovine bone mineral. A pilot study. *Journal of Periodontology*, Vol.80, pp.1599-1605, ISSN 0022-3492.

[24] Findikcioglu, K.; Findikcioglu, F.; Yavuzer, R.; Elmas, C. & Atabay, K. (2009). Effect of platelet-rich plasma and fibrin glue on healing of critical- size calvarial bone defects. *Journal of Craniofacial Surgery*, Vol.20, pp.34-40, ISSN 1049-2275.

[25] Garret, J.S. (1996). Periodontol regeneration around natural teeth. *Annals of Periodontology*, Vol.1, pp.621-666, ISSN 1553-0841.

[26] Giannobile, W.V. (1996). The potential role of growth and differentiation factors in periodontal regeneration. *Journal of Periodontology*, Vol.67, pp.545–553, ISSN 0022-3492.

[27] Giannopoulou, C.; Geinoz, A. & Cimasoni, G. (1999). Effects of nicotine on periodontal ligament fibroblasts in vitro. *Journal of Clinical Periodontology*, Vol.26, pp.49–55, ISSN 0303-6979.

[28] Gosain, A.K. & Lyon, V.B. (2002). The current status of tissue glues: II. For adhesion of soft tissues. *Plastic and Reconstructive Surgery*, Vol.110, pp.1581-1584, ISSN 0032-1052.

[29] Hamdan, A.A.; Loty, S.; Isaac, J.; Bouchard, P.; Berdal, A. & Sautier, J. (2009). Platelet-poor plasma stimulates the proliferation but inhibits the differentiation of rat osteoblastic cells in vitro. *Clinical Oral Implants Research*, Vol.20, pp.616-623, ISSN 0905-7161.

[30] Hanes, P.J.; Schuster, G.S. & Lubas, S. (1991). Binding, uptake, and release of nicotine by human gingival fibroblasts. *Journal of Periodontology*, Vol.62, pp.147-152, ISSN 0022-3492.

[31] Hanna, R.; Trejo, P.M. & Weltman, R.L. (2004). Treatment of intrabony defects with bovine-derived xenograft alone and in combination with platelet-rich plasma: A randomized clinical trial. *Journal of Periodontology*, Vol.75, pp.1668-1677, ISSN 0022-3492.

[32] Harnack, L.; Boedeker, R.H.; Kurtulus, I.; Boehm, S.; Gonzales, J. & Meyle, J. (2009). Use of platelet-rich plasma in periodontal surgery-a prospective randomized double blind clinical trial. *Clinical Oral Investigation*, Vol.13, pp.179-187, ISSN 1432-6981.

[33] Howell, T.H.; Fiorellini, J.P.; Paquette, D.W.; Offenbacher, S.; Giannobile, W.V. & Lynch, S.E. (1997). A phase I/II clinical trial to evaluate a combination of recombinant human-derived growth factor-BB and recombinant human insulin-like growth factor-I in patients with periodontal disease. *Journal of Periodontology*, Vol.68, pp.1186–1193, ISSN 0022-3492.

[34] James, J.A.; Sayers, N.M.; Drucker, D.B. & Hull, P.S. (1999). Effects of tobacco products on the attachment and growth of periodontal ligament fibroblasts. *Journal of Periodontology*, Vol.70, pp.518–525, ISSN 0022-3492.

[35] Jorgensen, L.N.; Kallehave, F.; Christensen, E.; Siana, J.E. & Gottrup, F. (1998). Less collagen production in smokers. *Surgery*, Vol.123, pp.450-455, ISSN 0039-6060.

[36] Kawase, T.; Okuda, K.; Wolff, L.F. & Yoshie, H. (2003). Platelet-rich plasma derived fibrin clot formation stimulates collagen synthesis in periodontal ligament and osteoblastic cells in vitro. *Journal of Periodontology*, Vol.74, pp.858-864, ISSN 0022-3492.

[37] Kawase, T.; Okuda, K.; Saito, Y. & Yoshie, H. (2005). In vitro evidence that the biological effects of platelet-rich plasma on periodontal ligament cells is not mediated solely by constituent transforming-growth factor-β or platelet-derived growth factor. *Journal of Periodontology*, Vol.76, pp.760-767, ISSN 0022-3492.

[38] Kersten, B.G.; Chamberlain, A.D.; Khorsandi, S.; Wikesjö, U.M.; Selvig, K.A. & Nilvéus, R.E. (1992). Healing of the intrabony periodontal lesion following root condition with citric acid and wound closure including an expanded PTFE membrane. *Journal of Periodontology*, Vol.63, pp.876-882, ISSN 0022-3492.

[39] Kuru, B.; Yılmaz, S.; Argın, K. & Noyan, U. (2006). Enamel matrix derivative alone or in combination with a bioactive glass in wide intrabony defects. *Clinical Oral Investigation*, Vol.10, pp.227-234, ISSN 1432-6981.

[40] Lekovic, V.; Camargo, P.M.; Weinlaender, M.; Vasilic, N. & Kenney, E.B. (2002). Comparison of platelet-rich plasma, bovine porous bone mineral, and guided tissue regeneration versus platelet-rich plasma and bovine porous bone mineral in the treatment of intrabony defects: A reentry study. *Journal of Periodontology*, Vol.73, pp.198-205, ISSN 0022-3492.

[41] Listgarten, M.A. & Rosenberg, M.M. (1979). Histological study of repair following new attachment procedures in human periodontal lesions. *Journal of Periodontology*, Vol.50, pp.333-344, ISSN 0022-3492.

[42] Lynch, S.E.; de Castilla, G.R.; Williams, R.C.; Kiritsy, C.P.; Howell, T.H.; Reddy, M.S. & Antoniades, H.N. (1991). The effects of short-term application of the combination of the platelet-derived growth factor and insulin-like growth factor on periodontal wound healing. *Journal of Periodontology*, Vol.62, pp.458–467, ISSN 0022-3492.

[43] Marx, R.E.; Carlsson, E.R.; Eichsteaedt, R.M.; Schimmele, S.R.; Strauss, J.E. & Georgeff, K.R. (1998). Platelet rich plasma: Growth factor enhancement for bone grafts. *Oral Surgery Oral Medicine Oral Pathology Oral Radiology and Endodontics*, Vol.85, pp.638-646, ISSN 1079-2104.

[44] Marx, R.E. (2001). Platelet rich plasma (PRP): What is PRP and what is not PRP? *Implant Dentistry*, Vol.10, pp.225-228, ISSN 1056-6163.

[45] Mellonig, J.T. (1999). Enamel matrix derivative for periodontal reconstructive surgery: technique and clinical and histologic case report. *International Journal of Periodontics and Restorative Dentistry*, Vol.19, pp.8-19, ISSN 0198-7569.

[46] Mühlemann, H.R. & Son, S. (1971). Gingival sulcus bleeding: a leading symptom in initial gingivitis. *Helvetica Odontologica Acta*, Vol.15, pp.107-112, ISSN 0018-0211.

[47] Nygaard-Ostby, P.; Bakke, V.; Nesdal, O.; Nilssen, H.K.; Susin, C. & Wikesjo, U.M. (2008). Periodontal healing following reconstructive surgery: effect of guided tissue regeneration using a bioresorbable barrier device when combined with autogenous bone grafting. A randomized controlled clinical trial. *Journal of Clinical Periodontology*, Vol.35, pp.37-43, ISSN 0303-6979.

[48] Nyman, S.; Gottlow, J.; Karring, T. & Lindhe, J. (1982). The regenerative potential of the periodontal ligament. An experimental study in the monkey. *Journal of Clinical Periodontology*, 9: 257-265, ISSN 0303-6979.

[49] Okuda, K.; Kawase, T.; Momose, M.; Murata, M.; Saito, Y.; Suzuki, H.; Wolf, L.F. & Yoshie, H. (2003). Platelet-rich plasma contains high levels of platelet-derived growth factor and transforming growth factor-β and modulates the proliferation of periodontally related cells in vitro. *Journal of Periodontology*, Vol.74, pp.849-857, ISSN 0022-3492.

[50] Okuda, K.; Tai, H.; Tanabe, K.; Suzuki, H.; Sato, T.; Kawase, T.; Saito, Y.; Wolff, L.F. & Yoshie, H. (2005). Platelet-rich plasma combined with a porous hydroxyapatite graft for the treatment of intrabony periodontal defects in humans: a comparative controlled clinical study. *Journal of Periodontology*, Vol.76, pp.890-898, ISSN 0022-3492.

[51] Ouyang, X.Y. & Qiao, J. (2006). Effect of platelet-rich plasma in the treatment of periodontal intrabony defects in humans. *Chinese Medical Journal*, Vol.119, pp.1511-1521, ISSN 0366-6999.

[52] Paolantonio, M. (2002). Combined regenerative technique in human intrabony defects by collagen membranes and anorganic bovine bone. A controlled clinical study. *Journal of Periodontology*, Vol.73, pp.158-166, ISSN 0022-3492.

[53] Papli, R. & Chen, S. (2007). Surgical treatment of intrabony defects with autologous platelet concentrate or bioabsorbable barrier membrane: A prospective case series. *Journal of Periodontology*, Vol.78, pp.185-193, ISSN 0022-3492.

[54] Peacock, M.E.; Sutherland, D.E.; Schuster, G.S.; Brennan, W.A.; O'Neal, R.B.; Strong, S.L. & Van Dyke, T.E. (1993). The effect of nicotine on reproduction and attachment of human gingival fibroblasts in vitro. *Journal of Periodontology*, Vol.64, pp.658-665, ISSN 0022-3492.

[55] Piemontese, M.; Domenico Aspriello, S.; Rubini, C.; Ferrante, L. & Procaccini, M. (2008). Treatment of periodontal intrabony defects with demineralized freeze-dried bone allograft in combination with platelet-rich plasma: A comparative clinical trial. *Journal of Periodontology*, Vol.79, pp.802-810, ISSN 0022-3492.

[56] Polson, A.M. & Proye, M.P. (1983). Fibrin linkage: A precursor for new attachment. *Journal of Periodontology*, Vol.54, pp.141-147, ISSN 0022-3492.

[57] Rutherford, R.B.; Niekrash, C.E.; Kennedy, J.E. & Charette, M.F. (1992). Platelet-derived and insulin-like growth factors stimulate periodontal attachment in monkeys. *Journal of Periodontal Research*, Vol.27, pp.285–290, ISSN 0022-3484.

[58] Scheyer, E.T.; Velasquez-Plata, D.; Brunsvold, M.A.; Lasho, D.J. & Mellonig J.T. (2002). A clinical comparison of a bovine-derived xenograft used alone and in combination with enamel matrix derivative for the treatment of periodontal osseous defects in humans. *Journal of Periodontology*, Vol.73, pp.423-432, ISSN 0022-3492.

[59] Sculean, S.; Pietruska, M.; Schwartz, F.; Willershausen, B.; Arweiler, N.B. & Auschill, T.M. (2005). Healing of human intrabony defects following regenerative periodontal therapy with an enamel matrix protein derivative alone and combined with a bioctive glass. A controlled clinical study. *Journal of Clinical Periodontology*, Vol.32, pp.111-117, ISSN 0303-6979.

[60] Silness, J. & Löe, H. (1964). Periodontal disease in pregnancy (II). Correlation between oral hygiene and periodontal conditioning. *Acta Odontologica Scandinavica*, Vol.22, pp.121-135, ISSN 0001-6357.

[61] Sporn, M.B. & Roberts, A.B. (1992). Transforming growth factor-β: Recent progress and new challenges. *Journal of Cell Biology*, Vol.119, pp.1017–1021, ISSN 0021-9525.

[62] Stavropoulos, A.; Mardas, N.; Herrero, F. & Karring, T. (2004). Smoking affects the outcome of guided tissue regeneration with bioresorbable membranes: a retrospective analysis of intrabony defects. *Journal of Clinical Periodontology*, Vol.31, pp.945–950, ISSN 0303-6979.

[63] Tamimi, F.M.; Montalvo, S.; Tresguerres, I. & Blanco Jerez, L. (2007). A comparative study of 2 methods for obtaining platelet rich plasma. *Journal of Oral and Maxillofacial Surgery*, Vol.65, pp.1084-1093, ISSN 0278-2391.

[64] Tipton, D.A, & Dabbous, M.K. (1995). Effects of nicotine on proliferation and extracellular matrix production on human gingival fibroblasts in vitro. *Journal of Periodontology*, Vol.66, pp.1056-1064, ISSN 0022-3492.

[65] Tonetti, M.S.; Pini-Prato, G. & Cortellini, P. (1995). Effect of cigarette smoking on periodontal healing following GTR in infrabony defects. A preliminary

retrospective study. *Journal of Clinical Periodontology*, Vol.22, pp.229-234, ISSN 0303-6979.

[66] Trombelli, L. (2005). Which reconstructive procedures are effective for treating the periodontal intraosseous defect? *Periodontology 2000*, Vol.37, pp.88-105, ISSN 0906-6713.

[67] Trombelli, L.; Kim, C.K.; Zimmerman, G.J. & Wikesjo, U.M.E. (1997). Retrospective analysis of factors related to clinical outcome of guided tissue regeneration procedures in intrabony defects. *Journal of Clinical Periodontology*, Vol.24, pp.366–371, ISSN 0303-6979.

[68] Trombelli, L. & Scabbia, A. (1997). Healing response of gingival recession defects following guided tissue regeneration procedures in smokers and non-smokers. *Journal of Clinical Periodontology*, Vol.24, pp.529–553, ISSN 0303-6979.

[69] Wachtel, H.; Schenk, G.; Bo''hm, S.; Weng, D.; Zuhr, O: & Hürzeler, M.B. (2003). Microsurgical access flap and enamel matrix derivative for the treatment of periodontal intrabony defects: A controlled clinical study. *Journal of Clinical Periodontology*, Vol.30, pp.496–504, ISSN 0303-6979.

[70] Weibrich, G.; Kleis, W.K. & Hafner, G. (2002). Growth factor levels in the platelet-rich plasma produced by 2 different methods: Curasan type PRP kit versus PCSS PRP system. *International Journal of Oral and Maxillofacial Implants*,Vol.17, pp.184-190, ISSN 0882-2786.

[71] Weibrich, G.; Kleis, W.K.; Buch, R.; Hitzler, W.E. & Hafner, G. (2003). The Harvest Smart PReP system versus the Friadent-Schütze platelet-rich plasma kit. Comparison of a semiautomatic method with a more complex method for the preparation of platelet concentrates. *Clinical Oral Implants Research*, Vol.4, pp.233-239, ISSN 0905-7161.

[72] Wikesjo, U.M.; Nilveus, R.E. & Selvig, K.A. (1992). Significance of early healing events on periodontal repair: A review. *Journal of Periodontology*, Vol.63, pp.158-165, ISSN 0022-3492.

[73] Yassıbag-Berkman, Z.; Tuncer, O.; Subasioglu, T. & Kantarci, A. (2007). Combined use of platelet-rich plasma and bone grafting with or without guided tissue regeneration in the treatment of anterior interproximal defects. *Journal of Periodontology*, Vol.78, pp.801-809, ISSN 0022-3492.

[74] Yılmaz, S.; Cakar, G.; Eren-Kuru, B. & Yıldırım, B. (2007). Platelet rich plasma in combination with bovine derived xenograft in the treatment of generalized aggressive periodontitis. *Platelets*, Vol.18, pp.535-539, ISSN 0953-7104.

[75] Yılmaz, S.; Cakar, G.; Kuru, B.; Dirikan, S. & Yıldırım, B. (2009). Platelet-rich plasma in combination with bovine derived xenograft in the treatment of deep intrabony periodontal defects: A report of 20 consecutively treated patients. *Platelets*, Vol.20, pp.432-440, ISSN 0953-7104.

[76] Yılmaz, S.; Cakar, G.; Dirikan Ipci, S.; Eren-Kuru, B. & Yıldırım, B. (2010a). Regenerative Treatment with Platelet Rich Plasma Combined with Bovine Derived Xenograft in Smokers and Non-smokers: 12 Month Clinical and Radiographic Results. *Journal of Clinical Periodontology*, Vol.37, pp.80-87, ISSN 0303-6979.

[77] Yılmaz, S.; Cakar, G.; Yildirim, B. & Sculean, A. (2010b). Healing of two and three wall intrabony periodontal defects following treatment with an enamel matrix derivative combined with autogenous bone. *Journal of Clinical Periodontology*, Vol.37, pp.544-550, ISSN 0303-6979.

[78] Yılmaz, S.; Kabadayı, C.; Dirikan Ipci, S.; Cakar, G. & Eren-Kuru, B. (2011). Treatment of intrabony periodontal defects with platelet rich plasma versus platelet poor plasma combined with a bovine derived xenograft: a controlled clinical trial. *Journal of Periodontology*, Vol.82, pp.837-844, ISSN 0022-3492.

[79] Zucchelli, G.; Bernardi, F.; Montebugnoli, L. & De, S.M. (2002). Enamel matrix proteins and guided tissue regeneration with titanium-reinforced expanded polytetrafluoroethylene membranes in the treatment of infrabony defects: a comparative controlled clinical trial. *Journal of Periodontology*, Vol.73, pp.3–12, ISSN 0022-3492.

13

Ocular Surface Reconstitution

Pho Nguyen, Shabnam Khashabi
and Samuel C Yiu
Doheny Eye Institute, Los Angeles
California

1. Introduction

1.1 The ocular surface – anatomy and pathology

The corneal epithelium, conjunctival epithelium, and the lacrimal system constitute the ocular surface. A healthy corneal epithelium is essential for corneal health and visual function. The corneal epithelium is a 5- to 6-cell-thick layer that provides a defensive barrier against pathologic organisms. It exists in a dynamic equilibrium, with superficial cells being constantly shed into the tear film. Populations of pluripotent stem cells reside in the palisades of Vogt at the human corneoscleral limbus and generate transient amplified cells that centripetally migrate toward the central cornea. These transient amplified cells undergo terminal differentiation into epithelial cells and repopulate the corneal epithelium, i.e. the XYZ hypothesis (Thoft et al., 1983). Severe ocular surface disorders, such as infection, keratoconjunctivitis sicca, Stevens-Johnson syndrome, ocular cicatricial pemphigoid or chemical/thermal injuries, can progress to corneal scarring, conjunctivalization, neovascularization, or stromal melts. Depletion of the limbal stem cells may follow, resulting in impaired vision or eventual corneal blindness. According to the World Health Organization, corneal disorders, e.g. trachoma or onchocerciasis, constitute a significant cause of loss of vision and blindness in the world (Thylefors et al., 1995).

The conjunctiva is a thin, transparent, mucus membrane, overlying a thin vascular stroma. It is divided into three geographic zones: bulbar, forniceal, and palpebral. The conjunctival nonkeratinized stratified epithelium contains mucin-producing goblet cells, which are important for tear film stability. Additionally, the conjunctiva participates in the ocular surface antimicrobial defense via the conjunctiva-associated lymphoid tissue, as well as secretory antimicrobial peptides, such as defensins (Haynes et al., 1999). Disorders of the conjunctiva include elastotic changes, fibrovascular proliferation, malignancies, and autoimmune conditions such as Stevens-Johnson syndrome or cicatricial pemphigoid. Complications include dysfunctional tear syndrome, keratinization, symblepharon formation, eyelid disfigurement, and eyelash misalignment. Patient discomfort, cosmetic imperfection, increased risk of infection, and visual impairment are some notable concerns.

A normal tear film is essential for maintenance of the corneal and conjunctival epithelia. Composed of three layers, mucin, aqueous and lipid layers, the human tripartite tear film has antimicrobial, epitheliotrophic, mechanical, and optical properties. A wide range of physiologic or pathologic conditions, such as biologic aging, hormonal changes, chemical or thermal injuries, chronic inflammation, or autoimmune disorders, may disrupt the tear film and trigger a deleterious cascade, injuring ocular surface epithelia. Furthermore, suboptimal lacrimal functions may result in poor surgical outcomes, especially after penetrating keratoplasty or limbal stem cell transplantation.

Traditionally, the eyelids and lacrimal gland were excluded from the definition of the ocular surface. It is evident that visual function and epithelial health would not be feasible without these structures. The eyelids are essential for ocular surface protection and tear film maintenance. Untreated eyelid deformities, lid malpositions, or eyelash misalignments can precipitate detrimental consequences to the integrity and function of the ocular surface epithelia. Thus, a functional ocular surface requires structurally and functionally intact eyelids and lacrimal gland.

2. Ocular surface reconstitution

In severe ocular surface disorders, the management strategies entail symptomatic relief, reconstitution of the anatomic and physiologic ocular surface, and treatment and prevention of recurrence of the causative condition. Here we will discuss strategies to restore the conjunctival epithelium, corneal epithelium, and lacrimal function. Figure 1 illustrates the management strategies. Injury or inflammation causes severe ocular surface disorder with conjunctival scarring, limbal stem cell deficiency, corneal opacity with neovascularization, lacrimal dysfunction, disorganized lashes, and lid malposition (a). Mainstay treatment options include antibiotics, anti-inflammatory agents, lubrication, and amniotic membrane transplantation, as well as removal of lashes and correction of lid changes (b). As progress is made in science and tissue bioengineering, tissue replacement and regeneration may be feasible to restore the ocular surface and vision (c).

2.1 Conjunctival tissue reconstitution
2.1.1 Suppression of cicatricial changes

Commonly, ocular surface diseases limited to the conjunctiva progress to excessive cicatricial changes and loss of normal epithelial anatomy. Cicatricial changes to the conjunctival epithelium generally result from poorly controlled fibroblastic activities, e.g. tissue injuries or persistent inflammation. In addition to disrupting the tear film, cicatricialization of the conjunctiva has important implications in glaucoma surgeries, where availablility of healthy conjunctiva is essential for good surgical outcomes. A widely adopted therapeutic strategy is pharmacologic suppression of the inflammatory cascade and the fibroblast activation pathway using corticosteroids and antimetabolites.

Recently, research efforts have been directed toward transforming growth factor beta (TGF-β) and its involvement in fibroblast proliferation. TGF-β is a multifunctional cytokine, which plays an important role in tissue repair and regeneration. After injury, TGF-β triggers a complex cascade involving monocyte and leukocyte chemotaxis, induction of angiogenesis, control of production of cytokines and inflammatory mediators, deposition of extracellular matrix materials, and prevention of their enzymatic degradation (Border & Ruoslahti, 1992; Massagué et al., 1992). Excessive TGF-β activity has been associated with exuberant fibrotic changes in the eye and other organs. In a murine model, TGF-β was associated with formation of granulation tissue (Roberts et al., 1986) and increased inflammatory cell activity, as well as with exuberant extracellular collagen type-III deposition (Siriwardena et al., 1999). Using immunohistochemistry, Razzaque et al (2003) found increased accumulations of type-I and type III collagens and heat shock protein 47, a collagen-binding protein in fibrotic conjunctiva of patients with ocular cicatricial pemphigoid compared to normal subjects. Up-regulation of these proteins was also detected when ex-vivo

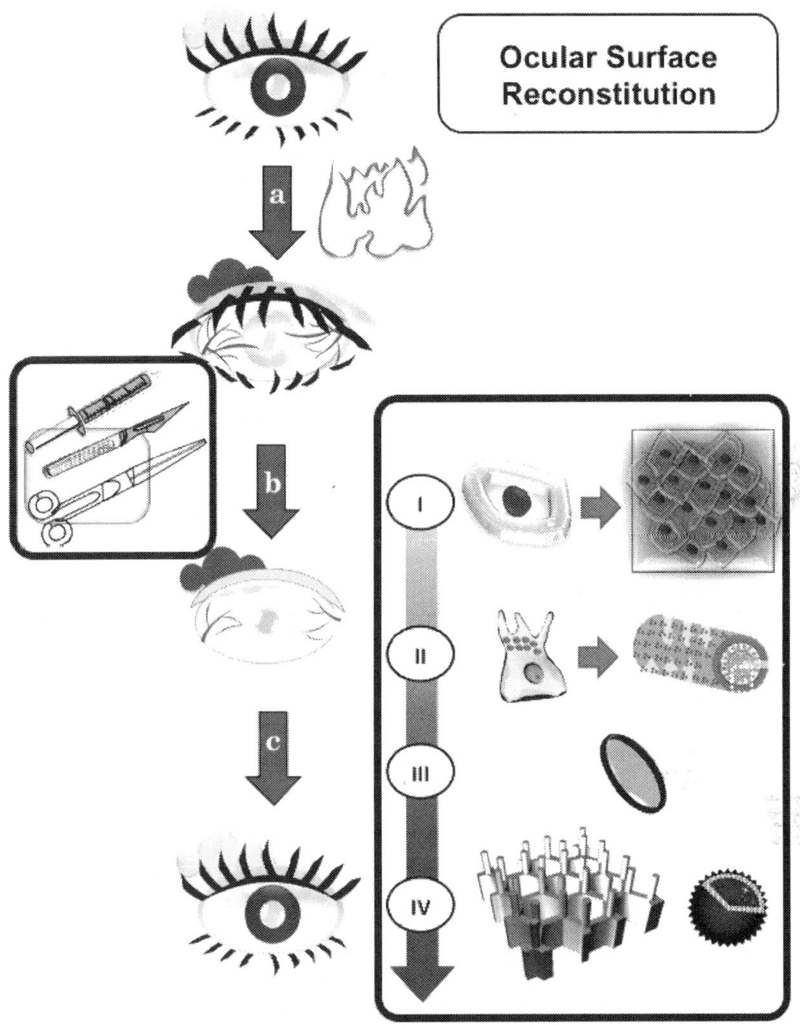

Fig. 1. Ocular Surface Reconstitution: (a) ocular surface destruction, (b) preliminary management – antibiotics; anti-inflammatory; amniotic membrane; eradication of microtrauma sources; etc. (c) ocular surface reconstitution – tissue transplant ± bioengineered substratum (I); lacrimal gland function restoration (II); corneal transplant or keratoprosthesis (III); regenerative medicine, targeted drug delivery, (IV); oculoplastic reconstruction, etc. (Adapted from Nguyen et al., 2008)

conjunctival fibroblasts were treated with TGF-β. Elevated levels of bone morphogenetic protein-6 and activin A, both part of the TGF- β superfamily, have been reported in scar tissue after glaucoma incisional surgeries, and their mRNA expression was immunolocalized to epithelial cells, vascular endothelia, stromal fibroblasts, and macrophage-like cells using real-time polymerase chain reaction (Andreev et al., 2006). It is suggested that these multifunctional growth factors participate in tissue wound healing, and altered expression of these factors may play a role in conjunctival scarring.

It is, therefore, hypothesized that modulation of TGF-β activity may help control the formation of scar tissue. Indeed, this has been demonstrated in animal models and clinical trials. CAT-152 is a human monoclonal antibody capable of neutralizing the growth factor TGF-β2. In in vitro and in vivo models, CAT-152 was shown to inhibit conjunctival scarring (Cordeiro et al., 1999; Thompson et al., 1999). In a rabbit model, SB-431542, a potent and selective inhibitor of the TGF-β1 activin receptor-like kinase, decreases subconjunctival deposition of collagen in rabbits undergoing glaucoma filtration surgery (Xiao et al., 2009). It is thought that SB-431542 inhibits phosphorylation of Smad2 stimulated by TGF-β, abrogating fibroblast transdifferentiation and type-I collagen synthesis. A prospective randomized placebo controlled clinical study of patients undergoing trabeculectomy showed that subconjunctival injection of CAT-152 appears to be safe in human undergoing trabeculectomy (Cordeiro 2003; Siriwardena et al., 2002). Successful and safe suppression of TGF- β superfamily activities may modulate fibroblast proliferation and cicatricial changes to the conjunctiva.

An alternative approach is stimulation of tissue regrowth and reduction of the conjunctival wound contraction mediated by modified contractile fibroblasts using biocompatible porous matrices (Gabbiani et al., 1971; Majno et al 1971). In a guinea pig skin model, highly porous, cross-linked collagen-glycosaminoglycan copolymer matrix had been reported to induce partial morphogenesis of skin when seeded with dermal and epidermal cells. After grafting on a full-thickness skin wound model (Yannas et al., 1982, 1989), the seeded matrix regenerated mature epidermis and near physiologic dermis, which was distinct from scar tissue and able to prevent wound contraction, compared to non-seeded matrix. For ophthalmic applications, porous collagen-glycosaminoglycan copolymer matrix was shown to decrease wound contraction in a full-thickness defect in a rabbit conjunctival model (Hsu et al., 2000). In animal glaucoma filtration surgery model, implantation of biodegradable porous collagen matrix in the subconjunctival space was associated with less initial scar formation and maintenance of long-term intraocular pressure control (Chen et al., 2006). In this study, Masson trichrome and β-smooth muscle actin immunocytochemistry established reduction of myofibroblast population in the eye having bioactive matrix implant compared to control. The authors concluded that the collagen matrix may simulate a more physiological structure for tissue regrowth (Chen et al., 2006; Hsu et al., 2000). Scleral buckle composed of biodegradable collagen-glycosaminoglycan copolymer had been reported to be safe and effective in a rabbit model (Wu et al., 2008). Similar results have been demonstrated with modified porous poly(lactide-co-glycolide) scaffold as well (Lee et al., 2003). These findings provide insight into cellular behaviors and assist in the understanding of conjunctival tissue bioengineering.

2.1.2 Conjunctival tissue substitutes

2.1.2.1 Autologous mucosal transplantation

Other therapeutic options for conjunctival tissue reconstitution include autologous tissue grafts and bioengineered conjunctival substitutes. Similar to any epithelia, the conjunctival epithelium is in a dynamic self-renewal equilibrium. Epithelial cells with high mitotic capacity are distributed predominantly in the bulbar and forniceal conjunctiva in humans (Pellegrini et al., 1999; Tsubota et al., 2002). In autologous conjunctival autografts, e.g. for pterygium excision, the conjunctival autograft typically successfully replenishes the donor bed. However, severe tissue deficiency necessitates transplantation of other mucosal tissues. Other mucosal tissues having structure analogous to the conjunctiva, i.e. nonkeratinized stratified squamous epithelium with goblet cells, include nasal, esophageal, vaginal, and tracheal mucosa. Gibson and colleagues (1986) demonstrated the feasibility of transplanting oral epithelium onto the ocular surface in a rabbit model. In a case report, oral mucosa was used in combination with a modified Gunderson conjunctival flap to treat ocular surface burn injury (Cheng & Chang, 2006). In a series of patients with severe bilateral mucus deficiency syndrome after severe lye, acid, or heat burns, radiation, or systemic mucosal disease, autologous nasal mucosal grafts from the nasal conchae were transplanted to cover the ocular surface (Naumann et al., 1986). Symptomatic and visual improvements were reported. Long-term follow-up, using impression cytology and hematoxylin and eosin and periodic acid-Schiff staining of biopsied tissue, demonstrated that functional goblet cells persisted in autologous nasal mucosa for up to 10 years after transplantation (Wenkel et al., 2000). Another study also reported similarly promising outcomes for patients with cicatricial ocular surface secondary to lye injury (Kim et al., 2010).

2.1.2.2 Cultivated bioengineered tissues

Currently, a variety of substrates are readily available for cultivation of conjunctival tissue, e.g. murine-derived 3T3 fibroblast feeder layer, human amniotic feeder layer, and serum-free media. Irradiated 3T3 murine feeder cells are widely utilized in a number of cell culture systems because these immortalized 3T3 cells secrete growth factors to support cellular proliferation (Todaro & Green, 1963). Although the murine-derived feeder layers provide satisfactory support for tissue growth and maintenance, the tissue is unsuitable for human use because of several considerations. The xenoantigenic nonhuman sialic acid Neu5Gc produced by the human embryonic stem cells using this culture system may trigger cell killing in vivo if exposed to human serum containing circulating antibodies specific for Neu5Gc (Martin et al., 2005; Rheinwald 1980). Other concerns include transmission of murine diseases or contamination with virus or prion agents (Boneva et al., 2001; Ramarli et al., 1995).

Consequently, alternatives are being explored. The human amniotic feeder layer, derived from human placentas, has been found to promote clonal growth of human limbal epithelial progenitors (Chen et al., 2007; Tanioka et al., 2006)). Such bioengineered tissues would be suitable candidates for replacement of native conjunctival tissue as the human amniotic membrane is a stable and elastic substratum that can be integrated into the ocular surface. The anti-inflammatory properties of human amniotic membrane would be especially advantageous in the case of inflammatory cicatricial changes, as any trauma to the recipient bed would reactivate or incite the inflammatory cascade that may inflict further injury to both the host and the graft. Other substrates include ultrathin poly(β-caprolactone) membrane, silicone, and others (Ang et al., 2006; Selvam et al., 2006; Yoon et al., 2007). To minimize the biohazard of using xenobiotic materials, serum-free media are being investigated as well. One study demonstrated good epithelization following transplantation of serum-free autologous cultivated conjunctival epithelium on human amniotic membrane in pterygium excision (Ang et al., 2005). Tissue cultivation will be discussed in detail below.

2.2 Corneal epithelium reconstitution

In humans, the known organs that demonstrate immunologic privilege are the eyes, brain, testicles, placenta, and fetus. The anterior chamber immunologic privilege and avascularity of the cornea significantly contribute to the high success rate of corneal allografts, e.g. penetrating keratoplasty, which can be performed without HLA-typing (Nguyen et al., 2007, 2008, 2010). However, the risks of allogeneic graft rejection increase significantly in patients with severe ocular surface disorders or limbal stem cell deficiency. In severe cases, management of corneal blindness involves treatment of the ocular surface disorder, restoration of limbal stem cells, development of corneal tissue equivalents, or use of a keratoprosthesis.

2.2.1 Human amniotic membrane transplantation

Now widely employed in burn management and ophthalmology, the human amniotic membrane found its first applications as surgical dressing and tissue graft in dermal burns and as a strategy to impede tissue adhesion in mucosal injuries (Davis, 1910; Stern, 1913). Subsequently, its utility was expanded to reconstruction of the oral cavity, bladder, and vagina, as well as tympanoplasty and arthroplasty. De Rotth (1940) first reported the suitability of amniotic membrane in ocular surface reconstruction where he applied the amniochorion to repair symblepharon. A suboptimal surgical outcome was observed because the chorion component can incite inflammation and scarring. The amniotic membrane enjoyed limited use until the 1990s when its properties were better understood and its ophthalmic applications became widely adopted.

In humans, the amniotic membrane, which lines the inner surface of the placenta, is composed of a basement membrane and a subjacent stromal matrix devoid of vascular and lymphatic vessels, as well as nerve fibers. It has many unique features, including suppression of inflammation and promotion of tissue healing. Its basement membrane assists in induction of adhesion and migration of epithelial cells for wound healing (Azuara-Blanco et al., 1999; Dua et al., 2004; John et al., 2002; Kim & Tseng 1995; Shimazaki et al., 1998). Modulation of the inflammatory cascade is realized by suppression of interleukin alpha and interleukin 1 beta in epithelial cells (Solomon et al., 2001). Release of growth factors such as epidermal growth factor, keratinocyte growth factor, or basic fibroblast growth factor (Koizumi et al., 2000) and inhibition of proteinase activity (Kim et al., 2000) promote tissue survival. The amniotic membrane also effectuates reduction of scar tissue formation by reducing polymorphonuclear infiltration (Park & Tseng, 2000) and suppressing the TGF-β signaling system and myofibroblast differentiation of normal fibroblasts (Lee et al., 2000; Tseng, 1999).

The available literature supports its use as an epithelial surrogate after excision of large ocular surface neoplasias (Espana et al., 2000; Gündüz et al., 2006; Tomita et al., 2006), reconstruction of conjunctival ocular surface (Solomon et al., 2003; Zhou et al., 2004; Oberhansli et al., 2005; Tseng et al., 2005), excision of pterygia (Ang et al., 2007; Nakamura et al., 2006), severe or refractory neurotrophic corneal ulcers (Chen et al., 2000; Ivekovic et al., 2002; Khokhar et al., 2005), bacterial keratitis (Barequet et al., 2008), and symptomatic alleviation following acute ocular burns (Tamhane et al., 2005). A prospective noncomparative interventional case series study reported that nonpreserved human amniotic membrane appears useful for ocular surface reconstruction following excision of extensive ocular surface neoplasia, such as squamous cell carcinoma, malignant melanoma and conjunctival-orbital lymphangioma (Gündüz et al., 2006). No immune graft rejection was encountered. For pterygium excision, complete epithelialization, early resolution of ocular inflammation, and no recurrence of pterygium had been reported following orthotopic transplantation of sterilized, freeze-dried amniotic membrane, over a follow-up period of 14 months (Nakamura et al., 2006). Caution is advised, however, in using amniotic membrane transplant for primary pterygium excision in patients with sufficiently recruitable conjunctiva. Randomized prospective studies comparing conjunctival autograft versus amniotic membrane transplant for pterygium showed a higher recurrence rate in the amniotic membrane transplant group (Essez et al., 2004; Luanratanakorn et al., 2006; Tananuvat & Martin, 2004). A recent study (Baraquet et al., 2008) demonstrated the use of human amniotic membrane in an animal model to assist in the healing of bacterial keratitis. In this rat model of Staphylococcus aureus keratitis, histopathologic examination revealed the least corneal haze and neovascularization in the cefazolin/amniotic membrane combination group, compared to the normal saline and amniotic membrane, and cefazolin without amniotic membrane transplantation groups.

Complete ocular surface coverage using amniotic membrane was performed in patients with severe acute chemical and thermal burns to protect and promote conjunctival regeneration and to prevent complications of burns, such as symblepharon formation and corneal melt (Joseph et al., 1991). However, Joseph and colleagues did not observe a positive association between amniotic membrane transplantation and overall success rate of ocular surface restoration or preservation of ocular integrity in patients with severe acute burns, whether used alone or in conjunction with other surgical procedures. Using laser scanning in vivo confocal microscopy, Nubile et al (2001) discovered two different mechanisms of adhesion: when the amniotic membrane behaved as a patch, instead of as a graft, the epithelial cells migrated under the membrane, culminating in membrane disintegration; when the amniotic membrane acted as a graft, it became the basement membrane matrix, integrating with the superficial corneal stroma and allowing epithelial cell migration and proliferation (Nubile et al., 2001; Resch et al., 2006). This behavior potentially explains the poor outcome observed by Joseph et al (1991). Accordingly, care should be taken during transplantation to allow the membrane to be a graft, thus promoting epithelial migration and proliferation. Zhou et al (2004) noted that severity of the symblepharon, severity of decrease in lacrimal function, and a decreased amount of remaining healthy conjunctiva are poor prognostic factors for long-term outcomes of amniotic membrane transplantation for conjunctival surface reconstruction. A higher rate of failure was associated with preoperative use of topical steroid in patients receiving amniotic membrane transplantation for persistent epithelial defect and conjunctival repair (Saw et al., 2007). However, the preoperative use of steroids may simply be an indication of the severity of the disease.

Traditionally, sutures are used to anchor the amniotic membrane to the ocular surface; however, there is a recent shift toward bioadhesive transfixation instead. There are a number of motivations for this shift: suture fixation requires a greater amount of surgical expertise and dexterity; sutures can cause corneal irritation and scarring; membrane shrinkage can result in graft loss; and sutures need to be extricated. For conjunctival autograft following pterygium excision, the biocompatibility, safety, and efficacy of fibrin bioadhesives was substantiated in a prospective randomized, interventional case series (Uy et al., 2005). In a rabbit model, Szurman et al (2006) demonstrated that fibrin glue provided a better surgical outcome than sutures for amniotic membrane transplantation. In the bioadhesive group, histology revealed a smooth fibrin layer at the graft-host interface and a continuous, stratified layer of cytokeratin-3 expressing corneal epithelial cells on the membrane surface. The authors found that suture fixation tends to promote membrane contraction, raised membrane edges, and epithelial ingrowth into the submembrane space (Szurman et al., 2006). These findings were essentially corroborated in another study (Sekiyama et al., 2007). The fibrin bioadhesive biodegrades within 2 weeks, whereas the amniotic membrane persists for at least 12 weeks, providing sufficient time for epithelialization (Sekiyama et al., 2007). Sutureless transplantation with fibrin glue or ProKera™ is reportedly a safe and easy method that avoids complications related to sutures and shortens the time of surgery (Kheirkhah et al., 2006).

2.2.1.1 Disadvantages of human amniotic membrane

Amniotic membrane only provides a supportive substratum for epithelial stem cells to proliferate and migrate to restore the ocular surface. Thus, transplantation of the amniotic membrane without cultivated epithelial cells for true limbal stem cell deficiency may not be curative. Without the limbal stem cells and corneal epithelial cells, the cornea may not maintain avascularity and clarity because of secondary conjunctivalization and neovascularization of the cornea. In a retrospective case study, Maharajan et al (2007) reported the correlation between a higher success rate of ocular surface reconstruction using amniotic membrane transplantation and the availability of stem cells.

There are some minor disadvantages of amniotic membrane transplantation. As already noted, suture fixation of amniotic membrane requires surgical dexterity to manipulate the membrane on the ocular surface. As a substrate for stem cell expansion, the surface with stem cells is transfixed face-up on the corneal surface. This position has some theoretical drawbacks: delayed migration of epithelial cells to the corneal surface before cellular repopulation, attrition of epithelial cells from exposure in patients with severe dry eye syndrome or ocular surface disorders, or mechanical displacement of the tissue. Another disadvantage is the association with corneal calcific opacification. In a large series of amniotic membrane transplantation for persistent epithelial defect, partial limbal stem cell deficiency and conjunctival reconstruction, Anderson and colleagues (2003) reported a significant (12.8%) rate of corneal calcification, occurring 3-17 weeks postoperatively. Risk factors were preoperative corneal calcification in either eye and postoperative use of phosphate-containing eye drops. A more recent study suggests that the use of phosphate-buffered eye drops, commonly found in artificial tear or topical corticosteroid formulations, favors the formation of insoluble crystalline calcium phosphate hydroxyapatite deposits in the stroma in the presence of epithelial keratopathy (Bernauer et al., 2006). Studies using an animal model confirmed these findings as well (Schrage et al., 2010). The authors discouraged the use of phosphate-buffered eye drops as excessive use of such eye drops may cause corneal calcification, which may necessitate corneal graft surgery for subsequent visual rehabilitation.

2.2.2 Epithelial cell transplantation

Treatment of limbal stem cell deficiency includes transplantation of cultivated limbal stem cells. The unique properties of amniotic membrane make it an attractive candidate as a substratum for stem cell expansion. Tsubota and colleagues (1996) were among the first to introduce the combined transplantation of allograft limbal tissue and amniotic membrane in patients with ocular cicatricial pemphigoid, where a short-term success rate of 86% was observed. Koizumi et al (2000) reported successful transplantation of cultivated corneal epithelium using acellular amniotic membrane as a substrate for ex vivo expansion of the epithelial cells. Nakamura et al (2004, 2006) proposed the application of sterilized, freeze-dried amniotic membrane as a substrate for cultivating autologous corneal epithelial cells for ocular surface reconstruction. In a rabbit model, the authors reported no significant difference between freeze-dried and cryopreserved amniotic membrane. Corneal epithelium cultivated well and reepithelialization of the grafted cornea was demonstrated postoperatively (Nakamura et al., 2004, 2006).

The conventional cultivation methods for mucosal epithelial tissues involve the use of xenobiotic materials such as fetal bovine serum (FBS) and murine-derived 3T3 feeder cells. Given the risks of zoonotic infection and xenoantigenic materials as discussed above, FBS-free culture systems have been developed, but these have reduced efficacy for cell propagation. Ang et al (2006) introduced the use of human serum for in vivo and ex vivo expansion of human conjunctival epithelial cells on amniotic membranes. Cultivated epithelial cells on amniotic membrane substrates using autologous serum showed complete corneal epithelialization, retained corneal clarity, and improved visual acuity in patients with severe limbal stem cell deficiency (Nakamura et al., 2006). Autologous serum expansion and FBS expansion appeared to have comparable proliferative capacities with equivalent morphologies (Nakamura et al., 2006). Autologous serum had also been used to cultivate oral epithelial cells for patients with severe ocular surface disease with promising outcomes (Ang et al., 2006).

Murine 3T3 feeder cells are typically used in these cultivation systems to maintain stemness of the stem cells. However, these xenobiotic feeders cannot be used for transplantation in humans. Accordingly, other culture systems, such as human amniotic epithelial cells, human embryonic fibroblasts or feeder cell-free, are being investigated for feasibility as the feeder cells. Chen and colleagues (2007) compared murine 3T3 cells and human amniotic epithelial cells, and found that, although murine 3T3 feeder cells demonstrated faster initial proliferative capacity, human amniotic epithelial feeder cells retained evidence suggestive of undifferentiated state for more generations. Lai et al. (2007) similarly concluded that human amniotic epithelial cells are suitable for cultivation of mouse embryonic stem cells, compared to mouse embryonic fibroblast feeder cells. Under both FBS-supplemented and serum-free conditions, human embryonic fibroblast feeder cells was found to support expansion of human limbal epithelial cells as well as 3T3 feeder cells (Notara et al., 2007). A feeder cell-free system had been introduced by Yokoo et al. in 2008. Of note, the authors found that, compared to conventional medium containing 10% FBS cocultured with 3T3 fibroblast feeder cells, cells grown in serum-, feeder cell- and bovine pituitary extract (BPE)-free culture medium demonstrated higher colony forming ability, equivalent proliferative capacity, and similar morphology. Two weeks after transplantation onto rabbit corneas, these cells showed histologic findings suggestive of normal corneal epithelium (Yokoo et al., 2008). These autologous cultivation or serum- feeder cell- and BPE-free systems would theoretically reduce the risk of allograft rejection and transmission of xenobiotic infectious agents, and the need for long-term corticosteroid and immunosuppressive therapy may be obviated. Adoption of a standardized protocol to ensure availability of tissues with consistent constituents and consistent clinical outcomes would be desirable (Hopkinson et al., 2006).

2.2.3 Other substrates for epithelial cell cultivation

Alternative materials are being investigated. Rama et al (2001) investigated the application of a fibrin substrate for cultivating autologous limbal stem cells. Good epithelization and regression of inflammation and vascularization were reported following transplantation onto corneas damaged by total limbal stem cell deficiency (Pellegrini et al., 1999; Rama et al., 2001). These findings were corroborated by another group (Han et al., 2002). A more recent study comparing fibrin bioadhesive substrate and amniotic membrane for cultivation of corneal epithelial sheets showed that both had similar levels of colony-forming progenitor cells, while more differentiation was observed in the fibrin group (Higa et al., 2007). A thermoresponsive surface, which allows easy harvest of intact transplantable corneal epithelial cell sheets from ex vivo expansion of limbal stem cells, has been developed (Nishida et al., 2004). Without transplantation of the substratum, e.g. amniotic membrane, theoretically, this epithelial sheet will rapidly repopulate the corneal surface, without the time delay for migration. Using this technique, Nishida et al (2004) performed ex vivo fabrication of autologous oral mucosal tissue and transplantation on a small group of four patients with bilateral total corneal stem cell deficiencies with reasonable success. Lai et al (2007) also demonstrated successful transplantation of bioengineered human corneal endothelial cell monolayer sheet grafts, cultivated on the thermoresponsive surface then attached to gelatin hydrogel discs, into denuded rabbit corneas. Restoration of corneal clarity was observed, compared to controls.

An innovative approach to cultivating cells on contact lens has recently been introduced (Di Girolamo et al., 2007). It was found that successful explant cultures could be made using either superior forniceal conjunctiva or superior limbal epithelial biopsies placed on a siloxane-hydrogel extended wear contact lens. These lenses were subsequently inserted onto the eyes of three patients with limbal stem cell deficiency. The patients were found to have a stable transparent corneal epithelium with no recurrence of conjunctivalization or corneal vascularization along with improvements in both best-corrected visual acuity and symptom scores. It is theorized that the close proximity between the therapeutic contact lens and the ocular surface facilitates cell migration from their artificial substratum to replenish the damaged ocular surface. Di Girolamo et al also postulated that the diffusion of secretory factors from the contact lens niche would promote corneal wound healing, inhibit angiogenesis, and rescue or activate any remaining limbal stem cells. Research is now being conducted to create a plasma polymer surface that can coat CL surfaces that would not only allow for appropriate attachment and growth of the epithelial cells but also transfer of the cultured cells onto the denuded corneal surface once inserted (Deshpande et al., 2010). This technology is similar to that being used in patients with extensive burn injuries and chronic nonhealing diabetic ulcers for transfer of autologous keratinocytes (Moustafa et al., 2004, 2007; Zhu et al., 2005). Both rabbit and human corneal cells were able to attach and proliferate well on acrylic acid-coated contact lens surfaces, and there could be reliable transfer of epithelial cells to rabbit corneas (Deshpande et al., 2010). Other substrates for ex vivo cell expansion being investigated are composite membranes of alginate polymeric membrane coated with chitosan, NaOH-surface-modified poly(epsilon-caprolactone), three-dimensional collagen-proteoglycan-based scaffolds, and human anterior lens capsule (Ang et al., 2006; Galal et al., 2007; Lai et al., 2007; Ma et al., 2007; Oztürk et al., 2006; Torbet et al., 2007). Though studies have not yet been performed in humans, these results are indeed exciting in the realm of corneal stem cell therapy.

2.2.4 Autologous non-ocular tissues

As discussed in the conjunctival section, when available, transplantation of autologous tissues is the treatment of choice to obviate the risks of allograft rejections despite the presence of the anterior chamber immunologic privilege. In cases of bilateral severe limbal stem cell deficiency, oral mucosal transplantation is an adequate alternative. Kinoshita and Nakamura proposed a two-step process for patients at increased risks for allograft rejection or patients having low tolerance for immunosuppressive therapies. Here, the autologous oral mucosal epithelial progenitor cells undergo ex vivo expansion prior to transplantation (Kinoshita & Nakamura, 2004). Ex vivo expansion of a carrier-free sheet is more beneficial than a direct transplantation of an autologous buccal mucosal graft because the latter typically contain opaque subepithelial fibrous tissue. A long-term study reported good epithelial coverage, regression or stabilization of corneal neovascularization, and improved visual acuity in a cohort of patients with severe ocular surface disorder (Inatomi et al., 2006; Nakamura et al., 2010). The main mechanisms of graft failure were loss of transplanted cultivated oral epithelial cells, leading to persistent epithelial defect, followed by invasion of neighboring conjunctival epithelial cells, infiltration of inflammatory cells, and neovascularization. And as expected, good graft survival, intact ocular surface integrity, and no inflammatory cell infiltration were observed in successful cases, (Nakamura et al., 2007). Another study also showed good results using human buccal mucosa cultivated on thermoresponsive surface with mitomycin-treated 3T3 feeder cells for patients with chronic conjunctival inflammation and limbal stem cell depletion (Nishida et al., 2004). Ex vivo expansion of embryonic stem cells or bone marrow-derived mesenchymal stem cells have been investigated with promising preliminary results (Homma et al., 2004; Ma et al., 2006; Oh et al., 2008).

The advantages of these autologous tissues are a significant reduction of xenobiotic materials in the transplant, a decreased risk of infection, a minimized risk of allograft immunologic rejection, and a reduced need for immunosuppressive therapy. However, some authors still used the 3T3 murine feeder cells. Also, it would be intriguing to look at long-term restoration of corneal avascularity and clarity as these tissues are not corneal epithelium and do not possess innate limbal stem cells.

2.2.5 Corneal tissue substitutes

Severe ocular surface diseases not only affect the epithelium but also involve the corneal stroma as well. Thus, reconstitution of the ocular surface may provide symptomatic relief but not visual rehabilitation. Tissue bioengineering is required to regenerate corneal tissue equivalents that integrate into the host and restore vision. A number of substrates have been proposed as three-dimensional growth scaffolds for ocular tissues: irradiated acellular non-human cornea disc (Zhang et al., 2007), fish scale-derived scaffold (Lin et al., 2010), collagen-copolymer extracellular matrix (Li et al., 2003; Liu et al., 2006), silk film biomaterials (Lawrence 2008), and electrospun chitosan-graft-poly (β-caprolactone)/poly (β-caprolactone) nanofibrous scaffolds for retinal progenitor cells (Chen et al., 2011).

Since corneal stroma is composed of lamella of collagen fibrils, many investigators developed collagen-surface modified scaffolds to promote host tissue integration, repair and regeneration. Of note, Myung and colleagues (2007, 2008) used photolithography to construct a three dimensional collagen type I - poly(ethylene glycol)/poly(acrylic acid) "core and skirt" corneal prosthesis, which can support surface reepithelization and stromal integration. In a rabbit model, the authors observed that corneal epithelium forms a confluent layer on the scaffold. The surface modification of collagen also assisted in the integration of the seeded corneal fibroblasts into the hydrogel skirt. Li et al (2003) performed lamellar keratoplasty using collagen-copolymer extracellular matrix scaffold in pigs. The authors reported increased procollagen synthesis and nerve regeneration, compared to control. Similar findings were reported by Liu et al (2006) using corneal substitutes made of cross-linked porcine type I collagen and water-soluble carbodiimides, in rabbit and minipig models. Using recombinant human collagens cross-linked with carbodiimide and hydroxysuccinimide, Merrett et al (2008) constructed bioengineered corneal equivalents for transplantation into minipigs. These synthetic corneal implants were found to be optically clear, with good epithelial coverage. Importantly, the authors also demonstrated repopulation of the synthetic stroma with host corneal stromal cells, as well as reinnervation with similar nerve density compared to allograft.

3. Ocular surface prosthetic devices

As discussed above, contact lenses are being investigated as a stem cell reservoir and delivery vehicle. Therapeutic applications of contact lenses extend to other uses as well, for symptomatic alleviation of aberrant ocular surface and drug delivery. Many individuals suffer severe eye discomfort and pain that is unresponsive to the mainstay treatment options. In these patients, the ocular surface can be improved by an ocular surface prosthetic device - not for limbal stem cell delivery but to create a microenvironment that supports the corneal surface, i.e. the Prosthetic Replacement of the Ocular Surface Ecosystem or PROSE, developed by Perry Rosenthal. The device itself is a fluid-ventilated, gas-permeable polymer with optic and haptic portions situated over the sclera that are linked to a transitional zone that suspends the lens over the cornea without contact. The unique design of this prosthesis provides a constant aqueous layer over the cornea surface, which contributes to healing of the corneal tissue and subsequent relief of symptoms (Rosenthal et al., 2000; Shepard et al., 2009)). Studies have demonstrated corneal healing as soon as 6 hours after device placement (Takahide et al., 2007). Stason and colleagues (2010) reported improved visual acuity and visual function in a significant number of these patients. Better quality of life has been reported as well (Romero-Rangel et al., 2000).
The most common route of drug delivery for the ocular surface is topical. This modality has some limitations, however, including rapid clearance from the ocular surface, short duration of therapeutic dosage, and systemic absorption (Bourlais et al., 1998; Ciolino et al., 2009; Ghate et al., 2008; Gulsen et al., 2004; Patton et al., 1978; Saettone et al., 2002). Accordingly, the use of contact lenses as drug delivery vehicles has been investigated. Sedlavek (1965) in Czechoslovakia first studied the use of soft contact lenses for drug delivery; since then many different drugs and technologies have been successfully tried using this technique (Sedlavek, 1965). The optimal design for a contact lens drug delivery system is one that allows for zero-order release kinetics, along with being comfortable and biocompatible (Ciolino et al., 2009). With these criteria in mind, the soft hydrogel lenses were the first lenses to be used in this capacity. The hydrogel lenses have high water content and large intermolecular pores, which allow them to absorb water soluble drugs and release them with a large initial pulse and then gradually thereafter (Jain, 1988). To maintain first-order kinetics, the hydrogel lens requires regular instillation of drops at intervals of 2-4 hours or reapplication of a soaked lens (Jain & Batra, 1970). This approach is suboptimal as drug

uptake and release kinetics are dependent on the equilibrium solubility of the drug present in the lens matrix (which is small for most of the hydrophobic drugs [Gulsen et al., 2004]), the inconvenience of frequent dosing, and the waste of drug remaining in the soaking solution.

Efforts to overcome these limitations focus on lens designs that allow for a sustained release of therapeutic dosage. One idea involves the use of nanoparticles in which the ophthalmic drug could be encapsulated and then dispersed in the lens material (Gulsen et al., 2004). These so called "drug-laden hydrogels" were found to release therapeutic levels of drugs for a few days with the ability to manipulate drug delivery rates by varying the load of nanoparticles in the gel (Gulsen et al., 2004). This discovery allowed for lenses that could release drugs over extended periods of time and thus behaved in a zero order kinetic fashion. However, although this was exciting, the development of a sustained, long-term drug delivery system that functions under the physiological temperature, pH and salinity of the eye continues to be problematic (Ciolino et al., 2009). Most recently, research is being conducted on the development of prototype lenses that incorporate a dual polymer system such that there is a polymer film containing the drug compounds; the film is then coated with a transparent polymer that is used in contact lenses (Ciolino et al., 2009). The results with this new lens showed that there could be controlled release with zero-order kinetics for over 4 weeks (Ciolino et al., 2009).

In addition to the soft hydrogel lens and their derivatives, collagen shields have also proven to be effective delivery modalities. The collagen shield contact lens is a therapeutic lens composed of non-cross linked porcine collagen with the capability of dissolving on the corneal surface over a period of around 12 hours (Phinney et al., 1988). Studies done using collagen shields soaked with gentamicin and vancomycin produced tear, cornea, and aqueous humor levels that were generally higher or comparable to those achieved through frequent drop administrations (Phinney et al., 1988). Further, collagen shields used for heparin delivery in postoperative rabbit eyes helped to prevent postoperative fibrin formation in eyes at risk for this complication (i.e. eyes requiring surgery for complications of proliferative diabetic retinopathy and glaucoma filtration surgery) (Murray et al., 1990). With time, as bioengineered materials continue to advance, so too will the quality of optimal delivery system contact lenses. Pharmacological contact lenses have the potential to revolutionize the current drug delivery modalities and conquer new ground in the long held battle of medication wastage and systemic absorption side effects.

4. Restoration of lacrimal function

Dry eye syndrome or dysfunctional tear syndrome is a chronic ailment, afflicting up to ten million people in the United States alone, and affecting women twice as often as men (Pflugfelder et al., 2000). As prevalent as it is, currently, there is only one medication (cyclosporine ophthalmic emulsion, 0.05%) approved by the Food and Drug Administration for decreased tear production due to ocular inflammation (Barber et al., 2005; Pflugfelder, 2004). Dry eye syndrome tends to coexist with ocular surface disorder. Importantly, poor lacrimal function correlates with poor surgical outcomes in ocular surface reconstructions, especially stem cell transplantation (Nguyen et al., 2008; Shimazaki et al., 2000, 2007). Temporary remedies, such as artificial tears, hydroxypropyl cellulose ophthalmic insert, punctal plugs, or moisture-chamber spectacles often provide only marginal relief and may interfere with the patient's quality of life. A long-term solution is the construction of an implantable lacrimal microbiosystem, which would house a colony of monolayer lacrimal acinar cells and produce tear fluid (Selvam et al., 2006, 2007, 2008).

4.1 Artificial lacrimal gland

The parenchyma of the lacrimal gland consists of specialized epithelia: ductal cells and acinar cells. The acinar cells form an acinus structure whose wall is composed of a polarized monolayer of cells. These lacrimal acinar epithelial cells are responsible for production and release of tear proteins such as lactoferrin, growth factors, secretory immunoglobulins, and lysosomal hydrolases into nascent tear fluid (Fullard, 1994). Their apices point toward a central lumen into which tear fluid is secreted. Many acini form a lobule. The ductal cells line the intralobular and interlobular ducts, bringing tear fluid into the eye. An implantable lacrimal microbiosystem would approximate the structure and function of the lacrimal gland. Figure 2 depicts a schematic artificial lacrimal gland using a dead-end tube concept.

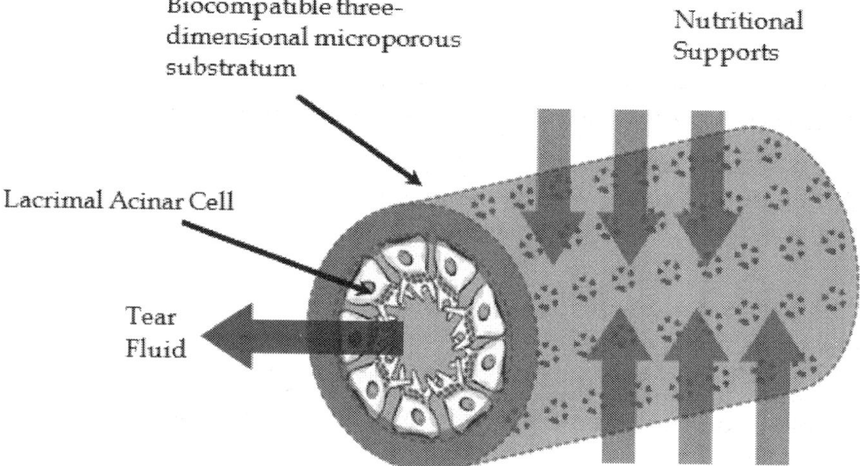

Fig. 2. Schematic of an artificial lacrimal gland where lacrimal acinar cells are seeded in a three-dimensional housing, allowing nutritional support from the environment and egress of tear fluid from the artificial gland.

Development of such devices faces many challenges. As with any implantable devices, a lacrimal microbiosystem must have a biologically compatible encapsulation to prevent rejection or toxicity to the host. However, the lacrimal microbiosystem encapsulation is much more intricate. Instead of a closed system, the enclosure must allow a dynamic bi-directional flux of ions, fluid, nutrients, and materials between the device and the host, while downregulating reactive fibroblastic proliferation in the surrounding environment that may diminish material transfer. Tear production is regulated by a variety of control modalities, such as regenerated intrinsic neurostimulation, extrinsic pharmacologic induction, or electronic excitation. Device construction includes ex-vivo seeding and maturation of the lacrimal acinar cells into a three-dimensional bioengineered scaffold. Strategic placement of pluripotent stem cells may be necessary. The stem cell clusters within the scaffold would repopulate the lacrimal acinar tissue, as needed. Precise stimuli for regeneration and terminal differentiation would be provided by extrinsic delivery or activation of previously implanted nanovesicles containing plasmids coding for growth factors, for example. As a stem cell device, safeguard mechanisms must be instituted to monitor and prevent autonomous unstructured proliferation and escape of the stem cells, i.e. tumorigenicity.

Recent studies demonstrated monolayer growth of lacrimal acinar cells on Matrigel®, a biocompatible preparation rich in extracellular matrix molecules, i.e. laminin, collagen IV, heparin-sulfate proteoglycans, and growth factors (Selvam et al., 2006, 2007, 2008). Compared to other biocompatible polymeric materials, poly-L-lactic acid (PLLA)-Matrigel® substrate showed superior expression of monolayer acinar cell morphology. Active exocytosis vesicle profile was evident on electron microscopy, and maintenance of protein secretory functions was confirmed by β-hexosaminidase catalytic activity assay. PLLA and PLGA copolymers have a wide range of applications: surgical sutures, scaffolds in bone augmentation and ligament restoration, osteoconductive materials in dental medicine, and components in vascular grafts, urethral stents, and nerve growth conduits. Growth of polarized lacrimal acinar cells on PLLA, which demonstrated the ability to maintain normal monolayer acinar cell morphology, secretory function, and EGF-dependent proliferation, suggests that PLLA may be a good candidate for the development of a tear secretory device. A dead-end tube construct using polyethersulfone has been reported (Li et al., 2006; Liu et al., 2009). After seeding the tube with cultured rat lacrimal acinar epithelial cells, the authors observed that a polyethersulfone dead-end tube may be suitable for attachment and proliferation of these cells, allowing its potential application as an artificial lacrimal gland.

4.2 Tear film measurements

Patients with ocular surface disorders typically have an underlying inflammatory process, such as ocular cicatricial pemphigoid, Stevens-Johnson syndrome, Sjögren syndrome, or post chemotherapy. The persistent inflammatory process and dysfunctional tear film lead to poor surgical outcomes. Good preoperative lacrimal function and a healthy conjunctival epithelium are important prognostic factors for limbal stem cell transplantation (Tsubota et al., 1999; Shimazaki et al., 2000). In fact, having a preoperative Schirmer's test > 10 mm correlates to a higher success rate for ocular surface reconstruction. Accordingly, preoperative characterization of tear film is important for ocular surface reconstruction.

Traditionally, tear production and quality are accessed using vital dye stain, tear break-up time, Schirmer's test, or the cotton-thread test. However, these tests have suboptimal repeatability, inter-class correlation, and correlation between the test results and patient symptom (Jordan & Baum, 1980; Lin et al 2005; Machado et al., 2009; Mainstone 1996; Nicols et al., 2004; Saleh et al., 2006; Savini et al., 2008; Schein et al., 1997; Yokoi & Komuro 2004). They are likely to also produce erroneous results by disrupting the natural tear film, affecting tear production, and modifying the tear film structure. Accordingly, recent efforts are directed toward non-invasive optical imaging techniques.

Anterior segment optical coherence tomography (OCT) is a noninvasive technology using a laser to quantitatively measure the tear film and tear meniscus without ocular surface contact or fluorescein dye instillation (Huang et al., 1991, 2008; Ibrahim et al., 2010; Shen et al., 2009; Wang et al., 2006; Zhou et al., 2009). Tear film measurement using OCT showed good correlation with Schirmer's test and patient symptom score. Figure 3 portrays application of anterior-segment OCT for measurement of lower tear meniscus height, depth, and cross-sectional area. OCT may also be used to measure the tear film or the upper tear meniscus. Other noninvasive technologies include slit-lamp equipped micrometer (Mainstone et al., 1996), slit-lamp photography (Kawai et al., 2007), and reflective meniscometry (Yokoi et al., 1999).

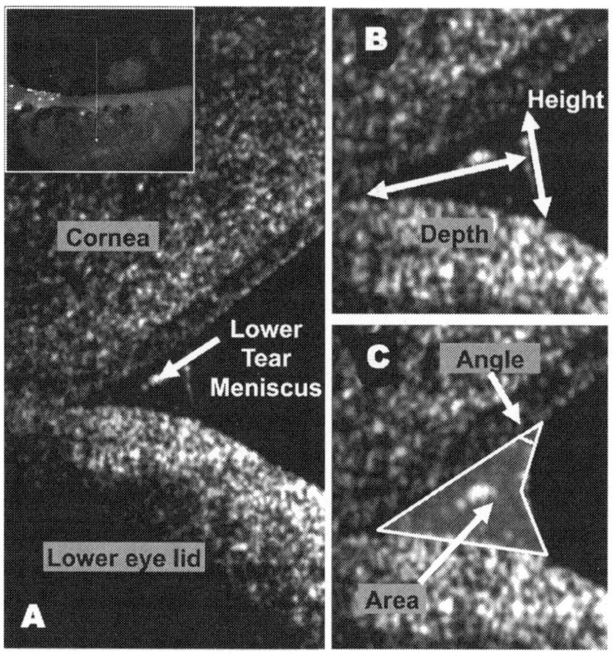

Fig. 3. Lower tear meniscus measurement using anterior segment optical coherence tomography. (A) Anatomy of the lower tear meniscus; the inset illustrates the scan position. (B) and (C) Caliper measurement protocol.

4. Conclusion

The corneal epithelium, conjunctival epithelium, tear film and lacrimal system, and the eyelids constitute the functional ocular surface. Management strategies for severe ocular surface disorders involve treatment and prevention of recurrence of the causative condition, symptomatic alleviation, and reconstitution of the anatomic and physiologic ocular surface. Human amniotic membrane transplantation with or without epithelial graft is a versatile modality for various injuries to the ocular surface epithelia. Other options for reconstitution of the epithelia include suppression of cicatricial changes, autologous mucosal transplantation, or cultivated bioengineered tissues. Motivated by the risk of zoonotic disease transmission or xenobiotic immunogenicity, researchers worldwide are investigating serum-free feeder cell-free culture systems for ex vivo expansion of epithelial tissues for ocular surface applications. The results of these investigations are promising.

Normal lacrimal function is not only important for symptomatic alleviation and optical pursuits; it is also exceedingly essential for surgical outcomes. Preoperative tear film is an important prognostic indicator for ocular surface reconstruction. Strategies for tear film restoration include lubrication, moisture chamber, and artificial lacrimal glands.

5. References

[1] Adeghate, E, Draper, C.E, Singh, J. Effects of ageing on changes in morphology of the rat lacrimal gland. Advanced Experimental Medical Biology, Vol. 506, No. Pt A, (2002), pp. 103–107.

[2] Ainsworth, G, Rotchford, A, Dua, H.S, King, A.J. A novel use of amniotic membrane in the management of tube exposure following glaucoma tube shunt surgery. British Journal of Ophthalmology, Vol. 90, No. 4, (2006), pp. 417–419.

[3} Anderson, S.B, de Souza, F., Hofmann-Rummelt, C., Seitz, B. Corneal calcification after amniotic membrane transplantation. British Journal of Ophthalmology, Vol. 87, No. 5, (2003), pp. 587–591.

[4] Andreev, K., Zenkel, M., Kruse, F., Jünemann, A., Schlötzer-Schrehardt, U. Expression of bone morphogenetic proteins (BMPs), their receptors, and activins in normal and scarred conjunctiva: Role of BMP-6 and activin-A in conjunctival scarring? Experimental Eye Research, Vol 83, No. 5 (2006), pp 1162-70.

[5] Ang, L.P, Cheng, Z.Y, Beuerman, R.W, Teoh, S.H, Zhu, X., Tan, D.T. The development of a serum-free derived bioengineered conjunctival epithelial equivalent using an ultrathin poly(epsilon-caprolactone) membrane substrate. Investigative Ophthalmology and Vision Science, Vol.47, No. 1, (2006), pp. 105–12.

[6] Ang, L.P, Chua, J.L, Tan, D.T. Current concepts and techniques in pterygium treatment. Current Opinions in Ophthalmology, Vol 18, No.4, (2007), pp. 308–313.

[7] Ang, L.P, Nakamura, T., Inatomi, T. et al. Autologous serum–derived cultivated oral epithelial transplants for severe ocular surface disease. Arch Ophthalmology, Vol 124, No. 11, (2006), pp. 1543–1551.

[8] Ang, L.P, Tan, D.T, Seah, C.J, Beuerman, R.W. The use of human serum in supporting the in vitro and in vivo proliferation of human conjunctival epithelial cells. British Journal of Ophthalmology, Vol. 89, No. 6, (2005), pp. 748–752.

[9] Ang, L.P.K, Cheng, Z.Y, Beuerman, R.W, et al. The development of a serum-free derived bioengineered conjunctival epithelial equivalent using an ultrathin poly (ε-caprolactone) membrane substrate. Investigative Ophthalmology and Vision Science,. Vol. 47, (2006), pp.105-112.

[10] Ang, L.P.K, Cheng, Z.Y., Beuerman, R.W, Teoh, S.H, Zhu, X., Tan, D.T.H.The Development of a Serum-Free Derived Bioengineered Conjunctival Epithelial Equivalent Using an Ultrathin Poly(ε-Caprolactone) Membrane Substrate. Investigative Ophthalmology and Vision Science, Vol. 47, No. 1 (January 2006). pp. 105-112.

[11] Arthur, B.W, Hay, G.J, Wasan, S.M, Willis, W.E. Ultra-structural effects of topical Timolol on rabbit cornea. Arch Ophthalmology, Vol. 10, (1983), pp. 1607-1610.

[12] Azuara-Blanco, A., Pillai, C.T, Dua, H.S. Amniotic membrane transplantation for ocular surface reconstruction. British Journal of Ophthalmology, Vol. 83, No. 4, (1999), pp. 399–402.

[13] Bacman, S.R, Berra, A, Sterin-Borda, L, Borda, E.S. Human primary Sjögren's syndrome autoantibodies as mediators of nitric oxide release coupled to lacrimal gland muscarinic acetylcholine receptors. Current Eye Research, Vol. 17, No. 12, (1998), pp. 1135–1142.

[14] Barber, L.D, Pflugfelder, S.C, Tauber, J., Foulks, G.N. Phase III safety evaluation of cyclosporine 0.1% ophthalmic emulsion administered twice daily to dry eye disease patients for up to 3 years. Ophthalmology, Vol. 112, No. 10, (2005), pp. 1790–1794.

[15] Barequet, I.S, Habot-Wilner, Z., Keller, N. et al. Effect of amniotic membrane transplantation on the healing of bacterial keratitis. Investigatiave Ophthalmology and Vision Science, Vol. 49, No. 1, (2008), pp. 163–167.

[16] Barequet, I.S, Habot-Wilner, Z., Keller, N. et al. Effect of amniotic membrane transplantation on the healing of bacterial keratitis. Investigative Ophthalmology and Vision Science, Vol 49, No.1, (2008), pp.163–167.

[17] Bawa R, inventor. Bausch & Lomb Inc. (Rochester, NY), assignee. Sustained-release formulation containing an amino acid polymer. US Patent 4 668 506. May 26, 1987.

[18] Beauregard, C, Brandt, P.C, Chiou, G.C.Y. Induction of nitric oxide synthase and over-production of nitric oxide by interleukin-1β in cultured lacrimal gland acinar cells. Experimental Eye Research, Vol. 77, No. 1, (2003), pp. 109-114.

[19] Bernauer, W., Thiel, M.A., Langenauer, U.M. and Rentsch, K.M. Phosphate concentration in artificial tears. Graefe's Archive for Clinical and Experimental Ophthalmology, Vol. 244, No. 8, (2006), pp. 1010-1014.

[20] Bernauer, W., Thiel, M.A, Kurrer, M. et al. Corneal calcification following intensified treatment with sodium hyaluronate artificial tears. British Journal of Ophthalmology, Vol. 90, No. 3, (2006), pp. 285–288.

[21] Bernauer,W., Thiel, M.A., and Rentsch, K.M. Phosphate concentration in ophthalmic corticoid preparations. Graefe's Archive For Clinical and Experimental Ophthalmology. Vol. 246, No. 7, (2008), pp. 975-978.

[22] Bisbee, C.A. Prolactin effects on ion transport across cultured mouse mammary epithelium. American Journal of Physiology, Vol. 240, No. 3, (1981), pp. C110–C115.

[23] Boneva R.S, Folks T.M, Chapman L.E. Infectious disease issues in xenotransplantation. Clinical Microbiology Review, Vol. 14, No. 1-1, (2001).

[24] Bourlais, C.L, Acar, L, Zia, H, et al. Ophthalmic drug delivery systems: recent advances. Progress in Retinal and Eye Research, Vol. 17, No. 1, (1998), pp. 33-58.

[25] Chen Y.T, Li W., Hayashida Y., et al. Human amniotic epithelial cells as novel feeder layers for promoting ex vivo expansion of limbal epithelial progenitor cells. Stem Cells, Vol. 25, No. 8, (2007), pp. 1995–2005.

[26] Chen, H., Fan, X., Xia, J., Chen, P., Zhou, X., Huang, J., Yu, J., Gu, P.. Electrospun chitosan-graft-poly (ε-caprolactone)/ poly (ε-caprolactone) nanofibrous scaffolds for retinal tissue engineering. International Journal of Nanomedicine. Vol. 6, (2011), pp. 453–461.

[27] Chen, H.J, Pires, R.T, Tseng, S.C. Amniotic membrane transplantation for severe neurotrophic corneal ulcers. Brithish Journal of Ophthalmology, Vol. 84, No.8, (2000), pp. 826–833.

[28] Chen, H.J, Pires, R.T, Tseng, S.C. Amniotic membrane transplantation for severe neurotrophic corneal ulcers. Britishin Journal of Ophthalmology, Vol 84, No. 8, (2000), pp. 826–833.

[29] Chen, H.S., Ritch, R., Krupin, T., Hsu, W. Efficacy and Safety of Biodegradable Collagen-Glycosaminoglycan Polymer as a Material for Scleral Buckling. Investigative Ophthalmology and Vision Science, Vol 49, No. 6, (June 2008), pp. 2673-2678.

[30] Chen, Y.T, Li, W., Hayashida, Y., He, H., et al. Human Amniotic Epithelial Cells as Novel Feeder Layers for Promoting Ex Vivo Expansion of Limbal Epithelial Progenitor Cells. Stem Cells, Vol. 25, No. 8, (2007), pp. 1995-2005.

[31] Cheng, K.C, Chang, C.H.Modified gunderson conjunctival flap combined with an oral mucosal graft to treat an intractable corneal lysis after chemical burn: a case report.Kaohsiung Journal of Medical Science, Vol. 22, No. 5, (2006), pp. 247-51.

[32] Ciolino, J.B, Hoare T.R, Iwata, N.G, Behlau, I., Dohlman, C.H, Langer, R., Kohane, D.S. A Drug-Eluting Contact Lens. Investigative Ophthalogy and Vision Science, Vol. 50, No.7, (2009), pp. 3346-52.

[33] Cordeiro, M.F. Role of transforming growth factor b in conjunctival scarring. Clinical Science (Lond), Vol 104, No. 2, (February 2003), pp 181-7.

[34] Cordeiro, M.F., Gay, J.A. and Khaw, P.T. Human anti-TGF-b2 monoclonal antibody: a new anti-scarring agent for glaucoma filtration surgery. Investigative Ophthalmology and Visual Science, Vol. 40, No.10, (September 1999), pp 2225–2234.

[35] Cordeiro, M.F., Reichel, M.B., Gay, J.A., D'Esposita, F., Alexander, R.A., Khaw, P.T. Transforming Growth Factor-β1, -β2, and -β3 In Vivo: Effects on Normal and Mitomycin C Modulated Conjunctival Scarring. Investigative Ophthalmology and Visual Science. Vol. 40, No. 9 (August 1999) pp. 1975-1982.

[36] Davis, J.W. Skin transplantation with a review of 550 cases at the Johns Hopkins Hospital. Johns Hopkins Medical Journal, Vol. 15, (1910), pp. 307-96.

[37] Daya, S.M. Living-related conjunctivo-limbal allograft (lr-CLAL) for the treatment of stem cell deficiency, an analysis of long-term outcome [abstract]. Ophthalmology, Vol. 106, No. supplement, (1999), pp. 243.

[38] De Rotth, A. Plastic repair of conjunctival defects with fetal membrane. Archives of Ophthalmology, Vol.23, No. 3, (1940), pp. 522–525.

[39] Deshpande, P., Notara, M., Bullett, N., Daniels, J., Haddow, D., Macneil, S. Development of a Surface-Modified Contact Lens for the Transfer of Cultured Limbal Epithelial Cells to the Cornea for Ocular Surface Diseases. Tissue Engineering: Part A. Vol. 15, No.10, (2009), pp. 2889-902.

[40] Di Girolamo, N., Chui, J., Wakefield, D., Coroneo, M.T. Cultured human ocular surface epithelium on therapeutic contact lenses. British Journal of Ophthalmology, Vol. 91, No. 4, (2007), pp. 459–464.

[41] Djalilian, A.R, Bagheri, M.M, Schwartz, G.S et al. Keratolimbal allograft for the treatment of limbal stem cell deficiency. Presented at: Castroviejo Cornea Society Annual Meeting, Orlando, FL, USA, 23 October 1999.

[42] Doane, M.G, Dohlman, C.H, Bearse, G. Fabrication of a keratoprosthesis. Cornea, Vol. 15, No.2, (1996), pp. 179–184.

[43] Dogru, M, Nakagawa, N, Tetsumoto, K, Katakami, C, Yamamoto, M. Ocular surface disease in atopic dermatitis. Japanese Journal of Ophthalmology, Vol. 43, No. 1, (1999), pp. 53–57.

[44] Dogru, M., Karakaya, H., Özçetin, H. et al. Tear function and ocular surface changes in keratoconus. Ophthalmology, Vol. 110, No. 6, (2003), pp. 1110–1118.

[45] Dogru, M., Katakami, C., Inoue, M. Tear function and ocular surface changes in noninsulin-dependent diabetes mellitus. Ophthalmology, Vol. 108, No. 3, (2001), pp. 586–592.

[46] Dua, H.S, Gomes, J.A, King, A.J, Maharajan, V.S. The amniotic membrane in ophthalmology. Survey in Ophthalmology, Vol. 49, No.1, (2004), pp. 51–77.

[47] Espana, E.M, Prabhasawat, P., Grueterich, M., Solomon, A., Tseng, S.C. Amniotic membrane transplantation for reconstruction after excision of large ocular surface neoplasias. British Journal of Ophthalmology, Vol. 86, No. 6, (2002), pp. 640–645.

[48] Essex, R.W, Snibson, G.R, Daniell, M., Tole, D.M. Amniotic membrane grafting in the surgical management of primary pterygium. Clinical and Experimental Ophthalmology, Vol. 32, No. 5, (2004), pp. 501–504.

[49] Fagerholm, P., Lagali, N.S., Carlsson, D.J., Merrett, K. and Griffith, M. Corneal Regeneration Following Implantation of a Biomimetic Tissue-Engineered Substitute. Clinical Translation, Vol. 2, No. 2, (2009), pp. 162-164.

[50] Fristrom, B. A 6-month, randomized, double-masked comparison of latanoprost with timolol in patients with open angle glaucoma or ocular hypertension. Acta Ophthalmol Scandanavia, Vol. 74, (1996), pp. 140-144.

[51] Fullard, R. (1994). Tear proteins arising from lacrimal tissue. In: Principles and Practice of Ophthalmology 2nd Edition, (Albert DM, Jacobiec FA, eds), 473-9, Saunders, Philadelphia, US.

[52] Gabbiani, G, Ryan, G.B, Majno, G. Presence of modified fibroblasts in granulation tissue and their possible role in wound contraction. Experientia. Vol. 27, No. 5, (1971), pp. 549–550.

[53] Galal, A., Perez-Santonja, J.J, Rodriguez-Prats, J.L, Abad, M., Alio, J. Human anterior lens capsule as a biologic substrate for the ex vivo expansion of limbal stem cells in ocular surface reconstruction. Cornea, Vol. 26, No. 4, (2007), pp. 473–478.

[54] Geerling, G., MacLennan, S., Hartwig, D. Autologous serum eye drops for ocular surface disorders. British Journal of Ophthalmology, Vol. 88, No. 11, (2004), pp. 1467–1474.

[55] Ghate, D., Edelhauser, H.F. Barriers to glaucoma drug delivery. Journal of Glaucoma, Vol. 17, No.2, (2008), pp. 147-156.

[56] Giambattista, B, Virno, M, Pecori, G, Pellegrino, N, Motolese, E. Possibility of isoproterenol therapy with soft contact lenses: ocular hypotension without systemic effects. Annex Ophthalmology, Vol.8, (1976), pp. 819-829.

[57] Gipson, I.K, Geggel, H.S, Spurr-Michaud, S.J. Transplant of oral mucosal epithelium to rabbit ocular surface wounds in vivo. Arch Ophthalmology, Vol. 104, No. 10, (1986), pp. 1529–1533.

[58] Girolamo, N.D, Bosch, M., Zamora, K., Coroneo, M T, Wakefield, D., Watson, S.L.A. Contact Lens-Based Technique for Expansion and Transplantation of Autologous Epithelial Progenitors for Ocular Surface Reconstruction. Transplantation, Vol. 87, No. 10, (2009), pp. 1571-78.

[59] Gomes, J.A, dos Santos, M.S, Cunha, M.C, Mascaro, V.L, Barros, J.N, de Sousa, L.B. Amniotic membrane transplantation for partial and total limbal stem cell deficiency secondary to chemical burn. Ophthalmology. Vol. 110, No. 3, (2003), pp. 466–473.

[60] Gomes, J.A.P., dos Santos, M.S., Cunha, M.C., Mascaro, VLuD, Barros, JdN, and de Sousa, L.B. Amniotic membrane transplantation for partial and total limbal stem cell

deficiency secondary to chemical burn. Ophthalmology, Vol. 110, No. 3, (2003), pp. 466-73.

[61] Grueterich, M., Espana, E.M., Tseng, S. C.G. Ex Vivo Expansion of Limbal Epithelial Stem Cells: Amniotic Membrane Serving as a Stem Cell Niche. Survey of Ophthalmology, Vol. 48, No. 6, (2003), pp. 631-646.

[62] Gulsen, D., Chauhan, Anuj. Ophthalmic Drug Delivery through Contact Lenses. Investigative Opthalmology and Vision Science, Vol. 45, No.7, (2004), pp. 2342-47.

[63] Gündüz, K., Uçakhan, Ö.Ö, Kanpolat, A., Günalp, I. Nonpreserved human amniotic membrane transplantation for conjunctival reconstruction after excision of extensive ocular surface neoplasia. Eye, Vol. 20, No. 3, (2006), pp. 351–357.

[64] Guo, Z., Song, D., Azzarolo, A.M et al. Autologous lacrimal-lymphoid mixed-cell reactions induce dacryoadenitis in rabbits. Experimental Eye Research, Vol. 71, No. 1, (2000), pp. 23–31.

[65] Haynesa, R.J, Tigheb, P.J., Dua, H.S. Antimicrobial defensin peptides of the human ocular surface. British Journal of Ophthalmology, Vol. 83, (1999), pp. 737-741.

[66] Hick, S., Demers, P.E, Brunette, I., La, C., Mabon, M., Duchesne, B. Amniotic membrane transplantation and fibrin glue in the management of corneal ulcers and perforations: a review of 33 cases. Cornea.Vol. 24, No.4, (2005), pp. 369–377.

[67] Hicks, C.R, Crawford, G.J, Dart, J.K et al. AlphaCor: Clinical outcomes. Cornea, Vol. 25, No. 9, (2006), pp. 1034–1042.

[68] Hicks, C.R, Crawford, G.J, Lou, X. et al. Corneal replacement using a synthetic hydrogel cornea, AlphaCor: device, preliminary outcomes and complications. Eye. Vol. 17, No. 3, (2003), pp. 385–392.

[69] Hicks, C.R, Crawford, G.J, Tan, D.T et al. Outcomes of implantation of an artificial cornea, AlphaCor: effects of prior ocular herpes simplex infection. Cornea, Vol. 21, No. 7, (2002), pp. 685–690.

[70] Higa, K., Shimmura, S., Kato, N., Kawakita, T., Miyashita, H., Itabashi, Y., Fukuda, K., Shimazaki, J., and Tsubota, K. Proliferation and Differentiation of Transplantable Rabbit Epithelial Sheets Engineered with or without an Amniotic Membrane Carrier. Investigative Ophthalmology and Vision Science, Vol. 48, No. 2, (February 2007), pp. 597-604.

[71] Hillman, J, Masters, J, Broad, A. Pilocarpine delivery by hydrophilic lens in the management of acute glaucoma. Trans Ophthalmol Soc UK. Vol. 95, (1975), pp. 79-84.

[72] Hillman, J.S. Management of acute glaucoma with pilocarpine-soaked hydrophilic lens. British Journal of Ophthalmology, Vol. 58, (1974), pp. 674-679.

[73] Holland, E.J, Schwartz, G.S. Epithelial stem-cell transplantation for severe ocular surface disease [editorial]. New England Journal of Medicine, Vol. 340, No. 22, (1999), pp. 1752-1753.

[74] Homma, R., Yoshikawa, H., Takeno, M. et al. Induction of epithelial progenitors in vitro from mouse embryonic stem cells and application for reconstruction of damaged cornea in mice. Investigative Ophthalmology and Vision Science, Vol. 45, No. 12, (2004), pp. 4320–4326.

[75] Hopkinson, A., McIntosh, R.S, Tighe, P.J, James, D.K, Harminder, S.D. Amniotic membrane for ocular surface reconstruction: donor variations and the effect of handling on TGF-ß content. Investigative Ophthalmology and Vision Science, Vol. 47, No. 10, (2006), pp. 4316–4322.

[76] Hsu, W.C, Spilker, M.H, Yannas, I.V, Rubin, P, A. Inhibition of conjunctival scarring and contraction by a porous collagen-glycosaminoglycan implant. Investigative Ophthalmology and Vision Sciece, Vol. 41, No. 9, (2000), pp. 2404–2411.

[77] Ilari, L., Daya, S.M. Long-term outcomes of keratolimbal allograft for the treatment of severe ocular surface disorders. Ophthalmology, Vol. 109, No. 7, (2002), pp. 1278–1284.

[78] Iveković, B., Tedeschi-Reiner, E., Petric, I., Novak-Laus, K., Bradić-Hammoud, M. Amniotic membrane transplantation for ocular surface reconstruction in neurotrophic corneal ulcers. Collegium Antropologicum, Vol. 26, No. 1, (2002), pp. 47–54.

[79] Iveković, B., Tedeschi-Reiner, E., Petric, I., Novak-Laus, K., Bradić-Hammoud, M. Amniotic membrane transplantation for ocular surface reconstruction in neurotrophic corneal ulcers. Collegium Antropologicum, Vol.26, No.1, (2002), pp. 47–54.

[80] Jain, M.R, Batra, V. Steroid penetration in human aqueous with Sauflon 70 lenses. Indian Journal of Ophthalmology, Vol. 11, No. 2, (1970), pp. 26-31.

[81] Jain, M.R. Drug delivery through soft contact lenses. British Journal of Ophthalmology. Vol. 72, (1988), pp. 150-154.

[82] John, T., Foulks, G., John, M.E., Cheng, K., Hu, D. Amniotic membrane in the surgical management of acute toxic epidermal necrolysis. Ophthalmology. Vol. 109, No. 2, (2002), pp. 351–360.

[83] Joseph, A., Dua, H.S, King, A.J. Failure of amniotic membrane transplantation in the treatment of acute ocular burns. British Journal of Ophthalmology, Vol. 85, No. 9, (2001), pp.1065–1069.

[84] Joseph, A., Raj, D., Shanmuganathan, V., Powell, R.J, Dua, H.S. Tacrolimus immunosuppression in high-risk corneal grafts. British Journal of Ophthalmology, Vol. 91, No. 1, (2007), pp. 51–55.

[85] Kheirkhah, A., Li, W., Casas, V., Tseng, S.C.G. Sutureless amniotic membrane transplantation. Expert Review of Ophthalmology, Vol. 1, No. 1, (2006), pp.49–62.

[86] Khokhar, S., Natung, T., Sony, P., Sharma, N., Agarwal, N., Vajpayee, R.B. Amniotic membrane transplantation in refractory neurotrophic corneal ulcers: a randomized, controlled clinical trial. Cornea.Vol. 24, No. 6, (2005), pp. 654–660.

[87] Kim, J.C, Tseng, S.C.G. Transplantation of preserved human amniotic membrane for surface reconstruction in severely damaged rabbit corneas. Cornea. Vol. 14, No. 5, (1995), pp. 473–484.

[88] Kim, J.H, Chun, Y.S, Lee, S.H, Mun, S.K, Jung, H.S, Lee, S.H, Son, Y., Kim, J.C. Ocular surface reconstruction with autologous nasal mucosa in cicatricial ocular surface disease. American Journal of Ophthalmology, Vol. 149, No.1, (January 2010), pp. 45–53.

[89] Kim, J.S, Kim, J.C, Na, B.K, Jeong, J.M, Song, C.Y. Amniotic membrane patching promotes healing and inhibits proteinase activity on wound healing following acute corneal alkali burn. Experimental Eye Research, Vol. 70, No.3, (2000), pp. 329–337.

[90] Kinoshita, S., Nakamura, T. Development of cultivated mucosal epithelial sheet transplantation for ocular surface reconstruction. The International Journal of Artificial Organs, Vol. 28, No. 1, (2004), pp. 22–27.

[91] Koizumi, N., Inatomi, T., Quantock, A., Fullwood, N.J, Dota, A., Kinoshita, S. Amniotic membrane as a substrate for cultivating limbal corneal epithelial cells for autologous transplantation in rabbits. Cornea. Vol. 19, No.1, (2000), pp. 65–71.

[92] Koizumi, N., Inatomi, T., Sotozono, C., Fullwood, N.J., Quantock, A.J., Kinoshita, S. Growth factor mRNA and protein in preserved human amniotic membrane. Current Eye Research, Vol. 20, No. 3, (2000), pp.173–177.

[93] Koizumi, N., Rigby, H., Fullwood, N., Kawasaki, S., Tanioka, H., Koizumi, K., Kociok, N., Joussen, A., and Kinoshita, S. Comparison of intact and denuded amniotic membrane as a substrate for cell-suspension culture of human limbal epithelial cells. Graefes Arch Clinical Experimental Ophthalmology, Vol. 245, No. 1, (2007), pp. 123-134.

[94] Lagnado, R., King, A.J, Donald, F., Dua, H.S. A protocol for low contamination risk of autologous serum drops in the management of ocular surface disorders. British Journal of Ophthalmology, Vol. 88, No. 4, (2004), pp. 464–465.

[95] Lai, D., Cheng, W., Liu, T., Jiang, L., Huang, Q., Liu, T. Use of Human Amnion Epithelial Cells as a Feeder Layer to Support Undifferentiated Growth of Mouse Embryonic Stem Cells. Cloning and Stem Cells. Vol. 11, No. 2, (June 2009), pp. 331-340.

[96] Lai, J.Y, Chen, K.H, Hsiue, G.H. Tissue-engineered human corneal endothelial cell sheet transplantation in a rabbit model using functional biomaterials. Transplantation, Vol. 84, No. 10, (2007), pp. 1222-1232.

[97] Lai, J.Y, Hsiue, G.H. Functional biomedical polymers for corneal regenerative medicine. Reaction and Functional Polymers, Vol. 67, No. 11, (2007), pp.1284-1291.

[98] Lam, D.S, Wong, A.K, Tham, C.C, Leung, A.T. The use of combined intravenous pulse methylprednisolone and oral cyclosporine A in the treatment of corneal graft rejection: a preliminary study. Eye, Vol. 12, Pt. 4, (1998), pp. 615–618.

[99] Lawrencea, B.D., Marchantb, J.K., Pindrusa, M.A, Omenettoa, F.G., Kaplana, D.L. Silk film biomaterials for cornea tissue engineering. Biomaterials, Vol. 30, No. 7, (March 2009), pp. 1299-1308.

[100] Lee ,S.Y, Oh, J.H, Kim, J.C, et al. In vivo conjunctival reconstruction using modified PLGA grafts for decreased scar formation and contraction. Biomaterials, Vol. 24, No. 27, (2003), pp. 5049-5059.

[101] Lee, S.B, Li, D.Q, Tan, D.T, Meller, D.C, Tseng, S.C.G. Suppression of TGF-beta signaling in both normal conjunctival fibroblasts and pterygial body fibroblasts by amniotic membrane. Current Eye Research, Vol. 20, No.4, (2000), pp. 325–334.

[102] Li, W., Hayashida, Y., Chen, Y., Tseng, S.C.G. Niche regulation of corneal epithelial stem cells at the limbus. Cell Research,Vol. 17, (2007), pp. 26-36.

[103] Li,F., Carlsson, D., Lohmann, C., Suuronen, E., Vascotto, S., Kobuch, K., Sheardown, H., Munger, R., Nakamura, M., Griffith, M. Cellular and nerve regeneration within a biosynthetic extracellular matrix for corneal transplantation. Proceedings of the National Academy of Sciences, Vol. 100, No. 25, (December 2003), pp. 15346–15351.

[104] Lin,C.C, Ritch, R., Ming Lin, S., Ni, M., Chang, Y., Lu, Y.L., Lai, H.J. and Lin,F. A New Fish Scale-derived Scaffold for Corneal Regeneration. European Cells and Materials, Vol. 19, (2010), pp. 50-57.

[105] Liu, Y., Gan, L., Carlsson, D.J., Fagerholm, P., Lagali, N., Watsky, M.A., Munger, R., Hodge, W.G., Priest, D. and Griffith, M. A Simple, Cross-linked Collagen Tissue Substitute for
Corneal Implantation. Investigative Ophthalmology and Vision Science, Vol. 47, (2006), pp. 1869–1875.

[106] Luanratanakorn, P., Ratanapakorn, T., Suwan-Apichon, O., Chuck, R.S. Randomised controlled study of conjunctival autograft versus amniotic membrane graft in

pterygium excision. Brithish Journal of Ophthalmology, Vol. 90, No.12, (2006), pp. 1476–1480.

[107] Luanratanakorn, P., Ratanapakorn, T., Suwan-Apichon, O., Chuck, R.S. Randomised controlled study of conjunctival autograft versus amniotic membrane graft in pterygium excision. British Journal of Ophthalmology, Vol. 90, No. 12, (2006), pp.1476–1480.

[108] Ma, L.H, Yu, W.T, Ma, X.J. Preparation and characterization of novel sodium alginate/chitosan two ply composite membranes. Journal of Applied Polymer Science, Vol. 106, No. 12, (2007), pp. 394–399.

[109] Ma, Y., Xu, Y., Xiao, Z. et al. Reconstruction of chemically burned rat corneal surface by bone marrow-derived human mesenchymal stem cells. Stem Cells, Vol. 24, No. 2, (2006), pp. 315–326.

[110] Maharajan, V.S, Shanmuganathan, V., Currie, A., Hopkinson, A., Powell-Richards, A., Dua, H.S. Amniotic membrane transplantation for ocular surface reconstruction: indications and outcomes. Clinical and Experimental Ophthalmology, Vol. 35, No. 2, (2007), pp. 140–147.

[111] Majno, G., Gabbiani, G., Hirchel, B.J, Ryan, G.B, Statkov, P.R. Contraction of granulation tissue in vitro: similarity to smooth muscle. Science. Vol 173, No 996, (1971), pp. 548–550.

[112] Marmion, V.J, Yarkakul, S. Pilocarpine administration by contact lens. Trans Ophthalmol Soc UK. Vol. 97, (1977), pp. 162-163.

[113] Martin M.J, Muotri A, Gage F, Varki A. Human embryonic stem cells express an immunogenic nonhuman sialic acid. Nature Medicine. Vol. 11, No. 2 (2005), pp. 228-232.

[114] Mathers, W.D, Stovall, D, Lane, J.A, Zimmerman, M.B, Johnson, S. Menopause and tear function: the influence of prolactin and sex hormones on human tear production. Cornea, Vol. 17, No. 4, (1998), pp. 353–358.

[115] Mills, R.A, Coster, D.J, Williams, K.A. Effect of immunosuppression on outcome measures in a model of rat limbal transplantation. Investigative Ophthalmology and Vision Science, Vol. 43, No. 3, (2002), pp. 647–655.

[116] Montague, R, Wakins, R. Pilocarpine dispensation for the soft hydrophilic contact lens. British Journal of Ophthalmology, Vol. 59, (1975), pp. 455-458.

[117] Moustafa, M., Bullock, A.J., Creagh, F.M., Heller S., Jeffcoate, W., Game, F., Amery, C., Tesfaye, S., Ince, Z., Haddow, D.B., and MacNeil, S. Randomized, controlled, single-blind study on use of autologous keratinocytes on a transfer dressing to treat nonhealing diabetic ulcers. Regenerative Medicine, Vol. 2, (2007), pp. 887-902.

[118] Moustafa, M., Simpson, C., Glover, M., Dawson, R.A., Tesfaye, S., Creagh, F.M., Haddow, D., Short, R. Heller, S., and MacNeil, S. A new autologous keratinocyte dressing treatment for non-healing diabetic neuropathic foot ulcers. Diabetic Medicine, Vol. 21, No. 7, (2004), pp. 786-789.

[119] Murray, T.G., Stern, W.H, Chin, D.H, MacGowan-Smith, E.A. Collagen Shield Heparin Delivery for Prevention of Postoperative Fibrin. Arch Ophthalmology, Vol. 108, (1990), pp. 104-106.

[120] Myung, D, Koh, W, Bakri, A, et al. Design and fabrication of an artificial cornea based on a photolithographically patterned hydrogel construct. Biomedical Microdevices, Vol. 9, No. 6, (Januaray 2007), pp. 911-922.

[121] Myung, D., Duhamel, P.-E., Cochran, J. R., Noolandi, J., Ta, C. N. and Frank, C. W. Development of Hydrogel-Based Keratoprostheses: A Materials Perspective. Biotechnology Progress, Vol. 24, No. 3, (October 2008), pp. 735-74.

[122] Myung, D., Farooqui, N., Waters, D., Schaber, S., Koh, W., Carrasco, M., Noolandi, J., Frank, C.W. and Ta, C.N. Glucose-Permeable Interpenetrating Polymer Network Hydrogels for Corneal Implant Applications: A Pilot Study. Current eye research, Vol. 33, No. 1, (2008), pp. 29-43.

[123] Myung, D., Koh, W., Bakri, A. et al. Design and fabrication of an artificial cornea based on a photolithographically patterned hydrogel construct. Biomedical Microdevices, Vol. 9, No. 6, (2007), pp. 911–922.

[124] Nakamura, T., Ang, L.P, Rigby, H. et al. The use of autologous serum in the development of corneal and oral epithelial equivalents in patients with Stevens-Johnson syndrome. Investigative Ophthalmology and Vision Science, Vol. 47, No. 3, (2006), pp. 909–916.

[125] Nakamura, T., Endo, K., Cooper, L.J, Fullwood, N.J., Tanifuji, N., Tsuzuki, M., Koizumi, N., Inatomi, T., Sano, Y., and Kinoshita,S. The Successful Culture and Autologous Transplantation of Rabbit Oral Mucosal Epithelial Cells on Amniotic Membrane. Investigative Ophthalmology and Vision Science, Vol. 44, No. 1, (January 2003), pp. 106-116.

[126] Nakamura, T., Inatomi, T., Cooper, L.J, Rigby, H., Fullwood, N.J., Kinoshita, S. Phenotypic investigation of human eyes with transplanted autologous cultivated oral mucosal epithelial sheets for severe ocular surface diseases. Ophthalmology, Vol. 114, No. 6, (2007), pp. 1080–1088.

[127] Nakamura, T., Inatomi, T., Sekiyama, E., Ang, L., Yokoi, N., Kinoshita, S. Novel clinical application of sterilized, freeze-dried amniotic membrane to treat patients with pterygium. Acta Ophthalmology Scand, Vol. 84, No.3, (2006), pp. 401–405.

[128] Nakamura, T., Inatomi, T., Sotozono, C. et al. Transplantation of autologous serum-derived cultivated corneal epithelial equivalents for the treatment of severe ocular surface disease. Ophthalmology, Vol. 113, No. 10, (2006), pp.1765–1772.

[129] Nakamura, T., Takeda, K., Inatomi, T., Sotozono, C., Kinoshita,S. Long-term results of autologous cultivated oral mucosal epithelial transplantation in the scar phase of severe ocular surface disorders. British Journal of Ophthalmology (November 2010). Epub ahead of print.

[130] Nakamura, T., Yoshitani, M., Rigby, H. et al. Sterilized, freeze-dried amniotic membrane: a useful substrate for ocular surface reconstruction. Investigative Ophthalmology and Vision Science, Vol. 45, No.1, (2004), pp. 93–94.

[131] Nakamura, T., Yoshitani, M., Rigby, H. et al. Sterilized, freeze-dried amniotic membrane: a useful substrate for ocular surface reconstruction. Investigative Ophthalmology and Vision Science, Vol.45, No.1, (2004), pp.93–94.

[132] Naumann, G.O, Lang, G.K, Rummelt, V., Wigand, M.E.Autologous nasal mucosa transplantation in severe bilateral conjunctival mucus deficiency syndrome.Ophthalmology, Vol. 97, No. 8, (August 1990), pp. 1011-7.

[133] Nguyen, P., Barte, F., Kang, K., Yiu, S.C, et al. (2008) A Novel Pharmaceutical Management of Immunogenic Rejection Following Repeat Penetrating Keratoplasty

in High-Risk Patients [Abstract] Presented at: Association for Research and Vision in Ophthalogy 2008 Annual Meeting, Fort Lauderdale, FL, USA, April 30-May4 2008.

[134] Nguyen, P., Yiu, SC. (2007). Multi-agent pharmaceutical therapy for modulation of corneal allograft immunologic rejection, In: Curr Insights Online publication. Available from www.omicsonline.org/2155-9570/2155-9570-1-103.php.

[135] Nishida, K., Yamato, M., Hayashida, Y. et al. Corneal reconstruction with tissue-engineered cell sheets composed of autologous oral mucosal epithelium. New England Journal of Medicine, Vol. 351, No. 12, (2004), pp. 1187–1196.

[136] Nishida, K., Yamato, M., Hayashida, Y. et al. Functional bioengineered corneal epithelial sheet grafts from corneal stem cells expanded ex vivo on a temperature-responsive cell culture surface. Transplantation, Vol. 77, No. 3, (2004), pp. 379–385.

[137] Notara, M., Haddow, D.B, MacNeil, S., Daniels, J.T. A xenobiotic-free culture system for human limbal epithelial stem cells. Regenerative Medicine, Vol. 2, No. 6, (2007), pp. 919-927.

[138] Nubile, M., Dua, H.S, Lanzini, T.E.M et al. Amniotic membrane transplantation for the management of corneal epithelial defects: an in vivo confocal microscopic study. British Journal of Ophthalmology, Vol. 92, No. 1, (2008), pp.54–60.

[139] Oberhansli, C., Spahn, B. Amniotic membrane transplantation for oculopalpebral and reconstructive surgery [French]. Journal of French Ophtalmology, Vol. 28, No. 7, (2005), pp. 759–764.

[140] Oh, J.Y, Kim, M.K, Shin, M.S, Lee, H.J, et al. The Anti-inflammatory and Anti-angiogenic Role of Mesenchymal Stem Cells in Corneal Wound Healing Following Chemical Injury. Stem Cells, Vol. 26, No. 4, (2008), pp. 1047-55.

[141] Osei-Bempong, C., Henein, C., Ahmad, S. Culture conditions for primary human limbal epithelial cells. Regenerative Medicine, Vol. 4, No. 3, (2009), pp. 461-470.

[142] Ozdemir, M., Buyukbese, M.A, Cetinkaya, A., Ozdemir, G. Risk factors for ocular surface disorders in patients with diabetes mellitus. Diabetes Research and Clinical Practice, Vol. 59, No. 3, (2003), pp. 195–199.

[143] Oztürk, E., Ergün, M.A, Oztürk, Z. et al. Chitosan-coated alginate membranes for cultivation limbal epithelial cells to use in the restoration of damaged corneal surfaces.International Journal of Artificial Organs, Vol. 29, No. 2, (2006), pp. 228-238.

[144] Park, W.C, Tseng, S.C.G. Modulation of acute inflammation and keratocyte death by suturing, blood, and amniotic membrane in PRK. Investigative Ophthalmology and Vision. Science, Vol. 41, No. 10, (2000), pp. 2906–2914.

[145] Parmar, D. N.,Alizadeh, H., Awwad, S., Bowman, R.W., Cavanagh, D.H., McCulley, J. P.Contact Lens-Based Expansion and Transplantation of Autologous Epithelial Progenitors for Ocular Surface Reconstruction: Crossover Control. Transplantation, Vo.. 89, No.4, (February 2010), pp. 483.

[146] Patton, T.F, Francoeur, M. Ocular bioavailability and systemic loss of topically applied ophthalmic drugs. American Journal of Ophthalmology, Vol. 85, No.2, (1978), pp. 225-229.

[147] Pellegrini G, Golisano O, Paterna P, et al. Location and clonal analysis of stem cells and their differentiated progeny in the human ocular surface. Journal of Cell Biology. Vol 145, (May 1999), pp. 769–782.

[148] Pellegrini, G., Traerso, C.E, Franzi, A.T, et al. Long-term restoration of damaged corneal surfaces with autologous cultivated corneal epithelium. Lancet, Vol. 329, No. 9057, (1997), pp.990-993.

[149] Pflugfelder, S.C, Solomon, A, Stern, M.E. The diagnosis and management of dry eye: a twenty-five-year review. Cornea, Vol. 19, No. 5, (2000), pp. 644–649.

[150] Pflugfelder, S.C, Solomon, A, Stern, M.E. The diagnosis and management of dry eye: a twenty-five-year review. Cornea, Vol. 19, No. 5, (2000), pp. 644–649.

[151] Pflugfelder, S.C. Antiinflammatory therapy for dry eye. American Journal of Ophthalmology, Vol. 137, No. 2, (2004), pp.337–342.

[152] Phinney, R.B, Schwartz, S.D, Lee, D.A, Mondino, B. Collagen-Shield Delivery of Gentamicin and Vancomycin. Arch Ophthalmology, Vol. 106, (1988), pp. 1599-1604.

[153] Powell-Richards,J., Shanmuganathan,A. O.R., Dua, H.S. Epithelial cell characteristics of cultured human limbal explants. British Journal of Ophthalmology, Vol. 88, No. 3, (2004), pp. 393-398.

[154] Rama, P., Bonini, S., Lambiase, A. et al. Autologous fibrin-cultured limbal stem cells permanently restore the corneal surface of patients with total limbal stem cell deficiency. Transplantation. Vol. 72, No.9, (2001), pp. 1478–1485.

[155] Ramarli D, Giri A, Reina S, et al. HIV-1 spreads from lymphocytes to normal human keratinocytes suitable for autologous and allogenic transplantation. Journal of Investigative Dermatology. Vol. 105, No. 5, (1995), pp. 644-647.

[156] Ramer, R, Gasset, A. Ocular penetration of pilocarpine. Annex of Ophthalmology, Vol. 6, (1974), pp. 1325-1327.

[157] Rao, S.K, Rajagopal, R., Sitalakshmi, G., Padmanabhan, P. Limbal allografting from related live donors for corneal surface reconstruction. Ophthalmology, Vol. 106, No. 4, (1999), pp. 822–828.

[158] Razzaque, M.S., Foster, S., Ahmed, R. Role of Collagen-Binding Heat Shock Protein 47 and Transforming Growth Factor-β1 in Conjunctival Scarring in Ocular Cicatricial Pemphigoid. Investigative Ophthalmology and Visual Science. Vol 44, No. 4, (April 2003), pp 1616-1621.

[159] Reinhard, T., Böhringer, D., Enczmann, J. et al. Improvement of graft prognosis in penetrating normal-risk keratoplasty by HLA class I and II matching. Eye, Vol. 18, No. 3, (2004), pp. 269–277.

[160] Reinhard, T., Spelsberg, H., Henke, L. et al. Long-term results of allogeneic penetrating limbo-keratoplasty in total limbal stem cell deficiency. Ophthalmology, Vol. 111, No.4, (2004), pp. 775–782.

[161] Reinhard, T., Sundmacher, R., Spelsberg, H., Althaus, C. Homologous penetrating central limbo-keratoplasty (HPCLK) in bilateral limbal stem cell insufficiency. Acta Ophthalmology Scandanavia, Vol. 77, No. 6, (1999), pp. 663-667.

[162] Resch, M.D, Schlotzer-Schrehardt, U., Hofmann-Rummelt, C. et al. Adhesion structures of amniotic membranes integrated into human corneas. Investigative Ophthalmology and Vision Science, Vol. 47, No. 5, (2006), pp.1853–1861.

[163] Rheinwald JG. Serial cultivation of normal human epidermal keratinocytes. Methods in Cellular Biology, Vol 21A, (1980), pp. 229-254.

[164] Roberts, A.B., Sporn, M.B., Assoian, R.K., Smith, J.M. and Roche, N.S. Transforming growth factor beta: rapid induction of fibrosis and angiogenesis in vivo and

stimulation of collagen formation in vitro. Proceedings of the National Academy of Sciences of the United States of America, Vol 83, No. 12, (June 1986), pp 4167–4171.

[165] Romero-Rangel, T, Stravrou, P, Cotter, J, Rosenthal, P, Baltatzis, S, Foster, C.S. Gas-permeable scleral contact lens therapy in ocular surface disease. American Journal of Ophthalmology. Vol. 130, No. 1, (2000), pp. 25-32.

[166] Rosenthal, P., Cotter, J.M., Baum, J. Treatment of persistent corneal epithelial defect with extended wear of a fluid-ventilated gas permeable scleral contact lens. American Journal of Ophthalmology, Vol.130, No. 1, (2000), pp. 33-38.

[167] Rosenwald PL, inventor. Ocular device. US Patent 4 484 922. November 27, 1984.

[168] Rumelt, S., Bersudsky, V., Blum-Hareuveni, T., Rehany, U. Systemic cyclosporin A in high failure risk, repeated corneal transplantation. British Journal of Ophthalmology, Vol. 86, No., 9, (2002), pp. 988–992.

[169] Saettone, M. Progress and problems in ophthalmic drug delivery. Pharmatechnology, Vol. 4, (2002), pp. 1-6.

[170] Samson, M., Nduaguba, C., Baltatzis, S., Foster, C.S. Limbal stem cell transplantation in chronic inflammatory eye disease. Ophthalmology, Vol. 109, No. 5, (2002), pp. 862–868.

[171] Saw, V.P.J, Minassian, D., Dart, J.K.G et al. Amniotic membrane transplantation for ocular disease: a review of the first 233 cases from the UK user group. British Journal of Ophthalmology, Vol. 91, No. 8, (2007), pp. 1042-1047.

[172] Schrage, N.F., Frentz, M., Reim, M. Changing the composition of buffered eye-drops prevents undesired side effects. British Journal of Ophthalmology, Vol. 94, No. 11, (2010), pp. 1519-1522.

[173] Schultz CL, Nunez IM, Silor DL, Neil ML, inventors; Johnson & Johnson Vision Products, Inc. (Jacksonville, FL), assignee. Contact lens containing a leachable absorbed material. US Patent 5 723 131. March 3, 1998.

[174] Schwab, I.R, Reyes, M., Isseroff, R.R. Successful transplantation of bioengineered tissue replacements in patients with ocular surface disease. Cornea. Vol.19, No.4, (2000), pp. 421–426.

[175] Schwartz, G.S, Gomes, J.A.P, Holland, E.J. (2002) Preoperative staging of disease severity. In: Ocular Surface Disease Medical and Surgical Management, Holland EJ, Mannis MJ (Eds.), 158-167, Springer-Verlag, New York, USA.

[176] Sedlavek, J. Possibilities of application of ophthalmic drugs with the aid of gel contact lens. Cesk Oftalmol. Vol. 21, (1965), pp. 509-14.

[177] Sekiyama, E., Nakamura, T., Kurihara, E. et al. Novel sutureless transplantation of bioadhesive-coated, freeze-dried amniotic membrane for ocular surface reconstruction. Investigative Ophthalmology and Vision Science, Vol. 48, No. 4, (2007), pp. 1528–1534.

[178] Selvam S., Thomas, P.B, Yiu S.C. Tissue engineering: current and future approaches to ocular surface reconstruction. Ocular Surface. Vol 4, No 3, (2006), pp. 120-136.

[179] Selvam, S, Nakamura, T, Samant, D et al. (2007) Impact of chronic stimulation with carbachol, histamine and serotonin on ion transport in a novel, chloride-secreting rabbit lacrimal acinar cell monolayer model. Presented at: Association for Research and Vision Ophthalmology 2007 Annual Meeting, Fort Lauderdale, FL, USA, April 30-May 4, 2007.

[180] Selvam, S, Thomas, P.B, Trousdale, M.D et al. Tissue-engineered tear secretory system: functional lacrimal gland acinar cells cultured on matrix protein-coated substrata. Journal of Biomedical Material Research Application Biomaterials, Vol. 80, No. 1, (2007), pp. 192–200.

[181] Selvam, S., Thomas, P.B, Yiu, S.C. Tissue engineering: current and future approaches to ocular surface reconstruction. Ocular Surface, Vol. 4, No.3, (2006), pp. 120–136.

[182] Shepard, D.S, Razavi, M., Stason, W.B, Jacobs, D.S, Suaya, J.A, Cohen, M., Rosenthal, P. Economic appraisal of the Boston Ocular Surface Prosthesis. American Journal of Ophthalmology, Vol 148, No.6, (2009), pp. 860-868.

[183] Shimazaki, J., Aiba, M., Goto, E., Kato, N., Shimmura, S., and Tsubota, K. Transplantation of human limbal epithelium cultivated on amniotic membrane for the treatment of severe ocular surface disorders. Ophthalmology, Vol. 109, No. 7, (2002), pp. 1285-1290.

[184] Shimazaki, J., Shinozaki, N., Tsubota, K. Transplantation of amniotic membrane and limbal autograft for patients with recurrent pterygium associated with symblepharon. British Journal of Ophthalmology, Vol 82, No. 3, (1998), pp. 235–240.

[185] Short, A.J, Secker, G.A, Notara, M.D, et al. Transplantation of ex vivo cultured limbal epithelial stem cells: A review of techniques and clinical results. Survey of Ophthalmology, Vol. 52, No. 5, (2007), pp.483.

[186] Siriwardena, D., Khaw, P.T., King, A.J., Donaldson, M.L., et al. Human antitransforming growth factor β2 monoclonal antibody—a new modulator of wound healing in trabeculectomy: A randomized placebo controlled clinical study. Ophthalmology. Vol. 109, No. 3, (March 2002), pp 427-431.

[187] Sloper, C.M, Powell, R.J, Dua, H.S. Tacrolimus (FK506) in the management of high-risk corneal and limbal grafts. Ophthalmology, Vol. 108, No. 10, (2001), pp. 1838–1844.

[188] Solomon, A, Ellies, P, Anderson, D.F et al. Long-term outcome of keratolimbal allograft with or without penetrating keratoplasty for total limbal stem cell deficiency. Ophthalmology, Vol. 109, No. 6, (2002), pp. 1159–1166.

[189] Solomon, A., Espana, E.M, Tseng, S.C. Amniotic membrane transplantation for reconstruction of the conjunctival fornices. Ophthalmology, Vol. 110, No. 1, (2003), pp. 93–100.

[190] Solomon, A., Rosenblatt, M., Monroy, D., Ji, Z., Pflugfelder, S.C., Tseng, S.C.G. Suppression of interleukin 1α and interleukin 1β in human limbal epithelial cells cultured on the amniotic membrane stromal matrix. British Journal of Ophthalmology, Vol. 85, No.4, (2001), pp. 444–449.

[191] Stason, W.B, Razavi, M., Jacobs, D.S, Shepard, D.S, Suaya, J.A, Johns, L., Rosenthal, P. Clinical Benefits of the Boston Ocular Surface Prosthesis. American Journal of Ophthalmology, Vol. 49, No.1, (2010), pp. 54-61.

[192] Stern, M. The grafting of preserved amniotic membrane to burned and ulcerated surfaces, substituting skin grafts. Journal of the American Medical Association, Vol. 60, (1913), pp.973-4.

[193] Sundmacher, R., Reinhard, T. Central corneolimbal transplantation under systemic ciclosporin A cover for severe limbal stem cell insufficiency. Graefes Arch Clinical and Experimental Ophthalmology, Vol. 234, No. Supplement 1, (1996), pp. S122–S125.

[194] Szurman, P., Warga, M., Grisanti, S. et al. Sutureless amniotic membrane fixation using fibrin glue for ocular surface reconstruction in a rabbit model. Cornea. Vol. 25, No. 4, (2006), pp. 460–466.

[195] Takahide, K, Parker, P.M, Wu, M, Hwang, W.Y.K, Carpenter, P.A, Moravec, C., Stehr, B., Martin, P.J, Rosenthal, P, Forman, S.J, Flowers, M.E.D. Use of fluid-ventilated, gas-permeable scleral lens for management of severe keratoconjunctivitis sicca secondary to chronic graft-versus-host disease. Biology of Blood Marrow Transplant, Vol. 13, No.9 (2007), pp. 1016-1021.

[196] Tamhane, A., Vajpayee, R.B, Biswas, N.R et al. Evaluation of amniotic membrane transplantation as an adjunct to medical therapy as compared with medical therapy alone in acute ocular burns. Ophthalmology, Vol. 112, No. 11, (2005), pp.1963–1969.

[197] Tananuvat, N., Martin, T. The results of amniotic membrane transplantation for primary pterygium compared with conjunctival autograft. Cornea. Vol. 23, No.5, (2004), pp. 458 – 463.

[198] Tanioka, H., Kawasaki, S., Yamasaki, K., Ang, LPK, Koizumi, N., Nakamura, T., Yokoi, N., Komuro, A., Inatomi, T., Kinoshita, S. Establishment of a Cultivated Human Conjunctival Epithelium as an Alternative Tissue Source for Autologous Corneal Epithelial Transplantation. Investigative Ophthalmology and Vision Science, Vol. 47, No. 9, (September 2006), pp. 3820-3827.

[199] Tatlipinar, S., Akpek, E. Topical ciclosporin in the treatment of ocular surface disorders. British Journal of Ophthalmology, Vol. 89, No. 10, (2005), pp. 1363–1367.

[200] Thomas, P. B., Liu, Y., Zhuang, F.F., Selvam, S., Song, S. W., Smith, R.E., Trousdale, M.D., Yiu, S.C. Identification of Notch-1 expression in the limbal basal epithelium. Molecular Vision, Vol. 13, (2007), pp. 337-44.

[201] Thompson, J.E., Vaughan, T.J., Williams, A.J. et al. A fully human antibody neutralising biologically active human TGFbeta2 for use in therapy. Journal of Immunological Methods, Vol 227, No. 1-2, (1999), pp 17–29.

[202] Thylefors, B, Negrel, A.D, Pararajasegaram, R, Dadzie, K.Y. Global data on blindness. Bull World Health Organ, Vol. 73, No. 1, (1995), pp. 115-121.

[203] Tomita, M., Goto, H., Muramatsu, R., Usui, M. Treatment of large conjunctival nevus by resection and reconstruction using amniotic membrane. Graefes Archives Clinical Experimental Ophthalmology, Vol. 244, No. 6, (2006), pp.761–764.

[204] Torbet, J., Malbouyres, M., Builles, N. et al. Orthogonal scaffold of magnetically aligned collagen lamellae for corneal stroma reconstruction. Biomaterials, Vol. 28, No. 29, (2007), pp. 4268–4276.

[205] Tsai, R.J, Li, L.., Chen, J. Reconstruction of Damaged Corneas by Transplantation of Autologous Limbal Epithelial Cells. The New England Journal of Medicine, Vol. 343, No. 2, (2000), pp. 86-93.

[206] Tsai, R.J, Li, L.M, Chen, J.K. Reconstruction of damaged corneas by transplantation of autologous limbal epithelial cells. New England Journal of Medicine, Vol. 343, No. 2, (2000), pp. 86–93.

[207] Tseng, S.C, Di Pascuale, M.A, Liu, D.T, Gao, Y.Y, Baradaran-Rafii, A. Intraoperative mitomycin C and amniotic membrane transplantation for fornix reconstruction in severe cicatricial ocular surface diseases. Ophthalmology, Vol. 112, No. 5, (2005), pp. 896–903.

[208] Tseng, S.C., Prabhasawat, P., Barton, K., Gray, T., and Meller, D. Amniotic membrane transplantation with or without limbal allografts for corneal surface reconstruction in patients with limbal stem cell deficiency. Arch Ophthalmology, Vol. 116, (1998), pp. 431-441.

[209] Tseng, S.C.G, Chen, J.J.Y, Huang, A.J.W, Kruse, F.E, Maskin, S.L, Tsai, R.J.F. Classification of conjunctival surgeries for corneal diseases based on stem cell concept. Ophthalmic Clinical North American, Vol. 3, (1990), pp. 595-610.

[210] Tseng, S.C.G, Li, D.Q, Ma, X. Suppression of transforming growth factor-beta isoforms, TGF-beta receptor type II, and myofibroblast differentiation in cultured human corneal and limbal fibroblasts by amniotic membrane matrix. Journal of Cellular Physiology, Vol. 179, No.3, (1999), pp.325–335.

[211] Tsubota K, Shimmura S, Shinozaki N, Holland EJ, Shimazaki J. Clinical application of living-related conjunctival-limbal allograft. American Journal of Ophthalmology. Vol. 133, No. 1, (2002), pp. 134–135.

[212] Tsubota, K., Satake, Y., Kaido, M., et al. Treatment of severe ocular-surface disorders with corneal epithelial stem-cell transplantation. New England Journal of Medicine, Vol. 340, No. 22, (1999), pp.1697-1703.

[213] Tsubota, K., Satake, Y., Ohyama, M. et al. Surgical reconstruction of the ocular surface in advanced ocular cicatricial pemphigoid and Stevens-Johnson syndrome. American Journal of Ophthalmology, Vol. 122, No. 1, (1996), pp.38–52.

[214] Tsubota, K., Shimazaki, J. Surgical treatment of children blinded by Stevens-Johnson syndrome. American Journal of Ophthalmology, Vol. 128, No. 5, (1999), pp. 573–581.

[215] Uy, H.S, Chan, P.S, Ang, R.E. Topical bevacizumab and ocular surface neovascularization in patients with Stevens-Johnson syndrome. Cornea, Vol. 27, No. 1, (2008), pp. 70–73.

[216] Uy, H.S., Reyes, J.M.G.,Flores, J.D., Lim-Bon-Siong, R. Comparison of Fibrin Glue and Sutures for Attaching Conjunctival Autografts After Pterygium Excision. Ophthalmology, Vol. 112, No. 4 , (April 2005), pp. 667-671.

[217] Wei, Z.G, Sun, T.T, Lavker, R.M. Rabbit conjunctival and corneal cells belong to two separate lineages. Investigative Ophthalmology and Vision Science, Vol. 37, (1996), pp. 523-33.

[218] Weisbach, V., Dietrich, T., Kruse, F.E, Eckstein, R., Cursiefen, C. HIV and hepatitis B/C infections in patients donating blood for use as autologous serum eye drops. British Journal of Ophthalmology, Vol. 91, No. 12, (2007), pp. 1724–1725.

[219] Wenkel ,H. , Rummelt, V., Naumann, G.O. Long term results after autologous nasal mucosal transplantation in severe mucus deficiency syndromes. British Journal of Ophthalmology, Vol. 84, No.3, (March 2000), pp. 279-84.

[220] Whitcher, J.P., Srinivasan, M., and Upadhyay, M.D. Corneal blindness: a global perspective. Bull World Health Organization, Vol. 79, No.3, (2001), pp. 214-221.

[221] Whitcher, J.P., Srinivasan, M., and Upadhyay, M.P. Corneal blindness: a global perspective. British World Health Organization, Vol. 79, No.3, (2001), pp. 214-221.

[222] Wilson, M.C, Shields, M.B. A comparison of clinical variations of the iridocorneal endothelial syndrome. Arch Ophthalmology, Vol. 107, (1989), pp. 1465-1468.

[223] Wylegala, E., Tarnawska, D. Amniotic membrane transplantation with cauterization for keratoconus complicated by persistent hydrops in mentally retarded patients. Ophthalmology.Vol. 113, No.4, (2006), pp. 561–564.

[224] Xiao, Y., Liu, K., Shen, J., Xu, G., Ye, W. SB-431542 Inhibition of Scar Formation after Filtration Surgery and Its Potential Mechanism. Investigative Ophthalmology and Visual Science, Vol. 50, No. 4, (April 2009), pp.1698-1706.

[225] Yannas, I.V, Burke, J.F, Orgill, D.P, Skrabut, E.M. Wound tissue can use a polymeric template to synthesize a functional extension of skin. Science.Vol 215, No. 4529, (1971), pp 174-`176.

[226] Yannas, I.V, Lee, E., Orgill, D.P, Skrabut, E.M, Murphy, G.F. Synthesis and characterization of a model extracellular matrix that induces partial regeneration of adult mammalian skin. Proceedings of the National Academy of Sciences of the United States of America, Vol. 86, (February 1989), pp. 933–937.

[227] Yao, Y.F, Inoue, Y., Miyazaki, D., Shimomura, Y., Ohashi, Y., Tano, Y. Ocular resurfacing and alloepithelial rejection in a murine keratoepithelioplasty model. Investigative Ophthalmology and Vision Science, Vol. 36, No. 13, (1995), pp. 2623–2633.

[228] Yokoo, S., Yamagami, S., Usui, T., Amano, S., Araie, M. Human Corneal Epithelial Equivalents for Ocular Surface Reconstruction in a Complete Serum-Free Culture System without Unknown Factors. Investigative Ophthalmology and Vision Science, Vol.49, No. 6, (June 2008), pp.2438-2443.

[229] Yoon, I., Lee, D., Rah. Conjunctival Expansion Using a Subtenon's Silicone Implant in New Zealand White Rabbits. Yonsei Medical Journal. Vol 48, No 6, (December 2007), pp. 955-962.

[230] Young, A.L, Rao, S.K, Cheng, L.L, Wong, A.K, Leung, A.T, Lam, D.S. Combined intravenous pulse methylprednisolone and oral cyclosporine A in the treatment of corneal graft rejection: 5-year experience. Eye, Vol. 16, No.3, (2002), pp. 304–308.

[231] Zhang,C., Nie, X., Hu, D., Liu, Y, Deng, Z., Dong, R., Zhang, Y., Jin, Y.. Survival and integration of tissue-engineered corneal stroma in a model of corneal ulcer. Cell Tissue Research, Vol. 329, (2007), pp. 249–257.

[232] Zhou, S.Y, Chen, J.Q, Chen, L.S, Liu, Z.G, Huang, T., Wang, Z.C. Long-term results of amniotic membrane transplantation for conjunctival surface reconstruction [Chinese]. Zhonghua Yan Ke Za Zhi, Vol. 40, No. 11., (2004), pp. 745–749.

[233] Zhou, S.Y, Chen, J.Q, Chen, L.S, Liu, Z.G, Huang, T., Wang, Z.C. Long-term results of amniotic membrane transplantation for conjunctival surface reconstruction [Chinese]. Zhonghua Yan Ke Za Zhi,Vol 40, No. 11, (2004), pp. 745–749.

[234] Zhu, N., Warner, R., Simpson, C., Glover, M., Hernon, C., Kelly, J., Fraser, S., Brotherston, T., Ralston, D., and Macneil, S. Treatment of burns and chronic wounds using a new cell transfer dressing for delivery of autologous keratinocytes. European Journal of Plastic Surgery, Vol. 28, No. 5, (2005), pp.319-330.

[235] Zoukhri, D. Effect of inflammation on lacrimal gland function. Experimental Eye Research, Vol. 82, No. 5, (May 2006), pp. 885–898.

14

A Liquid Ventilator Prototype for Total Liquid Ventilation Preclinical Studies

Philippe Micheau et al.[*]
Université de Sherbrooke
Canada

1. Introduction

1.1 Context

Mechanical ventilation is a life-saving procedure used for treating acute respiratory distress, when the respiratory system is no longer capable of regulating blood gases via pulmonary gas exchange. While conventional mechanical ventilation (CMV) is often sufficient to transiently replace lung function until recovery, the most severe respiratory distress syndromes must be treated either by non conventional mechanical ventilation such as high frequency ventilation or even non ventilator strategies such as extracorporeal gas exchange (Raoof et al., 2010).

Large literature data suggest a radical change in ventilator support by replacing the traditional gas mixture with a breathable liquid. This method, called liquid assisted ventilation, leads to the replacement of the air-liquid interface in the alveoli by a liquid-liquid interface. Since the 70s, perfluorocarbon liquids (PFC) have been identified as the best candidates to be used in liquid ventilation due to their high oxygen and carbon dioxide solubility (Wolfson & Shaffer, 2005). In addition, they are biochemically stable and bio-inert molecules, available as medical grade products including for respiratory use. Liquid assisted ventilation can be performed either as partial or total liquid ventilation. During partial liquid ventilation, only a fraction of the lungs are filled with perfluorocarbon liquid and a conventional mechanical gas ventilator ensures lung ventilation. In contrast, during total liquid ventilation (TLV), the lungs are completely filled with perfluorocarbon liquid while a dedicated device, called a liquid ventilator, must be used to periodically renew a liquid tidal volume in the lungs. A large number of preclinical studies involving various animal models of acute respiratory distress syndrome have demonstrated clear benefits from total liquid ventilation as compared to all other tested ventilation strategies, including partial liquid ventilation, conventional and high frequency gas ventilation (Hirschl et al., 1996; Wolfson et al., 2008). Among its several theoretical advantages over CMV, TLV is considered less aggressive for the lungs, due to lower positive inspiratory pressures and lower respiratory rates. This is felt to be beneficial in both pediatric and adult respiratory distress syndromes, where repeated alveolar overdistension during CMV contributes to

---

[*]Raymond Robert, Benoit Beaudry, Alexandre Beaulieu, Mathieu Nadeau, Olivier Avoine, Marie-Eve Rochon, Jean-Paul Praud and Hervé Walti
Université de Sherbrooke, Canada

acute and chronic lung injury (Chan et al., 2007; Hayes et al., 2010; Speer, 2009). Moreover, it offers a new means to clean the lung of inflammatory debris (Richman et al., 1992; Foust et al., 1996 Avoine et al., 2011). Consequently, a round table discussion of experts in liquid ventilation has unanimously recommended that a liquid ventilator must be developed for clinical applications (Costantino et al., 2009).

1.2 Problem

In TLV, minute ventilation has been previously reported to be a significant limiting factor for gas exchange (Bull et al., 2009; Matsuda et al., 2003), due to choked flows during expiration, which severely impede lung emptying and greatly decrease tidal volume (Koen et al., 1988; Baba et al., 2004; Robert et al., 2009). Even if the mechanical stresses of the collapses on the airway are not significant (Bagnoli et al., 2007), a decrease in minute ventilation can affect arterial blood gases. Previously, we have shown that the use of a pressure controlled mode prevents choked flows (Robert et al., 2010) and allows maximizing V_{min}. The location of the pressure sensor in the trachea is motivated by the fact that the expiratory collapses can occur in the first airway (Robert et al., 2009) or close to the carina (Bull et al., 2005). Hence, the pressure controlled mode maintains a constant acceptable negative pressure in the trachea during the expiratory flow, such that the limiting phenomena of collapse can be avoided (Robert et al., 2010). It is important to note that such control mode is new in TLV because all previously published liquid ventilator prototypes were controlled in volume (Baba et al., 1996; Corno et al., 2003; Hirschl et al., 1995; Polhmann et al., 2011; Sekins et al., 1999). However, the insertion of a pressure sensor in the mouth can be problematic in a clinical context, thus the main issue is to adapt this pressure controlled mode without a pressure sensor in the trachea.

Many efficient demonstrations of TLV have been performed with different prototypes (Corno et al. 2003; Cox et al. 2003; Degraeuwe et al., 2000a; Hirschl et al., 1995; Larrabe et al., 2001; Meinhardt et al., 2000; Parker et al., 2009; Pohlmann et al., 2011; Sekins et al., 1999; Tredici et al., 2004). With O_2 saturated PFC and limited to low minute ventilations (comparatively to CMV), they obtained acceptable ventilation. However, the use of O_2 saturated PFC has never been questioned, despite the fact that high concentrations of O_2 in the lungs are a well-known factor contributing to acute and chronic lung injury (Hayes et al. 2010; Speer 2009; Tasake et al., 2008). In addition, the use of O_2-saturated PFC during TLV often results in hyperoxia. This is particularly deleterious in neonatal medicine in which even a transitory hyperoxia can result in so-called neonatal oxygen radical disease. This disease affects multiple organs including lung, retina, gut and brain and results in acute and chronic morbidities with potential lifelong consequences (Bitterman, 2009; Dorfman et al., 2010; Gitto et al., 2009). Hence, an efficient liquid ventilator with a pressure regulated mode capable of reaching high V_{min} must include a device to control the oxygenation.

1.3 Objective

For the liquid ventilator prototype to reach the readiness level for clinical applications, the objective is dual. From an engineering perspective, it must ensure both targeted minute ventilation and oxygenation. From a medical perspective, it must resemble other conventional ventilators to be operated by clinicians in intensive care units.

Consequently, two distinct objectives are targeted:

i. Implement a pressure regulation mode, but with a pressure sensor located at the mouth and not in the trachea. Moreover, the proposition is to translate what has been done in

CMV to liquid ventilators: inspiration and expiration should be controlled in pressure (Simon et al., 2000). Finally, a control algorithm must prevent large lung volume which may compromise the cardiovascular system or induce a perfluorothorax (rupture of the lungs).

ii. Control the O_2 concentration in the inspired PFC, similarly to what is done in CMV with the control of oxygen concentration in inspired air (FiO_2).

2. Description of Inolivent-4

2.1 The prototype Inolivent-4

Following years of research in liquid ventilator development for animal experiments, this study presents the most advanced prototype of liquid ventilator, Inolivent-4 (figure 1). Its design is based on fundamental concepts developed for our third prototype, Inolivent-3 (Robert et al., 2006) and by including the most recent knowledge on flow dynamics in liquid ventilation (Bossé et al., 2010). Inolivent-3 comprises two independent piston pumps and an oxygenator unit which regroups the heating system, buffer reservoir and condenser (used to recuperate the PFC vapors emanating from the oxygenator columns). These concepts have demonstrated their efficiency during animal experiments and were maintained for our following prototype, Inolivent-4. The latter, compared to its predecessors, includes different ventilation control modes like those found on conventional ventilators (Robert et al., 2007b; Robert et al., 2010). From a clinician's point of view, volume controlled ventilation (VCV) during inspiration and pressure controlled ventilation (PCV) during expiration, greatly simplifies the use of liquid ventilator and minimizes the risk of incorrect ventilation parameter selection which could induce systematic airway collapses and lead to oxygen deprivation.

Parameter	Description	Unit
Fr	Respiratory frequency	RPM
V_t	Tidal volume	ml
$F_{gas}O_2$	Fraction of oxygen in the gas bubbled	%
T_i	Inspiration time	s
T_{eip}	End inspiration pause time	%
T_{eep}	End expiration pause time	%
$P_{ref,e}$	Expiration reference pressure	cmH$_2$O
$P_{ref,i}$	Inspiration reference pressure	cmH$_2$O
$P_{LIM,H}$	Upper inspiration limit pressure	cmH$_2$O
$P_{LIM,L}$	Lower expiration limit pressure	cmH$_2$O
$PEEP_{ref}$	Reference PEEP	cmH$_2$O

Table 1. Ventilation parameters available on the user interface.

The table 1 presents the basic ventilator parameters accessible on the touch screen user interface. A schematic of the ventilator circuit (figure 1) represents the PFC flow path. Its operation can be described by considering a typical liquid ventilation cycle. The cycle starts when the valve 1 is opened and the valve 2 is closed; a tidal volume of PFC is pumped from the buffer reservoir by the inspiration pump. When the inspiratory pump is ready, the valve 1 is closed, the valve 2 is opened. Then, the inspiratory pump pushes the tidal volume of PFC into the lungs. When the active inspiration phase is completed, the valve 2 is closed.

The inspiration pause is when both valves 2 and 4 are closed. Valve 4 is then opened and the active expiration starts: the expiratory pump removes the tidal volume of PFC from the lungs. Finally, the valve 4 is closed to operate the expiration pause. Simultaneously, the valve 3 is opened and the expiration pump pushes the PFC inside the filter which goes to the oxygenator. By overflow, the oxygenated PFC travels to the buffer reservoir. The time parameters (Fr and Ti) and the tidal volume serve as cycle limits. The end inspiration and expiration pauses are used to measure the positive-end inspiratory pressure (PEIP) and positive-end expiratory pressure (PEEP). The duration is proportional (in percentage) to the length of each phase. The upper and lower pressure limits stop the corresponding phase, in case the airway pressure goes beyond these values. The reference PEEP is used by the PEEP controller to correct the inspired or expired volume (see section *PEEP controller*).

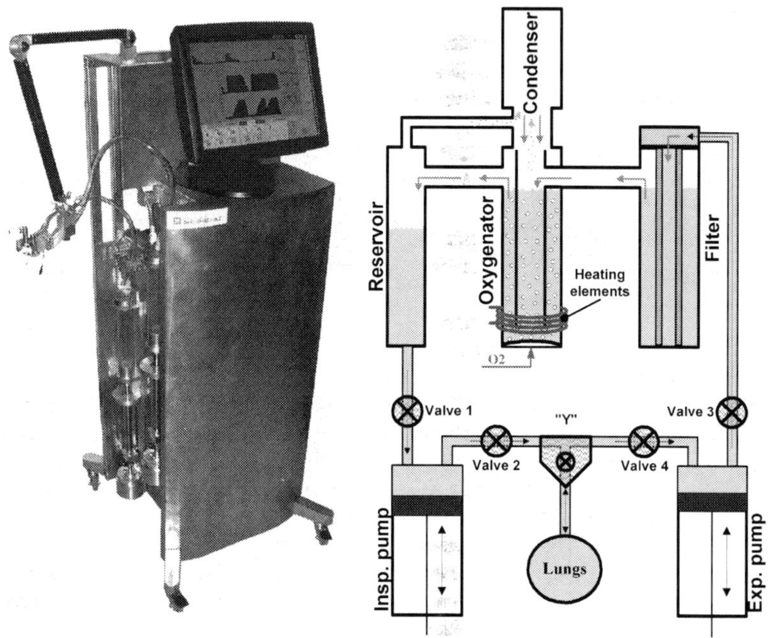

Fig. 1. Picture of Inolivent-4 (left) and the schematic of the ventilator circuit (right).

The "Y" piece (figure 2) is connected to the endotracheal tube (ET). It includes a mechanical 3 way valve to select the CMV port, the TLV circuit or the closed position. An airway pressure sensor can be located in the ET tube (via an epidural catheter inserted into the Y-piece) to monitor the tracheal pressure. A pressure sensor located in the Y-piece is used to measure the mouth pressure (Py). The new pressure controller regulates Py both during the inspiration and expiration phases.

All actuators and sensors are connected to a real-time control unit composed of a PC with two analog input boards (PCI-DAS1602, Measurement Computing, USA) and one analog output board (PCI-DAC6703, Measurement Computing, USA). The user interface is a touch screen PC which communicates with the real-time control unit.

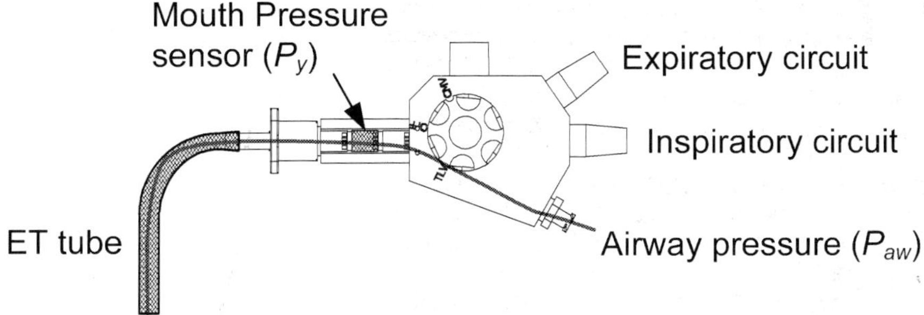

Fig. 2. "Y"-piece connected to the endotracheal tube.

2.2 Oxygenator description

Research teams in TLV have mainly used bubble oxygenators (Parker et al., 2009; Sekins et al., 1999; Tissier et al., 2009), commercial membrane oxygenators (Corno et al. 2003; Larrabe et al., 2001; Hirschl et al., 1995; Wolfson et al., 1999, 2008) or have developed membrane oxygenators specifically for TLV (Tredici et al., 2004). However, the requirements for extracorporeal blood oxygenation appear irrelevant in TLV, since there is no mechanical stress to the blood (hemolysis) or bubble infusion in the blood (Iwahashi et al., 2004). Hence, a bubble oxygenator seems acceptable for a clinical use.

The bubble oxygenator of Inolivent-4 is composed of 2 translucent, polycarbonate cylinders, which communicate with each other at the bottom of the inner cylinder (figure 3). When the PFC retrieved from the lung is pumped into the inner cylinder, an equivalent volume of PFC overflows into the outer cylinder and to the buffer reservoir. In this manner, the PFC coming from the lungs is not in direct contact with the PFC going to the lungs. Therefore the PFC residence time in the oxygenator is maximized. The gas to be bubbled flows through a perforated santoprene™ rubber membrane (McMaster, USA) at the bottom of the oxygenator. The latter has approximately 470 perforations made with a 1.2 mm diameter needle, (all equally spaced with a rounded pattern over the entire membrane surface) which generate bubbles into both cylinders (Beaudry, 2009). The stainless steel base contains three 100-watt cartridge heaters (Watlow, St-Louis, USA), which maintain the PFC at the targeted temperature. A condenser on top of the oxygenator liquefies the PFC vapors to minimize evaporative losses.

2.3 Oxygen concentration control

To control the O_2 concentration in the bubbled gas ($F_{gas}O_2$), medical air is blended with O_2 using a gas mixer system. The latter is composed of two proportional valves (ET-P-10-6025, Clippard, USA) controlled by two solenoid drives (B5950, Canfield Connector, USA). Each proportional valve controls the air and O_2 flow, measured by distinct flowmeter (41211, 41212, TSI, USA). The reference flow for each bubbled gas is computed based on the required $F_{gas}O_2$ and on the total gas flow requested by the user. These references and the measured flows are used by two separate feedback controllers (one for both gases), which were programmed using Simulink (MathWorks, MA, USA) and implemented with xPC target (MathWorks, MA, USA). An O_2 gas sensor (KE-25, Figaro, Japan) measures the $F_{gas}O_2$, allowing the latter to be modulated from 21% to 100%. Pressure regulators (MMR-1N-P60,

Clippard, USA) decrease line pressure while gas filters (C-02917-00, Parker, USA) remove unwanted particles, which could impair flowmeter measurements.

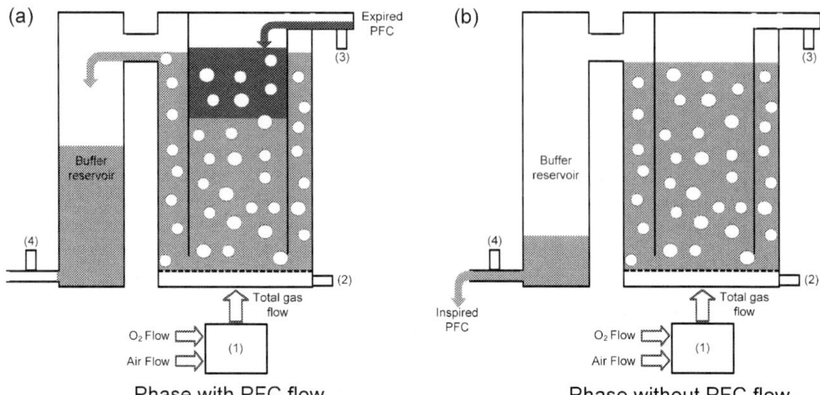

Fig. 3. Sketch of the oxygenator and the buffer reservoir with PFC flow (left) and without PFC flow (right). (1) Gas mixer system with O_2 gas sensor to measure $F_{gas}O_2$, (2) gas mixing chamber with a rubber membrane to generate bubbles, (3) circuit of the expired PFC with O_2 liquid sensor, (4) circuit of the expired PFC with O_2 liquid sensor, (a) when the expired PFC is inserted there is an overflow of PFC from the oxygenator to the buffer reservoir, (b) the inspired PFC is pumped from the buffer reservoir

In order to measure oxygen fraction concentration, a fluorescence sensor (Fibox 3 LCD, PreSens Precision Sensing GmbH, Germany) was used to determine the O_2 fraction in the PFC liquid, at the inspiratory ($Fi_{PFC}O_2$) and expiratory ($Fe_{PFC}O_2$) circuit alternately. Data were recorded with the software provided by PreSens (LCDPST3 V1.16, PreSens Precision Sensing GmbH, Germany). The 90th percentile response time of the sensor is 40 s and was measured in-vitro in our laboratory (Beaudry, 2009).

2.4 The pressure controller
2.4.1 The targeted pressure
The pressure controller commands the pumps during the inspiration and expiration to track desired pressure (figure 4) measured at the mouth, with the pressure sensor in the "Y-piece".

Before the inspiration, the airway pressure is equal to the PEEP which depends on the end-expiratory lung volume (V_{EELV}). At the start of the inspiration, the pressure (P_y) must increase from the PEEP level to the inspiratory reference pressure ($P_{ref,i}$) in less than 1 second. The pressure regulator regulates the inspiratory flow to maintain P_y within ± 1 cmH₂O of $P_{ref,i}$. When the tidal volume $V_{t,i}$ is completely inserted in the lungs or when the inspiration time T_i is reached, the inspiration is stopped. During the end inspiratory pause, the airway pressure is equal to the PEIP which depends on the end-inspiratory lung volume (V_{EILV}).

At the start of the expiration, P_y must decrease from the PEIP to the expiratory reference pressure ($P_{ref,e}$). The pressure regulator controls the expiratory flow which maintains P_y

within $\pm 1\,cmH_2O$ of $(P_{ref,e})$. When the tidal volume $V_{t,e}$ is completely removed from the lungs or when the expiratory time T_e is reached, the expiration is stopped. During the end-expiratory pause, the airway pressure must reach the desired PEEP, called $PEEP_{ref}$. If not, modifications are done on the volume to be inspired or expired in order to correct the PEEP. Those modifications are made by the PEEP follower.

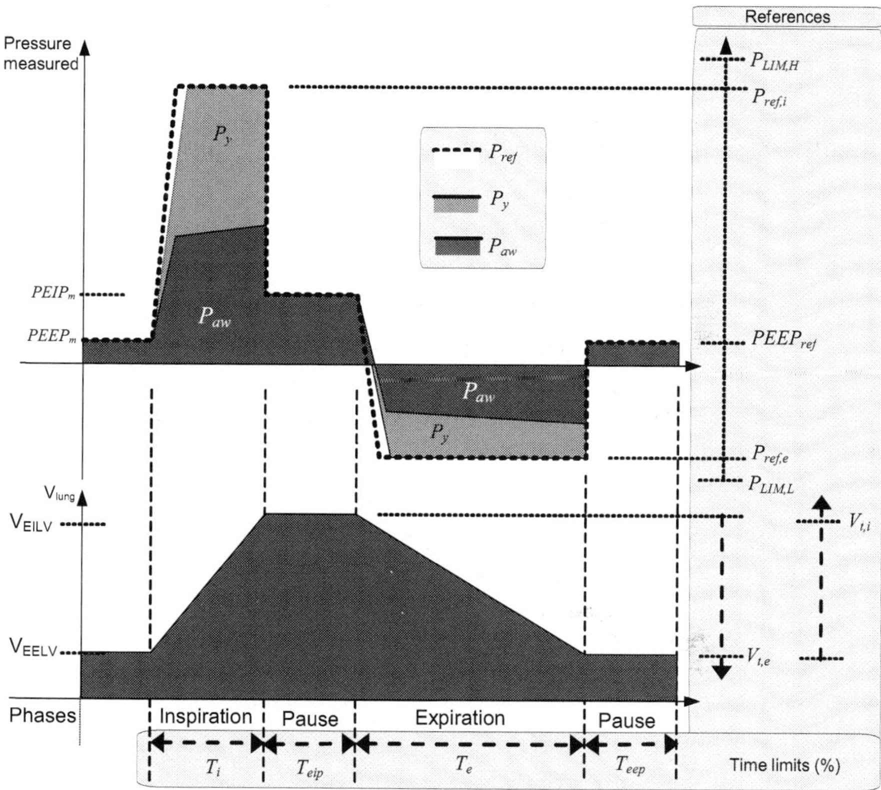

Fig. 4. Reference respiratory cycle in pressure regulated mode

2.4.2 The controller

Based on the work presented in a previous publication (Robert et al., 2010), the objective is to insert or remove the PFC liquid from the lungs at a specified pressure reference, $P_{ref,e}$ during expiration phase and $P_{ref,i}$ during inspiration. In control terms, the problem is to implement a feedback loop to command the pump, U, as a function of the error between the pressure reference, P_{ref}, and the Y pressure P_y as illustrated in figure 5.

The design of the pressure regulator for the inspiration and expiration was done by loop shaping, using the robust design toolbox of Matlab (Mathworks, USA). The pressure regulator bandwidths were set at 2 Hz. The controller design follows a pole placement procedure where the process plant *P(s)* with a regulator *C(s)* leads to an open loop transfer function, chosen to be an integrator with a low pass filter at 100 rad/s.

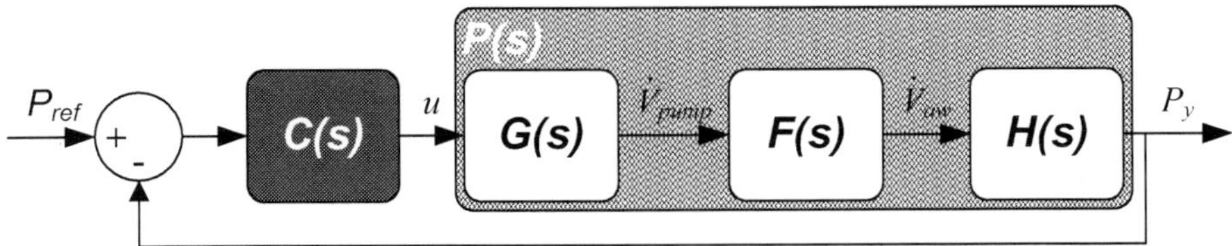

Fig. 5. Block diagram of the control system. C: Controller; G: Pump model; F: Tube dynamic; H: Lung model; P: Plant.

The robust control toolbox gave a lag filter coupled with a notch filter. The lag filter cancels the pole and zero of the motor, lungs and ET tube. The notch filter compensates for the tube resonance. Thus, the form of the controller is;

$$C(s) = K_c \left(\frac{1 + \tau_c s}{1 + \beta \tau_c s} \right) K_f^2 \left(\frac{s^2 + 2/c\omega_f s + \omega_f^2}{s^2 + 2K_f d/c\omega_f s + K_f^2 \omega_f^2} \right) \tag{1}$$

Where β and τ_c cancel the pole and zero of the lung, motor and ET tube, K_c is set to achieve the desired bandwidth, the gain K_f adjusts the filter gain in higher frequency (2.0), ω_f is the notch filter frequency (15 rad·s-1), c and d are the filter parameters (5.5). Gain scheduling was done in order to take into consideration all the ET tube diameters (from 4 to 6 mm) and different fluid properties (PFOB and PFDEC).

2.5 The PEEP controller
2.5.1 The specifications
The PEEP controller acts on the inspiration and expiration tidal volumes from cycle to cycle. It monitors the volumes inspired, the volume expired and the pressure measured during the end-expiration pause (PEEP). All these measures coupled with the PEEP reference set by the clinician ($PEEP_{ref}$), are used by the PEEP controller to adjust from cycle to cycle the volume pumped. In the end, the lung volume is corrected and the reference PEEP is tracked.

It is important to note the difference between the PEEP and the $P_{ref,e}$. In conventional mechanical ventilation, the PEEP set on the user interface is the pressure reference maintained in the airway by the pressure regulator. In TLV, the $P_{ref,e}$ is much lower than the PEEP in order to maximize the ventilation frequency. Since the PEEP is a clinical setting on all mechanical ventilators, a control scheme was developed for the liquid ventilator

2.5.2 The algorithm
PEEP regulation can be obtained by controlling the end-expiratory lung volume V_{EELV} with a supervisor (Robert et al., 2007a). Cycle after cycle, the balance between inserted and retrieved PFC volume can be used to increase or decrease V_{EELV}. The PEEP estimate takes into consideration the pumping errors that could occur during the expiration. The PEEP controller removes these volume errors from the PEEP measured during the expiratory pause ($PEEP_m$), since it is corrected by the supervisor.

As in a conventional mechanical ventilator, the clinician specifies a required PEEP ($PEEP_{ref}$). This value is compared to the PEEP estimated ($PEEP_{est}$) during the end-expiratory pause of the kth cycle. The step of increasing or decreasing V_{EELV} (the volume correction ΔV) is proportional to the PEEP error (limited to avoid too large variations and/or instabilities) and expiratory volume error E_e:

$$\Delta V[k] = K_{PEEP}\left(PEEP_{est}[k] - PEEP_{ref}[k]\right) + E_e[k] \tag{2}$$

A limit is imposed on the estimated PEEP variation but not on the expiratory volume error. The equation 3 is used to estimate the PEEP.

$$PEEP_{est}[k] = PEEP_m[k] - \frac{E_e[k]}{K_{PEEP}} \tag{3}$$

Where K_{Peep} is a gain and $E_e[k]$ is the volume error measured during the expiration. Theoretically, the product of lung elastance $E_L = C_L^{-1}$ ($cmH_2O \cdot ml^{-1}$) and expiratory volume error E_e provides an exact estimate of the PEEP. However, for improved robustness, K_{Peep} was fixed. If the volume correction is negative, the volume to be inspired for the next cycle ($V_{t,i}[k+1]$) is lower than the tidal volume V_t in order to increase the lung volume

$$V_{t,i}[k+1] = \begin{cases} V_t & if \quad \Delta V \geq 0 \\ V_t + \Delta V[k] & if \quad \Delta V < 0 \end{cases} \tag{4}$$

If the volume correction is positive, the volume to be expired for the next cycle is lower than the tidal volume V_t in order to decrease the lung volume

$$V_{t,e}[k+1] = \begin{cases} V_t - E_i[k+1] & if \quad \Delta V \leq 0 \\ V_t - E_i[k+1] - \Delta V[k] & if \quad \Delta V > 0 \end{cases} \tag{5}$$

Where $E_i[k+1]$ is the inspiratory volume error for the $[k+1]^{th}$ cycle.

3. Validation of the oxygenator

3.1 Oxygenator time-constant measurement

The oxygenator is modeled as a first order system characterized by a time constant, τ. To measure the oxygenator time constant, the method consisted to perform a step time test: $F_{PFC}O_2(t)$ was recorded when the oxygen concentration in the bubbled gas is abruptly switched from $F_{gas}O_2(t)=0$ for $t \leq 0$ to $F_{gas}O_2(t)=100\%$ for $t>0$, at null initial condition ($F_{PFC}O_2(0)=0\%$). To generate this system excitation, pure CO_2 was first bubbled in order to retrieve all dissolved O_2 from the PFC. Once $P_{PFC}O_2$ reached less than 5 mmHg, the gas brought under the membrane was switched to pure O_2 at a fixed flow rate. The $P_{PFC}O_2$ was recorded at a sampling time of 1 second (figure 6 presents a typical measured time response) to compute off-line the $F_{PFC}O_2$.

The data was fed to the System Identification Toolbox of Matlab (Mathworks, USA) to identify the time constant values (by selecting a first order linear system) and its 95% confidence interval. The procedure was repeated for different oxygen flows by using incremental values of 0.5, 1, 2, 4, 6 and 8 L/min at a PFC temperature of 39°C. As can be seen in the table 2, the smallest time constant was 14 seconds at an O_2 flow of 8 L/min. Thus, in the worst case scenario when there is no oxygen in the PFC, the concentration reaches 95% of the $F_{gas}O_2$ after 42 seconds. Fortunately, this case will not happen, since the expired PFC will always contain a certain amount of oxygen. A theoretical development can mathematically quantify this fact with the index α.

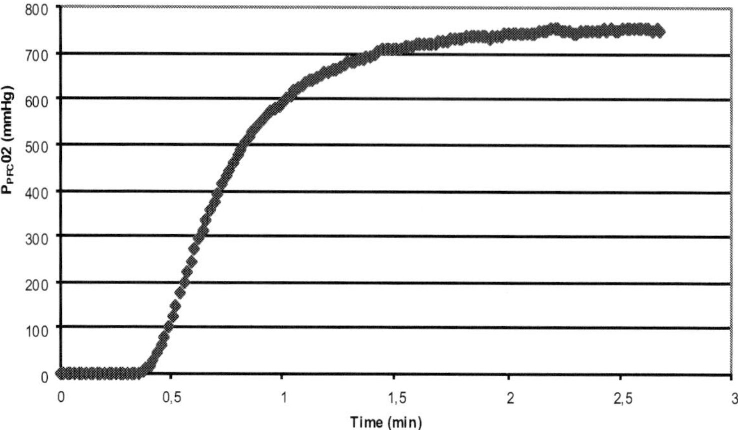

Fig. 6. Typical O_2 partial pressure variation measured in the PFC versus time at a flow rate of 6 l/min.

O_2 flow (L/min)	τ (s)	α (%)
0.5	167	38
1	77	58
2	41	73
4	22	85
6	17	89
8	14	91

Table 2. Identified time constants (in seconds) for different O_2 flow rates (L/min) at 39°C. The 95% confidence interval was equal or inferior to 1 second for all values. The α values are computed with equation (9) for t_c=10s and a=10%.

3.2 An index to evaluate the oxygenator efficiency

To quantify oxygenator efficiency throughout all of the experiments, the following index is computed:

$$\alpha = \left(Fi_{PFC}O_2 - Fe_{PFC}O_2 \right) / \left(F_{gas}O_2 - Fe_{PFC}O_2 \right) \tag{6}$$

An α equal to 0 signifies an inefficient oxygenator since $Fi_{PFC}O_2 = Fe_{PFC}O_2$. Conversely, an α equal to 1 signifies a perfect oxygenator since $Fi_{PFC}O_2 = F_{gas}O_2$.

Considering the dynamic response of the oxygenator as a first order system, it is possible to write the index of efficiency (6) as a function of the time constant and the ventilator parameters. In the following demonstration, the initial time refers to the instant when the tidal volume V_t is expired from the lungs and mixed with the PFC volume already present in the oxygenator. Considering the oxygenator as a first order system characterized by a time constant, τ, the initial oxygen concentration is assumed equal to the expired PFC mixed proportionally with the PFC volume in the oxygenator. Thus,

$$F_{PFC}O_2(0) = aFe_{PFC}O_2 + (1-a)Fi_{PFC}O_2 \qquad (7)$$

where $a=V_t/V_0$ is the volume ratio, V_0 is the PFC oxygenator volume and the PFC tidal volume, V_t. The final cycle time, $t=t_c$, is when a V_t goes through the oxygenator to the buffer reservoir by overflow (when the tidal volume of PFC expired from the lungs is pushed into the oxygenator). Since the PFC in the buffer reservoir will be pumped to the lungs during the next cycle, the inspired oxygen concentration $Fi_{PFC}O_2=F_{PFC}O_2(t_c)$. Considering these assumptions, it is obvious that:

$$Fi_{PFC}O_2 = \alpha F_{gas}O_2 + (1-\alpha)Fe_{PFC}O_2 \qquad (8)$$

Where the index defined by (6) is:

$$\alpha = \frac{1-\exp(-\frac{t_c}{\tau})}{1-\exp\left(-\frac{t_c}{\tau}\right)(1-a)} \qquad (9)$$

An α equal to 0 occurs if the cycle time t_c is much smaller than the time constant ($t_c/\tau \to 0$). On the other hand, an α equal to 1 signifies a perfect oxygenator since $Fi_{PFC}O_2=F_{gas}O_2$. This case occurs if the tidal volume is smaller than the oxygenator volume ($V_t/V_0 \to 0$). Finally, an α superior to 90% indicates that the oxygenator is almost independent of expired PFC gas concentration since the contribution of $Fe_{PFC}O_2$ to $Fi_{PFC}O_2$ is inferior to 10%.

3.3 In-vivo validation of the oxygenator
3.3.1 In-vivo protocol
The experimental protocol was approved by our institutional Ethics Committee for Animal Care and Experimentation. For the in vivo protocol, five healthy newborn lambs (<4 days old, 2.5-3.6 kg) were intubated, anaesthetized and paralyzed as detailed previously (Robert et al., 2010). The lambs were premedicated by intramuscular injection, orally intubated with a 5.0 or 5.5 mm cuffed endotracheal tube (Mallinckrodt, St. Louis, MO) and restrained in supine position under radiant heat to maintain a central temperature of 39±1°C. The lambs were ventilated with a conventional mechanical ventilator (Servo 300 ventilator, Siemens-Elema AB, Solna, Sweden) in pressure-regulated, volume controlled mode (positive end-expiratory pressure (PEEP) = 4 cmH$_2$O, V_t = 10 ml/kg, FiO_2 = 100%, respiratory frequency (Fr) = 50 breaths/minute, inspiratory time on expiratory time ratio (I:E) = 1:2). The baseline parameters were recorded after 30 minutes. Inolivent-4 was then used to initiate and perform TLV. Lungs were first filled (25 ml/kg) with warmed (39.0 ± 0.5°C) and pre-oxygenated perfluorodecalin with a PEEP of 7 cmH$_2$O. TLV was then initiated (Fr =

5.35/minute, V_t = 20 ml/kg, I:E = 1:3, $F_{gas}O_2$ = 100%) (Robert et al. 2010). Fifteen minutes after reaching each new setting, the PFC partial pressure of oxygen, $P_{PFC}O_{2,}$ was measured with the liquid sensor (Fibox 3 LCD, PreSens, Germany) in both the expired ($Pe_{PFC}O_2$) and inspired ($Pi_{PFC}O_2$) PFC. An arterial blood sample was also drawn for PaO_2 and $PaCO_2$ measurements.

Three different settings were tested using the following parameters. First a reduction in minute ventilation (V_{min}) from 180 to 120 ml/min/kg at an $F_{gas}O_2$ = 100% and constant gas flow of 6 l/min. Second, a reduction in gas flow from 8 to 0.5 l/min at an $F_{gas}O_2$ = 100% and constant V_{min} = 180 ml/min/kg. Finally, a reduction in $F_{gas}O_2$ from 100 to 65% at constant gas flow of 8 l/min and V_{min} = 180 ml/min/kg.

Prior to each experiment, the maximum O_2 partial pressure value in the PFC, $P_{PFC}O_{2,100\%}$, was noted in order to calculate both the O_2 concentration in the inspired PFC ($Fi_{PFC}O_2=Pi_{PFC}O_2/P_{PFC}O_{2,100\%}$) and the O_2 concentration in the expired PFC ($Fe_{PFC}O_2=Pe_{PFC}O_2/P_{PFC}O_{2,100\%}$). This value was obtained by bubbling pure O_2 in the oxygenator, during the TLV preparation phase (which is a closed-loop circulation of the PFC in the liquid ventilator). After 10 minutes, the measured O_2 partial pressure in the PFC was invariant and close to the atmospheric pressure. This measured value was noted as $P_{PFC}O_{2,100\%}$. The atmospheric pressure was measured with a precision dial barometer (4199, Control Company, USA).

Statistical values were obtained with the Statistics Toolbox provided with Matlab (Mathworks, USA). Statistical significance was assumed at p < 0.05. The Shapiro-Wilk test was first applied to each group to verify sample normality. The paired Student-t test was used to test a significant difference between two paired groups, after verification that homoscedacity was respected with the two samples F-test. The relationship between PaO_2 and $Fi_{PFC}O_2$ was established by using linear regression statistics (Seber, 1989) in order to compute the regression coefficients R^2. An F-test was finally performed on the slope.

3.3.2 In-vivo results

Upon transfer from conventional gas ventilation (V_{min} =602 ± 50 ml/min/kg; FiO_2=100%) to total liquid ventilation (V_{min}=180 ± 3 ml/min/kg; $F_{gas}O_2$=100%), blood gases were not significantly different (CMV: PaO_2=285 ± 87 mmHg, $PaCO_2$=39 ± 3 mmHg, pH=7.35 ± 0.05; TLV: PaO_2 = 217 ± 76 mmHg, $PaCO_2$=43 ± 5 mmHg, pH=7.25 ± 0.07).

Decreasing pure O_2 flow from 8 to 4 l/min (table 3) did not lead to a significant decrease in $Fi_{PFC}O_2$ and $Fe_{PFC}O_2$, the oxygenator efficiency index remaining in the range 80 to 73%. On the contrary, a significant decrease in $Fi_{PFC}O_2$ was observed for O_2 flow below 4 l/min. Despite this decrease in $Fi_{PFC}O_2$, no significant decrease in PaO_2 was observed for all O_2 flows tested and PaO_2 was maintained above 100 mmHg. Meanwhile, $PaCO_2$ and pH were significantly altered for pure O_2 flow at 0.5 l/min. Finally, the oxygenator efficiency index dramatically dropped at 54% for O_2 flow at 0.5 l/min, showing the inability of the oxygenator to saturate the PFC with O_2 at this flow level.

When $F_{gas}O_2$ was reduced from 100 to 65% (table 4), $Fi_{PFC}O_2$ decreased from 96 to 64% and $Fe_{PFC}O_2$ decreased from 80% to 31%. Between each decrement of $F_{gas}O_2$, significant differences were observed both in $Fi_{PFC}O_2$ and $Fe_{PFC}O_2$, although no significant differences were observed in oxygenator efficiency, which remained in the range of 96 to 80%. No significant differences were observed for either $PaCO_2$ or pH with the various $F_{gas}O_2$ tested. As expected, PaO_2 simultaneously decreased from 217 to 99 mmHg, a significant correlation being observed between PaO_2 and $Fi_{PFC}O_2$ (and the oxygenator efficiency index) in the 5 newborn lambs tested (Slope = 350 mmHg with R^2=0.37, Fisher test p<0.05).

	O_2 gas flow (l/min)					
	8	**6**	**4**	**2**	**1[1]**	**0.5[1]**
$Fi_{PFC}O_2$ (%)	96±1	96±1	95±2	94±1	93±2	88±2
$Fe_{PFC}O_2$ (%)	80±2	80±2	80±3	78±3	77±3	74±4
α (%)	80±7	77±4	77±9	73±7	69±7	54±7
PaO_2 (mmHg)	217±76	234±64	213±57	232±54	188±59	177±41[2]
$PaCO_2$ (mmHg)	43±5	41±3	44±5	47±4	50±6	60±10
pH	7.25±0.07	7.28±0.05	7.26±0.06	7.26±0.05	7.22±0.06	7.14±0.06

Table 3. In vivo results obtained in 5 newborn lambs while varying gas flow and maintaining $F_{gas}O_2$ at 100% and V_{min}=180 ml/min/kg. Data are presented as mean ± SD with underbraces indicating statistical significance (p<0.05). $Fi_{PFC}O_2$, oxygen fraction of inspired PFC; $Fe_{PFC}O_2$, oxygen fraction of expired PFC; α, oxygenator efficiency index; PaO_2, partial pressure of oxygen in arterial blood; $PaCO_2$, partial pressure of CO_2 in arterial blood; pH, arterial blood pH. [1]Data obtained with 4 lambs. [2]Measurements did not pass the Shapiro-Wilk normality test and were excluded from the significance test

	$F_{gas}O_2$ (%)					
	100	**90**	**80**	**75[1]**	**70[1]**	**65[1]**
$Fi_{PFC}O_2$ (%)	96 ± 1	89 ± 2	78 ± 1	74 ± 1	69 ± 1	64 ± 1
$Fe_{PFC}O_2$ (%)	80 ± 2	72 ± 5	61 ± 3	57 ± 3	52 ± 2	46 ± 4
α (%)	80 ± 7	96 ± 13	91 ± 7	93 ± 5	95 ± 5	95 ± 4
PaO_2 (mmHg)	217 ± 76	153 ± 53	133 ± 63	149 ± 53	100 ± 39	99 ± 47
$PaCO_2$ (mmHg)	43 ± 5	45 ± 6	44 ± 3[2]	43 ± 4	44 ± 5	45 ± 3
pH	7.25±0.07	7.23±0.07	7.25±0.05	7.25±0.06	7.26±0.06	7.25±0.06

Table 4. In vivo results obtained in 5 newborn lambs while varying $F_{gas}O_2$ at a constant gas flow of 8 l/min and V_{min} =180 ml/min/kg. Data are presented as mean ± SD with underbraces indicating statistical significance (p<0.05). V_{min}, minute ventilation; $Fi_{PFC}O_2$, oxygen fraction of inspired PFC; $Fe_{PFC}O_2$, oxygen fraction of expired PFC; a, oxygenator efficiency index; PaO_2, Partial pressure of oxygen in arterial blood; $PaCO_2$, Partial pressure of CO_2 in arterial blood; pH, arterial blood pH. [1]Data obtained with 4 lambs. [2]Measurements did not pass the Shapiro-Wilk normality test and were excluded from the significance test.

3.4 Discussion regarding the oxygenator

Results in the present and previous (Avoine et al., 2011; Beaudry 2009; Robert et al. 2010) studies show that our custom-designed bubble oxygenator can oxygenate PFC at a level up to $Fi_{PFC}O_2$=96%, which is similar to previously published data obtained with a silicone hollow fiber membrane oxygenator in rabbits (93%, when considering a standard atmospheric pressure of 760 mmHg) (Tredici et al., 2004) or commercial membrane oxygenator (84%) (Cox et al., 2003; Stavis et al., 1998). In addition, the maximal PaO_2 of 234 mmHg obtained herein is similar to the PaO_2 value of 201 mmHg previously reported in healthy newborn lambs by others (Hirschl et al., 1995; Larrabe et al., 2001; Stavis et al., 1998,

Cox et al. 2003). Beyond their low cost and simplicity of use, the present study clearly shows that bubble oxygenators can be highly efficient during TLV.

Our results highlight elevated PaO_2 values, which can be obtained in TLV in the presence of an O_2-saturated PFC, when minute ventilation is maximized. The present results show that it is possible to control PaO_2 during total liquid ventilation without altering alveolar ventilation or oxygenator efficiency. The PaO_2 can be controlled by adjusting the O_2 fraction in the gas flowing into the oxygenator, thus mimicking the method used in conventional gas ventilation. Reduction of gas flow to the oxygenator decreases the oxygenator efficiency index, with the most undesirable effect being the deterioration of the CO_2 removal before any effect on PaO_2 reduction. Similar results, on the high flow rate requirement in order to achieve an efficient CO_2 removal, were previously reported with a hollow-fiber oxygenator in a rabbit model (Tredici et al., 2004).

Finally, as expected, our method based on the control of $F_{gas}O_2$ with a gas mixer enabled us to adjust $Fi_{PFC}O_2$ and consequently to control PaO_2 down to safer values without significantly altering $PaCO_2$ and pH. The use of this particular method, which reproduces the manner in which FiO_2 is controlled in all existing conventional gas ventilators, has never been reported for TLV.

This study could be deemed as being limited to a specific PFC, a perfluorodecalin (PFDEC), since another PFC, the perfluoroctylbromide (PFOB), is usually considered as the PFC of choice for TLV. It was used in clinical studies I/II on partial liquid ventilation (Kacmarek et al., 2006; Hirschl et al., 1998, 2002). However, the conclusions should be the same, regardless of the PFC type, since the oxygenator is able to control the gas concentration, reach the desired oxygen fraction in the inspired PFC ($Fi_{PFC}O_2$) and remove the CO_2.

4. Validation of the pressure controller

4.1 In-vivo protocol

The experimentations were obtained with 10 healthy anaesthetized and paralyzed newborn lambs (age < 5 days, weight < 4 kg, 1 hours TLV trial with PFOB or PFDEC). The experimental protocol was approved by our institutional Ethics Committee for Animal Care and Experimentation. The lamb was placed in a supine position and an epidural catheter was inserted in the Y-piece such that the extremity extended 1 centimeter before the ET tube ending. The other end of the catheter was connected to a pressure sensor located at the same height as the ET tube in the trachea. The pressure sensor (Model 1620, Measurement Specialties, Hampton, VA) was used to measure airway pressure P_{aw} in the trachea.

After randomization, the lungs were filled at functional residual capacity (25 ml/kg) with warmed and oxygenated PFDEC or PFOB. After PFC instillation, total liquid ventilation was initiated in volume-controlled mode (during the first minute) at a rate of 5.35 breaths/minute and a V_t = 40 ml with an $F_{gas}O_2$ of 100%. Then, the ventilation modes were switched rapidly to pressure regulated modes for both inspiration and expiration and the pressure references $P_{ref,i}$ and $P_{ref,e}$ were adjusted to reach the desired tidal volume (25 ml/kg) at a frequency around 6.4 rpm.

4.2 In-vivo results

Figure 7 presents typical results for the pressure regulators during TLV using PFDEC and PFOB. The controllers were able to reach the pressure references, without oscillations. The airway pressure P_{aw} was much lower that the Y pressure P_y during the inspiration, since the

ET tube added a significant pressure loss in the fluid circuit. Those losses were highly influenced by the inside diameters of the ET tube (considered in the design of the controllers).

There was no airway collapse during the expiration even if the airway pressure was below -10 cmH$_2$O. The airway pressure decreased proportionally during the expiration. This can be explained as the pressure losses in the ET tube decrease with the flow which in turn, decreases exponentially with time, thus resulting in a proportional decrease of the airway pressure over time. The pressure references used during TLV with the PFOB are lower, because the PFOB viscosity is lower compared to the PFDEC, which affects directly the ET tube pressure losses.

The measured PEEP offset can be explained by looking closely at the volumes pumped in the lungs. With PFDEC, the requested tidal volume V_t was 2.5 ml/kg over the volume inspired $V_{t,i}$. This difference was directly reflected on the expiration volume error E_e (2.43 ± 0.69 ml/kg). The figure 8 presents typical results for the Peep controller during TLV using PFDEC and PFOB. For both PFC the $PEEP_{est}$ was near the $PEEP_{ref}$. There was a slight offset since the gain K_{Peep} used was not equal to the dynamic compliance measured. The table 5 indicates that the mean error between the $PEEP_{ref}$ and the $PEEP_{est}$ was around 0.3 ± 0.1 cmH$_2$O (PFDEC) and 0.4 ± 0.1 cmH$_2$O (PFOB). The measured PEEP was 2.7 ± 0.7 cmH$_2$O over the reference for the PFDEC and 1.4 ± 1.1 cmH$_2$O for the PFOB, once the ventilation parameters were stabilized. For both PFC, the PEEP controller was stable.

Using equation 3, it becomes obvious that an error on the expired volume will cause a PEEP offset. Since the PFOB is less viscous, it is easier to reach the requested tidal volume V_t. In consequence, the expiration error and the PEEP offset are smaller.

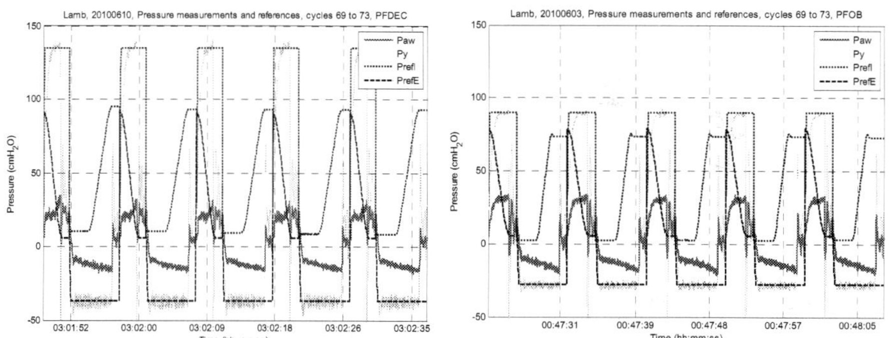

Fig. 7. Left graphic: Airway pressure (P_{aw}) and Y pressure (P_y) versus time for $P_{ref,i}$ = 135 cmH$_2$O and $P_{ref,e}$ = -37 cmH$_2$O using PFDEC. Right graphic: Airway pressure (P_{aw}) and Y pressure (P_y) versus time for $P_{ref,i}$ = 90 cmH$_2$O and $P_{ref,e}$ = -28 cmH$_2$O using PFOB

The table 5 presents the numerical results obtained on 5 newborn lambs (< 4 days old) per PFC group (PFOB and PFDEC). The inspired and expired volume for both groups were similar, but the frequency reached using PFOB was higher, even if the pressure reference and PEEP measured were lower. Again, the viscosity of the PFDEC is the explanation behind these observations. Adequate gas exchange and normal acid-base equilibrium were maintained during TLV.

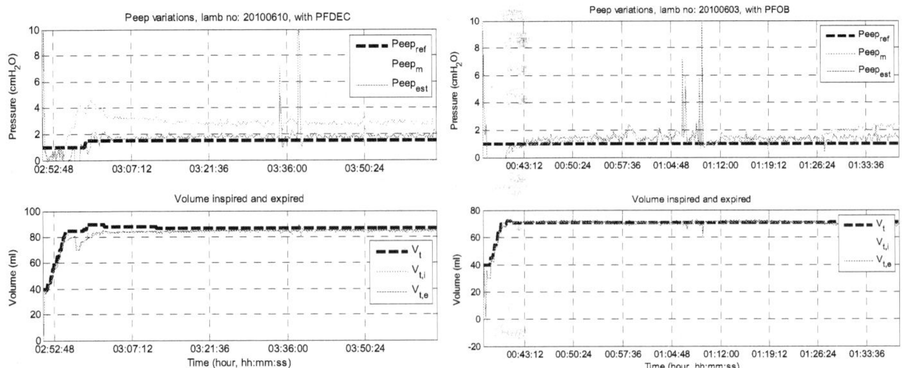

Fig. 8. Left top graphic: typical evolution of the PEEP measured ($PEEP_m$), estimated ($PEEP_{est}$) and reference ($PEEP_{ref}$) over a complete experiment with PFDEC. Left bottom graphic: typical evolution of the inspired ($V_{t,i}$) and expired volume ($V_{t,e}$) versus the tidal volume (V_t) for PFDEC. Right top graphic: typical evolution of the PEEP measured ($PEEP_m$), estimated ($PEEP_{est}$) and reference ($PEEP_{ref}$) over a complete experiment with PFOB. Right bottom graphic: typical evolution of the inspired ($V_{t,i}$) and expired volume ($V_{t,e}$) versus the tidal volume (V_t) for PFOB. References are shown in bold for the $PEEP_{ref}$ and V_t.

	PFDEC		PFOB	
V_t (ml/kg)	26.9	± 0.6	25.9	± 0.9
$V_{t,i}$ (ml/kg)	24.4	± 0.7	24.9	± 0.4
$V_{t,e}$ (ml/kg)	24.6	± 0.6	25.2	± 0.4
E_i (ml/kg)	-0.14	± 0.51	-0.41	± 0.05
E_e (ml/kg)	2.43	± 0.69	1.07	± 1.00
$P_{ref,i}$ (cmH₂O)	152.7	± 46.3	111.5	± 17.3
$P_{ref,e}$ (cmH₂O)	-39.1	± 8.02	-32.9	± 5.6
Fr (rpm)	6.14	± 0.13	6.43	± 0.08
V_{min} (ml/min/kg)	149.7	± 5.9	160.3	± 3.3
$PEEP_{ref}$ (cmH₂O)	1.8	± 0.6	1.0	± 0.1
$PEEP_{est}$ (cmH₂O)	2.1	± 0.7	1.3	± 0.2
$PEEP_m$ (cmH₂O)	4.5	± 1.0	2.3	± 1.1
PaO_2 (mmHg)	285.0	± 67.0	288.3	± 91.4
$PaCO_2$ (mmHg)	52.9	± 10.9	37.2	± 10.2
pH	7.29	± 0.11	7.39	± 0.11
$F_{gas}O_2$	1.0	± 0.0	1.0	± 0.0

Table 5. Results of 5 healthy newborn lambs for each TLV group (< 4 days old). (mean ± standard deviation)

4.3 Discussion regarding the PEEP-controller

The proposed PEEP controller manages directly the end-expiratory alveolar pressure (by controlling the PEEP) and all the measurable volume errors. However, all measurable volume errors do not include sensor non-linearities, machining tolerances and analog input precisions, so the lung volume can derive positively or negatively, if and only if such volume derivatives have no impact on the alveolar pressure. In the normal case, the control of the end-expiratory alveolar-pressure is equivalent to the control of the end-expiratory lung volume (EELV) because there is a direct relationship between these two variables (Degraeuwe et al., 2000b; Parker et al., 2009).

5. Conclusion

The oxygenator presented at the beginning of this chapter shows that we can control the oxygen fraction of the inspired PFC and removal of all the CO_2 contained in the expired liquid. This control of the oxygen concentration, $Fi_{PFC}O_2$, permits the adjustment of the arterial partial pressure of oxygen PaO_2 without any consequence on the arterial partial pressure of CO_2, $PaCO_2$. A decrease in the oxygen flow rate through the oxygenator is not a suitable solution, since they it compromises the $PaCO_2$, before any effect on PaO_2 reduction.

The choice of a bubble oxygenator could be seen as a limitation, since most research groups use membrane oxygenators with TLV (Corno et al. 2003; Cox et al. 2003; Hirschl et al., 1995; Larrabe et al., 2001; Parker et al., 2009; Pohlmann et al., 2011; Tredici et al., 2004). On the contrary, we strongly believe that bubble oxygenators are quite suitable for TLV. Beyond their low cost and simplicity of use, the present study shows that they can be highly efficient during TLV.

The presented pressure controlled ventilation modes seem similar to those used in conventional gas ventilators, although there are some differences. The PEEP follower maintains the end-expiratory alveolar pressure constant by commanding volume corrections based on the end-expiratory pressure. The pressure references $P_{ref,i}$ and $P_{ref,e}$ are higher compared to CMV. The pressure controlled mode eliminates the airway collapses and favors higher minute ventilation. Thus, it can be implemented successfully on liquid ventilators during the inspiration and expiration phases. When used in conjunction with the oxygenator presented, the arterial blood gases can be maintained in a normal range throughout the liquid ventilation. As for the PEEP controller, the latter compensates all the volume errors (measurable and non-measurable) from cycle to cycle, which secures the TLV by avoiding lung overdistension.

There is still room for improvement, as the observed results do not fully comply with the defined specifications. The PEEP controller is able to estimate quite well the real PEEP, but the measured PEEP can increase since the expired volume is not reached. Modifications will be performed to improve PEEP tracking. In some ventilation cycles, the volume expired is not achieved. Thus, the pressure limit $P_{ref,e}$ could be adapted from cycle to cycle to increase the expired volume. Nevertheless, the pressure regulators and PEEP controller have greatly improved the ventilation efficiency and simplified the interaction with the liquid ventilator, since all the volumes and errors are managed by controllers from cycle to cycle.

In the ultimate objective of transferring liquid ventilators into intensive care units (Costantino et al., 2009), we strongly recommend the addition of a gas mixer to the oxygenator (to adjust the O_2 concentration) coupled with pressure controlled ventilation mode as two efficient means to maintain normal blood gases during TLV.

6. Acknowledgment

This research was supported in part by Natural Sciences and Engineering Research Council of Canada, the Foundation of Stars and by the Fonds Québécois de la Recherche sur la Nature et les Technologies. Special thanks to Pulsion Medical System AG (Munich, Germany) for the PiCCO and VoLEF loan. Philippe Micheau, Jean-Paul Praud and Hervé Walti are members of the Centre de Recherche Clinique Étienne-Le Bel, funded by the Fonds de la Recherché en Santé du Québec.

7. References

Avoine, O., Bossé, D., Beaudry, B., Beaulieu, A., Albadine, R., Praud, J.-P., Robert, R., Micheau, P., & Walti, H. (2011). Total liquid ventilation efficacy in an ovine model of severe meconium aspiration syndrome. Crit Care Med, Vol. 39, No. 5, pp. 1097-1103

Baba, Y., Taenaka, Y., Akagi, H., Nakatani, T., Masuzawa, T., Tatsumi, E., Wakisaka, Y., Toda, K., Eya, K., Tsukahara, K., & Takano, H. (1996). A volume-controlled liquid ventilator with pressure-limit mode: Imperative expiratory control. Artif Organs, Vol. 20, pp. 1052-1056, ISSN 0160-564X.

Baba, Y., Brant, D., Brah, S.S., Grotberg, J., Bartlett, R.H., & Hirschl R.B. (2004). Assessment of the development of choked flow during total liquid ventilation. Crit Care Med., Vol. 32, pp. 201-208, ISSN 0090-3493.

Bagnoli, P., Tredici, S., Seetharamaiah, R., Brant, D. O., Hewell, L.A., Johnson, K., Bull, J.L.; Costantino, M.L., & Hirschl, R.B. (2007). Effect of repeated induced airway collapse during total liquid ventilation. ASAIO J, Vol. 53, pp. 549-555, ISSN 1538-943X.

Beaudry, B. (2009). L'évaluation et l'optimisation des échanges gazeux dans un oxygénateur de respirateur liquidien. Master's thesis (in French), Université de Sherbrooke, Canada.

Bitterman, H. (2009). Bench-to-bedside review: Oxygen as a drug. Critical Care, Vol. 13, No. 1, pp 205-212, ISSN 1466-609X.

Bosse, D., Beaulieu, A., Avoine, O., Micheau, P., Praud, JP, & Walti, H. (2010). Neonatal Total Liquid Ventilation: Is Low Frequency Forced Oscillation Technique Suitable for Respiratory Mechanics Assessment? J Appl Physiol, Vol. 109, pp. 501-510, ISSN 1522-1601.

Bull, J.L, Foley, D.S. Bagnoli, P., Tredici, S., Brant, D.O., & Hirschl, R.B. (2005). Location of Flow Limitation in Liquid-Filled Rabbit Lungs. ASAIO Journal, Vol. 51, No. 6, pp. 781-788, ISSN 1538-943X.

Bull, J., Tredici, S., Fujioka, H., Komori, E., Grotberg, J., Hirschl, R.B. (2009). Effects of respiratory rate and tidal volume on gas exchange in total liquid ventilation. ASAIO J, Vol. 55, pp. 373-81, ISSN 1538-943X.

Chan, K.P., Stewart, T.E., & Mehta, S. (2007). High-frequency oscillatory ventilation for adult patients with ARDS. Chest, Vol. 131, pp. 1907-1916, ISSN 0012-3692.

Chatburn, R.L., & Primiano, F.P. (2001). A new system for understanding modes of mechanical ventilation. Respir Care, Vol. 46, No. 6, pp. 604-621, ISSN 0020-1324.

Corno, C., Fiore, G.B., Martelli, E., Dani, C. & Costantino, M.L. (2003). Volume controlled apparatus for neonatal tidal liquid ventilation. ASAIO Journal, Vol. 49, No. 3, pp. 250-258, ISSN 1538-943X.

Corno, C., Fiore, G.B., & Costantino, M.L. (2004). A mathematical model of neonatal tidal liquid ventilation integrating airway mechanics and gas transfer phenomena. IEEE Trans Biomed Eng, Vol. 51, pp. 604-611, ISSN 0018-9294.

Costantino, M. L., & Fiore, G. B. (2001) A model of neonatal tidal liquid ventilation mechanics. Med Eng Phys, Vol. 23, pp. 457-471, ISSN 1350-4533.

Costantino, M.L., Micheau, P., Shaffer, T.H., Tredici, S., & Wolfson, M.R. (2009). Clinical Design Functions: Round table discussions on bioengineering of liquid ventilators, ASAIO J., Vol. 55, No. 3, pp. 206-207, ISSN 1538-943X.

Cox, C., Stavis, R.L., Wolfson, M.R., Shaffer, T.H. (2003). Long-Term Tidal Liquid Ventilation in Premature Lambs: Physiologic Biochemical and Histological Correlates. Biol. Neonate, Vol. 84, pp. 232-242, ISSN 0006-3126.

Degraeuwe, P.L., Vos, G.D., Geskens, G.G., Geilen, J.M., & Blanco, C.E. (2000a). Effect of perfluorochemical liquid ventilation on cardiac output and blood pressure variability in neonatal piglets with respiratory insufficiency. Pediatr Pulmonol, Vol. 30, pp. 114-124, ISSN 8755-6863.

Degraeuwe, P.L., Dohmen, L.R., Geilen, J.M., & Blanco, C.E. (2000b). A feedback controller for the maintenance of FRC during tidal liquid ventilation: theory, implementation, and testing, Int J Artif Organs, Vol. 23, No. 10, pp. 680-688, ISSN 0391-3988.

Dorfman, A.L., Chemtob, & S., Lachapelle, P. (2010). Postnatal hyperoxia and the developing rat retina: beyond the obvious vasculopathy. Doc Ophthalmol, Vol. 120, pp. 61-66, ISSN 0012-4486.

Foust, R., Tran, N.N., Cox, C., Miller, T.F., Greenspan, J.S., Wolfson M.R., & Shaffer T.H. (1996). Liquid Assisted Ventilation: An Alternative Ventilatory Strategy for Acute Meconium Aspiration Injury. Pediatr Pulmonol, Vol. 21, pp. 316-322, ISSN 8755-6863.

Gitto, E., Pellegrino, S., D'Arrigo, S., Barberi, I., & Reiter, R.J. (2009). Oxidative stress in resuscitation and in ventilation of newborns. Eur Respir J, Vol. 34, No. 6, pp. 1461-1469, ISSN 0903-1936.

Hayes D. Jr, Feola, D.J., Murphy, B.S., Shook, L.A., & Ballard, H.O. (2010). Pathogenesis of bronchopulmonary dysplasia. Respiration, Vol. 79, pp. 425-436, ISSN 1423-0356.

Hirschl, R. B., Merz, S. I., Montoya, J. P., Parent, A., Wolfson, M. R., Shaffer, T. H. & Bartlett, R. H. (1995). Development and application of a simplified liquid ventilator. Crit Care Med, Vol. 23, pp. 157-163, ISSN 0090-3493.

Hirschl, R.B., Tooley, R., Parent, A., Johnson, K., & Bartlett, R.H. (1996). Evaluation of gas exchange, pulmonary compliance, and lung injury during total and partial liquid ventilation in the acute respiratory distress syndrome. Crit Care Med, Vol. 24, pp. 1001-1008, ISSN 0090-3493.

Hirschl, R.B., Conrad, S., Kaiser, R., Zwischenberger, J.B., Bartlett, R.H., Booth, F. & Cardenas, V. (1998). Partial liquid ventilation in adult patients with ARDS: a multicenter phase I-II trial. Ann Surg, Vol. 228, pp. 692-700, ISSN 0003-4932.

Hirschl, R.B., Croce, M., Gore D., Wiedemann, H., Davis, K., Zwischenberger, J., & Bartlett, R.H. (2002). Prospective, randomized, controlled pilot study of partial liquid ventilation in adult acute respiratory distress syndrome. Am J Respir Crit Care Med, Vol. 165, No. 6, pp. 781-787, ISSN 1073-449X.

Iwahashi, H., Yuri, K., & Nosé, Y. (2004). Development of the oxygenator; past, present and future, J Artif Organs, Vol. 7, pp. 111-120.

Kacmarek, R.M., Wiedemann, H.P., Lavin, P.T., Wedel, M.K., Tütüncü, A.S., & Slutsky, A.S. (2006). Partial liquid ventilation in adult patients with acute respiratory distress syndrome. Am J Respir Crit Care Med, Vol. 173, pp. 882-889, ISSN 1073-449X.

Koen, P.A., Wolfson, M.R., & Shaffer T.H. (1988). Fluorocarbon ventilation: Maximal expiratory flows and CO_2 elimination. Pediatr Res, Vol. 24, pp. 291–296, ISSN 0031-3998.

Larrabe, J.L., Alvarez, F.J., Cuesta, E.G., Valls-i-Soler, A., Alfonso, L.F., Arnaiz, A., Fernandez, M.B., Loureiro, B., Publicover, N.G., Roman, L., Casla, J.A., & Gomez M.A. (2001). Development of a time-cycled volume-controlled pressure-limited respirator and lung mechanics system for total liquid ventilation. IEEE Trans Biomed Eng, Vol. 48, No. 10, pp. 1134-1144, ISSN 0018-9294.

Mancebo, J. (1992). PEEP, ARDS and alveolar recruitment. Intensive Care Medicine, Vol. 18, No. 7, pp. 383-385, ISSN 0342-4642.

Matsuda, K., Sawada, S., Bartlett, R., Hirschl, R.B. (2003). Effect of ventilatory variables on gas exchange and hemodynamics during total liquid ventilation in a rat model. Crit Care Med, Vol. 31, pp. 2034-2040, ISSN 0090-3493.

Meinhardt, J.P., Quintel, M., & Hirschl, R.B. (2000). Development and application of a double-piston configured, total-liquid ventilatory support device. Crit Care Med, Vol. 28, pp. 1483-1488, ISSN 0090-3493.

Munson, B.R., Young, D.F., Okiishi, T.H. (1998). Fundamentals of Fluid Mechanics. 3rd ed. John Wiley & Sons, New York, ISBN 0-471-17024-0.

Parker, J.C., Sakla, A., Donovan, F.M., Beam, D., Chekuri, A., Al-Khatib, M., Hamm, C.R., & Eyal, F.G. (2009). A microprocessor-controlled tracheal insufflation-assisted total liquid ventilation system. Med Biol Eng Comput, Vol. 47, pp. 931-939, ISSN 0140-0118.

Pohlmann, J.R., Brant, D.O., Daul, M.A., Reoma, J.L., Kim, A.C., Osterholzer, K.R., Johnson, K.J., Bartlett, R.H., Cook, K.E., & Hirschl, R.B. (2011). Total liquid ventilation provides superior respiratory support to conventional mechanical ventilation in a large animal model of severe respiratory failure. ASAIO J, Vol. 57, pp. 1-8, ISSN 1538-943X.

Raoof, S., Goulet, K., Esan, A., Hess, D.R., & Sessler, C.N. (2010). Severe hypoxemic respiratory failure: part 2-nonventilatory strategies, Chest, Vol. 137: pp. 1437-1448, ISSN 0012-3692.

Richman, P.S., Wolfson, M.R., & Shaffer, T.H. (1993). Lung lavage with oxygenated perfluorochemical liquid in acute lung injury. Crit Care Med, Vol. 21, pp. 768-774, ISSN 0090-3493.

Robert, R., Micheau, P., Cyr, S., Lesur, O., Praud, J.P. & Walti, H. (2006). A prototype of volume-controlled tidal liquid ventilator using independent piston pumps. ASAIO J, Vol. 52, pp. 638-645, ISSN 1538-943X.

Robert, R., Micheau, P., & Walti, H. (2007a). A supervisor for volume-controlled tidal liquid ventilator using independent piston pumps. Biomed Signal Process Control, Vol. 2, pp. 267-274, ISSN 1538-943X.

Robert, R., Beaudry, B., Gauthier, J.P., Beaulieu, A., Patenaude, B., Praud, J.P., Lesur, O., Walti H., Micheau, P. (2007b). Development of a tidal liquid ventilator for experimental research and towards clinical research, 6th Inter. Symp. on Perfluorocarbon App. and Liquid Ventilation, Montréal, Canada, October 11-12.

Robert, R. (2007c). Modélisation numérique et stratégies de commande du débit expiratoire pour éviter le collapsus des voies respiratoires en ventilation liquidienne totale, PhD Thesis, Université de Sherbrooke, Canada, 245 p.

Robert, R., Micheau, P., & Walti H. (2009). Optimal expiratory volume profile in tidal liquid ventilation under steady state conditions, based on a symmetrical lung model, ASAIO J, Vol. 55, No. 1, pp 63-72, ISSN 1538-943X.

Robert, R., Micheau, P., Avoine, O., Beaudry, B., Beaulieu, & A., Walti, H. (2010). A Regulator for Pressure Controlled Liquid Ventilation. IEEE Trans Biomed Eng, Vol. 57, No. 9, pp. 2267-2276, ISSN 0018-9294.

Seber, G.A.F., & Wild C.J. (1989). Nonlinear Regression. John Wiley & Sons Inc, New-York, ISBN 0471617601.

Sekins, K.M., Nugent, L., Mazzoni, M., Flanagan, C., Neer, L., Rozenberg, A., & Hoffman, J. (1999). Recent innovations in total liquid ventilation system and component design. Biomed Instrum Technol, Vol. 33, No. 3, pp. 251-284, ISSN 0899-8205.

Simon, F., Jenayeh, I., & Rake, H. (2000). Mechatronics in medical engineering: advanced control of a ventilation device. Microproc and Microsystems, Vol. 24, No. 2, pp. 63-69, ISSN 0141-9331.

Speer, C.P. (2009). Chorioamnionitis, postnatal factors and proinflammatory response in the pathogenetic sequence of bronchopulmonary dysplasia. Neonatology, Vol. 95, pp. 353-361, ISSN 1661-7800.

Stavis, R.L., Wolfson, M.R., Cox, C., Kechner, N., Shaffer, T.H. (1998). Physiologic, Biochemical, and Histologic Correlates Associated with Tidal Liquid Ventilation. Pediatr Res, Vol. 43, pp. 132-138, ISSN 0031-3998.

Tasaka, S., Amaya, F., Hashimoto, S., & Ishizaka, A. (2008). Roles of oxidants and redox signaling in the pathogenesis of acute respiratory distress syndrome. Antioxid Redox Signal, Vol. 10, pp. 739-753, ISSN 1523-0864.

Tissier, R., Couvreur, N., Ghaleh, B., Bruneval, P., Lidouren, F., Morin, D., Zini, R., Bize, A., Chenoune, M., Belair, M.F., Mandet, C., Douheret, M., Dubois-Rande, J.L., Parker, J.C., Cohen, M.V., Downey, J.M., & Berdeaux, A. (2009). Rapid cooling preserves the ischaemic myocardium against mitochondrial damage and left ventricular dysfunction. Cardiovasc Res, Vol. 83, pp. 345-53, ISSN 0008-6363.

Tredici, S., Komori, E., Funakubo, A., Brant, D.O., Bull, J.L., Bartlett, R.H., & Hirschl, R.B. (2004). A prototype of a liquid ventilator using a novel hollow-fiber oxygenator in a rabbit model. Crit Care Med, Vol. 32, No. 10, pp. 2104-2109, ISSN 0090-3493.

Venegas, J.G., Simon, R.S., & Harris B.A. (1998). A Comprehensive Equation for the Pulmonary Pressure-Volume Curve. J Appl Physiol, Vol. 84, No. 1, pp. 389-395, ISSN 1522-1601.

Wang, Y.Y.L., Chang C.C., Chen J.C., Hsiu H., & Wong W.K. (1997). Pressure Wave Propagation in Arteries: A model with radial dilatation for simulating the behavior of a real artery, IEEE Eng Med Biol Mag, Vol. 16, No. 1, pp. 51-54, ISSN 0739-5175.

Wolfson, M.R., Miller, T.F., Peck, G., & Shaffer, T.H. (1999). Multifactorial analysis of exchanger efficiency and liquid conservation during perfluorochemical liquid-assisted ventilation. Biomed Instrum Technol, Vol. 33, No. 3, pp. 260-267, ISSN 0899-8205.

Wolfson, M. R., & T. H. Shaffer (2005). Pulmonary applications of perfluorochemical liquids: ventilation and beyond. Paediatr Respir Rev, Vol. 6, No. 2, pp. 117-27, ISSN 1526-0542.

Wolfson, M.R., Hirschl R.B., Jackson, J.C., Gauvin, F., Foley, D.S., Lamm, W.J.E., Gaughan, J., & Shaffer, T.H. (2008). Multicenter comparative study of conventional mechanical gas ventilation to tidal liquid ventilation in oleic acid injured sheep. ASAIO Journal, Vol. 54, pp 256-269, ISSN 1538-943X.

Part 3

Molecular and Cellular Engineering: Industrial Application

15

Isolation and Purification of Bioactive Proteins from Bovine Colostrum

Mianbin Wu, Xuewan Wang, Zhengyu Zhang
and Rutao Wang
Department of Chemical and Biological Engineering
Zhejiang University
China

1. Introduction

Bovine colostrum is the milk secreted by cows during the first few days after parturition. It contains many essential nutrients and bioactive components, including growth factors, immunoglobulins (Igs), lactopcroxidase (Lp), lysozyme (Lys), lactoferrin (Lf), cytokines, nucleosides, vitamins, peptides and oligosaccharides, which are of increasing relevance to human health. Much research work has been done on the structure and function of bovine colostrum proteins. IgG was widely utilised in the immunological supplementation of foods, specifically in infant formulate, and yielded sales of approximately US$100 million in 2007 (Gapper, *et al.*, 2007). In the highly competitive and valuable international market for IgG-containing products, some of the products are usually priced based on IgG content. Another important protein from bovine colostrum is lactoferrin. Its diverse range of biological activities such as anti-infective activities toward a broad spectrum of species, antioxidant activities and promotion of iron transfer are expanding the demand in the market. It also exhibits the potential for chemoprevention of colon and other cancers as a natural gradient. Apart from the two kinds of bovine colostrum proteins, α-lactalbumin has been claimed as an important food additive in infant formula due to its high content in tryptophan and as a protective against ethanol and stress-induced gastric mucosal injury. β-Lactoglobulin is commonly used to stabilize food emulsions for its surface-active properties. Bovine serum albumin (BSA) has gelation properties and it is of interest in a number of food and therapeutic applications (Almecija, *et al.*, 2007). Therefore, fractionation for the recovery and isolation of these proteins has a great scientific and commercial interest.

As a result of this growth in the commercial use of bovine colostrum proteins, there is great interest in establishing more efficient, robust and low cost processes to purify them. Although great deals of studies have been done for the separation and purification of colostrum proteins due to their wide application in food industry, medicine and as supplements, large scale production system for the downstream processing of recombinant antibodies still represents the major issue. Lu (Lu, *et al.*, 2007) designed a two-step ultrafiltration process followed by a fast flow strong cation exchange chromatography to isolate LF from bovine colostrum in a production scale. A stepwise procedure for purification of the crude LF was conducted using a preparative-scale strong cation exchange chromatography. The purity and the recovery of the final LF product were 94.20% and 82.46%, respectively. The process developed in Lu's work was a significant improvement over the commercial practice for the fractionation of LF from bovine colostrum. Recently, Saufi et al. developed a cationic mixed matrix membrane for the recovery of LF from bovine whey, the absorbent was developed by embedding ground SP Sepharose cation exchange resin into an ethylene vinyl alcohol polymer base membrane (Saufi & Fee, 2011). The static LF binding capacity of the cationic Mixed Matrix Membrane (MMM) was 384 mg/mLmembrane or 155 mg/mL membrane, exceeding the capacity of several commercial adsorptive membranes. The membrane chromatography system was operated in cross-flow mode to minimize fouling and enhance LF binding, resulting in an LF recovery as high as of 91%, with high purity. The system was operated at a constant permeate flux rate of 100 Lm^{-2} h^{-1}, except during the whey loading step, which was run at 50 Lm^{-2} h^{-1}. This is the first time a cross-flow MMM process has been reported for LF recovery from whey.

The traditional protein fraction process usually included initial processes such as centrifugalization and membrane treatment, and polishing steps such as chromatographic procedures. To further utilize bioactive substance such as bovine colostrum sIgA and IgG, a procedure including salting out, ultra-filtration and gel chromatography in proper sequence on isolation and purification of bovine colostrum sIgA and IgG was reported (LIU& Y.Y.X.G.a.X. 2007). The purity and yield of bovine colostrum sIgA were 85.3% and 42.8%, respectively. The purity and yield of bovine colostrum IgG were respectively 97.2% and 64.4%.This preparative method provided theoretical and experimental foundation for sIgA and IgG industrial production. Depending on the market requirement, other procedures may be employed as the suitable steps for the products' commerciality, such as freeze-drying and crystallization. Therefore, the protocols for the purification of proteins should be designed according to the feed stock and final requirement.

Although a wide variety of protocols can be used to separate bioactive proteins from complex food stock, chromatographic procedure is the most prevalent form as high-resolution fractionation technique. In this section, we will discuss the use of chromatographic procedures and other techniques as high-resolution techniques for the fraction of bovine colostrum proteins. Special attention will be paid to the amount of bio-product denaturation or activity loss that occurs. Particular attention will also be paid to the quality of the separated bio-product. The understanding about processes that lead to these activity losses would then assist in minimizing these activity losses.

2. Precondition of bovine colostrum

2.1 Preparation of acid whey

In order to avoid the problems caused by high viscosity of bovine colostrum, researchers usually employ acid whey as the beginning feed stock. The method is as follows. Bovine colostrum samples were collected within the first day after cow parturition from the dairy plant and were immediately frozen and stored at −18°C. The frozen samples were thawed and the lipid fraction were removed by centrifugation at 8,000 r/min for 15~20 min at 4°C. Acid colostral whey was prepared by precipitation of the casein from skimmed colostrum with 1 mol/L HCl at pH 4.2 and the precipitated casein was removed by microfiltration. The whey was then adjusted to pH 6.8 with 1 mol/L NaOH and then went through centrifugation.

2.2 Membrane filtration

Membrane filtration provides promising results for the fractionation of whey proteins and it has traditionally been based solely on differences in molecular mass. Until recently, membranes were thought to achieve separation only between proteins differing in size by at least a factor of 10. Almecija (Almecija, *et al.*, 2007) investigated the potential of ceramic membrane ultrafiltration for the fractionation of clarified whey. They employed a 300 kDa tubular ceramic membrane in a continuous diafiltration mode. The effect of working pH was evaluated by measuring the flux-time profiles and the retentate and permeate yields of α-lactalbumin, β-lactoglobulin, BSA, IgG and lactoferrin. The study results showed that at pH 3, 9 and 10 permeate fluxes ranged from 68 to 85, 91 to 87 and 89 to 125 L/(m²h), respectively. On the other hand, around the isoelectric points of the major proteins (at pH 4 and 5), permeate fluxes varied from 40 to 25 and from 51 to 25 L/(m² h), respectively. For α-lactalbumin and β-lactoglobulin, the sum of retentate and permeate yields was around 100% in all cases, which indicates that no loss of these proteins occurred. After 4 diavolumes, retentate yield for alpha-lactalbumin ranged from 43% at pH 9 to 100% at pH 4, while for β-lactoglobulin, was from 67% at pH 3 to 100% at pH 4. In contrast, BSA, IgG and lactoferrin were mostly retained, with improvements up to 60% in purity at pH 9 with respect to the original whey. The results of this paper obtained were explained in terms of membrane–protein and protein–protein interactions.

2.3 Precipitation

Precipitation method is an effective way to concentrate the proteins due to their different pI, sensitivity to the ionic strength and other properties. Salting-out is widely used for the pretreatment of bovine whey to selectively precipitate the protein of interest or impurities. Lozano (Lozano, *et al.*, 2008) used an improved method successfully and rapidly separated β-lactoglobulin from bovine whey. Firstly, differential precipitation with ammonium sulfate was used to isolate β-lactoglobulin from other whey proteins using 50% ammonium sulfate. The precipitate was dissolved and separated again using 70% ammonium sulfate, leaving a supernatant liquid enriched in β-lactoglobulin. After dialysis and lyophilization, isolation of the protein was performed by ion-exchange chromatography. Comparison of physicochemical and immunochemical analysis showed that the identity and purity of the isolated protein was comparable with that of the Sigma standard. Spectroscopic results showed that the method used for protein isolation did not induce any changes in the protein native structural properties. Ammonium sulfate precipitation method played a vital role for this rapid, efficient and inexpensive two-step process that allowed high homogeneous protein yield.

3. Chromatographic procedures for the separation of bovine colostrum proteins

3.1 Ion exchange chromatography

3.1.1 Introduction

Proteins contain charged groups on their surfaces that enhance their interactions with solvent water and hence their solubility. Charged residues can be cationic or anionic and it is noteworthy that even polar residues can also be charged under certain pH conditions. These charged and polar groups are responsible for maintaining the protein in solution at physiological pH. Because proteins have unique amino acid sequences, the net charge on a protein at physiological pH is determined ultimately by the balance between these charges. This also underlies differing isoelectric points (pIs) of proteins (Himmelhoch, 1971). Therefore, bioactive proteins can be absorbed by different ion-exchange chromatography [Fig. 1] due to the different charge type and pI. The ion-exchange resins are then selectively eluted by slowly increasing the ionic strength (this disrupts ionic interactions between the protein and column matrix competitively) or by altering the pH (the reactive groups on the proteins lose their charge) (Dolman, et al., 2002)

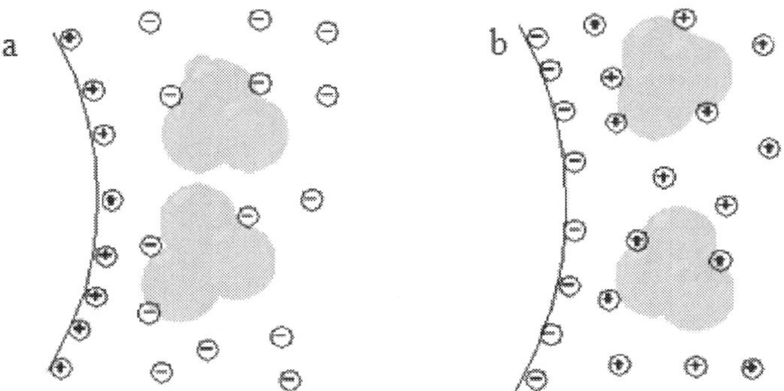

Fig. 1. a) Anionic (negatively charged) proteins exchange. b) Cationic (positively charged) proteins exchange.

3.1.2 Applications in Isolation and purification of bioactive proteins from bovine colostrum

The whey proteins can be fractionated and separated by different ion exchange chromatography. A water-jacketed chromatography column (XK 26/40, Amersham Biosciences) packed with SP Sepharose Big Beads cation exchanger was used to recover and fractionate whey proteins (Doultani, et al., 2004). The chromatographic procedure involved sequentially pumping different solutions into the column: (1) equilibration (EQ) buffer to adjust column pH; (2) whey; (3) EQ buffer to rinse unbound material from the column; and (4) different elution buffers to selectively desorbed different bound proteins.

The optimum conditions for initially separating the proteins such as α-lactalbumin, β-lactoglobulin, bovine serum albumin, immunoglobulin G and lactose from a sweet dairy whey mixture could be determined by a commercial anion-exchange resin (Gerberding & Byers, 1998). The separation was accomplished with simultaneous step elution changes in salt concentration and pH. It was found that the anion-exchange step was most effective in separating β-lactoglobulin from the feed mixture. Followed by the anion-exchange separation, the breakthrough curve was processed using a commercial cation-exchange resin to further recover the valuable immunoglobulin G.

A simple and useful method for β-lactoglobulin isolation from bovine whey was presented recently (Lozano, et al., 2008). Differential precipitation with ammonium sulfate was used to isolate β-lactoglobulin from other whey proteins using 50% ammonium sulfate. The precipitate was dissolved and separated again using 70% ammonium sulfate, leaving a supernatant liquid enriched in β-lactoglobulin. After dialysis and lyophilization, isolation of the protein was performed by ion-exchange chromatography. This was a rapid, efficient and inexpensive two step method that allows high homogeneous protein yield and has advantages over other methods since it preserves the native structure of β-lactoglobulin.

In 2006, Andrews reported a simple, rapid and cost-effective preparation of two milk peptide components in a high degree of purity, and in gramme quantities, for evaluation of such properties (Andrews, *et al.*, 2006). The purification process was more efficient if β-casein was used as starting material. In this work, we prepared 46 g of β-casein from sodium caseinate in a simple rapid DEAE-cellulose ion-exchange chromatography stage. This was followed by in vitro hydrolysis with plasmin and precipitation and gel filtration steps.

R. Hahn (Hahn, *et al.*, 1998) investigated a fractionation scheme for the economically interesting proteins, such as IgG, lactoferrin and lactoperoxidase, based on cation exchangers. In his work, S-Sepharose 2 FF, S-Hyper D-F and Fractogel EMD SO 650 (S) were considered as successful candidates for the large-scale purification of 3 bovine whey proteins.

Fweja (Fweja, *et al.*, 2010) isolated Lactoperoxidase (LP) from whey protein by cation-exchange using Carboxymethyl resin (CM-25C) and Sulphopropyl Toyopearl resin (SP-650C). The recovery was much greater with column procedures and the purity was higher than batch column.

Xiuyun Ye (Ye, *et al.*, 2002) described a mild and rapid method for isolating various milk proteins from bovine rennet whey. β-Lactoglobulin from bovine rennet whey was easily adsorbed on and desorbed from a weak anion exchanger, diethylaminoethyl-Toyopearl. However, α-lactalbumin could not be adsorbed onto the resin. α-Lactalbumin and β-lactoglobulin from rennet whey could also be adsorbed and separated using a strong anion exchanger, quaternary aminoethyl-Toyopearl. The rennet whey was passed through a strong cation exchanger, sulphopropyl-Toyopearl, to separate lactoperoxidase and lactoferrin. α-Lactalbumin and β -lactoglobulin were adsorbed onto quaternary aminoethyl-Toyopearl. α-Lactalbumin was eluted using a linear (0–0.15 M) concentration gradient of NaCl in 0.05 M Tris–HCl buffer (pH 8.5). Subsequently, β-lactoglobulin B and β-lactoglobulin A were eluted from the column with 0.05 M Tris–HCl (pH 6.8), using a linear (0.1–0.25 M) concentration gradient of NaCl. The disadvantage of this system may be the disappearance of Ig and bovine serum albumin (BSA).

3.1.3 New ion-exchange process and technology

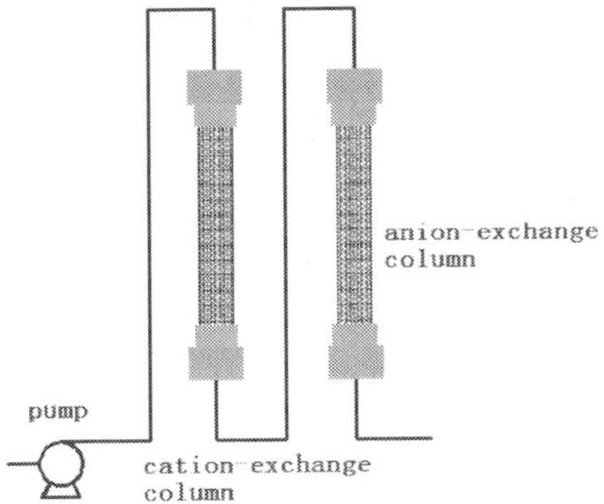

Fig. 2. The process of two ion-exchange columns in series for the isolation of Lf and IgG

Recently, the ion-exchange chromatography was improved to adapt the requirement of separation. It was combined with other ion exchange steps and with affinity chromatography to achieve complete purity in a wide range of biological systems and a wide variety of protein classes. Wu and Xu developed a novel process which could separate LF and IgG simultaneously from bovine colostrum by combining cation (CM-sepharose FF) and anion (DEAE-sepharose FF) ion exchange chromatography which showed in Fig.2.

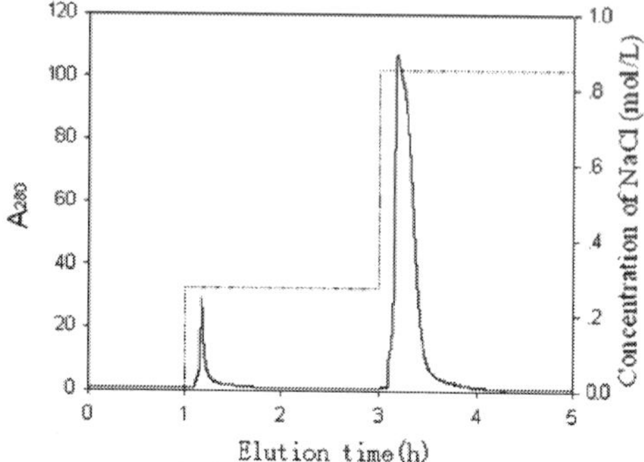

Fig. 3. Isolation of LF from of the ultrafiltrated colostrum whey by cation-exchange chromatography using CM-sepharose FF column (1.6 × 25 cm). Adsorption phase, 500 mL ultra-filtrated colostrum whey (pH 6.8); washing phase, 200 mL de-ironed water; eluting phase, 200 mL 0.27 mol/L and 200 mL 0.85 mol/L NaCl solution with sequential saline gradient.

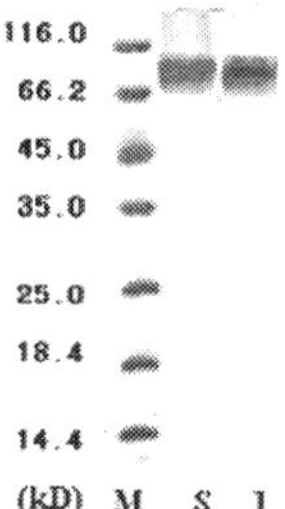

Fig. 4. SDS-PAGE profile of fractions obtained in ultrafiltrated whey by cation-exchange column using saline gradient. Lane M, protein markers; lane S, Lf standard; lane 1, elution peak with 0.85 mol/L NaCl.

After dilution, the ultra-filtrated whey was passed though a cation-exchange column of CM-sepharose FF followed by an anion-exchange column of DEAE-sepharose in series. When the whey (pH = 6.8) was passed through the CM-sepharose column, proteins with pI above 6.8 were adsorbed on the resin. Figure 3 showed the results of CM-sepharose FF cation-exchange chromatography. After the unabsorbed proteins were eluted from the column, the column was washed with sodium chloride solutions of increasing molarities (0.27 and 0.85 mol/L) in a stepwise manner. The fraction in the first peak (P1) was weakly adsorbed proteins which could not be retained on the resin during washing with 0.27 mol/L NaCl solution. The more strongly adsorbed proteins were eluted and formed the second peak (P2). The fraction in P2 was identified as Lactoferrin (LF) by SDS-PAGE (Fig. 4, Lane 1) and the purity of LF analysised by HPLC was 96.6%.

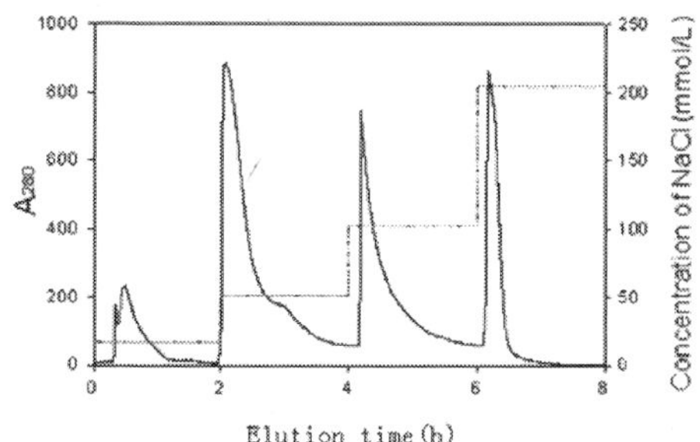

Fig. 5. Isolation of LF from the ultrafiltrated colostrum whey by an ion-exchange chromatography using DEAE-sepharose FF column (1.6 × 75 cm). Adsorption phase, 500 mL ultra-filtrated colostrum whey (pH 6.8); washing phase, 300 mL de-ironed water; eluting phase, 600 mL 17 mmol/L, 600 mL 51 mmol/L, 600 mL 103 mmol/L, and 600 mL 205 mmol/L NaCl solution in a stepwise manner.

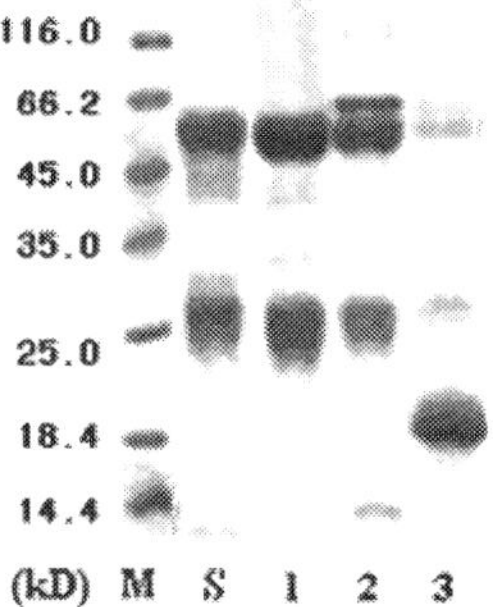

Fig. 6. SDS-PAGE profile of fractions obtained in ultrafiltrated whey by anion-exchange chromatography using saline gradient. Lane M, proteins marker; lane S, IgG standard; lane 1, elu-tion peak with 51 mmol/L NaCl; lane 2, elution peak with 103 mmol/L NaCl; lane 3, elution peak with 205 mmol/L NaCl with stepwise saline gradient.

When the colostrum whey was passed though the DEAE-Sepharose FF column, the proteins with pI below 6.8, including IgG were exchanged on the resin. After washed by de-ionized water, the column was eluted by sequential stepwise gradients with 17, 51, 103, and 205 mmol/L NaCl. The elution profiles were shown in Fig. 5. The second peak in Fig. 5, which was eluted by 51 mmol/L NaCl, was identified as IgG by SDS-PAGE (Fig. 6, lane 1) and it showed high IgG immune activity as measured by ELISA method. IgG was also detected in the third peak of Fig. 5, which was eluted with 103 mmol/L NaCl (Fig. 6, lane 2). Both SDS-

PAGE and ELISA methods shown that the fraction in the second peak had higher purity and IgG activity than that in the third peak.

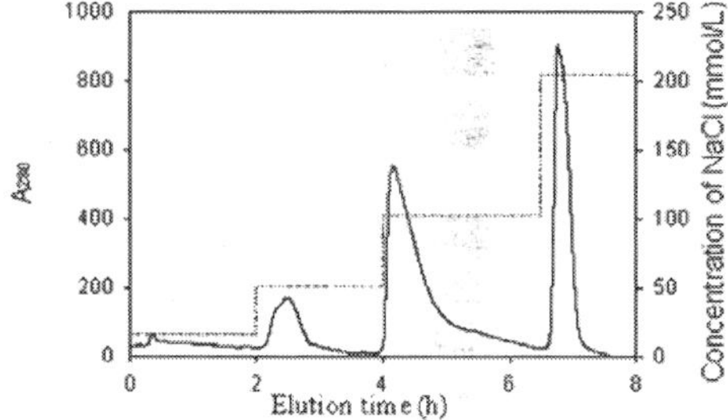

Fig. 7. Isolation of LF from of the un-ultrafiltrated colotrum whey by anion-exchange chromatography using DEAE-sepharose FF column (1.6 × 75 cm). Adsorption phase, 500 mL ultra-filtrated colostrum whey (pH 6.8); washing phase, 300 mL deironed water; eluting phase, 600 mL 17 mmol/L, 600 mL 51 mmol/L, 600 mL 103 mmol/L, and 600 mL 205 mmol/L NaCl solution in a stepwise manner.

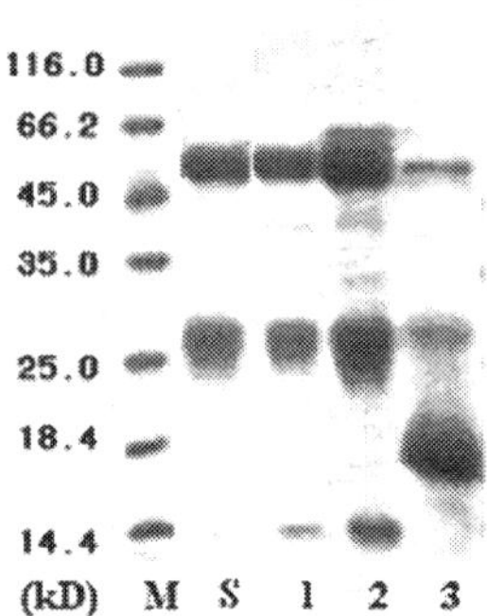

Fig. 8. SDS-PAGE profile of fractions obtained in un-ultrafiltrated whey by anion-exchange chromatography using saline gra-dient. Lane M, proteins marker; lane S, IgG standard; lane 1, elution peak with 51 mmol/L NaCl; lane 2, elution peak with 103 mmol/L NaCl; lane 3, elution peak with 205 mmol/L NaCl with sequential saline gradient.

Elution curves (Fig. 5 and Fig. 6) and SDS-PAGE profiles (Fig. 7 and Fig. 8) showed that four protein fractions could be separated by anion-exchange chromatography with the same saline gradient using both un-ultrafiltrated and ultrafil-trated samples. Compared Fig. 5 with Fig. 7, elution with 103 mmol/L and 205 mmol/L NaCl produced relatively both the same broad peaks with tailing, but the peaks washed by 17 mmol/L and 51 mmol/L NaCl showed that the fractions from ultrafiltrated whey sample had the higher protein concentration than those from the un-ultrafiltrated. Proteins in whey can be agglomerated and denaturized within ultrafiltra-tion process and the SDS-PAGE profiles also indicated that the peak contained other proteins in colostrum whey. From the results, it could be deduced that the higher concentration of the other proteins in the ultrafil-trated whey than that in un-ultrafiltrated whey was by reason that the other proteins which were exchanged nonspecifically on the resin could be desorbed at relatively low saline solu-tion such as 17 mmol/L NaCl.

The majority of IgG could be eluted with 51 mmol/L NaCl with both un-ultrafiltrated and ultrafiltrated colostrum whey and both fractions had the same IgG purity about 95% (w/w) by HPLC analysis, but the peak obtained in ultrafiltrated whey had higher IgG concentration than that obtained in un-ultrafiltrated whey. Small molecule such as salts (ions), sugars, and amino acids could be easily adsorbed on the sorbent and reduced IgG adsorption capacity of the resin. The ultrafiltrated whey, due to the fact that low molecular mate rials in the original whey were removed by ultrafiltration, showed higher ion exchange capacity for IgG and resulted in a higher concentration of IgG in the fraction. According to SDS-PAGE profiles (Fig. 6 and Fig. 8), the proteins eluted with 103 mmol/L contained IgG, BSA and α-lactalbumin. The concentration of α-lactalbumin in Fig. 6 was lower than in Fig. 8 for the reason that α-lactalbumin was mostly removed by ultrafiltration process. The major protein in the fractions eluted by 205 mmol/L NaCl for both ultrafiltrated and un-ultrafiltrated colostrum whey was β-lactoglobulin. Furthermore, there was no clear difference in the β-lactoglobulin concentration for both ultrafiltrated and un-ultrafiltrated colostrum whey samples. Although the molecular weight of β-lactoglobulin was 28 kD, under pH 7.0 the major portion of β-lactoglobulin could be polymerized into dimmer. Therefore, the major portion of β-lactoglobulin was retained, while the colostrum whey was ultrafiltrated with a 50 kD molecular weight cut-off polyethersulfone membrane.

	Bovine colostrum	Acid whey	Step 1[a]	Step 2[b]	Step 3[c]	Step 4[d]	Step 5[e]
Concentration of Lf (mg/mL)	1.02	0.94	0.8	0.03	0.01	0.7	-
Recovery (%)	100	92.16	78.43	-	-	68.63	-
Concentration of IgG (mg/mL)	25.00	18.25	17.75	17.24	0.11	-	11.34
Recovery (%)	100	73.00	71.00	68.96	-	-	45.38

[a]Acid whey was ultrafitrated with 50 kD molecular weight cut-off membrane.
[b]Whey was passed through cation- exchange column.
[c]Whey was passed through anion- exchange column.
[d]Fraction was eluted with 0.85 mol/L NaCl on cation exchange column.
[e]Fraction was eluted with 51 mmol/L NaCl on anion exchange column.

Table 1. Concentration and recovery yield of LF and IgG at each step of the over all separation process

Concentrations of LF and IgG at every step of the separation process were analyzed by ELISA method (Table 1). According to the results shown in Table 1, the activity of LF was only decreased by a little (about 7%), but the activity of IgG was lost severely (about 25%) during preparation of the acid colostrum whey. During ultrafiltration process, the activity of LF was lost a lot (about 14%), whereas that of IgG was lost a little (only 2%). On the other hand, 9.8% of LF activity and 23.6% of IgG activity were lost during cation-exchange chromatography and anion-exchange chromatog-raphy, respectively. In summary, the recovery yields for LF and IgG in the overall separation process were 68.83% and 45.38%, respectively.

In summary, a novel process for the isolation of the high value bovine LF and IgG from colostrum whey was developed. The LF and IgG were purified by two ion-exchange columns in series. The two resins had opposite polarity. Results showed that the proposed procedures were fast, reliable, and effective. Additionally, ultrafiltration can be used as a pretreatment method to remove small molecules and to increase both the product purity and recovery rate of LF and IgG. Furthermore, the serial ion-exchange chromatography need not use buffers to maintain pH of the whey samples and can be operated at high flow rates. In general, the purities of 96.6% (w/w) LF and 95.0% (w/w) IgG were obtained with respective recovery rates of 68.83% and 45.38% by serial cation-anion exchange chromatography from ultrafiltrated bovine colostrum. (Wu & Xu, 2009)

Isidra Recio (Recio & Visser, 1999) reported a membrane method for the rapid isolation of antibacterial peptides from lactoferrin (LF) which was more rapid and offers several economic advantages than exchange chromatography. Cheese whey was filtered through a cation-exchange membrane, and the selectively bound LF was directly hydrolysed in situ with pepsin. Inactive LF fragments were washed off the membrane with ammonia, and a fraction enriched in LFcin-B was obtained by further elution with 2 M NaCl.

Ulber (Ulber, et al., 2001) discussed the application of several membrane types for a crossflow filtration of sweet whey to remove insoluble particles and lipids from the whey with the aim of obtaining permeate which could be directly used for down-streaming the minor component via ion exchange membrane adsorber systems. Using a two-step downstream process consisting of a cross-flow filtration and a membrane adsorbent was possible to isolate bLF from sweet whey in a very suitable manner. The advantages of a membrane adsorbent system in direct comparison with ion exchange chromatographic support were to be found in its higher flow rates and, therefore, shorter cycle times as well as in easier handling and upscaling.

Saufi (Saufi & Fee, 2009) described the application of Mixed Matrix Membrane (MMM) chromatography for fractionation of β-Lactoglobulin from bovine whey. MMM chromatography was prepared using ethylene vinyl alcohol polymer and lewatit anion exchange resin to form a flat sheet membrane. The membrane was characterized in terms of structure and its static and dynamic binding capacities were measured. The optimum binding for β-Lactoglobulin was found to be at pH 6.0 using 20 mM sodium phosphate buffer. The MMM had a static binding capacity of 120 mg/g membrane (36 mg/mL membrane) and 90 mg/g membrane (27 mg/mL membrane) for β-Lactoglobulin and α-Lactalbumin, respectively. In batch fractionation of whey, the MMM showed selective binding towards β-Lactoglobulin compared to other proteins. The dynamic binding capacity of β -Lactoglobulin in whey solution was about 80 mg/g membrane (24 mg β-Lac/mL of MMM), which was promising for whey fractionation using this technology. The mixed matrix membrane showed excellent potential for a whey protein fractionation application, particularly for selective binding of β-Lac. The membrane had a defect-free structure and provided a high binding capacity for β-Lac in whey solution, compared with other proteins. The MMM had maximum equilibrium binding capacities of 150 mg β-Lac/g membrane and 90 mg α-Lac/g membrane in individual pure protein experiments. In batch fractionation of whey, the MMM had almost the same binding capacity for β-Lac as it did for pure β-Lac.

Anders Heebøll-Nielsen (Anders, et al., 2004) described the design, preparation and testing of superparamagnetic anion-exchangers, and their use together with cation-exchangers in the fractionation of bovine whey proteins as a model study for high-gradient magnetic fishing. Crude bovine whey was treated with a superparamagnetic cation-exchanger to adsorb basic protein species, and the supernatant arising from this treatment was then contacted with the anion-exchanger. In the initial cation-exchange step quantitative removal of lactoferrin (LF) and lactoperoxidase (LPO) was achieved with some simultaneous binding of immunoglobulins (Igs). The immunoglobulins were separated from the other two proteins by desorbing with a low concentration of NaCl (≤0.4 mol/L), whereas lactoferrin and lactoperoxidase were co-eluted in significantly purer form when the NaCl concentration was increased to 0.4-1 mol/L. The anion-exchanger adsorbed β-lactoglobulin selectively allowing separation from the remaining protein.

Compared with the other chromatographic methods, ion-exchanger chromatography has the advantages of low cost, reduced steps, continuous feed-in, and easy to scale-up. It has shown potential for commercial applications.

3.2 Affinity chromatography

Affinity chromatography is a prevailing procedure to isolate and purify the active substances. This technique is based on molecular recognition or bio-recognition which is widespread in many professional disciplines, such as biology, molecular biology and chemistry (Wilchek & Chaiken, 1968; Wilchek & Miron, 1999; Scopes, 1999)

3.2.1 Principles of affinity chromatography

Affinity chromatography primarily requires a group of proteins to have a reversible interaction with a specific ligand attached to a solid matrix; in addition, the effectiveness of affinity purifications relies on the ability of the protein to recognize specifically an affinity adsorbent. As for the procedure of affinity chromatography, when the compound is passed through the affinity column at a certain flow velocity, the desired active substances will be attached to an affinity adsorbent immobilized to the chromatography matrix. With the different solution passing through the affinity column, the binding between the absorbent and the active substances can be loosened by a change in buffer conditions, such as the pH, ionic strength or polarity, consequently the desired component are eluted relatively free of contaminants. Virtually, affinity chromatography always result in high selectivity, high resolution and high capacity for the proteins of interest. The key stages in an affinity chromatography are shown in Figure 9.

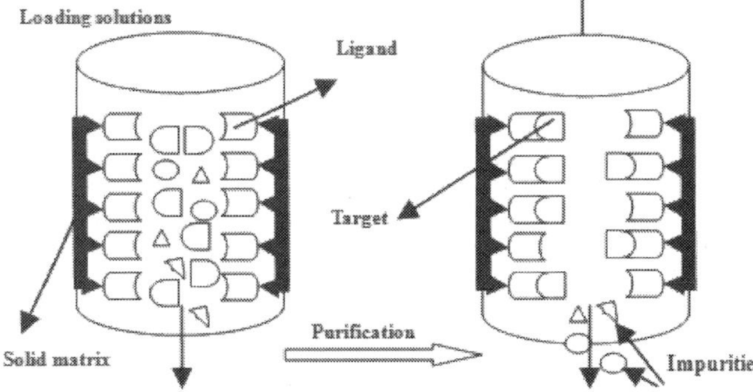

Fig. 9. The basic principle of affinity chromatography

The first protein which was purified by affinity chromatography was α-amylase in 1910. From then on, affinity chromatography was applied extensively. There are lots of various applications derived from affinity chromatography; we can see a series of those in table 2

1	Immunoaffinity chromatography	11	Centrifuged affinity chromatography
2	Lectin affinity chromatography	12	Affinity repulsion chromatography
3	Metal-chelate affinity chromatography	13	Theophilic chromatography
4	Covalent affinity chromatography	14	Membrane-based affinity chromatography
5	Perfusion affinity chromatography	15	Weak affinity chromatography
6	High performance affinity chromatography	16	Receptor affinity chromatography
7	Affinity precipitation	17	Molecular imprinting affinity
8	Filter affinity transfer chromatography	18	Library-derived affinity ligands
9	Dye-ligand affinity chromatography	19	Affinity partitioning
10	Affinity electrophoresis	20	Affinity capillary electrophoresis

Table 2. Various Techniques stemmed from Affinity Chromatography (Wilchek & Chaiken, 1968)

3.2.2 Application of affinity chromatography

In 1988, Lee applied the Cu-loaded immobilized metal affinity chromatography to separate of immunoglobulins from bovine blood which were pre-retreated by polyphosphate precipitation. The IgG gained by this procedure were almost pure after the residual polyphosphate (Lee, et al., 1966). In 1990, Timothy immobilized a DNA-aptamer specific for human L-selectin to a chromatography matrix to create an affinity column; meanwhile, they used this column to purify a recombinant human L-selectin-Ig fusion protein from Chinese hamster ovary cell-conditioned medium. A 1500-fold purification with an 83% single step recovery came out by the first-step purification, and this demonstrated that oligonucleotide aptamers could be effective affinity purification reagents (Romig et al., 1999). Roque immobilized the ligand 8/7 on to hexanediamine-modified agarose as affinity media, and applied this media to purify the immunoglobulins and Fab fragments by affinity chromatography. The finding shows the ligand 8/7 hibits the interaction of PpL with IgG and Fab by competitive ELISA and has negligible binding to Fc. The ligand 8/7 adsorbent is better than an artificial protein L to bind to immunoglobulins from different sources, in short, all this reflects the efficient isolation immunoglobulins from raw samples (Roque, et al., 2005)

In 1995, Bottomley isolated human IgG by using immobilized analogues of protein A for affinity chromatography. They applied a linear gradient from pH 5.0 to pH 3.0 of 0.5 M acetate buffer to elute the loaded column. In this study, the problems related to low pH elution could be decreased while the pH range for elution increased (Bottomley et al., 1995).

In 2001, Tu conducted a research of preparing LF bound sepharose 4B gel which was used as an affinity ligand. After the crude IgY or rabbit serum was loaded to the Lactoferrin bound sepharose 4B column, the column was washed and eluted by two kinds of buffer. All the collected fractions were treated and analyzed. This study revealed antibody specific against LF for affinity chromatography from crude IgY was more reliable than that from rabbit serum (Tu *et al.*, 2001). Analogously, Chen made a preparation of Lysozyme bound sepharose 4 fast flow gel which was applied to isolate IgY in the affinity chromatrography. This research showed the binding capacity was lower and the dissociation constant was higher than both of the monoclonal antibody immunoaffinity column chromatography; in addition, this Lysozyme bound sepharose 4 immunoaffinity column was competent in sparating IgY specific against Lysozyme from yolk (Chen *et al.*, 2002)

3.2.3 Novel affinity chromatography process for the purification of bioactive protein from Bovine colostrum

Ounis once used heparin affinity chromatography to separate the protein components from two whey protein solutions which produced by ion-exchange chromatography (IEC-WPI) and microfiltration / ultrafiltration (MF/UF-WPI) respectively (Ounis, *et al.*, 2008). After the column was equilibrated, WPI solution was passed through the column at a flow rate of 1 mL/min, then the column was washed by 0.01 M phosphate buffer , in the wake of this, sequential elution steps were executed with 0.01 M phosphate buffer containing 0.5, 1.0 or 2.0 M NaCl. The passed solutions were collected every step and determined by the bicinchoninic acid (BCA) protein assay, Enzyme-Linked ImmunoSorbent Assay, reversed-phase high-performance liquid chromatography and 2-dimensional gel electrophoresis respectively. The results from these determinations revealed that heparin affinity chromatography had not only the capacity to separate the major proteins contained in WPIs, but also the ability of concentrating the minor cationic proteins and some growth factors.

Affinity membrane chromatography is a technique which combines membrane chromatography with affinity interaction; the membranes contain biospecific ligands on their inner pore surface. As a result of convective flow of the solution through the pores, the mass transfer resistance is tremendously reduced, and binding kinetics dominates the adsorption process. Affinity membrane chromatography provides high selectivity and fast processing for the isolation and purification of proteins. In 2007, Wolman applied affinity membrane chromatography to purify lactoferrin from whey and colostrum in only one step. The study used a hollow fibres synthesized by grafting a glycidyl methacrylate or dimethyl acrylamide copolymer to polysulfone membranes and attaching the Red HE-3B dye to them. According to the comparison between the productivity produced by Red HE-3B hollow-fibre membranes and d-Sepharose, Red HE-3B hollow-fibre membranes showed a more acceptable chromatographic performance for Lf purification from bovine colostrum than the obtained with d-Sepharose. In addition, the Lf obtained from bovine colostrum by this one-step procedure contained the casein and immunoglobulin as the only contaminants, so it could be treated as a final product practically (F.J. Wolman, *et al.*, 2007; Dimartino, *et al.*, 2011; Zou, *et al.*, 2001)

Akita made an immunoaffinity column with specific egg yolk immunoglobulin (Ig) Y against bovine IgG1 and IgG2 and used this column to isolate the IgG1 and IgG2 from cheddar cheese whey of colostrun. The study revealed that the potential binding capacity of

IgY could come up to 38% after the immobilization by reductive amination. Meanwhile, this immunoaffinity column with specific egg yolk immunoglobulin (Ig) Y could be used to isolate the bovine immunoglobulin G subclasses from whey and colostrum specificly. (Akita & Chan, 1998)

In 1998, a study by Kim was based on application of affinity chromatography to separate the immunoglobulin G from Cheddar cheese whey. Initially, they make a preparation of IgY which is specific to IgG, then, biotinylation of IgY and immobilization of avidin columns were performed, after that, they coupled each other together and determined the binding capacity of avidin-biotinylated IgYIgG columns, finally the cheddar cheese whey was loaded and IgG was isolated. This study showed that IgG from Cheddar cheese whey could be isolated one step by the avidin biotinylated IgYIgG column chromatography. It's notable that the IgG binding capacity of this study was 50-55% and purity of the recovered IgG was 99%. There is possible for this avidin biotinylated IgYIgG column to be applied in high-purity IgG (Kim & Chan, 1998).

In 2007, Chen synthesized a micron-sized monodisperse superparamagnetic polyglycidyl methacrylate (PGMA) particles coupled with heparin (PGMA-heparin) and they isolated lactoferrin from bovine whey. In the main procedure, they made a preparation of magnetic affinity adsorbents and the whey which was going to be isolate firstly, then whey was incubated with magnetic affinity adsorbents at a certain proportion. After that, the adsorbents were eluted with the same butter respectively in different concentration sequentially. The results from analysis and determination indicate the potential application of magnetic PGMA-heparin particles for production of high purified LF from whey (Chen, *et al.*, 2007).

3.3 Hydrophobic Charge Induction Chromatography (HCIC)

The nutritional values and physiological benefits of Igs, a major whey protein in bovine colostrum, have received more and more attention in the last two decades. As a result, developing low cost and high efficiency purification process to fulfil the growing demand of Igs is significantly necessary. Traditionally, by taking the advantage of different isoelectric points of whey proteins, various kinds of ion-exchange sorbents have been synthesized for the purification of immunoglobulins. In practice, however, single or merely several ion-exchange chromate-graphic procedures can hardly obtain high purity protein of interest from acid whey of bovine colostrum. Hydrophobic charge-induction chromatography, or HCIC, is a novel chromatographic technique for separation of biological macromolecules, based on the pH-dependent behavior of ionizable, dual-mode ligands. Selectivity is orthogonal to ion exchange and other commonly employed chromatographic modes (Boschetti, *et al.*, 2000).

3.3.1 The mechanism of HCIC

HCIC binding is based on mild hydrophobic interaction and is achieved under near-physiological conditions, without the addition of lyotropic or other salts. Desorption is based on electrostatic charge repulsion and is accomplished by reducing the pH of the mobile phase. Under mild acidic conditions (pH4.0–4.5), the ligand and target molecule take on a net positive charge; binding is thus disrupted and elution occurs. Elution is conducted using dilute buffer (e.g., 50mM acetate). The new BioSepra MEP HyperCel sorbent from Life Technologies, Inc. (LTI; Rockville, MD) has been optimized for capture and purification of monoclonal and polyclonal IgG. The heterocyclic ligand, derived from 4-

mercaptoethylpyridine (4-MEP), provides efficient capture and purification of antibodies from a broad range of sources, such as animal sera, ascites fluid and a variety of cell culture supernatants, including protein-free, chemically defined, protein-supplemented and serum-supplemented media.

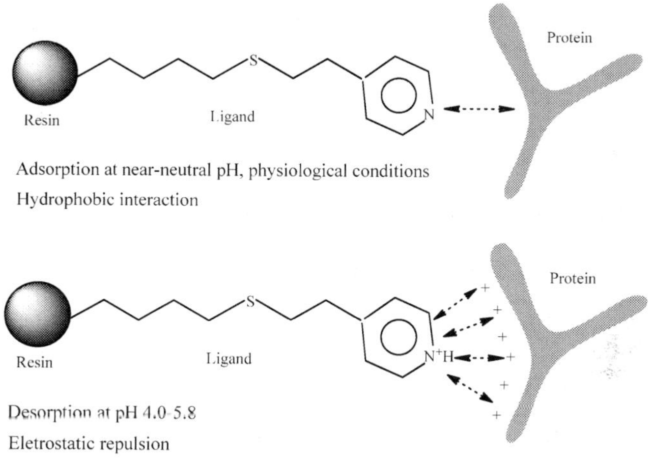

At neutral pH, (top) the ligand is uncharged and binds molecules through mild hydrophobic interaction. As the pH is reduced (bottom), the ligand becomes positively charged and hydrophobic binding is disrupted by electrostatic charge repulsion.

Fig. 10. Mechanism for hydrophobic charge-induction chromatography

3.3.2 The advantage of HCIC

a.   Independent of ionic strength

Compared with hydrophobic interaction chromatography (HIC), HCIC is also typified by adsorption of proteins to a moderately hydrophobic surface. However, HCIC could adsorb proteins without the presence of high concentrations of a lyotropic salt such as ammonium sulphate. HCIC matrices have a higher ligand density than HIC, therefore, it could bind proteins at low ionic strength. High ligand density (80 mmol/mL) matrices have been used for mixed mode hydrophobic ionic chromatography for the purification of chymosin, which resulted in high capacity (Burton, 1997). Furthermore, chymosin could be adsorbed at high and low ionic strength, therefore, a pretreatment step of salt addition, or removal by dialysis, dilution or ultrafiltration was not required. HCIC reduced sample preparation requirements. This method was simple, efficient, inexpensive and provided very good resolution of chymosin from a crude recombinant source.

b.   pH-dependent binding

At the beginning, the matrices of HCIC absorbents contained amine linkages or carboxyl groups, therefore, the absorbents were charged at pH 4-9 range. Adsorption to an uncharged surface was only possible at pH extremes. Nonspecific electrostatic interactions could result in lower capacity and product purity and/or matrix fouling problems. Furthermore, charged groups could interfere with adsorption of target proteins. If carboxyl/amine groups were replaced with weaker acids or bases such as imidazole, uncharged matrix form could be obtained within the pH 4–10 range. In its preferred form, adsorption is carried out under conditions which do not cause electrostatic repulsion between the protein and the matrix. However, by reducing the pH of the mobile phase, like charge are established on both ligand and protein. When pH of the mobile is reduced, the magnitude of the opposing charges depends on the pI of the target protein and the pKa of the ligand. Desorption is prompted by electrostatic charge repulsion by reducing the pH of the mobile phase.

3.3.3 Research progress and application of HCIC for protein purification

Recent efforts to improve hydrophobic interaction chromatography (HIC) for use in monoclonal antibody (mAb) purification have focused on two approaches: optimization of resin pore size to facilitate mAb mass transport, and use of novel hydrophobic charge induction (HCIC) mixed mode ligands that allow capture of mAbs under low salt conditions. Hydrophobic charge induction chromatography (HCIC), as a mixed-mode chromatography, achieves high adsorption capacity by hydrophobic interaction and facile elution by pH-induced charge repulsion between the Solute and ligand. In 2008, Chen (Chen, 2008) evaluated standard HIC and new generation HIC and HIC-related chromatography resins for mAb purification process efficiency and product quality both as isolated chromatography steps and in purification process trains. They found that the HCIC Mercapto-Ethyl-Pyridine (MEP) resin, which shows a different salt impact trend and impurity resolution pattern from standard HIC resin, can not only capture mAb from crude CHO fermentation supernatant but also substantially enhance mAb purification process flow efficiency when serving as a polishing role. Under the condition of 0.4 M NaCl, the binding capacity of MEP resin for IgG reached 30 mg/g resin near pH=7, higher than Butyl-650M resin 20.5 mg/g resin.

Large amount of study on the mechanism and optimization of HCIC resins have been conducted. Sun (Sun, 2008) reported a new medium, 5-aminoindole-modified Sepharose (Al-Sepharose) for HCIC. The adsorption equilibrium and kinetics of lysozyme and bovine serum albumin (BSA) to Al-Sepharose were determined by batch adsorption experiments at different conditions to provide insight into the adsorption properties of the medium. The results showed that the influence of salt type on protein adsorption to Al-Sepharose was corresponded with the trend for other hydrophobicity-related properties in literature. Both ligand density and salt concentration had positive influences on the adsorption of the two proteins investigated. The adsorption capacity of lysozyme decreased rapidly when pH decreased from 7 to 3 due to the increase of electrostatic repulsion, while BSA, an acidic protein, achieved maximum adsorption capacity around its isoelectric point. Dynamic adsorption experiments showed that the effective pore diffusion coefficient of lysozyme remained constant at different salt concentrations, while that of BSA decreased with increased salt concentration due to its greater steric hindrance in pore diffusion. High protein recovery by adsorption at pH 7.10 elution at pH 3.0 was obtained at a number of NaCl concentrations, indicating that the adsorbent has typical characteristics of HCIC and potentials for applications in protein purification.

In 2010, Wang (Wang, 2010) introduced the methods of molecular simulation to study the interactions between MEP and IgG. Firstly, molecular docking is used to identify the potential binding sites around the protein surface of Fc Chain A of IgG, and 12 potential binding sites were found. Then 6 sites were further studied using the molecular dynamics simulations. The results indicated that MEP ligand tends to bind on the hydrophobic area of Fc Chain A surface. At neutral conditions, MEP can bind stably on the site around TYR319 and LEU309 of Fc Chain A, which showed obviously a pocket structure with strong hydrophobicity. The analysis of trajectory revealed that hydrogen bonds exist between MEP and the former two amino acids around the simulation period. The binding of MEP to other sites were relatively unstable, and depends on the initial binding modes of MEP. When the pH lowered to 4.0, it could be found that MEP bound formerly on the Fe Chain A departed quickly due to the electrostatic repulsion, weaker hydrophobic interaction and the disappearance of hydrogen bonds. With the aids of molecular simulations, the separation mechanism of HCIC was verified from the view of molecular interactions-the binding with hydrophobic interactions at neutral condition and the desorption with electrostatic repulsion at acid condition.

Lin (Lin, 2010) used the immunoglobulin of egg yolk (IgY) to investigate the effects of salt on HCIC. The adsorption behavior of antibody IgY on several HCIC adsorbents as a function of salt concentration was studied using adsorption isotherms and adsorption kinetics. The hydrodynamic diameters and potentials of IgY at various salt concentrations were also determined. It was found that the saturated adsorption capacities increased linearly with increasing salt concentration because of the improvement of hydrophobic interactions between IgY and the HCIC ligands. The pore diffusion model was used to evaluate the dynamic adsorption process. The total effective diffusivity showed a maximum value at an ammonium sulfate concentration of 0.2 M. The results indicate salt-promoted adsorption under the appropriate concentration due to a reduction of protein size and the enhancement of hydrophobic interactions between IgY and the HCIC ligand. Therefore, the addition of a proper amount of salt is beneficial for antibody adsorption in the HCIC process. Although certain progress has been achieved in recent years, advanced study is still necessary for the wide and mature application of HCIC.

4. Conclusion

Bovine colostrum or whey is a mixture of lactose, protein, fat and minerals. Therefore, the isolation of specific bioactive proteins such as LF and Igs is still a challenge. The application of bovine active proteins should be considered when designing the isolation protocol. With the development of application scope in food industry and biomedicine, isolation of high purity bovine proteins has attracted more and more attentions. The criteria for separation of proteins from bovine colostrum and milk or their by-products should be 1) bioactive proteins retain a reasonable recovery rate and purity; 2) utilization of organic solvents and other non-food grade chemicals is avoided because of the potential application as nutraceutical and functional foods; and 3) the separation procedures have a potential for commercialization.

To get high purity bioactive proteins from bovine colostrum in commercial scale, chromatographic procedures are essential in the process. Compared with the other chromatography separation processes, IEC, HCIC and affinity chromatography have the potential to be utilized in purification of proteins from bovine colostrum in commercial scale. The protocol of selecting a certain chromatographic procedure is based on the characteristics of the proteins in the bovine colostrum, such as their size, shape, charge, hydrophobicity, solubility and biological activity.

IEC is one of the liquid chromatography techniques which based on electrostatic interactions. Different proteins in bovine colostrum have different charges and interact differently in ion exchange chromatography. As a main kind of bioactive protein, LF which has relative high isoelectric point (pI) compared with other milk proteins and is suitable to be isolated by this method. Many sorts of cation ion exchangers, such as CM and SP resins can be selected in purification of LF. Affinity chromatography which is based on molecular recognition or bio-recognition can be used in separation of antibodys in bovine colostrum or bovine whey with high purity, such as IgG and IgA. However, considering the production scale and cost, this technology is limited to be applied in commercial scale. Compared with affinity chromatography, hydrophobic charge-induction chromatography (HCIC) based on the pH-dependent behavior of ionizable, dual-mode ligands is a hopeful chromatographic technique for separation of biological macromolecules, especially antibodies in bovine colostrum with relative low cost and high efficiency such as high purity achieved in a single step, high protein capacity, and easy cleaning. Moreover, the small molecular substances in bovine colostrum or bovine whey, such as lactose, vitamins, and oligosaccharides, can be isolated by applying membrane filtration, especially nanofiltration and ultrafiltration.

5. References

Gapper, L.W., et al., Analysis of bovine immunoglobulin G in milk, colostrum and dietary supplements: a review. *Analytical and Bioanalytical Chemistry*, 2007. 389(1): p. 93-109.

Lu, R.R., et al., Isolation of lactoferrin from bovine colostrum by ultrafiltration coupled with strong cation exchange chromatography on a production scale. *Journal of Membrane Science*, 2007. 297(1-2): p. 152-161.

Saufi, S.M. and C.J. Fee, Recovery of lactoferrin from whey using cross-flow cation exchange mixed matrix membrane chromatography. *Separation and Purification Technology*, 2011. 77(1): p. 68-75.

LIU, Y.Y.X.G.a.X., Isolation and Purification of Bovine Colostrum sIgA and IgG. *Journal of Northeast Agricultural University* 2007. 15(1): p. 58-61.

Almecija, M.C., et al., Effect of pH on the fractionation of whey proteins with a ceramic ultrafiltration membrane. *Journal of Membrane Science*, 2007. 288(1-2): p. 28-35.

Lozano, J.M., G.I. Giraldo, and C.M. Romero, An improved method for isolation of beta-lactoglobulin. *International Dairy Journal*, 2008. 18(1): p. 55-63.

Himmelhoch S R (1971) Ion-exchange chromatography. *Methods Enzyme* vol 22, 273–290

Carl Dolman, Mark Page, and Robin Thorpe. Purification of IgG Using Ion-Exchange HPLC/FPLC. *The Protein Protocols Handbook (2nd Edition).* 2002:989

Shireen Doultani, K. Nazan Turhan, Mark R. Etzel. Fractionation of proteins from whey using cation exchange chromatography. *Process Biochemistry.* 2004(39):1737-1743.

S.J. Gerberding, C.H. Byers. Preparative ion-exchange chromatography of proteins from dairy whey. *Journal of Chromatography A.* 1998(808):141-151.

José Manuel Lozano, Gloria I. Giraldo, Carmen M. Romero. An improved method for isolation of β-lactoglobulin. *International Dairy Journal.* 2008(18):55-63.

A.T. Andrews et al. β-CN-5P and β-CN-4P components of bovine milk proteose-peptone: Large scale preparation and influence on the growth of cariogenic microorganisms. *Food Chemistry.* 2006(96):234-241.

Leonard WT Fweja, Michael J Lewis and Alistair S Grandison. Isolation of lactoperoxidase using different cation exchange resins by batch and column procedures. *Proprietors of Journal of Dairy Research.* 2010(77)357-367.

Xiuyun Ye, Shigeru Yoshida, T.B. Ng. Isolation of lactoperoxidase, lactoferrin, α-lactalbumin, β -lactoglobulin B and β -lactoglobulin A from bovine rennet whey using ion exchange chromatography. *The International Journal of Biochemistry & Cell Biology*. 2000(32):1143-1150.

Wu, M.B. and Y.J. Xu, Isolation and purification of lactoferrin and immunoglobulin G from bovine colostrum with serial cation-anion exchange chromatography. *Biotechnology and Bioprocess Engineering*, 2009. 14(2): p. 155-160.

Isidra Recio, Servaas Visser. Two ion-exchange chromatographic methods for the isolation of antibacterial peptides from lactoferrin In situ enzymatic hydrolysis on an ion-exchange membrane. *Journal of Chromatography A*. 1999(831):191-201.

Ulber, R. et al. Downstream Processing of Bovine Lactoferrin from Sweet Whey. *Acta Biotechnol*. 2001(21):1,27-34.

Syed M. Saufi, Conan J. Fee. Fractionation of β -Lactoglobulin From Whey by Mixed Matrix Membrane Ion Exchange Chromatography. *Biotechnology Bioengineering*. 2009(103):138-147.

Meir Wilchek, Irwin Chaiken. (1968). An Overview of Affinity Chromatography. *Methods*, 147.

Young-Zoon Lee, Jeoung S. Sim, Shalan Al-Mashikhi, and Shuryo Nakai. (1988). Separation of Immunoglobulins from Bovine Blood by Polyphosphate Precipitation and Chromatography. *Agricultural and food chemistry*, (1966), 922-928.

Timothy S. Romig, Carol Bell, Daniel W. Drolet. (1999). Aptamer affinity chromatography : combinatorial chemistry applied to protein purification. *Journal of Chromatography B*, 731, 275-284.

A. Cec´ilia A. Roque, M. ^Angela Taipa, Christopher R. Lowe. (2005). An artificial protein L for the purification of immunoglobulins and Fab fragments by affinity chromatography. *Journal of Chromatography A*, 1064, 157-167.

Stephen P.Bottomley,Brain J. Sutton,Michael G. Gore. (1995). Elution of human IgG from affinity columns containing immobilised variants of protein A. *Journal of Immunological Methods*, 182, 185-192.

Yann-Ying Tu, Chao-Cheng Chen, Hung-Min Chang. (2001). Isolation of immunoglobulin in yolk (IgY) and rabbit serum immunoglobulin G (IgG) specific against bovine lactoferrin by immunoanity chromatography. *Food Research international*,34, 783-789

F.J. Wolman, M. Grasselli, et al. (2007). One-step lactoferrin purification from bovine whey and colostrum by affinity membrane chromatography. *Journal of Membrane Science*, 288, 132-138.

Simone Dimartino, Cristiana Boi, Giulio C. Sarti. (2011). A validated model for the simulation of protein purification through affinity membrane chromatography. *Journal of Chromatography A*, 1218, 1677-1690.

Hanfa Zou, Quanzhou Luo, Dongmei Zhou. (2001). Affinity membrane chromatography for the analysis and purification of proteins. *Methods*.

Chao-Cheng Chen, Yann-Ying Tu, Tzy-Li Chen et al. (2002). Isolation and Characterization of Immunoglobulin in Yolk (IgY) Specific against Hen Egg White Lysozyme by Immunoaffinity Chromatography. *Agricultural and food chemistry*, 5424-5428.

Wassef Ben Ounis, Sylvie F. Gauthier, et al. (2008). Separation of minor protein components from whey protein isolates by heparin affinity chromatography. *International Dairy Journal*, 18, 1043-1050.

E.M. Akita and E.C.Y. Li-Chan, (1998), Isolation of Bovine Immunoglobulin G Subclasses from Milk, Colostrum, and Whey Using Immobilized Egg Yolk Antibodies, *Journal of Dairy Science*, Vol 81, Issue 1, 1998, Pp54-63.

H. Kim, E. C. Y. Li-Chan. (1998). Separation of Immunoglobulin G from Cheddar Cheese Whey by Avidin-Biotinylated IgY Chromatography. *Journal of Food Science*, 63(3), 429-434.

Lin Chen, Chen Guo, Yueping Guan, Huizhou Liu. (2007). Isolation of lactoferrin from acid whey by magnetic affinity separation. *Separation and Purification Technology*, 56, 168-174.

Sohel Dalal, Smita Raghava, M.N. Gupta. (2008). Single-step purification of recombinant green fluorescent protein on expanded beds of immobilized metal affinity chromatography media. *Biochemical Engineering Journal*, 42, 301-307.

Andreas Hassl, Horst Aspock. (1988). Purification of egg yolk immunoglobulins a two-step procedure using hydrophobic interaction chromatrography and gel filtration chromatography. *Journal of Immunological Methods*, 225-228.

Pavel Kovalenko, Hidehiko Fujinaka , Yutaka Yoshida, et al. (2004). Fc receptor-mediated accumulation of macrophages in crescentic glomerulonephritis induced by anti-glomerular basement membrane antibody administration in WKY rats. *The Japanese Society for Immunology*, 16(5), 625-634.

Ken Sugo, Tomohiko Yoshitake, Masahiro Tomita et al. (2011). Simplified purification of soluble histidine-tagged green fluorescent protein from cocoon of transgenic silkworm in metal affinity hydroxyapatite chromatography. *Separation and Purification Technology*, 76(3), 432-435.

Boschetti, E., et al., Hydrophobic charge-induction chromatography - Method has some advantages over traditional antibody production methods. *Genetic Engineering News*, 2000. 20(13): p. 34-+.

Burton, S.C., N.W. Haggarty, and D.R.K. Harding, One step purification of chymosin by mixed mode chromatography. *Biotechnology and Bioengineering*, 1997. 56(1): p. 45-55.

Chen, J., J. Tetrault, and A. Ley, Comparison of standard and new generation hydrophobic interaction chromatography resins in the monoclonal antibody purification process. *Journal of Chromatography A*, 2008. 1177(2): p. 272-281.

Sun, Y., et al., 5-Aminoindole, a new ligand for hydrophobic charge induction chromatography. *Journal of Chromatography A*, 2008. 1211(1-2): p. 90-98.

Wang, H.Y., et al., Molecular Simulation of the Interactions between 4-Mercaptoethyl-pyridine Ligand and IgG. *Acta Chimica Sinica*, 2010. 68(16): p. 1597-1602.

Lin, D.Q., et al., Salt-Promoted Adsorption of an Antibody onto Hydrophobic Charge-Induction Adsorbents. *Journal of Chemical and Engineering Data*, 2010. 55(12): p. 5751-5758.

16

Separation of Biosynthetic Products by Pertraction

Anca-Irina Galaction[1] and Dan Caşcaval[2]
[1]"Grigore T. Popa" University of Medicine and Pharmacy of Iasi,
Faculty of Medical Bioengineering, Dept. of Biotechnologies
[2]"Gheorghe Asachi" Technical University of Iasi,
Faculty of Chemical Engineering and Environmental
Protection, Dept. of Biochemical Engineering
Romania

1. Introduction

The industrial biotechnology has been considerably developed in the last years, especially for the fine chemicals production and food technologies (Caşcaval & Galaction, 2007). This evolution of the biotechnology at large-scale is supported by favorable political and social sentiments and leads to the gradually replace of the chemical technologies by sustainable biochemical technologies with significant benefits.

According to the Lisbon strategy, the improvement of the current technologies was the major objective until 2010 and remains an economic, technological and social challenge (Daugherty, 2006). This objective can be reached by defining an unitary vision concerning the world industrial biotechnology, by ensuring feasible framework programs for developing biotechnology, by increasing through knowledge and transparent information the public interest and support on industrial biotechnology, by establishing the partnerships between the public and private institutions. Thus, the new concept of "white biotechnology" is considered to be the "New Era" of biotechnology and joins all the initiatives dedicated to producing goods or services by sustainable biotechnologies. Being directed to the identification and utilization of the natural renewable sources of raw materials for biosynthesizing valuable bioactive compounds, by means of clean processes which will cut the waste generation and high energy consumption, the driving force of the white biotechnology is the sustainability by carefully managing of the finite resources. Therefore, according to the definition given by Gro Harlem Brundtland, the former Chair of the World Commission on Environment and Development, in its report *Our common future* (April 1987), the sustainable development imposes the equilibrium of three equally important requirements, of economic, ecologic and social types. This idea has been also underlined by Thomas Rachel, German Presidency of the Council of the European Union at the opening ceremony of the International Conference *European BioPerspectives - "En Route to a Knowledge-Based Bio-Economy"*(31 May - 1 June 2007, Cologne) (Caşcaval & Galaction, 2007). It is very important to think about the "white biotechnology" not only in terms of its potential economic benefits, but also in terms of environmental protection or of the starting-point for new business. The industrial biotechnology has became a hot topic especially among the manufacturers and companies using chemical synthesis technologies, because the biotechnology possesses the potential to improve and, then, to maintain the level of products competitiveness.

In this context, the actual trend to implement the "white biotechnology", defined as "the third wave of the biotechnology" too, is also dedicated to the design, optimization and application at macro-scale of new techniques for separation and purification. Compared to the chemical methods, the biosynthesis represents a very advantageous alternative for production of many compounds with biological activity, because of the reduction of the overall process stages number and of the advanced utilization of the low-cost raw materials. However, the undesirable particularity of industrial biotechnologies is the complexity of the separation from fermentation broths of the obtained products, especially due to their high dilution in broth, chemical and thermal lability and to the presence of secondary products. Therefore, the purification of biosynthetic compounds requires a laborious succession of separation stages with high material and energy consumption, the contribution of these stages to the overall cost being of 20 - 60%, or even more (Baird, 1991; Schugerl, 1994).

For these reasons, modern techniques have been developed or adapted for the separation of the biosynthetic products. Derived from the "classical" solvent extraction method, some new extraction techniques, namely as: reactive extraction, extraction and transport through liquid membranes, supercritical fluid extraction, two aqueous phases extraction, extraction by reverse micelles, have been experimented and applied at laboratory or industrial scale for bioseparations. One of the most attractive techniques is pertraction, defined as the extraction and transport through liquid membranes. Pertraction consists in the transfer of a solute between two aqueous phases of different pH or other chemical properties value, phases that are separated by a solvent layer of various sizes (Noble & Stern, 1995; Yordanov & Boyadzhiev, 2004; Kislik, 2010). The pertraction efficiency and selectivity could be significantly enhanced by adding a carrier, such as organophosphoric compounds, long chain amines or crown-ethers etc., into the liquid membrane, the separation process being called *facilitated pertraction* or *facilitated transport* (Li, 1978; Teramoto et al., 1990; Juang et al., 1998; Scovazzo et al., 2002; Luangrujiwong et al., 2007; Caşcaval et al., 2009).

The liquid membranes can be obtained either by emulsification, but their stability is poor, by including the solvent in a hydrophobic porous polymer matrix, or by using pertraction equipments of special construction, which allow to separate and easily maintain the three phases without adding surfactants (free liquid membranes) (Caşcaval et al., 2009).

Compared to the physical or reactive liquid-liquid extraction, the use of pertraction reduces the loss of solvent during the separation cycle, needs small quantity of solvent and carrier, owing to their continuous regeneration, and allows the solute transport against its concentration gradient, as long as the pH-gradient between the two aqueous phases is maintained (Baird, 1991; Schugerl, 1994; Fortunato et al., 2004; Kislik, 2010).

Beside the separation conditions and the physical properties of the liquid membrane, the pertraction mechanism and, implicitly, its performance are controlled by the solute and carrier characteristics, respectively by their ability to form products soluble in the liquid membrane. Among the mentioned factors, the pH-difference between the feed and stripping phase exhibits the most significant influence, this parameter controlling the yields and selectivities of the extraction and reextraction processes, on the one hand, and the rate of the solute transfer through the solvent layer, on the other hand.

Because of its generous offer in the field of biosynthetic compounds separation, pertraction represents a continuous challenge for bioengineering and biotechnology. Thus, this Chapter presents the main results of our experiments on individual or selective separation of some biosynthetic products (antibiotics, carboxylic acids, amino acids, vitamins) by free or facilitated pertraction, using carriers of long chain amines or organophosphoric acids types.

2. Selective pertraction of Penicillin V

The biosynthesis of beta-lactamic antibiotics (Penicillins G and V) by *Penicillium sp.* or *Aspergillus sp.* requires the use of precursors (phenylacetic acid, or phenoxyacetic acid, respectively). Due to their toxicity, the precursors are added in portions during the fermentation, their concentration being maintained at a constant level. Therefore, the acids final concentrations in the fermentation broth vary between 0.2 and 0.6 g/l, depending on the strain and biosynthesis conditions. For this reason, the selective separation is required for obtaining beta-lactamic antibiotics with high purity. Although this operation is difficult by using conventional separation techniques due to the similar physical and chemical characteristics, the antibiotics can be selectively separated from their precursors by facilitated pertraction with Amberlite LA-2 in 1,2-dichloroethane (Caşcaval et al., 2000).

For Penicillin V, the experiments emphasized the major role of pH on the permeability through liquid membrane and selectivity of separation of this antibiotic from phenoxyacetic acid. Thus, the permeability factor, P, is positively influenced by increasing the pH-gradient between the two aqueous phases (the permeability factor conveys the capacity of a solute transfer through liquid membrane, and has been defined as the ratio between the final mass flow and the initial mass flow of solute).

Contrary, Figure 1 indicates that the maximum values of selectivity factor, S, correspond to the minimum difference between the pH-values of the aqueous phases (the selectivity factor has been defined as the ratio between the final mass flow of antibiotic and the final mass flow of precursor). Thus, at a constant level of stripping phase pH of 10 and for a pH-value for feed phase of 6, S was 80.4. If the pH-value of feed phase is maintained at 3 and the pH-value of stripping phase is of 7, the value S = 24.2 was obtained.

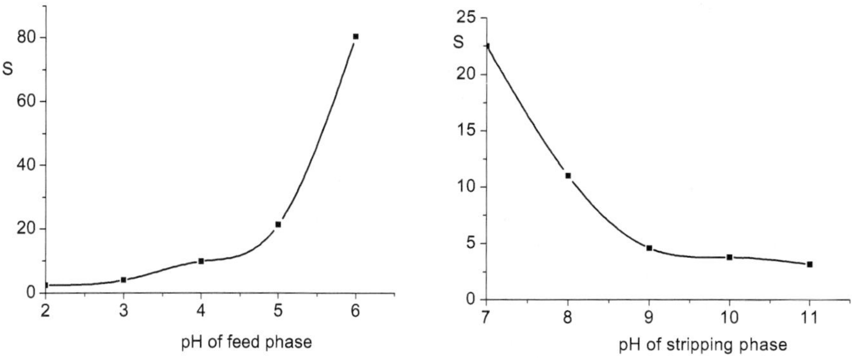

Fig. 1. Effect of feed phase and stripping phase pH-values on the selectivity factor (rotation speed = 500 rpm, carrier concentration = 80 g/l)

Another important factor is the concentration of Amberlite LA-2 inside the liquid membrane. Although the effect of this factor is quite similar for the two components of mixture, at lower carrier concentration the decrease of permeability factor of phenoxyacetic acid is more significant. By increasing the Amberlite LA-2 concentration inside the liquid membrane, the approaching of the permeability factors of the two compounds can be observed. This phenomenon, indicated in Figure 2 by the ratio of permeability factors, suggests that at low carrier concentrations it preferentially reacts with the compound of higher acidity, namely Penicillin V.

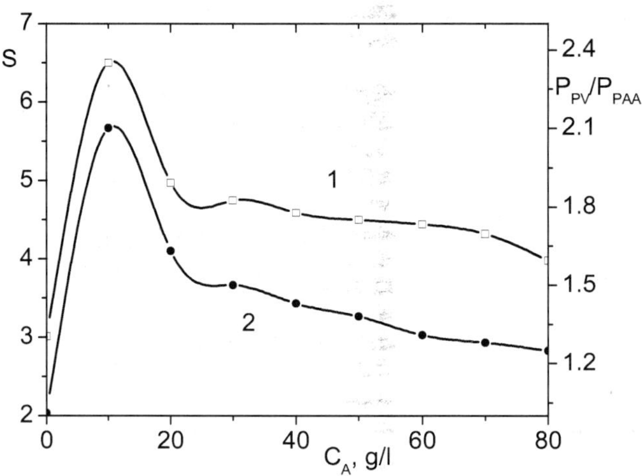

Fig. 2. Effect of carrier concentration on the ratio between permeability factors of Penicillin V and phenoxyacetic acid and on the selectivity factor (pH-value of feed phase = 3, pH-value of stripping phase = 10, rotation speed = 500 rpm; 1 - selectivity factor S, 2 - P_{PV}/P_{PAA})

At high Amberlite LA-2 concentrations, additional amounts of carrier will be available nearly the interface, means that the carrier will react also with the weaker acid, namely phenoxyacetic acid. These results have been found in the variation of selectivity factor, which reached the maximum value of 6.5 for 10 g/l Amberlite LA-2 inside the liquid membrane.

3. Direct pertraction of Erythromycin

Erythromycin is a macrolide antibiotic biosynthesized by *Streptomyces erythreus* on glucose substrate, being very active against the infections produced by *staphylococcus*, gram-positive *bacterium*, etc. Erythromycin exhibits a significant inhibitory effect, this leading to the diminution of microbial activity or cells lysis with the antibiotic accumulation in the broth (Galaction & Caşcaval, 2006). The phenomenon can be avoided by direct removal of antibiotic during the fermentation process.

At industrial scale, the antibiotic separation from fermentation broths is carried out by physical extraction with butyl acetate, with or without preliminary filtration of biomass, followed by its reextraction with diluted solutions of hydrochloric acid. Due to

Erythromycin dissociation and to the low polarity of butyl acetate, the physical extraction is possible only in alkaline pH-domain, the maximum extraction yields being reached for the pH-value of aqueous phase greater than 9 (Galaction & Caşcaval, 2006). In these conditions, some other components of fermentation broths, which are non-dissociated at the extraction pH-value, can be supplementary extracted, the ulterior purification of the antibiotic becoming more difficult.

For the above mentioned reason, a new separation method of Erythromycin from aqueous solutions by reactive extraction with di-(2-ethylhexyl) phosphoric acid (D2EHPA) has been investigated (Caşcaval & Galaction, 2004). This method has been developed and applied as facilitated pertraction, the addition of D2EHPA inside the liquid membrane allowing to increasing the antibiotic mass flow compared to the free pertraction without any carrier (Kawasaki et al., 1996; Caşcaval et al. 2007). But, in the case of media possessing rheological behavior and apparent viscosity similar to the *S. erythreus* broths, the efficiency of pertraction was strongly affected. The increase of apparent viscosity of feed phase from 1 to 30 cP led to the maximum decrease of antibiotic mass flows by the factors of 42.5 and 7.5 for free and facilitated pertraction, respectively (Galaction et al., 2009).

Similar to the direct extraction of other biosynthetic compounds from the fermentation broths (Katikaneni & Cheryan, 2002; Monteiro et al., 2005; Vijayakumar et al., 2008; Kang & Sim, 2008), the presence of biomass could supplementary affect the Erythromycin pertraction, owing to the following phenomena: the appearance of supplementary resistance to the antibiotic transfer from the feed phase to the liquid membrane due to the physical barrier induced by the cell adsorption to the interface; the increase of the apparent viscosity of the feed phase, and, consequently, the amplification of antibiotic diffusional resistance; the mechanical lysis of cells, as the result of the shear stress promoted by the impellers, with the release of the cytoplasmatic compounds which can be co-extracted (amino acids) or can precipitate (proteins).

The study on Erythromycin pertraction from aqueous solutions or simulated broths indicated that the free pertraction is not possible for the pH-value of feed phase, pH_f, lower than 4, due to the pronounced antibiotic ionization (Caşcaval et al. 2007; Galaction et al., 2009). By increasing the pH_f above this level, both the initial and final mass flows are strongly increased, as the result of the increase of physical extraction efficiency. This dependence between the mass flows and the pH of feed phase is respected also in the case of Erythromycin free pertraction from *S. erythreus* suspensions (Figure 3).

But, the accumulation of biomass led to the significant decrease of the initial mass flow (by increasing the biomass concentration from 0 to 20 g/l d.w., the initial mass flow has been reduced for about 7 times).

The increase of the stripping phase pH-value, pH_s, leads to the significant reduction of the antibiotic initial mass flow, this effect becoming more pronounced with the microorganism accumulation in the feed phase. In this case, the negative influence of the biomass is amplified by increasing pH_s, as the result of the supplementary effect of the neutral domain of pH_s (by increasing the biomass concentration from 0 to 20 g/l d.w., the initial mass flow of Erythromycin decreasing for about 5.8 and 19.2 times at pH_s of 2 and 7, respectively). The decreasing of the final mass flow is more important, this parameter reaching the value 0 for pH_s over 7 (at this value of pH_s the pH-gradient between the feed and stripping phases becomes 0). The increase of *S. erythreus* concentration induces the supplementary decrease of antibiotic final mass flow. Thus, comparatively with the pertraction from water, at 20 g/l d.w. *S. erythreus* the final mass flow of Erythromycin is reduced for about 2.8-5.8 times, this effect being amplified at lower values of pH_s.

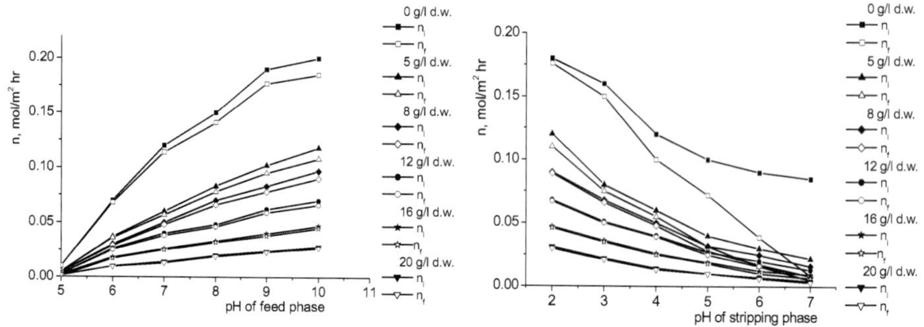

Fig. 3. Influence of pH-values of feed and stripping phases on initial, n_i, and final, n_f, mass flows of Erythromycin for free pertraction from *S. erythreus* broths (rotation speed = 500 rpm)

Independently on the biomass concentration in the feed phase, the permeability factor is continuously reduced by the increase of pH_f over 4, this suggesting that pH_f exhibits a more important influence on the initial mass flow than on the final one (Figure 4).

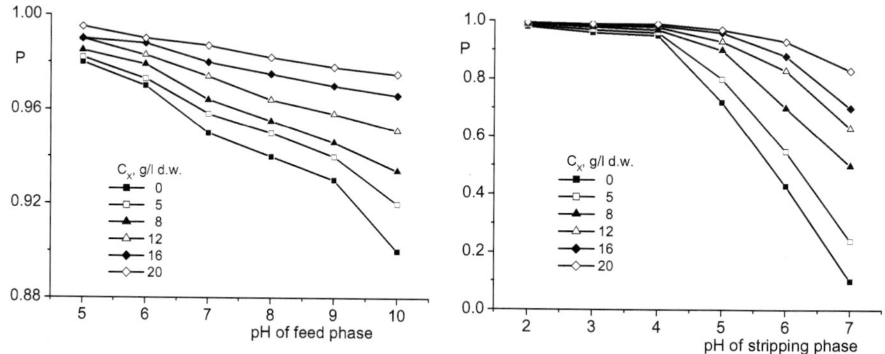

Fig. 4. Influence of pH-values of feed and stripping phases on permeability factor for free pertraction from *S. erythreus* broths (rotation speed = 500 rpm)

The magnitude of the effect of pH_f is diminished with the biomass accumulation, because the amount of antibiotic extracted in the liquid membrane becomes lower, thus facilitating its almost complete reextraction in the stripping phase.

For quantifying the effect of biomass presence, the *reduction factor*, F, has been defined as the ratio between the initial mass flows corresponding to the pertraction from *S. erythreus* broths and simulated ones (without biomass), F_i, respectively between the final mass flows recorded for the same two cases, F_f. The influence of the biomass is clearly underlined in the Figure 5, being recorded the reduction of over 3 times of the factors F_i and F_f with the accumulation of biomass to 20 g/l d.w. This effect is stronger for *S. erythreus* concentration increase up to 5 g/l d.w., thus emphasizing the important role of solid phase in hindering the pertraction.

The addition of the carrier, di-(2-ethylhexyl) phosphoric acid, D2EHPA, in the liquid membrane offers the possibility to carry out the pertraction also at pH-values lower than 4, due to the modification of the mechanism of Erythromycin extraction in the dichloromethane phase. Thus, the previously proposed and verified mechanism of antibiotic reactive extraction with D2EHPA occurs by means of an interfacial reaction of ionic exchange type controlled by the pH of aqueous phase (Caşcaval & Galaction, 2004):

$$Er - N^+H(CH_3)_{2\ (aq)} + HP_{(o)} \rightleftharpoons Er - N^+H(CH_3)_2\ P^-_{\ (o)} + H^+_{\ (aq)}$$

where HP is the carrier. According to the extraction mechanism, the carrier reacts only if Erythromycin exists in aqueous solution in its cationic forms, therefore at an acidic pH-value of the solution ($pH_f < 4$).

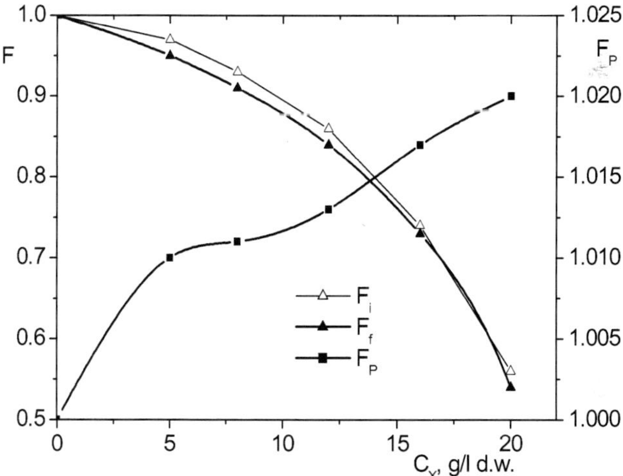

Fig. 5. Influence of biomass concentration, C_X, on factors F and F_P for facilitated pertraction from *S. erythreus* broths (pH of feed phase = 4, pH of stripping phase = 2, carrier concentration = 40 g/l, rotation speed = 500 rpm)

In the case of facilitated pertraction, the accumulation of biomass from 0 to 20 g/l d.w. led to the reduction for about 1.8 times of the factors F_i and F_f. Comparatively to the free pertraction, the magnitude of this effect is attenuated by D2EHPA addition, which increases the initial mass flows of Erythromycin. The dependence between the factor F_P and biomass concentration is opposite to those describing the variation of mass flows ratios. According to those concluded for the free pertraction, the accumulation of *S. erythreus* induces the equalization of the final and initial mass flows. For this reason, the permeability factors are greater for the facilitated pertraction from *S. erythreus* broths than those for the facilitated pertraction from simulated broths, thus leading to the increase of the factor F_P with the biomass concentration.

4. Selective pertraction of Gentamicins

Gentamicin is an aminoglycoside antibiotic, isolated in 1963 by Weinstein from the *Micromonospora purpurea* cultures. It was introduced in therapeutic practice in 1969 in USA. Gentamicin has a broad spectrum against the aerobic Gram positive and Gram negative bacteria, including the strains resistant to tetracycline, chloramphenicol, kanamycin, and colistin, namely *Pseudomonas, Proteus, Staphylococcus, Streptococcus, Klebsiella, Haemophilus, Aerobacter, Moraxella* and *Neisseria*. It was the first antibiotic efficient against *Pseudomonas*, being one of the most important members of the aminoglycoside antibiotics family (Korzybski, 1978; Williams & Lemke, 2002). This antibiotic is industrially obtained by *Micromonospora purpurea* or *echinospora* biosynthesis, the product being a complex mixture of some components of very similar structures. Among them, three are the most important: Gentamicins C_1, C_{1a} and C_2 (Gentamicin C_{2a} is considered also to be Gentamicin C_2, because it is its stereoisomer) (Isoherranen & Soback, 2000; Silverman, 2004). The biosynthetic complex contains also the active Gentamicin C_{2b}, but its concentration is very low. The chemical structures of the major Gentamicins are indicated in Figure 6.

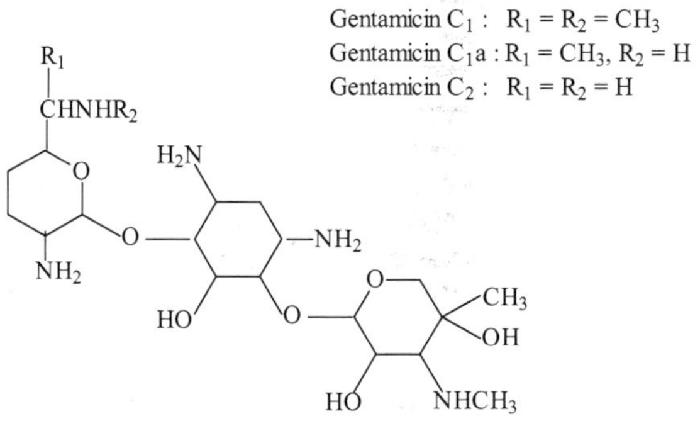

Fig. 6. Chemical structure of biosynthetic Gentamicins

The ratio of these components in the mixture varies from one biosynthetic product to another, the average values of their concentrations being: Gentamicin C_1 35%, Gentamicin C_{1a} 25%, Gentamicin C_2 (including Gentamicin C_{2a}) 40% (Yoshizawa et al., 1998). The antibacterial activities of the Gentamicins, respectively their affinities for the bacterial ribosomes, are different. Thus, the most efficient is Gentamicin C_{1a}, its activity being slightly higher than that of Gentamicin C_2. Gentamicin C_1 binds the ribosomal subunits with the lowest efficiency compared with the other two Gentamicins (there are no reports concerning the specific affinity of Gentamicin C_{2a}, probably due to its assimilation with Gentamicin C_2) (Rosenkrantz et al., 1980).

The separation of Gentamicin from the fermentation broths at industrial scale is achieved by sorption by cation-exchangers, followed by its desorption with a solution of 4-5% sulfuric acid. After the neutralization, the solution is purified and concentrated under vacuum, the antibiotic being precipitated as sulfate salt by acetone addition (Savitskaya et al., 1982). But, this technique doesn't allow the fractionation of the complex mixture of Gentamicins, the use only of Gentamicins C_{1a} and C_2 improving the specific biological activity per weight unit of antibiotic.

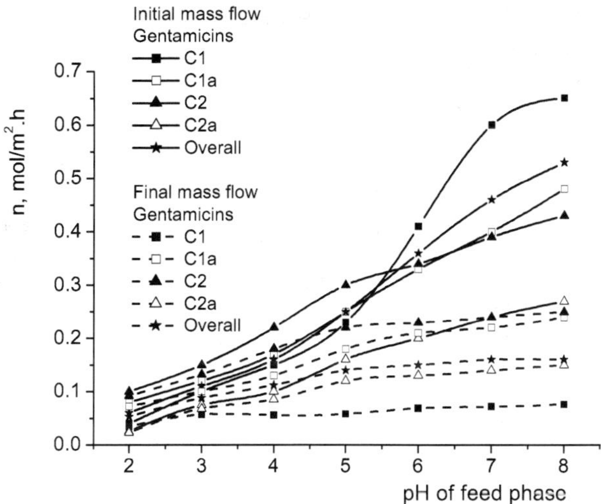

Fig. 7. Influence of pH-value of feed phase on mass flows of Gentamicins (pH of stripping phase = 1.5, D2EHPA concentration = 20 g/l, rotation speed = 500 rpm)

The investigations on the reactive extraction of Gentamicins (Caşcaval et al., 2007) have been developed by studying the possibility to fractionate the biosynthetic mixture of Gentamicins by facilitated pertraction with D2EHPA dissolved in dichloromethane as liquid membrane (Galaction et., 2008). The influence of pH-gradient on the Gentamicins pertraction is amplified by the ionization-protonation of these antibiotics in the two aqueous phases, these processes controlling the efficiency of extraction and re-extraction, as well as the rate of the transport through liquid membrane. Thus, from Figure 7 it can be observed that the initial and final mass flows of Gentamicins are continuously increased with the increase of pH-value. This variation is the result of the mechanism of reactive extraction of the Gentamicins. According to the previous studies, the reactive extraction with D2EHPA occurs by means of the formation of a strong hydrophobic compound by the following ionic exchange reaction (Caşcaval et al., 2007):

$$\text{Gentamicin}^{n+}_{(aq)} + n\,HP_{(o)} \rightleftharpoons \text{Gentamicin.}nHP_{(o)} + n\,H^{+}_{(aq)}$$

where Gentamicin^{n+} represents the antibiotic with protonated aminic groups, and HP the carrier, respectively (n = 1 - 5). The aminic groups of Gentamicins are involved in the reactive extraction, the interactions between the antibiotic and extractant being of ionic type. Gentamicins possess five aminic groups, which could react with the extractant, similar to the reaction with sulfuric acid in the desorption process from the cation-exchangers (Savitskaya et al., 1982). But, due to the voluminous molecules of the antibiotic and extractant, the steric hindrances appear, thus limiting the number of the aminic groups that can react. Furthermore, the basic character of the aminic groups is different and induces the competition between them in the reaction with D2EHPA. The substitutes, which differentiate the Gentamicins, control the basicity of the specific aminic groups and induce their different reactivity, respectively their different mass flows.

Although the effect of the pH-value of feed phase is similar for all Gentamicins, for the neutral pH domain there were recorded important modifications of the relative extraction rate of the four Gentamicins in the membrane phase. For pH-value below 5 the order of the increase of initial mass flows is due to the decrease of dissociation degree from Gentamicin C_1 to Gentamicin C_2, being as follows:

$$\text{gentamicin } C_2 > \text{gentamicin } C_{1a} > \text{gentamicin } C_1 > \text{gentamicin } C_{2a}.$$

This order also indicated significant difference between the initial mass flows of the Gentamicins C_2 and C_{2a}, both compounds having the same chemical structure and molecular weight. This phenomenon could be the result of the molecular conformation of Gentamicin C_{2a}, which alters the strength of the interactions of solvation type with the solvent molecules, and, consequently, its solubility in dichloromethane. For pH-values of feed phase over 5, the initial mass flows of Gentamicins C_1 and C_{1a} become superior to those of the other two Gentamicins. The reactive extraction with D2EHPA needs the protonation of Gentamicins in aqueous solution, this process being affected by the pH increase. Due to the different basicity of Gentamicins specific substituted aminic groups, the relative magnitude of the pH influence on initial mass flows is different. Thus, the increase of the mass flow for pH-value over 5 becomes more pronounced for the Gentamicin containing aminic groups with higher basicity, namely Gentamicin C_1. For the reasons above considered, the lowest mass flow was recorded for Gentamicin C_{2a}.

For describing the selectivity of pertraction, the selectivity factor, S, has been used, being defined as the ratio between the permeability factor of all Gentamicins and that of Gentamicin C_1. According to the above results, from Figure 8 it can be seen that the variation of pH of feed phase from 2 to 8 exhibits a favorable influence of the selectivity factor, this parameter increasing from 1 to 3.1 in the considered domain of pH.

The increase of stripping phase pH-value induces the significant reduction of initial and final mass flows of all Gentamicins (Figure 9). This variation is controlled by the re-extraction mechanism, which is based on the interfacial reaction between the Gentamicins-D2EHPA salts and five equivalents of sulfuric acid for each mole of Gentamicin (Savitskaya et al., 1982; Caşcaval et al., 2007):

$$\text{Gentamicin.HP}_{(o)} + 5/2\,H_2SO_{4(aq)} \rightleftharpoons \text{Gentamicin}^{5+}.5/2SO_4^{2-}{}_{(aq)} + HP_{(o)}$$

The reactivity of Gentamicins in the reaction with sulfuric acid is determined also by the basicity of their specific aminic groups, because they control the strength of the ionic interactions between the antibiotic and the carrier, and therefore the easiness of the antibiotic release from the membrane phase. Figure 9 indicates that at higher acidic pH-domain of stripping phase the highest initial mass flow corresponds to Gentamicin C_1. The decrease of the sulfuric acid concentration, respectively the increase of stripping phase pH-value, leads to the decrease of all Gentamicins initial mass flows, this variation being more pronounced for Gentamicin C_1. Therefore, for pH-value over 2 the initial mass flow of Gentamicin C_1 becomes lower than those of the other Gentamicins. This evolution is due to the different basicity of the specific aminic groups of Gentamicins, which induces different rates of re-extraction in the stripping phase, consequently different concentration gradients of Gentamicins between the two aqueous phases. At lower pH-value, the concentration gradients are maximum and, therefore, the extracted mass flows of all Gentamicins are high.

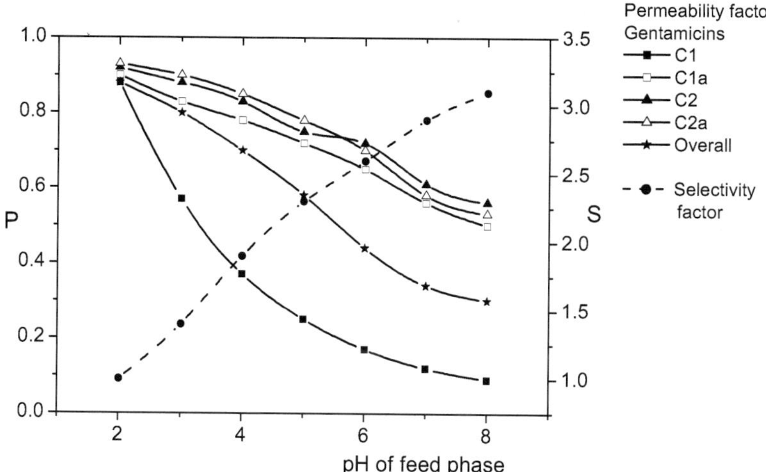

Fig. 8. Influence of pH-value of feed phase on permeability and selectivity factors (pH of stripping phase = 1.5, D2EHPA concentration = 20 g/l, rotation speed = 500 rpm)

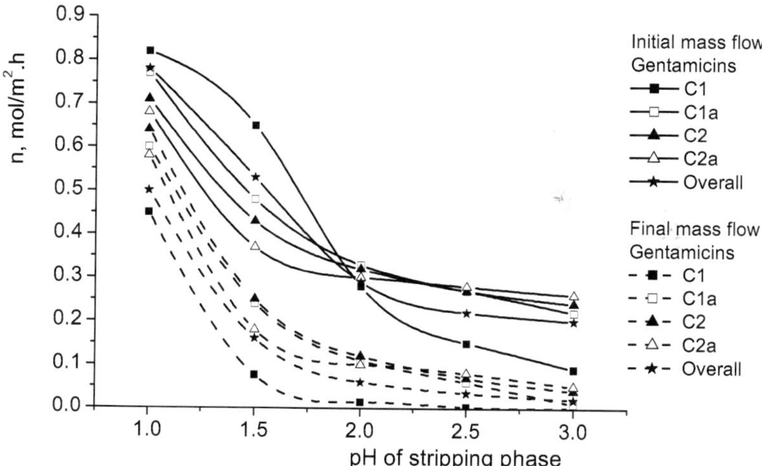

Fig. 9. Influence of pH-value of stripping phase on mass flows of Gentamicins (pH of feed phase = 8, D2EHPA concentration = 20 g/l, rotation speed = 500 rpm)

At higher pH-values of stripping phase, owing to the significant increase of the re-extraction efficiency of Gentamicins C_{1a}, C_2 and C_{2a} compared to Gentamicin C_1, the order of the decrease of the initial mass flows is changed, the Gentamicin C_1 becoming the poorer extracted compound. The selectivity factor increases with the increase of pH of stripping phase and reaches the highest values for pH=3 (S = 10.8). This variation is in concordance with the above results and indicates that the lowest permeability through liquid membrane and the most significant negative influence of stripping phase pH-value correspond to the pertraction of Gentamicin C_1, due to the above discussed reasons.

The increase of carrier concentration into the liquid membrane induces the increase of the initial and final mass flows of all Gentamicins, but the basicity of the specific aminic groups controls the magnitude of this influence. According to the results obtained for reactive extraction of Gentamicins (Caşcaval et al., 2007), if the carrier exists in a stoechiometric deficit related to the complete reaction with all Gentamicins, it will firstly reacts with Gentamicin having the characteristic aminic group with the highest basicity, consequently with Gentamicin C_1. For this reason, the maximum difference between the initial mass flow of Gentamicin C_1 and those of the other Gentamicins is reached for the D2EHPA concentration below 20 g/l. Moreover, contrary to the variation of Gentamicin C_1 initial mass flow, the mass flows of Gentamicins C_{1a}, C_2 and C_{2a} continuously increase without reaching any evident constant level in the domain 0- 60 g/l D2EHPA.

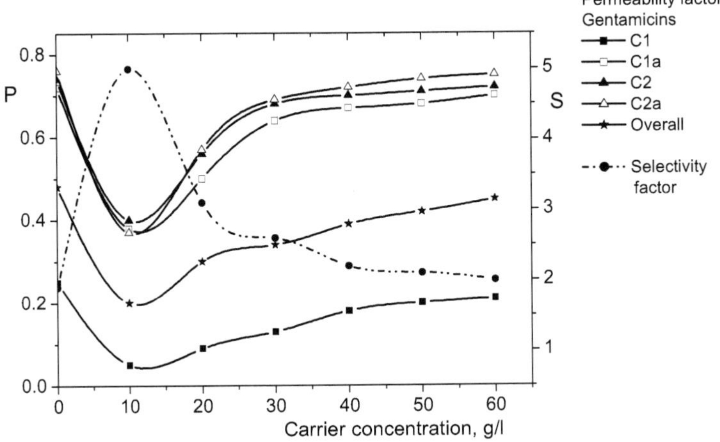

Fig. 10. Influence of carrier concentration on permeability and selectivity factors (pH of feed phase = 8, pH of stripping phase = 1.5, rotation speed = 500 rpm)

The variation of the selectivity factor with carrier concentration is opposite to that of the permeability factors (Figure 10). The maximum value of selectivity factor (S = 5) corresponds to the minimum of permeability factors, thus suggesting that at lower carrier concentration Gentamicin C_1 is less efficiently pertracted. The selectivity of pertraction is diminished for about 2.5 times by increasing the carrier concentration in the liquid membrane from 10 to 60 g/l.

Using the proper levels of the factors influencing the separation process (pH of feed phase of 8, pH of stripping phase of 3, rotation speed of the feed and stripping phases below 100 rpm and carrier concentration of 10 g/l), the most active Gentamicins (Gentamicins C_{1a}, C_2 and C_{2a}) can be selectively pertracted from the initial mixture. By removing Gentamicin C_1 from the biosynthetic mixture the biological activity of the antibiotic is enhanced and the therapeutic dose is reduced, its secondary effects being diminished.

5. Selective pertraction of carboxylic acids obtained by citric fermentation

Citric acid is one of the widely used carboxylic acids, having multiple applications in chemical, pharmaceutical, food and cosmetic industries. This compound is mainly obtained through a fermentation process by *Aspergillus niger* cultivated on molasses (Moo-Young et al., 1985). Due to the presence in the final broth of other carboxylic acids as secondary metabolic products, especially malic and succinic acids, the separation and purification technology of citric acid is quite complicated. The citric acid represents about 80 - 95% from the total amount of organic acids in the broth at the end of fermentation, its concentration being of 50 g/l. The rest are secondary acids, their concentration reaching 4 g/l. At industrial scale, the separation and purification of citric acid consist on carboxylic acids precipitation as calcium salts, solubilization of calcium citrate by heating the solution and citric acid release by treating with sulfuric acid (Moo-Young et al., 1985). This technology needs high amount of raw materials and energy consumption and produces large amounts of calcium sulfate as the by-product, without leading to high purity of citric acid.

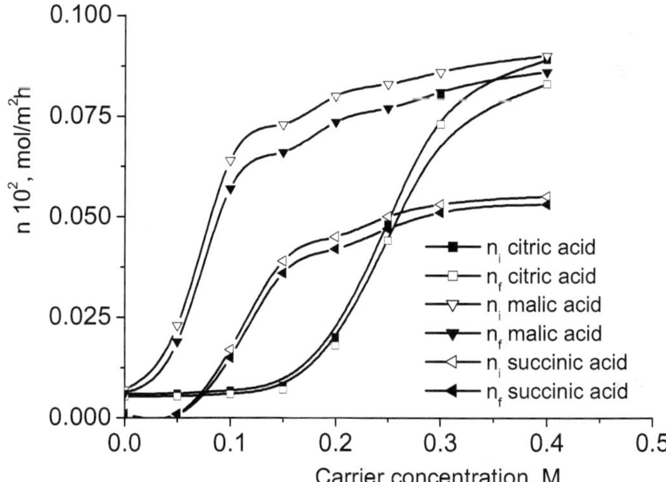

Fig. 11. Influence of carrier concentration on citric, malic and succinic acids mass flows (citric acid concentration in feed phase = 7.8×10^{-2} M, malic acid concentration in feed phase = 7.8×10^{-2} M, succinic acid concentration in feed phase = 7.8×10^{-2} M, rotation speed = 500 rpm, pH of feed phase = 3, pH of stripping phase = 11)

Based on the differences between the extraction mechanisms, acidity of these carboxylic acids and hydrophobicity of the compounds formed with the carrier, the selective removal of the malic and succinic acids from the final fermentation broth by pertraction with Amberlite LA-2 has been performed (Caşcaval et al., 2004a). In the case of these acids pertraction from a mixture, the dependence of their mass flows on the pH gradient has been correlated with their acidity, because the acidity controls the rate of interfacial reactions between solute and carrier. Thus, the obtained order of the pertraction efficiency, given as follows: succinic acid<citric acid<malic acid, was the result of the higher acidity of citric and

malic acids, on the one hand, and of the superior hydrophobicity of malic acid – Amberlite LA-2 complex, on the other hand.

The permeability factors of all studied acids tended to 1 with the increase of pH-gradient, underlining the approach between the acid extraction and re-extraction yields. Moreover, the values of permeability factors suggest an inverse proportionality between the transport capacity of liquid membrane and the acidity of transferred solute, the order of permeability factors diminution being: succinic acid>malic acid>citric acid.

This order could be explain by the similar variation of the rate of interfacial reaction between acid - carrier compound and sodium hydroxide, the increase of acidity leading to the appearance of a kinetic resistance to the re-extraction process.

Concentration of Amberlite LA-2 inside of the liquid membrane induces a different influence on pertraction efficiency of the carboxylic acids. The difference on carrier effects is due to the difference on acids extraction mechanisms, as well as to the difference on solutes acidity and hydrophobicity. As it can be seen from Figure 11, by increasing the carrier concentration the malic acid, succinic acid and citric acid are successively pertracted.

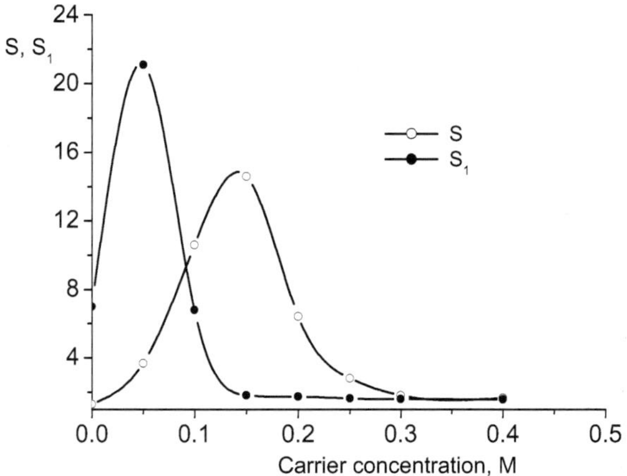

Fig. 12. Influence of carrier concentration on selectivity factors (citric acid concentration in feed phase = 7.8×10^{-2} M, malic acid concentration in feed phase = 7.8×10^{-2} M, succinic acid concentration in feed phase = 7.8×10^{-2} M, rotation speed = 500 rpm)

The succinic acid is extracted after the Amberlite LA-2 concentration exceeds the value stoechiometrically needed for the interfacial reaction with malic acid, respectively after it exceeds the molar ratio between carrier and malic acid of 1. The citric acid is extracted for carrier concentration higher than that corresponding to an equimolecular ratio with malic and succinic acids. Below the carrier concentrations that allow the reactive extraction of succinic and citric acids, their pertraction is possible only by physical solubilization in 1,2-dichloroethane, but the acids mass flows are very low. These results demonstrate the major influence of the Amberlite LA-2 concentration on pertraction selectivity.

The above discussed results suggested the possibility to selectively remove the malic and succinic acids, the citric acid remaining in the raffinate phase. For confirming this hypothesis and establishing the required conditions for reaching a high selectivity of separation, the influences of pH-gradient between the aqueous phases, carrier concentration and mixing intensity on pertraction selectivity have been studied. The selectivity of pertraction was described by means of the selectivity factors S and S_1. The selectivity factor S was introduced for the separation of malic and succinic acids from citric acid, and it was defined as the ratio between the sum of the final mass flows of malic and succinic acids and the final mass flow of citric acid. The factor S_1 was used for the separation of malic acid from succinic acid, being the ratio between the final mass flow of malic acid and that of citric acid.

The reduction of pH gradient leads to the increase of selectivity factors S and S_1, but the magnitude of this effect is rather different. The modification of pH value of feed phase induces a stronger effect on separation selectivity of secondary carboxylic acids from citric acid, while the modification of stripping phase pH exhibits a more pronounced effect on separation selectivity of malic acid from succinic acid.

The decisive influence of carrier concentration on pertraction selectivity is underlined by the dependence between the selectivity factors and this parameter (Figure 12).

Similar to the variation of acids mass flows with carrier concentration, the experimental data show that the maximum selectivity both for separation of secondary carboxylic acids from citric acid, and for separation of malic acid from succinic acid is reached for an equimolecular ratio between Amberlite LA-2 and the pertracted acids.

In order to verify the above conclusions, initially the pertraction of citric, malic and succinic acids from a mixture similar to that obtained by citric fermentation was performed. The concentrations of the carboxylic acids in the feed solution were as follows: 50 g/l (0.26 M) citric acid, 2.5 g/l (2.1×10^{-2} M) malic acid, and 2.5 g/l (1.9×10^{-2} M) succinic acid, respectively. In the second step, the malic acid was pertracted from a mixture containing 2.5 g/l (2.1×10^{-2} M) malic acid and 2.5 g/l (1.9×10^{-2} M) succinic acid. In both cases, the pertraction was carried out using the separation conditions that offer maximum selectivity and high rate of transport through liquid membrane: carrier concentration of 0.04 M, rotation speed of 500 rpm, pH of feed phase of 4 and pH of stripping phase of 11. The obtained results indicated that, by combining the favorable effects of pertraction parameters, superior values of selectivity factors can be obtained: S = 24.5, S_1 = 47.5.

6. Synergetic pertraction of p-aminobenzoic acid

p-Aminobenzoic acid (PABA), also called vitamin B_{10} or factor R, was found to be part of the folic acids. Because it is component of the pteroylglutamate, it is considered to act as provitamin for some bacteria and growth factor for some superior animals, in the human body possessing the capacity to synthesize folates. The most recent methods for PABA production are the chemical synthesis using methyl-4-formylbenzoate as the starting material or biosynthesis by mutant strains of *E. coli* (Amaratunga et al., 2000; Park et al., 2003). In both cases the separation stages are complex and require the consumption of a large amount of energy and materials.

Due to the insolubility of PABA in organic solvents immiscible with water, its separation by physical extraction is impossible. But, owing to the chemical structure of PABA which contains an acidic group, -COOH, and a basic one, $-NH_2$, the reactive extraction has been taken into consideration and has been performed using extractants of aminic and organophosphoric acid types, namely Amberlite LA-2 and (D2EHPA), respectively (Galaction et al., 2010). Because the formation of the third phase has been observed during the reactive extraction of PABA, the mechanisms and, consequently, the factors which control the mechanisms of acid extraction with the two extractants in presence of 1-octanol as phase modifier have been also investigated. On the basis of the experiments on the synergetic reactive extraction of PABA, the facilitated pertraction of this acid using a liquid membrane without and with 1-octanol has been comparatively analyzed from the viewpoint of the influences of the process parameters (pH gradient between the feed and stripping phases, carrier concentration, mixing intensity) (Kloetzer et al., 2010).

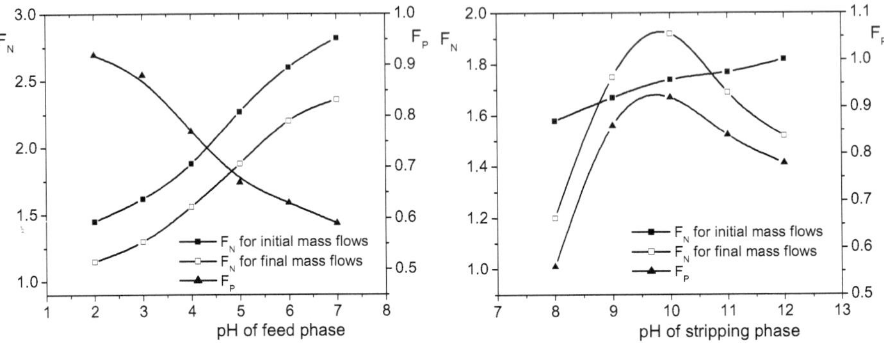

Fig. 13. Influence of pH-values of feed and stripping phases on factors F_N and F_P (Amberlite LA-2 concentration = 40 g/l, rotation speed = 500 rpm, 1-octanol concentration = 10% vol.)

In order to quantify the effect of 1-octanol addition inside the liquid membrane on the initial and final mass flows, the factor F_N has been considered and defined as the ratio between the mass flows recorded in presence and in absence of alcohol. Similarly, the factor F_P has been calculated as the ratio between the permeability factors reached for liquid membrane with and without 1-octanol.

The dependence of the values of factors F_N and F_P on the feed phase pH, plotted in Figure 13, suggests that the addition of 1-octanol exhibits two different effects. The factor F_N, calculated either for the initial mass flows or for the final ones, is greater than the unit for the entire considered pH_F-domain and increases with the increase of pH_F. Thus, for pH_F variation from 2 to 7, F_N increased from 1.5 to 2.9 for the initial mass flows, respectively from 1.2 to 2.4 for the final mass flows. These results are the consequence of the favorable effect of 1-octanol on the solubilization of PABA molecules, free or bounded to the carrier molecules, on the membrane phase. The increase of pH_F induces the dissociation of PABA in the feed phase, the presence of 1-octanol allowing also the extraction of the dissociated molecules of acid. The relative magnitude of the positive effect of alcohol addition is superior in the case of the initial mass flow, due to the supplementary kinetic resistance to the acid reextraction process from the membrane phase to the stripping solution.

Contrary, the values of factor F_P are lower than 1 for the entire experimented domain of the feed phase pH, the increase of pH_F inducing the reduction of this factor. In this case, the significant increase of the initial mass flow of PABA by adding 1-octanol inside the liquid membrane exceeds the membrane capacity to transport the acid and to release it into the stripping phase.

Similar to the influence of pH_F, the factors F_N are superior to 1 for all the considered pH_S-values, but the influence of pH_S has to be distinctly analyzed for the initial and final mass flows ratios. Thus, the factor F_N, calculated as the ratio between the initial mass flows, increased slowly with the increase of stripping phase pH, from 1.6 to 1.8. The variation of factor F_N related to the final mass flows with pH_S indicated its increase to a maximum level, corresponding to $pH_S = 10$, followed by its decrease. The maximum F_N (1.9) exceeded that obtained for the initial mass flows indifferent of pH_S-value, due to the more important influence of pH_S on the PABA reextraction step from the membrane phase.

The variation of F_P follows that of F_N calculated for the final mass flows, the two factors being directly correlated. Moreover, for the entire investigated domain of stripping phase pH, the value of F_P was lower than 1 (maximum F_P was 0.92).

Due to its favorable effect on PABA extraction from the feed phase into the liquid membrane, the addition of 1-octanol in dichloromethane induces the increase of the initial and final mass flows of the acid. Thus, for 40 g/l Amberlite LA-2 and alcohol concentration variation from 5 to 20% vol., the initial mass flow was amplified for about 1.4-2.2 times and the final mass flow for about 1.1-1.6 times compared with the corresponding mass flows in absence of 1-octanol (Caşcaval et al., 2009). This effect is more significant for the initial mass flow, because the reextraction rate tended to its maximum level for the given experimental conditions. For the same reason, the permeability factor was increased slowly from 0.4 to 0.7 by increasing the alcohol concentration inside the membrane phase.

7. Selective pertraction of cinnamic acids

Cinnamic acid, also known as phenylacrylic acid, is a natural compound derived from phenylalanine, its main vegetable sources being the cinnamon, the resin of *Liquidambar* tree, the storax, the balsam of tolu, and the balsam of Peru. The main utilization of cinnamic acid is in the cosmetic/perfumery industry, especially as methyl, ethyl or benzyl esters (the cinnamic acid and its volatile benzylic ester are responsible for the cinnamon flavor). The cinnamic acid itself, or the p-hydroxy- and p-methoxycinnamic acids, has different pharmaceutical applications, for pulmonary affections, cancer, lupus, infectious diseases (diarrhea, dysentery), possessing antibacterial and antifungal activity (Saraf & Simonyan, 1992; Tawata et al., 1996; Lee et al., 2004). It is also used in food, or for the synthetic ink, resins, elastomers, liquid crystalline polymers and adhesives production.

The cinnamic acids could be obtained by extraction from vegetable materials, by chemical synthesis or biosynthesis. New methods have been recently developed for cinnamic acid extraction (supercritical fluid extraction, vapor phase extraction, pressurized fluid extraction), but their applications are rather limited for high quantities of vegetable materials (Bartova et al., 2002; Palma et al., 2002; Smelz et al., 2004; Naczk & Shahidi, 2004). The cinnamic acid is synthesized from styrene and carbon tetrachloride, by oxidation of cinnamic aldehyde, or from benzyl dichloride and sodium acetate. The chemical methods are expensive due to the costs of the starting materials, the high number of required stages for product purification, and generated large amounts of unwanted secondary products. For these reasons, the production by fermentation or/and enzymatic methods of cinnamic acid and its main derivatives, the p-hydroxy- and p-methoxycinnamic acids, have been developed. Thus, *Saccharomyces cerevisiae*, *Escherichia coli*, *Pseudomonas sp.* have been cultivated on glucose, and *Cellulomonas galba* on n-paraffins with addition of alkylbenzenes (Parales et al., 2002). The glucose, fructose, lactose, sugar, cellulose and starch can be enzymatically transformed by phenylalanine ammonia lyase or tyrosine ammonia lyase in alkaline media. These enzymes are synthesized directly into the media by the mutant strains of *E. coli*, *Rhodotorula sp.*, *Rhodosporidium sp.*, *Sporobolomyces sp.*, *Rhizoctonia solani*, *Trichosporon cutaneum*, *Rhodobacter sp.* (Hanson & Havir, 1981; Naczk & Shahidi, 2004).

Excepting from our works, there are no reports on the possibility of separating cinnamic acid and its related acids from fermentation broths or enzymatic media by liquid-liquid extraction. This is probably due to the low solubility of these compounds in solvents immiscible with water. Their extraction became possible by adding an extractant of aminic type into the solvent, this compound reacting with the cinnamic acids and leading to the formation of hydrophobic derivatives (Camarut et al., 2006). This technique has been considered for developing the cinnamic and p-methoxycinnamic acids selective pertraction with Amberlite LA-2 (Galaction et al., 2007). Due to the methoxy group which differentiates the two studied acids, the influence of the feed phase pH is based on two different mechanisms. Thus, from Figure 14, plotted for pH of stripping phase of 10, it can be observed that the initial and final mass flows of the cinnamic acid are continuously reduced with the increase of pH-value. On the other hand, the mass flows of p-methoxycinnamic acid initially increase with the pH increase, reach a maximum level at pH=4, decreasing then. This variation is more pronounced for the initial mass flow.

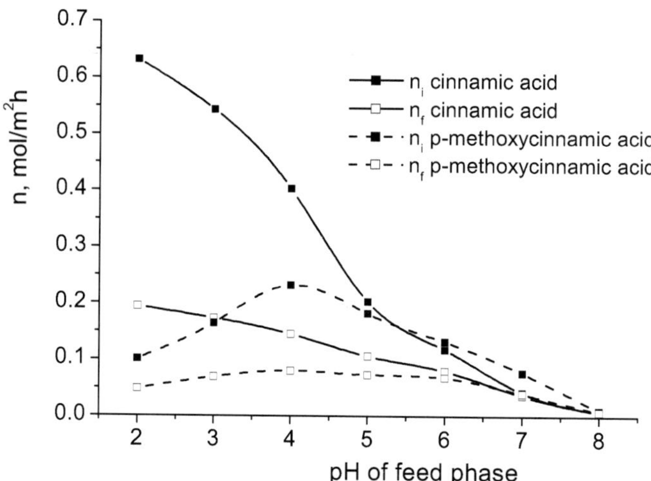

Fig. 14. Influence of pH-value of feed phase on mass flows of cinnamic and p-methoxy-cinnamic acids (pH of stripping phase = 10, Amberlite LA-2 concentration = 40 g/l, rotation speed = 500 rpm)

These variations are the result of the mechanism of reactive extraction of the two acids. The reactive extraction occurs by means of the interfacial interactions between the carboxylic groups of the cinnamic and p-methoxycinnamic acids and Amberlite LA-2. These interactions could be of hydrogen bonding type with the undissociated carboxylic groups, or of ionic type, if the acids dissociate in the aqueous solution. The initial mass flow of cinnamic acid continuously decreased with the pH increase due to its dissociation at higher pH-values. The existence of the maximum level of the initial mass flow of p-methoxycinnamic acid is the result of two opposite phenomena occurring with the pH increase: the diminution of the methoxy group protonation, this promoting the extraction, and the dissociation of the carboxylic group, with negative effect on extraction.

For pH-values bellow 5, the initial mass flow of cinnamic acid exceeded that of p-methoxycinnamic acid. Over pH=5, due to the superior hydrophobicity and acidity of p-methoxycinnamic acid, its initial mass flow becomes higher than that of cinnamic acid (pKa=4.44 for cinnamic acid, pKa = 4.28 for p-methoxycinnamic acid (Weast, 1974)). But, for the pertraction process, the differences between the mass flows of the two acids recorded for pH>5 are less pronounced than in the case of reactive extraction. This result is the consequence of the less intense mixing in the pertraction system, and, therefore, of the resistance to the diffusion through the boundary layers from liquid membrane interfaces, which is more important than that induced for the reactive extraction process, especially for the compounds with higher molecular weight. Among the two acids, the resistance to the diffusion of p-methoxycinnamic acid is higher, due to its more voluminous molecule.

The variations of the two acids final mass flows are similar to those of the initial mass flows, owing to their direct dependence to the acids concentrations in the organic layer.

The permeability factor of cinnamic acid increases with the pH increase, this variation suggesting that the reduction of its initial mass flow exhibits a positive effect on the permeability through liquid membrane, due to the diminution of the amount of acid accumulated into the organic phase (Figure 15). Thus, the maximum value of permeability factor for the considered experimental conditions was of 0.93, being reached at pH=8.

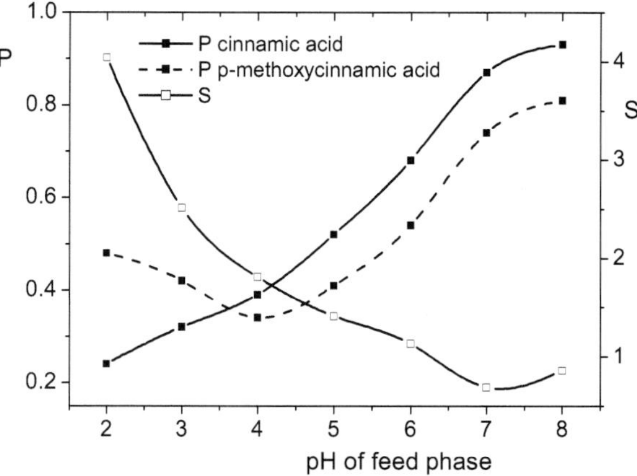

Fig. 15. Influence of pH-value of feed phase on permeability and selectivity factors (pH of stripping phase = 10, Amberlite LA-2 concentration = 40 g/l, rotation speed = 500 rpm)

The permeability factor of p-methoxycinnamic acid has a particular evolution with the pH increase. This parameter initially decreases and reaches a minimum level at pH=4, then increasing similarly as for cinnamic acid. For pH<4, the increase of the amount of p-methoxycinnamic acid extracted in organic layer exceeds the increase of its final mass flow, due to the high initial mass flow. Because the initial mass flow of p-methoxycinnamic acid is lower comparatively to cinnamic acid, its permeability factor is superior to that of cinnamic acid in this domain of pH. For higher pH-values, due to the resistance to the diffusion from the liquid membrane to the stripping phase, which is more important for p-methoxycinnamic acid, the permeability factor of this acid becomes lower than that of cinnamic acid.

For describing the selectivity of pertraction, the selectivity factor, S, has been defined as the ratio between the final mass flow of cinnamic acid and that of p-methoxycinnamic acid. From Figure 15 it can be seen that the maximum value of selectivity factor is reached at the pH of feed phase of 2, as the result of the highest difference between the acids extraction degree and, consequently, between their concentrations in the liquid membrane. The increase of pH induces a negative effect on the selectivity of cinnamic acid separation. Thus, for pH>6, the selectivity factor is less than 1, owing to the higher amount of p-methoxycinnamic acid in the liquid membrane and higher final mass flow compared to those of cinnamic acid.

The increase of carrier concentration into the liquid membrane induces the increase of the initial and final mass flows of both acids. At the concentration of Amberlite LA-2 bellow 10 g/l, the initial mass flow of p-methoxycinnamic acid is higher, due to its superior acidity compared with cinnamic acid. The increase of Amberlite LA-2 amount in the organic phase exhibits a more pronounced effect on cinnamic acid mass flow, because it compensates the lower acidity of this acid. This phenomenon cumulated with the slower diffusion of p-methoxycinnamic acid generates significant differences between the values of acids mass flows for carrier concentrations over 10 g/l. The initial mass flows of the acids reach a constant level at 40 g/l Amberlite LA-2. The variation of the final mass flows is similar, the constant level being reached for Amberlite LA-2 concentration of 60 g/l.

The evolution of the acids permeability factors are different. Similar to the pertraction of carboxylic acids obtained by citric fermentation, they initially decrease from a value corresponding to the absence of Amberlite LA-2 in the organic solvent to a minimum value for a concentration of 10 g/l Amberlite LA-2 and finally increase concomitantly with the carrier concentration.

The positive influence of the increase of carrier concentration is more important in the case of cinnamic acid, this leading to the increase of the selectivity factor from 0.6 for free pertraction to 2 for facilitated pertraction with 40 g/l Amberlite LA-2. For higher carrier concentration, the selectivity factor remains at the constant level.

8. Fractionation of amino acids mixture by pertraction

The amino acids can be obtained by biosynthesis, by protein hydrolysis or by extraction from natural sources. The most efficient methods are the first two, but the separation of amino acids from fermentation broths or protein hydrolysates is rather difficult. In the last decades a continuous and increasing interest has been observed in developing the techniques that can improve the selectivity and the yield of downstream processes for the separation and purification of amino acids (Liu & Dai, 2003). The separation techniques currently applied for removal and purification of amino acids from dilute aqueous solutions typically employ the ion exchange, crystallization at the isoelectric point or chromatography (Caşcaval et al., 2004b). But, these techniques are rather difficult to be transposed to the industrial scale, thus affecting the production of amino acids and increasing the cost of the used technology.

The reactive extraction became a very attractive method for amino acids separation, because it offers an advantageous alternative to the above mentioned separation techniques. Amino acids dissociate in aqueous solutions, forming characteristic ionic species as a function of the solution pH-value. This property makes amino acids hydrophilic at all pH-values and, thus, complicates their recovery by solvent extraction. For this reason, the amino acids solubility in conventional organic solvents is lower, their physical extraction being practically impossible. The liquid-liquid extraction of amino acids becomes possible only by adding extractants into the organic phase, namely derivatives of phosphoric acid (Kelly et al., 1998; Liu et al., 1999; Caşcaval et al., 2001; Juong & Wang, 2002; Lin et al., 2006; Lin & Chen, 2006), high molecular weight amines (Rehm & Reed, 1993; Schugerl, 1994; Tan et al., 2007) or some types of crown-ethers (Deblay et al., 1990).

The pertraction could be also used for amino acids separation, the proper carrier being chosen from the above listed extractants (organophosphoric acid, high molecular weight amines or crown-ethers). In this context, the separation of some amino acids of acidic character (L-aspartic acid, L-glutamic acid), basic character (L-histidine, L-lysine, L-arginine) or neutral character (L-glycine, L-tryptophan, L-cysteine, L-alanine) from their mixtures obtained either by fermentation or protein hydrolysis using the facilitated pertraction with D2EHPA in dichloromethane has been analyzed (Blaga et al., 2008).

In the case of amino acids pertraction, the influence of the pH-gradient between the phases is enhanced by the formation of the ionized forms of amino acids in the aqueous phases and controls both the efficiency of extraction/reextraction and the transport rate through the solvent layer. Thus, from Figure 16 it can be observed that for all studied amino acids the initial mass flows increase with the increase of feed phase pH, reach a maximum value followed by their strong decrease.

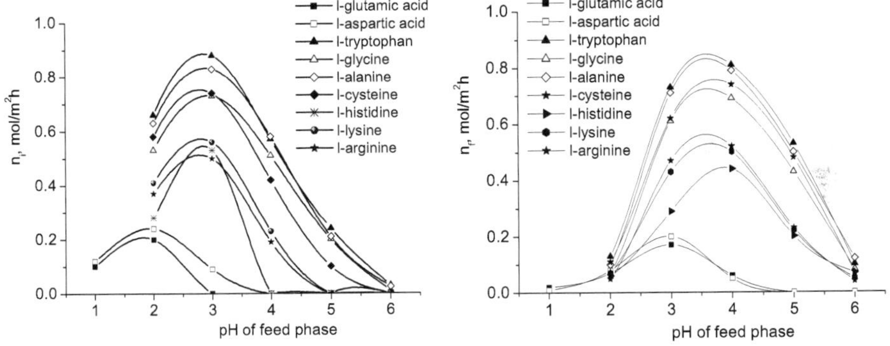

Fig. 16. Influence of pH value of feed phase on mass flows of amino acids (pH of stripping phase = 2, carrier concentration = 40 g/l, rotation speed = 500 rpm)

The value of the pH corresponding to the maximum initial mass flows is 2 for the acidic amino acids, and 3 for the other amino acids. This influence of the pH value on amino acids mass flows is the consequence of the reactive extraction mechanism of amino acids with D2EHPA, which occurs by means of an interfacial chemical reaction of the ion exchange type controlled by the pH of aqueous phase. According to the obtained results by studying the amino acids reactive extraction (Caşcaval et al., 2001), the carrier, D2EHPA, reacts only if the amino acids exist in aqueous solution in their cationic forms (pH of aqueous phase has to be below pH$_{isoelectric}$). The maximum of mass flow is the result of two opposite phenomena which occur with the pH increase: the increase of the concentration of extractant active form (in the strong acidic pH-domain the extractant is protonated and, consequently, becomes unable to react with the amino acid), and the decrease of the total amount of amino acid existing in cationic form. The further increase of the pH-value of feed phase leads to the increase of the concentration of the acidic and neutral amino acids zwitterions, and respectively, of the basic amino acids dication-anionic species or zwitterions, thus reducing significantly the initial mass flows of the amino acids (at the isoelectric point the reactive extraction of amino acids becomes impossible (Caşcaval et al., 2001)). Unlike the acidic or neutral amino acids, the pertraction of basic amino acids is not possible even if the pH values are lower than those corresponding to their isoelectric points, due to the formation of the dication-anionic species (L-histidine: $n_i = 0$ for $pH_i \geq 4$, L-lysine: $n_i = 0$ for $pH_i \geq 5$, L-arginine: $n_i = 0$ for $pH_i \geq 5$).

The recorded differences between the initial mass flows of the solutes are probably the result of the different hydrophobicity of the radicals R from the amino acids structures, this being in concordance with the previous conclusions regarding the reactive extraction yields of the same amino acids with D2EHPA (Caşcaval et al., 2001).

The final mass flows of amino acids initially increase with pH of the feed phase, owing to their accumulation in the liquid membrane, reaching the maximum values at $pH_i = 3$ for aspartic and glutamic acids, and at $pH_i = 4$ for the rest of amino acids, respectively. Because the amino acids are accumulated in the liquid membrane in different proportions, the differences between the final mass flows are rather similar to those between the initial mass flows. The further increase of pH_i to the neutral pH-domain leads to the decrease of the final mass flows, owing to the change of the direction of pH-gradient which controls the direction of solute transfer through the liquid membrane.

For all considered amino acids, the permeability factors strongly increase with the pH increase, becoming higher than 1 for $pH \geq 3$. This variation indicates that the final mass flows become larger than the initial ones, phenomenon that is possible due to the reextraction of the additional amount of amino acids accumulated into the organic layer.

The increase of the pH-value of the stripping phase caused the reducing of both initial and final mass flows of the amino acids that can be extracted at the prescribed pH of feed phase. For example, although at $pH_i = 2$ all the amino acids are extracted, at $pH_i = 4$ the initial mass flows of L-aspartic acid, L-glutamic acid and L-histidine are 0, for the above presented reasons.

A similar variation has been recorded also for the permeability factors as a function of the pH value of stripping phase (Figure 17). The maximum values of the permeability factors are reached for the pH-value of stripping phase of 1. This result, together with the variations of mass flows, indicated that by increasing the pH_f, the direction of the solutes transport through liquid membrane is inverted, and consequently the amount of the accumulated amino acids inside the solvent layer increases significantly.

According to the Figure 17, the maximum values of the permeability factors are higher for L-aspartic and L-glutamic acids, owing both to their lower initial mass flows, and to their lower hydrophobicity, which promote the reextraction in the stripping phase.

Therefore, by combining the feed phase pH-value, which strongly limits the amino acids transfer to the membrane phase, the pH-value of stripping phase, which controls the rate of the amino acids re-extraction from the liquid membrane and, consequently, their concentration gradients between the two aqueous phases, the carrier concentration, which controls the capacity of liquid membrane to transport the solute, and the mixing intensity, which can selectively diminish the resistance to the diffusion, the selective separation by facilitated pertraction becomes possible for different groups of amino acids with similar acidic properties. Therefore, for pH of feed phase over 5 only L-glycine, L-alanine, L-tryptophan and L-cysteine are pertracted, for pH of feed phase between 4 and 5 these amino acids and L-lysine and L-arginine, for pH of feed phase between 3 and 4 L-histidine can be added to the previous list of pertracted amino acids, and below pH of 3 L-aspartic acid and L-glutamic acid can be also separated.

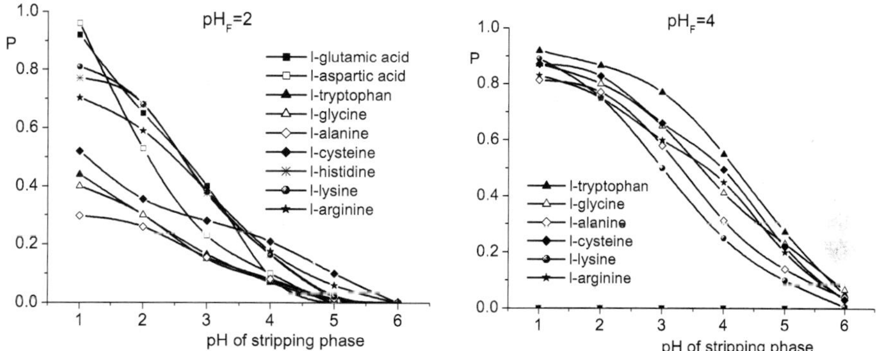

Fig. 17. Influence of pH-value of stripping phase on permeability factors (carrier concentration = 40 g/l, rotation speed = 500 rpm)

9. Conclusions

Extraction and transport through liquid membranes, also called pertraction, constitutes advantageous alternative to the conventional separation methods, because it reduces the number of stages required for an efficient separation and, therefore, the corresponding energy and material consumption. For these reasons, this technique has a considerable potential for biosynthetic products separation and purification, being required for further development of many biotechnologies, and represents very attractive research domain for chemical and biochemical engineers. In the actual context of the "white biotechnology", the studies in the field of bioseparations are dedicated to extending the application area of pertraction for including the separation of other biosynthetic or natural products and to scaling-up this technique at industrial level.

10. References

Amaratunga, M.; Lobos, J. H.; Johnson, B. F. & Wiliams, E. D (2000). Genetically engineered microorganisms and method for producing 4-hydroxybenzoic acid. Patent US 6030819/2000

Baird, M.H.I. (1991). Solvent extraction - the challenge of a "mature" technology. The Canadian Journal of Chemical Engineering, Vol.69, No.6, 1287-1301, ISSN 1939-019X

Bartova, M.; Opletal, L.; Chobot, V. & Sovova, H. (2002). Liquid chromatographic analysis of supercritical carbon dioxide extracts of Schizandra chinensis. Journal of Chromatography B: Analytical Technologies in the Biomedical and Life Science, Vol.770, No.1-2, pp. 283-288, ISSN 1570-0232

Blaga, A.C.; Galaction, A.I. & Caşcaval, D. (2008). Separation of amino acids from their mixture by facilitated pertraction with D2EHPA. Chemical and Biochemical Engineering Quarterly, Vol.22, No.4, pp. 439-446, ISSN 0352-9568

Camarut, M.; Galaction, A.I. & Caşcaval, D. (2006). Separation of trans-cinnamic acid by reactive extraction in low-polar solvent. Romanian Biotechnological Letters, Vol.11, No.5, pp. 2897-2903, ISSN 1224-5984

Caşcaval, D. & Galaction, A.I. (2004). New extraction techniques in bioseparation 1.Reactive extraction. Chemical Industry Journal, Vol.58, No.9, pp. 375-386

Caşcaval, D. & Galaction, A.I. (2007). The European colour of biotechnology is white. Romanian Biotechnological Letters, Vol.12, No.6, pp. 3489-3494, ISSN 1224-5984

Caşcaval, D.; Galaction, A.I. & Nicuţă, N. (2006). Separarea eritromicinei prin pertractie libera si facilitata. Revista de Chimie, Vol. 57, No.3, pp. 297-302, ISSN 0034-7752

Caşcaval, D.; Galaction, A.I. & Oniscu, C. (2004a). Selective pertraction of carboxylic acids obtained by citric fermentation. Separation Science and Technology, Vol.39, No.8, pp. 1907-1925, ISSN 0149-6395

Caşcaval, D.; Galaction, A.I. & Turnea, M. (2009). Study of the influence of solute and carrier characteristics on facilitated pertraction mechanism in pseudosteady-state conditions. Journal of Membrane Science, Vol.328, No.1-2, pp. 228-237, ISSN 0376-7388

Caşcaval, D.; Galaction, A.I.; Nicuţă, N. & Blaga, A.C. (2007). Selective separation of gentamicins from the biosynthetic mixture by reactive extraction. Separation and Purification Technology, Vol. 57, No.2, pp. 264-269, ISSN 1383-5866

Caşcaval, D.; Oniscu, C. & Caşcaval, C. (2000). Selective Separation of Penicillin V from Phenoxyacetic Acid Using Liquid Membranes. Biochemical Engineering Journal, Vol.5. No.1, pp. 45-50, ISSN 1369-703X

Caşcaval, D.; Oniscu, C. & Galaction, A.I. (2001). Selective Separation of Amino Acids by Reactive Extraction. Biochemical Engineering Journal. Vol. 7, No.3, pp. 171-176, ISSN 1369-703X

Caşcaval, D.; Oniscu, C. & Galaction, A.I. (2004b). Biochemical engineering and biotechnology. 3. Bioseparations, Performantica, ISBN 973-99683-3-3, Iasi, Romania

Daugherty, E. (2006). Biotechnology: Science for the new millennium, EMC Paradigm Publishing, ISBN 0-763822828, New York, USA

Deblay, P.; Minier, M. & Renon, H. (1990). Separation of L-valine from fermentation broths using a supported liquid membrane. Biotechnology and Bioengineering, Vol.35, pp. 123-131, ISSN 1097-0290

Fortunato, R.; Afonso, C.A.M.; Reis, M.A. & Crespo, J.G. (2004). Supported liquid membranes using ionic liquids: study of transport mechanisms and stability. Journal of Membrane Science, Vol.242, No.1, pp. 197-204, ISSN 0376-7388

Galaction A.I.; Kloetzer, L. & Caşcaval, D. (2010). Separation of p-aminobenzoic acid by reactive extraction in the presence of 1-octanol as phase modifier. Chemical and

Biochemical Engineering Quarterly, Vol.24, No.2, pp. 149-157, ISSN 0352-9568

Galaction, A.I. & Caşcaval, D. (2006). Secondary metabolites with pharmaceutical, cosmetic and food applications, Casa de Editura Venus, ISBN 973-756-000-0, Iaşi, Romania

Galaction, A.I.; Camarut, M. & Caşcaval, D. (2007). Selective separation of cinnamic and p-methoxycinnamic acids by facilitated pertraction. Separation Science and Technology, Vol.42, No.16, pp. 3727-3740; ISSN 0149-6395

Galaction, A.I.; Caşcaval, D. & Nicuţă, N. (2008). Selective removal of Gentamicin C_1 from biosynthetic Gentamicins by facilitated pertraction for increasing antibiotic activity. Biochemical Engineering Journal, Vol.42, No.1, pp. 28-33, ISSN 1369-703X

Galaction, A.I.; Nicuţă, N. & Caşcaval, D. (2009). Separarea eritromicinei prin pertracţie directă din lichide de fermentaţie simulate. Revista Medico-Chirurgicala, Vol.113, No.2, pp. 619-624, ISSN 0048-7848

Hanson, K.R. & Havir, E.A. (1981). The Biochemistry of Plants, Vol. 7, Academic Press, ISBN 0-12-67407-1, New York, USA

Isoherranen, N. & Soback, S. (2000). Determination of gentamicin after trimethylsilylimidazole and trifluoroacetic anhydride derivatization using gas chromatography and negative ion chemical ionization ion trap mass spectrometry. Analyst, Vol.125, pp. 1573-1576, ISSN 0003-2654

Juang, R.S.; Lee, S.H. & Shiau, R.C. (1998). Carrier-facilitated liquid membrane extraction of penicillin G from aqueous streams. Journal of Membrane Science, Vol.146, No.1, pp. 95-101, ISSN 0376-7388

Juong, R.-S., Wang, Y.-Y.: Amino acids separation with D2EHPA by solvent extraction and liquid surfactant membranes. Journal of Membrane Science. Vol.207, No.2, pp. 241-246, ISSN 0376-7388

Kang, C.D. & Sim, S.J. (2008). Direct extraction of astaxanthin from Haematococcus culture using vegetable oils. Biotechnology Letters Vol.30, No.3, pp. 441-444, ISSN 1573-6776

Katikaneni, S.P.R. & Cheryan, M. (2002). Purification of fermentation-derived acetic acid by liquid-liquid extraction and esterification. Industrial & Engineering Chemistry Research, Vol.41, No.11, pp. 2745-2752, ISBN 1520-5045

Kawasaki J.; Egashira R.; Kawai T.; Hara H. & Boyadzhiev L. (1996). Recovery of erythromycin by a liquid membrane. Journal of Membrane Science Vol.112, No.2, pp. 209-217, ISSN 0376-7388

Kelly, N.A.; Lukhezo, M.; Reuben, B.G.; Dunne, L.J. & Verrall, M.S. (1998). Reactive solvent extraction of amino acids with cationic extractants. Journal of Chemical Technology & Biotechnology, Vol.72, pp. 347-355, ISSN 0268-2575

Kislik, V.S. (2010). Liquid membranes - principles and applications in chemical separations and wastewater treatment, Elsevier, ISBN 978-0-444-53218-3, London, UK

Kloetzer, L.; Galaction, A.I. & Caşcaval, D. (2010). Facilitated pertraction of p-aminobenzoic acid with Amberlite LA-2 in presence of 1-octanol. Separation Science and Technology Vol.45, No.10, pp. 1440-1447, ISSN 0149-6395

Korzybski, T. (1978). Antibiotics: origin, nature and properties, American Society for Microbiology, ISBN-091482614X, Washington, USA

Lee, S.; Han, J.M.; Kim, H.; Kim, E.; Jeong, T.S.; Lee, W.S. & Cho, K.H. (2004). Synthesis of cinnamic acid derivatives and their inhibitory effects on LDL-oxidation, acyl-CoA: cholesterol acyltransferase-1 and -2 activity, and decrease of HDL-particle size. Bioorganic & Medicinal Chemistry Letters, Vol.14, No.18, pp. 4677-4684, ISSN 0960-894X

Li, N.N. (1978). Facilitated transport through liquid membranes. An extended abstract. Journal of Membrane Science, Vol.3. No.3, pp.265-279, ISSN 0376-7388

Lin, S.H. & Chen, C.-N. (2006). Simultaneous reactive extraction separation of amino acids from water with D2EHPA in hollow fiber contactors. Journal of Membrane Science, Vol.280, No.1-2, pp. 771-780, ISSN 0376-7388

Lin, S.H.; Chen, C.N. & Juang, R.S. (2006). Extraction equilibria and separation of phenylalanine and aspartic acid from water with di(2-ethylhexyl)phosphoric acid. Journal of Chemical Technology and Biotechnology. Vol.81, pp. 406-412, ISSN 0268-2575

Liu, Y.S. & Dai, Y.Y. (2003). Distribution behavior of alpha-amino acids and aminobenzoic acid by extraction with trioctylamin. Separation Science and Technology, Vol.38, No.5, pp. 1217-1228, ISSN 0149-6395

Liu, Y.S.; Dai, Y.Y. & Wang, H.D. (1999). Distribution behavior of l-phenylalanine by extraction with di(2-ethylhexyl) phosphoric acid. Separation Science and Technology, Vol.34, pp. 2165-2176, ISSN 0149-6395

Luangrujiwong, P.; Sungpet, A.; Jiraratananon, R. & Way, J. D. (2007). Investigation of the carrier saturation in facilitated transport of unsaturated hydrocarbons. Journal of Membrane Science, Vol.250, No.2, pp. 277-285, ISSN 0376-7388

Monteiro, T.I.R.C.; Porto, T.S.; Carneiro-Leão, A.M.A.; Silva, M.P.C. & Carneiro-Da-Cunha, M.G. (2005). Reversed micellar extraction of an extracellular protease from Nocardiopsis sp. fermentation broth. Biochemical Engineering Journal, Vol.24, No.1, pp. 87-90, ISSN 1369-703X

Moo-Young, M.; Cooney, Ch.L. & Humphrey, A.E. (1985). Comprehensive Biotechnology, Vol. 3, 254-270, Pergamon Press, ISBN 9780080262048, Oxford, UK

Naczk, M. & Shahidi, F. (2004): Extraction and analysis of phenolics in food. Journal of Cromatography A, Vol.1054, No.1-2, pp. 95-102, ISSN 0021-9673

Noble, r.d. & Stern, s.a. (1995). Membrane separations technology. Principles and applications, Elsevier, ISBN: 978-0-444-81633-7, London, UK

Palma, M.; Pineiro, Z. & Barroso, C.J. (2002). In-line pressurized-fluid extraction–solid-phase extraction for determining phenolic compounds in grapes. Journal of Cromatography A, Vol.968, No.1-2, pp. 1-5, ISSN 0021-9673

Parales, R.E.; Bruce, N.C.; Schmid, A. & Wackett, L.P. (2002). Biodegradation, Biotransformation, and Biocatalysis (B3). Applied and Environmental Microbiology, Vol.68, No.10, pp. 4699-5005, ISSN 1098-5336

Park, S.S.; Park, J.H.; Kim, S.H. & Hwang, S.H. (2003). Method for preparing p-aminobenzoic acid. Patent WO 072534/2003.

Rehm, H.J. & Reed, G. (1993). Biotechnology, VCH, ISBN 3-527-25764.0, Weinheim, Deutschland

Rosenkrantz, B.E.; Greco, J.R.; Hoogerheide, J.G. & Oden, E.M. (1980). Gentamicin, In: Analytical profiles of drug substances, K. Florey (Ed.), Vol. 9, 295-340, Academic Press, ISBN: 0122608267, Orlando, USA

Saraf, A.S. & Simonyan, A.V. (1992). Synthesis and antiallergic activity in a series of cinnamic acid. Pharmaceutical Chemistry Journal, Vol.26, No.7-8, pp. 598-605, ISSN 0091-150X

Savitskaya, E.M.; Yakhontova, L.F. & Nys, P.S. (1982). Sorption of organic substances by ion exchangers of various nature. Pure Applied Chemistry, Vol.54, pp. 2169-2180, ISSN 1365-3075

Schuegerl, K. (1994). Solvent Extraction in Biotechnology, Springer, ISBN: 978-3-540-57694-5, Berlin, Deutschland

Scovazzo, P.; Visser, A.E.; Davis Jr. J.H.; Rogers, R.D.; Koval, C.A.; DuBois, D.L. & Noble, R.D. (2002). Supported ionic liquid membranes and facilitated ionic liquid membranes, In: Ionic liquids: industrial applications to green chemistry, Rogers, R.D., Seddon, K.R., (Ed.), 69-87, ACS Symposium Series 818, ISBN 978-084123789-6, Washington, USA

Silverman, R.B. (2004). The organic chemistry of drug design and drug action, Second edition, Elsevier Academic Press, ISBN 978-0-12-643732-4, London, UK

Smelz, E.A.; Engelberth, J.; Tumlinson, J.H.; Block, A. & Alborn, H.T. (2004). The use of vapor phase extraction in metabolic profiling of phytohormones and other metabolites. The Plant Journal, Vol.39, No.5, pp. 790-796, ISSN 1365-313X

Tan, B.; Luo, G. & Wang, J. (2007). Extractive separation of amino acid enantiomers with co-extractants of tartaric acid derivative and aliquat-336. Separation and Purification Technology, Vol.53, No.3, pp. 330-336, ISSN 1383-5866

Tawata, S.; Taira, S.; Kobamoto, N.; Zhu, J.; Ishihara, M. & Toyama, S. (1996). Synthesis and antifungal activity of cinnamic acid esters. Bioscience, Biotechnology and Biochemistry, Vol. 60, No.5, pp. 909-1004, ISSN 1347-6947

Teramoto, M.; Matsuyama, H. & Yonehara, T. (1990). Selective facilitated transport of benzene across supported and flowing liquid membranes containing silver nitrate as a carrier. Journal of Membrane Science, Vol.50, No.3, pp. 269-277, ISSN 0376-7388

Vijayakumar, J.; Aravindan, R. & Viruthagiric, T. (2008). Recent Trends in the Production, Purification and Application of Lactic Acid. Chemical and Biochemical Engineering Quarterly, Vol.22, No.2, pp. 245-264, ISSN 0352-9568

Weast, R.C. (1974). Handbook of Chemistry and Physics, 54th edition, CRC Press, ISBN: 0849304814, Cleveland, USA

Williams, D.A. & Lemke, T.L. (2002). Foye's Principles of Medicinal Chemistry, Fifth edition, Lippincot Williams & Wilkins, ISBN 978-0-7817-6879-5, New York, USA

Yordanov, B. & Boyadzhiev, L. (2004). Pertraction of citric acid by means of emulsion liquid membranes. Journal of Membrane Science, Vol.238, No.2, pp. 191-198; ISSN 0376-7388

Yoshizawa, S.; Fourmy, D. & Puglishi, J.D. (1998). Structural origin of gentamicin antibiotic action. The EMBO Journal, Vol.17, pp. 6437-6448, ISSN 0261-4189

17

Screening of Factors Influencing Exopolymer Production by *Bacillus licheniformis* Strain T221a Using 2-Level Factorial Design

Nurrazean Haireen Mohd Tumpang[1],
Madihah Md. Salleh[1] and Suraini Abd-Aziz[2]
*[1]Industrial Biotechnology Department, Faculty of
Biosciences and Bioengineering,
Universiti Teknologi Malaysia,
[2]Bioprocess Technology Department, Faculty of
Biotechnology and Biomolecular Sciences,
Universiti Putra Malaysia,
Malaysia*

1. Introduction

The term exopolymer can be defined as biopolymer that is secreted by microorganisms outside the cell. Exopolymer is also known as exopolysaccharides (EPS) (Verhoef, 2005; Beech *et al.*, 1999). The exopolymer consist of polysaccharides that are either found associated with the microbial cell wall in the form of capsules or completely dissociated from the microbial cell. Besides, nucleic acids, proteins and also amphiphilic compounds including (phospho-) lipids can also be present in exopolymer. The contents of carbohydrate, protein and nucleic acid in exopolymer have a substantial effect on the flocculation of bacteria (Verhoef, 2005; Beech *et al.*, 1999, Sheng *et al.*, 2005). Generally bacteria exopolysaccharides have unique rheological properties because of their high purity and regular structure. These bacteria produce wide variety of exopolymers that have been used to cope in various ways with the external environment (Kornmann *et al.*, 2003). The exopolymer has advantages to enhance the viscosity of the solution thus making it applicable as thickeners, emulsifiers, and suspending agents in food, dairy product, pharmaceutical and petroleum industries (Lungmann *et al.*, 2007; Gandhi *et al.*, 1997, Yakimov *et al.*, 1997; Lee *et al.*, 1999; Degeest *et al.*, 2001; Torino *et al.*, 2005).

In Microbial Enhanced Oil Recovery (MEOR) application, the exopolymer-producing bacteria are highly dependent on both physical and chemical environment (Yakimov *et al.*, 1997; Haghighat *et al.*, 2008; Duta *et al.*, 2006, Lin *et al.*, 2007). Physical environment include the fermentation conditions such as temperature, pH environment and incubation time. The chemical environments include the carbon and nitrogen source of the medium (Liu *et al.*, 2009; Degeest *et al.*, 2001; Cerning *et al.*, 1994). Temperature and pH are important factors that affect the performance of cells and exopolymer production (Lee *et al.*, 1997, Degeest *et al.*, 2001; Tharek *et al.*, 2006). The acidity and alkalinity could result in the inhibition of bacterial growth, thus retarding its metabolic production (Norell and Messley, 2003). Ghaly *et al.*, 2007 reported that bacteria *Bacillus licherniformis* produced exopolymer (levan) which has potential applications as a selective plugging agent in microbial enhanced oil recovery when grown in sucrose. The bacteria was able to grow in sucrose, glucose, and fructose, but produced exopolymer only in the presence of sucrose. Exopolymer production will be very low if one of these environmental factors is not controlled at the proper levels (Turimin, 2003). Therefore, intensive research on optimization of exopolymer production should be conducted at effective low cost medium and little manpower.

Recently, the results analyzed by a statistical planned experiment are better acknowledged than those carried out by the traditional one-variable-at-a-time method. Statistical experimental designs have been used for many decades and can be adopted on several steps of an optimization strategy, such as for screening experiments or searching for the optimal conditions of a targeted response (Duta *et al.*, 2006; Cui *et al.*, 2006; Lungmann et al., 2007; Reddy *et al.*, 1999; Casas *et al.*, 1997). The 2-level factorial design can be considered to be a multivariable sequential search technique in which the effects of two or three factors are studied together and the responses are analyzed statistically to arrive at a decision (Tunga *et al.*, 1998, Duta et al., 2006; Anbu *et al.*, 2006). It is useful to identify the important nutrients and interactions between two or more nutrients in relatively few experiments as compared to the one-factor-at-time technique (Ooijkaas *et al.*, 1998, Luo et al., 2009).

In this study, five statistically significant parameters which are temperatures, pH, sucrose, NaCl and peptone concentrations were used for screening process using *DESIGN EXPERT 6.0.4 software*. Each independent variable was investigated at a high (+1) and a low (-1) level. The design consists of 37 experiments which include five replicates at center points.

2. Materials and methods

2.1 Microorganism

Bacillus licheniformis strain T221a was used in the experiments. The bacteria was locally isolated from Tiong A27 Petroleum Reservoir in Sarawak (Tharek *et al.*, 2006). The stock culture was maintained in cryopreservation beads and stored at -80°C.

2.2 Culture media

This microorganism was grown and maintained on Modified Reinforced Clostridia (RGM1) agar medium. This medium contained NaCI 5.0 g/L, peptone 10.0 g/L, yeast extract 3.0 g/L, glucose 5.0 g/L, NH_4NO_3 2.0 g/L, meat extract 10.0 g/L, soluble starch 1.0 g/L, L-Cystein HCl 0.5 g/L and sodium acetate 3.0 g/L. The pH of the media was adjusted to approximately 8.5. *B. licheniformis* was grown on this media at 50°C for 12 hours. After growth, the cultures were stored at 4°C.

The stock culture was used for preparing the inoculum in 150 mL serum bottles containing 100 mL of RGM1 broth medium. The preparation of the RGM1 broth was conducted under anaerobic condition. 1 mL of trace mineral and 1 mL of vitamin was added into the medium. Resazurin solution (0.1% w/v) was added as indicator of anaerobiosis. The medium was sparged for 10 minutes to remove oxygen gas. The culture medium was incubated at 50°C for 12 hours without agitation until the optical density at 660nm was above 0.6.

2.3 Exopolymer production medium

The enhancement of exopolymer production was performed in Modified Exopolymer Production (MM2) medium which consisted of the following composition (g/L): KH_2PO_4

(0.5), K_2HPO_4 (0.5), cystein HCl (0.5), NH_4NO_3 (2.0), and sodium bicarbonate (10.0). The concentration of sucrose, peptone and NaCl was added in the range according to the experimental design; sucrose (30.0 - 70.0 g/L), peptone (4.0-8.0 g/L) and NaCl (10.0 – 30.0 g/L). The initial pH of the medium was adjusted in the ranged accordance to the experimental design which was 7 - 10. Preparation of production medium was performed in 150 mL serum bottles under aerobic condition. All the culture media were sterilized for 15 minutes at 121°C.

2.4 Production of exopolymer in batch culture

Batch culture was initiated by inoculating approximately 10% (v/v) of culture from serum bottles that were prepared previously and transferred into 100 ml sterile MM2 medium. The experiments were done in duplicate. Culture broth was incubated at their respective temperatures for 16 hours without agitation. Withdrawn samples were centrifuged at 5,000 rpm for 30 minutes. The supernatant was used for the determination of exopolymer concentration and sucrose concentration. The pellet (cells) was used for the determination of cell concentration.

2.5 Exopolymer quantification

Exopolymer concentration was determined via product-dry weight method by centrifuging culture sample (5 mL) at 5000 rpm for 30 minutes. The supernatant was kept and two volumes of chilled (4°C) ethanol (99.8%, v/v) were added to one volume of supernatant in order to precipitate out the soluble exopolymers. After the exopolymers total precipitation, the suspended material was filtered through pre-weighed 0.2um nylon membrane filter (Whatman) and was washed twice with chilled (4°C) ethanol (99.8%, v/v). Finally, the filtered soluble exopolymers was dried in an oven at 70°C until a constant weight was produced (Duta *et al.*, 2006).

2.6 Determination of sucrose

The procedures described by Mokrasch (1954) were followed to determine the sucrose content. The Anthrone solution was prepared in a mixture of concentrated sulphuric acid and water at ratio of 5:2 and was chilled on ice until it becomes cold. The Anthrone solution was stored at 4°C in the dark. 1 mL of sample was spun down and 5 mL of chilled Anthrone solution was added and mixed. The mixture was kept on ice about 5 minutes to form blue-green colour before reading on spectrophotometer at 620nm.

2.7 Cell concentration determination

The cell concentration was determined by dry-cell weight method (Soni *et al.*, 1987). The culture sample 5 mL was centrifuged at 5,000 rpm for 30 minutes and the supernatant was decanted. The cells was washed twice with distilled water by filtering it through a pre-weighed 0.2 μm cellulose nitrate membrane filter (Whatman) prior to drying it in an oven at 95°C until a constant reading was produced.

2.8 Experimental design

The factors that influenced the exopolymer production were screened using 2-level factorial design created by *DESIGN EXPERT software* (State-Ease Inc., Statistic made easy, Minneapolis, MN, USA, Version 6.0.4). Five variables factors, which are temperature, pH,

NaCl, sucrose (carbon source) and peptone (nitrogen source) concentration were expected to have a significant effect on exopolymer production. The design contains a total of 37 experimental trials involving five replicates at centre points. Each independent variable was investigated duplicate at superior a high (+1) and a low (-1) level (Table 1). Runs of center points were included in the matrix and statistical analysis was used to identify the effect of each variable on exopolymer production. The runs were randomized for statistical reasons.

* Factor	Unit	Low Level (-1)	High Level (+)
Temperature (A)	oC	40	60
pH (B)	-	7	10
NaCl concentration (C)	g/l	10	30
Sucrose concentration (D)	g/l	30	90

* The data of the factors and the low level (-1) and high level (+1) according to the conventional factors that give the significant influenced towards exopolymer production.

Table 1. Factors in real value, for screening by the 2-level fractional factorial design.

3. Results & discussions

3.1 Factors significantly affecting exopolymer production

The medium components and fermentation condition have played an important role for exopolymer production in batch culture. In order to find out the key ingredients significantly affecting the production of exopolymer, the relative significance of five variable factors (temperature, pH, sucrose, peptone and NaCl concentration) were investigated by the 2-level factorial design. The design consisted of 37 experiments in duplicate plus five center point (Table 2). All the experiments were conducted in static flask culture.

The results of the 2-level factorial design model in the form of analysis of variance (ANOVA) are shown in Table 3. ANOVA is a statistical technique that subdivides the total variation of a set of data into component associated to specific sources of variation for the purpose of testing hypotheses for the modeled parameters (Duta et $al.$, 2006). According to the ANOVA, the Fisher's F-test with a very low probability value [$(P_{model} > F) < 0.005$] indicated the model was highly significant on exopolymer production. The larger the magnitude of t-test and smaller the P-value, the more significant is the corresponding coefficient. Among the variables screened, the concentration of sucrose (D), temperature (A) and pH (B) were determined as the most significant variables influencing exopolymer production. Concentration of peptone (E) and NaCl (C) in exopolymer production did not result in significant variation due to the P-value is greater than 0.100.

The goodness of fit of the model was examined by determination coefficient $R^2 = 0.9360$, which implied that the sample variation with more than 93.6% was attributed to the variables. However, only 6.40% of the total variance could not be explained by the model. The adjusted determination coefficient (Adj $R^2 = 0.8879$) was also satisfactory to confirm the significance of the model. Also, the model has an "adequate precision value" of 15.392, which suggests that the model can be used to navigate the design space.

The predicted optimum levels of tested variables (temperature (A), pH (B), NaCl (C), sucrose (D) and peptone (E)) were obtained from ANOVA. The optimal levels for the variables were as follows: 90.0 g/L sucrose, 4.00 g/L peptone, 29.98 g/L NaCl, 40oC temperature and initial pH 10 with the corresponding Y = 4.22 g/L. To validate this model, these predicted parameters were tested in the laboratory and the samples were taken at

Sd	Run	Block 1	Temp (°C)	pH	[NaCl] (g/L)	[sucrose] (g/L)	[peptone] (g/L)	*Exopolymer Production (g/L)	
								Act. Value	Pred. Value
1	32	Block 1	40	7	10	30	8	2.48	2.38
2	12	Block 1	40	7	10	30	8	2.28	2.38
3	35	Block 1	60	7	10	30	4	0.22	0.24
4	9	Block 1	60	7	10	30	4	0.26	0.24
5	18	Block 1	40	10	10	30	4	2.36	2.75
6	27	Block 1	40	10	10	30	4	3.14	2.75
7	36	Block 1	60	10	10	30	8	1.40	1.58
8	4	Block 1	60	10	10	30	8	1.74	1.58
9	24	Block 1	40	7	30	30	4	2.40	2.36
10	1	Block 1	40	7	30	30	4	2.32	2.36
11	33	Block 1	60	7	30	30	8	0.34	0.32
12	17	Block 1	60	7	30	30	8	0.30	0.32
13	21	Block 1	40	10	30	30	8	1.72	1.60
14	14	Block 1	40	10	30	30	8	1.48	1.60
15	3	Block 1	60	10	30	30	4	1.26	1.03
16	7	Block 1	60	10	30	30	4	0.80	1.03
17	19	Block 1	40	7	10	90	4	1.54	1.58
18	10	Block 1	40	7	10	90	4	1.62	1.58
19	2	Block 1	60	7	10	90	8	0.24	0.61
20	25	Block 1	60	7	10	90	8	0.98	0.61
21	30	Block 1	40	10	10	90	8	3.50	3.67
22	26	Block 1	40	10	10	90	8	3.84	3.67
23	29	Block 1	60	10	10	90	4	1.98	1.16
24	13	Block 1	60	10	10	90	4	0.34	1.16
25	31	Block 1	40	7	30	90	8	0.28	0.32
26	15	Block 1	40	7	30	90	8	0.36	0.32
27	5	Block 1	60	7	30	90	4	1.88	1.66
28	11	Block 1	60	7	30	90	4	1.44	1.66
29	28	Block 1	40	10	30	90	4	3.84	4.24
30	8	Block 1	40	10	30	90	4	4.64	4.24
31	37	Block 1	60	10	30	90	8	2.50	2.85
32	6	Block 1	60	10	30	90	8	3.20	2.85
33	16	Block 1	50	8.5	20	70	6	1.38	1.36
34	22	Block 1	50	8.5	20	70	6	1.26	1.36
35	20	Block 1	50	8.5	20	70	6	1.28	1.36
36	34	Block 1	50	8.5	20	70	6	1.58	1.36
37	23	Block 1	50	8.5	20	70	6	1.32	1.36

* Exopolymers were determined after cultivation of the bacteria in MM2 medium. The actual values of exopolymer production were compared to the predicted values given by 2-level factorial design. The actual experimental results were in agreement with the prediction.

Table 2. Experimental designs for the screening of significant factors that influences exopolymer production using 2-level factorial design.

Source	Sum of Squares	Mean Square	F Value	Prob > F	
* Model	42.96	2.86	19.49	< 0.0001	*Significant*
A (Temperature)	11.16	11.16	75.94	< 0.0001	
B (pH)	11.07	11.07	75.30	< 0.0001	
C (NaCl)	0.021	0.021	0.14	0.7093	
D (Sucrose)	1.83	1.83	12.47	0.0021	
E (Peptone)	0.36	0.36	2.43	0.1348	
AB	0.42	0.42	2.85	0.1070	
AC	2.13	2.13	14.51	0.0011	
AD	0.71	0.71	4.86	0.0394	
AE	2.24	2.24	15.22	0.0009	
BC	0.063	0.0693	0.43	0.5201	
BD	4.64	4.64	31.54	< 0.0001	
BE	0.93	0.93	6.34	0.0205	
CD	1.70	1.70	11.58	0.0028	
CE	5.63	5.63	38.29	< 0.0001	
DE	0.060	0.060	0.40	0.5318	
Curvature	0.72	0.72	4.89	0.0387	*Significant*
Pure Error	2.94	0.15			
Cor Total	46.62				
Std. Deviation	0.38	**Adj R-squared**		0.8879	
Mean	1.72	**Pred R-quared**		0.7513	
R-squared	0.960	**Adeq Precision**		15.392	

* The data were shown that the model was highly significant towards exopolymer production by *Bacillus licheniformis.*

Table 3. Regression analysis (ANOVA) for the production of exopolymer

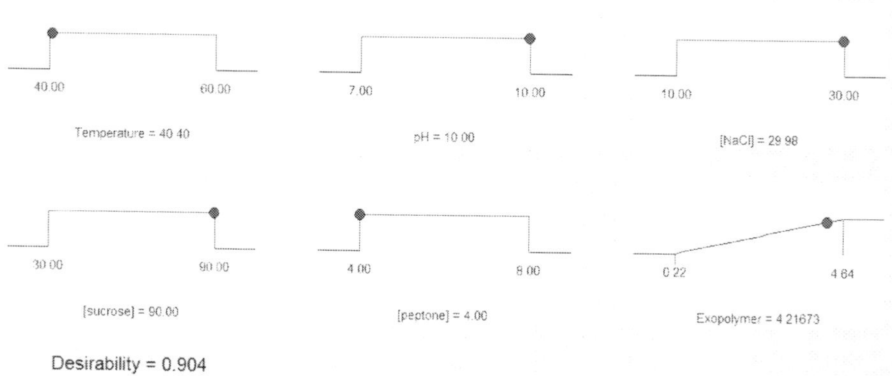

Fig. 1. Predicted optimum levels of five independent variables which indicate the highest exopolymer production.

certain interval during fermentation hour for exopolymer production and other analysis. The final exopolymer concentration obtained was 4.04 g/L, which is almost reaching the predicted value under same condition. This result corroborated the validity and the effectiveness of this model. Figure 1 shows the predicted optimum levels of five independent variables with desirability 90.04%.

3.2 Time course and kinetic evaluation of exopolymer production in batch culture

The suggested optimal values for each variable obtained from 2-level factorial design, were tested in the laboratory and the kinetic was studied. The pattern of exopolymer production by microorganisms may either be growth-associated, non-growth-associated or mixed (Gandhi et al., 1997; Desai and Banat, 1997). Time course of exopolymer production by *B. licheniformis* strain T221a in static culture is presented in Figure 2. The growth of *B. licheniformis* strain T221a increased rapidly during the first 16 hours of incubation. After 16 hours incubation, the bacteria entered death phase and the cell mass decreased. However, exopolymer production kept increasing after 16 hours of incubation even though the cell mass entered death phase. The profile of exopolymer production by *B. licheniformis* strain T221a assigned it as growth-associated. This result is similar to other bacteria such as *Ralstonia eutropha* ATCC 17699 (Wang and Yu, 2007), *Lactobacillus sakei 0-1* (Degeest *et al.,* 2001), *Lactobacillus salivarius* BCRC 14759 and *Bifidobacterium bifidum* BCRC 14615 (Liu *et al.,* 2009) which produced same trends of exopolymer production. Maximum exopolymer production ($P_{max} = 7.12$ g/L) produced by *B. licherniformis* was observed at 24 hours.

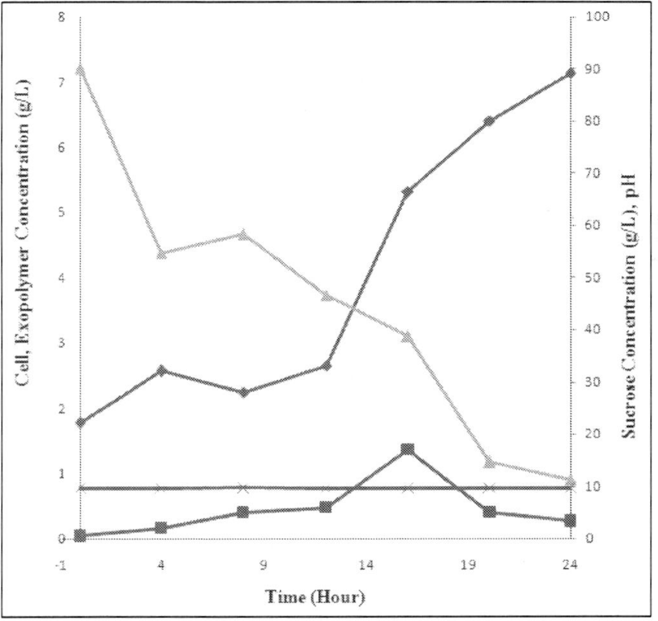

Fig. 2. The time course of exopolymer production in batch culture by *Bacillus licheniformis*. Symbols represent: (▶) Sucrose concentration, (■) Cell mass concentration, (♦) Exopolymer Concentration, (x) pH

The maximum concentration of cells mass (X_{max} = 1.36 g/L) was obtained at 16 hours of incubation. During active cell growth and exopolymer production, sucrose was actively utilized. The concentration of sucrose decreased exponentially reaching about 11.26 g/L after 24 hours. Depletion of sucrose caused a decrease in bacterial cell mass but not in exopolymer production. According to Desai and Banat (1997), the exopolymer can also be produced under nitrogen-limiting conditions. A number of investigators have demonstrated that an overproduction of biopolymer by *Pseudomonas* spp. was obtained when the culture reaches the stationary phase of growth due to the limitation of nitrogen and iron.

The kinetic evaluation of exopolymer production by *B. licheniformis* strain T221a showed the yield of exopolymer formation per gram of substrate utilized (Yp/s) was two times higher than yield of cell formation per gram of substrate utilized (Yx/s), 0.025 g/g. The maximum and overall productivities of exopolymer were 0.64 g/L/h and 0.297 g/L/h, respectively. This result indicated that the predicted parameters in medium composition are more suitable for production of exopolymer rather than the cell mass concentration.

4. Conclusions

2-level factorial design is useful to screen the effects of five variables factors that influenced the exopolymer production. A fitted model obtained showed suitable prediction response that indicates improvement of a model. Based on the data obtained, it was proven that pH, temperature and sucrose concentration was highly significant towards exopolymer production. This design suggested that the optimal value for each variable are; 90.00 g/L sucrose, 4.00 g/L peptone, 29.98 g/L NaCl, 40°C temperature and initial pH 10 with predicted exopolymer of 4.22 g/L. This predicted value was performed in laboratory and 4.04 g/L of exopolymer yield was obtained. The actual experimental results were in agreement with the prediction. This statistical design proved to locate the optimum levels of the most significant parameters for exopolymer production, with minimum effort and time.

5. Acknowledgement

Authors wish to thank Petronas Research Scientific and Services (PRSS) for samples supplied and Faculty of Bioscience and Bioengineering, Universiti Teknologi Malaysia for the analyses services.

6. References

Anbu P., Gopinath S.C.B., Hilda A., Lakshmipriya T. and Annadurai G. (2006). Optimization of extracellular keratinase production by poultry farm isolate *Scopulariopsis brevicaulis*. *Bioresource Technology*. Vol. 98, pp. 1298-1303, ISSN 0960-8524

Beech I., Hanjagsit L., Kalaji M., Neal A. L. and Zinkevich V. (1999). Chemical and structural characterization of exopolymers produced by *Pseudomonas* sp. NCIMB 2021 in continuous culture. *Microbiology*. Vol. 145, pp. 1491-1497. ISSN 1465-2080

Casas, J. A., de Lara, S. G. and Ochoa, F. G. (1997). Optimization of a synthetic medium for *Candida bombicola* growth using factorial design of experiments. *Enzyme Microbiology Technology*. Vol. 21, pp. 221-229, ISSN 0141-0229

Cui F. J., Li, Y., Xu, Z. H., Sun, K. and Tao, W. Y. (2006). Optimization of the medium composition for production mycelial biomass and exo-polymer by *Grifola frondosa* GF9801 using response surface methodology. *Bioresourse Technology*. Vol. 97, pp. 1209-1216, ISSN 0960-8524

Cerning, J., Renard, C. M. G. C., Thibault, J. F. and Bouillanne, C. (1994). Carbon source requirements for exopolysaccharide production by *Lactobacillus casei* CG11 and partial structure analysis of the polymer. *Applied and Environmental Microbiology*. Vol 60, No 11, pp. 3914-3919, ISSN 0099 -2240

Degeest, B., Janssens, B. and Vuyst, L. D (2001). Exopolysaccharide (EPS) Biosynthesis by *Lactobacillus sakei* 0-1: Production kinetics, enzyme activities and EPS yields. *Applied Microbiology*. Vol. 91, pp. 470 – 477, ISSN 1365 - 2672

Desai, J. D. and Banat, I. M. (1997). Microbial production of surfactants and their commercial potential. *Microbiology and Molecular Biology Reviews*. Vol 61, No 1, pp. 47-64, ISSN 1092 - 2172

Duta, F. P., de Franca, F. P. and Lopes, L.M.D.A. (2006). Optimization of culture conditions for exopolysaccharides production in *Rhizobium* sp. using the response surface method. *Electronic Journal of Biotechnology*. Vol. 9, No 4, pp. 391-399, ISSN 0717 -3458

Gandhi, H. P., Ray, R. M. and Patel R. M. (1997). Exopolymer production by *Bacillus* species. *Carbohydrate Polymers*. Vol. 34, pp. 323-327, ISSN 0144-8617

Ghaly, A. E., Arab, F., Mahmoud N. S. and Higgins, J. (2007). Production of levan by *Bacillus licheniformis* for use as a soil sealant in earthen manure storage structures. *American Journal Biotechnology and Biochemistry*. Vol. 3, No 2, pp. 47-54. ISSN 1553-3468

Haghighat, S., Sepahy, A. A., Assadi, M. M. and Pasdar, H. (2008). Ability of indigenous *Bacillus licheniformis* and *Bacillus subtilis* in microbial enhanced oil recovery. *Environmental Science Technology*. Vol 5, No 2, pp. 385-390. ISSN 0013- 936X

Kornmann, H., Duboc P., Marison I., and Stockar U. V. (2003). Influence of nutritional factors on the nature, yield, and composition of exopolysaccharides produced by *Gluconacetobacter xylinus* I-2281. *Applied and Environmental Microbiology*. Vol. 69, No 10, pp. 6091–609, ISSN 0099 -2240

Lee, J. W., Yeomans, W. G., Allen, A. L., Deng, F., Gross, R. A., and Kaplan, D. L. (1999). Biosynthesis of novel exopolymers by *Aureobasidium pullulans*. *Applied and Environmental Microbiology*. Vol. 65, No 12, pp. 5265-5271, ISSN 0099 -2240

Lin, Y, Zhang, Z and Thibault, J. (2007). *Aureobasidium pullulans* batch cultivations based on a factorial design for improving the production and molecular weight of exopolysaccharides. *Process Biochemistry*. Vol. 42, pp. 820–827, ISSN 0032-9592

Liu, C. T., Hsu, I. T., Chou C. C., Lo, P. R and Yu, R. C. (2009). Exopolysaccharide production of *Lactobacillus salivarius* BCRC 14759 and *Bifidobacterium bifidum* BCRC 14615. *World Journal Microbiology Biotechnology*. Vol.25: 883–890. ISSN 0959-3993

Luo, J., Liu, J., Ke, C., Qiao, D, Ye, H., Sun, Yi. And Zeng, X. (2009). Optimization of medium composition for the production of exopolysaccharides from Phellinus baumii Pilát in submerged culture and the immuno-stimulating activity of exopolysaccharides. *Carbohydrate Polymer*. Vol 78, No. 3, pp 409-415, ISSN 0144-8617

Lungmann, P., Choorit, W. and Prasertsan, P. (2007). Application of statistical experimental methods to optimize medium for exopolymer production by newly isolated

Halobacterium sp. SM5. *Electronic Journal Biotechnology*. Vol. 10, No 1, pp 1-11, ISSN 0717 -3458

Mokrasch, L. C. (1953). Analysis of hexose phosphate and sugar mixtures with the anthrone reagent. *Biological Chemistry*. Vol. 208, pp 56-59, ISSN 0021-9258.

Norell, S. A and Messley, K. E (2003). Principles and Application: Microbiological Laboratory Manual. Pearson Education, ISBN 10:013100293, Inc. United States of America.

Ooijkaas, L. P., Wilkinson, E. C., Tramper, J. and Buitelaar, R. M. (1998). Medium optimization for spore production of coniothyrium minitans using statistically-based experimental designs. *Biotechnology and Bioengineering*. Vol 64, No 1, pp 92-100, ISSN 0006-3592

Reddy, P. R. M, Reddy, G. and Seenayya, G. (1999). Production of thermostable pullunase by *Clostridium Thermosulfurogenes* SV2 in solid-state fermentation: Optimization of nutrients levels using response surface methodology. *Bioprocess Engineering*, Vol. 21, pp 497-503, ISSN 0178-515X

Sheng, G. P., Yu, H. Q. and Yu, Z. (2005). Extraction of extracellular polymeric substances from the photosynthetic bacterium *Rhodopseudomonas acidophila*. *Applied Microbiology Biotechnology*. Vol. 67, pp 125–130, ISSN 1432-0614

Soni, B. K., Soucaille, P. And Goma, G. (1987). Continuous acetone-butanol fermentation: global approach for the improvement in the solvent productivity in synthetic medium. *Applied Microbiology Biotechnology*. Vol. 25, pp 317-321, ISSN 1432-0614

Tharek, M., Ibrahim. Z., Hamzah, S. H., Markum, N., Aris, A. M., Daud, F. N., Salleh, M. M., Yahya, A., Wai, L. C., Khairuddin, N., Illias, R., Omar, M. I., Foo, K. S., Elias, E. J. and Bailey, S. (2006). Isolation, screening and characterization of soluble exopolymer-producing bacteria for enhanced oil recovery. *Proceeding of Regional Postgraduate Conference on Engineering and Science (RPCES 2006)*, pp 26-27, ISBN 978-953-307-109-1, Johore, Malaysia, July 17-22 2006.

Torino, M. I., Mozzi, F. and Valdez, G. Font de. (2005). Exopolysaccharide biosynthesis by *Lactobacillus helveticus* ATCC 15807. *Applied Microbiology Biotechnology*. Vol. 68, pp 259 – 265, ISSN 1432-0614

Tunga, R., Banerjee, R. and Bhattacharyya, B. C. (1998). Optimization of variable biological experiments by evolutionary operation-factorial design technique. *Bioscience and Bioengineeering*. Vol. 88, No 2, pp 224-230, ISSN 1389 - 1723

Verhoef, R. P. (2005). *Structural characterisation and enzymatic degradation of exopolysaccharides involved in paper mill slime deposition*. Ph.D. Thesis, Wageningen University, Netherlands.

Wang, J. and Yu, Q. (2007). Biosynthesis of polyhydroxybutyrate (PHB) and extracellular polymeric substances (EPS) by *Ralstonia eutropha* ATCC 17699 in Batch Cultures. *Applied Microbiology Biotechnology*. Vol. 75, pp 871–878. ISSN 1432-0614

Wei, O. S. (2000). Isolation and characterization of indigenous microorganisms in Malaysian oil fields. Master Thesis, Universiti Teknologi Malaysia.

Yakimov, M. M., Amro, M. M., Bock M., Boseker, K., Fredrickson, H. L., Kessel, D. G., and Timmis, K. N. (1997). The potential of Bacillus Licheniformis strains for in situ enhanced oil recovery. *Petroleum Science and Engineering*. Vol 18, pp 147-160, ISSN 0920-4105

18

Biocatalysts in Control of Phytopatogenic Fungi and Methods for Antifungal Effect Detection

Cecilia Balvantín–García, Karla M. Gregorio-Jáuregui, Erika Nava-Reyna,
Alejandra I. Perez-Molina, José L. Martínez-Hernández,
Jesús Rodríguez-Martínez and Anna Ilyina
Department of Biotechnology, Chemistry School,
Universidad Autónoma de Coahuila
México

1. Introduction

The diseases in plants can be reduced with chemical control, in the case of plant fungi illness can be controlled using fungicides. Chemical fungicides are toxic substances that are used to prevent growth or kill fungi harmful to plants, animals or humans. Their use presents many environmental problems such as: the development of resistance to insecticides in pest populations, the resurgence of the populations treated, chemical waste accumulation, risks and legal complications, destruction of beneficial species. Moreover, it is need to be taking in consideration high cost of fumigants, equipment, labor work and material. Thus, to minimize an environmental damage it is very important to substitute chemical control by the biological control.

Biocontrol means to use natural enemies or their metabolites against pathogens causing diseases. This method provides a decrease of pathogenic microorganisms' population. The main advantages are: little or no adverse side effects to the other biological systems different from pathogenic microorganisms; rare resistance of pathogenic forms to biological control; not harmful to environment; the favorable relation cost *vs* benefit; prevention of food contamination by chemicals. Biocontrol methods apply to regulate phytopathogenic fungi growth by using an antagonistic microorganism (fungi or bacteria) and / or its derivatives as an active ingredient. These biofungicides, called like that due to its biological origin, are easily assimilated into the environment and they are alternative techniques to replace the use of chemicals in agriculture. Development of new biocontrol approaches and biofungicides, as well as methodologies for detection of antifungal activity are the one of the important goals of biotechnology.

A fungal disease can be described as polycyclic whether the causative agent is capable of producing spores and re-infect plants during a growing season or monocyclic when the causative agent must wait for a new season. This classification applies to regions with four seasons where pathogens must produce specific structures to survive the winter. Phytopathogenic fungi can also be distinguished by the types of produced spores and the method by which they penetrate into the plant. Once the pathogen has penetrated, it produces a haustorium and grows inside the plant (biotrophic power), or kills target cells and feeds on dead tissue (necrotrophic power). The identification of pathogens is performed based on the signs and symptoms of the disease. Signs refer to the observation of some of the structures of the pathogen (such as sporulation). Symptoms are secondary evidence produced by the plant where a pathogen is present (such as wilting leaves).

The fungal cell wall contains different chemical constituents such as polysaccharides, proteins, chitin and other substances. The cell wall formation varies among species, also varies with age of the fungus, since substances may be present in young hyphae, disappear in the older or deposit other materials to mask the presence of initial constituents. Also the composition of the medium, the pH and temperature influence composition of fungi walls.

Enzymes such exo-1,3- β -D-glucosidase (laminarinase; EC 3.2.1.6) and β -N-acetyl-D-glucosaminidase (chitinase; EC 3.2.1.14) are hydrolytic enzymes produced by *Trichoderma spp* wich are strong inhibitors of many important plant pathogens, mainly of the genus *Phytophthora*, *Rhizoctonia*, *Sclerotium*, *Pythium* and *Fusarium* among others. *Trichoderma* species are most commonly used as antagonists to control plant diseases caused by fungi and they do not attack plants. There are essentially three types of mechanisms by which *Trichoderma* strains influence the plant pathogen: direct competition for space or nutrients, production of antibiotic metabolites, volatile or nonvolatile nature, and direct parasitism of certain species of *Trichoderma* on plant pathogenic fungi. *Trichoderma* species that act as competitive hyperparasites, produce antifungal metabolites and hydrolytic enzymes. Their activity is related to structural changes at the cellular level, such as vacuolation, granulation, disintegration of cytoplasm and cell lysis, which found in organisms they interact with. The chitinases and laminarinases hydrolyze the fungi cell wall and are able to degrade chlamydospores, conidia and polysaccharides of the mature hyphae.

Trichoderma enzymes (chitinase and laminarinase) are substantially more antifungal against wider range of pathogens (e.g. *Fusarium oxysporum*). They are effective and non-toxic in comparison to other purified enzymes from any other source when assayed under the same conditions (Lorito et al., 1998).

F. oxysporum is a pathogenic plant fungus. It causes severe chlorosis, defoliation, desiccation, and wilt in leaves of plants which can lead to plant death. This generates considerable economic loss in many important crops (Jimenez-Gasco et al., 2004).

The use of mycolytic enzymes as antifungal treatments in the protection of some commercially important crops is promising. Nevertheless, the application of enzyme *in situ* needs to increase enzyme stability and protection against environmental factors (Wang & Chio, 1998) that may be achieved by means of immobilization. The immobilization on solid supports can affect the enzyme mobility to fungi in soil contained system and increases its heterogeneity. Therefore, the immobilization in liposomes possibly can help to avoid the problems related to union of enzyme on solid support. However, gradual release of enzymes is desirable in the hydroponic systems with frequent fluid current. The immobilization in solid support may be useful. Recently, in our laboratory a research focused on the immobilization of chitinase and laminarinase on liposomes and brown seaweed bagasse was performed in order to analyze their stability and activity against *Fusarium oxysporum* during tomato growth under greenhouse conditions and hydroponic green fodder (HGF) production, respectively.

The objectives of this study are: to describe the bioluminescence assay approach for detection of mycolytic activity of biocatalysts (chitinase and laminarinase) using *Fusarium oxysporun* as model of phytopathogenic fungus; to demonstrate the effect of chitinase and laminarinase immobilized on brown seaweed bagasse on *Fusarium oxysporum* growth in HGF system; to compare the partitioning behavior of chitinase and laminarinase in soya lecithin liposomes, using a thermodynamic approach based on the variation of partitioning with temperature as well as to define the synergetic activity of microencapsulated enzymes against *Fusarium oxysporum in vitro* and *in vivo* testing in the presence and absence of chemical fungicide thiabendazole.

2. Methods for antifungal effect detection

2.1 Common methods

Fungal growth inhibition could be measured by different methods, which are affected by several *in vitro* factors (Table 1). These variables should be considered when a susceptibility test is designed or interpreted.

Organism specific factors	Drug specific factor	Variables that influence results
Variable growth characteristics Pleomorphism (in yeasts) Type of Metabolism (aerobic or anaerobic) Medium, pH and incubation temperature can affect growth and pleomorphism	Limited aqueous solubility of some agents Partial inhibition of growth over a wide concentration range giving trailing end points Buffer and pH effects on activity Interaction with media components and buffer	Inoculum Medium formulation and pH Agar versus broth Type of buffer Temperature and duration of incubation Minimal inhibitory concentration (MIC) end point criteria

Table 1. Variables related to the drug, organism and technique used in antifungal testing.

Broth dilution test was established as the standard reference for antifungal susceptibility testing, serving as the basis for comparison, for the development of novel tools for antifungal susceptibility testing (Rex *et al.*, 2008a; Rex *et al.*, 2008b). These alternative methods require correlation with MIC results in broth dilution one. At first, National Committee for Clinical Laboratory Standards (NCCLS) recommended broth macrodilution methods, but broth microdilution tests were later determined, having the same effect (Espinel-Ingroff et al., 1992). The methodology is useful for testing common filamentous fungi or yeasts, including the dermatophytes. The fungi encompass *Aspergillus* spp., *Fusarium* spp., *Rhizopus* spp., *P. boydii*, *S. prolificans*, *S. chenckii* (Espinel-Ingroff et al, 1995), *Trichophyton, Microsporum, Epidermophyton* spp. (Ghannoum *et al.*, 2004), *Candida* spp. and *C. neoformans* (Rex *et al.*, 2008b). Difference between methodologies is test volume: for macrodilution is 1 ml and test is done in test tubes, while microdilution is 200 µl and the test is performed in a 96-multiwell microdilution plate. Medium, antifungal substance and inoculum are added to the test tubes or wells and are incubated at 35°C from 21 to 74 hours depending of the fungus. Results are expressed as MIC or MEC (minimal effective concentration). The MIC is the lowest concentration of an antifungal agent that inhibits organism's growth and the MEC is the lowest concentration of drug that leads to the growth of compact hyphal forms as compared to the hyphal growth seen in the growth control assay. MEC evaluation is more appropriated than MIC reading for testing echinocandin antifungal agents (Espinel-Ingroff et al., 2003).

NCCLS, now Clinical and Laboratory Standard Institute (CLSI), recommended some alternative methodologies to the conventional broth dilution tests to probe yeasts and molds susceptibility, which provide reproducible results: YeastOne, Alamar Blue, MTT–test, E-test and Disc diffusion.

Disk diffusion testing is a simple and economic alternative to broth dilution tests. It has been probed with yeasts inhibition. Furthermore there have been identified parameters for testing the antifungal effect over filamentous fungi to five agents (amphotericin B, caspofungin, itraconazole, posaconazole, and voriconazole) by this method (Diekema et al., 2003). Results are provided between 8 to 24 hours, quicker than broth dilution test and the use of nonsupplemented Mueller-Hinton agar instead supplemented one should make this method more available to conventional laboratories at a less cost. There is a good correlation between minimal inhibitory concentration (MIC) and diameter of inhibition in disk diffusion testing.

The E-test (stable agar gradient method) is an alternative procedure to test antifungal susceptibility of yeast (Espinel-Ingroff *et al.*, 1996) and molds (Espinel-Ingroff *et al.*, 2001; Szekely *et al.*, 1999). The method is based on a combination of the concepts of dilution and diffusion tests. It quantifies antifungal susceptibility directly as MIC values, like dilution methods. E-test also consists of a predefined and continuous concentration gradient, making this methodology more precise than conventional procedures based on discontinuous two-fold serial dilutions, and it is not affected by antifungal agent properties (such as molecular weight, diffusion characteristics and aqueous solubility) or by different growth rates of fungus as disk diffusion testing (AB BIODIS, 2000). This method involves placing a plastic strip containing a gradient of an antifungal agent on the surface of an inoculated agar plate (plates are inoculated with a suspension of yeast or mold, turbidity equal to 0.5 McFarland standard (1 McFarland standard for *C. neoformans*)), across the entire surface of agar in three directions. The drug diffuses into the agar and establishes a stable concentration gradient. Inhibition of fungal growth produces an ellipse, and the MIC is read where the ellipse intersects the test strip. Plates are incubated at 35 °C between 24 to 72 h.

MICs determination can be facilitated for a method which quantifies the hyphal growth of filamentous fungi and overcomes observer bias, which can be getting by colorimetric methods based on the measurement of metabolic activity. Alternative methods use different colorimetric growth indicators and they take at least 24 h before reading. The commercially available YeastOne (Trek Diagnostics Systems) consists of a microtiter plate with dried antifungal drugs (Table 2). Every well includes an oxidation-reduction indicator (Alamar Blue) that changes from blue to pink in the presence of microbial growth. The first well to show a change from pink (growth) to purple or blue (growth inhibition) is recorded as the MIC. Easy set-up procedures eliminate time-consuming broth dilution alternative and results are ready after 24 hours of incubation. Several multicenter studies found good correlation between microbroth dilution and Alamar Blue colorimetric susceptibility tests among *Candida sp.* and *C. neoformans* (Eraso et al., 2008).

Another preliminary colorimetric test used for filamentous fungi and yeast isolates utilized the yellow tetrazolium salt dye 3-(4,5-dimethyl-2-thiazyl)-2,5-diphenyl-2H-tetrazolium bromide (MTT). This salt is cut by dehydrogenases to form its purple formazan derivative, which can be measured spectrophotometrically at 550 nm (Levitz & Diamond, 1985). All living and metabolically active fungi can cleave MTT. This method has demonstrated a good agreement with MICs of standard broth dilution tests for the fungal inhibition test of yeasts (Clancy & Nguyen, 1997) and some molds (Meletiadis *et al.*, 2000). Initial inoculum and the dye MTT are incubated for 48 h or more to get results.

The methods described above require a long time to perform. Thus, the development of the faster methods for antifungal activity detection is obvious.

Antifungal Agent	Dilution Range µg/ml
Amphotericin B	0.12 - 8
5-Flucytosine	0.060 - 64
Anidulafungin	0.015 - 8
Caspofungin	0.008 - 8
Micafungin	0.008 - 8
Fluconazole	0.120 - 256
Itraconazole	0.015 - 16
Posaconazole	0.008 - 8
Voriconazole	0.008 - 8

Table 2. Antifungal agents and its dilution range on a YeastOne plate.

2.2 Bioluminescence ATP assay as antifungal susceptibility testing

Due to conventional antifungal methods need long periods of time to evaluate inhibitory effect of the antifungal compounds, a research focused in the development of a faster and reliable technique using luciferase catalyzed bioluminescent reaction for detection of ATP released from the cells destroyed by mycolytic compounds was carried out in our laboratory.

The luciferase catalyzed reaction allows to selective ATP detection (Ansehn & Nilsson, 1984)). However, ATP of dead cells is very unstable metabolite due to ATPases activity, which in this case catalyzes its hydrolysis. To improve detection of ATP released from dead fungal cells, there were applied vanadate ions as inhibitor of a wide spectrum of ATP-ases (Angelis & Gobbetti, 2004). Assay was carried out testing the susceptibility of *Fusarium oxysporum* to such micolytic agents as chitinase (C), laminarinase (L) and syringomycin-E (SR-E) (Cano *et al.*, 2008; De Lucca *et al.*, 1999).

Firstly, vanadate concentration that doesn't affect fungi's viability has been determined by evaluation of the *F. oxysporum* radial growth in agar poisoned with different concentration of ammonium vanadate. Results led to choose 0.75 mM as right vanadate concentration for next assays.

The effect of the mycolitic agents on extracellular and intracellular ATP levels in the presence and absence of vanadate (applied at 0.75 mM) was evaluated (Fig. 1). These assays were performed using enzymes at 20 µg/ml (Cano *et al.*, 2008) and syringomycin E at 2 µg/ml (Espinel-Ingroff et al., 1995) as final concentrations. Results confirm the hypothesis that vanadate addition leads to keep ATP of dead cells due to ATPases inhibition. The increase of extracellular ATP could be considered as the measurable parameter that demonstrates the effect of mycolytic antifungal compounds because it has linear correlation with the CFU (colony forming unit) change rate (Fig. 2).

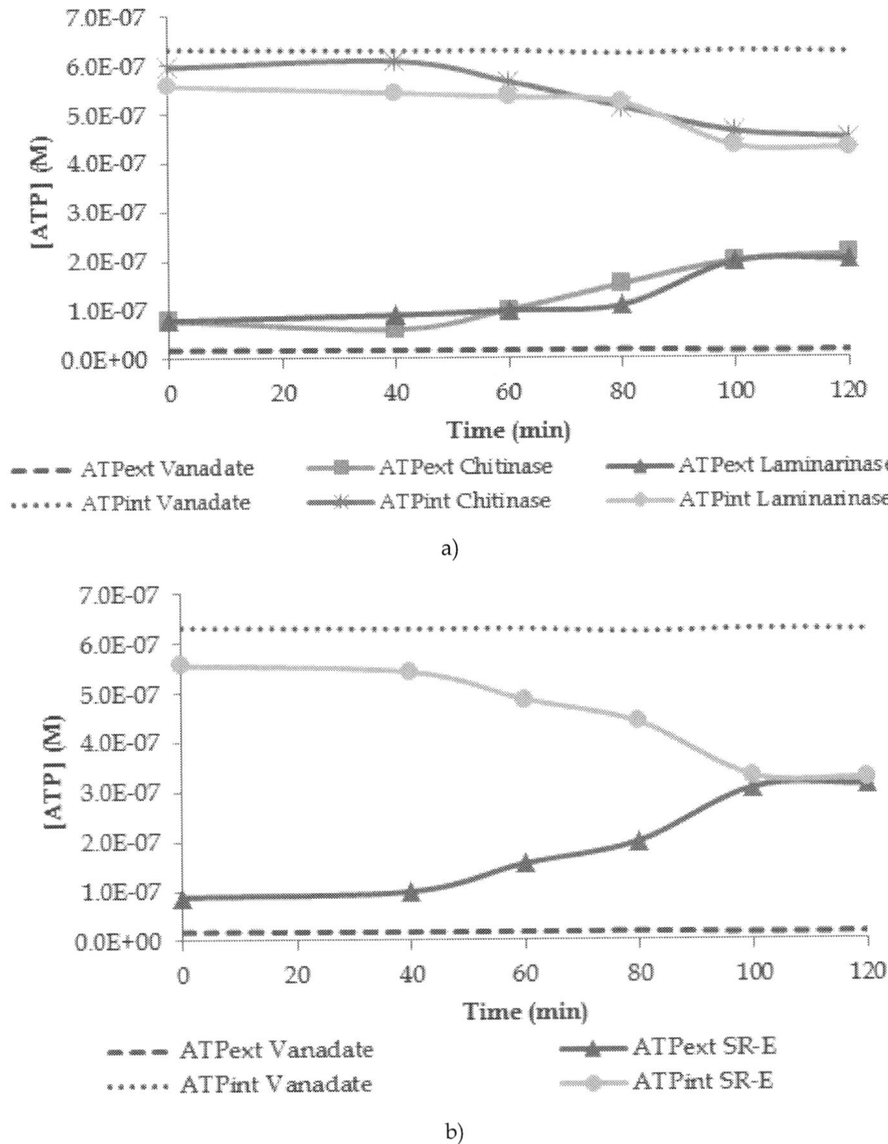

Fig. 1. Intracellular ATP (ATPint) and extracellular ATP (ATPext) detected after 120 min of incubation with antifungal biochemical compounds applied to *Fusarium oxysporum*: a) Chitinase and laminarinase; b) Syringomycin E. The dotted lines indicates the extracellular ATP y intracellular ATP levels in control without antifungal agents.

The decline of intracellular ATP occurs simultaneously with the decrease of fungus viability: the CFU concentration remained constant in the absence of biofungicides, and decreased in their presence. This parameter also may be indicative to measure the inhibitory effect of antifungal compounds. According to the relationship between the CFU/ml and ATP concentration (Fig. 2), is defined that each CFU of *Fusarium oxysporum* affected by mycolitic activity of SR-E corresponds to 10^{-11} mol of ATP. However, it should be reported that ATP level in cells is affected by multiple factors such as crop growth stage, culture media and the presence of metabolic regulators, which may influence the relationship between the number of CFU and the amount of ATP per cell (Stanley, 1986).

Selection of change in intracellular ATP level as indicative parameter of antifungal effect is confirmed by results of kinetic study performed with different concentrations of the tested substances (Fig. 3). The increase of biofungicides concentration leads to greater decrease of intracellular ATP level as well as fungus viability, i.e. a good correlation is observed.

The obtained results led to conclusion that the bioluminescent assay may be considered as the fast and reliable method to antifungal activity evaluation.

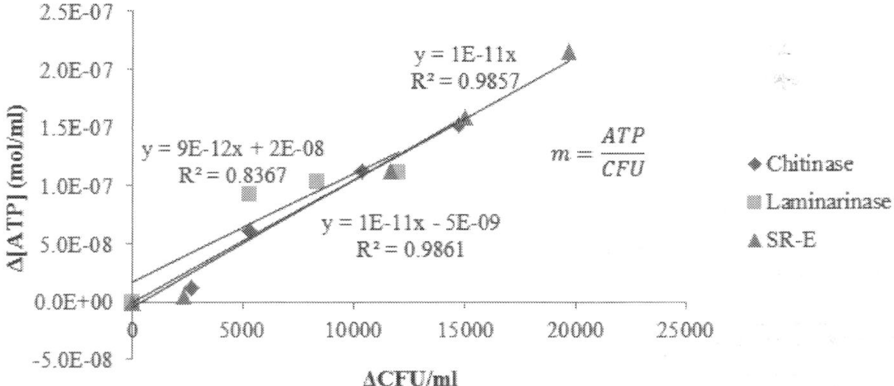

Fig. 2. Correlation between ΔCFU/ml and extracellular ΔATP to estimate the ATP quantity on a CFU of *Fusarium oxysporum*.

3. Immobilization of chitinase and laminarinase in seaweed bagasse as an alternative for fungal control in HGF system

The effect of chitnase and laminarinase (Sigma, USA) on *Fusarium oxysporum* viability was evaluated in the HGF production system. Free and immobilized on seaweed bagasse enzymes were applied. Enzymes were immobilized by adsorption on bagasse at 4°C and 250 rpm. The protein adsorption kinetics and their activity were evaluated at different contact times of immobilization. Immobilized protein was calculated by measuring initial and final protein concentration in the medium by Bradford method (1976). Chitinase and laminarinase activities were determined spectrophotometrically according to the methods described by Pantoom (2008) and Lethbridge (1978), respectively. The results are presented in Table 3. The greater enzymatic activity of chitinase and laminarinase immobilized on seaweed bagasse was detected at 120 min.

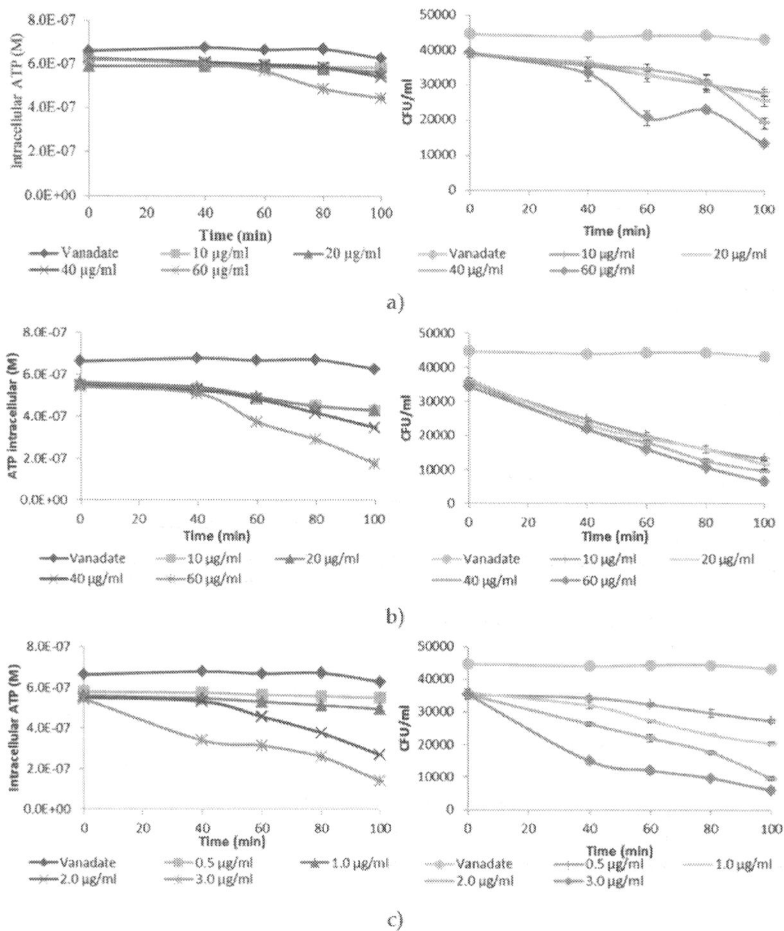

Fig. 3. Kinetics of intracellular ATP (left) and *F. oxysporum* viability (CFU/ml) (right) in the presence of vanadate and different biofungicides concentrations: a) Chitinase; b) Laminarinase; c) Syringomicin E.

After immobilization, antifungal properties of these preparations were compared with free enzymes on wheat HGF. This experiment was carried out under controlled conditions (on an environmental chamber at 21°C). Five grams of wheat seeds were used for each test. Seeds were irrigated with 20 ml of nutritious solution per day and inoculated with 10 4 *F. oxysporum* spores/seed gram. The first day of assay was added 2.5 ml of chitinase or laminarinase (at 20 µg/ml) or 2.5 g of immobilized enzymes. At the harvest day (10th day) fungal viability expressed as CFU/ HGF g as well as height, wet and dry mass were evaluated. Table 4 summarizes the effect of free and immobilized on seaweed bagasse chitinase and laminarinase on HGF production and fungus viability.

Time (min)	Chitinase		Laminarinase	
	Immobilized protein (%)	*IU/g of bagasse*	*Immobilized protein (%)*	*IU/g of bagasse*
0	0	0	0	0
60	38.87	0.017	31.5	0.13
90	62.71	0.018	35.26	0.14
120	65.7	0.02	82.76	0.17
180	66.48	0.019	83.75	0.14

Table 3. Kinetics of chitinase and laminarinase immobilization.

Treatment	Wet mass (g)	Dry mass (g)	Height (cm)	Fungal viability (CFU/g of HGF)
Control without F. oxysporum	38.23	5.74	19.33	5.5×10^5
Control	39.63	6.07	19.66	2.8×10^6
Chitinase	49.86	5.85	20.33	8×10^4
Laminarinase	47.33	5.7	20.16	1.5×10^6
Immobilized chitinase	62.5	5.5	24.5	6×10^4
Immobilized laminarinase	50.06	5.2	24.3	1×10^4

Table 4. Effect of chitinase and laminarinase free and immobilized on seaweed bagasse in HGF characteristics.

Table 4 shows that treatments with chitinase and laminarinase immobilized were effective to inhibit *F. oxysporum* viability compared with the controls. Moreover it was demonstrated that the addition on seaweed bagasse led to increase HGF wet mass and height. Also, the addition of seaweed bagasse on HGF system led to increase plants height and yield.

Thus, with immobilized chitinase and laminarinase on seaweed bagasse led to better control of phytopathogenic fungi under HGF system.

4. Liposomes as chitinase and laminarinase carriers

4.1 Properties of liposomes

Liposomes are vesicles of colloidal dimensions and spherical shape, with a membrane composed of a lipid bilayer in which (phospho) lipid bilayer sequesters part of the solvent (Lasic, 1995). Liposomes were discovered in 1961 by Alec D. Bangham who was studying phospholipids and blood clotting, and since then they became very versatile tools in biology, biochemistry and medicine.

Due to their chemical composition, structure and small size, liposomes exhibit several properties which may be useful in various applications (Table 5). The most important properties include bilayer phase behavior, its mechanical properties and permeability, charge density, presence of surface bound or grafted polymers, or attachment of special ligands. Additionally liposomes exhibit many special biological characteristics, including specific interactions with biological membranes and various cells (Lasic, 1995).

Liposomes are broadly classified by their structure, composition and size (Table 6). The size, lamellarity (unilamellar or multilamellar) and lipid composition of the bilayers influence

Use	Application area	Reference
Containing drugs or markers. As a model, tool, or reagent in the basic studies of cell interactions. Recognition processes.	Pharmacology and medicine	Lasic, 1995
Used to improve the therapeutic efficacy of the encapsulated drug molecules. Site specific targeting.	Pharmacokinetics and biodistribution	Lasic, 1995
As carriers of nystatin and amphotericin *B*, particularly as a means of overcoming toxic side effects and low aqueous solubility of these antibiotics.	Pharmacology	Yamskov *et al.*, 2008
Antimicrobial effect of liposome-encapsulated polymyxin B. formulations against a *P. aeruginosa*.	Microbiology	McAllister *et al.*, 1999
To yield better correlations with partitioning and solvation of ketoprofen.	Pharmacology	Lozano & Martínez, 2006
To deliver exogenous genetic material intra cellularly via fusion with the cell	Pharmacology	Ravichandiran *et al.*, 2011
The antimicrobial and antiviral activity of liposomes as carriers of essential oils.	Microbiology	Martin *et al.*, 2010
Drug delivery systems such antifungal, local anesthetics and Retinoids.	Pharmacology	Granda & Diduk, 1996
Encapsulation of chitinase and laminarinase on soya lecithin liposomes against *Fusarium oxysporum*.	Biotechnology	Joublanc *et al.*, 2010

Table 5. Applications of liposomes in different science areas.

many of the important properties like the fluidity, permeability, stability and structure, these can be controlled and customized to serve specific needs. The properties are also influenced by external parameters like the temperature, ionic strength and the presence of certain molecules nearby.

Liposomes are under investigation both as models for biological membranes and as carriers for various bioactive agents such as drugs, diagnostic and genetic materials, and vaccines. The thermodynamics of molecules transfer can be studied by measuring the partition coefficient as a function of temperature. Such data were used for the prediction of absorption, membrane permeability, and *in vivo* distribution in the case of various drugs (Bangham, 1993; Ávila *et al.*, 2003).

To obtain liposomes, different phospholipids may be applied such soya lecithin, which is the popular and commercial name for a naturally occurring mixture of phospholipids (also called phosphatides or phosphoglycerides). Different soya lecithin samples vary in color from light tan to dark reddish brown. Lecithin is the gummy material contained in crude vegetable oils and removed by degumming. Soybeans are by far the most important source of commercial lecithin and lecithin is the most important by-product of the soy oil processing industry because of its many applications in foods and industrial products. Three main phospholipids in this mixture called "commercial soy lecithin" are 33.0% of phosphatidylcholine (also called "pure" or "chemical" lecithin to distinguish it from the natural mixture), 14.1% of phosphatidylethanolamine (popularly called "cephalin"), and 16.8% of phosphatidylinositols (also called inositol phosphatides) as well as 0.4% of phosphatidylserine. Commercial soy lecithin also typically contains unrefined soy oil as well as additives insoluble in organic solvents (Beare- Rogers *et al.* 1992).

Structure	* Multilamellar: Spherically concentric multilamellar (many bilayers) structures. * Unilamellar: Spherical concentric unilamellar (one bilayer) structures.
Composition	* Phospholipids, cholesterol, phosphatidylethanolamine, free fatty acids, divalent cations. * Conventional, pH-sensitive, cationic, immune and long-circulating.
Size (nm)	* Small unilamellar 20-50 * Large unilamellar, 200-1000 * Multilamellar, 400-3500

Table 6. Classification of liposomas according to structure, composition and size.

Liposomes are widely used to deliver drugs for cancer and other diseases, as well as physiologically active substances in cosmetic products. Their application for immobilization of mycolytic enzymes such as chitinase and laminarinase, and the effect of microencapsulation on their antifungal properties are studied by our scientific group.

4.2 Comparison of the partitioning behavior of chitinase and laminarinase in soya lecithin liposomes, using a thermodynamic approach based on the variation of partitioning with temperature

Liposomes were prepared similar to Bangham method (1993). This resulted in the formation of multilamellar vesicles (MLVs), which was verified by microscopy according to Ávila *et al.*

2003. The molal partition coefficients ($K_{o/w}$), were calculated by Lozano & Martínez (2006) reported method. The standard free energy of transfer ($\Delta G_{w\to o}$), from aqueous media to organic system was calculated in agreement Ávila & Martínez (2003) approach. The temperature dependence of partitioning (van't Hoff method) was employed to obtain data on the enthalpy of transfer ($\Delta H_{w\to o}$). The entropy of transfer ($\Delta S_{w\to o}$) was quantified by means equation $\Delta S_{w\to o} = (\Delta H_{w\to o} - \Delta G_{w\to o})/ T$, and van't Hoff linearization.

The 'partition' means, in this case, that the enzyme is distributed between two phases in a dynamic equilibrium. It is a heterogeneous equilibrium since the 'solute' is distributed between two distinct phases: water and liposomes lipids. As the evidence confirming the distribution process of the enzyme might be considered the decrease of its concentration in aqueous phase after liposome formation related to the partition process between these two phases. Fig. 4 shows the temperature dependence of the partition coefficients for laminarinase and chitinase in studied systems. The $K_{o/w}$ values diminish with rising temperature in chitinase contained systems and increase for laminarinase microencapsulation (Fig. 4). The partition coefficients of enzymes laminarinase and chitinase ($K_{o/w}$) are greater than 1 indicating affinity of enzymes for microencapsulation in liposomes. However, the mechanisms of microencapsulation are different for each enzyme that may be related with differences of their primary structure and amount of lipophilic nature aminoacids (Nobe *et al.*, 2004).

The thermodynamic functions related to the transfer of laminarinase and chitinase from aqueous media to soya lecithin liposomes are summarize in Table 7. In both cases values of $\Delta G_{w\to o}$ at 25 °C are similar and negative. This indicates the preference of each enzyme for the organic phase confirming that enzyme transfer from aqueous media to organic system is spontaneous.

The enthalpic changes imply to energetic requirements and the entropic changes the molecular randomness (increase or decrease in the molecular disorder), resulting in the net transfer of enzyme from water to organic phase. The $\Delta S_{w\to o}$ values defined for chitinase and laminarinase microencapsulation in soy lecithin liposomes are differed to the sign: positive for laminarinase immobilization and negative for chitinase microencapsulation (Table 7). The enthalpy of chitinase transfer ($\Delta H_{w\to o}$) is negative and that of laminarinase is positive. Therefore, the process is exothermic and endothermic, respectively. Negative enthalpy indicates the presence of significant interaction between molecules of chitinase and soya lecithin phospholipids. Phospholipids can establish hydrogen bonding as donor or acceptor of hydrogen (Ávila & Martínez, 2003). On the other hand, after a certain number of enzyme molecules have migrated from the aqueous to the liposome organic phase, the original cavities occupied by the protein in the aqueous phase now are occupied by water molecules. This event is accompanied by release of energy due to water-water interactions. However, depending on enzyme's molecular structure, it is also necessary to keep in mind that the water molecules can organize around the enzyme hydrophobic aminoacids (hydrophobic hydration). This event is accompanied by an intake of energy in addition to a local entropy increase which is related to the separation of some water molecules.

Table 7 shows that for the laminarinase, transfer processes from water to lecithin liposomes were endothermic, and imply high increments in the system net entropy. The entropies of transfer ($\Delta S_{w\to o}$) are positive only for laminarinase contained system. The increase in entropy at the transfer of laminarinase to lecithin liposomes is possibly due to the disorder produced in the hydrophobic core of the lipid layers during separating the phospholipids hydrophobic tails to accommodate the protein molecules in liposomes. The obtained results indicate that the transfer of laminarinase is entropy driven due to positive value of entropy, while chitinase transfer is enthalpy driven due to its negative value.

Thus, laminarinase and chitinase microencapsulation performed by means of thermodynamically different mechanism that might be taken in account for process optimization.

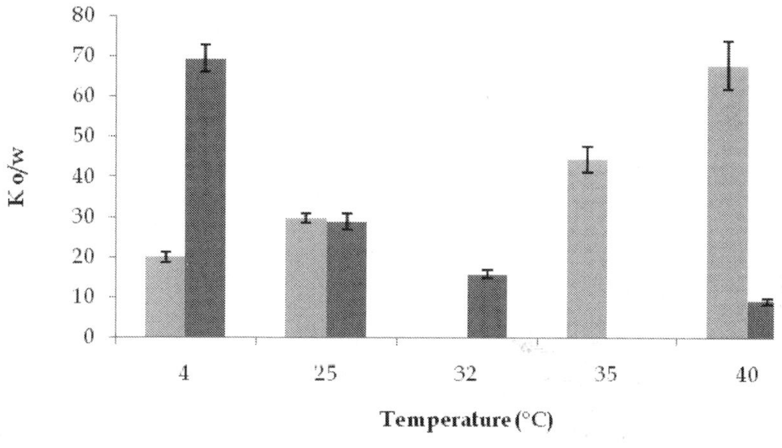

Fig. 4. Partition coefficients of enzymes (laminarinase and chitinase) in soya lecithin liposomes system as a function of temperature (±0.1 °C), in molality (± standard deviation).

Enzyme, mg/Ml	Laminarinase at 0.01 mg/mL	Chitinase at 0.01 mg/Ml
$\Delta G_{w \to o}$, kJ/mol	-8.4	-8.3
$\Delta H_{w \to o}$, kJ/mol	19.4	-39.3
$\Delta S_{w \to o}$, J/(mol x K) from equation	93.4	-103.9

Table 7. Free energy, enthalpy and entropy for the transfer of enzymes (laminarinase and chitinase) from aqueous media to soya lecithin liposomes.

4.3 Liposome storage stability

The liposomes number were measure by optical light microscopy (40X) immediately after their preparation and each tenth day during their storage at 4° and 25°C (Ávila et al., 2003).

The enzyme presence and rising temperature for liposome formation led to decrease in their number and storage stability over long period of time (Table 8). In the presence of chitinase significant decreasing was observed in twentieth day under both storage temperatures and disappearance at 40th and 30th days, respectively for 4° and 25°C. The number chitinase contained liposomes obtained at 25°C (Table 8) was significantly lower than number of liposomes at the same temperature without enzymes or in the presence of laminarinase (Table 8) that seems to relate to a different interaction mechanisms. In the presence of laminarinase the concentration was similar to quantified in the system without enzymes at 4°C and it was greater than detected at 25°C. Significant decreasing was observed in 10th and 20th day and disappearance after 50 and 30 days, respectively, for 4° and 25°C. In this case, at 25°C liposomes number did not decrease as drastically as in liposomes without enzymes. Effects of enzymes on the stability of liposomes could be related to the interaction of proteins and lipids, which in the case of chitinase destabilized the liposomes, while laminarinase stabilized them for short periods of interaction followed by destabilization.

Enzyme concentration (mg/mL)	Temperature (°C)	Time (days)							
		0	10	20	30	40	50	60	70
		Liposomes/Ml							
Without enzyme	4	9 E+09	9 E+09	9 E+09	9 E+09	6 E+09	5 E+09	5 E+09	5 E+09
	25	6 E+09	8 E+08	6 E+08	9 E+07	7 E+06	5 E+06	5 E+06	4 E+06
Chitinase 0.005	4	4.63 E+09	4.81 E+09	2.76 E+09	2.78 E+09	n.d.	n.d.	n.d.	n.d.
	25	6.83 E+06	6.08 E+06	2 E+06	n.d.	n.d.	n.d.	n.d.	n.d.
Chitinase 0.01	4	2.88 E+09	2.43 E+09	2.12 E+09	1.68 E+09	n.d.	n.d.	n.d.	n.d.
	25	3 E+06	3.17 E+06	1.25 E+06	n.d.	n.d.	n.d.	n.d.	n.d.
Laminarinase 0.005	4	8.65 E+09	8.68 E+09	7.75 E+09	2.22 E+09	2.17 E+07	n.d.	n.d.	n.d.
	25	7.1 E+09	5.53 E+09	5.83 E+08	n.d.	n.d.	n.d.	n.d.	n.d.
Laminarinase 0.01	4	8.44 E+09	8.68 E+09	9.02 E+09	1.67 E+09	3.17 E+07	n.d.	n.d.	n.d.
	25	6.3 E+09	5.4 E+09	4.9 E+08	n.d.	n.d.	n.d.	n.d.	n.d.

(n.d.- undetectable by means of optical microscopy)

Table 8. Liposome storage stability at 4°C and 25°C

There are two aspects that affect stability of liposome systems: 1) the liposome component may degrade by hydrolysis and oxidation; chemical changes in the layer-forming molecules may affect physical stability; e.g., if phospholipids lose one of their acyl chains (turn into their lysoforms), the liposome structure is affected; and 2) the physical structure of the liposomes may be affected by changes within the lipid-layer, aggregation, or fusion. The storage stability may be increased by the use of purified phospholipids (Bangham, 1993; Ávila *et al.*, 2003). Thus, the soya lecithin liposomes are sensible to the enzyme presence and are better stored at low temperature.

4.4 Enzyme activity and storage stability

Chitinase activity was quantitatively determined by colorimetric measuring the nitrophenyl group of p-nitrophenyl-β-D-N-acetyl-glucosamide served as substrate, as described previously by Li *et al.* (2004). The laminarinase (β-1, 3-glucanase) activity was measured according to the method of Singh (1999) using laminarine from *Laminaria digitata* as substrate. The activity of laminarinase was determined spectrophotometrically by measuring the release of reducing sugar by the method of Somogyi-Nelson (Nelson, 1944).

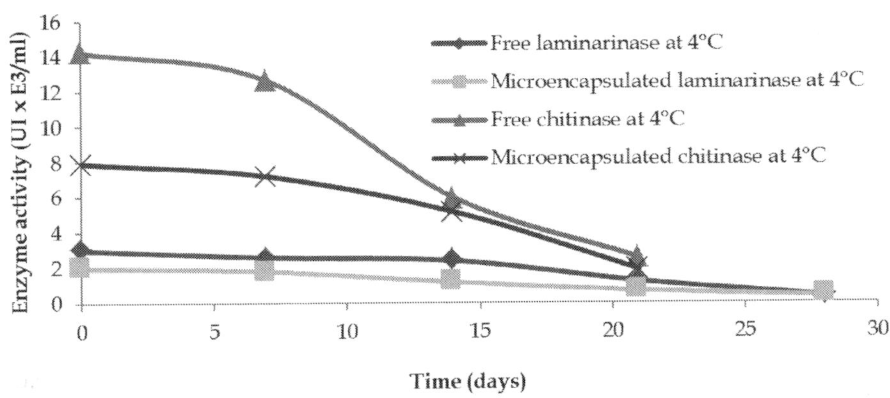

Fig. 5. Chitinase and laminarinase activity during storage at 4°C for microencapsulated in liposomes and free enzyme using at 0.01 mg/mL.

Results described activity of free and encapsulated chitinase and laminarinase during their storage at 4°C are shown in Fig. 5. Microencapsulated enzymes are less active than free enzymes, possibly due to encapsulation effect (Fig. 5). In the presence of liposome the chitinase activity was twice less than free enzyme, while laminarinase was significantly less active after encapsulation. The lower activity is related to partly enzyme encapsulation. The enzyme concentration is less in microencapsulated form than in free form.

Moreover, Chaize (2004) reports that once encapsulated, the enzymes have encountered another problem. The permeability barrier of the lipid membrane drastically diminishes the activity of the enzyme trapped in the liposome by reducing the entrance rate of the substrate molecules and then reducing the substrate concentration inside the liposome. It could be the reason for decreasing of chitinase and laminarinase activity after their microencapsulation. Microencapsulated enzymes lost activity slower than free enzymes. Thus, immobilization of laminarinase and chitinase led to increase the enzymes stability.

4.5 Definition of the activity of microencapsulated enzymes against *Fusarium oxysporum in vitro* and *in vivo* testing

The inhibitory effects of free and microencapsulated chitinase and laminarinase were estimated by using the radial growth inhibition assay as described previously by Prapagdee *et al.* (2007). Fungal growth inhibition was expressed as the percentage of radial growth inhibition relative to the control assay. The means and standard deviations of all obtained results were calculated. Data were analyzed by one-way analysis of variance (ANOVA). Significant differences (p ≤ 0.05) between the means were determined by the Duncan multiple range tests (Bewick *et al.*, 2004).

4.5.1 *In vitro* antifungal activity of free and immobilized enzymes, chemical fungicide and their mixtures

Previously we reported that chitinase and laminarinase from *Trichaderma spp.* applied at 0.02 mg/ml in solid agar media totally inhibited the growth of *Fusrium oxysporum* (Joublanc *et al.*, 2010). Free enzymes (laminarinase and chitinase) applied at 10 μg/ml each partially inhibited the *F. oxysporum* growth and spore production. In these cases the inhibition increased with increasing fungus growth time. In contrast, separately applied microencapsulated enzymes inhibit the fungus growth only during the first four days. The difference may be explained by significantly lesser activity of enzymes encapsulated in liposomes. Moreover, the inhibition of *F. oxysporum* mycelia growth in the presence of liposomes without enzymes is very similar to the one observed in the presence of separately applied microencapsulated enzymes. This leads to conclusion, that the inhibition of *F. oxysporum* growth in these cases is also related to lipid presence. This result is consistent with other reports (Ment *et al.*, 2010) that discussed similar effect observed with other fungi types and other lipid extracts.

The concentration (spores/ml) was decreased in comparison with one detected in the presence of liposomes without enzyme. Conidia concentration decrease was greater in the presence of free enzymes that related to their higher activity and greater growth inhibition.

However, the growth of *Fusarium oxysporum* was totally inhibited in the presence of free and microencapsulated chitinase and laminarinase mixtures applied at 0.005 and 0.01 mg/ml, concentrations lower than 0.02 mg/ml. Our results also are in agreement with the findings of Lorito *et al.* (1993), who indicated that mixtures of hydrolytic enzymes with a complimentary mode of action may benefit an organism by improving its antifungal activity through mycoparasitism or survival in an antagonistic environment.

Use of free or microencapsulated laminarinase and chitinase at 0.01 mg/mL allows at least double decrease of chemical fungicide concentration to obtaining complete inhibition of fungal growth. It can be useful in agriculture practice to reduce levels of chemical fungicides. Thus, the results demonstrate a synergistic effect on the *Fusarium oxysporum* growth: the effect of two treatments (laminarinase/chitinase, or enzyme/thiabendazole) is greater than the effect of each treatment applied individually. Considering the synergistic effects of microencapsulated enzymes demonstrated in the present study, it may be supposed an advantage of its potential application in agro-industry.

In vitro testing performed using soil contained system was also carried out. One milliliter of free or microencapsulated enzymes , as well as their micture at 0.005 and 0.01 mg/mL was added at each tube contained 1 g of soil. Next assay was performed using chemical fungicide (thiabendazole) solution added to soil at final concentration 0.006 mg/g of soil followed by addition of 1 ml of free or microencapsulated enzymes solution at 5 or 10 µg/mL. Control assays were carried out using soil without treatments, adding 1 mL of acetate buffer at pH 5.4, as well as using 1 mL of liposomes suspension prepared without enzymes or with thiabendazole at 6 or 12.5 µg/g of soil without enzyme addition.

All tubes were inoculated with *F. oxysporum* using 1.3 E+05 conidia/mL suspension obtained from a 12 day old culture. Immediately after tube inoculation fungus viability (CFU/g of soil) was evaluated (Madigan *et al.*, 2003). The same measurements were carried out after tube incubation at 25°C at 20th, 30th and 40th day.

It has been demonstrated that the enzymes addition led to decrease of the fungus proliferation that was expressed in lower values of CFU/(g of soil). Free and immobilized laminarinase significantly controlled the fungus viability only at 10 or 20 µg/mL, while chitinase applied at 5 µg/mL also provoked the inhibitory effect. The effect was greater with microencapsulated chitinase than free enzyme, while it was not observed in the case of laminarinase. The results obtained in assays carried out in soil demonstrate fungistatic effect (an inhibiting effect upon the growth and reproduction of fungi without its total destruction) which is more pronounced in the presence of higher enzyme concentrations.

The low fungistatic effect also was demonstrated in assay carried out with liposomes without enzymes that demonstrated the lecithin capacity to inhibit conidia germination. The difference between free and microencapsulated enzyme activities decreased during the assay, possibly due to greater storage stability of enzymes immobilized in soya lecithin liposomes. The relative activity of microencapsulated enzymes was superior to free enzymes. It may be considered as higher enzymatic stability under storage in the soil. Microencapsulation of chitinase and laminarinase protects the protein active structure from these inactivation factors.

The synergic effect to control fungus viability was demonstrated. A mixture of chitinases and β- 1,3-glucanase was significantly more effective against phytopathogenic fungus than either of these enzymes used individually.

With thiabendazole applied at 12.5 µg/mL the fungicide effect was observed, while at 6 µg/mL the effect was lower. The enzyme addition led to greater inhibition of *F. oxysporum* viability. The greater inhibition was observed in the presence of 10 µg/mL of microencapsulated chitinase, as well as laminarinase. Each mixture provoked a fungicide effect detected at 30th day of assay. The effect of free enzyme applied in the presence of thiadendazole was slightly lower than of microencapsulated enzyme. This difference may be related to an inactivation of enzymes in the presence of chemical fungicide. Free enzymes were inactivated after 40 days. Due to inhibitory effect of chemical fungicide, the initial activity of immobilized chitinase was greater than of free enzyme activity in soil. Initially, free laminarinase was more active than microencapsulated preparation but drastically lost its activity in the next days of assay. Thus, the use of microencapsulated enzymes has advantages in comparison to free enzyme application due to their greater enzyme stability as well as higher inhibitory effect on *F. oxysporum* viability with and without chemical fungicide.

4.5.2 *In vivo* testing under greenhouse conditions

Microencapsulated laminarinase and chitinase were applied in the soil under greenhouse condition against *Fusarium oxysporum* to the tomato crops (*L. esculentum Mill*). *Fusarium oxysporum* is one of the most important pathogens of the tomato plants and the other vegetable crops in the greenhouse and field conditions (Larkin *et al*, 1996). Tomato diseases are normally controlled by means of the selected fungicides, e.g. thiabendazole. Due to fungi adaptability and population diversity, the pathogen frequently overcomes all of these currently used disease control strategies. The use of broad-spectrum fungicides results in imbalances within the microbial community creating unfavorable conditions for the activity of beneficial organisms, contaminates the environment, affects ozone layer, and must be applied every season due to its null residual activity and the rapid re-colonization of soils by the phytopathogens. The enzymes of *Trichoderma* are strong inhibitors of many important plant pathogens. Chitinases and laminarinases are able to lyse polysaccharides of mature hyphae, conidia, chlamydospores, and sclerotia.

Three different treatments included encapsulated enzymes were applied: the mixture of laminarinase and chtinase at 10 µg/g of soil and 5 µg/ g of soil, respectively; immobilized laminarinase at 10 µg/g of soil with thiabendazole at 6 µg/g of soil, and immobilized chitinase at 10 µg/g of soil in the presence of thiabendazole at the same concentration. The assays with thiabendazole at 12.5 µg/g of soil and with liposomes without enzyme were carried out. Two controls inoculated and not inoculated with *F. oxysporum* and treated with the buffer without additional substances were performed. Thus, 7 different groups of plants were employed with 10 plants in each group.

Tomato seedlings (*L. esculentum* Mill) were grown according to conventional techniques (Montealegre *et al.*, 2010) during 14 days. Then, they were transplanted into previously sterilized 30 g of the soil / peatmoss mixture (weight ratio 1:1) using as substrate for plant growth. The corresponding treatments were added at total volume of 15 mL to reach the mentioned final concentration. The pots were inoculated with 15 mL of *F. oxysporum* conidias at 1E+05 conidias/mL. The pots were kept at random complete block under controlled greenhouse conditions (controlled range of temperature: 20°C (minimum) and 30°C (maximum); 12 h natural light; irrigation was applied once at three days).

The group of plants grown on the substrate infected with *F. oxysporum* lost 90% of its population due to fungus attack. In all cases the damage at the root of the plant caused by fungus was observed. In other blocks corresponding to other treatments the dead plants were not observed.

Obtained results demonstrated that application of all treatments contained enzymes or chemical fungicide led to control of CFU/(g of soil). The best results were obtained with thiabendazole at 12.5 µg/(g of soil) and at 6 µg/(g of soil) in the presence of chitinase microencapsulated in liposomes at 10 µg/mL. With three other treatments, a similar effect to fungus viability was observed. The CFU/(g of soil) measurements were significantly greater in the untreated infected pots than in the pots with applied treatments, although the increase of CFU/(g of soil) was observed in all cases. The activity maintained at the initial level for the first 14 days followed by its decrease for the next two weeks.

The growth of plants under treatments contained enzymes and/or liposomes with and without thiabendazole was greater than in the case of non-treated and non-infected control. The lowest growth was detected on the infected group without treatments that indicated positive effect of microencapsulated enzymes applied in different treatments to control of phytopathogenic fungus *F. oxysporum* and protection of tomato plants (*L. esculentum* Mill).

5. Conclusion

To determine antifungal activity of biochemical micolytic compounds, the bioluminescent assay may be considered as the fast and reliable method of its evaluation.

It can be concluded that immobilization of chitinase and laminarinase can be a good alternative for enzyme stabilization. Mycolitic enzymes immobilized on seaweed bagasse can be used for the control of phytopathogenic fungi in HGF system.

Chitinase and laminarinase have affinity to soya lecithin liposomes. The findings on the thermodynamic properties of enzymes microencapsulation on liposomes can be considered for process optimization in future studies and applications. Stability of enzyme preparations was increased. Finally, the possibility of using of mycolytic enzymes immobilized in liposomes for the control of some pathogens was confirmed. This finding may provide the alternative means of reducing the dependency on synthetic chemical fungicides. The synergistic effect on viability of *F. oxysporum* was demonstrated in the presence of mixture of encapsulated enzymes and enzymes with thiabendazole. Thus, it confirmed the original idea regarding of use the mycolytic enzyme immobilized in soya lecithin liposome for control of some pathogens.

6. Acknowledgements

The authors would like to thank SEP–CONACYT 2006-57118 project for the financial support.

7. References

AB BIODIS. (2000). Etest for Antifungal Susceptibility Testing. Available at http://www.aspergillus.org.uk/secure/diagnosis/Etest.pdf

Angelis M., Gobbetti M. (2004). Environmental stress responses in *Lactobacillus*: a review. *Proteomics*. 4:106-122.

Ansehn S., Nilsson L. (1984). Direct Membrane-Damaging Effect of ketoconazole and tioconazole on Candida albicans demonstrated by bioluminescent assay of ATP. *Antimicrob Agents Chemother* 26(1):22-25.

Ávila, C.M. Gómez, A. Martínez, F. (2003). Estudio termodinámico del reparto de algunas sulfonamidas entre liposomas de lecitina de huevo y sistemas acuosos. *Acta Farm Bonaerense*, vol.22, pp. 119-126.

Bangham, A.D. (1993). Liposomes: the babraham connection. *Chem Phys Lipids*, vol.64, pp. 275-285.

Beare-Rogers, J.L. Bonekamp-Nasner, A. Dieffenbacher, A. (1992). Determination of the phospholipid profile of lecithins by high performance liquid chromatography. *Pure Appl Chem*, vol.64, pp. 447-454.

Bewick, V. Cheek, L. Ball, J. (2004). Statistics review 9: One-way analysis of variance. *Crit Care*, vol.8, pp. 130-136.

Bradford, M.M. (1976). A rapid and sensitive method for the quantization of microgram quantities of protein utilizing the principle of protein–dye binding. *Anal. Biochem*, vol.72 pp. 248-254.

Cano L., León E., Pérez A., Iliná A., Martínez J.L. (2008). Encapsulación de enzimas hidrolíticas en liposomas: perspectivas de aplicación en agricultura. *Ciencia Cierta*. 15:22-25.

Chaize, B. Colletier, J.P. Winterhalter, M. Fournier, D. (2004). Encapsulation of enzymes in liposomes: high encapsulation effi ciency and control of substrate permeability. *Art Cells Blood Substit Biotechnol*, vol.32 pp. 67-75.

Clancy C.J., Nguyen M.H. (1997). Comparison of a photometric method with standardized methods of antifungal susceptibility testing of yeasts. *J Clin Microbiol*. 35:2878-2882.

De Lucca A.J., Jacks T.J., Takemoto J., Vinyard B., Peter J., Navarro E., Walsh T.J. (1999). Fungal lethality, binding, and cytotoxicity of syringomycin-E. *Antimicrob Agents Chemother*. 43(2):371-373.

Diekema D.J., Messer S.A., Hollis R.J., Jones R.N., Pfaller M.A. (2003). Activities of Caspofungin, Itraconazole, Posaconazole, Ravuconazole, Voriconazole, and Amphotericin B against 448 Recent Clinical Isolates of Filamentous Fungi. *J Clin Microbiol*. 41(8): 3623-3626.

Eraso E., Ruesga M., Villar-Vidal M., Carrillo-Muñoz A.J., Espinel-Ingroff A., Quindós G. (2008). Comparative evaluation of ATB Fungus 2 and Sensititre YeastOne panels for testing in vitro Candida antifungal susceptibility. *Rev Iberoam Micol*, 25:3-6.

Espinel-Ingroff A. (2001). Comparison of the E-test with the NCCLS M38-P method for antifungal susceptibility testing of common and emerging pathogenic filamentous fungi. *J Clin Microbiol.* 39:1360–1367.

Espinel-Ingroff A. (2003). Evaluation of broth microdilution testing parameters and agar diffusion e-test procedure for testing susceptibilities of *Aspergillus* spp. to caspofungin acetate (MK-0991). *J Clin Microbiol.* 41:403-409.

Espinel-Ingroff A., Dawson K., Pfaller M., Anaissie E., Breslin B., Dixon D., Fothergill A., Paetznick V., Peter J., Rinaldi M., Walsh T. (1995). Comparative and collaborative evaluation of standardization of antifungal susceptibility testing for filamentous fungi. *Antimicrob Agents Chemother.* 39(2):314–319.

Espinel-Ingroff A., Kish C.W., Kerkering T.M., Fromtling R.A., Bartizal K., Galgiani J.N., Villareal K., Pfaller M.A., Gerarden T., Rinaldi M.G., Fothergill A. (1992). Collaborative comparison of broth macrodilution and microdilution antifungal susceptibility tests. *J Clin Microbiol.* 30(12):3138–3145.

Espinel-Ingroff A., Pfaller M., Erwin M.E., Jones R.N. (1996). Interlaboratory evaluation of Etest method for testing antifungal susceptibilities of pathogenic yeasts to five antifungal agents by using Casitone agar and solidified RPMI 1640 medium with 2% glucose. *J Clin Microbiol.* 34:848–852.

Ghannoum M.A., Chaturvedi V., Espinel-Ingroff A., Pfaller A., Rinaldi M.G., Lee-Yang W., Warnock W. (2004). Intra- and interlaboratory study of a method for testing the antifungal susceptibilities of dermatophytes. *J Clin Microbiol.* 42:2977-2979.

Granda-Cortada D. Diduk, N.A. (1996). Los liposomas en dermatoterapia. Rev Cubana Farm, vol.30, ISSN 0034-7515.

Jimenez-Gasco, M.M. Navas-Cortes, J.A. Jimenez-Diaz, R.M. (2004). Evolución de *Fusarium oxysporum f.sp. ciceris*, el agente de la fusariosis vascular del garbanzo, en raza patogénicas y patotipos. *Biol Sag Veg Plagas*, vol.31, pp. 59–69.

Joublanc, E.L. Vazquéz, B.B. Ramírez, G. Martínez, J.L. Iliná, A. (2010). Potential of mycolytic enzymes encapsulated on soya lecithin liposomes as biocontrol agents against *Fusarium oxysporum, Innovations in food science and food biotechnology in developing countries*, Regalado G.C. & García A.B.E., pp. 73-84, Asociación Mexicana de Ciencia de los Alimentos, Mexico.

Larkin RP, Hopkins DL, Martin FN. (1996). Suppression of Fusarium wilt of watermelon by nonpathogenic Fusarium oxysporum and other microorganisms recovered from a disease-suppressive soil. *Phytopathology.* 86(812-819), 0031-949X

Lasic D.D. (1995). Applications of Liposomes, Handbook of Biological Physics, Lipowsky R. & Sackmann E., pp. 491, Elsevier Science B.V., Inc. 1050 Hamilton Court.

Lethbridge, G. Bull, A.T. Burns, R.G. (1978). Assay and properties of 1,3-β- glucanase in soil. *Soil Biol. Biochem*, vol.10 pp. 389-391.

Levitz S.M., Diamond R.D. (1985). A rapid colorimetric assay of fungal viability with the tetrazolium salt MTT. *J Infect Dis.* 152:938–945.

Li, H.M. Sullivan, R. Moy, M. Kobayashi, D.Y. Belanger, F.C. (2004). Expression of a novel endophytic fungal chitinase in the infected host grass. *Mycol.* vol.96, pp. 526-536.

Lorito, M. Harman, G.E. Hayes, C.K. Broadway, R.M. Tronsmo, A. Woo, S.L. Di-Pietro, A. (1993). Chitinolytic enzymes produced by Trichoderma harzianum: antifungal activity of purified endochitinase and chitobiosidase. *Phytopathology*, Vol.83, pp. 302–307.

Lozano, H.R. Martínez, F. (2006). Thermodynamics of partitioning and solvation of ketoprofen in some organic solvent/buffer and liposome systems. *Brazil J Pharm Sci.* vol.42, pp. 601-613.

Madigan MT, Martinko JM, Parker J. (2003). *Pearson Educación S.A.* (10ª ed.), Publisher, pp. 1096, Madrid

Martín, A. Varona, S. Navarrete, A. Cocero, M.J. (2010). Encapsulation and co-precipitation processes with supercritical fluids: applications with essential oils. *The Open Chem Engin J,* vol.4, pp. 31-41.

McAllister, M., Alpar, H.O. Brown, M.R.W. (1999). Antimicrobial properties of liposomal polymyxin B S. *Journal of Antimicrobial Chemotherapy,* vol.43, pp. 203-210.

Meletiadis J., Meis J.F.G.M., Mouton J.W., Donnelly J.P,. Verweij P.E. (2000). Comparison of NCCLS and 3-(4,5-Dimethyl-2-Thiazyl)-2,5-Diphenyl-2H-Tetrazolium Bromide (MTT) Methods of in vitro susceptibility testing of filamentous fungi and development of a new simplified method. J Clin Microbiol. 38(8):2949-2954.

Ment, D. Gindin, G. Soroker, V. Glazer, I. Rot, A. Samish, M. (2010). Metarhizium anisopliae conidial responses to lipids from tick cuticle and tick mammalian host surface. *J Invertebr Pathol,* vol.103 pp. 132–139.

Montealegre J, Valderrama L, Sánchez S, Herrera R, Besoain X, Pérez L. (2010). Biological control of Rhizoctonia solani in tomatoes with Trichoderma harzianum mutants, In: *Electronic Journal of Biotechnology,*
http://www.ejbiotechnology.info/content/vol13/issue2/full/6/6.pdf

Nelson, N. (1944). A photometric adaptation of the Somogyi method for the determination of glucose. *J Biol Chem,* vol.153, pp. 375-376.

Nobe, R. Sakakibar,a Y. Ogawa, K. Suiko, M. (2004). Cloning and expression of a novel *Trichoderma viride* laminarinase AI gene (lamAI). *Biosci Biotechnol Biochem,* vol.68, pp. 2111–2119.

Pantoom, S. Songsiriritthigul, C. Suginata, W. (2008). The effects of the surface- exposed residues on the binding and hydrolytic activities of *Vibrio carchariae* chitinase A, *BMC Biochem,* vol.9 doi: 10.1186/1471-2091-9-2.

Prapagdee, B. Kotchadat, K. Kumsopa, A. Visarathanonth, N. (2007). The role of chitosan in protection of soybean from sudden death syndrome caused by *Fusarium solani f. sp.* glycines. *Biores Tech,* vol.7, pp. 1353-1358.

Ravichandiran, V. Masilamani, K. Senthilnathan, B. (2011). Liposome- A Versatile Drug Delivery System *Pelagia Research Library Der Pharmacia Sinica,* vol.2, pp. 19-30.

Rex J.H, Alexander B.D., Andes D., Arthington-Skaggs B., Brown S.D, Chaturveli V., Espinel-Ingroff A., Ghannoum M.A, Knapp C.C., Motyl M.R., Ostrosky-Zeichner L., Pfaller M., Sheehan D.J., Walsh T.J. (2008). Reference method for broth dilution antifungal susceptibility testing of filamentous fungi; Approved standard-Second edition. National Committee for Clinical Laboratory Standards. M38-A2. Wayne, PA.

Singh, P.P. Shin, Y.C. Park, C.S. Chung, Y.R. (1999). Biological control of *Fusarium* wilt of cucumber by chitinolytic bacteria. *Phytopathol,* vol.89, pp. 92-99.

Stanley, P.E. (1986). Extraction of adenosine triphosphate from microbial and somatic cells. *Methods Enzymol.* 133: 14–22.

Szekely, A., E. M. Johnson, and D. W. Warnock. 1999. Comparison of E-test and broth microdilution methods for antifungal drug susceptibility testing of molds. *J Clin Microbiol*, 37:1480–1483.

Wang, S.L. Chio, S.H. (1998). Reversible immobilization of chitinase via coupling to reversibly soluble polymer. *Enzyme Microb Technol*, vol.22, pp. 634–640.

Yamskov, I.A. Kuskov, A.N. Babievsky, K.K. Berezin, B.B. Krayukhina, M.A. Samoylova, N.A. Tikhonov, V.E. Shtilman, M.I. (2008). Novel liposomal forms of antifungal antibiotics modified by amphiphilic polymers. *Appl Biochem and Microbiol*, vol.44, pp. 624-628.

19

Cofactor Engineering Enhances the Physiological Function of an Industrial Strain

Liming Liu and Jian Chen
State Key Laboratory of Food Science and Technology
Jiangnan University
China

1. Introduction

Microorganisms are able to produce a wide range of valuable chemicals and materials, and microbial fermentation is widely used as an alternative route for the production of chemicals in industry[1]. The key elements that determine the efficiency of a fermentation process are high titer, high yield, high productivity and process robustness[2]. These parameters are highly dependent on the host microorganism. In order to enhance the metabolic capabilities of the host microorganism, early research focused on screening appropriate microorganisms that naturally overproduce target products and improving their performance by random mutagenesis and by optimizing the fermentation processes. With the advent of metabolic engineering, many different genetic or metabolic engineering strategies have been adopted to improve the metabolic capabilities of the host strains, including relief of feedback inhibition, deletion of competing pathways, up-regulation of primary biosynthetic pathways, re-direction of central metabolism towards the target pathway, over-expression of export processes and insertion of new metabolic pathways. More recently, the emergence of systems biology integrated with metabolic engineering has provided a comprehensive understanding of microbial physiology, followed by a more global-wide identification of the target genes to be manipulated[3]. Those approaches have been proven to be powerful in developing microbial strains for the commercial production of organic acids[4], amino acids, biofuels and pharmaceuticals[5,6,7]. Nevertheless, problems such as the accumulation of toxic intermediates or metabolic stress resulting in decreased cellular fitness are still far from being solved. Over-expression, deletion or introduction of heterologous genes in target metabolic pathways does not always result in the desired phenotype. A good example is the attempts to increase the glycolytic flux, which cannot be increased by individual or combinational over-expression of genes encoding the key enzymes in either a eukaryotic or prokaryotic microorganism[8]. The essence of the problems listed above lies in the fact that, in addition to the modification of key genes by metabolic engineering, the researcher needs to study the effects of the internal environment (e.g. the intracellular energy charge and the interior redox potential and intracellular pH) on the phenotype, based on an accurate analysis of the metabolic network structure. If such an approach is adopted, manipulation of the form and level of intracellular cofactors can potentially be an efficient strategy for obtaining a desired phenotype.

In 1998, Hugenholtz from Delft University of Technology introduced the *nox-2* gene, which encodes the H_2O-forming NADH oxidase, into *Lactococcus lactis* resulting in a shift from homolactic to mixed-acid fermentation during aerobic glucose catabolism. The magnitude of the shift was directly dependent on the level of NADH oxidase overproduced[9]. This result is different from that of metabolic engineering. Most current metabolic engineering strategies have focused on manipulating enzyme levels through the amplification, interruption or addition of a metabolic pathway. The cofactors involved in microbial physiology include ATP / ADP / AMP, NADH / NAD^+, NADPH / $NADP^+$, acetyl coenzyme A and its derivatives, vitamins and trace elements. As illustrated in KEGG (www.kegg.com)[10], cofactors are essential to a large number of biochemical reactions; for example, NADH is involved in 740 biochemical reactions with 433 enzymes and NADPH is involved in 887 biochemical reactions with 462 enzymes (Table 1, updated in Mar 30, 2011). Their manipulation is expected to have substantial effects on metabolic networks. Cofactor engineering, therefore, has potential as a tool for metabolic manipulation.

	ATP	ADP	NADH	NAD+	NADPH	NADP+	CoA	Acetyl-CoA
Number of reactions	496	347	740	750	887	889	480	169
Number of enzymes	454	350	433	455	462	462	250	119

Table 1. The reactions and enzymes involved with nucleotide cofactors, as listed in KEGG

2. Strategies and applications of ATP manipulation

ATP, a kind of nucleotide, widely serves as substrate, product, activator or/and inhibitor in metabolic networks. Based on these four basic functions, the demand and supply of ATP could affect active transportation, peptide folding, subunit assembly, protein relocation and phosphorylation, cell morphology, signal transduction, and stress response. Through these complicated physiology process, ATP is involved in many metabolic pathways and production of almost all of the metabolites by industrial strains. Therefore, the manipulation of ATP supply and demand could be a powerful tool to increase the metabolic performance of industrial strains. Substrate-level phosphorylation (anaerobic conditions) and oxidative phosphorylation (aerobic conditions) were two different ATP regeneration pathways. It seems that manipulation of oxidative phosphorylation was a more efficient way to regulate the intracellular ATP concentration, because under aerobic conditions, most ATP production origin from oxidative phosphorylation pathway. It is conceivable that NADH availability, electron transfer chain (ETC), proton gradient, F_0F_1-ATPase and oxygen supply could all be regulatory candidates for manipulating the intracellular ATP availability.

2.1 Strategies for manipulation ATP availability

Intracellular NADH, produced from glycolysis, fatty acid oxidation, and the citric acid cycle, can be converted to NAD in three separate ways. Under aerobic growth, NADH oxidation occurs through ETC, in which oxygen is used as the final electron acceptor, and a large amount of ATP is produced. Under anaerobic growth and in the absence of an alternate oxidizing agent, the oxidation of NADH can occur by fermentative pathways, such as aldehyde dehydrogenase[11], or lactate dehydrogenase[12]. In this case, energy production is mainly from substrate-level phosphorylation. NADH can also be directly oxidized into water and NAD through NADH oxidase[13]. Therefore, manipulating the availability and oxidation pathway of NADH may be an efficient strategy to manipulate the intracellular ATP level.

There are three different strategies to manipulate the NADH availability to adjust the intracellular ATP content, based on NADH-related metabolic pathways. Firstly, manipulating NADH availability through over-expression or deletion of the key NADH related enzymes, such as *ackA* (acetate kinase)[14], *aldA* (aldehyde dehydrogenase)[11], *ldh* (lactate dehydrogenase), and *pfl* (pyruvate formate-lyase). Secondly, supplement the culture medium with specific substrates for NAD-dependent dehydrogenase, such as formate, citrate and oxalate. Finally, overexpression of NADH oxidase genes, such as *noxE* from *Lactococcus lactis* or *nox* from *Streptococcus pneumoniae*, that oxidize NADH into NAD and water without ATP regeneration[13].

Complex I, II, III and IV are the key components of ETC and play the vital role in ATP production. Focusing on those four different complexes, three separate strategies have been used to disrupt the ETC's capacity to reduce energy production. To decrease ATP content by disrupting ETC, specific inhibitors of ETC components were added to the culture broth and a reduced ATP level was observed[15]. For yeast *T. glabrata*, when 10 mg·L^{-1} rotenone or antimycin A was added to the culture broth at the mid-exponential growth phase, the intracellular ATP level decreased 43% and 27.7%, respectively[8]. The second strategy was deficiency of key components by mutagenesis or genetic operation. The deficiency of cytochrome aa$_3$ and b in yeast led to an energy level decrease of 25%. The third method was disruption of ETC by ectopic expression of some enzymes, such as alternative oxidase (*AOX1*) from *Histoplasma capsulatum*[16], which disrupted ETC through oxidation of electrons and decreased ATP supply.

In aerobic growth, when oxygen is used as the final electron acceptor of the ETC, the abundance of oxygen in culture broth is the decisive environmental factor of ATP production, especially for some fermentation processes, which are high-density, high-viscous and high-energy requiring. Many studies have demonstrated that an increased ATP supply can be achieved by increasing oxygen supply. In the past decades, the strategies of process control and genetic modification have been applied to enhance the ATP production efficiency through increasing oxygen supply. The first strategy can be further divided into two different approaches. One is controlling the aeration rate through the agitation speed in bioreactors, or aeration with pure oxygen[17,18,19]. Many complicated oxygen-supply control strategies were developed based on these simple methods. Another approach is adding oxygen vectors to the culture broth, such as *n*-hexane, *n*-heptane and *n*-dodecane, result in high oxygen solubility in culture broth. In an example of a genetic strategy, hemoglobin from *Vitreoscilla* (*vgb*) was ectopically expressed in different industrial strain to improve oxygen transfer by binding oxygen at low extracellular oxygen content[20].

F$_0$F$_1$-ATPase, the final component of oxidative phosphorylation, plays the central role in ATP production. Three different methods have been performed to reduce the intracellular ATP by decreasing F$_0$F$_1$-ATPase activity. The first was supplementing the culture medium with an external and specific inhibitor of F$_0$F$_1$-ATPase, such as oligomycin, neomycin and

N′,N′-dicyclohexylcarbodiimide[21]. The second was genetic manipulation and traditional mutation of F_0F_1-ATPase[22]. In prokaryotic microorganisms, *Bacillus subtilis* and *Corynebacterium glutamicum* mutants defective in the activity of F_0F_1-ATPase have a decreased ATP supply and intracellular ATP level, thus increasing ATP demand through the glycolytic pathway[23]. For yeast *Torulopsis glabrata*, mutagenesis to decrease F_0F_1-ATPase activity by about 65% decreased the intracellular ATP level by 24%[24]. In *Saccharomyces cerevisiae*, F_0F_1-ATPase may be under strict regulation by autologous ATPase inhibitor peptides, such as IF1, which could affect the oligomerization of F_0F_1-ATPase and thus affect its activity[25]. All three methods described above used down-regulation of F_0F_1-ATPase. Over-expression of F_0F_1-ATPase was believed to be a most direct method for up-regulation of activity of the enzyme complex, which is not achieved for a long period. A most recent study has shown that over-expression of a mitochondrial *ATP6* gene from *Arabidopsis thaliana* in *S. cerevisiae* and *A. thaliana* could effective enhance the activity of F_0F_1-ATPase and ATP regeneration, thus enhance the tolerance to several kind of common stress[26].

2.2 Applications of ATP manipulation

The strategies to enhance the concentration, the yield, and the productivity of the target metabolites with ATP-based manipulation could be divided into three groups: (1) decreasing ATP supply; (2) increasing ATP supply; (3) multi-phase ATP-supply regulation strategies. The ultimate objective of ATP manipulation is to achieve the highest product concentration, the highest yield and the highest productivity, singly or in combination. In the past decades, ATP-oriented bioprocess optimization has developed expeditiously, and has successfully extended the boundaries of metabolic engineering. Here we present some representative works to further illustrate the concept of bioprocess optimization based on the regulation of ATP availability.

A higher target metabolite concentration in the fermentation broth increases the bioreactor utility and reduces the expense for the subsequent extraction process. Regulation of the ATP availability in industrial strain could further increase the target metabolite concentration. Three examples are presented to illustrate the feasibility of further increasing target metabolite concentration by increasing ATP supply.

Studies had demonstrated that a continuous and abundant supply of ATP was essential for glutathione (GSH) synthesis and secretion. A direct, efficient, but costly method to further increase GSH production is to supplement the culture broth with pure ATP, although this is too expensive to use on an industrial scale. Since 1978, researchers have been attempting to establish a coupled system for GSH production using genetically engineered *Escherichia coli* with *gshI/gshII* for GSH synthesis and permeabilized *S. cerevisiae* for ATP regeneration. Recently, using an improved coupling system, a high GSH concentration of 8.92 mM was achieved without supplement of ATP[27]. Another example is the enhancement of xanthan gum concentration by increasing ATP supply through two different methods. The first concentrated on the improvement of oxygen supply to increase ATP production. The second was to feed cells with an extra energy source, such as citric acid. It was interesting that the average molecular weight of xanthan gum was also improved by increasing ATP supply. With pulse-feeding of citric acid, similar results for the synthesis of poly-γ-glutamate (PGA) from glutamate by *Bacillus subtilis* (35 g·L⁻¹ PGA concentration and 1 g·L⁻¹·h⁻¹ PGA productivity) was also achieved[28].

Increasing the fermentation productivity is an efficient way to increase the economy of bioprocess, because a high productivity decreases the fermentation period, the cost of equipment and energy expenditure. For this aim, it is extremely important to increase the rate of carbon flux through central metabolic pathways, *e.g.* the glycolytic pathway and the citric acid cycle. It has been demonstrated that the flux through the glycolytic pathway towards energy production is enhanced by decreasing the energy level of prokaryotic and eukaryotic cells. It has been shown that mutants with a deficiency in ATP synthase have a decreased intracellular ATP level, resulting in an accelerated glycolytic flux and enhanced productivity[19,23,24].

On the other hand, an elevated intracellular ATP level can also improve the productivity of some metabolites. For hyaluronic acid (HA) production by *Streptococcus zooepidemicus*, it has been shown that cell growth and HA production are closely associated with ATP level in fermentation processes by metabolic flux analysis. The continuous production of HA by increasing the supply of ATP through glucose limitation and increased yeast extract supply, could reduce the time spent on repeated bioreactor cleaning processes, thus improving the total productivity.

During fine or bulk chemicals production by industrial strain, many byproducts, such as acetic acid, lactic acid and glycerol, are secreted into the culture broth. The accumulation of byproducts results in a decrease in the yield of product on substrate and an increase in the bioprocess cost, and the environmental burden. The following examples illustrate how to decrease the byproduct formation by manipulating ATP-related metabolic pathways[29,30].

For production of penicillin and its derivatives by *Penicillium chrysogenum*, a higher ATP supply is required for the fast growth of *P. chrysogenum*. Moreover, both the synthesis (73 mol of ATP per mol of penicillin-G) and secretion of penicillin are high ATP-requiring processes. The ATP supply during these processes is very important and should be under strict regulation. Otherwise, organic acids and other intermediate metabolites accumulate in the culture broth due to the deficiency of ATP. Previous research showed that a low oxygen supply can result in a decrease in the yield of penicillin on glucose. In order to further increase the penicillin G yield on glucose, 200 mM of formate was co-fed to penicillin G cultures as an energy source. As a result, the yield of biomass or penicillin G on glucose increased from 49% to 62%[31,32,33]. Lactic acid production is an anaerobic process, in which all of ATP is generated from glycolysis. A high yield of lactic acid on glucose was achieved by increasing ATP demand and accelerating the glycolytic flux through deficiency of pyruvate-formate-lyase (*pfl*) and oxygen limitation. Increasing ATP demand promoted the rate of glycolysis and inhibited the synthesis of byproducts under the oxygen limitation condition. Moreover, the deficiency of *pfl* efficiently eliminated the accumulation of formate. By this strategy, the yield of lactic acid on glucose was improved by 72.5%[12,34].

During the bioprocess, industrial strain may encounter a series of environmental stresses, such as acid, cold, oxidative and osmotic changes. As a consequence, the survival, growth, and metabolic function of industrial strain are affected by those stresses. A number of environmental stress resistance mechanisms have been identified and characterized. It was hypothesized that the supply of ATP plays significant roles in facilitating the stress response of industrial strain, through active transport and signaling pathways[35,36]. The primary mechanism by which industrial strain survive high stress is to control the intracellular environment by membrane-bound ATPases, which translocate specific ions to the environment at the expense of ATP hydrolysis. A deficiency in those ATPases greatly weakens the cells' resistance to environmental challenges, resulting in the cessation of growth and target metabolite accumulation. For instance, a mutant of *S. cerevisiae* deficient in the activity of vacuolar proton-translocating ATPase has chronic oxidative stress. Both the aluminum and NaCl tolerance were sharply decreased in an *S. cerevisiae* strain deficient in the H+-ATPase. In contrast, it is well documented that the tolerance of *S. cerevisiae* to high ethanol content is improved with increased ATP level. An enhanced ATPase system could well facilitate the cells in dealing with harsher conditions[37,38,39].

The ATP-based stress-induced signaling pathways have been widely studied in industrial strain. ATP was an essential substrate for signal pathways. Several signal transduction nodes in the high osmotic glycerol (HOG) pathway were shown to use ATP as an energy source in protecting against high osmotic stress. Similarly, ATP also facilitated signaling in other stress response networks, such as the signal of cold stress, heat stress and oxidative stress. In turn, some signaling pathways could also affect ATP synthesis under stress. In *Streptococcus*, mutants deficient in the minimally conserved bacterial signal recognition particle (SRP) elements remain viable but are more sensitive to environmental stress because the SRP deficiency decreases ATPase activity and limits the ATP supply[40,41]. It was found that industrial strain grown on high environmental stress have a high ATP demand[42] and increase ATP supply[43] could well facilitate the resistance to stress. However, most researches are focused on ATP-related mechanisms or phenomenon description during resistance to different stresses. Few reports are available on the deliberately up-regulating ATP levels to enhance the ability of industrial strain to deal with stress challenges.

3. Strategies and applications of NADH manipulation

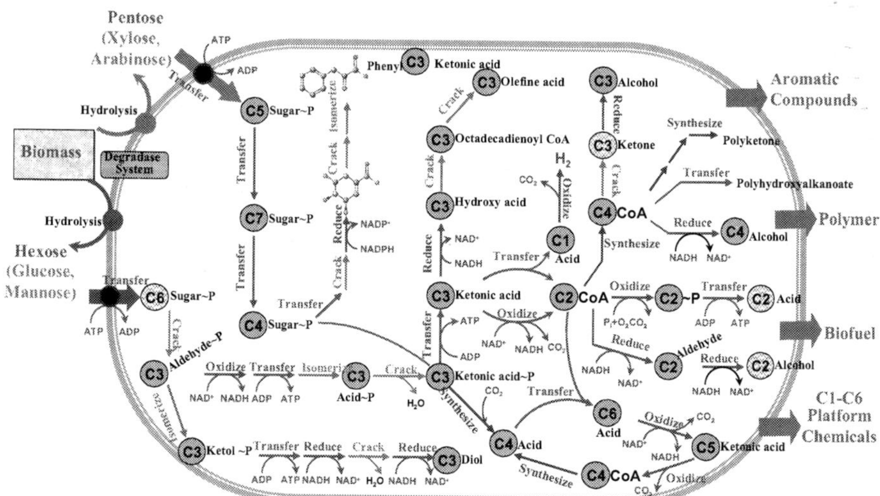

Fig. 1. An illustration of the broad use of the cofactor couple NADH/NAD+ in different metabolic reactions involved in the production of key metabolic products identified as platform chemicals

As the predominant redox product of catabolism, NADH has been found to be involved in more than 700 biochemical reactions in the microbial metabolic network (Table 1). Its

Manipulation strategy	Results/conclusion	Ref.
Biochemical engineering		
Feeding external electron acceptors		
Acetaldehyde	Decreased NADH/NAD$^+$ ratio	[47]
Fumarate or nitrate	Decreased NADH/NAD$^+$ ratio	[48]
Acetoin	Decreased NADH level	[49,50]
Pyruvate, citrate, O$_2$ or fructose	Decreased NAD(P)H level	[51]
Furfural	Decreased NADH level	[52]
Adding carbon sources with different oxidation states		
Sorbitol	Increased NADH availability	[44]
Gluconate	Decreased NADH/NAD$^+$ ratio	[53]
Adding a NAD$^+$ precursor		
Nicotinic acid	Increased NAD$^+$ level	[47]
Altering culture conditions		
Lower temperature	Increased NADH/NAD$^+$ ratio	
Increased dissolved oxygen level	Increased NADH availability	[54]
Extracellular oxidoreduction potential	Decreased NAD$^+$/NADH ratio in a relatively oxidative environment	[55]
Metabolic engineering		
Over-expressing enzymes association with NADH metabolism		
Nicotinic acid phosphoribosyltransferase	Increased NAD$^+$ levels and decreased NADH/NAD$^+$ ratio	[56]
Eliminating NADH competition pathways		
Inactivating aldehyde dehydrogenase	Increased NADH availability	[57]
Deactivating *adh*E, *ldh*A and *ack-pta*, simultaneously	Increased NADH availability	[58]
Introducing heterogeneous NADH metabolic pathways		
H$_2$O-NADH oxidase	Decreased NADH level and NADH/NAD$^+$ ratio	[59]
Alternative oxidase	Decreased NADH/NAD$^+$ ratio	[60]
NAD$^+$-dependent formate dehydrogenase	Increased NADH availability	[53,61]

Table 2. Manipulation strategies of NADH

physiological roles can be divided into five aspects (Fig. 1): (1) regulation of energy metabolism – NADH uses oxygen as the final electron acceptor to produce a large quantity of ATP through the electron transport chain in mitochondria; (2) adjustment of the microbial intracellular redox state – NADH/NAD+ is the main index of redox potential; (3) manipulation of carbon flux – NADH can redistribute carbon flux by activating or inhibiting key enzymes in the target metabolic pathway; (4) modification of mitochondrial function – NADH can modify mitochondrial function by affecting mitochondrial permeability, controlling the mitochondrial membrane anion channel and increasing the mitochondrial membrane potential; (5) regulation of cell life cycle. Based on the above, it is conceivable that NADH/NAD+ could potentially act as an efficient tool to manipulate microbial growth and phenotype. In general, there are two different manipulation strategies for NAD(H/+) availability (Table 2): (1) biochemical engineering approaches that include feeding external electron acceptors, adding carbon sources with different oxidation potentials or NAD synthesis precursors to the fermentation broth and controlling the culture conditions, such as the dissolved oxygen content, temperature and extracellular oxidoreduction potential[44,45]; (2) metabolic engineering methods such as over-expressing enzymes associated with NAD+(NADH) metabolism, eliminating NAD+(NADH) competition pathways and introducing an NAD+(NADH) regeneration system[46]. These strategies have been proven to provide efficient control of the intracellular NAD+(NADH) content.

3.1 Manipulation of NADH availability through biochemical engineering approaches
Many reports have demonstrated that aldehydes, ketones, organic acids, molecular nitrogen or nitrate can be used as internal electron acceptors to enhance NADH oxidation to maintain an optimum oxidoreduction level (using the $NADH/NAD^+$ ratio as the index) in industrial microorganisms. For example, during the heterofermentative lactic acid fermentation by *Lactobacillus* strains with hexose as the substrate, fructose, pyruvate, citrate, and O_2 were separately fed to the fermentation broth as external electron acceptors to accelerate the oxidation rate of NADH, to increase the growth rate of the lactic acid bacteria and to decrease the production of the byproducts erythritol and glycerol. When the external electron receptor acetoin was supplemented to the process of ethanol fermentation by *Saccharomyces cerevisiae* TMB3001 and *Fusarium oxysporum* with xylose as the substrate, the intracellular NAD^+ content and the yield of ethanol were efficiently increased. Another example was 4 mM acetyldehyde fed to the culture broth of *T. glabrata*, which led to a decrease in the $NADH/NAD^+$ ratio to 0.22 and improvement in the glucose consumption rate and the pyruvate titer of 26.3% and 22.5%, respectively.

When glucose, sorbitol and gluconate were compared as carbon sources in microbial glycolysis, sorbitol produced more NADH than glucose while gluconate was transformed directly to pyruvate with no NADH production. Therefore, the oxidation states of these three different carbon sources were -1 (sorbitol), 0 (glucose) and +1 (gluconate). It is thus conceivable that these carbon sources will have a pronounced effect on the intracellular $NADH/NAD^+$ ratio and subsequently on the carbon flux distribution. In a series of chemostat experiments under anaerobic conditions, San et al used three different carbon sources as a simple way of manipulating the cellular $NADH/NAD^+$ ratio from 0.51 (gluconate) to 0.75 (glucose) to 0.94 (sorbitol). The changes in the $NADH/NAD^+$ ratio increased the ethanol to acetate ratio from 1.00 with glucose to 3.62 with sorbitol and decreased it to 0.29 with gluconate. This result provided a simple method for manipulating the distribution of metabolic flux to the desired metabolites. In the case of succinate production by an engineered strain of *E. coli*, supplementation of sorbitol made the intracellular NADH content increase to 0.33 mmol/DCW and, as a result, the titer and yield of succinate increased by 96% and 81%, respectively.

In microbial cells, there are two different NAD synthesis pathways for maintaining the total $NADH/NAD^+$ intracellular pool: the de novo pathway and the pyridine nucleotide salvage pathway. For the de novo pathway, NAD is synthesized from aspartate and dihydroxyacetone phosphate. The pyridine nucleotide salvage pathway recycles intracellular NADH breakdown products, such as nicotinamide mononucleotide (NMN), as well as other preformed pyridine compounds from the environment, such as nicotinamide and nicotinic acid (NA). As the NA concentration (8 mg/L) increased in the fermentation medium of *T. glabrata*, the glucose consumption rate and the pyruvate concentration increased by 48.4% and 29%, respectively.

External environmental conditions, such as the dissolved oxygen concentration and temperature, also modulate the intracellular NADH, NAD^+ and ATP levels, thus shifting the metabolic pattern. Under oxygen limited conditions, oxidation of NADH in *Aspergillus niger* mainly depends on mannitol-phosphate dehydrogenase. When the dissolved oxygen concentration increased from 1% to 10%, the intracellular NADH content in *T. glabrata* IFO 0005 increased by 50%; however, the intracellular NADH content did not continue to increase with further increases in the dissolved oxygen concentration. Under conditions of high dissolved oxygen, the specific enzyme activities and gene expression levels of NAD^+-related glucose-6-phosphate dehydrogenase, malate dehydrogenase and isocitrate dehydrogenase were significantly increased, but the activities of NAD^+-related enzymes in the TCA cycle were significantly inhibited. Singh et al demonstrated that *Pseudomonas fluorescens* could adjust NAD^+ kinase and $NADP^+$ phosphatase activity to supply enough NADPH while limiting NADH synthesis. According to the fact that microbial cells demand different NADH availabilities at different dissolved oxygen concentrations, Liu et al have shown that the $NADH/NAD^+$ ratio was efficiently reduced by adding acetaldehyde to the *T. glabrata* fermentation broth at 20% dissolved oxygen content. The reduction in $NADH/NAD^+$ led to increases in the pyruvate titer, yield and productivity of 68%, 44% and 45%, respectively. For the case of temperature effects on NADH availability, Cuihua Wang et al demonstrated that lower temperatures are beneficial to higher $NADH/NAD^+$ ratios in *T. glabrata*.

The extracellular oxidoreduction potential (ORP) of the fermentation medium is a comprehensive index of environmental conditions, essentially depending on the chemical composition, pH, temperature and dissolved oxygen (DO) concentration of the culture medium. Many reports have revealed that ORP plays a major role in the distribution of carbon flux through changes in the activities of key enzymes that have a metal cofactor in the active site. Du suggested that ORP manipulates the NADH level by affecting the activities of some NADH- or NAD^+-related enzymes that participate in electron transcription. Another example was presented by Qin et al, when potassium ferricyanide was added to the *T. glabrata* culture broth at 20% DO concentration, leading to the NADH content, $NADH/NAD^+$ ratio and ATP level decreasing by 45.3%, 60.3% and 15.2%, respectively. As a consequence, the specific glucose consumption rate increased by 45.5%.

3.2 Manipulation of NADH availability through metabolic engineering strategies

The second strategy that can be used to manipulate NADH availability is the metabolic engineering approach. The application of genetic and metabolic engineering has the potential to considerably affect NADH availability through the amplification, addition or deletion of NAD-related metabolic pathways. Two distinct genetic engineering methods can be used: the first approach aims at increasing the total $NAD(H/^+)$ pool while the second approach focuses on changing the $NADH/NAD^+$ ratio. It is also conceivable that a combination of these two approaches may lead to both an increased NAD(H/+) pool and an increased ratio. As previously mentioned, nicotinic acid can be used to directly synthesize nicotinate mononucleotide, a direct precursor of NAD, catalyzed by nicotinic acid phosphoribosyltransferase (NAPRTase; EC2.4.2.11). This enzyme is encoded by the *pncB* gene. San et al has found that the total level of NAD(H/+) in *E. coli* was increased by 26% but the NADH/NAD+ ratio did not exhibit a consistent trend when the *pncB* gene was over-expressed. Similarly, Heuser et al over-produced the *pncB* and *nadE* genes, encoding nicotinic acid phosphoribosyltransferase and NAD synthetase in *E. coli*, which led to increases in the total NAD(H) and NADP(H) pools of 7-fold and 2-fold, respectively. For the NAD+-dependent dehydrogenase, Cordier et al manipulated NADH regeneration and decreased NADH consumption through deletion of ADH1 and over-expression of ALD3, encoding, respectively, the major NAD+-dependent alcohol dehydrogenase and a cytosolic NAD+-dependent aldehyde dehydrogenase. As a metabolic response to the changing levels of intracellular NADH/NAD+, the engineered *S. cerevisiae* secreted 0.46 g glycerol/g glucose at a rate of 3.1 mmol/g dry mass/h in aerated batch cultures. Another example in *S. cerevisiae* is over-expression of malic enzyme in the mitochondria or in the cytosol, which has a pronounced effect on the intracellular NAD(P)(H) pool. It was found that the total levels of NAD(H/+) remained constant when the mitochondrial malic enzyme was over-expressed but decreased by 34% with over-expression of the cytosolic malic enzyme; both mitochondrial and cytosolic malic enzymes efficiently decreased the NADH/NAD+ ratio. For NADP(H), the over-expression of mitochondrial malic enzyme decreased the total levels of NADP(H/+) (by 17%) and the ratio of NADPH/NADP (by 6%).

The second approach to manipulating NADH is the deletion or weakening of the NADH competition pathways (Fig.2), to redirect NADH to the target metabolic pathway to enhance the production of the desired metabolites. By inactivating aldehyde dehydrogenase (ALDH) in *Klebsiella pneumonia*, an enzyme that competes with 1,3-PD oxidoreductase for NADH, the final titer, the productivity of 1,3-PD and the yield of 1,3-PD relative to glycerol reached 927.6 mmol L^{-1}, 14.05 mmol L^{-1} h^{-1} and 0.699 mol mol^{-1}, respectively. During the production of glycerol from glucose with *S. cerevisiae*, Geertman adopted a series of metabolic engineering strategies, which included: (1) maintaining flexibility at fructose-1,6-bisphosphatase and triosephosphate isomerase; (2) deleting pyruvate decarboxylases, NADH dehydrogenases and glycerol-3-phosphate dehydrogenase; (3) feeding formate, to make the glycerol yield from glucose reach 1.08 mol mol^{-1}. Similarly, inactivation of competing NADH pathways (alcohol dehydrogenase and lactate dehydrogenase) and heterologous production of pyruvate carboxylase (PYC) in *E. coli*, to provide maximum quantities of NADH for succinate synthesis, led to achievement of a succinate yield from glucose of 1.31 mol/mol. Based on the above result, Sanchez et al reconstructed a recombinant *E. coli* strain with deactivated adhE, ldhA and ack-pta and, by activating the glyoxylate pathway through the inactivation of iclR, reduced the NADH demand for production of a mole of succinate from 2 moles to 1.25 moles.

The third approach for manipulating NADH is heterologous production of oxido-reduction related enzymes, to change the ways NADH is regenerated or oxidized, thus changing the ratio NADH/NAD+. In microbial cells, cytosolic NADH needs to shuttle to the mitochondria and be oxidized. It is conceivable that an increase in the efficiency and rate of NADH oxidation may be achieved through over-expression of water-forming NADH oxidase, which directly oxidizes NADH to NAD+ in the cytoplasm. In *L. lactis*, the heterologous expression of water-forming NADH oxidase led to a significantly decreased NADH/NAD$^+$ ratio and to a shift from homolactic to mixed-acid fermentation. Over-expression of the *noxE* gene, encoding the water-forming NADH oxidase in *S. cerevisiae*, resulted in decreases in the level of NADH and the NADH/NAD$^+$ ratio of 5- and 6-fold, respectively. As a consequence, the glucose consumption rate increased by 10%. In mitochondria, the over-expression of NADH alternative oxidase redirects the NADH oxidation pathway from the oxidative phosphorylation pathway to alternative oxidation and could effectively decrease the NADH/NAD$^+$ ratio and ATP content. In addition to accelerated NADH oxidation, some specific metabolite syntheses require a large quantity of NADH. Therefore, increasing the NADH availability is the limiting step for improving the production efficiency of the target metabolites. To this aim, Zhao et al significantly increased NADH availability through over-expression of phosphite dehydrogenase (PTDH), which catalyzes phosphate to phosphite and reduces NAD+ to NADH, from *Pseudomonas stutzeri*. Similarly, Berríos-Rivera et al reconstructed an efficient NADH regeneration pathway in *E. coli* through over-expression of NAD$^+$-dependent formate dehydrogenase from *Candida boidinii*, replacing the corresponding enzyme, which does not depend on any cofactor. The mutant exhibited high NADH synthesis ability, from 1 mole to 4 moles (under aerobic conditions) and 3 moles (under anaerobic conditions) per mole glucose.

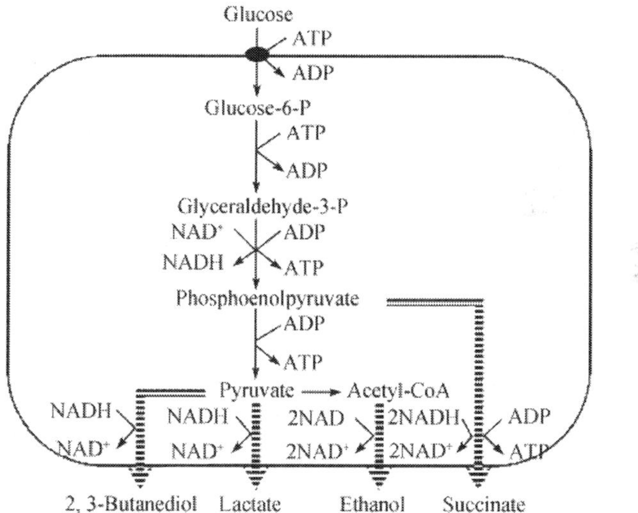

Fig. 2. The NADH competing pathway at the pyruvate node

4. Strategies and applications of CoA manipulation

As acetyl carriers, coenzyme A (CoA) (Table 2) and its derivatives, acetyl-CoA, succinyl coenzyme A (succinyl-CoA) and malonyl coenzyme A (malonyl-CoA), are involved in more than 600 biochemical reactions in microbial cell metabolism. Acetyl-CoA is an essential intermediate in many energy-yielding metabolic pathways and is a substrate in enzymatic

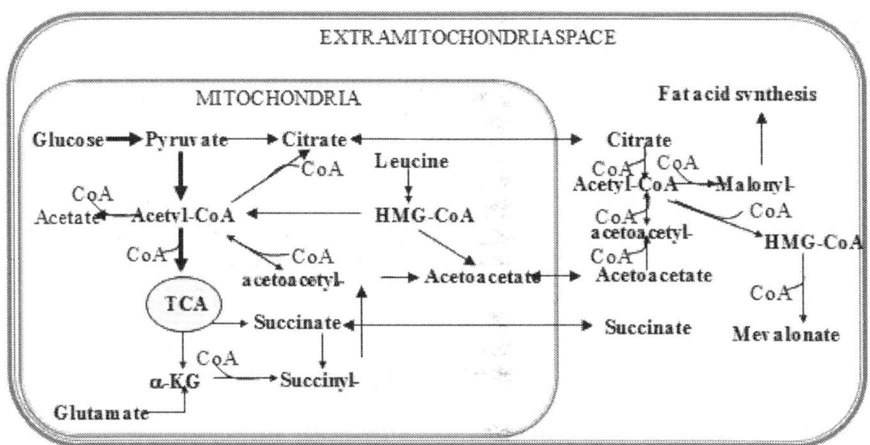

Compound	R	Formula	Exact mass
CoASH	H	$C_{21}H_{36}N_7O_{16}P_3S$	767.1152
Acetyl-CoA		$C_{23}H_{38}N_7O_{17}P_3S$	809.1258
3-Hydroxy-3-methyl-glutaryl-CoA		$C_{27}H_{44}N_7O_{20}P_3S$	911.1575
β-Hydroxybutyryl-CoA		$C_{25}H_{42}N_7O_{18}P_3S$	853.1520
Malonyl-CoA		$C_{24}H_{38}N_7O_{19}P_3S$	853.1156
Succinyl-CoA		$C_{25}H_{40}N_7O_{19}P_3S$	867.1313
Propionyl-CoA		$C_{24}H_{40}N_7O_{17}P_3S$	823.1414
Isobutyryl-CoA		$C_{25}H_{42}N_7O_{17}P_3S$	837.1571

Table 3. CoA and its thioester derivatives

Fig. 3. Involvement of CoA and its derivates in metabolic networks

production of industrially useful compounds such as esters and lipid molecules. As
illustrated in Fig. 3, CoA and its derivatives take part in a variety of metabolic functions,

such as the citric acid cycle, fatty acid synthesis and decomposition, macromolecule fat synthesis, amino acid metabolism, ketogenesis, sterol synthesis and as regulators to control some key enzymes in specific metabolic pathway[62,63].

4.1 Strategies of CoA manipulation

The total levels of CoA and its derivatives are dependent on the kinds and amounts of carbon sources in the fermentation medium. When *E. coli* were grown on a medium with glucose as the sole carbon source, the content of CoA reached a maximum value (400 umol/L) with acetyl-CoA as a derivative. However, the content of CoA decreased to 100 umol/L when the *E. coli* culture medium contained the amino acid mixture from casein hydrolysis. For pantothenic acid auxotrophic strains, the intracellular CoA content is determined by the concentration of acetate (or carboxylic acid salt). Due to the fact that the auxotrophic strain does not synthesize pantothenic acid when grown on a limited medium with glucose as the sole carbon source, this leads to a deficiency of succinyl-CoA for the synthesis of amino acids and proteins; as a consequence, the growth of the microbial cells is very slow. For the acetic acid auxotrophic *E. coli* strain, the intracellular acetyl-CoA level is low, and the total CoA level decreases with the hydrolysis and further metabolism of CoA. In contrast, the intracellular CoA level significantly increases with over-expression of CoA synthase in *E. coli*, and supplementation of pantothenic acid in the medium results in an increase in acetate production. Furthermore, it was found that the addition of acetate does not increase the acetyl-CoA content, but decreases the acetate concentration in the medium quickly reduce the level of intracellular acetyl-CoA. The kinds of carbon sources also have a pronounced effect on the CoA/acetyl-CoA ratio. Many studies have shown that the level of intracellular acetyl-CoA rapidly increases when microbial cells use D-glucose, D-fructose, D-mannose, glycerol or sorbitol as the sole carbon source, but remains constant with L-glucose, sucrose, maltose, succinate or acetate as the carbon source. Compared with that of glucose as a carbon source, acetate as the carbon source has a high CoA/acetyl-CoA ratio[64]. When microbial cells switched from acetate to glucose, the intracellular CoA level increased 50%. Consistent with this result, an increase in the CoA/acetyl-CoA ratio was observed. Furthermore, it was found that the specific activity of CoA synthase decreased with a shift from glucose to acetate[65]. This result demonstrated that CoA synthase is not sensitive to acetyl-CoA. Based on these findings, Lee et al enhanced PHB production by increasing intracellular acetyl-CoA and acetoacetyl-CoA levels in *E. coli* by adding a synthetic nitrogen-based amino acid and oleic acid[66].

4.2 Applications of CoA manipulation

From the above discussion, it can be concluded that the ratio of acetyl-CoA/CoA is a key index that reflects the metabolic state of carbon and energy metabolism during the fermentation process. San et al over-expressed pantothenate kinase in *E. coli*, by a combination of pantothenic acid addition, to significantly increase the intracellular CoA level. As a metabolic response to the change in CoA level, the carbon flux redistributed to isoamyl acetate, resulting in a significantly increased titer and yield of isoamyl acetate[67,68]. Similarly, the over-expression of pantothenate kinase gene also increased the CoA level (10-fold), acetyl-CoA level (5-fold) and the acetyl-CoA/CoA ratio, which channeled more carbon flux into acetate formation and to excessive accumulation of acetate[69].

Apart from metabolic engineering strategies, the biochemical engineering strategy has been proven to be an effective way to manipulate CoA and acetyl-CoA levels and the acetyl-

CoA/CoA ratio. An example is the manipulation of thiamine, biotin and Ca^{2+} levels as a tool for redistributing carbon flux from pyruvate to α-ketoglutarate in *T. glabrata*: (1) the carbon flux was blocked at the pyruvate node (69 g/L) with the sub-optimization of vitamins in the fermentation medium; (2) the titer of α-ketoglutarate reached 10.3 g/L by selectively opening the valve of carbon flux from pyruvate to the pyruvate dehydrogenase complex in the pyruvate carboxylase pathway by increasing the concentrations of thiamine and biotin; (3) the concentration of α-ketoglutarate (43.7 g/L) was further increased through increasing the pyruvate carboxylase level with Ca^{2+} present in the fermentation medium[70].

5. Concluding remarks and future directions

Recently, many studies have demonstrated that the fermentation process from sugar to the target product is not just a simple biochemical reaction but rather a comprehensive network, including the gene regulatory network, protein-protein interaction network, signal transduction network and metabolic network, dependent on the physical and chemical interactions of genes, proteins and metabolites. Therefore, interest in cofactor engineering in the future should be concerned with: (1) identification of the active site of the cofactor in the biochemical reaction, metabolic pathway and metabolic network; (2) evaluation of the effect of the cofactor on the metabolic reaction, pathway and network; (3) finding the threshold values at which the metabolic networks and regulatory networks respond to cofactor changes; (4) development of directed, precise strategies for cofactor manipulation. The increasing availability of genome sequences and accumulation of high-throughput biological data allow us to understand the physiological functions of cofactors and to propose precise strategies for cofactor manipulation of microbial physiology.

6. References

[1] Nielsen J, Otero JM (2010) Industrial Systems Biology. Biotechnology and Bioengineering 105: 439-460.

[2] Nielsen J, Liu LM, Agren R, Bordel S (2010) Use of genome-scale metabolic models for understanding microbial physiology. FEBS Letters 584: 2556-2564.

[3] Lee SY, Park JH (2008) Towards systems metabolic engineering of microorganisms for amino acid production. Current Opinion in Biotechnology 19: 454-460.

[4] Lin H, Bennett GN, San KY (2005) Fed-batch culture of a metabolically engineered Escherichia coli strain designed for high-level succinate production and yield under aerobic conditions. Biotechnology and Bioengineering 90: 775-779.

[5] Liao JC, Atsumi S (2008) Metabolic engineering for advanced biofuels production from Escherichia coli. Current Opinion in Biotechnology 19: 414-419.

[6] Wendisch VF, Bott M, Eikmanns BJ (2006) Metabolic engineering of Escherichia coli and Corynebacterium glutamicum for biotechnological production of organic acids and amino acids. Current Opinion in Microbiology 9: 268-274.

[7] Xian M, Yu C, Cao YJ, Zou HB (2011) Metabolic engineering of Escherichia coli for biotechnological production of high-value organic acids and alcohols. Applied Microbiology and Biotechnology 89: 573-583.

[8] Chen J, Liu LM, Li Y, Li HZ (2006) Significant increase of glycolytic flux in Torulopsis glabrata by inhibition of oxidative phosphorylation. FEMS Yeast Research 6: 1117-1129.

[9] Lopez de Felipe F, Kleerebezem M, de Vos WM, Hugenholtz J (1998) Cofactor engineering: a novel approach to metabolic engineering in Lactococcus lactis by controlled expression of NADH oxidase. Journal of Bacteriology 180: 3804-3808.

[10] Aoki-Kinoshita KF (2006) Overview of KEGG applications to omics-related research. Journal of Pesticide Science 31: 296-299.

[11] Cao Z, Zhang YP, Li Y, Du CY, Liu M (2006) Inactivation of aldehyde dehydrogenase: A key factor for engineering 1,3-propanediol production by Klebsiella pneumoniae. Metabolic Engineering 8: 578-586.

[12] Zhu J, Shimizu K (2004) The effect of pfl gene knockout on the metabolism for optically pure D-lactate production by Escherichia coli. Applied Microbiology and Biotechnology 64: 367-375.

[13] Nielsen J, Vemuri GN, Eiteman MA, McEwen JE, Olsson L (2007) Increasing NADH oxidation reduces overflow metabolism in Saccharomyces cerevisiae. Proceedings of the National Academy of Sciences of the United States of America 104: 2402-2407.

[14] Underwood SA, Zhou S, Causey TB, Yomano LP, Shanmugam KT, et al. (2002) Genetic changes to optimize carbon partitioning between ethanol and biosynthesis in ethanologenic Escherichia coli. Applied and Environmental Microbiology 68: 6263-6272.

[15] Miyoshi H, Abe M, Kubo A, Yamamoto S, Hatoh Y, et al. (2008) Dynamic function of the spacer region of acetogenins in the inhibition of bovine mitochondrial NADH-ubiquinone oxidoreductase (complex I). Biochemistry 47: 6260-6266.

[16] Johnson CH, Prigge JT, Warren AD, McEwen JE (2003) Characterization of an alternative oxidase activity of Histoplasma capsulatum. Yeast 20: 381-388.

[17] Huang H, Qu L, Ji XJ, Ren LJ, Nie ZK, et al. (2011) Enhancement of docosahexaenoic acid production by Schizochytrium sp. using a two-stage oxygen supply control strategy based on oxygen transfer coefficient. Letters in Applied Microbiology 52: 22-27.

[18] Huang H, Peng C, Ji XJ, Liu X, Ren LJ, et al. (2010) Effects of n-Hexadecane Concentration and a Two-Stage Oxygen Supply Control Strategy on Arachidonic Acid Production by Mortierella Alpina ME-1. Chemical Engineering & Technology 33: 692-697.

[19] Li Y, Hugenholtz J, Chen J, Lun SY (2002) Enhancement of pyruvate production by Torulopsis glabrata using a two-stage oxygen supply control strategy. Applied Microbiology and Biotechnology 60: 101-106.

[20] Tang KX, Zhang L, Li YJ, Wang ZN, Xia Y, et al. (2007) Recent developments and future prospects of Vitreoscilla hemoglobin application in metabolic engineering. Biotechnology Advances 25: 123-136.

[21] Johnson KM, Cleary J, Fierke CA, Opipari AW, Jr., Glick GD (2006) Mechanistic basis for therapeutic targeting of the mitochondrial F1F0-ATPase. ACS Chem Biol 1: 304-308.

[22] Yokota A, Henmi M, Takaoka N, Hayashi C, Takezawa Y, et al. (1997) Enhancement of glucose metabolism in a pyruvic acid-hyperproducing Escherichia coli mutant defective in F-1-ATPase activity. Journal of Fermentation and Bioengineering 83: 132-138.

[23] Sekine H, Shimada T, Hayashi C, Ishiguro A, Tomita F, et al. (2001) H+-ATPase defect in Corynebacterium glutamicum abolishes glutamic acid production with enhancement of glucose consumption rate. Appl Microbiol Biotechnol 57: 534-540.

[24] Liu LM, Li Y, Du GC, Chen J (2006) Increasing glycolytic flux in Torulopsis glabrata by redirecting ATP production from oxidative phosphorylation to substrate-level phosphorylation. Journal of Applied Microbiology 100: 1043-1053.

[25] Garcia JJ, Morales-Rios E, Cortes-Hernandez P, Rodriguez-Zavala JS (2006) The inhibitor protein (IF1) promotes dimerization of the mitochondrial F1F0-ATP synthase. Biochemistry 45: 12695-12703.

[26] Zhang X, Liu S, Takano T (2008) Overexpression of a mitochondrial ATP synthase small subunit gene (AtMtATP6) confers tolerance to several abiotic stresses in Saccharomyces cerevisiae and Arabidopsis thaliana. Biotechnol Letter 30: 1289-1294.

[27] Liao X, Deng T, Zhu Y, Du G, Chen J (2008) Enhancement of glutathione production by altering adenosine metabolism of Escherichia coli in a coupled ATP regeneration system with Saccharomyces cerevisiae. Journal of Applied Microbiology 104: 345-352.

[28] Ahn WS, Park SJ, Lee SY (2000) Production of Poly(3-hydroxybutyrate) by fed-batch culture of recombinant Escherichia coli with a highly concentrated whey solution. Applied Environmental Microbiology 66: 3624-3627.

[29] Karakashev D, Thomsen AB, Angelidaki I (2007) Anaerobic biotechnological approaches for production of liquid energy carriers from biomass. Biotechnology Letter 29: 1005-1012.

[30] Suwannakham S, Huang Y, Yang ST (2006) Construction and characterization of ack knock-out mutants of Propionibacterium acidipropionici for enhanced propionic acid fermentation. Biotechnology and Bioengineering 94: 383-395.

[31] Harris DM, van der Krogt ZA, van Gulik WM, van Dijken JP, Pronk JT (2007) Formate as an auxiliary substrate for glucose-limited cultivation of Penicillium chrysogenum: impact on penicillin G production and biomass yield. Applied Environmental Microbiology 73: 5020-5025.

[32] Garrido JM, van Benthum WA, van Loosdrecht MC, Heijnen JJ (1997) Influence of dissolved oxygen concentration on nitrite accumulation in a biofilm airlift suspension reactor. Biotechnology and Bioengineering 53: 168-178.

[33] van Gulik WM, de Laat WT, Vinke JL, Heijnen JJ (2000) Application of metabolic flux analysis for the identification of metabolic bottlenecks in the biosynthesis of penicillin-G. Biotechnology and Bioengineering 68: 602-618.

[34] Neves AR, Pool WA, Kok J, Kuipers OP, Santos H (2005) Overview on sugar metabolism and its control in Lactococcus lactis - the input from in vivo NMR. FEMS Microbiol Review 29: 531-554.

[35] Lokanath NK, Matsuura Y, Kuroishi C, Takahashi N, Kunishima N (2007) Dimeric core structure of modular stator subunit E of archaeal H+ -ATPase. J Mol Biol 366: 933-944.

[36] Salinas P, Ruiz D, Cantos R, Lopez-Redondo ML, Marina A, et al. (2007) The regulatory factor SipA provides a link between NblS and NblR signal transduction pathways in the cyanobacterium Synechococcus sp. PCC 7942. Mol Microbiol 66: 1607-1619.

[37] Shima J, Ando A, Takagi H (2008) Possible roles of vacuolar H+-ATPase and mitochondrial function in tolerance to air-drying stress revealed by genome-wide screening of Saccharomyces cerevisiae deletion strains. Yeast 25: 179-190.

[38] Milgrom E, Diab H, Middleton F, Kane PM (2007) Loss of vacuolar proton-translocating ATPase activity in yeast results in chronic oxidative stress. Journal of Biological Chemistry 282: 7125-7136.

[39] Stewart CM, Cole MB, Legan JD, Slade L, Schaffner DW (2005) Solute-specific effects of osmotic stress on Staphylococcus aureus. Journal of Applied Microbiology 98: 193-202.

[40] Hasona A, Crowley PJ, Levesque CM, Mair RW, Cvitkovitch DG, et al. (2005) Streptococcal viability and diminished stress tolerance in mutants lacking the signal recognition particle pathway or YidC2. Proc Natl Acad Sci U S A 102: 17466-17471.

[41] Hasona A, Zuobi-Hasona K, Crowley PJ, Abranches J, Ruelf MA, et al. (2007) Membrane composition changes and physiological adaptation by Streptococcus mutans signal recognition particle pathway mutants. Journal of Bacteriology 189: 1219-1230.

[42] Canovas M, Bernal V, Sevilla A, Iborra JL (2007) Salt stress effects on the central and carnitine metabolisms of Escherichia coli. Biotechnology and Bioengineering 96: 722-737.

[43] Sanchez C, Neves AR, Cavalheiro J, dos Santos MM, Garcia-Quintans N, et al. (2008) Contribution of citrate metabolism to the growth of Lactococcus lactis CRL264 at low pH. Appl Environ Microbiol 74: 1136-1144.

[44] Lin H, Bennett GN, San KY (2005) Effect of carbon sources differing in oxidation state and transport route on succinate production in metabolically engineered Escherichia coli. J Ind Microbiol Biotechnol 32: 87-93.

[45] Ma B, Pan SJ, Zupancic ML, Cormack BP (2007) Assimilation of NAD(+) precursors in Candida glabrata. Mol Microbiol 66: 14-25.

[46] Cordier H, Mendes F, Vasconcelos I, Francois JM (2007) A metabolic and genomic study of engineered Saccharomyces cerevisiae strains for high glycerol production. Metabolic Engineering 9: 364-378.

[47] Liu LM, Li Y, Shi ZP, Du GC, Chen J (2006) Enhancement of pyruvate productivity in Torulopsis glabrata: Increase of NAD(+) availability. J Biotechnol 126: 173-185.

[48] de Graef MR, Alexeeva S, Snoep JL, Teixeira de Mattos MJ (1999) The steady-state internal redox state (NADH/NAD) reflects the external redox state and is correlated with catabolic adaptation in Escherichia coli. J Bacteriol 181: 2351-2357.

[49] Panagiotou G, Christakopoulos P (2004) NADPH-dependent D-aldose reductases and xylose fermentation in Fusarium oxysporum. J Biosci Bioeng 97: 299-304.

[50] Panagiotou G, Christakopoulos P, Villas-Boas SG, Olsson L (2005) Fermentation performance and intracellular metabolite profiling of Fusarium oxysporum cultivated on a glucose-xylose mixture. Enzyme Microb Technol 36: 100-106.

[51] Zaunmuller T, Eichert M, Richter H, Unden G (2006) Variations in the energy metabolism of biotechnologically relevant heterofermentative lactic acid bacteria during growth on sugars and organic acids. Applied Microbiology Biotechnology 72: 421-429.

[52] Wahlbom CF, Hahn-Hagerdal B (2002) Furfural, 5-hydroxymethyl furfural, and acetoin act as external electron acceptors during anaerobic fermentation of xylose in recombinant Saccharomyces cerevisiae. BiotechnologyBioengineering 78: 172-178.

[53] San KY, Bennett GN, Berrios-Rivera SJ, Vadali RV, Yang YT, et al. (2002) Metabolic engineering through cofactor manipulation and its effects on metabolic flux redistribution in Escherichia coli. Metabolic Engineering 4: 182-192.

[54] Qiang H, Shimuzu K (1999) Effect of dissolved oxygen concentration on the intracellular flux distribution for pyruvate fermentation. Journal of Biotechnology 68: 135-147.

[55] Du CY, Yan H, Zhang YP, Li Y, Cao ZA (2006) Use of oxidoreduction potential as an indicator to regulate 1,3-propanediol fermentation by Klebsiella pneumoniae. Applied Microbiology Biotechnology 69: 554-563.

[56] Berrios- Rivera SJ, San KY, Bennett GN (2002) The effect of NAPRTase overexpression on the total levels of NAD, the NADH/NAD(+) ratio, and the distribution of metabolites in Escherichia coli. Metabolic Engineering 4: 238-247.

[57] Zhang YP, Li Y, Du CY, Liu M, Cao Z (2006) Inactivation of aldehyde dehydrogenase: A key factor for engineering 1,3-propanediol production by Klebsiella pneumoniae. Metabolic Engineering 8: 578-586.

[58] Sanchez AM, Bennett GN, San KY (2005) Novel pathway engineering design of the anaerobic central metabolic pathway in Escherichia coli to increase succinate yield and productivity. Metabolic Engineering 7: 229-239.

[59] Heux S, Cachon R, Dequin S (2006) Cofactor engineering in Saccharomyces cerevisiae: Expression of a H2O-forming NADH oxidase and impact on redox metabolism. Metabolic Engineering 8: 303-314.

[60] Vemuri GN, Eiteman MA, McEwen JE, Olsson L, Nielsen J (2007) Increasing NADH oxidation reduces overflow metabolism in Saccharomyces cerevisiae. PNAS 104: 2402-2407.

[61] Saanchez AM, Bennett GN, San KY (2005) Effect of different levels of NADH availability on metabolic fluxes of Escherichia coli chemostat cultures in defined medium. Journal Biotechnologyl 117: 395-405.

[62] Jackowski S, Rock CO (1986) Consequences of reduced intracellular coenzyme A content in Escherichia coli. Journal of Bacteriology 166: 866-871.

[63] Chohnan S, Furukawa H, Fujio T, Nishihara H, Takamura Y (1997) Changes in the size and composition of intracellular pools of nonesterified coenzyme A and coenzyme A thioesters in aerobic and facultatively anaerobic bacteria. Applied Environmental Microbiology 63: 553-560.

[64] Vallari DS, Jackowski S, Rock CO (1987) Regulation of pantothenate kinase by coenzyme A and its thioesters. Journal of Biological Chemistry 262: 2468-2471.

[65] Chohnan S, Izawa H, Nishihara H, Takamura Y (1998) Changes in size of intracellular pools of coenzyme A and its thioesters in Escherichia coli K-12 cells to various carbon sources and stresses. Bioscience Biotechnolology and Biochemical 62: 1122-1128.

[66] Lee SY, Chang HN (1995) Production of poly(3-hydroxybutyric acid) by recombinant Escherichia coli strains: genetic and fermentation studies. Candia Journal Microbiology 41 Suppl 1: 207-215.

[67] Vadali RV, Bennett GN, San KY (2004) Enhanced isoamyl acetate production upon manipulation of the acetyl-CoA node in Escherichia coli. Biotechnol Prog 20: 692-697.

[68] Vadali RV, Bennett GN, San KY (2004) Applicability of CoA/acetyl-CoA manipulation system to enhance isoamyl acetate production in Escherichia coli. Metabolic Engineering 6: 294-299.

[69] Vadali RV, Bennett GN, San KY (2004) Cofactor engineering of intracellular CoA/acetyl-CoA and its effect on metabolic flux redistribution in Escherichia coli. Metabolic Engineering 6: 133-139.

[70] Liu L, Li Y, Zhu Y, Du G, Chen J (2007) Redistribution of carbon flux in Torulopsis glabrata by altering vitamin and calcium level. Metabolic Engineering 9: 21-29.

20

The Bioengineering and Industrial Applications of Bacterial Alkaline Proteases: the Case of SAPB and KERAB

Bassem Jaouadi[1], Badis Abdelmalek[2], Nedia Zaraî Jaouadi[1]
and Samir Bejar[1]
[1]Laboratory of Microorganisms and Biomolecules, Centre de Biotechnologie de Sfax
University of Sfax, Road of Sidi Mansour Km 6
[2]Laboratory of Biochemistry and Industrial Microbiology,
Department of Industrial Chemistry,
University Saad Dahlab of Blida
[1]Tunisia
[2]Algeria

1. Introduction

Enzymes have long been used as alternatives to chemicals to improve the efficiency and cost-effectiveness of a wide range of industrial systems and processes. They are currently used in basic and applied arenas of research as well as in a wide range of product design and manufacturing processes, such as those pertaining to the food, beverage, pharmaceutical, detergent, leather processing, and peptide synthesis industries (Gupta et al., 2002). Of particular interest to the aims of the present work, proteases have often been reported to constitute a resourceful class of enzymes with promising industrial applications. According to recent estimates, these enzymes account for nearly 65% of total worldwide enzyme sales (Anonyme, 2007; Rao et al., 1998). They are widely distributed in nature and play a vital role in life processes. They are particularly known for their capacity to hydrolyze peptide bonds in aqueous environments and to synthesize peptide bonds in non-aqueous biocatalysis.

Proteases have been employed in a wide array of applications for many years with satisfactory results. They constitute a large family of enzymes present in a wide range of living organisms, such as plants, animals and microorganisms. In biotechnologically oriented systems and processes, however, proteases from microbial origins have often been reported to have distinct advantages when compared to plant or animal proteases, particularly because they possess almost all the characteristics desired for biotechnological applications. Among these biocatalysts, high-alkaline proteases, which alone account for about 40% of the total worldwide enzyme sales (Kirk et al., 2002), proved particularly suitable for industrial use. This is mainly due to their high stability and activity under harsh conditions.

Nowadays, the use of alkaline protease-based detergents is preferred over the conventional synthetic ones. This is partly because of their better cleaning properties, higher performance efficiency at lower washing temperature, and safer dirt removal conditions (Gupta et al., 2002). Typically, a detergent protease needs to be active, stable, and compatible with the alkaline environment encountered under harsh washing conditions: pH 9 - 11, temperature of 20 - 60°C, as well as high concentrations of salt, bleach, and surfactant. Some of the alkaline proteases that are particularly preferred in contemporary detergent formulations include Savinase™ (Subtilisin 309), Subtilisin Novo (BPN'), Alcalase™ (Subtilisin Carlsberg; SC), Maxacal™ (Novozymes A/S, Denmark), BLAP S^b (Henkel, Germany) and Properase™ (Genecor Int. USA). They are often reported to be stable at conditions of elevated temperatures and pH. Most of them have, however, been criticized for their limited efficiency in the presence of liquid or solid laundry detergents wherein their stability decreases (Beg and Gupta, 2003; Maurer, 2004). Therefore, the search for and screening of alternative microorganisms that produce detergent-stable enzymes and preserve their high activity and stability at extreme conditions would be highly desired, particularly within the framework of the persistent aspirations that consumers, industrialists and, by extension, researchers, have towards improved laundry detergents with powerful, safe and healthy cleansing abilities.

Various alkaline proteases have been reported to constitute appropriate additives for a variety of detergent, laundry and cleansing supplies as well as other leather processing, dyeing, and finishing applications. Keratinases are a group of mostly extracellular serine-proteases that have often been reported for their excellent potency to degrade keratins, a group of fibrous, insoluble and abundant structural proteins that constitute the major components of structures growing from the skin of vertebrates, such as hair, wool, nails, hooves, horns and feather quills. In fact, due to their high degree of cross-linking to disulphide bonds, hydrogen bonds, and hydrophobic interactions, these proteins show high stability and resistance to proteolytic hydrolysis (Coulombe and Omary, 2002).

Large amounts of keratin containing wastes are discharged every year from poultry, leather and meat processing industries. Current estimates indicate that the global annual discharge of feather from the poultry processing industry alone reaches millions of tons (C.A.S.T., 1995; Freeman et al., 2009). This keratinous poultry waste is degraded very slowly in nature and is, therefore, considered hazardous to the environment. Seeing that keratinous waste represents a valuable source for proteins and amino acids, several steam pressure and chemical treatment processes have been developed to convert feathers into feather meal for animals (Hess and FitzGerald, 2007). These physico-chemical conversion methods have, nevertheless, been reported to involve costly treatments under harsh temperature and pressure conditions that result in the loss of essential amino acids (Onifade et al., 1998). Alternatively, feather biodegradation processes have been proposed as viable substitutes (Ignatova et al., 1999; Xie et al., 2010).

Keratinolytic microorganisms can be employed in the manufacture of nutritious, cost-effective, environmentally safe feather meal for poultry, as well as in the enhancement of drug delivery, hydrolysis of prions, construction of biodegradable films, and production of biofuels (Brandelli et al., 2010). Additionally, these keratinolytic enzymes have a variety of current and potential applications in a wide range of biotechnological processes that involve keratin hydrolysis, including the enzymatic dehairing and catalysis for leather and cosmetic industries, the breaking down of recalcitrant matter for the laundry and detergent industries, the slowing down of nitrogen release for fertilizer and pesticide industries, and the production of biohydrogen and rare amino acids for animal feed and foodstuff industries (Bertsch and Coello, 2005).

Several microorganisms that possess keratinolytic activity have been reported to accede to the biodegradation of keratin waste by secreting keratinolytic peptidases into the culture medium and to offer valuable tools for the development of efficient and cost-effective keratin waste bioconversion methods (Onifade et al., 1998). In this respect, various keratinases have been purified from different microorganisms, namely fungi, such as *Microsporum* (Essien et al., 2009) and *Chryseobacterium indologenes* TKU014 (Wang et al., 2008), and bacteria, such as *Bacillus* (Pillai and Archana, 2008; Radha and Gunasekaran, 2008) and *Streptomyces* (Syed et al., 2009; Tatineni et al., 2008). As corresponds to their habitat, these bacteria are nutritionally quite versatile, and most of them produce extracellular hydrolytic enzymes that permit the use of high-molecular-weight biopolymers, such as proteins, polysaccharides, fats, and a variety of other substrates (Gupta et al., 1995). Among these enzymes, several serine peptidases have so far been isolated, purified, and characterized from various species, such as *S. griseus* (Awad et al., 1972; Johnson and Smillie, 1974), *S. fradiae* (Kitadokoro et al., 1994), *S. thermoviolaceus* SD8 (Chitte et al., 1999), and *S. graminofaciens* (Szabo et al., 2000).

Despite this large flow of data on keratinases, however, little information has so far been reported on the characterization and purification of keratinases from *Streptomyces*. Moreover, and particularly due to the relatively poor levels of stability and catalytic activity obtained for the *Streptomyces* enzymes so far investigated under the specific operational conditions required by current industrial applications, namely high temperature and pH values, as well as the presence of detergents or non-aqueous solvent, their practical application still remained very limited. Accordingly, the isolation and screening of new keratinolytically active *Streptomyces* strains from natural habitats could open new pathways for the discovery and use of novel keratinases.

The present chapter aim to provide an overview on the current quest for novel natural bacterial alkaline proteases with special emphasis on the purification and characterization of two enzymes, namely SAPB and KERAB, from isolated alkaline proteinase and keratinase producing microbial strains, whose promising properties and attributes are likely to open new pathways in current and future research and new possibilities for the improvement of current detergent formulations and leather processing industries. In fact, both SAPB and KERAB showed valuable operational characteristics that made them strong potential candidates for future application as additives in biotechnological applications and processes, particularly in detergent formulations and in dehairing during leather processing. They also showed relatively high stability in the presence of organic solvents, a feature which is highly desired in applications involving the biocatalysis of non-aqueous peptides. Accordingly, this chapter intends to report on the screening, identification, and phylogenetic analysis of the *Bacillus pumilus* strain CBS producing SAPB and the *Streptomyces* sp. strain AB1 producing KERAB. It also aims to describe the laundry detergent compatibility and high dehairing capacity of both enzymes, and to report on the ability of each strain or enzyme (SAPB or KERAB) alone to accomplish the whole keratin-degradation process of various keratinacious biowastes.

2. Screening and identification of alkaline proteinase and keratinase producing microbes

The isolation and screening of micro-organisms from naturally occurring alkaline habitats and keratinacious biowaste is likely to help identify potential microbial strains capable of producing active and stable enzymes that can resist the aforementioned harsh substances and conditions present in detergent formulations and leather dehairing processes.

2.1 Screening of alkaline protease and keratinase producing strains

A recent work by the authors (Jaouadi et al., 2009; Badis et al., 2009) involved the screening of about 125 bacterial strains (Bacilli and Actinomyces), originating from a collection of bacterial strains at the CBS and other strains that were previously isolated from surface soil samples at the Mitidja plain, North of Algeria (Badis et al., 2010), for protease and keratinase activities. Based on the ratio of the diameter of the clear zone (onto skimmed milk or keratin-containing medium agar plates at pH 9.) and that of the colony, only 24 isolates, which exhibited the highest ratio (> 3 mm), were selected for further assays pertaining to protease or keratinase production in liquid media. The two bacterial strains that displayed the highest extracellular protease and keratinase activity were termed as strain CBS (from the CBS bacterial strain collection) and strain AB1 (from Algerian soil samples) and retained for all subsequent experimental assays.

2.2 Identification and molecular phylogeny of the microorganisms

The two newly isolated bacterial strains, CBS and AB1, were submitted to identification and typing by molecular and catabolic techniques. The data from the morphological, biochemical and physiological characterization tests, performed on the isolates in accordance with the methods described in the Bergey's Manual of Systematic Bacteriology, showed that the CBS and AB1 strains appeared in a bacilli and filamentous form, respectively, that are aerobic, endospore-forming, Gram-positive, catalase+, oxydase+ and motile rod-shaped. The findings from API 50 CH gallery tests revealed that the CBS isolate metabolized L -arabinose, D-tagatose, ribose, and mannitol in addition to several other simple sugars. The AB1 strain, on the other hand, could use galactose, sucrose, maltose, cellobiose, fucose, raffinose, D-xylose, L-arabinose, and D-ribose, but not lactose, starch, L-rhamnose, erythritol, adonitol, and inositol. The results from API ZYM tests revealed that strain AB1 also exhibited alkaline phosphatase, esterase lipase (C8), leucine arylamidase and valine arylamidase activities, but no lipase (C14), trypsin, α-chymotrypsin, N-acetyl-β-lucosamidase, β -glucuronidase, α-mannosidase, and α-fucosidase ones. Taken together, the data obtained with regard to the physiological and biochemical properties of the two isolates strongly confirmed that the strains CBS and AB1 belonged to the *Bacillus* and *Streptomyces* genera, respectively.

A molecular approach was used to establish further support for the identification of the CBS and AB1 isolates. Two 16S rRNA gene fragments, namely 1,497 bp (Jaouadi et al., 2009) and 1541 bp (Jaouadi et al., 2010a), were amplified from the genomic DNA of the CBS and AB1 isolates, respectively, and then cloned and sequenced on both strands. The 16S rRNA gene sequences obtained were subjected to GenBank BLAST search analyses, which yielded strong homologies of up to 98 and 99% with those of several cultivated strains of *Bacillus* and *Streptomyces*, respectively. The nearest *Bacillus* and *Streptomyces* strains identified by the BLAST analysis were the *Bacillus pumilus*, with the accession numbers of DQ988522, AM292995, AY548955, AB195283, and EF173329, and the *Streptomyces rochei* strains of A-1 (GQ392058) and NRRL B-1559 (EF626598) as well as the *Streptomyces* sp. Strain B5W22-2 (EF114310), respectively. Those sequences were imported into the ARB and MEGA software packages, respectively, and then aligned. After that, the phylogenetic trees were constructed using neighbour-joining methods and Jukes-Cantor distance matrices (Fig. 1). Phylogenetic analyses confirmed that the CBS and AB1 strains were closely related to the five isolated *Bacillus* and three isolated *Streptomyces* strains mentioned earlier. In conclusion, all the results obtained strongly supported the assignment of the CBS and AB1 isolates to the *Bacillus pumilus* strain CBS and *Streptomyces* sp. strain AB1, respectively.

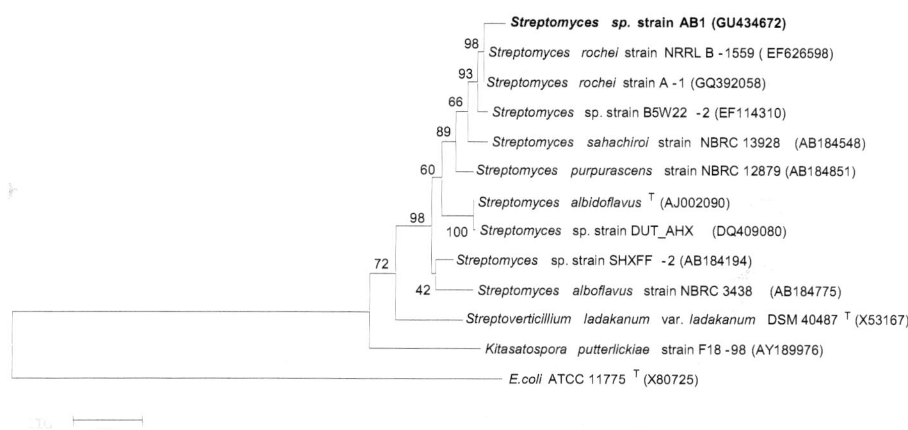

Fig. 1. Example of the phylogenetic tree of *Streptomyces sp.* strain AB1. Phylogenetic and molecular evolutionary analyses were conducted using MEGA version 4.1. Reference type-strain organisms are included and sequence accession numbers are given in parentheses. Bootstrap values, expressed as percentage of 100 replications, are shown in branching points and bar indicated 2 substitutions per 100 nt. The out-group used in the analysis, *E. coli* (X80725), was chosen arbitrarily.

3. Production, purification and biochemical characterization of SAPB and KERAB enzymes

3.1 SAPB and KERAB production

Different carbon and nitrogen sources and trace elements were assayed to optimize the culture growth conditions for the production of the enzymes. In the medium containing (g/l): gelatin 10, yeast extract 5, $CaCl_2$ 1, K_2HPO_4 1, and KH_2PO_4 1, the addition of 0.1% (v/v) trace elements [composed of (g/l): $ZnCl_2$, 0.4; $FeSO_4 \cdot 7H_2O$, 2; H_3BO_3, 0.065; and $MoNa_2O_4 \cdot 2H_2O$, 0.135] at pH 10 was noted to bring about a significant enhancement of 1.32 folds in SAPB production, which reached 6,500 U/ml under the optimal conditions used (pH 10.6 and 65°C), after 24 h of incubation at 37°C and 250 rpm (Jaouadi et al., 2009). In medium containing trace salts with feather as carbon and nitrogen source (g/l): NaCl, 0.5; KH_2PO_4, 0.5; K_2HPO_4, 0.5; KCl, 0.1; $MgSO_4 \cdot 7H_2O$, 1; and chicken feather meal, 10; at pH 9, KERAB production was observed to undergo a significant improvement, reaching a maximum of 9,500 U/ml under the optimal conditions used (pH 11.5 and 75°C) after 96 h of incubation at 30°C and 200 rpm (Jaouadi et al., 2010a). Under these particular conditions, the production of the SAPB and KERAB enzymes started after a 6- and 10-h lag phase, respectively. These productions were then noted to increase exponentially and concomitantly with the increase of cellular growth and to reach the maxima within 24 h of cultivation for SAPB (Fig. 2) and 96 h for KERAB (data not shown).

Compared to the production yields obtained in flask cultivations, the use of a 7-litre fermentor containing the optimized medium after 24-h cultivation at 37°C, an aeration of 1.5 vvm, and an agitation of 600 rpm was noted to improve SAPB production by about 4-folds. It is worth noting here that the cell densities obtained in both cases (Rotary flask and fermentor) were almost the same (about O.D. = 10.9). Based on this particular finding, it was possible to infer that the improvement of enzyme production was related not only to the cell's growth but also to the stability of fermentation parameters (pH and pO_2).

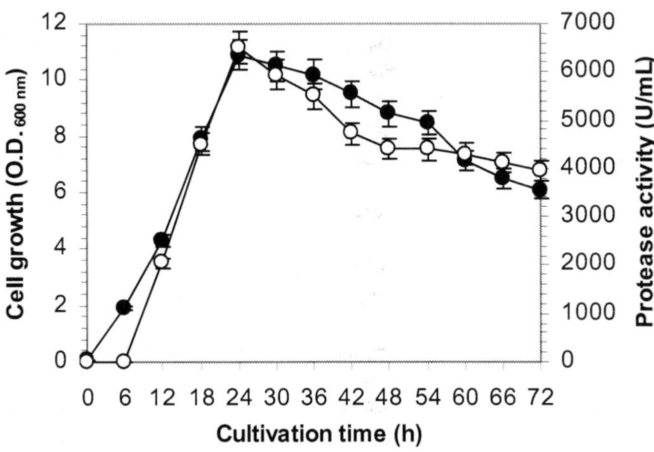

Fig. 2. Time course of *B. pumilus* strain CBS cell growth (●) and SAPB production (○). The culture was carried out under the submerged shaking flask conditions at 37°C for 72 h with an agitation rate of 250 rpm in broth medium containing (g/l): gelatin 10, yeast extract 5, $CaCl_2$ 1, K_2HPO_4 1, KH_2PO_4 1, and trace elements 0.1% (v/v) at pH 10. Cell growth was monitored by measuring the O.D. at 600 nm.

3.2 SAPB and KERAB purification and characterization

The purification protocols used for the purification of each enzyme were conducted at temperatures not exceeding 4°C. Five-hundred ml of 24 h and 96 h cultures of *B. pumilus* strain CBS and *Streptomyces* sp. strain AB1, respectively, were centrifuged to remove microbial cells. Ammonium sulfate was added to each supernatant to a final concentration of 270 g/l. In the case of SAPB, the precipitate formed was collected by centrifugation, dissolved in a minimum amount of 50 mM Tris-HCl (pH 7.5) supplemented with 2 mM $CaCl_2$ and 0.05% Triton X-100 (Buffer A). In the case of KERAB, the precipitate was suspended in 50 mM bicarbonate-NaOH buffer and supplemented with 5 mM $MgSO_4$ at pH 11.5 (Buffer B) containing 10 mM NaCl (Buffer C), and then dialyzed overnight against repeated changes of the buffer A and C, respectively.

Purification to homogeneity was achieved for SAPB by HPLC using Shodex Protein WK 802-5 column. The analysis indicated that enzyme achieved a degree of purity that was about 38-fold greater than that of the crude extract. Under the optimal assay conditions used, the purified enzyme preparation exhibited a yield of about 12% with a specific activity of 25,500 U/mg (Jaouadi et al., 2008). As far as KERAB was concerned, the insoluble material was then removed by centrifugation. The supernatant obtained was incubated for 1 h at 50°C and insoluble material was removed by centrifugation. The supernatant was loaded on a Sephacryl S-200 column equilibrated with buffer B. The elution of protease was performed with the same buffer. The fractions containing keratinase activity were then pooled and applied to a Q-Sepharose column equilibrated in buffer D. The column was rinsed with 500 ml of the same buffer and the adsorbed material was eluted with a linear NaCl gradient. At the final purification step, Keratinase activity was eluted between 0.15 and 0.3 M NaCl. The purity of the enzyme was estimated to be about 86-fold greater than that of the crude extract. The purified enzyme preparation contained about 24% of the total activity of the crude enzyme and had a specific activity of 67,000 U/mg (Jaouadi et al., 2010a). These preparations were homogeneous enzymes with high purity as they exhibited single protein bands on native PAGE and unique elution symmetrical peaks on gel filtration chromatography.

For determination of their molecular weight, enzyme preparations were treated with 1 mM PMSF prior to electrophoresis to inhibit possible autolysis during electrophoresis. Electrophoresis under denaturing conditions (SDS-PAGE) also revealed single bands with molecular masses estimated as 34 kDa for SAPB (Jaouadi et al., 2008) and 30 kDa for KERAB (Jaouadi et al., 2010a). The exact molecular masses obtained for the purified SAPB and KERAB were confirmed by MALDI-TOF mass spectrometry as being 34598.19 and 29850.17 Da, respectively. Zymogram activity staining also revealed two clear zones of proteolytic activity at 34 and 30 kDa for the SAPB and KERAB, respectively. These observations indicated that SAPB extracted from the newly isolated bacterium B. pumilus CBS was a monomeric holoenzyme comparable to those previously reported for other proteases from B. pumilus strains (Han and Damodaran, 1998; Huang et al., 2003; Kumar, 2002; Miyaji et al., 2006; Yasuda et al., 1999). They also showed that KERAB was a monomeric protein comparable to those previously reported for other proteases from Streptomyces strains (Syed et al., 2009; Tatineni et al., 2008).

The molecular mass of SAPB determined by SDS-PAGE (~ 34000 Da) and conducted by MALDI-TOF mass spectrometry (34598.19 Da) were not close to that calculated from the primary sequence of the mature polypeptide (27789 Da), which strongly suggested that the protein underwent noteworthy post-translational changes that were presumably pertaining to glycosylation. Similar differences between experimental and theoretical determinations were previously observed for several B. pumilus proteases, including those from B. pumilus TYO-67 (Yasuda et al., 1999), B. pumilus UN-31-C-42 (Huang et al., 2003), and B. pumilus MS-1 (Miyaji et al., 2006).

3.3 Physico-chemical and kinetic properties of SAPB and KERAB

Phenylmethanesulfonyl fluoride (PMSF) and diiodopropyl fluorophosphates (DIFP) were noted to strongly inhibit SAPB and KERAB, which indicated that both enzymes belonged to the serine proteases family. While the optimal pH and temperature values of 10.6 and 65°C were determined for SAPB using casein as a substrate, those obtained for KERAB were 11.5 and 75°C with keratin azure as substrate. The thermoactivity and thermostability of KERAB were also demonstrated to be enhanced in the presence of 5 mM Mg^{2+} against 2 mM Ca^{2+} for

SAPB. One of the distinguishing properties of SAPB was its catalytic efficiency (k_{cat}/K_m) which was 4.77, 2.73, and 2.11 times higher than those of Subtilisin Carlsberg, Subtilisin BPN', and Subtilisin 309, respectively. The catalytic efficiency of KERAB was higher than those of SAPB, nattokinase and subtilisin Carlsberg.

3.4 Substrate specificity of SAPB and KERAB

The activity of the purified SAPB and KERAB enzymes towards various natural and modified protein substrates is summarized in Table 1. Among the proteinaceous substrates tested, casein and keratin were most efficiently hydrolyzed by SAPB and KERAB, respectively. When SAPB and KERAB activities against casein and keratin were taken as 100%, the hydrolysis rates of gelatine and casein were 95 and 92%, respectively. Poor BSA hydrolysis rates were, however, noted in both cases. A similarly low hydrolysis level was also observed with gluten and egg albumin. Using modified proteins as substrates, the highest activities observed for SAPB and KERAB were with azocasein and keratin azure, respectively. Previous reports also showed that alkaline serine proteases from *B. stearotermophilus* FI (Rahman et al., 1994) and *Bacillus pumilus* A1 (Fakhfakh-Zouari et al., 2010) exhibited highest activities towards casein and keratin, respectively. Interestingly, no collagenase activities were detected for SAPB and KERAB on collagen types I and II, which suggests the potential utility of both enzymes for hair removal in the leather industry.

The cleavage specificities of SAPB and KERAB toward various oligopeptidyl and ester substrates were also investigated. The findings revealed that SAPB exhibited both esterase and amidase activities on oligopeptides, with Tyr or Phe at position P_1 (the amino acid residue at the N-terminal side of the scissile peptide bond). This included N-benzol-L-arginine ethyl ester (BTEE) or N-acetyl-L-tyrosine ethyl ester monohydrate (ATEE) and N-succinyl-L-Ala-L-Ala-L-Pro-L-Phe-p-nitroanilide (AAPF), which are specific substrates for chymotrypsin-like proteases (DelMar et al., 1979; Walsh, 1970). In fact, however, the activity of SAPB on AAPF did not necessarily mean that it was a chymotrypsin-like enzyme. Firstly, most of the microbial members of the Subtilisin family are reported to have specificity that is somewhat similar to that of chymotrypsin (Rawlings and Barrett, 1977). Moreover, SAPB was not observed to show sensitivity to Nα-p-tosyl L-phenylalanine chloromethyl ketone (TPCK), which is an inhibitor of chymotrypsin-like enzymes (Schoellman and Shaw, 1963). Last but not least, SAPB showed neither esterase nor amidase activity on synthetic substrates with P_1 = Arg, such as N-benzol-L-arginine ethyl ester (BAEE) and a-benzoyl-L-tyrosine p-nitroanilide (BAPNA), which are substrates for trypsin-like proteases (Rick, 1995). In contrast, the purified KERAB exhibited esterase and amidase activities on BAEE and BAPNA, but not on BTEE and ATEE.

In the same way, the purified KERAB was noted to exhibit a preference for aromatic and hydrophobic amino acid residues, such as Phe, Leu, Ala, and Val, at the carboxyl side of the splitting point in the P1 position. KERAB was, therefore, active against leucine peptide bonds. When Suc-(Ala)$_n$-pNA was used as the synthetic oligopeptide substrate, a minimum length of three residues was necessary for hydrolysis. Enzymatic activity was observed to depend mainly on secondary enzyme substrate contacts with amino acid residues (P2, P3, etc.) more distant from the scissile bond, as illustrated by the differences observed between the kinetic parameters of Suc-(Ala)$_2$-Val-pNA and those of Suc-Tyr-Leu-Val-pNA. The highest hydrolysis levels achieved by KERAB and SAPB were 100% for AAPF and Suc-Tyr-Leu-Val-pNA, respectively.

Substrate	Concentration	Relative activity (%)[*]	
		SAPB	KERAB
Natural protein[a]			
Keratin	10 g/l	65 ± 1.4	100 ± 3.0
Casein	20 g/l	100 ± 2.5	92 ± 2.0
Gelatine	20 g/l	95 ± 2.4	79 ± 1.4
BSA	20 g/l	52 ± 1.3	66 ± 1.4
Albumin (egg)	20 g/l	15 ± 0.7	26 ± 0.9
Gluten (wheat)	20 g/l	20 ± 0.8	11 ± 0.8
Modified protein[b]			
Keratin azure	10 g/l	63 ± 1.4	100 ± 3.0
Azo-casein	20 g/l	100 ± 2.5	94 ± 2.5
Collagen type I[c]	1 mg/ml	0 ± 0.0	0 ± 0.0
Collagen type II[c]	1 mg/ml	0 ± 0.0	0 ± 0.0
Ester[d]			
BAEE	4 mM	0 ± 0.0	100 ± 3.0
BTEE	3 mM	71 ± 1.4	10 ± 1.5
ATEE	3 mM	100 ± 2.5	20 ± 0.9
Synthetic peptide[e]			
Suc-Tyr-Leu-Val-pNA	2 mM	30 ± 1.1	100 ± 3.0
Suc-(Ala)$_2$-Pro-Phe-pNA	3 mM	100 ± 2.5	17 ± 0.9
Suc-(Ala)$_2$-Pro-Leu-pNA	3 mM	45 ± 1.3	13 ± 0.8
Suc-(Ala)$_2$-Val-Ala-pNA	3 mM	39 ± 1.2	50 ± 1.8
Suc-(Ala)$_2$-Val-pNA	3 mM	25 ± 0.9	56 ± 2.0
Suc-(Ala)$_3$-pNA	2 mM	10 ± 0.4	67 ± 2.2
Suc-(Ala)$_2$-Phe-pNA	2 mM	17 ± 0.5	89 ± 2.5
BAPNA	2 mM	0 ± 0.0	66 ± 1.3

[a]The activity of these natural protein substrates were assessed by measuring absorbance at 660 nm following the previously reported Folin-Ciocalteu method (Jaouadi et al., 2010b).
[b]The activity of these modified protein substrates were determined by measuring absorbance at 440 nm following the method of Riffel and Brandelli, (2002).
[c]The collagenolytic activity was determined by measuring absorbance at 490 nm as described in the protocol of Sigma Co.
[d]The esterase and amidase activities of these substrates were determined by measuring absorbance at 253 nm as described in the method of Walsh (Walsh, 1970).
[e]The activity of these synthetic oligopeptide substrates was determined by measuring absorbance at 410 nm according to the method of DelMar et al., (1979).
[*]Values represent the mean of four replicates and standard errors are reported.

Table 1. Substrate specificity of SAPB and KERAB.

4. Molecular cloning of *sapB* gene and engineering of more efficient SAPB mutant enzymes

The *sap*B gene encoding SAPB was cloned, sequenced, and over-expressed in *Escherichia coli*. The purified recombinant enzyme, called rSAPB, exhibited the same biochemical properties of the native enzyme (Jaouadi and Bejar, 2008). An additional study by the authors further investigated the implications of five amino acid residues (L31, T33, N99, F159, and G182) on the pH and temperature behavior as well as kinetic parameters of the enzyme using site-directed mutagenesis and 3D-modeling approaches (Jaouadi et al., 2010b). Seven more efficient SAPB mutant enzymes, particularly L31I/T33S/N99Y, were generated. The latter had an optimal of pH of 12 and an optimal temperature of 70°C. It was also noted to exhibit a high specific activity that was approximately 2-fold higher than that of the wild-type enzyme and a prominent increase in its k_{cat}/K_m value that was 42-fold higher than that of the wild-type enzyme.

5. Potential and prospects for SAPB and KERAB in detergent formulations

5.1 Effect of detergents on the activity and stability of SAPB and KERAB

With the aim of evaluating the performance of the purified proteases in real life-like detergents, SAPB and KERAB were pre-incubated at 40°C and in the presence of several commercially available laboratory non-ionic surfactants, denaturing agents or anionic surfactants, and bleach agents for 24 and 72 h, respectively. The residual activity was determined at pH 10.6 and 65°C (for SAPB) and pH 11.5 and 75°C (for KERAB). The findings revealed that the SAPB enzyme exhibited high stability at 10% of oxidizing agents (Tween 60 or Triton X-100) as well as against strong anionic surfactants, particularly sodium dodecyl sulphate(SDS) and linear alkylbenzene sulfonate (LAS) (Jaouadi et al., 2008). In fact, SAPB retained its activity upon treatment with 0.8% SDS and 0.5% LAS. In addition, 80 and 65% residual activity were obtained after incubation with 1.5% SDS and 1% LAS, respectively. The SAPB and KERAB enzymes were also highly stable against bleaching agents for they retained 110 and 115% of their initial activity after treatment with 15% hydrogen peroxide, respectively. This is an important behaviour of SAPB and KERAB because oxidant-, surfactant-, and bleach-stable wild-type enzymes are rarely reported.

By way of comparison, the alkaline protease from alkalophilic *Bacillus* sp. JB-99 lost 25% activity during treatment with 0.5% SDS for only 1 h of incubation at 40°C (Johnvesly and Naik, 2001) while two other alkaline proteases (FI and FII) from *Vibrio fluvialis* TKU005 were activated by 1% SDS (Wang et al., 2007b). The present native SAPB and KERAB enzymes showed inherent stability in the presence of high concentrations of detergent compounds, especially Tween 60 at 10%, SDS at 1.5%, and hydrogen peroxide at 15%. In addition, their enzymatic activity and stability were observed to improve in the presence of high concentrations of 1% perfume and anti-redeposition agents, particularly 100 mM Na_2 CMC, and of cationic (TTAB, CTAB) and zwitterionic (Zwittergent 3-12, CHAPS) detergent agents (Table 2). This stability is of interest since only few wild-type proteases have so far been reported to be oxidant, surfactant and bleach stable. These include those reported by Gupta et al., (1999) and Haddar et al., (2009). Bleach stability was also attained through protein engineering (Pillai and Archana, 2008; Radha and Gunasekaran, 2008). These findings suggest the potential strong candidacy of SAPB and KERAB for application as cleaning additives in detergent formulations to facilitate the release of proteinacious materials in tough stains caused by blood, chocolate, grime, milk, etc.

Detergent additive	Concentration	Relative activity (%)		Residual activity (%)	
		SAPB	**KERAB**	**SAPB**	**KERAB**
None	–	100 ± 2.5	100 ± 2.5	100 ± 2.5	100 ± 2.5
H_2O_2	15%	140 ± 3.7	155 ± 3.8	110 ± 2.6	115 ± 2.6
Sodium perborate	2% (w/v)	85 ± 2.2	110 ± 2.6	55 ± 2.0	85 ± 2.2
SDS	1.5%	110 ± 2.6	125 ± 3.0	80 ± 2.2	109 ± 2.6
LAS	1% (w/v)	79 ± 2.2	120 ± 3.2	65 ± 2.1	103 ± 2.5
Sulfobetaine	30 mM	105 ± 2.5	130 ± 3.3	90 ± 2.3	113 ± 2.6
Tween 40	5% (v/v)	111 ± 2.6	135 ± 3.4	101 ± 2.5	119 ± 2.8
Tween 60	10% (v/v)	120 ± 3.0	126 ± 3.3	105 ± 2.5	111 ± 2.6
Triton X-100	10% (v/v)	101 ± 2.5	132 ± 3.5	94 ± 2.3	112 ± 2.7
TAED	10% (w/v)	115 ± 2.8	128 ± 3.5	103 ± 2.5	117 ± 2.7
Na_2·CMC	5% (w/v)	109 ± 2.6	137 ± 3.5	101 ± 2.5	112 ± 2.6
Zeolithe	1% (w/v)	99 ± 2.5	100 ± 2.5	94 ± 2.3	95 ± 2.3
STPP	1% (w/v)	88 ± 2.3	90 ± 2.3	80 ± 2.2	82 ± 2.2
Perfume	1% (v/v)	115 ± 2.6	116 ± 2.6	103 ± 2.5	104 ± 2.5
Na_2·CO$_3$	100 mM	50 ± 2.0	113 ± 2.4	42 ± 1.8	100 ± 2.5
Zwittergent 3-12	10 mM	107 ± 2.5	116 ± 2.6	100 ± 2.5	109 ± 2.5
CHAPS	15 mM	121 ± 3.0	133 ± 3.1	106 ± 2.5	115 ± 2.6
CTAB	25 mM	104 ± 2.5	107 ± 2.3	95 ± 2.4	100 ± 2.5
TTAB	25 mM	99 ± 2.4	105 ± 2.6	90 ± 2.3	98 ± 2.3

Sulfobetaine: *N*-dodecyl-N-N'-dimethyl-3-ammonio-1-propane sulfonate; Tween: poly (oxyethylene) sorbitan monolaurate; Triton: octyphenolpoly (ethylene glycolether); TAED: tetraacetylethylenediamine; STPP: sodium tripolyphosphate; CHAPS: 3-[(3-cholamidopropyl) dimethylammonio]-1-propane sulfonate; CTAB: hexadecyltrimethylammonium bromide; TTAB: tetradecyl trimethylammonium bromides.

Table 2. Effect of various detergents on SAPB and KERAB activity and stability. The non-incubated enzymes were considered as 100%. The activity is expressed as a percentage of the activity level in the absence of additives. Values represent the mean of three replicates and standard errors are reported.

5.2 Compatibility of SAPB and KERAB enzymes with various commercial laundry detergents

To check the compatibility and stability of the alkaline proteases towards detergents, the enzymes were pre-incubated in the presence of various commercial laundry detergents of different compositions for 1 h at 40°C. The laundry detergents were diluted in tap water to a final concentration of 7 mg/ml to simulate washing conditions. The endogenous proteases were inactivated by incubating the diluted detergents for 1 h at 65°C, prior to the addition of the SAPB and KERAB enzymes, or the SB 309 commercial enzyme, which was used for comparison (Table 3). The findings showed that SAPB and KERAB were relatively more stable and compatible with some commercial liquid detergents than the commercial enzyme. In fact, while SB 309 retained 100, 85, 70 and 90% of its initial activity in the presence of Axion, Dinol, Nadhif, and Lav+, SAPB retained about 100, 95, 94, and 85% and KERAB about 87, 90, 75, and 95% of their initial activities, respectively. The SAPB and SB 309 enzymes were, however, less stable in the presence of Axion, where they were totally active. Furthermore, SAPB and KERAB showed excellent stability and compatibility in the presence of some commercial solid detergents, namely OMO, New Det, and Skip, with SAPB retaining about 96, 82, and 69% of its initial activity, and KERAB about 88, 93, and 95%, respectively. SAPB and KERAB were, however, less stable in the presence of Ariel, retaining about 55 and 51% of their initial activities, respectively. Nevertheless, the compatibility and stability exhibited by SAPB and KERAB were much more significant than that of SB 309, which retained only 70, 68, and 84% of its initial activity in the presence of OMO, New Det, and Skip, respectively. Incubated in the same conditions in the presence of New Det, the NH1 protease was reported to retain 60% of its initial activity (Hadj-Ali et al., 2007) and, in the presence of Ariel, the VM10 (Venugopal and Saramma, 2006) and SSR1 (Singh et al., 2001) proteases were reported to retain only 42 and 37% of their initial activities, respectively. Overall, the results obtained clearly indicated the superior performance of SAPB and KERAB enzymes in detergents compared to currently commercialized or previously described proteases. A minor discordance was, however, reported as present with regards this performance, which was presumably correlated to the nature and concentration of the laundry detergent compounds used.

Laundry detergent (7 mg/ml)	Relative activity (%)			Residual activity (%)		
	SAPB	KERAB	SB 309	SAPB	KERAB	SB 309
None	100 ± 2.5	100 ± 2.5	100 ± 2.5	100 ± 2.5	100 ± 2.5	100 ± 2.5
Liquid detergent						
Dinol	95 ± 2.4	90 ± 2.2	85 ± 2.2	81 ± 1.4	80 ± 2.2	77 ± 2.0
Lav+	85 ± 2.2	95 ± 2.4	90 ± 2.2	75 ± 2.0	81 ± 2.2	75 ± 2.0
Nadhif	94 ± 2.4	75 ± 2.0	70 ± 2.0	77 ± 2.0	62 ± 1.8	60 ± 1.7
Axion	100 ± 2.5	87 ± 2.1	100 ± 2.5	91 ± 2.3	66 ± 1.9	85 ± 2.2
Solid detergent						
New Det	99 ± 2.5	94 ± 2.4	68 ± 2.0	82 ± 2.2	93 ± 2.4	58 ± 1.7
Skip	75 ± 2.0	100 ± 2.5	84 ± 2.2	69 ± 2.0	95 ± 2.4	72 ± 2.2
Ariel	65 ± 1.8	60 ± 1.7	61 ± 1.7	55 ± 1.6	51 ± 1.5	50 ± 1.5
OMO	100 ± 2.5	95 ± 2.4	70 ± 2.2	96 ± 2.4	88 ± 2.1	61 ± 1.7

Values represent the mean of three replicates and standard errors are reported.

Table 3. Stability of the purified SAPB and KERAB proteases in the presence of various commercial laundry detergents. The non-incubated enzyme was considered as 100%. The activity is expressed as a percentage of the activity level in the absence of organic solvent.

5.3 Wash performance analysis of SAPB

In order to evaluate the performance of SAPB in terms of ability to remove harsh stains, namely those caused by chocolate or human blood, several pieces of stained cotton cloth were incubated at different conditions (Fig. 3). The findings from these assays revealed that the blood and chocolate stain removal levels achieved with the use of SAPB alone were more effective than the ones obtained with detergent (Det) alone. In fact, SAPB facilitated the release of proteinacious materials in a much easier way than the commercialized SB 309 protease (Jaouadi et al., 2009). Furthermore, the combination of SAPB and the Det detergent resulted in complete stain removal (Fig. 3). In fact, a similar study has previously reported on the usefulness of alkaline proteases from *Spilosoma obliqua* (Anwar and Saleemuddin, 1997) and *B. brevis* (Banerjee et al., 1999) in the assistance of blood stain removal from cotton

cloth both in the presence and absence of detergents, but, in terms of reported results, the
SAPB enzyme was more effective.

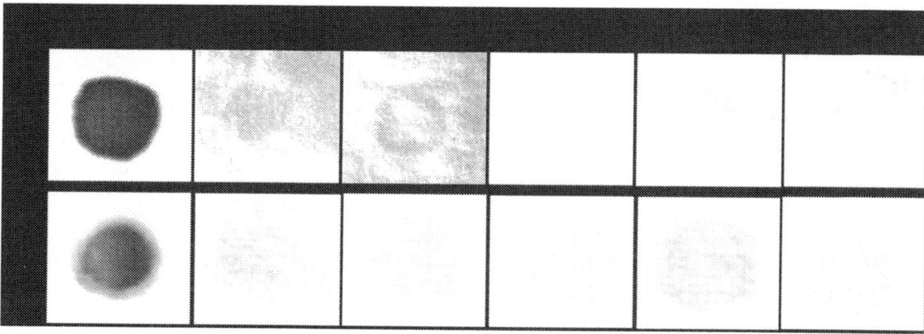

Fig. 3. Example of washing performance analysis test of SAPB. Stained cloth pieces with
blood (I) or chocolate (II). (A) Control: untreated stained cloth pieces; or stained cloth pieces
washed with: (B) distilled water, (C) Det detergent (7 mg/ml), (D) SAPB (500 U/ml), (E) SB
309 (commercial enzyme, 500 U/ml), and (F) SAPB (500 U/ml) + Det detergent (7 mg/ml).

5.4 Storage stability of the spray-dried and lyophilized SAPB

	Condition	t = 2 months	t = 12 months
		Residual activity (%)	
Spray-died	SAPB alone	76	55
	SAPB + Xylitol	88	70
	SAPB + Det	64	50
	SAPB + Det + Xylitol	78	70
Lyophilized	SAPB alone	74	55
	SAPB + Xylitol	80	68
	SAPB + Det	61	50
	SAPB + Det + Xylitol	75	65

Values are the means of three independent experiments.

Table 4. Stability of the spray-dried and lyophilized SAPB with and without xylitol at 1%
during storage at room temperature and during prolonged storage within Det solid
detergent. The activity of each treated SAPB before incubation was taken as 100% and the
residual activity was determined at regular intervals.

The findings indicated that spray-dried SAPB, from fermentor culture, lost about 3% of its
original activity; lyophilized SAPB lost about 10% (Jaouadi et al., 2009). Several of the
additives used during the spray drying and lyophilizing processes were noted to improve
SAPB stability (Table 4). However, the best results were actually obtained with 1% of xylitol,
maltodextrin, and PEG 8000, which preserved about 100, 99 and 97% of its proteolytic
activity, respectively. The stability of the spray-dried and lyophilized SAPB during
subsequent storage in the presence of 1% xylitol showed that, after incubation at room

temperature for 12 months, the enzymes lost only about 20 and 25% of their original activity, respectively, against 35% for the control without additives. The non-treated enzyme was rapidly inactivated, losing about 50% of its initial activity after 2 months of incubation. Moreover, compared to the treated enzyme and in the absence of additives, 1% xylitol clearly enhanced SAPB stability during storage within the Det solid detergent (Jaouadi et al., 2009). In fact, after being incubated for 12 months at room temperature in the presence and absence of xylitol, the spray-dried enzyme retained 68 and 55% of its initial activity, respectively. The level of stability enhancement achieved for the lyophilized SAPB by xylitol was, on the other hand, less pronounced, since the enzyme retained only 65% of its initial activity.

6. Potential and prospects for SAPB and KERAB in the leather processing industry

6.1 Keratin-degradation profile of *Bacillus pumilus* CBS and *Streptomyces* sp. AB1

Keratinacious substrates, such as keratin and keratin azure, were previously reported to be significantly hydrolyzed by SAPB (Jaouadi et al., 2008) and KERAB (Jaouadi et al., 2010a). It was also demonstrated that the *B. pumilus* strain CBS was able to grow in an optimized medium containing 10 g/l of feather-meal, chicken feather (Fig. 4A), goat hair, bovine hair, and sheep wool (as a sole carbon and nitrogen source) instead of gelatin and yeast extract, reaching an absorbance at 600 nm of 6 to 10 after 48 h-culture (Jaouadi et al., 2009). Of the 5 keratin substrates tested, feather-meal was the most strongly degraded (98.5%), followed by chicken feather (92%), goat hair (80%), and bovine hair (68%), with sheep wool showing a relatively low degradation rate (12%).

The feather-meal degradation rate achieved by *B. pumilus* CBS was higher than those of *B. pumilis* F3-4 (97%) (Son et al., 2008) and *Streptomyces albidoflavus* (67%) (Bressollier et al., 1999). The maximum release of protein obtained with the *B. pumilus* strain CBS occurred in the feather-meal medium, which was followed by the chicken feather medium. Moreover, while feather-meal and chicken feather gave the best SAPB production yields of 4,800 and 4,512 U/ml, respectively, sheep wool supported very low keratinolytic activity (1250 U/ml) (Jaouadi et al., 2009). Hence, its full-grown and intense Feather-Degrading (FD) activity could be achieved, in 24 h, at the range of 30 - 37°C, and with initial pH adjusted from 8 to 9. This profile contrasts with previously reported results stipulating that *B. pumilus* FH9 solubilize feather in 72 h at 55°C with pH 9 (El-Refai et al., 2005) while *B. pumilis* F3-4 show intense FD activity in 168 h at 30°C with pH 7.5 (Son et al., 2008).

An increase simultaneous to keratin degradation was noted in protein levels and sulfhdryl groups (Jaouadi et al., 2009). Higher levels of keratin degradation resulted in high sulfhydryl group formation. The results obtained, therefore, suggested that *B. pumilus* CBS had a disulfide bond-reducing ability. Moreover, the processing of data from amino acid analysis following keratin degradation revealed a marked increase in the release of free amino acids after 12 h of incubation. The profile suggested that phenylalanine, tryptophan, leucine, isoleucine, valine, and alanine were the major amino acids liberated, whereas the untreated keratin (control) did not release any free amino acids. In fact, this amino acid profile matched well with the one described for the keratinolytic serine-enzyme produced by *B. licheniformis* PWD-1 (Williams et al., 1990). When SAPB was shaking-incubated with a white feather, a partial degradation was observed after 24 h with a simultaneous increase in protein concentration and sulfhydryl group formation, whereas no degradation was noticed

with the control (Fig. 4B). These results confirmed that SAPB alone could accomplish the
whole dehairing process.

 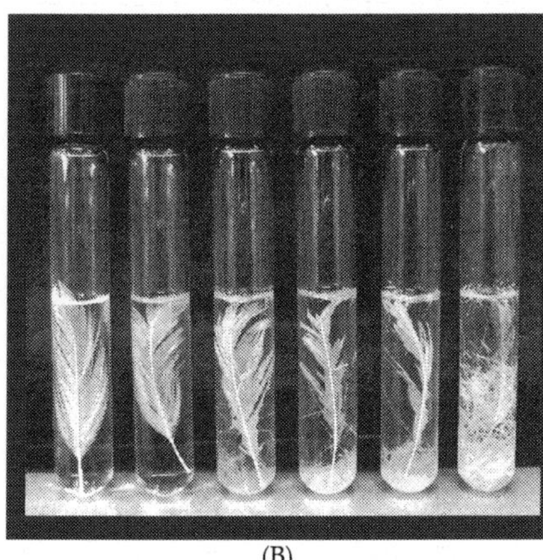

(A)   (B)

Fig. 4. Keratin(feather)-degradation by *B. pumilus* strain CBS and SAPB. (A) Feathers were
incubated for 24 h at 37°C under shake culture condition with 2.8×10^8 cells/ml as an initial
inoculum density of the strain CBS (right flask) and with autoclaved inoculum as control
(left flask). (B) SABP was incubated for 24 h at 37°C with chicken feather.

The *Streptomyces* sp. strain AB1 was able to grow after 2 days of culture in a mineral salt
medium containing 30 g/l of intact chicken feathers as sole carbon, nitrogen, and sulfur
sources instead of 10 g/l of feather meal. Intense feather-degrading activity was achieved at
30 °C and initial pH 9 (Jaouadi et al., 2010a). Interestingly, a nearly complete feather
degradation was achieved, including the delamination of the rachis. A simultaneous
increase of protein concentration and sulfhydryl group formation followed by a higher
disulfide bond-reducing activity of KERAB were also observed. In contrast, no degradation
was noted with the control. These results, therefore, suggested that the *Streptomyces* sp.
strain AB1 had a disulfide bond-reducing ability. Furthermore, when KERAB was incubated
with native chicken feathers, total degradation was observed after 24 h with a simultaneous
increase of protein concentration and sulfhydryl group formation; whereas no degradation
was noted with the control (Jaouadi et al., 2009).
The use of enzymatic and/or microbiological methods for the hydrolysis of feathers is an
attractive alternative to the currently used methods of feather meal preparation which
involve high temperature and pressure treatments that result in the loss of essential amino
acids (Hess and FitzGerald, 2007). The ability of the *B. pumilus* strain CBS and *Streptomyces*
sp. strain AB1 to grow and produce appreciable levels of protease and keratinase using
feather as a substrate could open new opportunities for the achievement of efficient
biodegradation and valorization processes of keratin-containing wastes and, thereby, help
reduce the environmental impact of such biowaste.

6.2 Dehairing utility of SAPB

The incubation of the SAPB protease with skin from goat (Fig. 5), bovine (Jaouadi et al.,
2009), and sheep (Jaouadi et al., 2009) for dehairing showed that after 24 h-incubation at

37°C, hair was removed very easily for all skins, compared to the corresponding controls, with no observable damage on the collagen. Therefore, the dehaired skins obtained exhibited clean hair pore and clear grain structure (data not shown). Again, these results confirmed that SAPB alone could accomplish the whole dehairing process.

The dehairing operation in leather processing is generally carried out under a relatively high pH value of about 8 - 10 (Dayanandan et al., 2003). This criterion was also satisfied by SAPB. In fact, approximately similar results were reached with the *A. tamarri* alkaline protease on goat skin at pH 9 - 11 and temperature of 30-37°C (Dayanandan et al., 2003). Likewise, similar results were obtained with the *Vibrio* sp. strain Kr2 but at pH values ranging from 6 to 8 and temperature of 30°C (Grazziotin et al., 2007). Other alkaline proteases from *B. pumilus* with high keratinolytic activity were also reported to accomplish alone the dehairing process on bovine hair (Kumar et al., 2008), cowhides (Wang et al., 2007a) and goatskins (Huang et al., 2003). However, with the higher dehairing ability and FD activities reported for SAPB, the latter could be considered a potential strong candidate for application in biotechnological bioprocesses involving the dehairing of hides or skins or the conversion of feather-rich wastes into economically useful feather-meal.

Fig. 5. Dehairing function of SAPB. SABP was incubated for 24 h at 37°C with, goat hair. (Left = control, Right = test).

7. Effect of organic solvents on the activity and stability of SAPB and KERAB

In addition to the key areas of application discussed for proteases above, the latter constitute a highly resourceful class of enzymes for various industrial sectors. They are, for instance, necessary in the biocatalysis of various peptide coupling reactions, which are of an extremely pharmaceutical and nutritional interest, namely those involved in the synthesis of several drug precursors such as the enkephalin (Kimura et al., 1990) and aspartame precursors (Nakanishi et al., 1990). However, the ultimate application of proteases in the synthesis of peptides has often been curtailed by the poor levels of specificity and instability in the presence of organic solvents so far reported in the literature. Accordingly, various water-miscible organic solvents and alcohols at final concentrations of 50% were assayed for their effect on SAPB and KERAB activity at pH 10 and 60°C. Buthanol, acetonitrile, and ethyl acetate had significant inhibitory effects on the activity of both enzymes (Table 5). By contrast, dimethylformamide (DMF), DMSO, and hexane were noted to enhance the activity and stability of both enzymes while isopropanol and ethanol enhanced those of SAPB and KERAB, respectively (Table 5). Hence, good stability rates of 115, 97, 90 and 85% were exhibited by SAPB in the presence of DMF, hexane, isopropanol and DMSO, respectively. Equally good stability rates of 150, 125, 115 and 105% were displayed by KERAB in the presence DMSO, DMF, ethanol, and haxane, respectively. Acetonitrile, however, exerted a considerably negative effect on enzyme stability. Compared to SAPB, NH1 (Hadj-Ali et al., 2007) seemed less efficient for it exhibited only 181.5 and 94.5% of its initial activity and stability in the presence of 25% DMSO, respectively. The exception was observed with the organic solvent-tolerant protease BG1 (Ghorbel-Frikha et al., 2005; Ghorbel et al., 2003), which showed a half-life of 50 days of its activity in the presence of 25% DMSO. While the only report available to date on organic solvent protease from *B. pumilus* 115b (Rahman et al., 2007) showed that it exhibited 134% of its initial activity in the presence of 25% hexane as opposed to the 190% for SAPB and 145% for KERAB.

Organic solvent (50%)	Relative activity (%)		Residual activity (%)	
	SAPB	KERAB	SAPB	KERAB
None	100 ± 2.5	100 ± 2.5	100 ± 2.5	100 ± 2.5
Methanol	100 ± 2.1	80 ± 2.1	85 ± 2.2	75 ± 2.0
Ethanol	75 ± 2.0	132 ± 3.2	55 ± 2.6	115 ± 2.6
Buthanol	50 ± 1.4	79 ± 2.0	38 ± 1.4	63 ± 1.5
Isopropanol	115 ± 2.6	25 ± 0.5	90 ± 2.3	15 ± 0.8
Acetonitrile	25 ± 1.0	20 ± 1.0	10 ± 0.8	0 ± 0.1
Ethyl acetate	85 ± 2.2	66 ± 1.5	72 ± 2.0	58 ± 1.5
DMF	200 ± 5.0	155 ± 3.7	115 ± 3.0	125 ± 3.0
DMSO	150 ± 3.7	195 ± 4.9	85 ± 2.2	150 ± 3.7
Hexane	170 ± 4.0	160 ± 3.8	97 ± 2.5	105 ± 2.5

Values represent the mean of three replicates and standard errors are reported.

Table 5. Effect organic solvents on SAPB and KERAB activity and stability. The non-incubated enzyme was considered as 100%. The activity is expressed as a percentage of the activity level in the absence of organic solvent.

A combination of high esterase and low amidase activities is necessary for several synthetic applications of proteases, including peptides coupling (Plettner et al., 1999). In addition to demonstrating its organic tolerance, the findings presented above show that both SAPB and KERAB exhibited powerful esterase activities on BTEE and on BAEE. Furthermore, no amidase activity was detected for SAPB and KERAB on BAEE with P_1= Arg and ATEE with P1 = Tyr, respectively. These findings, in addition of the observed activity and stability in certain organic solvents strongly suggested that SAPB and KERAB are potential strong candidates for use in peptide synthesis reactions in low water systems.

8. Conclusion

This chapter described the valuable advantages inherent in proteases and the promising opportunities they offer for the enhancement of a variety of industrial and consumer product applications. This was illustrated by an overview on the purification and characterization of two extracellular extremozyme serine alkaline proteinases, namely SAPB and KERAB, which were isolated from *B. pumilus* strain CBS and *Streptomyces* sp. strain AB1, respectively. These pure enzymes were significantly tolerant and stable in the presence of the various laundry detergents tested, which strongly supported their suitablity for liquid and solid laundry detergents. Furthermore, and in comparison with the standard enzyme, namely SB 309, both SAPB and KERAB turned to be more effective under alkaline and high temperature conditions. Furthermore, the *B. pumilus* strain CBS and *Streptomyces* sp. strain AB1 proved suitable for the degradation of avian feathers and feather-meal, showing strong potential for application in future biotechnological processes. More interestingly, SAPB demonstrated powerful dehairing abilities against various skins with minimal damage on collagen. Last but not least, these enzymes showed high esterase and low amidase activities as well a good tolerance for several organic solvents. Overall, the findings presented in this chapter strongly suggest that both enzymes, SAPB and KERAB, offer new and promising opportunities for prospective application in biotechnological bioprocesses, particularly those involving the synthesis of detergent formulations, dehairing during leather processing, and peptide biocatalysis in non-aqueous environments.

9. Acknowledgments

This work was funded by the Tunisian Ministry of Higher Education and Scientific Research (contract program CBS-LEMP, grant no. RL02CBS01) and the Algerian Ministry of Higher Education and Scientific Research (CNEPRU project grant no. JO100420070004). The authors wish to express their sincere gratitude to Pr. Anouar Smaoui, from the English department at the Sfax Faculty of Science for carefully structuring, proofreading, and polishing the language and format of the present book chapter.

10. References

Anonyme (2007). World enzymes to 2011 (2229). *Focus on Catalysts* 2007, 2-2.

Anwar, A. & Saleemuddin, M. (1997). Alkaline-pH-acting digestive enzymes of *Polyphagous brevis* and its characterization as a laundry detergent additive. *Process Biochem.* 35, 213-216.

Awad, W. M., Jr., Soto, A. R., Siegel, S., Skiba, W. E., Bernstrom, G. G. & Ochoa, M. S. (1972). The proteolytic enzymes of the K-1 strain of *Streptomyces griseus* obtained from a commercial preparation (Pronase). I. Purification of four serine endopeptidases. *J Biol Chem.* 247, 4144-4154.

Badis, A., Ferradji, F. Z., Boucherit, A., Fodil, D. & Boutoumi, H. (2010). Characterization and biodegradation of soil humic acids and preliminary identification of decolorizing actinomycetes at Mitidja plain soils (Algeria). *Afr J Microbiol Res.* 3, 997-1007.

Banerjee, U., Sani, R., Azmi, W. & Sani, R. K. (1999). Thermostable alkaline protease from *Bacillus brevis* and its characterisation as a laundry detergent additive. *Process Biochem.* 35, 213-219.

Beg, Q. & Gupta, R. (2003). Purification and characterization of an oxidation stable, thiol-dependent serine alkaline protease from *Bacillus mojavensis*. *Enzyme Microb Technol.* 32, 294-304.

Bertsch, A. & Coello, N. (2005). A biotechnological process for treatment and recycling poultry feathers as a feed ingredient. *Bioresour Technol.* 96, 1703-1708.

Brandelli, A., Daroit, D. J. & Riffel, A. (2010). Biochemical features of microbial keratinases and their production and applications. *Appl Microbiol Biotechnol.* 85, 1735-1750.

Bressollier, P., Letourneau, F., Urdaci, M. & Verneuil, B. (1999). Purification and characterization of a keratinolytic serine proteinase from *Streptomyces albidoflavus*. *Appl Environ Microbiol.* 65, 2570-2576.

Chitte, R. R., Nalawade, V. K. & Dey, S. (1999). Keratinolytic activity from the broth of a feather-degrading thermophilic *Streptomyces thermoviolaceus* strain SD8. *Lett Appl Microbiol.* 28, 131-136.

Coulombe, P. A. & Omary, M. B. (2002). 'Hard' and 'soft' principles defining the structure, function and regulation of keratin intermediate filaments. *Curr Opin Cell Biol.* 14, 110-122.

Dayanandan, A., Kanagaraj, J., Sounderraj, L., Govindaraju, R. & Rajkumar, G. S. (2003). Application of an alkaline protease in leather processing: An ecofriendly approach. *Journal Clean Prod.* 11, 533-536.

DelMar, E. G., Largman, C., Brodrick, J. W. & Geokas, M. C. (1979). A sensitive new substrate for chymotrypsin. *Anal Biochem.* 99, 316-320.

El-Refai, H. A., AbdelNaby, M. A., Gaballa, A., El-Araby, M. H. & Abdel Fattah, A. F. (2005). Improvement of the newly isolated *Bacillus pumilus* FH9 keratinolytic activity. *Process Biochem.* 40, 2325-2332.

Essien, J. P., Umoh, A. A., Akpan, E. J., Eduok, S. I. & Umoiyoho, A. (2009). Growth, keratinolytic proteinase activity and thermotolerance of dermatophytes associated with alopecia in Uyo, Nigeria. *Acta Microbiol Immunol Hung.* 56, 61-69.

Fakhfakh-Zouari, N., Hmidet, N., Haddar, A., Kanoun, S. & Nasri, M. (2010). A novel serine metallokeratinase from a newly isolated *Bacillus pumilus* A1 grown on chicken feather meal: biochemical and molecular characterization. *Appl Biochem Biotechnol.* 162, 329-344.

Ghorbel-Frikha, B., Sellami-Kamoun, A., Fakhfakh, N., Haddar, A., Manni, L. & Nasri, M. (2005). Production and purification of a calcium-dependent protease from *Bacillus cereus* BG1. *J Ind Microbiol Biotechnol.* 32, 186-194.

Ghorbel, B., Sellami-Kamoun, A. & Nasri, M. (2003). Stability studies of protease from *Bacillus cereus* BG1. *Enzyme Microb Technol.* 32, 513-518.

Grazziotin, A., Pimentel, F. A., Sangali, S., de Jong, E. V. & Brandelli, A. (2007). Production of feather protein hydrolysate by keratinolytic bacterium *Vibrio* sp. kr2. *Bioresour Technol.* 98, 3172-3175.

Gupta, R., Beg, Q. K. & Lorenz, P. (2002). Bacterial alkaline proteases: Molecular approaches and industrial applications. *Appl Microbiol Biotechnol.* 59, 15-32.

Gupta, R., Gupta, K., Saxena, R. & Khan, S. (1999). Bleach-stable alkaline protease from & sp. *Biotechnol Lett.* 21, 135-138.

Gupta, R., Saxena, R. K., Chaturvedi, P. & Virdi, J. S. (1995). Chitinase production by *Streptomyces viridificans*: Its potential in fungal cell wall lysis. *J Appl Bacteriol.* 78, 378-383.

Haddar, A., Agrebi, R., Bougatef, A., Hmidet, N., Sellami-Kamoun, A. & Nasri, M. (2009). Two detergent stable alkaline serine-proteases from *Bacillus mojavensis* A21: Purification, characterization and potential application as a laundry detergent additive. *Bioresour Technolo.* 100, 3366-3373.

Hadj-Ali, N. E., Agrebi, R., Ghorbel-Frikha, B., Sellami-Kamoun, A., Kanoun, S. & Nasri, M. (2007). Biochemical and molecular characterization of a detergent stable alkaline serine-protease from a newly isolated *Bacillus licheniformis* NH1. *Enzyme Microb Technol.* 40, 515-523.

Han, X.-Q. & Damodaran, S. (1998). Purification and Characterization of Protease Q: A detergent- and urea-stable serine endopeptidase from *Bacillus pumilus*. *J Agric Food Chem.* 46, 3596-3603.

Hess, J. F. & FitzGerald, P. G. (2007). Treatment of keratin intermediate filaments with sulfur mustard analogs. *Biochem Biophys Res Commun.* 359, 616-621.

Huang, Q., Peng, Y., Li, X., Wang, H. & Zhang, Y. (2003). Purification and characterization of an extracellular alkaline serine protease with dehairing function from *Bacillus pumilus*. *Curr Microbiol.* 46, 169-173.

Ignatova, Z., Gousterova, A., Spassov, G. & Nedkov, P. (1999). Isolation and partial characterisation of extracellular keratinase from a wool degrading thermophilic actinomycete strain *Thermoactinomyces candidus*. *Can J Microbiol.* 45, 217-222.

Jaouadi, B., Abdelmalek, B., Fodil, D., Ferradji, F. Z., Rekik, H., Zaraî, N. & Bejar, S. (2010a). Purification and characterization of a thermostable keratinolytic serine alkaline proteinase from *Streptomyces* sp. strain AB1 with high stability in organic solvents. *Bioresour Technol.* 101, 8361-8369.

Jaouadi, B., Aghajari, N., Haser, R. & Bejar, S. (2010b). Enhancement of the thermostability and the catalytic efficiency of *Bacillus pumilus* CBS protease by site-directed mutagenesis. *Biochimie.* 92, 360-369.

Jaouadi, B. & Bejar, S. (2008). Characterization of an original serine alkaline proteinase from *Bacillus pumilus* CBS. *J Biotechnol.* 136 (Suppl.) S305.

Jaouadi, B., Ellouz-Chaabouni, S., Ben Ali, M., Ben Messaoud, E., Naili, B., Dhouib, A. & Bejar, S. (2009). Excellent laundry detergent compatibility and high dehairing ability of the *Bacillus pumilus* CBS alkaline proteinase (SAPB). *Biotechnol Bioprocess Eng.* 14, 503-512.

Jaouadi, B., Ellouz-Chaabouni, S., Rhimi, M. & Bejar, S. (2008). Biochemical and molecular characterization of a detergent-stable serine alkaline protease from *Bacillus pumilus* CBS with high catalytic efficiency. *Biochimie.* 90, 1291-305.

Johnson, P. & Smillie, L. B. (1974). The amino acid sequence and predicted structure of *Streptomyces griseus* protease A. *FEBS Lett.* 47, 1-6.

Johnvesly, B. & Naik, G. (2001). Studies on production of thermostable alkaline protease from thermophilic and alkaliphilic *Bacillus* sp. JB-99 in a chemically defined medium. *Process Biochem.* 37, 139-144.

Kimura, Y., Nakanishi, K. & Matsuno, R. (1990). Enzymatic synthesis of the precursor of Leu-enkephalin in water-immiscible organic solvent systems. *Enzyme Microb Technol.* 12, 272-280.

Kirk, O., Borchert, T. V. & Fuglsang, C. C. (2002). Industrial enzyme applications. *Curr Opin Biotechnol.* 13, 345-351.

Kitadokoro, K., Tsuzuki, H., Nakamura, E., Sato, T. & Teraoka, H. (1994). Purification, characterization, primary structure, crystallization and preliminary crystallographic study of a serine proteinase from *Streptomyces fradiae* ATCC 14544. *Eur J Biochem.* 220, 55-61.

Kumar, A. G., Swarnalatha, S., Gayathri, S., Nagesh, N. & Sekaran, G. (2008). Characterization of an alkaline active-thiol forming extracellular serine keratinase by the newly isolated *Bacillus pumilus. J Appl Microbiol.* 104, 411-420.

Kumar, C. G. (2002). Purification and characterization of a thermostable alkaline protease from alkalophilic *Bacillus pumilus. Lett Appl Microbiol.* 34, 13-17.

Maurer, K. H. (2004). Detergent proteases. *Curr Opin Biotechnol.* 15, 330-334.

Miyaji, T., Otta, Y., Nakagawa, T., Watanabe, T., Niimura, Y. & Tomizuka, N. (2006). Purification and molecular characterization of subtilisin-like alkaline protease BPP-A from *Bacillus pumilus* strain MS-1. *Lett Appl Microbiol.* 42, 242-247.

Nakanishi, K., Takeuchi, A. & Matsuno, R. (1990). Long-term continuous synthesis of aspartame precursor in a column reactor with an immobilized thermolysin. *Appl Microbiol Biotechnol.* 32, 633-636.

Onifade, A. A., Al-Sane, N. A., Al-Musallam, A. A. & Al-Zarban, S. (1998). A review: Potentials for biotechnological applications of keratin-degrading microorganisms and their enzymes for nutritional improvement of feathers and other keratins as livestock feed resources. *Bioresour Technol.* 66, 1-11.

Pillai, P. & Archana, G. (2008). Hide depilation and feather disintegration studies with keratinolytic serine protease from a novel *Bacillus subtilis* isolate. *Appl Microbiol Biotechnol.* 78, 643-650.

Plettner, E., DeSantis, G., R. Stabile, M. & Jones, J. B. (1999). Modulation of esterase and amidase activity of subtilisin bacillus lentus by chemical modification of cysteine mutants. *J Am Chem Soc.* 121 4977-4981.

Radha, S. & Gunasekaran, P. (2008). Sustained expression of keratinase gene under PxylA and PamyL promoters in the recombinant *Bacillus megaterium* MS941. *Bioresour Technol.* 99, 5528-5537.

Rahman, R. N., Mahamad, S., Salleh, A. B. & Basri, M. (2007). A new organic solvent tolerant protease from *Bacillus pumilus* 115b. *J Ind Microbiol Biotechnol.* 34, 509-517.

Rahman, R. N. Z. A., Razak, C. N., Ampon, K., Basri, M., Zin, W. M., Yunus, W. & Salleh, A. B. (1994). Purification and characterization of a heat-stable alkaline protease from *Bacillus stearothermophilus* F1. *Appl Microb Biotechnol.* 40, 822-827.

Rao, M. B., Tanksale, A. M., Ghatge, M. S. & Deshpande, V. V. (1998). Molecular and biotechnological aspects of microbial proteases. *Microbiol Mol Biol Rev.* 62, 597-635.

Rawlings, D. N. & Barrett, A. J. (1994). Families of serine peptidases. Methods Enzymol. 244, 19-61.

Rick, W. Trypsin. In: H.U. Bergmeyer, Editor, (2nd English edition.), Methods of Enzymatic Analysis. Vol. 2, Verlag Chemie, Weinheim (1974), pp. 1013-1024 Academic Press, New York, NY USA.

Riffel, A. & Brandelli, A. (2002). Isolation and characterization of a feather-degrading bacterium from the poultry processing industry. *J Ind Microbiol Biotechnol.* 29, 255-258.

Schoellman, G. & Shaw, E. (1963). Direct evidence for the presence of histidine in the active center of chympotrypsin. *Biochemistry*. 2, 252-255.

Singh, J., Batra, N. & Sobti, R. C. (2001). Serine alkaline protease from a newly isolated *Bacillus* sp. SSR1. *Process Biochem*. 36, 781-785.

Son, H. J., Park, H. C., Kim, H. S. & Lee, C. Y. (2008). Nutritional regulation of keratinolytic activity in *Bacillus pumilis*. *Biotechnol Lett*. 30, 461-465.

Syed, D. G., Lee, J. C., Li, W. J., Kim, C. J. & Agasar, D. (2009). Production, characterization and application of keratinase from *Streptomyces gulbargensis*. *Bioresour Technol*. 100, 1868-1871.

Szabo, I., Benedek, A. & Mihaly Szabo, I. B., G.Y. (2000). Feather degradation with a thermotolerant *Streptomyces graminofaciens* strain. *World J Microbiol Biotechnol*. 16, 153-255.

Tatineni, R., Doddapaneni, K. K., Potumarthi, R. C., Vellanki, R. N., Kandathil, M. T., Kolli, N. & Mangamoori, L. N. (2008). Purification and characterization of an alkaline keratinase from *Streptomyces* sp. *Bioresour Technol*. 99, 1596-1602.

Venugopal, M. & Saramma, A. V. (2006). Characterization of alkaline protease from *Vibrio fluvialis* strain VM10 isolated from a mangrove sediment sample and its application as a laundry detergent additive. *Process Biochem*. 41, 1239-1243.

Walsh, K. A. (1970). Trypsinogens and trypsins of varoius species. *Methods Enzymol*. 19, 41-63.

Wang, H. Y., Liu, D. M., Liu, Y., Cheng, C. F., Ma, Q. Y., Huang, Q. & Zhang, Y. Z. (2007a). Screening and mutagenesis of a novel *Bacillus pumilus* strain producing alkaline protease for dehairing. *Lett Appl Microbiol*. 44, 1-6.

Wang, S. L., Chio, Y. H., Yen, Y. H. & Wang, C. L. (2007b). Two novel surfactant-stable alkaline proteases from *Vibrio fluvialis* TKU005 and their applications. *Enzyme Microb Technol*. 40, 1213-1220.

Wang, S. L., Hsu, W. T., Liang, T. W., Yen, Y. H. & Wang, C. L. (2008). Purification and characterization of three novel keratinolytic metalloproteases produced by *Chryseobacterium indologenes* TKU014 in a shrimp shell powder medium. *Bioresour Technol*. 99, 5679-5686.

Williams, C. M., Richter, C. S., Mackenzie, J. M. & Shih, J. C. (1990). Isolation, identification, and characterization of a feather-degrading bacterium. *Appl Environ Microbiol*. 56, 1509-1515.

Xie, F., Chao, Y., Yang, X., Yang, J., Xue, Z., Luo, Y. & Qian, S. (2010). Purification and characterization of four keratinases produced by *Streptomyces* sp. strain 16 in native human foot skin medium. *Bioresour Technol*.101, 344-350.

Yasuda, M., Aoyama, M., Sakaguchi, M., Nakachi, K. & Kobamoto, N. (1999). Purification and characterization of a soybean-milk-coagulating enzyme from *Bacillus pumilus* TYO-67. *Appl Microbiol Biotechnol*. 51, 474-479.

21

Bioengineering Recombinant Diacylglycerol Acyltransferases

Heping Cao
U.S. Department of Agriculture, Agricultural Research Service
Southern Regional Research Center
U.S.A.

1. Introduction

The complete genomes of many organisms including human, mouse, *Arabidopsis*, and rice have been sequenced. However, the functions of the proteins encoded by a large percentage of the genes in these organisms have not been determined. The immediate challenge of the post-genomic biology is to determine the biological functions of proteins coded for by those unknown genes. Many endogenous proteins occur in extremely low abundance (such as the anti-inflammatory protein tristetraprolin, TTP) (Cao et al., 2004) and are labile (such as omega-3 fatty-acid desaturase, FAD3) (O'Quin et al., 2010), which are major problems inherent to characterization of those proteins.

Recombinant proteins can be used as an alternative source to endogenous proteins. Production of active proteins in large quantities is necessary for the study of protein structure and function (Cao et al., 2003). Purified recombinant proteins are also important for the production of antibodies (Cao 2004; Cao et al., 2008; Cao et al., 2004) and pharmaceutical reagents. Unfortunately, a great number of proteins are difficult to express and purify. Those proteins include membrane proteins, lipid-associated proteins, and low-abundance proteins. The causes of the difficulties in protein expression and purification are various, among which are protein insolubility, protein degradation, and low-level protein expression (Cao 2010). Therefore, production of high-quality recombinant proteins requires optimization of protein expression and purification procedures in each case.

Diacylglycerol acyltransferases (DGATs) catalyze the last and rate-limiting step of triacylglycerol (TAG) biosynthesis in eukaryotic organisms. DGAT genes have been isolated from many organisms. At least two forms of DGATs are present in mammals (Cases et al., 1998; Cases et al., 2001) and plants (Lardizabal et al., 2001; Shockey et al., 2006) with additional forms reported in burning bush (*Euonymus alatus*) (Durrett et al., 2010), peanut (Saha et al., 2006), and *Arabidopsis* (Rani et al., 2010). Plants and animals deficient in DGATs accumulate less TAG (Smith et al., 2000; Stone et al., 2004; Zou et al., 1999). Animals with reduced DGAT activity are resistant to diet-induced obesity (Chen et al., 2004; Smith et al., 2000) and lack milk production (Smith et al., 2000). Over-expression of DGAT enzymes increases TAG content in plants (Andrianov et al., 2010; Bouvier-Nave et al., 2000; Burgal et al., 2008; Durrett et al., 2010; Jako et al., 2001; Lardizabal et al., 2008; Xu et al., 2008), animals (Kamisaka et al., 2010; Liu et al., 2009; Liu et al., 2007; Roorda et al., 2005), and yeast (Kamisaka et al., 2007). DGATs have nonredundant functions in TAG biosynthesis in species such as mice (Stone et al., 2004) and tung tree (*Vernicia fordii*) (Shockey et al., 2006). Mice deficient in DGAT1 are viable, have modest decreases in TAG, and are resistant to diet-induced obesity (Chen et al., 2002; Smith et al., 2000). In contrast, mice deficient in DGAT2 have severe reduction of TAG and die shortly after birth (Stone et al., 2004). The fact that DGAT1 is unable to compensate for the deficiency in DGAT2 indicates the nonredundant functions of each DGAT isoform in TAG biosynthesis during animal development. Therefore, understanding the roles of DGATs in plants and animals will have tremendous implications in creating new oilseed crops with value-added properties and in providing clues for therapeutic intervention in obesity and related diseases.

Over-production of DGATs has been the subject of a number of studies, but progress has been slow in the characterization of the enzymes because DGATs are integral membrane proteins (Shockey et al., 2006; Stone et al., 2006) and difficult to express and purify (Cheng et al., 2001; Weselake et al., 2006). Information regarding the expression of DGAT genes in *E. coli* is limited. The expression of DGAT1 and DGAT2 as full-length proteins in *E. coli* had not been reported. We recently developed a reliable procedure for the expression and purification of tung DGATs in *E. coli* (Cao et al., 2010; Cao et al., 2011).

2. Bioengineering recombinant diacylglycerol acyltransferases

2.1 DGAT genes have been identified in a wide range of organisms

Database search identified at least 115 DGAT sequences from 69 organisms including plants (such as *Arabidopsis*, barley, caster bean, cauliflower, corn, rape, rice, sorghum, soybean, tobacco, tung tree), animals (such as bird, chimpanzee, cow, dog, fish, fly, frog, monkey, mosquito, mouse, pig, rabbit, rat, sheep, worm), fungi (such as yeast), and human. The names of organisms, the subfamilies of DGATs (DGAT1 and DGAT2) and the GenBank accession numbers are listed in Table 1. Although more than two isoforms of DGATs are found in some species, most of them could be classified into the DGAT1 or DGAT2 subfamily according to their sequence similarities and phylogenetic analysis (data not shown). However, DGAT3 (Saha et al., 2006) and DGAT4 (Rani et al., 2010) were reported recently which have very different sequences with those of DGAT1 and DGAT2. DGAT1 and DGAT2 subfamilies have many conserved residues among the diverse species. However, addition of DGAT3 and DGAT4 from *Arabidopsis* (GenBank accession number: *AAN31909.1)*, caster bean (GenBank accession number: XP_002519339.1), peanut (GenBank accession number: AY875644.1), and yeast (GenBank accession number: DG315417.1) to the multiple sequence alignment completely destroyed all the conserved residues (data not shown), which is contrary to the general belief that the active sites of the enzymes should have certain degree of conservation during the evolution because all are supposed to catalyze the same/similar biochemical reaction.

No.	Organism	DGAT	GenBank accession number	No.	Organism	DGAT	GenBank accession number
1	*Aedes aegypti* (A)	1	XP_001658299	59	*Medicago truncatula* (P)	2	ACJ84867.1
2	*Ajellomyces capsulatus* (F)	1	EGC41804.1	60	*Nicotiana tabacum* (P, tobacco)	1	AAF19345.1

3	*Anolis carolinensis* (A)	2	XP_003225477.1	61	*Nematostella vectensis* (A, worm)	2a	XP_0016304 35.1
4	*Ashbya gossypii* (F)	2	NP_983542.1	62	*Nematostella vectensis* (A, worm)	2b	XP_0016333 22.1
5	*Arthroderma otae* (F)	1	EEQ31683.1	63	*Nematostella vectensis* (A, worm)	2c	XP_0016355 48.1
6	*Arabidopsis thaliana* (P)	1	NP_179535.1	64	*Ovis aries* (A, sheep)	1	NP_0011036 34.1
7	*Arabidopsis thaliana* (P)	2	NP_566952	65	*Ovis aries* (A, sheep)	2	XP_0015188 99.1
8	*Bubalus bubalis* (A, buffalo)	1	AAZ22403.1	66	*Oryctolagus cuniculus* (A, rabbit)	1	XP_0027244 27.1
9	*Brassica juncea* (P)	1a	AAY40784.1	67	*Olea europaea* (P, tree)	1	AAS01606.1
10	*Brassica juncea* (P)	1b	AAY40785.1	68	*Olea europaea* (P, tree)	2	ADG22608.1
11	*Brassica napus* (P)	1a	AAD45536.1	69	*Oryza sativa* (P, rice)	1	NP_0010548 69.2
12	*Brassica napus* (P)	1b	AAD40881.1	70	*Oryza sativa* (P, rice)	2a	NP_0010479 17
13	*Brassica napus* (P)	2	ACO90187	71	*Oryza sativa* (P, rice)	2b	NP_0010575 30
14	*Brassica napus* (P)	2	ACO90188	72	*Ostreococcus tauri* (algae)	2	XP_0030835 39.1
15	*Bos taurus* (A, cow)	1	NP_777118.2	73	*Pongo abelii* (A)	2	XP_0028223 04.1
16	*Bos taurus* (A, cow)	2a	DAA21853.1	74	*Paracoccidioides brasiliensis* (F)	1	EEH17170.1
17	*Bos taurus* (A, cow)	2b	XP_875499.3	75	*Perilla frutescens* (P)	1	AAG23696.1
18	*Bos taurus* (A, cow)	2c	XP_002683800.1	76	*Polysphondylium pallidum* (F)	1	EFA85004.1
19	*Caenorhabditis elegans* (A, worm)	2a	NP_505413.1	77	*Polysphondylium pallidum* (F)	2	EFA83646.1
20	*Caenorhabditis elegans* (A, worm)	2b	NP_872180.1	78	*Physcomitrella patens* (P, moss)	1	XP_0017709 29.1
21	*Canis familiaris* (A, dog)	1b	XP_849176.1	79	*Physcomitrella patens* (P, moss)	1	XP_0017587 58.1
22	*Canis familiaris* (A, dog)	1c	XP_858062.1	80	*Physcomitrella patens* (P, moss)	2b	XP_0017777 26.1
23	*Capra hircus* (A, sheep)	1	ABD59375.1	81	*Picea sitchensis* (P, tree)	2	ABK26256.1

24	*Ciona intestinalis* (A)	2	XP_002120879.1	82	*Pan troglodytes* (A, chimpanzee)	1	XP_520014.2
25	*Chlamydomonas reinhardtii* (algae)	2a	XP_001694904.1	83	*Pan troglodytes* (A, chimpanzee)	2	XP_527842.2
26	*Chlamydomonas reinhardtii* (algae)	2b	XP_001693189.1	84	*Phaeodactylum tricornutum* (F)	1	XP_0021777 53.1
27	*Chlorella variabilis* (algae)	1	EFN50697.1	85	*Populus trichocarpa* (P, tree)	1a	XP_0023082 78.1
28	*Chlorella variabilis* (algae)	2	EFN51306.1	86	*Populus trichocarpa* (P, tree)	1b	XP_0023305 10.1
29	*Dictyostelium discoideum* (mold)	1	XP_645633.2	87	*Populus trichocarpa* (P, tree)	2	XP_0023176 35.1
30	*Dictyostelium discoideum* (mold)	2	XP_635762.1	88	*Ricinus communis* (P, castor bean)	1	XP_0025141 32.1
31	*Drosophila melanogaster* (A, fly)	1a	NP_609813.1	89	*Ricinus communis* (P, castor bean)	1	XP_0025285 31.1
32	*Drosophila melanogaster* (A, fly)	1d	NP_995724.1	90	*Rattus norvegicus* (A, rat)	1	NP_445889.1
33	*Danio rerio* (A, zebrafish)	1a	NP_956024.1	91	*Rattus norvegicus* (A, rat)	2	NP_0010123 45.1
34	*Danio rerio* (A, zebrafish)	1b	NP_00100245 8.1	92	*Sorghum bicolor* (P, sorghum)	1a	XP_0024371 65.1
35	*Danio rerio* (A, zebrafish)	2	NP_00102536 7.1	93	*Sorghum bicolor* (P, sorghum)	1b	XP_0024394 19.1
36	*Euonymus alatus* (P)	1	AAV31083.1	94	*Sorghum bicolor* (P, sorghum)	2	XP_0024526 52.1
37	*Euonymus alatus* (P)	2	ADF57328.1	95	*Saccharomyces cerevisiae* (F, yeast)	2	NP_014888.1
38	*Elaeis oleifera* (P)	2	ACO35365.1	96	*Saccoglossus kowalevskii* (A, worm)	1	XP_0027361 60.1
39	*Echium pitardii* (P)	1	ACO55635.1	97	*Selaginella moellendorffii* (P)	1	XP_0029641 65.1
40	*Glycine max* (P, soybean)	1a	AAS78662.1	98	*Selaginella moellendorffii* (P)	2	XP_0029720 54.1

41	*Glycine max* (P, soybean)	1b	BAE93461.1	99	*Spirodela polyrhiza* (P)	2	AAQ89590.1
42	*Glycine max* (P, soybean)	2	ACU20344.1	100	*Schizosaccharo myces pombe* (F, yeast)	2	XP_0017131 60.1
43	*Helianthus annuus* (P)	2	ABU50328.1	101	*Sus scrofa* (A, pig)	1	NP_999216.1
44	*Homo sapiens* (human)	1	NP_036211.2	102	*Tribolium castaneum* (A)	1	XP_975142.1
45	*Homo sapiens* (human)	2a	AAQ88896.1	103	*Tribolium castaneum* (A)	2	XP_975146.1
46	*Homo sapiens* (human)	2b	NP_835470.1	104	*Toxoplasma gondii* (A)	1	AAP94209.1
47	*Hordeum vulgare* (P, barley)	2	BAJ85730.1	105	*Taeniopygia guttata* (A, bird)	2	XP_0021876 43.1
48	*Ictalurus punctatus* (A, catfish)	2b	NP_00118800 5.1	106	*Tropaeolum majus* (P)	1	AAM03340. 2
49	*Jatropha curcas* (P)	1	ABB84383.1	107	*Vernicia fordii* (P, tung tree)	1	DQ356680.1
50	*Lotus japonicas* (P)	1	AAW51456.1	108	*Vernicia fordii* (P, tung tree)	2	DQ356682
51	*Metarhizium acridum* (F)	1a	EFY86774.1	109	*Vernonia galamensis* (P)	1	ABV21945.1
52	*Metarhizium anisopliae* (F)	1b	EFY97444.1	110	*Vernonia galamensis* (P)	2	ACV40232.1
53	*Monodelphis domestica* (A)	1	XP_001371565 .1	111	*Vitis vinifera* (P, grape)	1	XP_0022793 45.1
54	*Monodelphis domestica* (A)	2	XP_001365685 .1	112	*Vitis vinifera* (P, grape)	2	XP_0022636 26
55	*Mus musculus* (A, mouse)	1	NP_034176.1	113	*Xenopus tropicalis* (A, frog)	2	NP_989372.1
56	*Mus musculus* (A, mouse)	2	NP_080660.1	114	*Zea mays* (P, corn)	1b	EU039830
57	*Macaca mulatta* (A, monkey)	1	XP_001090134 .1	115	*Zea mays* (P, corn)	2	NP_0011501 74.1
58	*Medicago truncatula* (P)	1	ABN09107.1				

Table 1. DGAT1 and DGAT2 sequence information (DGAT3 and DGAT4 are not included in the Table because of their divergent sequences). A: animal, F: fungus, P: plant.

2.2 Literature survey of DGAT expression

A literature survey was performed to find out how many publications related to DGATs have been collected by the two most popular databases, PubMed and Scopus. The data in Table 2 indicate that approximately 1000 papers had been collected by the two databases during the past 28 years when using DGAT and diacylglycerol acyltransferase as search terms in title/abstracts/keywords. Approximately four times of publications were obtained when using the full name of the enzyme "diacylglycerol acyltransferase" as a search term instead of using the abbreviation "DGAT" in the database search. More than half of the publications were from animals and approximately one quarter of the publications were from plants. Less than half of those publications dealt with expression of DGATs at the RNA and protein levels. Some of the publications reported of using more than one organism in the same paper, resulting in the total number of publications less than the number of publications from plants, animals, and human adding together (Table 2). Similarly, the total expression papers are less than the combination because more than one expression methods were used in the same paper. Approximately 5% of the publications were related to heterologous expression. However, only a few papers were from *E. coli* expression system.

Database	PubMed	PubMed	Scopus	Scopus
Search terms in title/abstracts/keywords	DGAT	diacylglycerol acyltransferase	DGAT	diacylglycerol acyltransferase
Total publications	216	817	255	1102
Plant	57	118	60	137
Human	74	203	72	316
Animal	138	588	164	760
Total expression papers	90	225	122	322
Plant expression	31	50	34	62
Human expression	31	85	42	131
Animal expression	53	144	78	220
E. coli expression	4	8	1	6
Yeast expression	17	32	17	33
Insect expression	5	12	7	15

Table 2. Literature survey of publications related to DGAT expression in PubMed and Scopus databases (1982-2010).

2.3 Recombinant DGAT expression update

Expression and purification of recombinant DGATs from any source represents a challenge because DGATs are integral membrane proteins (Hobbs et al., 1999; Siloto et al., 2008; Weselake et al., 2006). In addition, more than 40% of the total amino acid residues are hydrophobic (Table 3). Yeast was the preferred host for DGAT expression (Bouvier-Nave et al., 2000; Burgal et al., 2008; Cao et al., 2010; He et al., 2004; Kalscheuer et al., 2004; Kalscheuer & Steinbuchel 2003; Kroon et al., 2006; Liu et al., 2011; Liu et al., 2010; Manas Fernandez et al., 2009; Mavraganis et al., 2010; Milcamps et al., 2005; Nykiforuk et al., 2002; Quittnat et al., 2004; Shockey et al., 2006; Siloto et al., 2009; Wagner et al., 2010; Xu et al., 2008; Yu et al., 2008) followed by insect cells (Buszczak et al., 2002; Cases et al., 1998; Cases et al., 2001; Lardizabal et al., 2001). A limited number of reports used other host cells including *E. coli* (Saha et al., 2006; Siloto et al., 2008; Weselake et al., 2006) and human cells (Cheng et al., 2001). The great majority of the yeast and insect cell expression studies were designed to confirm the functions of full-length cloned genes. A few studies were directly related to the expression and purification of recombinant DGATs using *E. coli* expression system for functional and structural studies. The recombinant N-terminal region of *Brassica napus* DGAT1 was purified from *E. coli* with a predicted molecular mass of 13,278 Da which was confirmed by MALDI-TOF mass spectrometry. However, the apparent molecular mass on SDS-PAGE was doubled and the native size was four times of the size of the monomer due to self-association (Weselake et al., 2006). The N-terminal region of mouse DGAT1 was also studied in a similar way (Siloto et al., 2008). Full-length DGAT1 or DGAT2 from any organism was, however, not successfully expressed in *E. coli* (Hobbs et al., 1999; Weselake et al., 2006). The exceptional case was that expression of soluble peanut DGAT (DGAT3) in *E. coli* resulted in high levels of DGAT activity and the formation of labeled TAG (Saha et al., 2006), although its sequence is very different from those of DGAT1 or DGAT2.

	Tung tree DGAT1	Tung tree DGAT2	DGAT1 – DGAT2
Length (aa)	526	322	204
Molecular weight	59773.84	36726.20	23047.64
Isoelectric point (PI)	8.91	9.24	- 0.33
Charge at pH 7	11.78	8.44	3.34
Charged (RKHYCDE) (%)	27.00	23.60	3.40
Acidic (DE) (%)	7.98	7.14	0.84
Basic (KR) (%)	10.08	9.63	0.45
Polar (NCQSTY) (%)	25.86	21.74	4.12
Hydrophobic (AILFWV) (%)	41.06	43.48	-2.42

Table 3. Tung DGATs properties and amino acid composition.

2.4 Bioengineering recombinant DGAT for expression in bacteria

We recently described a procedure for over-expression of recombinant full-length DGAT1 and DGAT2 in a bacterial expression system (Cao et al., 2010; Cao et al., 2011). DGAT1 is much larger than DGAT2, although they are similar in other properties and amino acid composition (on % of frequency basis) (Table 3). The two DGAT isoforms have only limited sequence identity and similarity (Figure 1). We were able to express both proteins in *E. coli* as full-length recombinant proteins. In our study, we engineered a maltose binding protein (MBP) tag at the amino terminus and 6 histidine residues (His-tag) at the carboxyl terminus of full-length tung DGATs (Table 4).

Primer	Sequence (5′ to 3′)	Comments
DGAT1 forward	AATATTGGTACCCTGTTTCAGGGTCC GACAATCCTTGAAACGCCG	*Kpn*I site underlined Codons for PreScission protease site Colored
DGAT1 reverse	CGATTAACTAGTAGCTAGCTCAATG ATGATGATGATGATGTCTTGATTCGG TAGTCCC	*Spe*I site underlined Codons for 6 His Colored
DGAT2 forward	AATATTGGTACCCTGTTTCAGGGTCC GGGGATGGTGGAAGTTAAG	*Kpn*I site underlined Codons for PreScission protease site Colored
DGAT2 reverse	CGATTAACTAGTAGCTAGCTCAATG ATGATGATGATGATGAAAAATTTCA AGTTTAAG	*Spe*I site underlined Codons for 6 His Colored

Table 4. Primers for PCR-amplification of the full-length DGAT1 and DGAT2 insert sequences.

We engineered plasmids pMBP-DGAT1-His and pMBP-DGAT2-His for expressing the full-length tung tree type 1 and type 2 diacylglycerol acyltransferases (DGAT1 and DGAT2, GenBank Accession No. DQ356680 and DQ356682, respectively (Shockey et al., 2006) as fusion proteins in *E. coli*. The recombinant proteins contained MBP at the amino terminus and His-tag at the carboxyl terminus. The cloning vector pMBP-hTTP (Figure 2) was

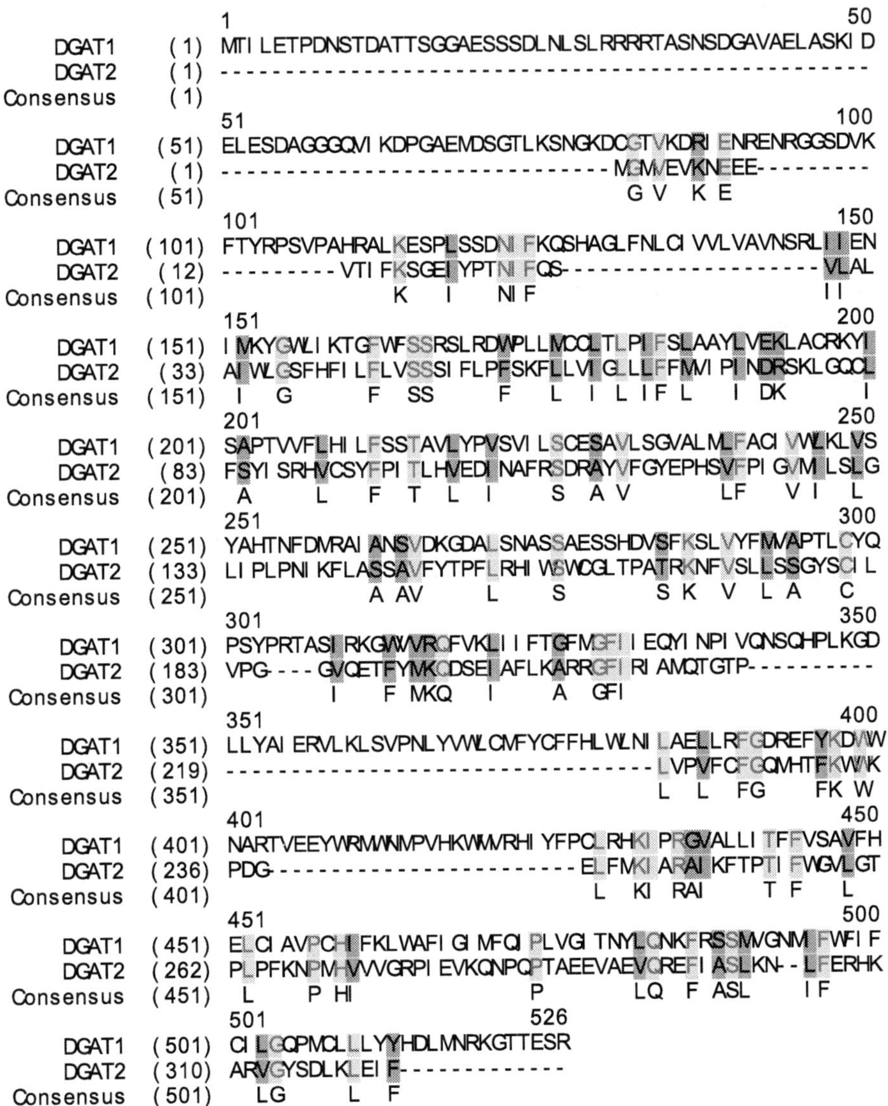

Fig. 1. Alignment of tung tree DGAT1 and DGAT2 amino acid sequences.

reported previously (Cao et al., 2003). Plasmids pMBP-DGAT1-His (Figure 3) and pMBP-DGAT2-His (Figure 4) were constructed by replacing the hTTP fragment in plasmid pMBP-hTTP (Figure 2) with the PCR-amplified DGAT1 and DGAT2 fragments at the *Kpn*I and *Spe*I sites (Table 4). Existing DGAT plasmid DNAs were used as the templates for PCR-amplification of the DGAT DNA open reading frames (Shockey et al., 2006). DGAT forward primers contained DNA sequence for a *Kpn*I/*Asp718*I restriction enzyme recognition site followed by a PreScission protease cleavage site (5'-CTGTTTCAGGGTCCG-3') (Cao et al., 2003) which codes for 5 amino acid residues (LFQGP) between MBP and DGAT protein sequences (Table 4). DGAT reverse primers contained sequence for a His-tag (5'-ATGATGATGATGATGATG-3') coding for 6 histidine residues at the carboxyl terminus of DGATs (Table 4).

The successful expression of full-length recombinant DGATs was probably due to the fusion to MBP, which was shown to increase the solubility of target proteins such as human and mouse TTP (Cao et al., 2003; Cao et al., 2008; Kapust & Waugh 1999). Although we engineered double affinity tags for facilitating purification of recombinant DGAT from *E. coli*, recombinant DGATs were only partially purified from the extract by either type of affinity beads [amylose resin and nickel-nitrilotriacetic agarose (Ni-NTA) beads] or both kinds of beads in tandem. Our data, together with the various published reports cited in the previous section, underline the tremendous challenges that exist for the purification of recombinant full-length DGAT proteins.

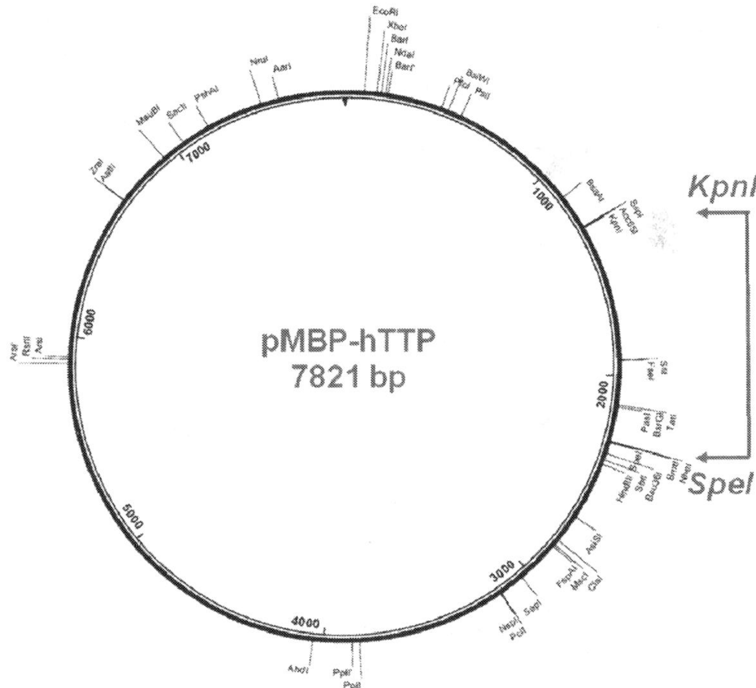

Fig. 2. Plasmid map of *E. coli* expression vector pMBP-hTTP.

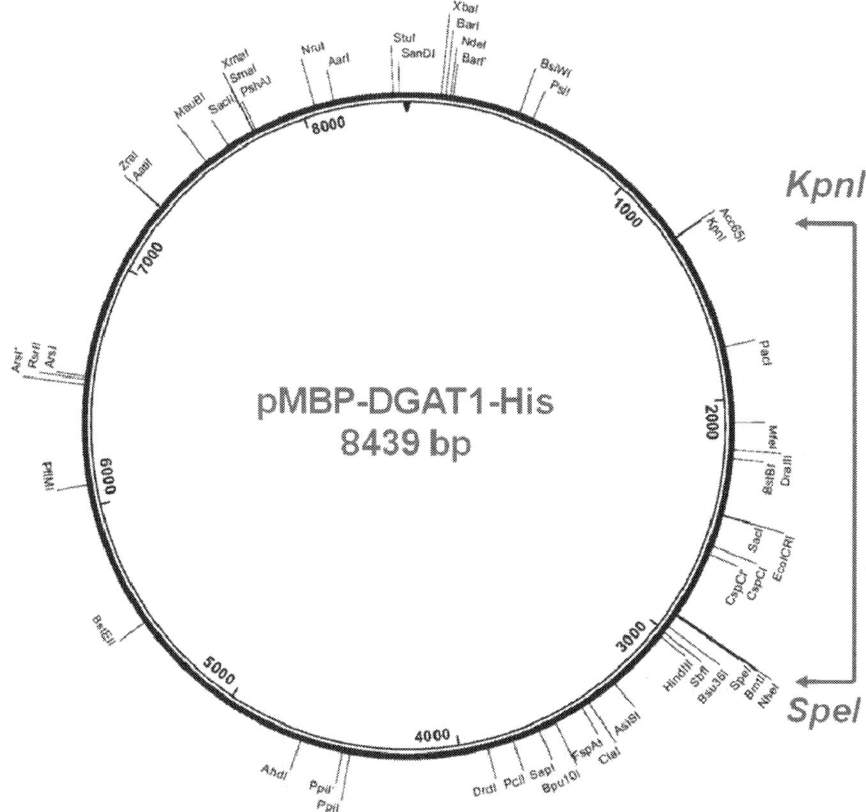

Fig. 3. Plasmid map of *E. coli* expression vector pMBP-DGAT1-His.

3. Conclusion

Diacylglycerol acyltransferases (DGATs) catalyze the last and rate-limiting step of triacylglycerol (TAG) biosynthesis in eukaryotic organisms. At least 115 DGAT sequences are identified from 69 organisms in the GenBank databases. Only a few papers have been published in the last 28 years on the expression of the recombinant DGAT proteins in a bacterial expression system. None of the full-length DGAT1 or DGAT2 had been expressed in *E. coli* expression system. The difficulties in DGAT expression and purification are due to the nature of these proteins being integral membrane proteins with more than 40% of the total amino acid residues being hydrophobic. Therefore, progress in characterization of the enzymes has been slow. We recently developed a procedure for full-length DGAT expression in *E. coli*. Expression plasmids were engineered to express tung DGATs fused to maltose binding protein and poly-histidine. The development of the technique should help to purify full-length DGATs for further studies such as raising high-titer antibodies and studying the structure-function relationship. Understanding the roles of DGATs in plant oil

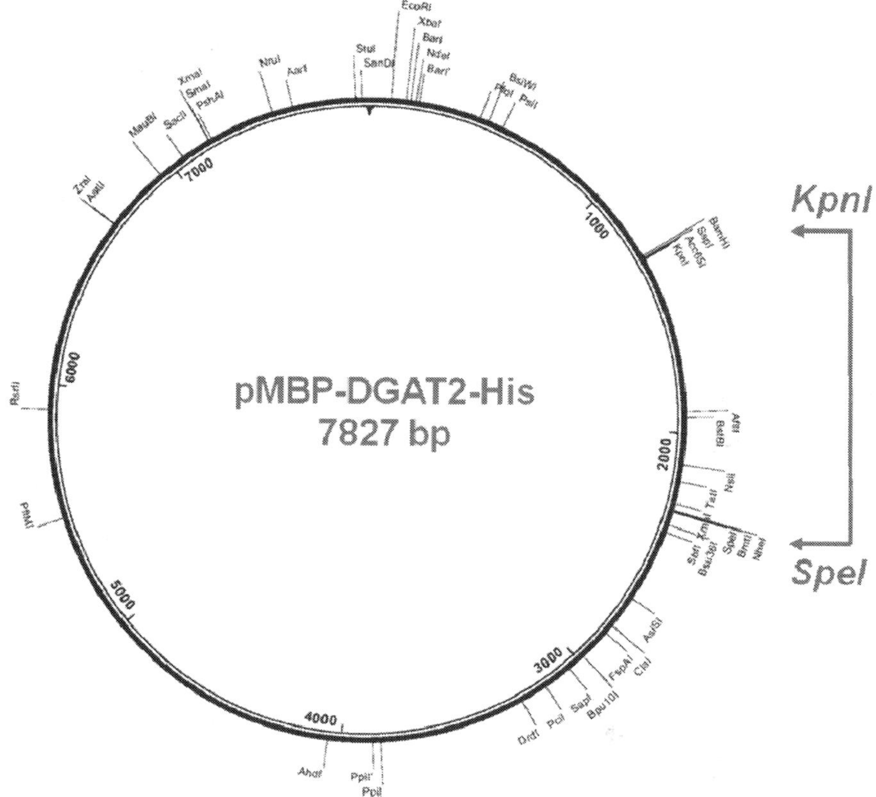

Fig. 4. Plasmid map of *E. coli* expression vector pMBP-DGAT2-His.

biosynthesis will help to create new oilseed crops with value-added properties. The elucidation of the precise roles of DGATs in animal and human fat synthesis and deposition may provide clues for nutritional and therapeutic intervention in obesity and related diseases.

4. Abbreviations

DGAT, diacylglycerol acyltransferase; FAD3, omega-3 fatty-acid desaturase; His, poly histidine; MBP, maltose binding protein; Ni-NTA, nickel-nitrilotriacetic agarose; TAG, triacylglycerol; TTP; tristetraprolin.

5. References

Andrianov, V.; Borisjuk, N.; Pogrebnyak, N.; Brinker, A.; Dixon, J.; Spitsin, S.; Flynn, J.; Matyszczuk, P.; Andryszak, K.; Laurelli, M.; Golovkin, M., & Koprowski, H.

(2010) Tobacco as a production platform for biofuel: overexpression of Arabidopsis DGAT and LEC2 genes increases accumulation and shifts the composition of lipids in green biomass. *Plant Biotechnol J* 8: 277-287

Bouvier-Nave, P.; Benveniste, P.; Oelkers, P.; Sturley, S. L., & Schaller, H. (2000) Expression in yeast and tobacco of plant cDNAs encoding acyl CoA:diacylglycerol acyltransferase. *Eur J Biochem* 267: 85-96

Burgal, J.; Shockey, J.; Lu, C.; Dyer, J.; Larson, T.; Graham, I., & Browse, J. (2008) Metabolic engineering of hydroxy fatty acid production in plants: RcDGAT2 drives dramatic increases in ricinoleate levels in seed oil. *Plant Biotechnol J* 6: 819-831

Buszczak, M.; Lu, X.; Segraves, W. A.; Chang, T. Y., & Cooley, L. (2002) Mutations in the midway gene disrupt a Drosophila acyl coenzyme A: diacylglycerol acyltransferase. *Genetics* 160: 1511-1518

Cao, H. (2004) Expression, purification, and biochemical characterization of the antiinflammatory tristetraprolin: a zinc-dependent mRNA binding protein affected by posttranslational modifications. *Biochemistry* 43: 13724-13738

Cao, H. (2010) Recombinant protein production technology [Review]. *J Jiangxi Agric Univ* 32: 1018-1031

Cao, H.; Chapital, D. C.; Shockey, J. M., & Klasson, T. K. (2011). Expression of tung tree diacylglycerol acyltransferase 1 in *E. coli. BMC Biotechnol*, 11:72.

Cao, H.; Chapital, D. C.; Howard, O. D.; Jiang, X. N.; Shockey, J. M., & Klasson, K. T. (2011) Purification of recombinant tung tree diacylglycerol acyltransferases from *E. coli. The FASEB Journal* 25: 765.8

Cao, H.; Dzineku, F., & Blackshear, P. J. (2003) Expression and purification of recombinant tristetraprolin that can bind to tumor necrosis factor-alpha mRNA and serve as a substrate for mitogen-activated protein kinases. *Arch Biochem Biophys* 412: 106-120

Cao, H.; Lin, R.; Ghosh, S.; Anderson, R. A., & Urban, J. F., Jr. (2008) Production and characterization of ZFP36L1 antiserum against recombinant protein from *Escherichia coli. Biotechnol Prog* 24: 326-333

Cao, H.; Tuttle, J. S., & Blackshear, P. J. (2004) Immunological characterization of tristetraprolin as a low abundance, inducible, stable cytosolic protein. *J Biol Chem* 279: 21489-21499

Cases, S.; Smith, S. J.; Zheng, Y. W.; Myers, H. M.; Lear, S. R.; Sande, E.; Novak, S.; Collins, C.; Welch, C. B.; Lusis, A. J.; Erickson, S. K., & Farese, R. V., Jr. (1998) Identification of a gene encoding an acyl CoA:diacylglycerol acyltransferase, a key enzyme in triacylglycerol synthesis. *Proc Natl Acad Sci U S A* 95: 13018-13023

Cases, S.; Stone, S. J.; Zhou, P.; Yen, E.; Tow, B.; Lardizabal, K. D.; Voelker, T., & Farese, R. V., Jr. (2001) Cloning of DGAT2, a second mammalian diacylglycerol acyltransferase, and related family members. *J Biol Chem* 276: 38870-38876

Chen, H. C.; Rao, M.; Sajan, M. P.; Standaert, M.; Kanoh, Y.; Miura, A.; Farese, R. V., Jr., & Farese, R. V. (2004) Role of adipocyte-derived factors in enhancing insulin signaling in skeletal muscle and white adipose tissue of mice lacking Acyl CoA:diacylglycerol acyltransferase 1. *Diabetes* 53: 1445-1451

Chen, H. C.; Smith, S. J.; Ladha, Z.; Jensen, D. R.; Ferreira, L. D.; Pulawa, L. K.; McGuire, J. G.; Pitas, R. E.; Eckel, R. H., & Farese, R. V., Jr. (2002) Increased insulin and leptin sensitivity in mice lacking acyl CoA:diacylglycerol acyltransferase 1. *J Clin Invest* 109: 1049-1055

Cheng, D.; Meegalla, R. L.; He, B.; Cromley, D. A.; Billheimer, J. T., & Young, P. R. (2001) Human acyl-CoA:diacylglycerol acyltransferase is a tetrameric protein. *Biochem J* 359: 707-714

Durrett, T. P.; McClosky, D. D.; Tumaney, A. W.; Elzinga, D. A.; Ohlrogge, J., & Pollard, M. (2010) A distinct DGAT with sn-3 acetyltransferase activity that synthesizes unusual, reduced-viscosity oils in Euonymus and transgenic seeds. *Proc Natl Acad Sci U S A* 107: 9464-9469

He, X.; Turner, C.; Chen, G. Q.; Lin, J. T., & McKeon, T. A. (2004) Cloning and characterization of a cDNA encoding diacylglycerol acyltransferase from castor bean. *Lipids* 39: 311-318

Hobbs, D. H.; Lu, C., & Hills, M. J. (1999) Cloning of a cDNA encoding diacylglycerol acyltransferase from Arabidopsis thaliana and its functional expression. *FEBS Lett* 452: 145-149

Jako, C.; Kumar, A.; Wei, Y.; Zou, J.; Barton, D. L.; Giblin, E. M.; Covello, P. S., & Taylor, D. C. (2001) Seed-specific over-expression of an Arabidopsis cDNA encoding a diacylglycerol acyltransferase enhances seed oil content and seed weight. *Plant Physiol* 126: 861-874

Kalscheuer, R.; Luttmann, H., & Steinbuchel, A. (2004) Synthesis of novel lipids in Saccharomyces cerevisiae by heterologous expression of an unspecific bacterial acyltransferase. *Appl Environ Microbiol* 70: 7119-7125

Kalscheuer, R. & Steinbuchel, A. (2003) A novel bifunctional wax ester synthase/acyl-CoA:diacylglycerol acyltransferase mediates wax ester and triacylglycerol biosynthesis in Acinetobacter calcoaceticus ADP1. *J Biol Chem* 278: 8075-8082

Kamisaka, Y.; Kimura, K.; Uemura, H., & Shibakami, M. (2010) Activation of diacylglycerol acyltransferase expressed in Saccharomyces cerevisiae: overexpression of Dga1p lacking the N-terminal region in the Deltasnf2 disruptant produces a significant increase in its enzyme activity. *Appl Microbiol Biotechnol* 88: 105-115

Kamisaka, Y.; Tomita, N.; Kimura, K.; Kainou, K., & Uemura, H. (2007) DGA1 (diacylglycerol acyltransferase gene) overexpression and leucine biosynthesis significantly increase lipid accumulation in the Deltasnf2 disruptant of Saccharomyces cerevisiae. *Biochem J* 408: 61-68

Kapust, R. B. & Waugh, D. S. (1999) Escherichia coli maltose-binding protein is uncommonly effective at promoting the solubility of polypeptides to which it is fused. *Protein Sci* 8: 1668-1674

Kroon, J. T.; Wei, W.; Simon, W. J., & Slabas, A. R. (2006) Identification and functional expression of a type 2 acyl-CoA:diacylglycerol acyltransferase (DGAT2) in developing castor bean seeds which has high homology to the major triglyceride biosynthetic enzyme of fungi and animals. *Phytochemistry* 67: 2541-2549

Lardizabal, K.; Effertz, R.; Levering, C.; Mai, J.; Pedroso, M. C.; Jury, T.; Aasen, E.; Gruys, K., & Bennett, K. (2008) Expression of Umbelopsis ramanniana DGAT2A in seed increases oil in soybean. *Plant Physiol* 148: 89-96

Lardizabal, K. D.; Mai, J. T.; Wagner, N. W.; Wyrick, A.; Voelker, T., & Hawkins, D. J. (2001) DGAT2 is a new diacylglycerol acyltransferase gene family: purification, cloning, and expression in insect cells of two polypeptides from Mortierella ramanniana with diacylglycerol acyltransferase activity. *J Biol Chem* 276: 38862-38869

Liu, L.; Shi, X.; Bharadwaj, K. G.; Ikeda, S.; Yamashita, H.; Yagyu, H.; Schaffer, J. E.; Yu, Y. H., & Goldberg, I. J. (2009) DGAT1 expression increases heart triglyceride content but ameliorates lipotoxicity. *J Biol Chem* 284: 36312-36323

Liu, L.; Zhang, Y.; Chen, N.; Shi, X.; Tsang, B., & Yu, Y. H. (2007) Upregulation of myocellular DGAT1 augments triglyceride synthesis in skeletal muscle and protects against fat-induced insulin resistance. *J Clin Invest* 117: 1679-1689

Liu, Q.; Siloto, R. M.; Snyder, C. L., & Weselake, R. J. (2011) Functional and topological analysis of yeast acyl-coa:diacylglycerol acyltransferase 2, an endoplasmic reticulum enzyme essential for triacylglycerol biosynthesis. *J Biol Chem* 286: 13115-13126

Liu, Q.; Siloto, R. M., & Weselake, R. J. (2010) Role of cysteine residues in thiol modification of acyl-CoA:diacylglycerol acyltransferase 2 from yeast. *Biochemistry* 49: 3237-3245

Manas-Fernandez, A.; Vilches-Ferron, M.; Garrido-Cardenas, J. A.; Belarbi, E. H.; Alonso, D. L., & Garcia-Maroto, F. (2009) Cloning and molecular characterization of the acyl-CoA: diacylglycerol acyltransferase 1 (DGAT1) gene from Echium. *Lipids* 44: 555-568

Mavraganis, I.; Meesapyodsuk, D.; Vrinten, P.; Smith, M., & Qiu, X. (2010) Type II diacylglycerol acyltransferase from Claviceps purpurea with ricinoleic acid, a hydroxyl fatty acid of industrial importance, as preferred substrate. *Appl Environ Microbiol* 76: 1135-1142

Milcamps, A.; Tumaney, A. W.; Paddock, T.; Pan, D. A.; Ohlrogge, J., & Pollard, M. (2005) Isolation of a gene encoding a 1,2-diacylglycerol-sn-acetyl-CoA acetyltransferase from developing seeds of Euonymus alatus. *J Biol Chem* 280: 5370-5377

Nykiforuk, C. L.; Furukawa-Stoffer, T. L.; Huff, P. W.; Sarna, M.; Laroche, A.; Moloney, M. M., & Weselake, R. J. (2002) Characterization of cDNAs encoding diacylglycerol acyltransferase from cultures of Brassica napus and sucrose-mediated induction of enzyme biosynthesis. *Biochim Biophys Acta* 1580: 95-109

O'Quin, J. B.; Bourassa, L.; Zhang, D.; Shockey, J. M.; Gidda, S. K.; Fosnot, S.; Chapman, K. D.; Mullen, R. T., & Dyer, J. M. (2010) Temperature-sensitive post-translational regulation of plant omega-3 fatty-acid desaturases is mediated by the endoplasmic reticulum-associated degradation pathway. *J Biol Chem* 285: 21781-21796

Quittnat, F.; Nishikawa, Y.; Stedman, T. T.; Voelker, D. R.; Choi, J. Y.; Zahn, M. M.; Murphy, R. C.; Barkley, R. M.; Pypaert, M.; Joiner, K. A., & Coppens, I. (2004) On the biogenesis of lipid bodies in ancient eukaryotes: synthesis of triacylglycerols by a Toxoplasma DGAT1-related enzyme. *Mol Biochem Parasitol* 138: 107-122

Rani, S. H.; Krishna, T. H.; Saha, S.; Negi, A. S., & Rajasekharan, R. (2010) Defective in cuticular ridges (DCR) of Arabidopsis thaliana, a gene associated with surface cutin formation, encodes a soluble diacylglycerol acyltransferase. *J Biol Chem* 285: 38337-38347

Roorda, B. D.; Hesselink, M. K.; Schaart, G.; Moonen-Kornips, E.; Martinez-Martinez, P.; Losen, M.; De Baets, M. H.; Mensink, R. P., & Schrauwen, P. (2005) DGAT1 overexpression in muscle by in vivo DNA electroporation increases intramyocellular lipid content. *J Lipid Res* 46: 230-236

Saha, S.; Enugutti, B.; Rajakumari, S., & Rajasekharan, R. (2006) Cytosolic triacylglycerol biosynthetic pathway in oilseeds. Molecular cloning and expression of peanut cytosolic diacylglycerol acyltransferase. *Plant Physiol* 141: 1533-1543

Shockey, J. M.; Gidda, S. K.; Chapital, D. C.; Kuan, J. C.; Dhanoa, P. K.; Bland, J. M.; Rothstein, S. J.; Mullen, R. T., & Dyer, J. M. (2006) Tung tree DGAT1 and DGAT2 have nonredundant functions in triacylglycerol biosynthesis and are localized to different subdomains of the endoplasmic reticulum. *Plant Cell* 18: 2294-2313

Siloto, R. M.; Madhavji, M.; Wiehler, W. B.; Burton, T. L.; Boora, P. S.; Laroche, A., & Weselake, R. J. (2008) An N-terminal fragment of mouse DGAT1 binds different acyl-CoAs with varying affinity. *Biochem Biophys Res Commun* 373: 350-354

Siloto, R. M.; Truksa, M.; Brownfield, D.; Good, A. G., & Weselake, R. J. (2009) Directed evolution of acyl-CoA:diacylglycerol acyltransferase: Development and characterization of Brassica napus DGAT1 mutagenized libraries. *Plant Physiol Biochem* 47: 456-461

Smith, S. J.; Cases, S.; Jensen, D. R.; Chen, H. C.; Sande, E.; Tow, B.; Sanan, D. A.; Raber, J.; Eckel, R. H., & Farese, R. V., Jr. (2000) Obesity resistance and multiple mechanisms of triglyceride synthesis in mice lacking Dgat. *Nat Genet* 25: 87-90

Stone, S. J.; Levin, M. C., & Farese, R. V., Jr. (2006) Membrane topology and identification of key functional amino acid residues of murine acyl-CoA:diacylglycerol acyltransferase-2. *J Biol Chem* 281: 40273-40282

Stone, S. J.; Myers, H. M.; Watkins, S. M.; Brown, B. E.; Feingold, K. R.; Elias, P. M., & Farese, R. V., Jr. (2004) Lipopenia and skin barrier abnormalities in DGAT2-deficient mice. *J Biol Chem* 279: 11767-11776

Wagner, M.; Hoppe, K.; Czabany, T.; Heilmann, M.; Daum, G.; Feussner, I., & Fulda, M. (2010) Identification and characterization of an acyl-CoA:diacylglycerol acyltransferase 2 (DGAT2) gene from the microalga O. tauri. *Plant Physiol Biochem* 48: 407-416

Weselake, R. J.; Madhavji, M.; Szarka, S. J.; Patterson, N. A.; Wiehler, W. B.; Nykiforuk, C. L.; Burton, T. L.; Boora, P. S.; Mosimann, S. C.; Foroud, N. A.; Thibault, B. J.; Moloney, M. M.; Laroche, A., & Furukawa-Stoffer, T. L. (2006) Acyl-CoA-binding and self-associating properties of a recombinant 13.3 kDa N-terminal fragment of diacylglycerol acyltransferase-1 from oilseed rape. *BMC Biochem* 7: 24

Xu, J.; Francis, T.; Mietkiewska, E.; Giblin, E. M.; Barton, D. L.; Zhang, Y.; Zhang, M., & Taylor, D. C. (2008) Cloning and characterization of an acyl-CoA-dependent diacylglycerol acyltransferase 1 (DGAT1) gene from Tropaeolum majus, and a study of the functional motifs of the DGAT protein using site-directed mutagenesis to modify enzyme activity and oil content. *Plant Biotechnol J* 6: 799-818

Yu, K.; Li, R.; Hatanaka, T., & Hildebrand, D. (2008) Cloning and functional analysis of two type 1 diacylglycerol acyltransferases from *Vernonia galamensis*. *Phytochemistry* 69: 1119-1127

Zou, J.; Wei, Y.; Jako, C.; Kumar, A.; Selvaraj, G., & Taylor, D. C. (1999) The *Arabidopsis thaliana* TAG1 mutant has a mutation in a diacylglycerol acyltransferase gene. *Plant J* 19: 645-653

22

Microalgal Biotechnology and Bioenergy in *Dunaliella*

Mansour Shariati[1] and Mohammad Reza Hadi[2]
[1]Department of Biology, University of Isfahan, Isfahan
[2]Department of Biology, Sciences and Research Branch of Fars, Islamic Azad University,
Iran

1. Introduction

Dunaliella is a halotolerant green alga that now belongs to the phylum Chlorophyta and family Polyblepharidaceae (Avron and Ben-Amotz 1992; Garcia et al., 2007). It lacks a rigid cell wall nevertheless it can grow in aquatic environments varied salinities from 0. 5 to 5.0 M NaCl (Shariati & Hadi, 2000, Phadwal & Singh, 2003; Jahnke & White, 2003). Under stress conditions, *Dunaliella* species can accumulate significant amounts of valuable chemical matters such as carotenoids (Hosseini Tafreshi & Shariati, 2006; Hadi et al., 2008), glycerol (Hadi et al., 2008), vitamins and proteins (Ghoshal et al., 2002). The mechanism by which *Dunaliella* cells can adapt to this wide range of salt concentrations was shown to be based on the ability of the alga to change its intracellular concentration of glycerol (Raja et al., 2007). In fact, the accumulation of glycerol in this alga is regulated by external water activity rather than the specific solute effect (Shariati & Lilley, 1994). In this condition, glycerol acts as a 'compatible solute' that protects enzymes against both inactivation and inhibition (Telfer, 2002). It was also shown that both the glycerol synthesis under hypertonic conditions and its elimination under hypotonic condition are independent of protein synthesis and occur in the light or dark (Shariati & Lilley, 1994). Therefore recently, algal biotechnology has made major advances, and microalgae like *Dunaliella* sp. are cultivated for the production of carotenoids and glycerol (Hosseini Tafreshi & Shariati, 2009; Spolaore et al., 2006). The potential ability of carotenoids to act as antioxidants and immunomodulatory agents has led to more active research investigating their application in the prevention of human cancers (Chidambara Murthy et al. 2005). Also, this alga accumulates large amounts of β-carotene (up to 14% of dry weight) in conditions such as high light intensity (Coesel et al., 2008), increased temperature (Ben-Amotz, 1996; Gomez & Gonalez, 2005), high salinity (Hadi et al., 2008) and nutrient deficiency (Marin et al., 1998) such as sulfate deficiency (Aghaii & Shariati, 2007) and nitrate deficiency (Shariati & Zoofan, 2003). *Dunaliella* β-carotene is used in the food (Dufosse et al., 2005), cosmetic, and pharmaceutical industries as a colorant, antioxidant (Chidambara Murthy et al., 2005), anti-tumor agent, and heart disease preventive (Tornwall et al., 2004), in addition to its characteristic as precursor of vitamin A. Among the parameters that can significantly influence the production of biomass and β-carotene in open ponds is the selection of the area (Hosseini Tafreshi & Shariati, 2006). *Dunaliella* production plants are located in areas having a hot and dry climate with minimal cloudiness and commonly situated at, or near a suitable source of brine. The climatic conditions in most of regions of world make it into one of the most suitable areas not only for mass culture of *Dunaliella* but also for other algae (Hosseini Tafreshi & Shariati, 2009). On the other hand, microalgae biofuel is necessary for economic sustainability, because the continued use of fossil fuels is not sustainable and resources are finite (Tsukahara & Sawayama, 2005). Also, the oil productivity of many microalgae exceeds the best producing oil crops (Chisti, 2007). In particular, some species have been identified as promising producers of useful lipids for biofuels production with cells containing about 37% oils. *Dunaliella salina* is one such species, which shows lipid accumulation in response to high environmental salinities with content of the cell up to 70% (Takagi et al., 2006). Therefore, it seems that mass culture of *Dunaliella* at a commercial level in world as pilot ponds can be suitable to promote economical productions from this microalga for bioenergy via direct conversion of algal biomass to liquid fuel (Tsukahara & Sawayama, 2005).

2. Biology, morphology and taxonomy of *Dunaliella*

At first in 18 centuries, scientists thought that *Dunaliella* is same *Haematococcus*, but after in the first of 19 centuries, they reported that this genus clearly differed from *Haematococcus* and erected the new generic name *Dunaliella* (Avron and Ben-Amotz 1992). So far, twenty-eight species of *Dunaliella* are recognized (fig.1). They comprise five species (shapes of 1-5 in fig.1), all of which occur in freshwater and appear to be very rare whereas 23 species of these (shapes of 6-28 in fig.1) occur in saline environments (Avron & Ben-Amotz, 1992). The cell shape in species of *Dunaliella* varies from ellipsoid, ovoid, cylindrical and pyriform to almost spherical (Borowitzka & Siva, 2007). Cells of a given species may change shape with changing conditions, often becoming spherical under unfavorable conditions (Ben-Amotz et al., 2009). Cell size may also vary to some degree with growth conditions and light intensity (Coesel et al., 2008). The general cell organization has been studied in most detail in *D. salina*, both with the light microscope and the electron microscope. In the following survey, reference is given primarily to *D. salina*, and other species are only mentioned when major differences are evident. A rigid wall is lacking, but there is a distinctive mucilaginous cell coat. The cell coat can be visualized in the light microscope with Indian ink, whereas in thin sections, it is seen as irregular electron-dense material covering the plasmalemma (Avron & Ben-Amotz, 1992). It consists of 25 to 200 nm long fibrils and appears to be largely glycoproteic in nature. The two flagella are apically inserted, equal in length, and usually exhibit a homodynamic pattern of beating (Vismara et al., 2004). The two basal bodies are displaced against each other and carry microtubular flagellar roots. The single chloroplast occupies most of the cell body. It is cup-, dish-, or bell-shaped and has a thickened basal portion containing a pyrenoid. Anteriorly, the chloroplast is sometimes incised into several lobes. The thylakoids of the chloroplast are sometimes arranged in dense stacks of up to 10 units. Stacking of thylakoids was found to be particularly pronounced in cells grown at high light intensity and high salt concentration. Starch grains usually surround the pyrenoid, but may also be found at other places of the chloroplast. In some species (*D. salina, D. parva*), the chloroplast may also accumulate large quantities of β-carotene within oily globules in the interthylakoid spaces, so that the cells appear orange-red rather than green (Avron & Ben-Amotz, 1992). The β-carotene globules of *D. salina* were found to be composed of practically only neutral lipids, more than half of which were β-carotene. Most of the reddish forms may lose their red color when grown at low light intensities (Sarmad et al., 2006).

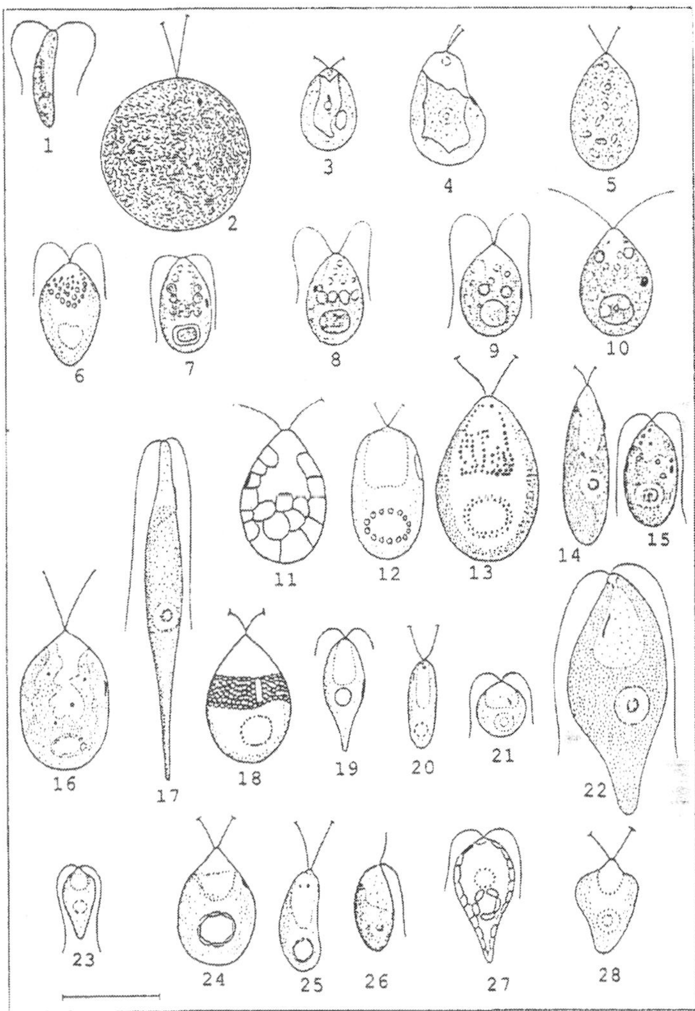

Fig. 1. 1. *D. acidophila;* 2. *D. flagellata;* 3. *D. lateralis;* 4. *D. obliqua;* 5. *D. paupera,* 6. *D. maritima,* 7. *D. polymorpha,* 8. *D. primolecta,* 9. *D. quartolecta,* 10. *D. tertiolecta,* 11. *D. parva,* 12. *D. pseudosalina,* 13. *D. salina,* 14. *D. baas-beckingii,* 15. *D. bioculata,* 16. *D. carpatica,* 17. *D. gracilis.* 18. *D. granulata,* 19. *D. media,* 20. *D minuta,* 21. *D. minutissima,* 22. *D. ruineniana,* 23. *D. terricola,* 24. *D. viridis,* 25. *D. asymmetrica,* 26. *D. jacobae,* 27. *D. peircei,* 28. *D. turcomanica.* Scale bar: 10 μm (Avron & Ben-Amotz, 1992).

The eyespot (stigma) has an anterior peripheral location in the chloroplast. In some species (especially in *D. salina),* the eyespot may be hardly visible in the light microscope (Avron & Ben-Amotz, 1992). The nucleus is generally obscured in life by a number of granules. It occupies most of the anterior part of the cell and is often surrounded by anterior lobes of the chloroplast. Ultrastructural studies show that it has a porous envelope and a single prominent nucleolus, which is often surrounded by clumped heterochromatin. Mitochondrial profiles can be seen in various parts of the cell in thin sections. It is showed that the number and size of mitochondria may vary among cells at different stages of growth. Dictyosomes (Golgi bodies) occur in numbers of 2 to 4, each consisting of 10 to 15 cisternae. The endoplasmic reticulum (ER) typically underlies the plasmalemma over most parts of the cell. During hyperosmotic stress periods, there may be marked increases in ER. It appears that ER serves as a temporary reservoir for membrane material in temporary excess during stress periods when major cellular compartments shrink. Vacuoles of different types occur in *Dunaliella*. Vacuoles containing portions of membrane and vesicles as well as granular or thread-like material are often prominent ultrastructural constituents of the cytoplasm in *Dunaliella*. Large lipid globules or vacuoles containing smaller lipid globules similar to those of the chloroplast may also occur at various places in the cell. Asexual cysts may be formed under extreme conditions such as drastic dilution of the medium or drying up of the environment. Sexual reproduction is by isogamy, with gametic fusion proceeding in a manner similar to that in *Chlamydomonas*. The gametes have the same size and the same structural features as growing cells of the same species. Several species of *Dunaliella* appear to be homothallic, whereas *D. salina* has been reported to be heterothallic. The zygote is green or red and is surrounded by a thick, smooth wall. After a resting stage, the zygote nucleus divides forming up to 32 cells, which are liberated through a rupture in the mother cell wall. Meiosis takes place during the germination of the zygote. Cell morphology may be influenced to some degree by environmental growth conditions. Conspicuous changes may occur between logarithmic and stationary growth phases. It has also been clearly shown that salt concentration, light intensity, and temperature may have some effects on, e.g., thylakoid structure, appearance of pyrenoid, and proliferation of the endoplasmic reticulum. Originally, *Dunaliella* and other wall-less green flagellates were all classified in the Polyblepharidaceae within the Volvocales. Subsequently, detailed studies on many of these green flagellates led to the reclassification of certain genera in other taxa. It is placed *Dunaliella* in a separate order of the Chlorophyceae (Dunaliellales) (Borowitzka & Siva, 2007). Except for the lack of a rigid wall, *Dunaliella* also has many characteristics in common with members of the Chlamydomonadales (Avron & Ben-Amotz, 1992).

3. Osmoregulation and glycerol production

The genus *Dunaliella* contains several species which stand out as being the only eukaryotic and photosynthetic organisms which are able to grow in media containing an extremely wide range of salt concentrations, from 0.05 to 5.5 M NaCl (Avron & Ben-Amotz, 1992), but, optimal growth is different in various species and in order to Hadi et al., (2008) reported that optimal growth of *D. salina* (Iranian strain) was obtained at 2 M NaCl (Fig. 2 and Fig. 6 partition of D). *Dunaliella* was already recognized as a major constituent of saline lakes. It has been shown to be present in all natural hypersaline environments. In contrast to other green algae, cells of the genus *Dunaliella* do not contain a rigid cell wall (Avron & Ben-Amotz, 1992) but have a thin elastic plasma membrane that responds rapidly to changes in osmotic pressure by changes in cell volume (Shariati & Lilley, 1994). Therefore, *Dunaliella* can physically withstand three- to fourfold increases or decreases in osmotic pressure, shrinking or swelling in response, respectively (that know as osmoregulation). However, much larger changes will lead to cell-bursting during a hypoosmotic stress, and to an irreversible shrinkage under hyperosmotic stress (Avron & Ben-Amotz, 1992).

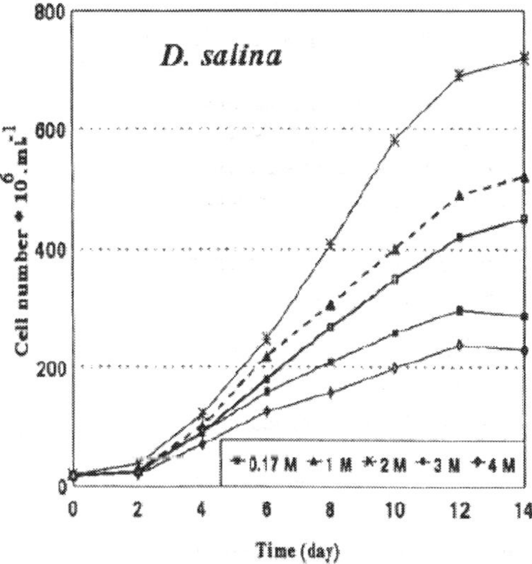

Fig. 2. Growth of *Dunaliella salina* in media containing different NaCl concentrations. The algae can grow in media containing an extremely wide range of salt concentration, from 0.17 M to 4.0 M NaCl (Hadi et al., 2008).

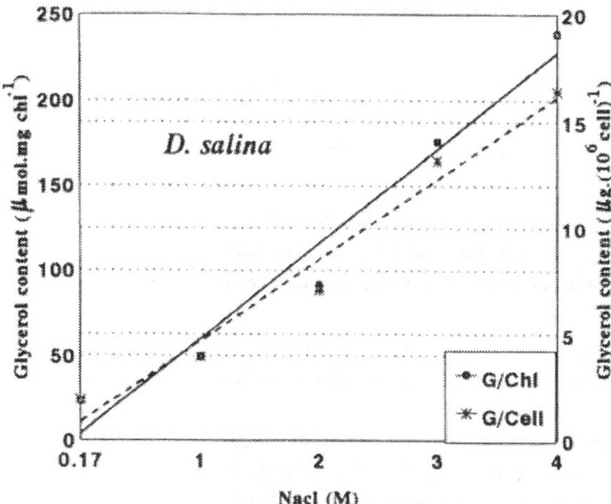

Fig. 3. Intracellular glycerol content as a function of the extracellular NaCl oncentration in *Dunaliella salina*. Intercellular glycerol and chlorophyll content was determined in cells grown in media containing the indicated salt concentration (Hadi et al., 2008).

It seems that in *Dunaliella*, osmolyte of main for balancing the extracellular osmotic stress is glycerol, because its intracellular concentration intensively accumulates in response to any osmotic stress such as high medium salinity (Shariati, 2003; Hadi et al., 2008). Also, it shown that its intracellular concentration was linearly related to the medium osmotic pressure (Fig. 3). When cells of *Dunaliella* grown at high salinity, the intracellular glycerol concentration exceeds 50% and is sufficient to account for essentially all of the osmotic pressure required to balance the extracellular osmolarity (Avron & Ben-Amotz, 1992). Glycerol production in *D. salina* grown in different salt concentrations is shown in Fig. 3. Also, glycerol production in *D. salina* grown under saltshock conditions (by transfer of algae from medium containing 1 to 3 M NaCl) is shown in Fig. 4. However glycerol accumulates in *Dunaliella* in response to osmotic stress but it seems that physiological role and function of glycerol may be different in each algae. Another possible physiological role of glycerol is that of a compatible solute, that is, a substance that at high concentrations protects enzyme activity.

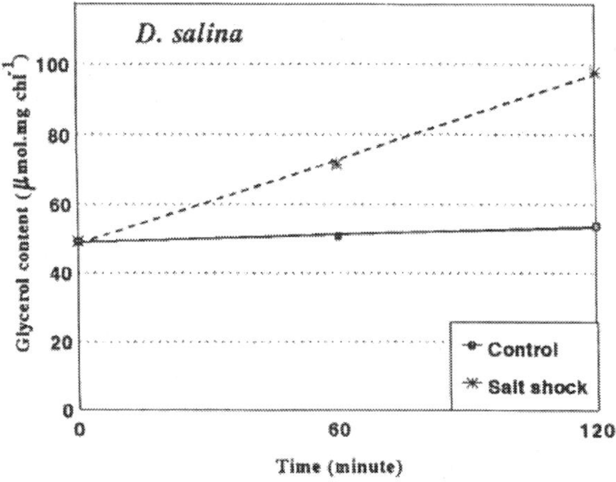

Fig. 4. The response in glycerol production of *Dunaliella salina* grown in 1.0 M NaCl and subjected to dense stress by transfer to 3.0 M NaCl. Time was measured from the moment of transfer to the higher salt concentration (Hadi et al., 2008)

Glycerol in *Dunaliella* parva can accumulate to the maximum content under saturated NaCl conditions, which is about 8 Mor 55% of cell weight (Alkayal et al., 2010). When cells of *Dunaliella* expose an osmotic shock, the initial reaction is physical in nature. Within seconds water enters or leaves the cell osmotically, causing marked volume changes which bring the cells back into osmotic equilibrium with the medium. Thereafter, depending on the direction and extent of the osmotic shock and on the metabolic conditions of the cells, glycerol is synthesized or eliminated via temperature-dependent enzymatic pathways, accompanied by water reentry or efflux, respectively, so that the cells regain approximately their original volume (Avron & Ben-Amotz, 1992). Glycerol is produced in *Dunaliella* either by photosynthetic CO_2 fixation or by starch degradation. The contribution of these metabolic pathways to glycerol synthesis depends on the availability of light, the starch reserve pool, and the size of the salt stress. In the dark, *Dunaliella* produces glycerol exclusively by degradation of starch, and the capacity of the cells to recover from hyperosmotic shock depends on their starch reserve pools. Hyperosmotic shock in the light greatly stimulates the rate of glycerol production and in parallel enhances starch degradation, indicating that starch degradation also has a significant contribution to glycerol production in the light. This complex interplay between photosynthesis and starch degradation probably results from the excessive demand for glycerol synthesis on one hand, and from the inhibition of photosynthesis by hyperosmotic shock on the other hand. Starch synthesis is also strongly inhibited following hyperosmotic shock. Hypoosmotic shock induces in *Dunaliella* a decrease in glycerol content and a parallel increase in starch content, indicating metabolic conversion of glycerol to starch. Hypoosmotic shocks also induce a transient inhibition of photosynthesis and a substantial, but not complete, inhibition of glycerol synthesis. These and other observations suggest that *Dunaliella* utilizes a dynamic interconversion between glycerol and starch, the two major carbon pools, as well as photosynthesis, to meet the osmotic requirements which are set by the external salt concentration (Avron & Ben-Amotz, 1992). Also, Chen & Jiang (2009) reported that changes of shape and volume of *D. salina* cell cultured chronically at various salinities were minor, but when the salinity was changed rapidly, the variations of cell shape and cell volume of *D. salina* were significant, which were recovered basically after 2 h except treating by high salinity. In addition, they indicated that, it was found some lipid globules in the surface of *D. salina* cells when the salinity increased from 2.0 to 4.0–5.0 M NaCl rapidly. In this researchs when *D. salina* was cultured chronically at various salinities, the accumulation of single cell glycerol increased with increased salinity, and *D. salina* also could rapidly decrease or increase single cell glycerol contents to adapt to hypoosmotic or hyperosmotic shock (Chen et al., 2011). Biosynthesis of glycerol in *Dunaliella* is carried out mostly within the chloroplast and partly in the cytosol and may be broadly divided into the production of dihydroxyacetone phosphate and its conversion to glycerol, as is summarized in Fig. 5. Four enzymes catalyze the interconversions between dihydroxyacetone phosphate and glycerol in *Dunaliella*: two reversible steps, namely a glycerophosphate dehydrogenase and an NADPH-specific dihydroxyacetone reductose, and two irreversible steps, namely a glycerol phosphate phosphatase (GPase) and a dihydroxyacetone kinase. Dihydroxyacetone reductose which catalyzes the interconversion of glycerol to dihydroxyacetone, is $NADP^+$-specific, has an exceptionally low affinity for glycerol, and an almost absolute specificity towards both glycerol and dihydroxyacetone phosphate. Dihydroxyacetone kinase, mediating the phosphorylation of dihydroxyacetone by ATP, has an absolute specificity towards dihydroxyacetone. The chloroplastic glycerophosphate dehydrogenase, catalyzing conversion of dihydroxyacetone phosphate to glycerol-3-phosphate with either NADH or NADPH, has unusual catalytic and regulatory properties and is distinct from the cytosolic isoenzyme in its stability towards detergents and high salt. These enzymes compose the so-called glycerol cycle (Avron & Ben-Amotz, 1992).

4. β-carotene

β-carotenes are aliphatic-alicyclic components that composed of five carbon isoprene groups and belong to carotenoid components. Those containing only hydrogen and carbon. β-carotenes are the pigments included in many plants and algae. Some of plants and red-orange foods owe their typical color principally to β-carotenes. β-carotenes are very widely distributed in nature; their various roles include provitamin A activity, absorption of light energy, triplet chlorophyll and singlet oxygen quenching, antioxidation activity, oxygen

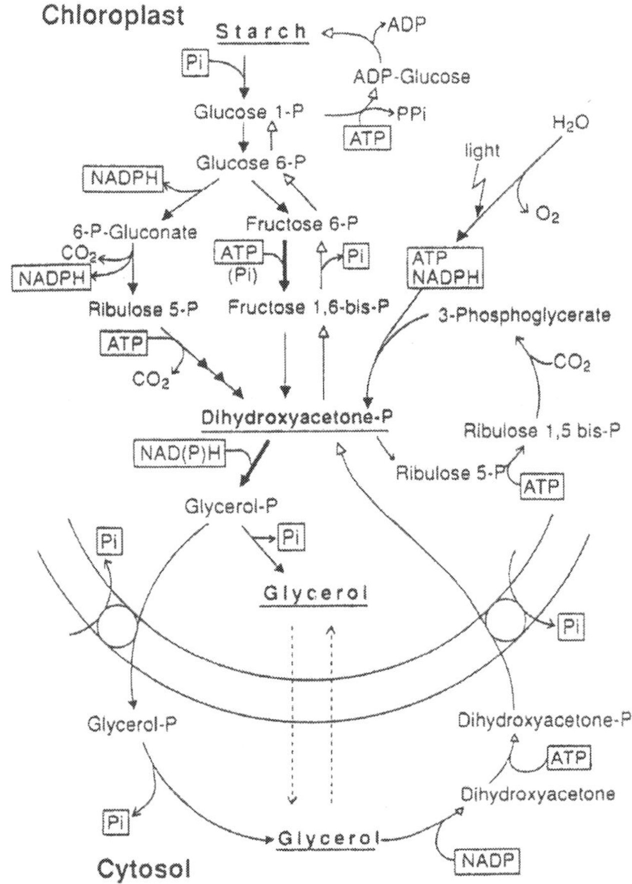

Fig. 5. Biosynthesis pathway of glycerol in *Dunaliella* (Avron & Ben-Amotz, 1992).

transport, and general coloration of many different organisms. β-carotene has the formula of $C_{40}H_{56}$, a molecular weight of 536.9, eleven conjugated double bonds and a typical violet-red crystalline color and it forms a solution in oil. β-carotene shows an absorption maxima in petroleum ether at 453 nm and 481 nm. The content of β-carotene in plants varies considerably in the range of 0.01 to 10 mg/100 g. The most common β-carotene-rich plants are green leafy plants such as parsley, spinach, and broccoli; yellow-orange fruits such as mangos, peaches, and red palm; and certain vegetables such as carrots, sweet potatoes, and pumpkin. A few microbial organisms accumulate β-carotene to a high extent. The fungus *Phycomyces blakesleanus* and the yeast *Rhodotortila* are examples of relatively high β-carotene content of 5 and 0.5 mg per g dry weight, respectively. *Dunaliella* has been shown to be capable of producing extraordinarily large amounts of β-carotene within oily globules in the interthylakoid spaces of the chloroplast. The β-carotene-rich *Dunaliella* strains are widely distributed in salt water bodies which contain more than 10% salt and most dominantly in high salt habitats approaching NaCl saturation (Shariati & Maddadkar Haghjoo, 1998). The orange-reddish color of many salt-lanes in high light intensity ecological niches is usually due to the color of the β-carotene rich *Dunaliella* (Fig. 6 partition B). Under non-inducing, non-accumulating conditions (optimum conditions, 1-2 M NaCl), *D. salina* is green (Table 1 and Fig. 6 partition of D) and contain only about 0.3 % β-carotene, similar to the content in plant leaves and other algae. Following induction and growth under appropriate cultivation conditions, the β-carotene is accumulated within oily globules in the interthylakoid space of the chloroplast to more than 10% of the algal dry weight, which is the highest content of β-carotene of any known alga, plant, or other microorganism. The extent of β-carotene accumulation and the rate of synthesis depend on certain physiological growth parameters, namely light intensity, salt concentration, temperature (Madadkar Haghjou & Shariati, 2007; Madadkar Haghjou et al., 2009), and nutrient deficiency. Indeed, the higher the stress intensity and as a result the slower the growth rate of the alga, the greater is the total amount of the light absorbed by the cell during one division cycle. This situation can lead to higher accumulation of β-carotene per cell. However, these conditions at the same time decrease the cell number per unit culture volume by affecting cell viability. Therefore, it is recommended by one group of authors that adjusting light and salinity likely is one of the best strategies to achieve optimal b-carotene production in mass cultures of *D. salina* (Marin et al. 1998). The Iranian strain of *D. salina* grown in medium containing 1-2 M NaCl appeared green in color (Table 1 and Fig. 6 partition of D), while at 4 M NaCl an orangered color was observed (Table 1 and Fig. 6 partition of A and C). The high accumulation of carotenoids is the main reason for the change to this orange-red color in *D. salina* at 4 M NaCl.

Alga Color	Car./Chl. Ratio	Total Chl. (µg ml-1)	Total Car. (µg ml-1)	NaCl (M)
Green	0.39	3.25	1.25	0.17
Green	0.61	4.17	2.57	1
Green	0.82	6.01	4.57	2
Orange	1.53	2.14	3.32	3
Orange-red	2.58	1.10	2.84	4

Table 1. Total carotenoid (Car.), total chlorophyll (Chl.) contents, carotenoid/chlorophyll ratio, and alga color of *Dunaliella salina* grown at the indicated concentrations of NaCl (M) at one week following inoculation (Hadi et al., 2008).

The amount of carotenoid accumulated in *D. salina* after 7 days of growth are depicted in Table 1. *D. salina* had the highest carotenoid/chlorophyll ratio (2.58) at 4 M NaCl (Table 1). Prieto et al., (2011) reported that highest carotenoid production was achieved with this culture system operated following the two-stage strategy. Also they indicated that closed tubular photo-bioreactor provided the highest carotenoid contents (10% of dry weight) in

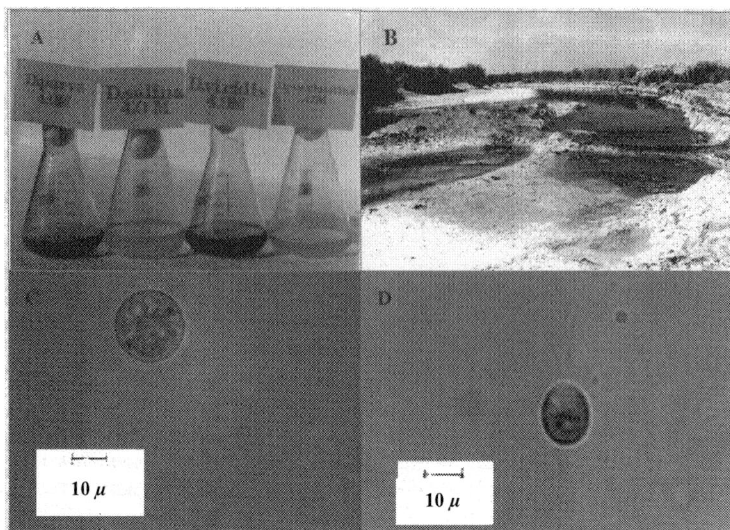

Fig. 6. Four species of *D. Parva, D. salina, D. viridis* and *D. pseudosalina* grown in 4.0 M NaCl, the liquid medium of *D. salina, D. Parva* and *D. pseudosalina* cause cells have rich β-carotene so, those are about orange colour (A), the pools of salt evaporation in Gave-Khooni Salt Marsh, Iran that cause present *D. salina* containing β-carotene have orange colour (B), shape of *D. salina* grown in 4 M NaCl (stress conditions) that cell have rich β-carotene so, it is orange colour (C) and shape of *D. salina* grown in 2 M NaCl at optimum growth that cell have rich chlorophyll so, it is green colour (D) (Hadi, 1996).

Dunaliella biomass and β-carotene abundance (90% of total carotenoids) as well as the highest 9-cis to all-trans β-carotene isomer ratio (Prieto et al., 2011). In addition, Hosseini Tafreshi & Shariati (2006) reported that Iranian strain G of *Dunaliella salina* had the highest potential for β-carotene accumulation and suitable for outdoor cultivation (Fig. 7). The higher the light intensity and the slower the growth rate of the alga, the higher the cellular β-carotene content. Highest β-carotene content per cell can be obtained by exposing the nitrogen-deficient cells to high light intensity in a short period of 1 to 2 d. The effect of light quality on carotenogenesis indicates that β-carotene biosynthesis and accumulation in *D. bardawil* is independent of light quality within the photo-synthetically active radiation region. The commercial production of large quantities of the β-carotene-rich *Dunaliella* provides natural β-carotene for different nutritional, dietary, and clinical studies for evaluation in comparison with the synthetic β-carotene. A few possible reasons for the increased synthesis of β-carotene and its possible function in *Dunaliella* have been suggested and investigated (Avron & Ben-Amotz, 1992):

1. Carbon storage: β-carotene is accumulated and stored as an extra-photosynthetic product for later use under limited growth rate. This hypothesis was studied by a "carbon sink" utilization which revealed that the *Dunaliella* cells do not consume the accumulated β - carotene on transfer to darkness or to a CO_2-free medium in the light.

2. Singlet oxygen quencher: β-carotene protects against chlorophyll catalyzed singlet oxygen and possibly other excited chlorophyll damaging agents. This hypothesis was analyzed by

electron micrographs of the β-carotene-rich *Dunaliella*. Due to the large distance between the β-carotene globules and the thylakoid located chlorophyll, and due to the short lifetime of these damaging compounds, the massively accumulated β-carotene cannot be effective through this mechanism.

3. Absorption effect: β-carotene protects the cell against injury by high intensity radiation under limited growth conditions by acting as a screen to absorb excess radiation. This hypothesis is well accepted through the observation that the β-carotene-rich *Dunaliella* shows maximal photoprotection against high irradiation with blue light, and minimal photoprotection against light irradiation with red light. Strains unable to accumulate β-carotene and the β-carotene-poor *D. bardawil* show low photo-protection and die when exposed to the extreme irradiation while the β-carotene-rich *D. bardawil* survives and flourishes. Thus, the function of β-carotene in *Dunaliella* seems to be photo-protection through its absorption properties. The oily nature of the globules at the periphery of the cup shaped chloroplast functions structurally most efficiently for this purpose.

In mammals, the structure of β-carotene possessing two β-ionone rings, one at either end, and a long one plane polyene chain, provides the optimal structure for enzymatic cleavage to two molecules of vitamin A and thus, the highest efficiency as a vitamin A precursor. Many factors affect the biological activity of β-carotene metabolism and its potency as a vitamin A precursor, namely, physical form, oily solubility, isomerization, dietary fat level, state of oxidation, presence of dietary antioxidants, the diet composition, animal studied, age, sex, and many more. Acetyl-CoA, and in a few organisms valine and leucine, are considered the starting compounds in the biosynthesis of β-carotenes. These early steps of the pathway, leading to C5 isoprenoid units and the subsequent prenyl diphosphate intermediates, are common to all classes of terpenoids. Only the later stages after geranylgeranyl, diphosphate (GGDP) and after phytoene, which is a product of condensation of two molecules of GGDP, are unique to the formation of β-carotenes. In most order, the general biosynthesis of β-carotene can be divided into four stages (Fig. 8): (1) formation of GGDP from mevalonic acid; (2) condensation to form phytoene; (3) desaturation of phytoene to lycopene; and (4) cyclization of lycopene to form β-carotene (Ye et al., 2008). The ability of β-carotene in *Dunaliella* to protect the algae from death by high blue irradiation was utilized to select mutants of *D. bardawil* which accumulate a higher content of β-carotene. Both the production rate of phytoene and the conversion rate of phytoene to lycopene and β-carotene are accelerated in the isolated mutants. By the analysis of the effect of protein synthesis inhibitors, it was suggested that the mutants are affected in the mechanism which regulates the activation of the carotene biosynthetic pathway, most probably at the metabolic steps which precede GGPP and which allow enhanced production of both β-carotene and chlorophyll, and possibly at a later site after phytofluene. Recent epidemiological and oncological studies suggest that normal to high levels of β-carotene in the body may protect it against cancer. Humans and animals fed a diet high in carotenoid-rich vegetables and fruits and who maintain higher than average levels of serum β-carotene have a lower incidence of several types of cancer. The interest in a natural source of β-carotene is increasing with the buildup of information relating carotenoids to preventive medicine. Of special interest in this regard is the observation that natural β-carotene, as found in *Dunaliella* and in most fruits and vegetables, contains a mixture of all-trans β-carotene and 9-cis β-carotene together with a few other stereoisomers, as discussed above. The requirement for better absorbed β-carotene for disease prevention purposes has created a new market of high commercial potential for the *Dunaliella* β-carotene stereoisomeric mixture. Large scale cultivation of *Dunaliella* for β-carotene production is based on autotrophic growth in media containing inorganic nutrients with carbon dioxide as a carbon source. Commercial attempts are being made to apply the basic biological information of β-carotene optimization to mass production of β-carotene in outdoor ponds of *Dunaliella* in areas located where the solar light output is maximal and the high salt concentration can eliminate foreign grazers. Two modes of cultivation are being used in large-scale bioreactors of *Dunaliella*. In the more common, the intensive mode, attempts are made to control all factors affecting cell growth and chemistry. The growth limitation is usually provided by controlling the availability of the nitrogen supply on continuously grown induced cells of *Dunaliella*. In the other mode, the extensive growth, *Dunaliella* grows very slowly in nearly saturated brine where the high salt concentration is used to control consistent production of β-carotene. In addition, since commercial activity in the microalgae extractable chemical sector is currently limited to two main products, *Dunaliella*-derived carotenoid pigments as a human nutritional supplement and genetic modification of *Dunaliella* strains, transgenic strains should also be explored (Jin & Melis, 2003). It seems that β-carotene accumulation protects cells against the deleterious effects of high intensity irradiation by absorbing light in the blue region of the spectrum (Ben-Amotz, 1993).

Ben-Amotz (1995) cultivated *D. salina* by a new two-phase growth strategy for β-carotene production. In this mode, the cells were firstly cultivated in small nursery ponds to attain optimal biomass and then transferred to large production ponds and diluted by adding medium deficient in nitrate and / or higher concentration of salt to approximately one third for carotenoid induction (Hosseini Tafreshi and Shariati 2006). At much higher aerial density the amount of light absorbed by a cell is low (due to the shading effect) and hence the resident time of the cell in ponds required to reach the maximal β-carotene content is longer. Consequently, optimization of the aerial density in which the maximal biomass and carotenoid content would be obtained is an important step both in ponds and photobioreactors. Garcia-Gonzalez et al. (2003) reported that the optimal values of population density, which yield the highest output rate in semi-continuous regime, were between 0.7 and 0.9 × 10⁶ cell ml⁻¹. The operation parameters of culture systems like mixing rate, depth of the culture, etc. can also affect the output rate and will be considered later. From a biotechnological point of view, it is desirable to increase the 9-cis to all-trans β-carotene ratio in the cell because 9-cis isomer has shown to be a better antioxidative and cancer-preventive than another (Chidambara Murthy et al. 2005). The information about the conditions that trigger synthesis of 9-cis isomer as well as β-carotene accumulation is also controversial. Garcia-Gonzalez et al. (2005) also found that a suitable approach for the production of high quality β-carotene with high 9-cis isomer content is the cultivation of *Dunaliella* in closed tubular photobioreactors, which have low mutual shading. Exposure to low temperature in the range of 10–15 °C could also induce the 9-cis isomer synthesis in *D. bardawil* (Ben-Amotz, 1996). Consequently, there is a great physiological variability in response to different carotene induction factors among different strains of *D. salina* (Hosseini Tafreshi & Shariati, 2009). The intrinsic response of each strain to each inductive factor alongside the complex interactions among various environmental conditions demonstrate that there is no predictable unique condition for reaching the maximum carotenoid and 9-cis β-carotene contents per unit time and per unit volume (Hosseini Tafreshi & Shariati, 2009). The optimization procedure should be done by testing the best strains and the most effective strategies under optimal conditions. Recently, Mojaat et al. (2008) studied the effects of Fe^{2+} ions and organic carbon source on growth and carotenogenesis of *Dunaliella salina*. In their study, a significant increase in β-carotene contents per cell was observed, with a maximum value of 70 pg cell^{-1} when the culture was supplemented with acetate and FeSO4. The approach might be a good alternative method for production of carotenoids by alga in photobioreactors after optimization (Hosseini Tafreshi & Shariati, 2006).

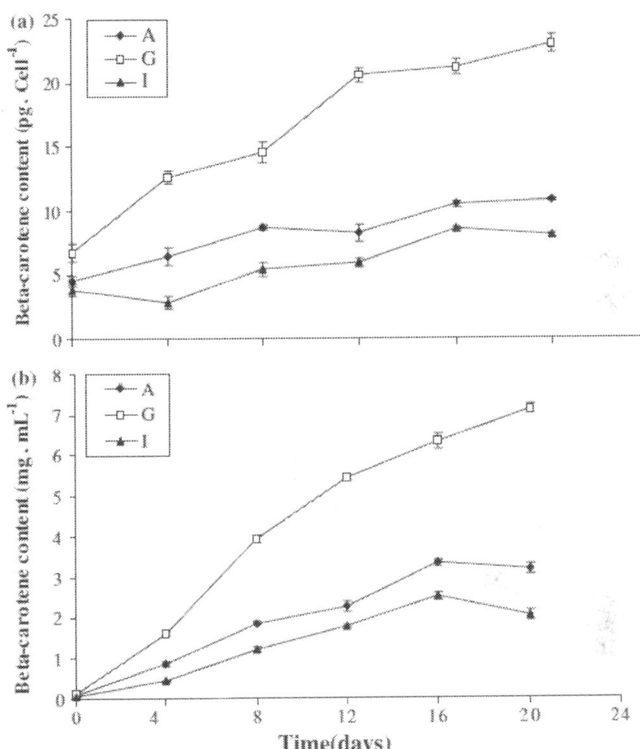

Fig. 7. Quantitative changes in β-carotene as a function of time per cell (a) and per unit volume (b) of three strains of *Dunaliella salina* (A, G and I) in open ponds during the stage 2 (nutrient-poor medium, containing 2.5 M NaCl). The values are means of three replicates ± SD (Hosseini Tafreshi and Shariati 2006).

5. Production of biofuel from *Dunaliella*

The global economy literally runs on energy. An economic growth combined with a rising population has led to a steady increase in the global energy demands. If the governments around the world stick to current policies, the world will need almost 60% more energy in 2030 than today (IEA, 2007; Vishwanath et al., 2008). The oil productivity of many microalgae exceeds the best producing oil crops (Vishwanath et al., 2008). Past research in the use of hydrothermal technology for direct liquefaction of biomass was very active. Only

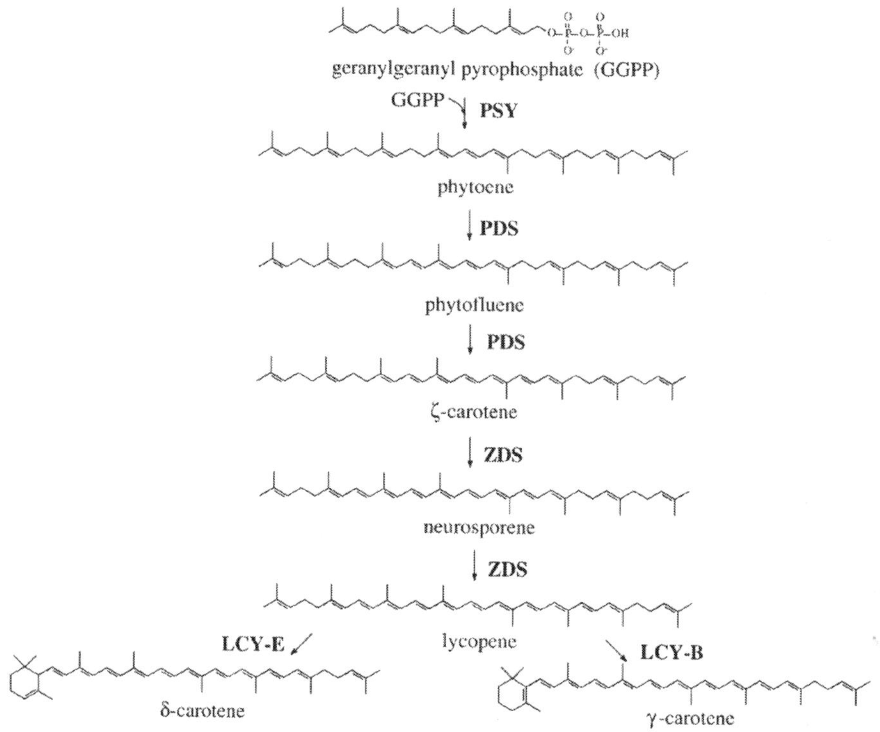

Fig. 8. A hypothetic pathway for carotenogenesis in *Dunaliella*. Enzymes for the relative conversions are in bold. PSY: phytoene synthase; PDS: phytoene desaturase; ZDS: ζ-carotene desaturase; LYC-E: lycopene ε-cyclase; LYC-B: lycopene β-cyclase (Ye et al., 2008).

a few of them, however, used algal biomass as feedstock for the technology. Minowa et al., (1995) report an oil yield of about 37% (organic basis) by direct hydrothermal liquefaction at around 300°C and 10 MPa from *Dunaliella tertiolecta* with a moisture content of 78.4 wt%. The oil obtained at a reaction temperature of 340°C and holding time of 60 min had a viscosity of 150–330 mPas and a calorific value of 36 kJ g−1, comparable to those of fuel oil. The liquefaction technique was concluded to be a net energy producer from the energy balance (Vishwanath et al., 2008).

The key for large scale production of biofuels is to grow suitable biomass species in an integrated biomass production conversion system (IBPCS) at costs that enable the overall system to be operated at a profit. The illustration in Figure 9 is a conceptual model for integrated biomass production (Klass, 1997) that can be adopted for microalgal biodiesel production.

The production of microalgal biodiesel requires large quantities of algal biomass. Most of algal species are obligate phototrophs and thus require light for their growth. Several cultivation technologies that are used for production microalgal biomass have been developed by researchers and commercial producers. The phototropic microalgae are most

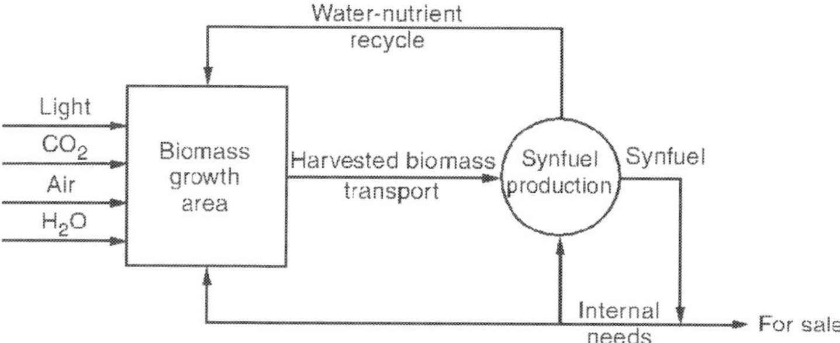

Fig. 9. A conceptual model for integrated biomass production and conversion integration system (Vishwanath et al., 2008).

commonly grown in open ponds and photobioreactors (Patil et al., 2005). The open pond cultures are economically more favorable, but raise the issues of land use cost, water availability, and appropriate climatic conditions. Further, there is the problem of contamination by fungi, bacteria and protozoa and competition by other microalgae. Photobioreactors offer a closed culture environment, which is protected from direct fallout, relatively safe from invading microorganisms, where temperatures are controlled with an enhanced CO_2 fixation that is bubbled through culture medium. This technology is relatively expensive compared to the open ponds because of the infrastructure costs. An ideal biomass production system should use the freely available sunlight. It is reported the best annual averaged productivity of open ponds was about 24 g^{-1} dry weight m^{-2} d^{-1} (Weisz, 2004). A productivity of 100 g^{-1} dry weight m^{-2} d^{-1} was achieved in simple 300 l culture systems (Patil et al., 2005). This level has been viewed as deriving from the light saturation effect. The light requirement coupled with high extinction coefficient of chlorophyll in algae has necessitated the design and development of novel system for large scale growth. Experiments have also elucidated that algal biomass production can be boosted by the flashing light effect (Matthijs et al., 1996), namely by better matching photon input rate to the limiting steps of photosynthesis. Indeed, the best annual averaged productivity has been achieved in closed bioreactors. Tridici (2004) has reviewed mass production in photobioreactors. Many different designs of photobioreactor have been developed, but a tubular photobioreactor seems to be most satisfactory for producing algal biomass on the scale needed for biofuel production. Closed, controlled, indoor algal photobioreactors driven by artificial light are already economical for special high-value products such as pharmaceuticals, which can be combined with production of biodiesel to reduce the cost (Vishwanath et al., 2008).

6. *Dunaliella*-based bioenergy options

In recent years, biofuel production from algae has attracted the most attention among other possible products. This can be explained by the global concerns over depleting fossil fuel reserves and climate change. Furthermore, increasing energy access and energy security are seen as key actions for reducing poverty thus contributing to the Millennium Development Goals. Access to modern energy services such as electricity or liquid fuels is a basic requirement to improve living standards. One of the steps taken to increase access and reduce fossil fuel dependency is the production of biofuels, especially because they are currently the only short-term alternative to fossil fuels for transportation, and so until the advent of electromobility. The so-called first generation biofuels are produced from agricultural feedstocks that can also be used as food or feed purposes. The possible competition between food and fuel makes it impossible to produce enough first generation biofuel to offset a large percentage of the total fuel consumption for transportation. As opposed to land-based biofuels produced from agricultural feedstocks, cultivation of algae for biofuel does not necessarily use agricultural land and requires only negligible amounts of freshwater (if any), and therefore competes less with agriculture than first generation biofuels. Combined with the promise of high productivity, direct combustion gas utilization, potential wastewater treatment, year-round production, biochemical content of algae and chemical conditions of their oil content can be influenced by changing cultivation conditions. Since they do not need herbicides and pesticides, algae appear to be a high potential feedstock for biofuel production that could potentially avoid the aforementioned problems. On the other hand, microalgae, as opposed to most plants, lack heavy supporting structures and anchorage organs which pose some technical limitations to their harvesting. The real advantage of microalgae over plants lies in their metabolic flexibility, which offers the possibility of modification of their biochemical pathways (e.g. towards protein, carbohydrate or oil synthesis) and cellular composition. Algae-based biofuels have an enormous market potential, can displace imports of fossil fuels from other countries (hence reduce a country's dependence), and is one of the new, sustainable technologies which can count on ever-increasing political and consumer support. The reasons for investigating algae as a biofuel feedstock are strong but these reasons also apply to other products that can be produced from algae. There are many products in the agricultural, chemical or food industry that could be produced using more sustainable inputs and which can be produced locally with a lower impact on natural resources. Co-producing some of these products together with biofuels, can make the process economically viable, less dependent from imports and fossil fuels, locally self sufficient and expected to generate new jobs, with a positive effect on the overall sustainability (Mata et al. 2010). A wave of renewed interest in algae cultivation has developed over the last few years' scientific research, commercialization initiatives and media coverage have exploded since 2007. In most cases, the main driver of the interest in algae is its high potential as a renewable energy source, mainly algae-based biofuels (ABB) for the transport sector. In 2009 FAO published a report detailing various options for algae cultivation, multiple biofuels that can be produced and the environmental benefits and potential threats associated with ABB production. One of the main conclusions of this report is that the economic feasibility of producing a (single) low-price commodity like biofuels from algae is not realistic, at least in the short term. This chapter summarizes some of the technology key findings of the aforementioned report and gives a brief overview of how algae can be cultivated and which biofuels can be produced. The following chapter investigates which other products can be produced from algae, and tries to asses the viability of co-production with bioenergy. Ethanol is commonly produced from starch-containing feedstocks; some algae have been reported to contain over 50% of starch. Algal cell walls consist of polysaccharides which can be used as a feedstock in a process similar to cellulosic ethanol production, with the added advantage that algae rarely contain lignin and their polysaccharides, are generally more easily broken down than woody biomass. Coproducts can potentially be derived from the non-carbohydrate part of the algal biomass. There are a variety of ways to produce biofuel with algae. Figure 10 provides an overview of the options, which are explained in detail in FAO (2009). In this section only the requirements of the algal biomass needed to produce various biofuels are briefly discussed in order to facilitate the selection of different coproduction options further in the report.

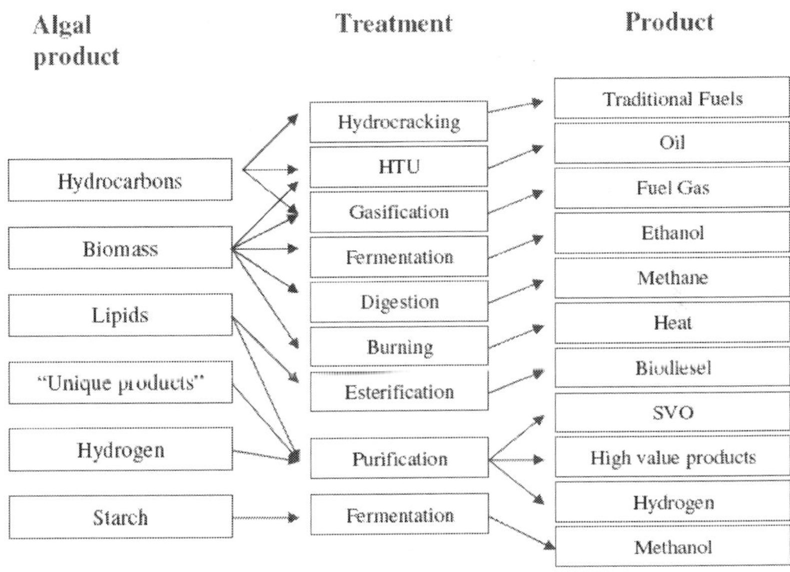

Fig. 10. Overview of algae-to-energy options (FAO, 2009).

Biodiesel production from algal oils has received most attention since algae can contain potentially over 80% total lipids, (while rapeseed plants, for instance, contain about 6% lipids). Under normal growth conditions the lipid concentration is lower (<40%) and high oil content is always associated with very low yields. The various lipids production can be stimulated under stress conditions, e.g. insufficient nitrogenavailability. Under such conditions, biomass production is not optimal though, reducing the non-lipid part of the biomass that can be further used as a source for co-products. Biomass has attracted more and more interests as an alternative energy source, since it is a renewable and environmentally friendly source and it fixes CO_2 in the atmosphere through photosynthesis (Yang et al., 2011). Biomass resources mainly include agricultural crops and their waste byproducts, forestry products, marine products and wastes. Among these biomass resources, microalgae are seen as being a future source of third generation bio-fuels and chemicals, due to their fast growth rate in an aquatic medium and high lipid content. Furthermore, microalgae are not lignocellulosic in composition but are comprised of proteins, lipids, non-cellulosic carbohydrates, and nucleic acids, which can be decomposed and hydrolyzed more easily than lignocellulosic biomass. Liquefaction of biomass had received extensive research. Microalgae as feedstock were seldom employed to liquefy. Yang et al., (2004) had investigated the liquefaction of Microcystis viridis using the same liquid alkali catalyst (sodium carbonate), and obtained the maximum oil yields of 33–40 wt%. Zou et al., (2009) described the liquefaction of microalgae in the presence of liquid acid catalyst. Ross et al. studied the hydrothermal processing of microalgae using alkali and organic acids, and the yields of biocrude on an organic basis were higher in the presence of organic acids compared with alkali catalysts (Ross et al., 2010). There was a high demand on the liquefaction equipment to avoid the corrosion effects of liquid alkali and acid catalysts. In order to solve these problems, conventional homogeneous catalysts were expected to be replaced in the near future by environmentally friendly heterogeneous catalysts (Perego and Bianchi, 2010). Many efforts had been devoted to the search of solid acid and alkali catalysts in the transesterification of vegetable oils to bio-diesel (Singh and Fernando, 2008), and the upgrading of bio-oils (Peng et al., 2009). The solid heterogeneous catalysts were almost fully recovered from reaction products, were normally easy and safe to dispose of, more selective (Yang et al., 2011). In the industry, microalgae have been used as source for a wide variety of practical and potential metabolic products, such as food supplements, pharmacological substances, lipids, enzymes, biomass, polymers, toxins, pigments, tertiary wastewater treatment, and "green energy". Microalgae are also important in aquiculture as they are a source of nutrients and have great importance in production of oxygen, in consumption of carbon dioxide, and in consumption of nitrogen-based compounds such as ammonium. The most common procedure for cultivation of microalgae is autotrophic growth. Because all microalgae are photosynthetic, and many microalgae are especially efficient solar energy convertors, microalgae are cultivated in illuminated environments naturally or artificially. Under autotrophic cultivation, the cells harvest light energy and use CO_2 as a carbon source. The main driving force to grow microalgae commercially is harvesting metabolic products, feed for marine and terrestrial organisms, food supplements for humans, or to use the microalgae for environmental processes, such as wastewater treatment, fertilization of soils, biofuels, and phytoremediation of toxic wastes (Perez-Garcia et al., 2011). Microalgae have been recognized as a promising alternative source for oil production. Several species of microalgae can be induced to overproduce specific lipids and fatty acids through relative simple manipulations of the physical and chemical properties of their culture medium. By manipulating fatty acid content, microalgae represent a significant source of unusual and valuable lipids and fatty acids for numerous industrial applications. Microalgae can accumulate substantial amounts of lipids – up to 50% of dry cell weight in certain species. Many microalgae species can grow in brackish water or seawater, thereby avoiding demand for fresh water, a limited resource in many parts of the world. Several species grow very fast; doubling their mass in 24 h. Bioprospection of strains is important to select the best strains that can produce higher amounts of desired metabolic products. Several studies have evaluated the use of several microalgae (Mutanda et al., 2011) but more work still need to be done given the number of existing microalgae. In microalgae, lipids have as a basic function the synthesis of lipoproteic membranes and are important in floating and as an energetic reserve. Accumulation of lipids can be attributed to consumption of sugars at a rate higher than the rate of cell generation, which would promote conversion of excess sugar into lipids. The lipids extracted from microalgae may be used in human nutrition as source of Omega-3 (El-Baky et al., 2004). Accumulation of lipids in the microalgae cells, as well as for other oleaginous microorganisms (high oil producers), depends on diverse factors, such as growth temperature, pH, availability of micronutrients, salinity and other factors. Production of bioactive compounds and bio fuels recently, microalgae have been considered for producing biofuels, especially biodiesel (Huntley and Redalje 2007). Properties like rapid growth and high accumulation of oils (exceeding 80% by weight of dry biomass) make microalgae an attractive potential source for biodiesel, a replacement for fossil diesel. Some species of *Dunaliella*, like *D. tertiolecta*, may share this potential, with cells containing about 37% oils. *Dunaliella tertiolecta* is a fast-growing species which means it has a high rate of absorption of CO_2. The pyrolysis characteristics of *D. tertiolecta* were studied by thermogravimetric analysis. In addition, Park et al. (1998) proved that the hydrocarbon productivity of *D. salina* 1650 was similar to that from Botryococcus braunii, which was known to economically produce liquid fuels. The effect of salt concentration on lipid and triacylglyceride contents of *Dunaliella* cells was also tested (Takagi et al. 2006).

7. Conclusion

The alga *Dunaliella* includes many important scientific aspects and application on the physiology (osmoregulation, function of H^+-ATPase in *D. acidophila*), biotechnology (β-carotene and glycerol production) and bioenergy (bioreactor and biofuel production). Carcinogenic effects of β-carotene and particularly the beneficial properties of *Dunaliella* natural β-carotene will promote demand for the natural product. This will lead to further development by the traditional commercial manufacturers and is likely to attract new producers into the mass culture of *Dunaliella* for β-carotene production and other biotechnological purposes. The ability to induce, modify and scale up *Dunaliella* to produce a series of uncommon carotenoids of high nutritional and medical value, like phytoene and phytofluene, also opens a new field in the area of *Dunaliella* biotechnology. The cultivation of *Dunaliella* in photobioreactors in diverse autotrophic, heterotrophic and mixotrophic culture modes as well as in two-phase systems are other promising approaches requiring further development to make them more economically competitive in future. Exploitation of reliable approaches for genetic transformation and metabolic engineering of *Dunaliella* combined with its use as a biological source for mass-producing high-value proteins such as vaccines, antibiotics and enzymes, seriously under consideration by several research groups, could open an interesting new facet of microalgal biotechnology in future. Finally, it seems that the enormous potentialities of different species of this fantastic alga for exploitation in various biotechnological areas such as wastewater management programmes, designing of biosensors, production of new antibiotic substances and production of biofuels will make *Dunaliella* a main topic for many future microalgal investigations.

8. Reference

Aghaii, P., Shariati, M. (2007). The effect of sulfate deficiency on cell division, β-carotene, chlorophyll accumulation, and photosynthesis in green alga *Dunalliella salina* (isolated from salt marsh of Gavkhoni, Isfahan). *Iranian J. Biol*, 2 (20): 1-11.

Alkayal, F., Albion, R. L., Tillett, R. L., Hathwaik, L. T., Lemos, M. S., Cushman, J. C. (2010). Expressed sequence tag (EST) profiling in hyper saline shocked *Dunaliella salina* reveals high expression of protein synthetic apparatus components. *Plant Sci.*, 179: 437– 449.

Avron, M, Ben-Amotz, A. (eds) (1992). *Dunaliella*: Physiology, Biochemistry, and Biotechnology. CRC Press, Boca Raton.

Ben-Amotz, A., Polle, J. E. W., Subba Rao, D. V., Ed. (2009). The Alga *Dunaliella*: Biodiversity, Physiology, Genomics and Biotechnology. Enfield, Science Publishers. USA.

Ben-Amotz, A. (1993). Production of b-carotene and vitamin by the halo-tolerant algae *Dunaliella*. In Marine Biotechnology. ed. Ahaway, A. and Zabrosky, O. pp. 411–417.

Ben-Amotz, A. (1995). New mode of *Dunaliella* biotechology: two-phase growth for b-carotene production. *J. Appl. Phycol.*, 7: 65–68.

Ben-Amotz, A. (1996). Effect of low temperature on the stereoisomer composition of β-carotene in the halo-tolerant alga *Dunaliella bardawil* (Chlorophyta). *J. Phycol.*, 32, 272–275.

Ben-Amotz, A. (2009). Bioactive compounds: glycerol production, carotenoid production, fatty acids production. In: Ben-Amotz, A., Polle, J. E .W., Subba Rao, D.V. (Eds.), The Alga *Dunaliella*, Biodiversity, Physiology, Genomics and Biotechnology. Science Publishers, Enfield, USA, pp. 189–208.

Borowitzka, M. A., Siva, C. J. (2007). The taxonomy of the genus *Dunaliella* (Chlorophyta, Dunaliellales) with emphasis on the Marine and halophilic species. *J. Appl. Phycol.*, 19: 567–590.

Chen, H., Jiang, J. G. (2009). Osmotic responses of *Dunaliella* to the changes of salinity. *J. Cell Physiol.*, 219:251–8.

Chen, H., Lao, Y. M., Jiang, J. G. (2011). Effects of salinities on the gene expression of a (NAD+)-dependent glycerol-3-phosphate dehydrogenase in *Dunaliella salina*. *Sci. Total Environ.*, 409 :1291–1297.

Chidambara Murthy, K. N., Vanitha, A., Rajesha, J., Mahadeva Swamy, M., Sowmya, P.R., Ravishankar, G.A. (2005). In vivo antioxidant activity of carotenoids from *Dunaliella salina* – a green microalga. *Life Sci.*, 76: 1381–1390.

Chisti, Y. (2007). Biodiesel from microalgae. *Biotechnol Adv.*, 25: 294–306.

Coesel, S. N., Baumgartner, A. C., Teles, L. M., Ramos, A. A., Henriques, N. M., Cancela, L., Varela, J. C. S. (2008). Nutrient limitation is the main regulatory factor for carotenoid accumulation and for Psy and Pds steadystate transcript levels in *Dunaliella salina* (Chlorophyta) exposed to high light and salt stress. *Mar. Biotechnol* 10: 602–611.

Dufosse, L., Galaup, P., Yaron, A., Arad, S.M., Blanc, P., Chidambara Murthy, K.N. and Ravishankar, G.A. (2005). Microorganisms and microalgae as sources of pigments for food use: a scientific oddity or an industrial reality? *Trends Food Sci. Technol.*, 16: 389–406.

El-Baky A., H.H., El-Baz, F.K. and El-Baroty, G.S. (2004). Production of lipids rich in omega 3 fatty acids from the halotolerant alga *Dunaliella salina*. *Biotechnology*, 3:102– 108.

FAO, Food and Agriculture Organization of the United Nations, (2009). website of FAO: www.fao.org/docrep/012/i1704e/i1704e01.pdf

Garcia, F., Freile-Pelegrin, Y., Robledo D. (2007). Physiological characterization of *Dunaliella* sp. (Chlorophyta, Volvocales) from Yucatan, Mexico. *Bioresour. Technol.* 98: 1359-1365.

Garcia-Gonzalez, M., Moreno, J., Canavate, J. P., Anguis, V., Prieto, A., Manzano, C., Folrencio, F. J., Guerrero, M. G. (2003). Condition for open-air outdoor of *Dunaliella salina* in souther spain. *J. Appl. Phycol.*, 15: 177–184.

Garcia-Gonzalez, M., Moreno, J., Manzano, J. C., Canavate, J. P., Folrencio, F. J., Guerrero, M. G. (2005). Production of *Dunaliella salina* biomass rich in 9-cis β-carotene and lutein in a closed tubular photobioreactor. *J. Biotechnol.*, 115: 81–90.

Ghoshal, D., Mach, D., Agarwal, M., Goyal, A., Goyal, A. (2002). Osmoregulatory isoform of dihydroxyacetone phosphate reductase from *Dunaliella tertiolecta*: Purification and characterization. *Protein Expr. Purif.*, 24: 404-411.

Gomez, P., Gonalez, M., (2005). The effect of temperature and irradiance on the growth and carotenogenic capacity of seven strains of *Dunaliella salina* (Chlorophyta) cultivated under laboratory conditions. *Biol. Res.*, 38: 151–162.

Hadi, M. R. (1996). Isolation, purification and identification of some species of green alga *Dunaliella* from salt marsh of Gave Khoni of Isfahan. M. Sc. thesis in biology (plant physiology). Isfahan University – Iran. pp, 1-107

Hadi, M. R., Shariati M., Afsharzadeh S. (2008). Microalgal biotechnology: carotenoid and glycerol production by Dunaliella sp. algae isolated from the Gave khooni salt marsh, Iran. *Biotech. Bioproc. Eng.*, 13(5): 540-544.

Hosseini Tafreshi, A., Shariati, M. (2006). Pilot culture of three strains of *Dunaliella salina* for b-carotene production in open ponds in the central region of Iran. *World J Microbiol Biotechnol*, 22: 1003–1009.

Huntley, M.E., Redalje, D.G. (2007). CO_2 mitigation and renewable oil from photosynthetic microbs: a new appraisal. *Mitig. Adapt. Strat. Glob. Change*, 12: 573–608.

International Energy Agency (IEA). World Energy Outlook (2007). China and India Insights. International Energy Agency Publications: Paris, France, 2007.

Jahnke, L. S., White, A. L. (2003). Long-term hyposaline and hypersaline stresses produce distinct antioxidant responses in the marine alga *Dunaliella tertiolecta*. *J. Plant Physiol.*, 160: 1193-1202.

Jin, E. S., Melis, A. (2003). Microalgal biotechnology: carotenoid production by the green algae *Dunaliella salina*. *Biotechnol. Bioprocess Eng.*, 8: 331-337.

Klass. D. L. (1998). Biomass for Renewable Energy, Fuels, and Chemicals; Academic Press: San Diego, USA, pp. 651.

Madadkar Haghjou, M., Shariati, M., Smirnoff, N. (2009). The effect of acute high light and low temperature stresses on the ascorbate-glutathione cycle and superoxide dismutase activity in two *Dunaliella salin*a strains. *Physiol. Plant.*, 135:272-280.

Madadkar Haghjou, M., Shariati. M. (2007). Photosynthesis and respiration responses to short-term low temperature stress in two *Dunaliella salina* strains, *World Appl. Sci. J.*, 2(4): 276-282.

Marin, N., Morales, F., Lodeiros, C., Tamigneaux, E. (1998). Effect of nitrate concentration on growth and pigment synthesis of *Dunaliella salina* cultivated under low illumination and preadapted to different salinities. *J. Appl. Phycol.*, 10: 405-411.

Mata, T.M., Martins, A.A., Caetano, N.S., (2010). Microalgae for biodiesel production and other applications: a review. *Renew. Sust. Energy Rev.*, 14: 217–232.

Matthijs, H. C. P., Balke, H.,Van Hes, U. M., Kroon, B. M. A., Mur, L. R., Binot, R. A. (1996). Application of Light-emitting Diodes in Bioreactors: Flashing Light Effects and Energy Economy in Algal Culture (Chlorella pyrenoidosa). *Biotechnol. Bioeng.*, 50:98-107.

Minowa, T., Yokoyama, S.-Y., Kishimoto, M., Okakura, T. (1995). Oil Production from Algal Cells of *Dunaliella tertiolecta* by Direct Thermochemical Liquefaction. *Fuel*, 74:1735-1738.

Mojaat, M., Pruvost, J., Foucault, A. and Legrand, J. (2008). Effect of organic carbon sources and Fe2+ ions on growth and β-carotene accumulation by *Dunaliella salina*. *Biochem. Eng. J.*, 39: 177–184.

Mutanda, T., Ramesh, D., Karthikeyan, S., Kumari, S., Anandraj, A., Bux, F. (2011). Bioprospecting for hyper-lipid producing microalgal strains for sustainable biofuel production. *Bioresour. Technol.*, 102: 57–70.

Park, D. H., Ruy, H. W., Lee, K. Y., Kang, C. H., Kim, T. H., Lee, H. Y. (1998). The production of hydrocarbons from photoautotrophic growth of *Dunaliella salina* 1650. *Appl Biochem Biotechnol.*, 70-71:739–746.

Patil, V., Reitan, K. I., Knudsen, G., Mortensen, L., Kallqvist, T., Olsen, E., Vogt, G., Gislerød, H. R. (2005). Microalgae as Source of Polyunsaturated Fatty Acids for Aquaculture. *Curr. Topics Plant Biol.*, 6: 57-65.

Peng, J., Chen, P., Lou, H., Zheng, X. M., (2009). Catalytic upgrading of bio-oil by HZSM-5 in sub- and super-critical ethanol. *Bioresour. Technol.*, 100: 3415–3418.

Perego, C., Bianchi, D. (2010). Biomass upgrading through acid–base catalysis. *Chem. Eng. J.*, 161: 314–322.

Perez-Garcia, O., Escalante, F. M. E., de-Bashan, L. E., Bashan, Y. (2011). Heterotrophic cultures of microalgae: metabolism and potential products. *Water Res.*, 45: 11– 36.

Phadwal, K., Singh, P. K. (2003). Isolation and characterization of an indigenous solate of *Dunaliella* sp. for β-carotene and glycerol production from a hypersaline lake in India. *J. Basic Microbiol.*, 43 (5): 423–429.

Prieto, A., Canavate, J. P., Garcia-Gonzalezb, M. (2011). Assessment of carotenoid production by *Dunaliella salina* in different culture systems and operation regimes. *J. Biotechnol.*, 151: 180–185.

Raja, R., Hemaiswarya, S., Rengasamy, R. (2007). Exploitation of *Dunaliella* for β-carotene production. *Appl. Microbiol. Biotechnol.*, 74: 517-523.

Ross, A. B., Biller, P., Kubacki, M. L., Lin, H., Lea-Langton, A., Jones, J. M. (2010). Hydrothermal processing of microalgae using alkali and organic acids. *Fuel*, 89: 2234–2243.

Sarmad, J., Shariati. M., Hosseini Tafreshi A. (2006). Preliminary assessment of β-carotene accumulation in four strains of *Dunaliella salina* cultivated under the different salinities and low light intensity. *Pakistan J Biol. Sci.*, 9 (8): 1492-1496.

Shariati, M., Hadi, M. R. (2000). Isolation, purification and identification of three unicellular green alga species of *Dunaliella salina*, *D. parva and D. pseudosalina* from salt marsh of Gave Khooni of Isfahan. *Iranian J. Biol.*, 9(1- 4): 45-54.

Shariati, M. (2003). Characterization of three species of *Dunaliella salina, Dunaliella parva and Dunaliella pseudosalina* isolated from salt marsh of GaveKhoni of Isfahan-IRAN. *Iranian J. Sci. Technol.*, 27: 185-190.

Shariati, M., Maddadkar Haghjoo, M. (1998). Study of β-carotene and chlorophyll content of green alga *D. salina* in response to high light intensity. *Iranian Journal of Biology*, 7(3, 4): 112-132.

Shariati, M., Lilley, R. M. (1994). Loss of intracellular glycerol from *Dunaliella* by electroporation at constant osmotic pressure: Subsequent restoration of glycerol content and associated volume changes. *Plant Cell Environ.*, 17: 1295-1304.

Shariati, M., Zoofan, P. (2003). The effect of nitrate deficiency on cell division and beta-carotene and chlorophyll synthesis in unicellular green alga *Dunaliella salina* (isolated from salt marsh of Gavkhoni, Isfahan-Iran). *J. of Pajouhesh-va-Sazandegi (Iran)*. 59: 7-13.

Singh, A. K., Fernando, S. D. (2008). Transesterification of soybean oil using heterogeneous catalysts. *Energy Fuels*, 22: 2067–2069.

Spolaore, P., Joannis-Cassan, C., Duran, E., Isam bert, A. (2006). Commercial applications of microalgae. *J. Biosci. Bioeng.*, 101: 87-96.

Takagi, M., Karseno Yoshida, T. (2006). Effect of salt concentration on intracellular accumulation of lipids and triacylglyceride in marine microalgae *Dunaliella* cells. *J. Biosci. Bioeng.*, 101: 223–226.

Telfer, A. (2002). What is β-carotene doing in the photo system II reaction centre? *Philos. Trans. R. Soc. Lond. B Biol. Sci.*, 357: 1431-1439.

Tornwall, M. E., Virtamo, J., Korhonen, P. A., Virtanen, M. J., Taylor, P. R., Albanes, D., Huttunen, J. K. (2004). Effect of a-tocopherol and β-carotene supplementation on coronary heart disease during the 6-year post-trial follow-up in the ATBC study. *Eur. Heart J.* 25: 1171–1178.

Tredici, M. R. (2004). Mass production of microalgae: photobioreactors. In: Richmond, A. (Ed.), Handbook of Microalgal Culture: Biotechnology and Applied Phycology. Blackwell, Oxford, UK, pp. 178–214.

Tsukahara, K. and Sawayama, S. (2005). Liquid fuel production using microalgae. *J. Jpn. Petrol. Inst.* 48:251–259.

Vishwanath, P., Tran, K. Q., Giselrod, H. R., (2008). Towards Sustainable Production of Biofuels from Microalgae. *Int. J. Mol. Sci.*, 9:1188-1195.

Vismara, R., Verni, F., Barsanti, L., Evangelista, V., Gualtieri, P. (2004). A short flagella mutant of *Dunaliella salina* (Chlorophyta, Chlorophyceae). *Micron.*, 35: 337– 344.

Weisz, P.B. (2004). Basic Choices and Constraints on Long-term Energy Supplies. *Physics Today*, 57: 47-52.

Yang, C., Jia, L., Chen, C., Liu, G., Fang, W. (2011). Bio-oil from hydro-liquefaction of *Dunaliella salina* over Ni/REHY catalyst. *Biores. Technol.* 102 : 4580–4584.

Yang, Y. F., Feng, C. P., Inamori, Y., Maekawa, T. (2004). Analysis of energy conversion characteristics in liquefaction of algae. *Resour. Conserv. Recycl.*, 43, 21–33.

Ye, Z. W., Jiang, J. G., Wu, G. H. (2008). Biosynthesis and regulation of carotenoids in *Dunaliella*: progresses and prospects. *Biotechnol Adv.*, 26: 352–360

Zou, S. P., Wu, Y. L., Yang, M. D., Li, C., Tong, J. M. (2009). Thermochemical catalytic liquefaction of the marine microalgae *Dunaliella salina* and characterization of bio-oils. *Energy Fuels* 23: 3753–3758.

23

New Trends for Understanding Stability of Biological Materials from Engineering Prospective

Ayman H. Amer Eissa[1,2] and Abdul Rahman O. Alghannam[1]
[1]Department of Agriculture Engineering, Faculty of Agricultural, Minoufiya University
[2]Department of Agriculture Systems Engineering, College of Agricultural and Food Sciences, King Faisal University,
[1]Egypt
[2]Saudi Arabia

1. Introduction

Economic loss due to egg shell breakage is an important problem in intensive egg production systems (Oosterwoud, 1987). Eggs are subjected to mechanical impacts at the moment of lay, during collection, in the sorting equipment and during transport in trays. Bain (1991) reported that 6% to 8% of all eggs laid are broken during handling from the production unit to the consumer, and in monetary terms this gives rise to losses of at least $600 million on a world wide basis. The value of egg production was $3.389 billion in 1992, compared to $ 3.209 billion in 1987 (Madison and Perez, 1994). Therefore, egg shell breakage continues to be a costly problem for the egg industry. In addition, people are at high risk when eating eggs which might be contaminated after being damaged, i.e., with cracks or checks (Amer, 1998; Bain, 1990).

Egg shell breakage depends both on the strength of the egg shell and magnitude of the mechanical load applied. Variables associated with shell strength are biological in nature (Hamilton, 1982). They involve the material and structural properties of the different layers which comprise the egg shell. The material properties depend on the type and the association between the mineral and organic components of the shell. The structural properties depend on egg shell thickness, size and shape of the egg and the distribution of shell over the egg surface.

In most of the methods used for measuring egg shell strength these basic physical properties are not quantified separately because of their complexity. Indeed, the shell curvature and its brittle nature make it difficult to measure the material properties of the shell by classical means (Amer Eissa & Gamea, 2003; Bain, 1990). Therefore, more practical techniques have to be developed. Such methods, however, describe the behaviour in terms of a superposition of several material and structural properties of the egg. One such commonly used method is the non-destructive, quasi static compression test. In this test the elastic stiffness properties of the whole egg shell structure are measured. An egg is placed horizontally between 2 flat parallel steel plates and then compressed at a constant compression speed until a predefined non-destructive load is applied. The slope of the force-deformation curve is an indicator of the mechanical stiffness of the shell. This mechanical stiffness depends on the thickness of the egg shell, the curvature of the egg shell, the diameter of the egg and the Young's modulus of the material in each layer of the shell (Amer, 1998; Bain, 1990).

A more precise method of measuring the impact strength of an egg is to record the force on the shell throughout the impact. Using this technique, Voisey and Hunt (1967 a,b) developed an instrument to measure the impact force required to fracture an egg. With this instrument the egg is supported in an adjustable plastic (nylon) cradle directly below a suspended aluminium rod 1900 mm in length, this cradle is adjusted, using a dial gauge, so that the upper surface of the egg is at a constant distance (5.0 mm) from the end of the aluminium rod for all measurements. The rod is suspended by an upper spherical end that fits into a socket connected to a vacuum pump. By opening a solenoid valve the vacuum is released and the rod falls freely until it strikes the egg. The impact of the rod on the egg is measured by a piezoelectric transducer attached to the lower end of the rod; the rod and transducer weigh 695 g. The transducer can be calibrated in either force or acceleration units. The electrical output of the transducer during the impact is recorded by a peak-shock meter and the maximum value displayed by a 4-digit voltmeter. The impact strength of about 150 eggs can be measured per hour with this instrument. Force-time plots can also be obtained using an oscilographa and camera to record the graph for each egg. It is very important that shock waves generated within the rod during the impact with an egg do not affect the measurement. This effect may be overcome by changing the length of the rod or the material from which it is made (Voisey and Hunt, 1968).

During handling of eggs, several impacts may be imparted to the egg shells, if not limited, will cause considerable damage. Several researchers have investigated the effects of impacts on agricultural products. Zapp et al., (1990) investigated the impact on apples by simulating an apple with an 'instrumented sphere'. Jindal and Mohsenin (1976) analysed a pendulum impacting device with which apples and corn kernels were tested. Finney and Massie (1975) also investigated a pendulum impacting device for testing the response of fruits to impacts. Other authors who have investigated impact devices for agricultural products include Amin (1995), Tennes et al., (1988), Siyami et al., (1986), Hughes et al., (1985), Simpson and Rehkugler (1972) and Bilanski (1964).

In the case of a chicken egg, the interpretation of the impact response of an egg is even more complex than for apples. An egg consists of several types of material: a shell, an air-chamber, egg white and yolk. The egg shell structure itself contains several distinct layers that are penetrated by pores. Within 1 layer, the material properties are not homogeneous. The material properties of the egg content are also variable. Nevertheless, it is worthwhile to asses the experimental relationships between the dynamic impact, mechanical properties of an egg and commonly measured physical egg and egg shell variables.

The values of the material property of eggshell were reported over 25 years ago. These values might be obsolete due to significant changes in chickens' species, feed and management practices. Therefore, it is necessary to measure the key material properties of present day shell eggs.

The aim of this chapter is to describe the different approaches used to evaluate mechanical properties and damage to the chicken eggshell. Furthermore we   describe a pendulum which is portable and which will impact eggshell under a wide range of velocities and energies.

2. Finite element methods applications in biological materials

Finite element approach for simulating the dynamic mechanical behaviour of a chicken egg
Perianu, et al., (2010) noticed that A numerical investigation of the structural acoustic
behaviour of a chicken egg was carried out. A three-dimensional finite element (FE) model
was developed to simulate both the dynamic behaviour of the eggshell and the fluid loading
of the inside fluid. The aim was to analyse the effects of variations in certain geometrical and
material parameters of the egg on the structural acoustic frequency response functions. It
was found that geometrical modifications (eggshell thickness, size of the egg) had a
considerable influence on the fluid structure coupled natural frequencies. In general,
variations of material characteristics did not have much influence on dynamic behaviour.
However, the Young's modulus of the eggshell strongly affected the natural frequencies of
the coupled system. The results obtained were used to interpret experimentally observed
relationships. Nowadays, consumers are primarily concerned with the quality and safety of
the food. In the case of the consumption of eggs, shell integrity is one of the main issues.
During the last decade, several methods were developed for rapid assessment of shell
integrity and strength (De Ketelaere, et al., 2004). Eggshell strength is regulated by a certain
number of variables such as genetic origin, the age of the laying hen, environmental factors
such as feed composition, diseases, climatic conditions and management by the farmer
(Solomon, 1991).

Classically, eggshell strength this evaluated by means of a nondestructive, quasi-static
compression test (Voisey & Hunt, 1974). The egg is placed horizontally between two parallel
steel plates and a compression load is exerted on the object. Force and displacement are
recorded throughout the test and used to calculate the static stiffness (k_{stat}), being defined as
the slope of the force e displacement curve. However, this method is time-consuming and
requires expensive test equipment. Hence, it cannot be used as an industrial tool for real-
time assessment of shell strength. As an alternative to traditional techniques, Coucke, et al.,
(1994) introduced a dynamic test method for eggshell stiffness assessment. The dynamic
behaviour of a chicken egg was characterised using an experimental modal analysis. Several
spherical modes were identified in the frequency range between 3 and 8 kHz. The mode
shapes in this frequency range all showed maximum deformation at the equator of the egg,
whilst the sharp and blunt poles were immobile. The mode shape of the first, flexural
spherical mode (S20) produced an oblate-prolate deformation at the equator. The damped
natural frequency of this mode and the total egg mass were used to calculate the dynamic
eggshell stiffness (k_{dyn}). Modeling the egg as a mass-spring system, the dynamic stiffness
k_{dyn} is given as: $k_{dyn} = cte.\ m.\ RF^2$, with m the mass of the egg, cte a constant (set to 1) and RF
the resonant frequency of the first spherical mode (S20). Later, Coucke, et al., (1999) and De
Ketelaere, et al., (2000) described a laboratory scale device for the measurement of the
dynamic stiffness of the eggshell based on the identification of the mode shapes, the
corresponding resonant frequencies and damping ratios of these modes using the
interpretation of the vibration response of an egg excited at its equator with a
nondestructive impact. This technique can also be used to detect cracks in the eggshell. The
relationships between some of the physical and mechanical egg and eggshell quality
parameters and the dynamic, mechanical properties of a chicken egg have been evaluated in
many studies. However, little information is available about the contribution of the basic
material and geometrical properties to the results of the dynamic tests. No full explanation
could be found for the different effects of egg size and eggshell thickness on the mechanical

behaviour. Coucke (1998) utilised simple structural models to simulate the dynamic mechanical behaviour of the egg using the finite element (FE) method. The analysis was incomplete since the egg content, i.e. the interior fluid, was not incorporated in the model. Since the content effect was neglected, the analysis results showed several deficiencies when comparing numerical and experimental data. In spite of some important observations achieved using the above-mentioned structural models, there is still a gap between models including only the eggshell and highly detailed structural acoustic models incorporating both eggshell and fluid content In the light of the above, the object of this paper is to set up a realistic model for an egg structural acoustic analysis, that is useful to assess, and can be visualised and compared with the structural acoustic behaviour of the egg for different material and geometrical properties. Using this parametric model, the influence of gradual changes in different parameters of the egg on structural acoustic frequency response functions is investigated. It was expected that the approach and the obtained results would be of interest to specialists working on acoustic-based egg grading machines. It was also expected that this approach could allow an improved interpretation of the experimentally observed correlations between egg and eggshell parameters.

The avian egg is a biological structure of high complexity. It contains an air chamber and a viscous liquid surrounded by two membranes and an external covering called the eggshell. The base numerical model used in this study represents a simplified replica of a chicken egg, a fluid filled shell, yielding a coupled structural acoustic problem. Here, the eggshell is modeled as a single layer shell structure of uniform thickness. The acoustic content includes the air chamber and water, the major component of albumen (w90%) and yolk (w50%). The shell membranes are not incorporated in the model. Vibrating structures induce acoustic pressure waves in a connected fluid and, vice versa, acoustic pressure waves act as external loads yielding structural vibrations. This fully coupled structural acoustic problem description is a thoroughly investigated field of research (Fahy, 1985). The numerical approach used in this paper for the representation of the coupling effects between fluid and structure is based on a FE representation of the structure as well as the interior fluids (Zienkiewicz, et al., 2005). The main advantage of such a method is that it is possible to represent in one model cavities with different types of fluid, e.g. water and air (Stavrinidis, et al., 2001). In the following section, the formulation of the coupled structural acoustic problem using the finite element method (FEM) is described.

2.1 Description of the base model

The base model represents a simplified replica of a chicken egg. The egg-shaped geometry was approximated by a half ellipsoid fused to a half-sphere. The overall dimensions of the egg model are 4.6, 5.8 and 4.6 cm, respectively in X (longitudinal), Y (vertical) and Z (lateral) direction. Eggshell thickness is assumed to be uniform over the shell surface. A default value of 0.38mm was applied. The material parameters of the eggshell are as follows: Young's modulus, $E = 3 \times 10^{10}$ Nm^{-2}, Poisson's ratio $\upsilon = 0.307$ and the mass density $\rho = 2400$ kgm^{-3} (Coucke, 1998). The egg content was represented by an air chamber and a water domain. The height of the air chamber for the default configuration is 4 mm. The acoustic parameters of the air are: speed of sound 343 ms^{-1} and the mass density 1.25 kgm^{-3}. The default values for the acoustic parameters of water are: speed of sound 1500 ms^{-1} and the mass density 997 kgm^{-3}. The FE meshes (see Fig. 1) for both structural and acoustic domain were generated using MSC. Patran (MSC Software, Santa Ana, CA, USA). All uncoupled structural results were obtained with the MSC. Nastran software, while the acoustic and coupled vibro-acoustic results are obtained with the LMS. Sysnoise software (LMS International, Leuven Belgium). Using mesh morphing techniques, the base meshes were modified for the various analyses (see below). Therefore all models comprised exactly the same number of nodes and elements. For the structural part of the analysis, bilinear four-noded quadrilateral and three-noded triangular shell elements (6 degrees of freedom per node) were used, whereas for the acoustic part linear eight-noded hexahedral and six-noded wedge elements (1 degree of freedom per node) were employed. The applied element size was consistent with the rule of thumb that states that at least six linear elements should be used per wavelength to assure sufficient prediction accuracy. The maximum frequency of interest was selected as 5000 Hz yielding an element size of 0.0025 m, which results in 4920 acoustic and 1050 structural elements.

The structural acoustic model involved in the simulations was a free boundary condition model excited by a unit normal point force exerted at the egg equator (Fig. 1).

Perianu et al., (2010) studied that, the first 50 structural modes were calculated as a first step. The eigenfrequencies of these uncoupled structural modes are in general higher than the experimentally observed values. This could be expected since only the eggshell is modeled and the egg content, which represents 85-90% of the total egg mass, is not yet incorporated in the structural model. However, for a fluid filled egg, the eigenfrequencies of the coupled modes were close to the experimental results. The mode shapes and the sequence of appearance of the calculated modes were very similar to the experimentally observed modes (see Coucke, 1998). Fig. 2 represents a top view (left) and a front view (right) of the mode shape of the first, flexural spherical mode. The first flexural mode has an eigenfrequency of 4190 Hz; this is called the oblate-prolate mode. As can be seen in Fig. 2, all deformation is concentrated towards the equator zone of the egg. An elliptic shape can be recognised at the equator ring. The magnitude of the deformation decreases gradually towards the poles of the egg. In the subsequent analysis each single geometrical and material property was varied within a reasonable range. The effect of these changes on the eigenfrequency of the first flexural mode (S20) was evaluated.

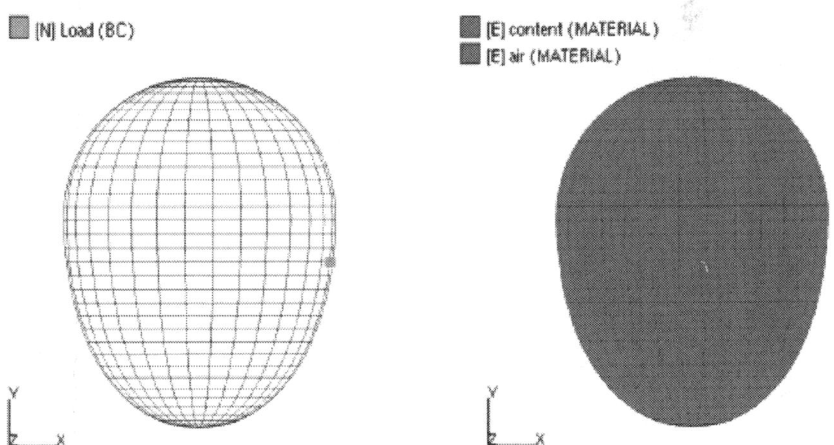

Fig. 1. FE mesh of the eggshell (left) and of its acoustic content (right).

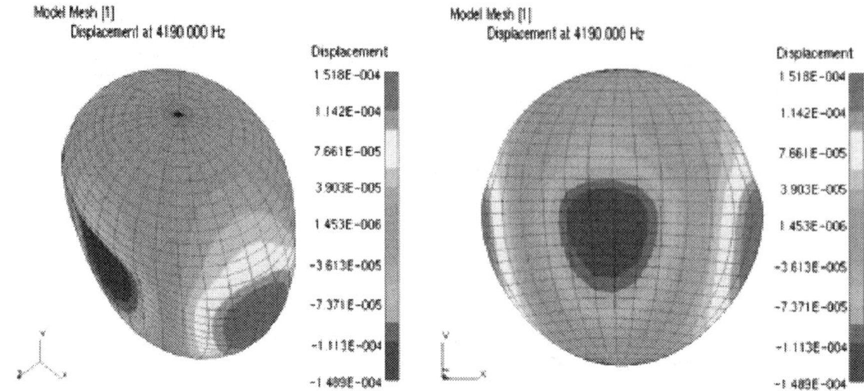

Fig. 2. Top and front view of the mode shape of S20 mode.

3. Operational modal analysis

Amer Eissa and Gomaa (2007) showed that the presence of cracks in eggshells is a common problem in high speed commercial egg grading machines. Non-destructive physical quality assessment of foods is mainly based on vibration analysis with resonant frequency and damping of the vibration being the main parameters. The focus of this research was to investigate the capability of the operational modal analysis as a non-destructive tool to characterise and to quantify the fracture behaviour of eggshells. This was achieved by studying the response of modal testing to the variation in strength as the main factor affecting fracture. Two different methods were applied: 1- traditional modal analysis using transfer function between actuator and sensor and 2- operational modal analysis using transmittance function between pairs of sensors in which input excitation is provided through a piezoelectric force transducer bonded to the centre of the cup used for egg installation in order to extract the corresponding modal parameters and damping ratio. The test for strength was performed on 200 eggs deriving from two genotypes of hens that were correlated to modal parameters and the behaviour of the fracture. Damping loss factor enabled to distinguish the strength for two types of eggs. Natural frequency showed a greater response to egg type and strength increased as frequency response function (FRF) increased for the two types of eggs (R= 0.979 & 0.984). These methods allow for an evaluation of the suitability of modal testing predicting fracture using an empirical formula. The damage index based on changes in transmittance function is very sensitive to a change in crack length. Structural information obtained from the biomaterial at different length scales is important in relation to the functional properties of the structure. This knowledge and the principles behind the formation of biomaterials could be used in an attempt to develop new systems of bioengineering.

Cracking of eggshells is a common problem in commercial egg production resulting in downgrading of eggs and substantial economic losses. Also cracked eggs are more vulnerable to Salmonella and other bacterial infections leading to health hazards. Therefore, detection and removal of cracked eggs continues to be very important for quality assurance in the egg industry. Candling is a reliable and most often used technique for egg quality assessment. However, egg candling is an immensely difficult task and requires a combination of great skill, practice and concentration by the human operators, especially when linking it with high speed grading machines. As a result, attempts have been made to replace the human operators with the so-called machine vision systems to detect cracks and other defects on the egg surface (Elster & Goodrum, 1991; Goodrum & Elster, 1992; Patel et al., 1994). It has been shown that several problems associated with misclassified images may affect the overall accuracy of such systems considerably (Garcia-Alegre et al., 2001). Recently it has been reported that acoustic impulses resulting from a soft mechanical impact could be used for online inspection of eggshell cracks non-destructively in real time (Cho et al., 2000; De Ketelaere et al., 2000). Cho et al. (2000) developed classification criteria for detecting surface cracks in egg shells based on the frequency spectrum of acoustic impulses using multiple regression and multivariate discriminate analysis. The errors associated with incorrect identification of cracked and intact eggs were found at 4 and 6%, respectively. In another approach, De Ketelaere et al. (2000) used Pearson correlation coefficients between the frequency spectra of acoustic responses to classify eggs. They showed an accuracy of about 90% for cracked egg detection with a false reject of less than 0.5%. Finally, Bain et al., (2006) and Mertens et al. (2006) showed that dynamic stiffness provides a good estimation of eggshell strength in relation to the likelihood of breakage in practice. Artificial neural networks (ANNs) have been widely applied to classification problems in many fields. Several successful implementations of ANN classifiers reveal their capability to extract trends and patterns from large data sets (Das & Evans, 1992; Chen et al., 1998; Ghazanfari et al., 1996; Jindal & Chauhan, 2001). They offer significant advantages when dealing with noisy or obscure patterns from statistical pattern classifiers (Jindal and Chauhan, 2001). Due to higher speeds in commercial egg grading machines (which grade up to 120,000 eggs/h), an automated quality sorting device is of interest to assure a consistent egg quality. One of the main physical quality parameters for the consumption of eggs is the absence of eggshell cracks. Very recent research shows that it is possible to detect cracks in eggshells on-line using vibration analysis (De Ketelaere et al. 1997; De Ketelaere et al. 2000; Coucke, 1998; Moshou et al., 1997). For this purpose, the egg is hit four times around its equator and the similarity (correlation) between the four measurements is used as a sorting criterion. In this way, up to 90% of the cracks can be removed while the number of false rejects (the percentage of intact eggs that are classified as broken) remains well below 0.5% (De Ketelaere et al., 2000). Because of the complexity of the eggs' structure and transport conditions, no clear guideline and methodology for the experimental analysis of the dynamic egg characteristics has emerged, especially damping on line. Egg shell strength provides a proper estimation of eggshell life. To avoid sudden or unexpected fractures of eggshells, non-destructive tests and modal parameters are investigated and correlated to strength. Two methods for estimating modal damping to verify the accuracy of the estimated modal damping ratio were investigated by Brincker et al. (2002).

Our research focuses on the question of whether it is possible to use vibration measurements to assess eggshell strength online as another potential quality parameter towards an integrated on-line egg quality assessment. Different techniques can be found for eggshell strength determination in the literature. In general, these can be split up into direct and indirect methods (Hamilton, 1982). Indirect methods measure a parameter that is related to the eggshell strength. The correlation between the different methods is moderate, and the method of choice often depends on the application, although most methods are destructive. Measuring eggshell thickness is one of the most frequent used indirect methods to get an indication of eggshell strength. Using a micrometer, it is possible to determine the thickness up to 0.02 mm. Another indirect measure for the strength of the egg is provided by the calculation of the eggshell percentage. Abdallah et al. (1993) showed that 80% of the breakage percentage of a batch of consumption eggs can be explained by this percentage. A third widely used indirect method makes use of the quasi-static compression of the eggs between two parallel plates. By measuring the force deformation curve, it is possible to determine the static stiffness of the egg. In the present study, this method is used as a reference.

Various direct methods are described in the literature. The most widely used is the compression fraction force measured during quasi-static compression (Voisey & Hunt, 1967 a,b; Abdallah et al., 1993). Other methods include puncture tests and impact tests. All direct methods are destructive. From the literature it appears that their behaviour has been simulated as a mass–spring system from which the stiffness of the product is the factor describing its quality. The stiffness of the product is hence a function of both the mass of the object and its resonant frequency given by Coucke (1998) with the dynamic stiffness, the mass of the egg and the resonant frequency. This is clearly an invariable model, linking only one vibration parameter (the dynamic stiffness) with a reference quality parameter. The dynamic stiffness of the eggshell is used to estimate the static stiffness. This invariable model is not applicable as such for eggshell strength assessment because of the moderate correlation between the kdyn and kstat (Coucke et al., 1999). The current study focuses on the expansion of the invariable model to obtain more accurate estimates of the eggshell strength. Additional and improved information will be provided by:

1. A very accurate estimation of the resonant frequency (note that this is quadratically related to the dynamic stiffness and hence plays a crucial role).
2. Expanding the model by incorporating the damping of the vibration, which was ignored in all research found in the literature; including the mode shape, and provision of an empirical formula to correlate between modal parameters (natural frequency, damping ratio and magnitude of frequency response function) and strength. Therefore, the objective of this study was to develop a method for detecting eggshell cracks based on transmittance function of frequency response of eggs on line.

2.1 Methods of damage detection

Various methods of detecting damage have been proposed. One of the approaches that have received considerable attention in the technical literature is vibration–based damage detection. The fundamental idea behind vibration-based damage detection techniques is that changes in the physical properties will alter a system's modal properties such as natural frequencies, modal shapes and damping. The discussion herein focuses on feature selection for damage detection (Doebling et al., 1998) summarizing many features that have been proposed for vibration-based damage detection.

Comprehensive literature reviews of subject structural health monitoring can be found (Doebling et al., 1996; Farrar & Doebling, 1997; Zou et al., 2000; Amer Eissa and Gomaa, 2009). Todd et al. (2001) used chaotic input signature and a state-space method for damage detection. A novel feature called the local attractor variance ratio was developed using chaos theory. They showed how a properly tuned chaotic excitation could be used to robustly detect structural changes. Techniques based on neural networks require a model to train the system to detect damage (Wang & Huang, 2000). Zubaydi et al. (2002) investigated the damage detection of composite ship hulls using neural networks. They developed a finite element model for a stiffened plate to stimulate dynamic response of the structure with and without damage. Very small damage in composite materials, such as cracks, was successfully found using wavelet analysis (Yan & Yam, 2002). Ganguli (2001) used a fuzzy logic system to locate damage on helicopter rotor blades. Salawu (1997) presented a review of various investigations on the effects of structural damage on natural frequencies. Many damage location methods use the change in resonant frequencies because frequency measurements can quickly be conducted and are often reliable. However, changes in ambient condition can cause a significant frequency change in composite materials, and findings suggest that detection of damage using frequency measurements might be unreliable when the damage is located at regions of low stress. Kuo & Jayasuriya (2002) used transfer functions to determine the extent of joint loosening in automobile vehicle frames with high mileage. The method was successful but did neither give specifics of the frequency range investigated nor about the used type of FRFs .

The successful transmittance function testing for wind turbine blade damage analysis was presented by (Ghoshal et al. (2000) and Schulz et al. (1999). Caccese et al. (2004) used three different monitoring techniques at low frequency modal analysis and for high frequency transfer functions between the actuator and sensors. These experiments demonstrated that the transmittance function is very sensitive to changed bolt load. To confirm the characteristics of the transmittance function technique, this paper focuses on the detection of fractures in eggshells.

2.2 Operational modal analysis using transmittance function

Transmittance testing procedures are similar to the procedures using the transfer function (Richardson & Potter, 1974; Caccese et al., 2004). For the transmittance test the response frequency domains of the two sensors are compared with each other. In contrast to this test, the response of the sensor is compared to the excitation signals. Thus, for transmittance testing sensor A was connected to input channel A and sensor B was connected to reference channel B. The transmittance function between two response points, a and b is given by:

$$T_{ab}(f) = \frac{G_{ab}(f)}{G_{bb}(f)}$$

where
G_{bb}: is an auto spectral density function.
G_{ab}: cross spectral density function, one side auto spectral density function and cross-spectral can be computed from real data:

$$G_{ab}(f) = \frac{2}{(n_d \Delta T)} \sum_{i=1}^{n_d} A(f,T)B^*(f,T).....(f > 0)$$

The spectral densities are a function of frequency that can be averaged across (n_d), where n_d: is a distinct sub-record of duration ΔT , * : Complex conjugate.

The cross-spectral density function is the Fourier transformation of the cross correlation function. It represents the frequency domain, and characterization of similarity of the magnitude and phase of two signals, therefore it can accurately detect damage over small distances on the structure. Furthermore, measured transmittance data include complete information on the dynamic behaviour of the test structure, in terms of vibration modes and damping, at many frequency points. So it is much easier to display ODS's from a set of ODS FRF and observe mode shapes at resonance frequencies. The magnitude of the scale factor is calculated by Vold et al. (2000):

$$Scale\ factor(i) = \frac{\sum_{i=1}^{No.of.Meas.sets} ARM(i)}{No.of.Meas.setXARM(i)}$$

where ARM (i) =Average value of the reference response APs for measurements set (i). This scale factor corrects each of ODS FRF magnitude according to the average level of all reference response signals and the average value can be calculated for any desired range of frequency samples (Mohanty & Rixen, 2004). Modal analysis is used to analyze the effect of aging on the dynamic properties of an egg. The egg is made up of several types of materials such as shell, yolk and egg white and these material properties are non homogeneous.

Amer Eissa and Gomaa (2007) found that, the response of the eggshell sensed by a light piezoelectric accelerometer, weight (2.2 gm), mounting surface flatness, charge sensitivity (0.318) pc / ms-2, voltage sensitivity (0.415) mv/ms-2 which was bonded to the opposite direction of impact hammer using a wax for free suspension that represent the inherent properties of egg. Without regard to the external constraint condition at frequency (0 - 1.6 KHz), both signals from impact hammer and accelerometer were supplied to FFT analyzer as in Fig. (3 a, b), and estimation FRF with a narrow band of 800 Hz, centered around the fundamental frequency a sample of FRF and coherence function are in Fig. (4) Because of complexity of eggs structure and transporting condition there is no clear guide line and

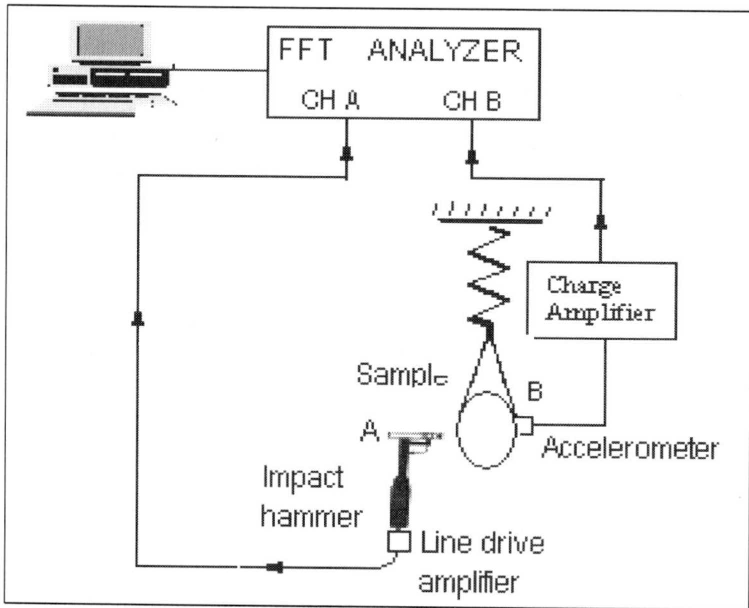

Fig. 3. (a) Schematic layout of the system used for (TMA) modal analysis.

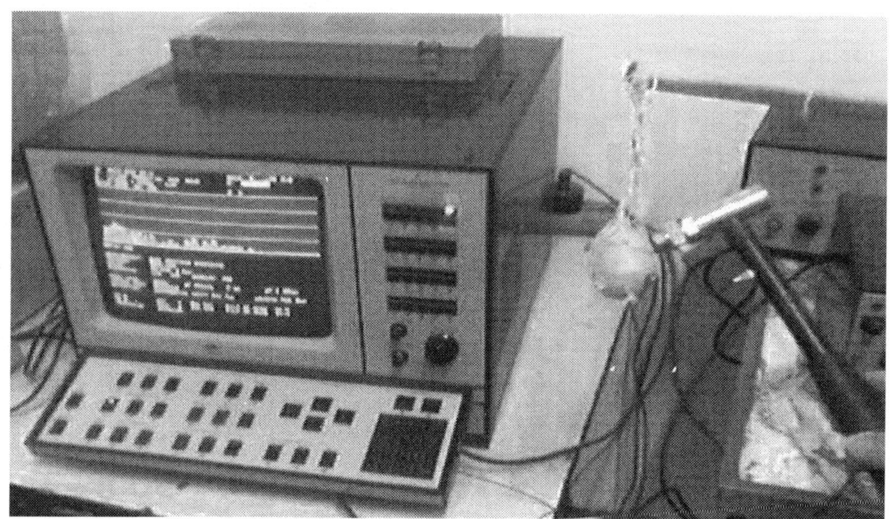

Fig. 3. (b) Free suspension for modal analysis.

methodology for experimental analysis to be emerged. As mentioned this work focused on two different methods for vibration measurements traditional Modal Analysis (TMA) and transmittance function (T.F). In order to verify the accuracy of the estimated modal damping, the same set of data were presented using operating moving average ODS Brian Schwarz et al (2000). Fixed installation using cup and transmittance function were obtained experimentally, as in Fig. (5 a, b) [The support of the egg is chosen in such a way that they coincide with the nodal points of the first elliptic]. The system excitation was using exciter control through generator with variable force but did not measure. For transmittance testing sensor A was connected to input channel A and sensor B was connected to reference channel B of FFT analyzer (Amer Eissa and Gomaa, 2007).

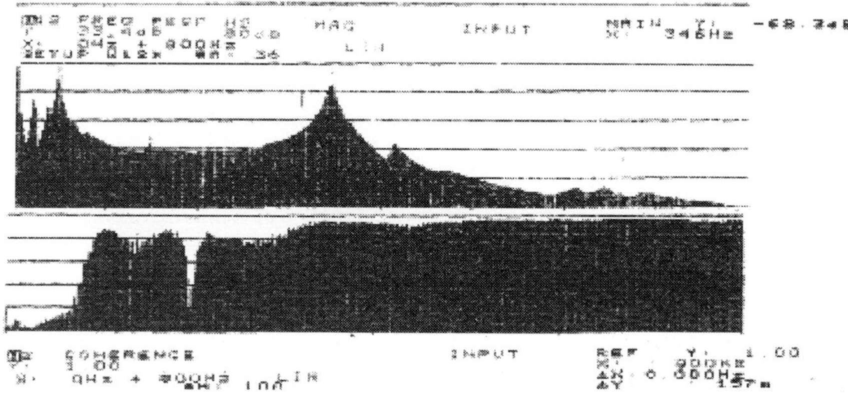

Fig. 4. A sample of FRF and coherence function.

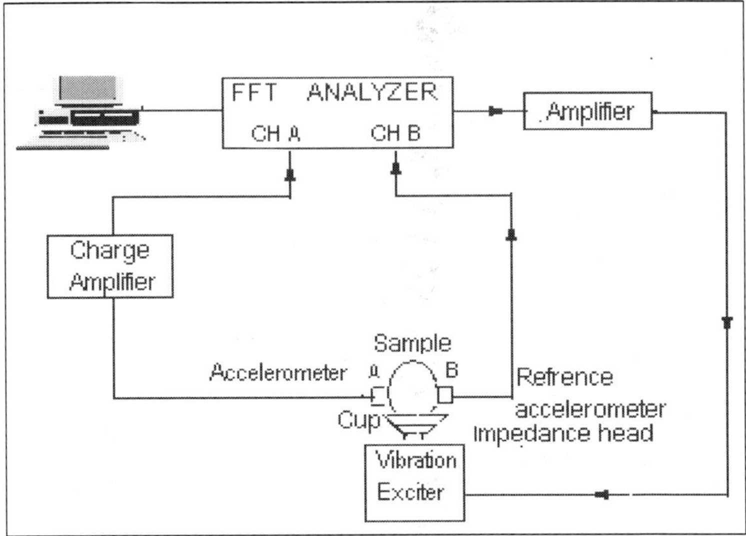

Fig. 5. (a) Schematic layout of the system used for operation modal analysis.

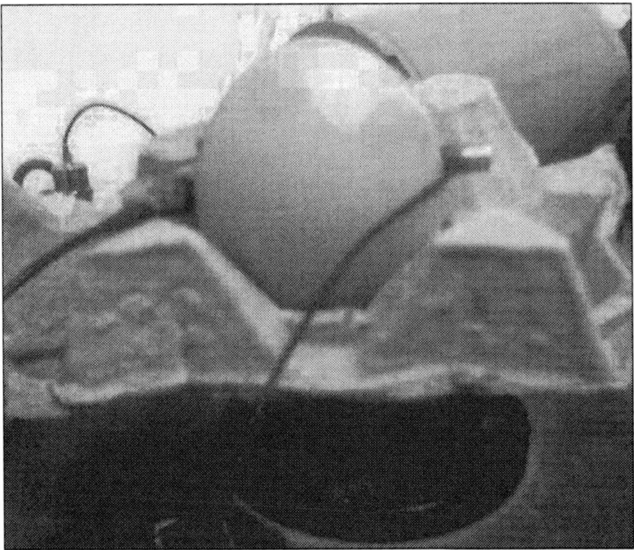

Fig. 5. (b) Fixed installation using a cup for operation modal analysis.

4. Pendulum for impact characterization

Amer Eissa (2005) suggested that, a mechanical breakage egg is cumulative and can occur each time when eggs are handled. It is a major problem in Egypt, Europe and world leads to direct losses in intensive egg production systems and increased labours costs during grading, sorting and processing. Eggs are subjected to mechanical impacts at the moment of lay, during collection, in the sorting equipment and during transport in trays.

A chicken egg is an already packaged food. An important quality aspect of the packaging material is the mechanical strength of the egg shell. A commonly used technique for the measurement of the shell strength is the quasi-static, non-destructive compression of an egg. The slope of the force-deformation curve is a measure for the stiffness of the shell and for the egg shell strength. However this method is tedious and the required equipment is expensive. An alternative method has been developed based on the impact behaviour of the egg. A portable pendulum for measuring the extent of dynamic shell breakage in eggs. It can be used to measure energy absorbed during impact. Measurements are made using an angular displacement transducer and interpreted by means of logic circuits. A coefficient of restitution parameter is defined which is based of the dynamic behaviour of the egg. Both impact velocity and rebound velocity showed promise as an indicator for evaluating the coefficient of restitution of egg shell during impact.

The new impact test method is a promising alternative for evaluating mechanical egg shell properties because it's destructive and non-destructive nature.

Economic loss due to egg shell breakage is an important problem in intensive egg production systems (Oosterwuud, 1987). Eggs are subjected to mechanical impacts at the moment of lay, during collection, in the sorting equipment and during transport in trays. Bain (1991) reported that 6% to 8% of all eggs laid are broken during handling from the production unit to the consumer, and in monetary terms this gives rise to losses of at least $600 million on a world wide basis. The value of egg production was $3.389 billion in 1992, compared to $ 3.209 billion in 1987 (Madison and Perez, 1994). Therefore, egg shell breakage continues to be a costly problem for the egg industry. In addition, people are at high risk when eating eggs which might be contaminated after being damaged, i.e., with cracks or checks (Amer, 1998; Bain, 1990).

Egg shell breakage depends both on the strength of the egg shell and magnitude of the mechanical load applied. Variables associated with shell strength are biological in nature (Hamilton, 1982). They involve the material and structural properties of the different layers which comprise the egg shell. The material properties depend on the type and the association between the mineral and organic components of the shell. The structural properties depend on egg shell thickness, size and shape of the egg and the distribution of shell over the egg surface.

In most of the methods used for measuring egg shell strength these basic physical properties are not quantified separately because of their complexity. Indeed, the shell curvature and its brittle nature make it difficult to measure the material properties of the shell by classical means (Amer Eissa & Gamea, 2003; Bain, 1990). Therefore, more practical techniques have to be developed. Such methods, however, describe the behaviour in terms of a superposition of several material and structural properties of the egg. One such commonly used method is the non-destructive, quasi static compression test. In this test the elastic stiffness properties of the whole egg shell structure are measured. An egg is placed horizontally between 2 flat parallel steel plates and then compressed at a constant compression speed until a predefined non-destructive load is applied. The slope of the force-deformation curve is an indicator of the mechanical stiffness of the shell. This mechanical stiffness depends on the thickness of the egg shell, the curvature of the egg shell, the diameter of the egg and the Young's modulus of the material in each layer of the shell (Amer, 1998; Bain, 1990).

A more precise method of measuring the impact strength of an egg is to record the force on the shell throughout the impact. Using this technique, Voisey and Hunt (1967 a,b) developed an instrument to measure the impact force required to fracture an egg. With this instrument the egg is supported in an adjustable plastic (nylon) cradle directly below a suspended aluminium rod 1900 mm in length, this cradle is adjusted, using a dial gauge, so that the upper surface of the egg is at a constant distance (5.0 mm) from the end of the aluminium rod for all measurements. The rod is suspended by an upper spherical end that fits into a socket connected to a vacuum pump. By opening a solenoid valve the vacuum is released and the rod falls freely until it strikes the egg. The impact of the rod on the egg is measured by a piezoelectric transducer attached to the lower end of the rod; the rod and transducer weigh 695 g. The transducer can be calibrated in either force or acceleration units. The electrical output of the transducer during the impact is recorded by a peak-shock meter and the maximum value displayed by a 4-digit voltmeter. The impact strength of about 150 eggs can be measured per hour with this instrument. Force-time plots can also be obtained using an oscilographa and camera to record the graph for each egg. It is very important that shock waves generated within the rod during the impact with an egg do not affect the measurement. This effect may be overcome by changing the length of the rod or the material from which it is made (Voisey and Hunt, 1968).

During handling of eggs, several impacts may be imparted to the egg shells, if not limited, will cause considerable damage. Several researchers have investigated the effects of impacts on agricultural products. Zapp et al., (1990) investigated the impact on apples by simulating an apple with an 'instrumented sphere'. Jindal and Mohsenin (1976) analysed a pendulum impacting device with which apples and corn kernels were tested. Finney and Massie (1975) also investigated a pendulum impacting device for testing the response of fruits to impacts. Other authors who have investigated impact devices for agricultural products include Amin (1995), Tennes et al., (1988), Siyami et al., (1986), Hughes et al., (1985), Simpson and Rehkugler (1972) and Bilanski (1964).

In the case of a chicken egg, the interpretation of the impact response of an egg is even more complex than for apples. An egg consists of several types of material: a shell, an air-chamber, egg white and yolk. The egg shell structure itself contains several distinct layers that are penetrated by pores. Within 1 layer, the material properties are not homogeneous. The material properties of the egg content are also variable. Nevertheless, it is worthwhile to asses the experimental relationships between the dynamic impact, mechanical properties of an egg and commonly measured physical egg and egg shell variables.

The values of the material property of eggshell were reported over 25 years ago. These values might be obsolete due to significant changes in chickens' species, feed and management practices. Therefore, it is necessary to measure the key material properties of present day shell eggs.

The objective of this work has been to develop apparatus able to evaluate shell dynamic strength, stiffness and relate their measurements with the egg shell quality purposes. This paper describes a pendulum which is portable and which will impact eggshell under a wide range of velocities and energies.

4.1 Construction of the pendulum

A pendulum impactor Figure (6) was used to apply the preselected amount of energy to the eggshell during impact. The portable folding pendulum a dynamic shell egg tester. It consists of two main parts: (1) a pendulum with an angular displacement transducer and arm release mechanism, length of the pendulum arm is 40 cm (2) a box containing the battery powered electronics, including control and display units, connected by cable to the angular displacement transducer. An angular displacement transducer attached to one end of the spindle is used to determine the rebound angle of the arm after impact for absorbed energy calculation. A simple pendulum, consisting of a steel ball cylindrical and spherical specimen on a lightweight (6.83 and 5.4 5 g, respectively), was used as the basic dropping apparatus as shown in Figure (6). This provided a simple system for negligible frictional losses during the drop. The main body and the platform of the pendulum are constructed from metal and its arm oscillates along a pivot supported by bearing. The duration of sample impact by the indentor and the time taken for the indentor to return to the point of initial contact with the egg are shown on individual displays. Signal from the angle meter first goes to a synchronic converter and then to a structural dynamic analyser. Impact parameters such as rebound angle can be monitored and recorded. These data are displayed by a single digital read out and selector switch. The height of the pendulum arm can be adjusted by moving the release catch mechanism on the metal plate (calibrated in cm drop height of the indentor). The various drop weights of the indentor are interchangeable and can be screwed to the end of the pendulum arm.

4.2 Dynamic tests

They differed in size, shape and mechanical properties from those tested in the fall under static loading will affect the impact characteristics to that from drop height, one of these two parameters must be known or estimated. By mathematical dependency, the known or estimated parameter can be derived from one of the following: impact velocity, rebound velocity, velocity ratio (coefficient of restitution), drop height and rebound height. The impact loads were expressed in terms of total energy, energy absorbed, total momentum and momentum absorbed. For user convenience, maximum and minimum values for each of these parameters are displayed for each impact. Once any one of the above parameter is estimated Amer Eissa, (2004). These properties may all be displayed simultaneously. These were defined by.

$E_{imp} = W\ h_{drop}$ = total energy or energy of impact, (N.m=J.106= μ J)

E_{abs} = W (hdrop-hreb) = energy absorbed by the egg, (N.m=J.106= μ J)

M = (W / 9.81) V1 = total momentum, (N.s.106=μN. s)

M_{abs}= (W / 9.81) (V1-V2) = momentum absorbed, (N.s.106=μN. s)

E_{reb} = Eimp (h_{reb} / h_{drop}) = energy of rebound of the pendulum arm, (J)

Where: V1 = (19.62 hdrop / 100)1/2   = impact velocity, m/s

V2 = (19.62 h_{reb} /100)1/2   = rebound velocity, m/s

W = Weight of the ball or rigid object (indentor), kg

h_{drop} = The height of drop, m = L (1- cos θ)

h_{reb} = The height of rebound, m = L (1- cos α)

L = The height of arm (0.40 m)

θ = The angle of drop, in degrees

α = The angle of rebound, in degrees.

The coefficient of restitution (e) which describes the rebound characteristics was calculated from.

$$e = \left(h_{reb} / h_{drop} \right) 1/2 = V2\ /\ V1$$

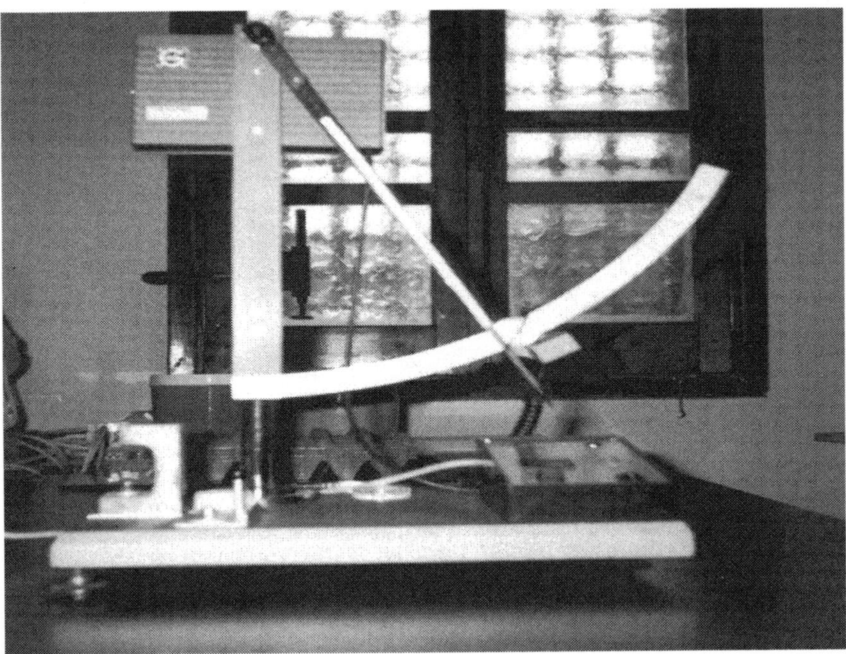

Fig. 6. The portable pendulum instrument.

This coefficient is usually defined as the ratio of final to initial relative velocity components of the striking bodies in the direction normal to the contact surfaces.

Mechanical breakage during collection of egg operations can affect both the egg and egg shell quality. The slope of the force deformation curve is a measure for the static stiffness (k_{stat}) of the egg, as:

$$F = k_{stat}\delta$$

where: F = force on the egg shell
    k_{stat} = static stiffness
    δ = deformation of the egg shell.

4.3 Strength under impact loading

Hen's eggshell strength under impact loading Šarka Nedomova (2009) found that An experimental method for evaluation of an eggshell's mechanical characteristics under impact loading is discussed. Proposed experimental set up enables recording of time history of the force at the contact area between the rod and eggshell, as well as the vibration response of the tested egg. By the gradual increasing of the rod impact velocity, a rupture force of the eggshell was determined. The value of this force depends on the position of the rod impact. This dependence is more significant than that of static loading of the eggs. The preliminary results also show the non-negligible dependence of rupture force on the loading rate. An introductory numerical simulation of the given test was also performed. The LS

DYNA finite element code was used. Numerical results exhibit reasonable agreement with experimental results.

For the egg industry and its consumers, an intact shell is the first and most important egg quality criterion. Since egg handling became more and more mechanized, eggshell quality became the topic of extensive research, probably because the balance between eggshell strength and the handling load applied to the shell became less favorable for the eggshell. The shell must be free of cracks and highly resistant to dynamic impact loads and static compression. Eggs are exposed to mechanical impacts at the moment of lay, during collection, in the sorting equipment, and during transport in trays. Static compression is the most critical consequence of the egg's packing. There are many methods for the study of the eggshell strength. Total shell strength is influenced by material and structural strength. In most of the methods used for measuring eggshell strength these basic physical properties are not quantified separately because of their complexity. Indeed, the shell curvature and its brittle nature make it difficult to measure the material properties of the shell by classical means (Bain, 1992). This is probably the main reason why eggshell strength is determined by the entire egg. For example, static stiffness (Ks; Voisey and Hunt, 1967) and dynamic stiffness (Kd; Coucke et al., 1998) are a measure for total shell strength. To determine Ks, an egg is compressed between parallel plates and, therefore, it is a measure for the interaction between structure and material characteristics. To determine Kd, the egg is excited by a small impact, and the vibration behavior is registered. Subsequently, from the resonant frequency and the mass of the egg, the Kd is calculated. These methods are non-destructive. To study the egg's resistance to impact, many other methods have been developed – see e.g. Tyler and Geake (1963), Voisey and Hunt (1968). These methods use balls or rods, which are dropped on the eggshell. The height of the fall, the size of ball, or a number of blows are used to estimate shell strength. The aim of the given paper is to further develop these experimental methods of dynamic strength evaluation. The loading has been performed by the impact of the free falling rods. The record of the force at the point of rod–eggshell contact enables evaluation of the rupture force at a definite impact velocity. The obtained results have been compared with results of the static compression of the eggs in order to find some evidence of the loading rate influence.

The values of the eggshell strength in terms of the rupture force are affected by many factors (egg specific gravity, egg mass, egg volume, egg surface area, egg thickness, shell weight, shape, and shell percentage). In order to avoid the simultaneous influence of these factors, the numerical simulation of these experiments should be performed. This procedure has been successfully used in Macleod et al. (2006) for the description of static loading. In the given paper this analysis has been performed for the dynamic loading.

4.3.1 Static loading

The test device (TIRATEST 27025, Germany) used for performing the measurements has three main components: a stationary and a moving platform, and a data acquisition system. Compression force was measured by the data acquisition system. The egg sample was placed on a block of polyurethane foam positioned on the stationary plate. The egg has been loaded by the moving rod (6 mm in diameter) at a speed of 20 mm/min. Two compression axes (X and Z) for the egg were used in order to determine the rupture force and deformation. The X-axis was the loading axis through the length dimension, and the Z-axis was the transverse axis containing the width dimension. Along the X-axis, two other orientations were considered. The eggs were loaded at the sharp end and at the blunt end see Fig. 7. For each orientation, 30 eggs were tested. The geometry of the eggshell was described using the shape index, SI. The average value of the shape index has been 76.18 ± 2.35 (%). In Fig. 9, the example of experimental record of the force vs. rod displacement is shown. One can see that the shape of this curve is different from those obtained at the eggs compression between two plates see Altuntas and Saekerog lu (2008).

4.3.2 Dynamic loading

The experimental set-up is shown in Fig. 8 and consists of three major components; the egg support, the loading device, and the response- measuring device (laser vibrometers POLITEC CLV 2000, USA). (1) The egg support used is a cube of soft polyurethane foam. The stiffness of this foam is significantly lower than the eggshell stiffness; therefore, there is very little influence of this foam on the dynamic behavior of the egg. (2) A bar of the circular cross-section with strain gauges (semi conducting, 3 mm in length) is used as a loading device. The bar is made from aluminum alloy. It is 200 mm long with a diameter of 6 mm. The bar is allowed to fall freely from a pre selected height. The instrumentation of the bar by the strain gauges enables to record time history of the force at the area of bar–eggshell contact. (3) The response of the egg to the impact loading, described above, was measured using the laser vibrometer. This device enables one to obtain the time history of the eggshell surface displacement. The eggs were impacted on the sharp end, on the blunt end, and on the equator. For each loading orientation, 20 eggs were used. The height of the bar fall was increased up to a value at which

the eggshell damage was observed. The displacement was recorded on the equator of the egg. The displacement was measured in the normal direction to the eggshell surface. Measuring the conductance of eggshells using the acoustic resonance technique and optical transmission spectra Bamelis, et al., (2008) found that , during the incubation of an avian egg, water vapour, oxygen and carbon dioxide are exchanged through the porous shell of the incubated egg. Due to the high variability of the eggshell conductance (G), large variation in exchange rates are present and hence a significant number of eggs are incubated in suboptimal conditions for humidity and partial pressures of carbon dioxide. Because there is no reliable technique to measure G in a non-destructive and fast way, the direct adaptation of the ambient conditions during incubation in relation to the G of the incubated eggs is not yet possible and this has repercussions on both the hatchability and chick quality. In the subject, two non-destructive and fast techniques, the Acoustic Resonance Technique (ART) and the measurement of light transmission through the egg, are used to estimate G. It was found that the dynamic stiffness of the egg ($kdyn$) and the optical transmission at 611nm are the parameters with the highest predictive power when estimating G. Although this model is highly significant ($P < 0.0001$), the R-value for the best model is only moderate ($R = 0.67$).

This indicates that there are still other parameters involved in the eggshell conductance that are not measurable by the ART and transmission of light. However, with the presented combination of non-destructive techniques, different classes of eggs based on their shell conductance could be created and incubated separately. During the incubation of chicken eggs, water vapour, oxygen and carbon dioxide are exchanged through the porous eggshell. Theoretically, the exchange rate of each gas is determined by the difference in the vapour pressure of each gas between the inside and the outside of the egg and the eggshell conductance for each gas ($GH2O$, $GO2$ or $GCO2$) (Paganelli, 1980; Tullett, 1981). Since $GH2O$, $GO2$ and $GCO2$ are closely related (Rahn, 1981), $GH2O$ is often used as a general

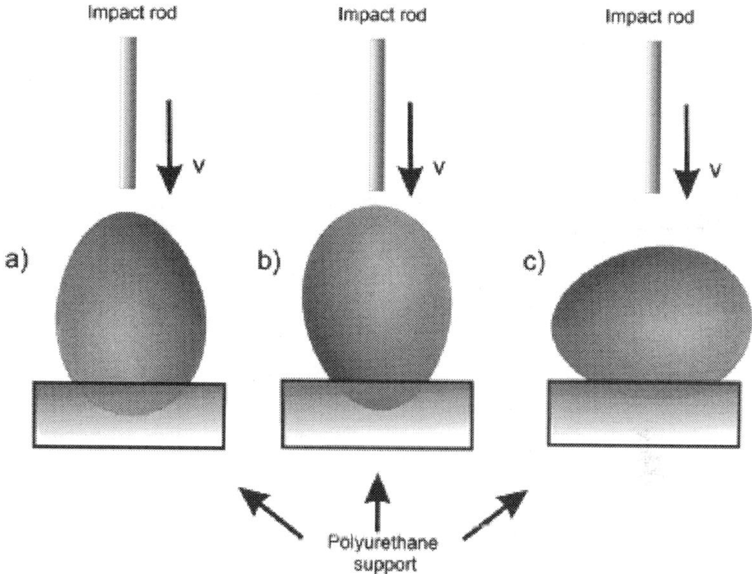

Fig. 7. Schematic of impact loading; (a) along X-axis (sharp end), (b) along X-axis (blunt
end), (c) along Z-axis (equator).

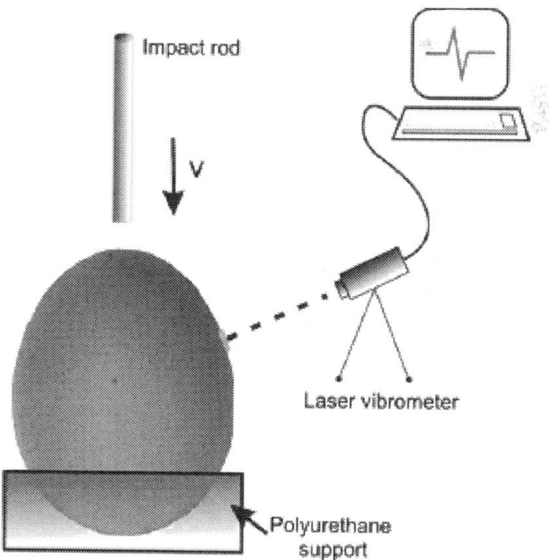

Fig. 8. Experimental set-up – dynamic loading.

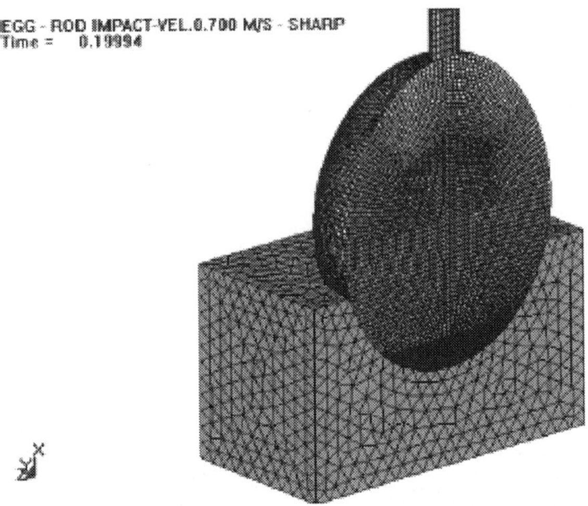

Fig. 9. Numerical model of the experimental set-up.

measure for the eggshell conductance and noted as G. The ambient environment of eggs during artificial incubation is usually kept constant and therefore the eggshell conductance and the inside partial gas pressure in the egg are the major players in the exchange process of water vapour, oxygen and carbon dioxide.

The air in the air chamber and between the fibres of the shell membranes is totally saturated with water vapour, making the exchange rate of water vapour through the eggshell towards the outside only dependent on G for fixed incubator humidity. Since the respiration quotient of the developing chick inside is about 0.772, the mass of the exchanged CO_2 equals the mass of the O_2 taken up (Tullett, 1981). Thus, the measurement of the mass loss of the egg during the incubation is a suitable estimator for the water loss and hence for G. The developing embryo produces more CO_2 and requires more O_2 towards the end of the 21-day incubation period. This change causes the gradient of the partial pressure of CO_2 and O_2 over the eggshell to increase, and as a consequence, the diffusion rate of those two gases will increase during the ongoing incubation. At a certain moment, the difference in partial pressures over the eggshell for both gases cannot increase more for physiologic reasons, and hence, a maximal diffusion rate of CO_2 and O_2 is reached. The period in which a maximal exchange rate is established is called the plateau phase (Rahn et al., 1979). During the plateau phase, the amount of O_2 that can reach the embryo and the amount of CO_2 that can leave from the embryo are limited by G. A high variability is reported on G by French and Tullett (1991), even in eggs from chickens of the same line and age. Therefore the duration and the level of the plateau phase are variable and hence the physiology of the developing embryo will be affected in different ways concerning the conductance of the eggshell (Ar and Rahn, 1980; Tazawa, 1980).

Until now, there has been no suitable technique to measure or closely estimate G besides the measurement of mass loss over a fixed period (mostly 4 days) in a controlled humidity environment with at least two different measurements of the mass of the egg. Online measurements of weigh losses of egg trays in modern incubators are valuable for adapting the mass losses of the mean of all setted eggs, not for differentiating eggs based on G before the incubation starts. If there should be a single measurement technique that is not destructive and fast before the incubation starts, then it could become possible to adapt the environmental factors in the incubator (humidity, CO2 andO2 partial pressures). This adjustment could make the conditions for different classes of eggs more adapted to their physiological needs, and hence, an increased hatchability may be expected. In the research presented here, two non-destructive and fast measurement techniques were tested for their ability to estimate G. The first technique, the Acoustic Resonance Technique (ART) is used to estimate the dynamic stiffness of the eggshell (k_{dyn}) from the resonant frequency of the S20 vibrational mode after an excitation of the egg (Coucke et al., 1999). During this S20 vibrational mode, the equator of the shell describes an elliptic, spherical deformation with two nodal lines passing over the poles of the egg. Since k_{dyn} is correlated with the eggshell thickness, in the same manner as G, k_{dyn} might be an estimator for G. Moreover, the ART technology is fast and non-destructive (Coucke, 1998). In the second technique, the light transmission (200–900 nm) through the egg was measured. Since it was assumed that light may travel more easily through a more porous medium, the transmitted spectrum of visible light through the egg may include some information about the porosity of the eggshell. Also this type of measurement is non-destructive and fast (Williams and Norris, 1987).

5. References

Abdallah A G; Harms R H; El-Husseiny O (1993). Various methods of measuring shell quality in relation to percentage of cracked eggs. Poultry Science, 72, 2038–2043

Altuntas, E., and Sekeroglu, A. (2008). Effect of egg shape index on mechanical properties of chicken eggs. Journal of Food Engineering 85: 606-612.

Amer Eissa A. H. and G. R. Gamea. (2003). Physical and mechanical properties of bulb onion. Misr J. Ag. Eng., 20(3): 661 – 676.

Amer Eissa A. H., and Gomaa. F. R. (2009). Operational Modal Analysis in Fruit Quality Assessment Using Different Methods of Packaging. Proceedings of the 8th Fruit, Nut and Vegetable Production Engineering Symposium, Concepcion Chile, 5-9-Jan., 2009.

Amer Eissa A.H. (2004). A portable pendulum for impact characterization of whole eggshell. Misr j. Agr. Eng., 21(1):1-13.

Amer Eissa A.H. (2005). Effects of age on some physical and rheological properties of eggshell quality in chickens. Minofiya J. Agric. Res. Vol. 30 No. 51-76.

Amer Eissa A.H. (2009). Comparative eggshell stability assessment using three different non-destructive sensing instruments and breakage force strength. Journal of Food Engineering, (93) 444-452.

Amer Eissa A.H. and Gomaa. F. R., (2007). "Prediction of Fracture of Eggshell Using Operational Modal Analysis". Engineering Research Journal, Minoufiya University, Vol.30, No.1, January 2007.

Amer Eissa, A.H. (1998). The engineering factors affecting handling and loss reduction of egg production. Ph.D Thesis, University of Minoufiya, Shibin El-Kom, Egypt.

Amin, E. E. A. A., (1995). Development of a grading machine for some horticultur farm crops. J. Ag. Sci. Mansoura Univ. 19(7):2399-2411.

Ar, A. and H. Rahn. (1980). Water in the avian egg: Overall budget of incubation. Am. Zool. 20:373–384.

Bain, M., I. C. Dunn, P. W. Wilson, N. Joseph, B. De Ketelaere, J. De Baerdermaeker, and D. Waddington. (2006). Dynamic stiffness: A novel method of determining eggshell quality in domestic hens that predicts the likelihood of damage in the field. Br. Poult. Sci. (accepted) Bell, D. Egg breakage—from the hen to the consumer. Calif. Poult. Lett. (Apr.):2–6.

Bain, M., I. C. Dunn, P. W. Wilson, N. Joseph, B. De Ketelaere, J. De Baerdermaeker, and D. Waddington (2006). Probability of an egg cracking during packing can be predicted using a simple non-destructive acoustic test. British Poultry Science, 47, 462-469

Bamelis, F.R., De Ketelaere, B., Mertens, K., Kemps, B.J., Decuypere, E.M., De Baerdemaeker, J.G., 2008. Measuring the conductance of eggshells using the acoustic resonance technique and optical transmission spectra. Computers and Electronics in Agriculture 62 (1), 35–40.

Blevins R D (1979). Formulas for natural frequency and mode shape. Litton Educational Publishing Inc.

Bowman, J. C., and N. I. Challender. (1963). Egg shell strength. A comparison of two laboratory tests and field results. Br. Poult. Sci. 4:103–116.

Brian Schwarz & Mark Richardson (2000). (Measuring operating Deflection shapes under non- stationary conditions. 18th International Conference of Modal Analysis. February 7-10 IMAG 2000.

Brincker R; Zhang L M; Andersen P (2002). Modal identification from ambient responses using frequency domain decomposition. Proceedings of the 18th International Modal Analysis Conference, San Antonio, Texas, USA

Brooks, J., and H. P. Hale. (1955). Strength of the shell of the hen's egg. Nature 175:848.

Caccese V; Mewer R; Vel S S (2004). Detection of bolt loss in hybrid composite/metal bolted connections. Engineering Structures, 26(7), 895-906

Charles, F., and J. R. Strong. (1988). Research note: Relationship between several measures of shell quality and egg-breakage in a commercial processing plant. Poult. Sci. 68:1730–1733.

Chen Y R; Park B; Huffman RW; Nguyen M (1998). Classification of on-line poultry carcasses with backpropagation neural networks. Journal of Food Process Engineering, 21(1), 33-48

Chen, C., Y. P. Chiang, Y. Pomeranz. (1998). Image analysis and characterization of cereal grains with a laser range finder and camera contour extractor. Cereal Chemistry 66 (6), 466–470.

Cho, H. K., Choi, W. K., & Paek, J. H. (2000). Detection of surface in shell eggs by acoustic impulse method. Transactions of ASAE, 43(6), 1921–1926.

Coucke P; Dewil E; Decuypere E; De Baerdemaeker J (1999). Measuring the mechanical stiffness of an eggshell using resonant frequency analysis. British Poultry Science, 40 (2) 227-232

Coucke, P. (1998). Assessment of some physical egg quality parameters based on vibration analysis. PhD Thesis, Katholieke Univ. Leuven, Belgium.

Coucke, P., Langenakens, J., Sas, P., & De Baerdemaeker, J. (1994). Experimental modal analysis on chicken eggs. Proceedings of the 12th International Modal Analysis Conference, 12, 1258-1263.

Damir A N; Elkhatib A; Nassef G (2004). Prediction of fatigue life using modal analysis for grey and ductile iron. International Journal of Fatigue, 29(3), 499-507

Das K; Evans M D (1992). Detecting fertility of hatching eggs using machine vision II: Neural network classifiers. Transaction of the ASAE, 35(6), 2035-2041

De Ketelaere B; Moshou D; Coucke P; De Baerdemaeker J (1997). A hierarchical self-organizing map for classification problems. In: Proceedings WSOM '97: International Workshop on Self-Organizing Maps, Helsinki University of Technology, Espoo, Finland, 86–90

De Ketelaere, B., Bamelis, F., Kemps, B., Decuypere, E., & De Baerdemaeker, J. (2004). Nondestructive measurements of the egg quality. World's Poultry Science Journal, 60(3), 289-302.

De Ketelaere, B., Coucke, P., & De Baerdemaeker, J. (2000). Eggshell crack detection based on acoustic resonance frequency analysis. Journal of Agricultural Engineering Research, 76(1), 157–163.

De Ketelaere, B., T. Govaerts, P. Coucke, E. Dewil, J. Visscher, E. Decuypere, and J. De Baerdemaeker. (2002). Measuring the eggshell strength of 6 different genetic strains of laying hens: Techniques and comparisons. Br. Poult. Sci. 43:238–244.

Doebling S W; Farrar C R; Prime M B (1998). A summary review of vibration- based damage identification methods. The Shock and Vibration Digest, 30 (2), 91-105.

Doebling S W; Farrar C R; Prime M B; Shevitz DW (1996). Damage identification and health monitoring of structural and mechanical systems from changes in their vibration characteristics: a literature review. Los Alamos National Laboratory Report LA-13070-MS, New Mexico, USA

Elster R T; Goodrum J W (1991). Detection of cracks in eggs using machine vision. Transaction of the ASAE, 34(1), 307-312

Fahy, F. (1985). Sound and structural vibration eradiation, transmission and response. London: Academic Press.

Farrar C R; Doebling S W (1997). An overview of modal-based damage identification methods. In: Proceeding of DAMAS Conference, Sheffield, UK

Ganguli R (2001). A fuzzy logic system for ground based structural health monitoring of a helicopter rotor using modal data. Journal of Intelligent Material System and Structures, 12(6), 397-407

García-Alegre M C; Ribeiro A; Guinea D M; Cristóbal G (2001). Color index analysis for automatic detection of eggshell defects. URL:http://www.iv.optica.csic.es/papers/huevos.pdf. [Download: December 18, 2001]

Ghazanfari A; Irudayaraj J; Kusalik A (1996). Grading pistachio nuts using a neural network approach. Transaction of the ASAE, 39(6), 2319-2324

Ghoshal A; Sundaresan M J; Schulz M J; Pai P F (2000). Structural health monitoring techniques for wind turbine blades. Journal of Wind Engineering and Industrial Aerodynamics, 85(3), 309-324

Goodrum J W; Elster R T (1992). Machine vision for crack detection in rotating eggs. Transaction of the ASAE, 35(4), 1323-1328

GRUNDER, A.A., FAIRFULL, R.W., HAMILTON, R.M.G. and THOMPSON, B.K. (1991) Correlations between measures of eggshell quality or percentage of intact eggs and various economic traits. Poultry Science, 70: 1855–1860.

Hamilton R M G (1982) Methods and factors that affect the measurement of eggshell quality. Poultry Science, 61 (10), 2022–2039

Hunton, P. (1993). Understanding the architecture of the eggshell. Proc. 5th Eur. Symp. Qual. Eggs and Egg Prod., Tours, France. World's Poult. Sci. Assoc., Cedex, France. Pages 141–147

Hunton, P., 1995. Understanding the architecture of the eggshell. World's Poultry Science Journal 51, 140–147.

Jindal V K; Chauhan V (2001). Neural networks approach to modelling food processing operations. In: Food Processing Operations Modelling: Design and Analysis (Irudayaraj J, ed). Marcel Dekker, Inc, New York, 305-323

Kemps, B.J., T. Govaerts, B. De Ketelaere, K. Mertens, F.R. Bamelis, M.M. Bain, E.M. Decupere and J.G. De Baerdemaeker (2006). The influence of Line and Laying Period on the Relationship Between Different Eggshell and Membrane Strength Parameters. Poultry Science (85): 1309-1317.

Kuo E Y; Jayasuriya A M M (2002) A high mileage vehicle body joint degradation estimation method. International Journal of Materials and Product Technology, 17(5-6), 400-410

MacLeod, N.; M. M. Bain; and J. W. Hancock (2006). The mechanics and mechanisms of failure of hens Eggs. Int J Fract 142: 29-41.

Mertens K; Bamelis F; Kemps B; Kamers B; Verhoelst E; De Ketelaere E; Bain M; Decuypere E; De Baerdemaeker J (2006). Monitoring of eggshell breakage and eggshell strength in different production chains of consumption eggs. Poultry Science, 85, 1670-1677

Mohanty P; Rixen D J (2004). Operational modal analysis in the presence of harmonic excitation. Journal of Sound and Vibration, 270(1-2), 93-109

Mohsenin, N. N. (1986). physical properties of plant and animal materials. Gordon of Breach science publishers, New York.

Moshou D; De Ketelaere B; Coucke P; De Baerdemaeker J; Ramon H (1997). A hierarchical self-organizing map for egg breakage classification. In: Third IFAC/ISHS International Workshop on Mathematical and Control Applications in Agriculture and Horticulture, Hanover, Germany, 125–129

Olsson, N. (1934). Studies on Specific Gravity of Hen's Egg. A Method for Determining the Percentage of Shell on Hen's Eggs. Otto Harrassowitz, Leipzig, Germany.

Patel V C; McClendon R W; Goodrum J W (1994). Crack detection in eggs using computer vision and neural networks. AI Applications, 8(2), 21-31

Perianu C., B. De Ketelaere , B. Pluymers , W. Desmet , J. DeBaerdemaeker , E. Decuypere (2010). Finite element approach for simulating the dynamic mechanical behaviour of a chicken egg. Biosystems engineering 106 (2010) 79-85.

PREISINGER, R. and FLOCK, D.K. (2000) Genetic changes in layer breeding: historical trends and future prospects. Occasional Publications of the British Society of Animal Science, 27: 20–28.

Rehkugler, G. E. (1963). Modulus of elasticity and ultimate strength of the hen's egg shell. Journal of Agricultural Engineering Research, 8, 352-354.

Richardson M; Potter R (1974) Identification of modal properties of an elastic structure from measured transfer function data. 20th International Instrumentation Symposium (I.S.A), Albuquerque, New Mexico, USA

Salawu O S (1997). Detection of structural damage through changes in frequency: a review. Engineering Structures, 19(9), 718 -723

Sarka Nedomova, Jan Trnka , Pavla Dvorakov , Jaroslav Buchar , Libor Severa .(2009). Hen's eggshell strength under impact loading. Journal of Food Engineering 94 (2009) 350–357.

Schulz M J; Pai P F, Inman D J (1999) Health monitoring and active control of composite structures using piezoceramic patches. Composites Part B Engineering, 30(7),713 - 725

Seide P (1975). Small elastic deformation of thin shells. Noordhoff International, Leyden, The Netherlands, 615-620

Solomon, S. E. (1991). Egg and eggshell quality. London, United Kingdom: Wolfe Publishing Ltd.

Stavrinidis, C., Witting, M., & Klein, M. (2001). Advancements in vibroacoustic evaluation of satellite structures. Acta Astronautica, 48(4), 203e210.

Thompson, B. K., and R. M. G. Hamilton. (1986). Relationship between laboratory measures of shell strength and breakage of eggs collected at a commercial grading station. Poult. Sci. 65:1877–1885.

Thompson, B. K., R. M. G. Hamilton, and P. W. Voisey. (1981). Relationships among various traits relating to shell strength, among and within five avian species. Poult. Sci. 60:2388–2394.

THOMPSON, B.K., HAMILTON, R.M.G. and GRUNDER, A.A. (1985) Relationships between laboratory measures of eggshell quality and breakage in commercial egg washing and candling equipment. Poultry Science, 64: 901–909.

Todd M D; Nichols J M; Pecora L M; Virgin L N (2001). Vibration-based damage assessment utilizing state space geometry changes: local attractor variance ratio. Smart Materials and Structures,10(5),1000-1008

Voisey P W; Hunt J R (1967) Relationship between applied force deformation of egg shells and fracture force. Journal of Agricultural Engineering Research, 12 (1) 1–4

Voisey, P. W., & Hunt, J. R. (1974). Measurement of eggshell strength. Journal of Texture Studies, 5, 135-182.

Voisey, P. W., and J. R. Hunt. (1967a). Physical properties of egg shells. Stress distribution in the shell. Br. Poult. Sci. 8:263–271.

Voisey, P. W., and J. R. Hunt. (1967b). Relationship between applied force, deformation of eggshells and fracture force. J. Agric. Eng. Res. 12:1–4.

Voisey, P. W., and R. M. G. Hamilton. (1976). Factors affecting the non-destructive and destructive methods of measuring eggshell strength by the quasi-static compression test. Br. Poult. Sci. 17:103–124.

Vold H I; Schwarz B; Richardson M (2000). Measuring operating deflection shapes under non- stationary conditions. 18th International Conference of Modal Analysis (IMAC), San Antonio, Texas, USA

Wang D H; Huang S L (2000). Health monitoring and diagnosis for flexible structures with PVDF piezoelectric film sensor array. Journal of Intelligent Material System and Structures, 11(6), 482-491

Wang, J., Teng, B., & Yu, Y. (2004). Pear dynamic characteristics and firmness detection. European Food Research and Technology, 218, 289–294.

Wells, R. G. (1967). Egg shell strength. 1. The relationship between egg breakage in the field and certain laboratory assessments of shell strength. Br. Poult. Sci. 8:131–139.

Yan Y J; Yam L H (2002). Online detection of crack damage in composite plates using embedded piezoelectric actuators/sensors and wavelet analysis. Composite Structures, 58(1), 29-38

Zienkiewicz, O. C., Taylor, R. L., Zhu, J. Z., & Nithiarasu, P. (2005). The finite element method. The three volume set (6th ed.). United Kingdom: Butterworth-Heinemann.

Zou Y; Tong L; Steven G P (2000). Vibration-based model-dependent damage (delamination) identification and health monitoring for composite structures- a review. Journal of Sound and Vibration, 230(2), 357-78

Zubaydi A; Haddara M R; Swamidas A S J (2002). Damage identification in a ship's structure using neural networks. Ocean Engineering, 29(10), 1187-1200.

Morphology Control of Ordered Mesoporous Carbon Using Organic-Templating Approach

Shunsuke Tanaka[1] and Norikazu Nishiyama[2]
[1]Department of Chemical, Energy and Environmental Engineering
Faculty of Environmental and Urban Engineering, Kansai University
[2]Division of Chemical Engineering
Graduate School of Engineering Science, Osaka University
Japan

1. Introduction

The discovery of nanostructured carbon materials such as fullerenes (Kroto et al., 1985) and carbon nanotubes (Iijima, 1991) has led to a considerable interest in the development of various carbonaceous materials. In particular, porous carbonaceous materials have been attracting much attention because of their high surface area, large pore volume, chemical inertness, and high mechanical stability. Porous carbons show promise in the fields of hydrogen-storage, catalysis, separation, nanoreactors, electrochemistry, and biochemical engineering. Traditional synthesis methods, which involve carbonization of activated carbon (Marsh et al., 1971; Tamai et al., 1996; Hu et al., 2000), produce only disordered materials. To date, fabrication of highly ordered structure remains challenging. Research efforts to produce porous carbon materials with well-tailored pore systems have focused on the use of various inorganic template materials such as porous anodic aluminum oxide (AAO) films (Kyotani et al., 1995, 2006), zeolites (Kyotani et al., 1997; Ma et al., 2000; Nishihara et al., 2009), siliceous opals (Zakhidov et al., 1998; Yu et al., 2002), and mesoporous silicas (Ryoo et al., 1999; Lee et al., 1999; Kaneda et al., 2002; Kleitz et al., 2003; Xia et al., 2006) to template the carbon.

The use of zeolites, which have 3-dimensionaly connected framework structures constructed from corner-sharing TO_4 tetrahedra, where T is any tetrahedrally-coordinated cation such as Si and Al, as templates for the synthesis of carbon deposits on the micropore walls has been successful. The resulting materials have ordered and uniform angstrom-sized pores. However, long-range ordered microporous carbon replicas require repetitive carbonization steps to completely fill the template pores.

Silica opals, also called colloidal crystals, which are made by the self-assembly of uniform submicrometre-sized silica spheres, have been used as templates for the synthesis of ordered macroporous carbons. The porosity and contact sites between the silica spheres provided walls and interconnected spherical pores, respectively, in the resulting carbons.

Similarly, synthesis of ordered mesoporous carbons has focused on the use of ordered mesoporous silicas with interconnected pore structures as templates. Ordered structures of mesoporous silicas are derived from the self-assembly of surfactants and silica precursors.

The development of the M41S family (Kresge et al., 1992; Beck et al., 1992; Zhao et al., 1998) triggered the synthesis of a wide variety of mesoporous materials with diverse symmetries using various surfactants. As a result, various mesoporous carbon nanostructures with different pore systems have been synthesized using a variety of different mesoporous silica templates (Fig. 1). Pore size is controllable by selecting silica templates of different lengths and adjusting the silica wall thickness, though there is no report of tailoring only the silica wall thickness of the mesoporous silicas with constant pore diameter.

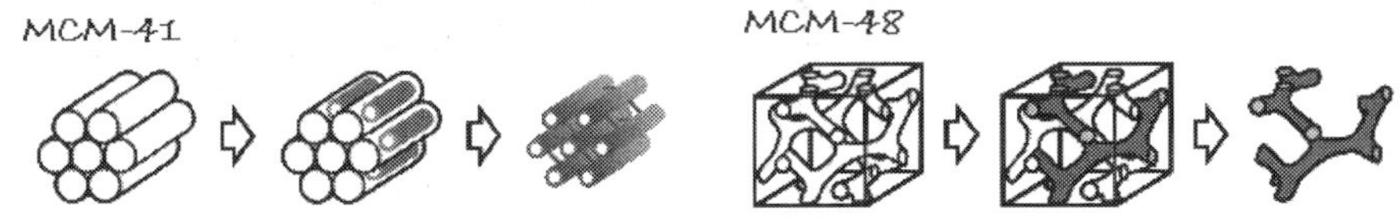

Fig. 1. Schematic illustration of the concept of M41S-template synthesis for ordered mesoporous carbons.

All the above inorganic-templating techniques require (1) preparation of inorganic templates, (2) impregnation of template pores with appropriate carbon precursors, (3) carbonization, and (4) subsequent selective removal of the templates using hydrofluoric acid or sodium hydroxide (Fig. 2). It is interesting to note that the pore system of carbons is inversely replicated from the silica templates, and thus, various mesostructured carbons with different pore systems have been synthesized using a variety of different mesoporous silica templates. Often, this time-consuming and costly process requires multiple infiltrations to complete the filling of the template pores. It is difficult to perform pore filling of the carbon precursors into the mesopores of silicas with low accessible pore systems.

Alternative methods, which eliminate the need for an inorganic template, have recently been developed to synthesize highly ordered mesoporous carbons, by directly assembly of organic templates with the carbon precursors (Tanaka et al., 2005, 2007, 2009; Meng et al., 2005; Zhang et al., 2005, 2006; Jin et al., 2009, 2010; Simanjuntak et al., 2009). This strategy uses an organic–organic interaction between a thermosetting resin and a thermally-decomposable copolymer to form a periodic ordered nanocomposite. The thermosetting

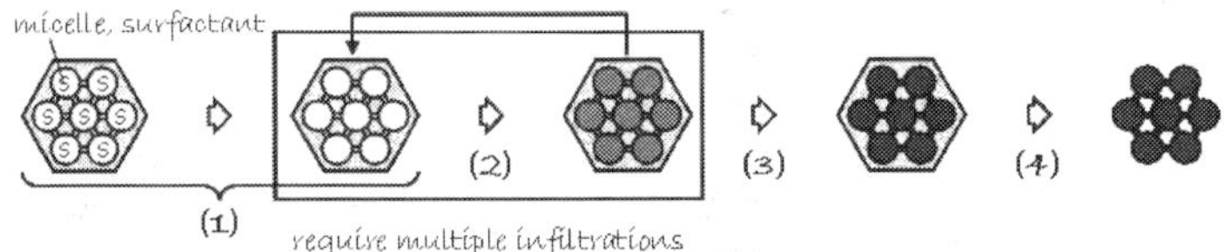

Fig. 2. Schematic illustration of the synthesis routes for ordered mesoporous carbons.

resin remains as the carbonaceous pore walls, while thermally-decomposable copolymer decomposes to form the mesopores (Fig. 2).

Simple techniques to control morphology and configuration of ordered mesoporous carbons are required for the development of practical applications. In this chapter, an advantageous organic-templating method for morphology control of ordered mesoporous carbons is introduced.

2. Self-assembly of organic–organic nanocomposites

In the inorganic-templating method, variable carbon precursors, e.g., sucrose, furfuryl alcohol, naphthalene, acetylene, polyacrylonitrile, and phenolic resin, can be utilized. On the other hand, in the organic-templating method, the main carbon precursors have been phenolic polymer resins prepared using phenolic resin monomer and formaldehyde (Fig. 3). The major reactions between phenolic resin monomer and formaldehyde include an addition reaction to form methylene and hydroxymethyl derivatives to form methylene and methylene ether bridged compounds. Interestingly, it has been pointed out that the polymerization mechanism and structure of resorcinol/formaldehyde are analogous to that described for the sol–gel processing of silica (Pekala 1989).

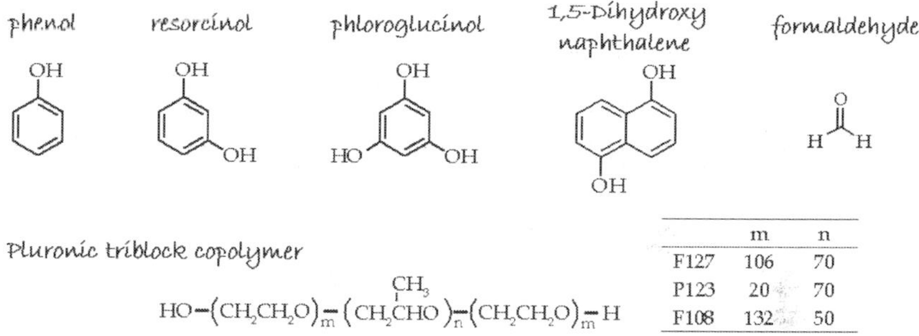

Fig. 3. Typical molecular structures of carbon precursors and thermally-decomposable polymer templates.

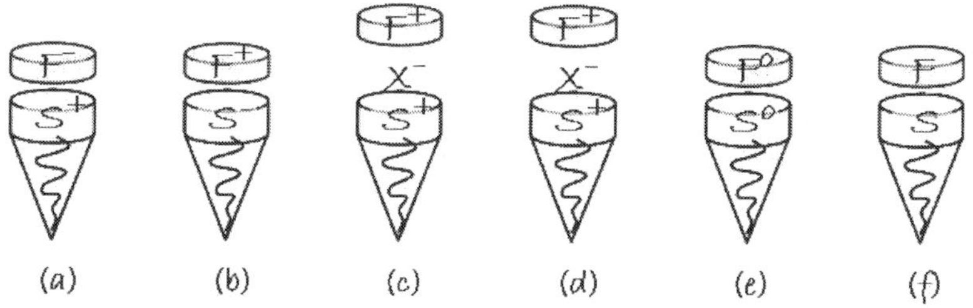

Fig. 4. Schematic representation of the various types of framework precursor–surfactant head group interactions: electrostatic S+F- (a), S-F+ (b), S+X-F+ (c), and S-X+F- (d) hydrogen bonding S⁰F⁰ (e), and covalent bonding S–F (f).

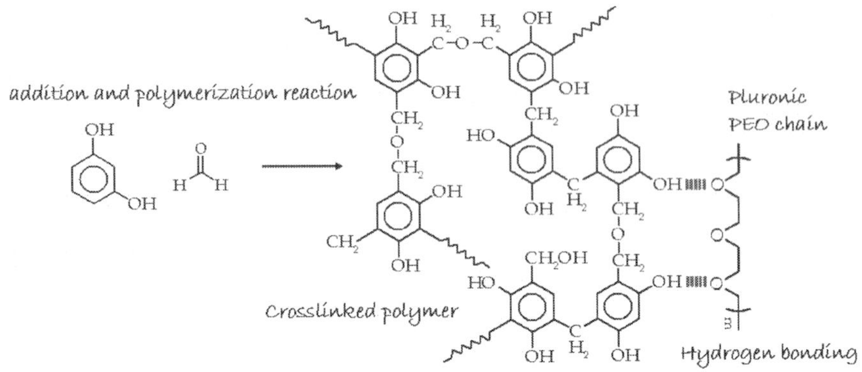

Fig. 5. Schematic representation of the reaction of resorcinol with formaldehyde and hydrogen-bonding interaction between PEO-containing block-copolymers and the hydroxyl-group-containing organic precursors.

The commercially available Pluronic block copolymers, e.g., F127, P123, and F108, have been used as templates (Fig. 3). To fabricate mesostructure, it is important to adjust the chemistry of the template head groups that can fit the requirement of the carbon precursors. The molecular interaction between the template head group and framework precursor can be expected using conventional reaction schemes. Six different possible molecular reaction pathways which use the principle of surfactant liquid crystal templating have been identified: (S^+F^-), (S^-F^+), $(S^+X^-F^+)$, $(S^-X^+F^-)$, (S^0F^0), and $(S–F)$, where S is the surfactant (cationic S^+, anionic S^-, neutral S^0), F is the soluble framework precursor (cationic F^+, anionic F^-, neutral F^0), and X is the intermediated (cationic X^+, anionic X^-) molecular species (Fig. 4). S–F indicates systems where the framework specie is covalently bonded to the template. The pathway applicable to a particular synthesis will be dictated by the reagents and synthesis conditions and can be influence the physical and chemical properties of the product. Hydrogen-bonding interaction between the copolymer template and the phenolic polymer resin is an efficient route to prepare mesoporous carbons (Fig. 5).

2.1 Synthesis of ordered mesoporous carbons using organic templates

In a typical synthesis, phenolic resin monomers were completely dissolved in a mixture composed of deionized water, ethanol and hydrochloric acid. Pluronic F127 was then added, and after it was completely dissolved, formaldehyde (37 wt.%) was added to the solution. The final molar composition of the solution was 4 phenolic resin monomers : 1 : 0.005–0.05 Pluronic F127 : 9 formaldehyde : 0.1 HCl : 20–100 ethanol : 40 water. The solutions were left at room temperature, during which they separated into two phases. The transparent upper phase was ethanol–water rich and the lower dark brown phase was polymer-rich. The upper clear phase was discarded; the lower dark brown phase was preheated at 100 °C for 1 h in air. Subsequently, the resultant brown sample was carbonized under a nitrogen atmosphere at 200–800 °C.

On the basis of the thermosetting resorcinol/formaldehyde resins, ordered mesoporous carbons, designated as COU, have been synthesized via the triblock copolymer F127-

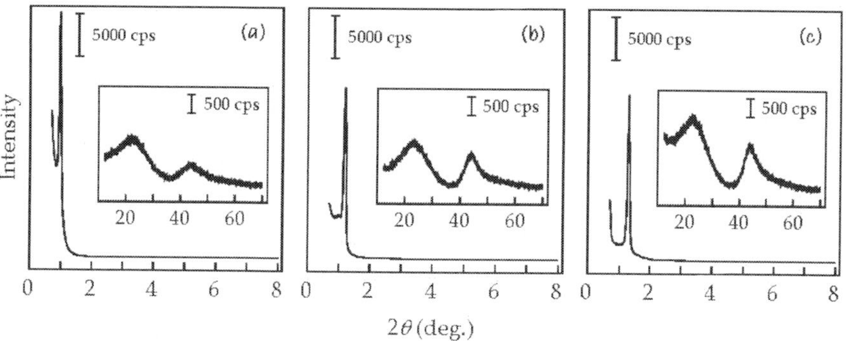

Fig. 6. XRD patterns of carbonized COU-1. The carbonization temperatures were (a) 400 °C, (b) 600 °C, and (c) 800 °C. (Tanaka et al., 2005)

templating route with the addition of triethyl orthoacetate as a co-carbon source. Fig. 6 shows X-ray diffraction (XRD) patterns of ordered mesoporous carbons COU-1. XRD pattern revealed a sharp reflection peak at a 2θ angle of 0.9–1.3°, demonstrating the periodically ordered structure of the carbons. The key to the success of their synthetic procedure was the formation of a periodically ordered organic–organic nanocomposite composed of thermosetting polymeric carbon precursors and the use of a thermally decomposable triblock copolymer Pluronic F127. Ordered mesoporous carbons COU-1 carbonized at different temperatures show typical type-IV N_2 adsorption/desorption curves with hysteresis loops, ascribed to the uniform mesopores inside the carbons (Fig. 7). The pore diameters of COU-1 carbonized at 400, 600 and 800 °C were estimated to be 7.4, 6.2 and 5.9 nm, respectively. The field-emission SEM images clearly show the hexagonally arranged channel pores and strongly support the results of the XRD and N_2 adsorption measurements (Fig. 8). Ordered straight channels have never before been seen in ordered mesoporous carbons synthesized by the inorganic-templating method.

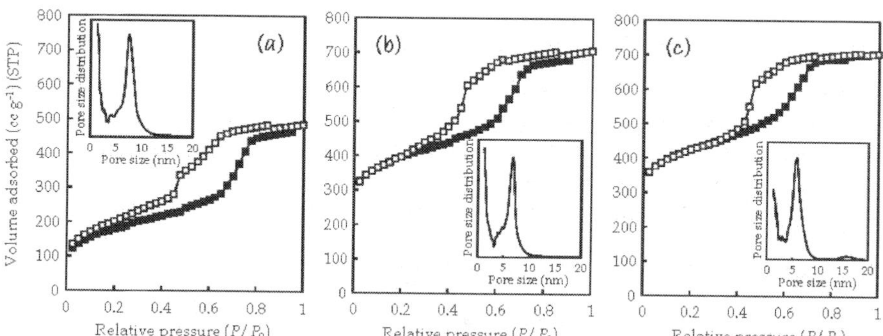

Fig. 7. N_2 adsorption/desorption isotherms and pore size distribution (inset) for carbonized COU-1. The carbonization temperatures were (a) 400 °C, (b) 600 °C, and (c) 800 °C. (Tanaka et al., 2005)

Fig. 8. FESEM images of carbonized COU-1. The carbonization temperatures were (a) 400 °C, (b) 600 °C, and (c) 800 °C. (Tanaka et al., 2005)

The surfactant F127 must have decomposed below 400 °C, because mesopores were already generated in the COU-1 carbonized at 400 °C. The thermosetting RF polymer remained as the carbonaceous pore walls, while the Pluronic F127 decomposed to form the mesopores. The increase in the micropore volume and the BET surface area at 600 °C must be due to the generation of gases from the decomposition of the RF polymer. The molar ratios of C/H were increased with increasing the carbonization temperature. The large increase in the C/H ratio at 800 °C indicates that the decomposition of the RF polymer still continued at 600–800 °C, although the micropore volume and the BET surface area showed no further increase above 600 °C.

2.2 Synthesis of hierarchically ordered mesoporous carbons

Although the inorganic-templating method is quite attractive, one should keep it in mind that this technique requires both the use of an expensive template and its removal by a severe treatment, which hampers the practical use of the template technique. On the other hand, because the template can be water-soluble in the organic-templating technique, the process would be much simpler and could be performed at low cost. Furthermore, it is noted that the organic-templating approach is advantageous for controlling the morphology and configuration and one can be use the inorganic template, which have larger template structure than that of Pluronic triblock copolymer, for preparation of mesoporous carbons with hierarchical porous structures.

The use of AAO films, which have uniform straight channels with 10–250 nm diameters, as templates for the synthesis of carbon deposits on the channel walls has been successful (Kyotani et al., 1995, 2006). The resulting tubular carbon materials have tunable diameters, lengths, and wall thickness. Additionally, the uniformity of the carbon nanostructures had never been seen before in carbon nanotubes when compared with carbon nanostructures prepared by conventional arc-evaporation or catalytic chemical vapor deposition (CVD) techniques.

On the other hand, a method to fabricate mesostructured silica within columnar pores of the AAO membranes via the surfactant-templating method has been developed (Yamaguchi et al., 2004). When ordered mesoporous silica is synthesized in film morphology, the mesostructure tends to orient with a specific (*hkl*) plane parallel to the solid–liquid interface (Hillhouse et al., 2001). When nonporous smooth substrates are used, the channel direction of the resulting mesoporous films is oriented parallel to the substrate, and transportation of molecules across the film is not possible. From the standpoint of molecular accessibility, the channel direction of the mesoporous films should be oriented perpendicular to the film surface. Macroscopic structures of silica–surfactant mesophases grown at the interfacial region depend on the shape of the interfaces. When the silica–surfactant nanocomposite is grown inside the columnar pores, the pore wall is expected to assist the self-assembly of the silica–surfactant nanocomposite, and the resulting mesophase might be oriented along the interface.

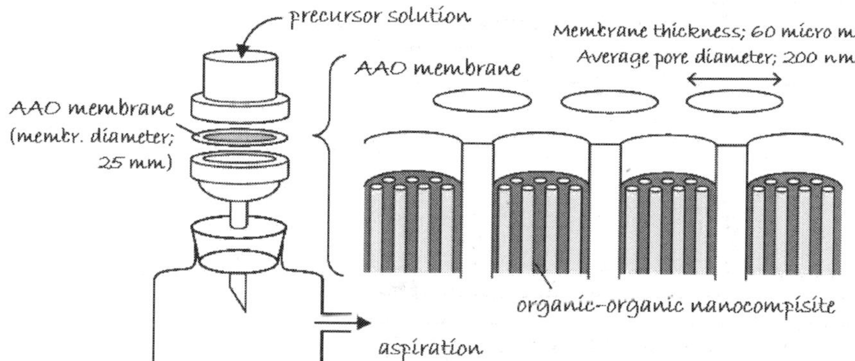

Fig. 9. Scheme to impregnate the columnar pores with triblock copolymer–templated phenolic polymer resin. Schematic illustration of the assembly of organic–organic nanocomposite formed inside the columnar alumina pores.

The procedure is quite simple and rapid; the films of triblock copolymer F127–templated phenolic polymer resin can be deposited at the columnar pore surface of the AAO membrane by simply immersing the membrane in the precursor solution. AAO membrane (average pore diameter = 200 nm, thickness = 60 µm, membrane diameter = 25 mm) was set in an ordinary membrane filtration apparatus, and the precursor solution described above was dropped onto it (Fig. 9). Moderate aspiration was applied so that the solution penetrated into the columnar alumina pores. The membrane including the precursor solution was dried in air. Carbonization was performed as described above. The AAO scaffold was completely removed by immersing the composite membrane in alkali solution. Fig. 10 shows TEM images of mesostructured carbon nanofibers embedded within the pores of AAO membranes and released by dissolving the AAO membrane. A well-ordered structure can be obtained and the mean diameter of the mesopores in the carbon nanofibers is approximately 8 nm. The results of N_2 adsorption/desorption analysis of the mesoporous carbon nanofibers attached to the AAO membrane revealed that the pore diameter is 7.6 nm, which is in good agreement with TEM observations.

Hydrothermal stability and alkaline resistance of the nanofibers were investigated by TEM observation before and after immersing in 5 M NaOH solution at 100 °C for 24 h. For nanofibers carbonized at 400 °C, the ordered mesostructure collapses completely after alkaline hydrothermal treatment, because the framework is still composed of an intermediate between a polymer and carbon. In contrast, for nanofibers carbonized at 600 and 800 °C, the mesostructures show no difference before and after treatment. Thus,

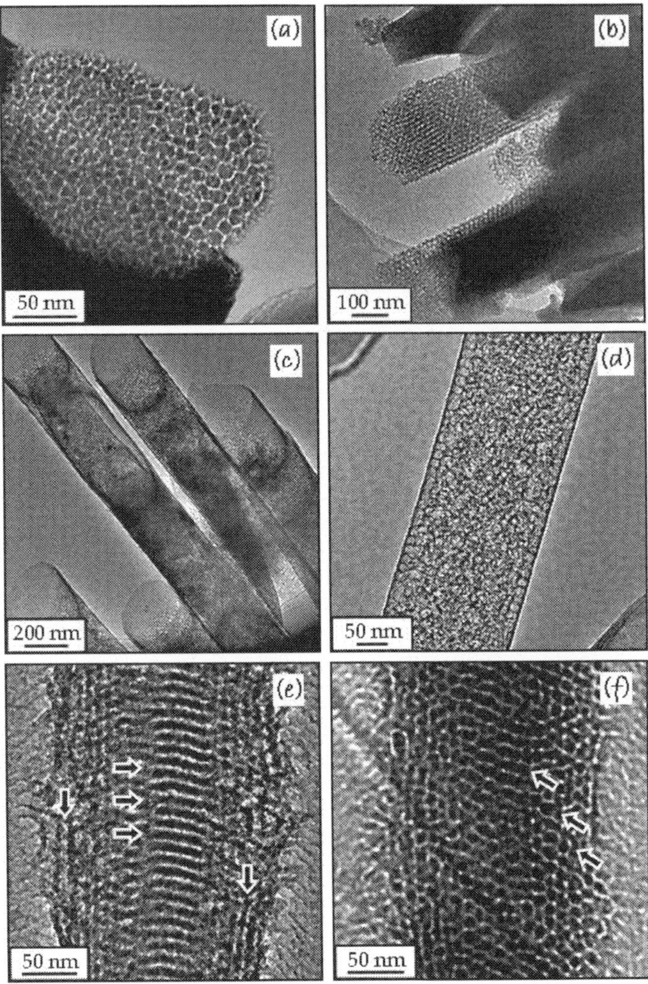

Fig. 10. (a) Top- and (b) side-view TEM images of mesoporous carbon nanofibers embedded within the pores of AAO membranes. (c) Low-magnification and (d–f) high-magnification TEM images of mesoporous carbon nanofibers after dissolving the AAO scaffolds. (Tanaka et al., 2009)

nanofibers carbonized at high temperatures (above 600 °C) show high hydrothermal stability and alkaline resistance. The longitudinal dimension and diameter of the nanofibers are about 60 µm and 200 nm, respectively, which are consistent with the pore dimension of the AAO membranes. The structural order seems to become distorted in the interior region. At the fiber–air interface, ordered layers with a short-range-order structure seem to be oriented along the interface in different directions, as indicated by the arrows.

2.3 Synthesis of ordered mesoporous carbon films and membranes

Development of well-ordered mesoporous carbon films will lead to new applications such as separation membranes and electronic devices. In the inorganic-templating method, the use of mesoporous silica thin films, rather than bulk powders, are unsuitable as a template because of the lack of pore accessibility in the silica films. Then, it is difficult to maintain the continuous film and film adhesion to the substrate may become poor.

In this section, a simple synthesis of mesoporous carbons in thin film and membrane (thick film) morphologies by means of organic–organic self-assembly using phenolic resin monomers and Pluronic F127 is introduced. Completely continuous films composed of ordered mesoporous carbon can be obtained by dip-coating or spin-coating.

Fig. 11 shows an FESEM image of a cross-section of ordered mesoporous carbon film. The color of the film turned black after carbonization at 400 °C. A continuous and flat film about 600 nm thick was grown from the silicon substrate. The carbon films were tightly adhered to the substrate even after carbonization at 800 °C. Increasing ethanol/water molar ratio and consequent decreasing concentration of the component decreases film thickness. In addition, film thickness can be controlled by adjusting the withdrawal rate during dip-coating.

When ordered mesoporous materials are synthesized in film morphology, the mesostructure tends to orient with a specific (*hkl*) plane parallel to the substrate. As such, film lattice constants are notoriously difficult to identify from XRD patterns alone, because of the limited number of observed peaks. Additionally, it was demonstrated that refraction effects of XRD are not negligible for many of the mesostructured films (Tanaka et al., 2006). To remedy this, the mesophase topology, order, and orientation of the films should be characterized by the combination of grazing-incidence small angle X-ray scattering (GISAXS), FESEM, and TEM measurements.

The molar ratio of phenolic resin monomers to F127 was changed from 115 to 800 by varying the concentration of the F127. The films prepared at phenolic resin monomers/F127 molar ratio of 200 and 160 are referred to as CKU-F69 and CKU-C12, respectively. GISAXS patterns were collected from CKU-F69 films carbonized at temperatures from 200 to 800 °C (Fig. 12). Interpretation of the GISAXS patterns was aided by NANOCELL (Tate et al., 2006), a program which simulates quantitatively the positions of Bragg diffraction peaks based on the distorted-wave Born approximation (DWBA) to account for the effects of refraction and reflection at the film–substrate and film–air interfaces. The experimental results were fitted to a face-centered orthorhombic *Fmmm* structure with the (010) planes parallel to the substrate, but where other planes were free to rotate about the substrate normal. The mesophase with cell parameters and orientation with respect to the substrate is shown schematically in Fig. 12. On the other hand, the CKU-C12 is (10)-oriented and possesses a rectangular *c2mm* symmetry, which results from uniaxial contraction of 2D hexagonal *p6mm* symmetry. The mesostructure were controlled by simply adjusting the molar ratio of RP/F127.

Upon the initial assembly, it is conjectured that the mesostructure is described by the body-centered lattice, likely a (110) oriented *Im3m* cubic close packing of micellar aggregates. The *Fmmm* mesostructure results from uniaxial shrinkage of *Im3m* symmetry along the substrate normal. In addition to the Bragg diffraction peaks in the GISAXS patterns, there is a diffuse ring present in the pattern. A diffraction ring superimposed on the octagon-shaped spot pattern indicates the presence of some polyoriented domains in the film. In other words some domains are not perfectly aligned about the substrate normal. Furthermore, the critical angle for X-ray scattering from the mesoporous carbon film was measured by GISAXS. The critical angle decreased from 0.16° to 0.15° after carbonization at 400 °C, indicating the reduction of the average electron density of the film due to removal of the template. In addition, at carbonization above 600 °C further reduction of the critical angle to 0.14° indicates a decrease in the density of the carbonaceous framework.

The ordered mesostructures were also preserved during this high temperature carbonization process. However, the interplanar distance, $d010$, did decrease, and at the same time there was a decrease in the film thickness as measured by using FESEM that followed a similar trend. The shrinkage percentage of the film carbonized at 800 °C was calculated to be 68%. The majority of the decrease in the $d010$ value was observed at a carbonization temperature of 400 °C, which corresponds to 66% of total contraction. This result implies that the majority of the residual hydroxyl groups in the carbonaceous walls condense at elevated temperatures. This temperature also corresponds to the decomposition of the majority of the organic template, as described in detail below using nitrogen sorption and thermogravimetric analyses. In contrast to the changes observed in the b lattice constant, the other parameters, a and c, did not change during the carbonization process, indicating that the shrinkage in the directions parallel to the substrate was hindered by the adhesion of the coating.

The pores have an ellipsoidal shape due to anisotropic contraction upon drying and the carbonization process, in contrast to isotropic contraction for bulk powders without a support medium. Besides the uniformity of the pore size, the pore shape may be useful for limiting the sizes or orientations of guest molecules in separation, catalysis, and sensor applications. From TEM observations, it was found that many domains exist and are oriented parallel to the substrate with different rotational directions. Highly ordered patterns of cage-like pores support the conclusion that CKU-F69 products possess orthorhombic $Fmmm$ symmetry.

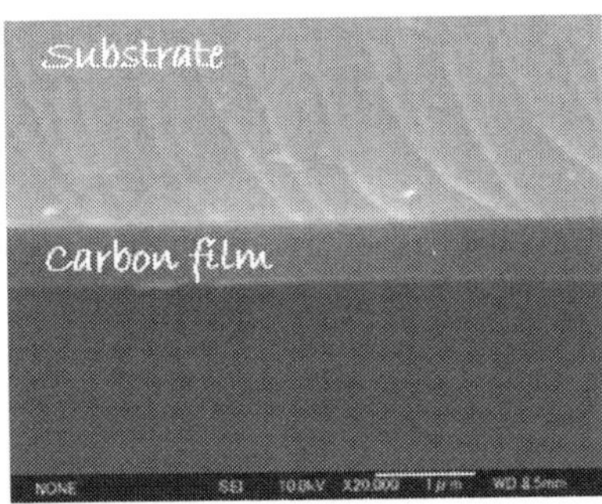

Fig. 11. FESEM image for the cross-section of CKU-F69. (Tanaka et al., 2007)

Microporous and mesoporous inorganic membranes have been investigated mainly with respect to silica membrane prepared by the sol–gel method (Park et al., 2001; Nishiyama et al., 2003; Sakamoto et al., 2007). However, silicate materials dissolved in water and alkaline solutions, which decrease the possibility of practical use. On the other hand, carbon membranes have attracted increasing interest because of their advantages, such as high surface area, high hydrothermal stability, and chemical inertness. Ordered mesoporous carbon is a promising material in the field of membrane filtration technologies, such as nanofiltration and ultrafiltration.

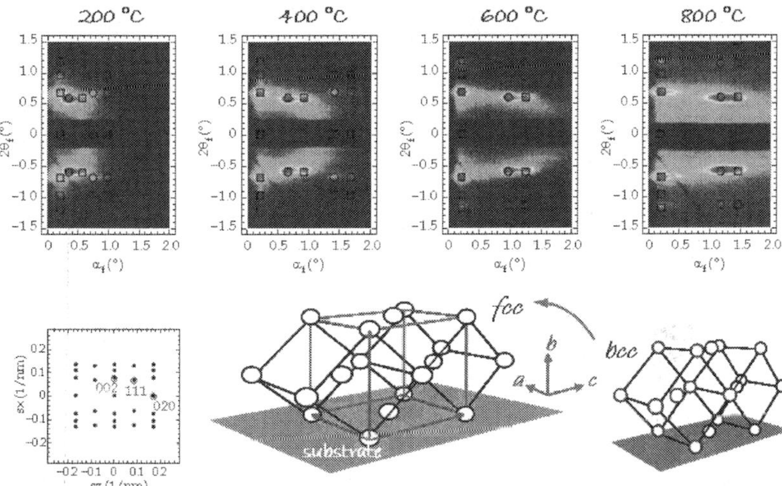

Fig. 12. GISAXS patterns of CKU-F69 carbonized at 200–800 °C. The overlay of simulated spots is from a NANOCELL simulation. The circles and squares identify transmitted and reflected Bragg peaks, respectively. In addition, NANOCELL simulated reciprocal space of an (010) oriented *Fmmm* with many domain orientations about the substrate normal. Schematic showing the cell parameters and orientation of the *Fmmm* structure with respect to the substrate. (Tanaka et al., 2007)

Next, mesoporous carbon membranes were prepared on the porous α-alumina support. α-Alumina porous tubular supports (outer diameter: 10 mm, inner diameter: 7 mm, length: 35mm, average pore size: 100 nm, average porosity: 40%) were purchased from Noritake Co., Ltd., Japan. To prepare cylindrical porous α-alumina tubes with dead-end structure, a 20 mm section of the porous α-alumina tube was joined to a dense α-alumina tube (outside diameter: 10 mm, inner diameter: 7 mm, length: 250 mm) and an α-alumina disk (diameter: 12 mm, thickness: 2 mm) with a SiO_2–BaO–CaO sealant and then calcined at 1100 °C. Membranes were prepared by dip-coating the α-alumina porous tubular support in the coating solution.

The lattice d-spacing of ordered mesoporous carbon membrane carbonized at 600 °C was 5.6 nm. The BET surface area and the total pore volume were 670 m^2/g and 0.58 cc/g, respectively. The pore diameter calculated from N_2 adsorption branch was estimated as 4.2 nm. The pore accessibility can be determined by the pore entrance diameter, which is useful for the limiting the sizes or orientations of guest molecules in separation, catalysis, and sensor applications.

A continuous and smooth layer about 5 µm thick was formed on the support surface. From the EDX analysis, the membrane had a carbon/alumina composite layer with a thickness of about 3.5 µm because a coating solution penetrated into the alumina pores. The ordered mesostructures were also preserved during this high temperature carbonization process, although the carbonization of the membrane shrunk the ordered structure, which corresponds to 5% contraction. Gas permeation measurements were carried out using N_2 to confirm the compactness of the membranes. In the permeation test, a total pressure on the feed side was kept constant in the range of 75–180 kPa and the permeate side was kept constant at atmospheric pressure. The permeance of N_2 through the alumina support increased with increasing the pressure drops, suggesting that the contribution of viscous flow cannot be negligible for the alumina support. In contrast, the permeances of N_2 through the mesoporous carbon membranes were independent of the pressure drops across the membranes. The N_2 gas permeance through the mesoporous carbon membranes is predominantly governed by the Knudsen diffusion (Uhlhorn et al., 1989). These results strongly indicate that there are no important pinholes and/or defects in the membranes.

The moisture and alkaline resistance is one of the most important factors in the filed of membrane separation (Park et al., 2001; Nishiyama et al., 2003). The membranes were placed under extremely severe hydrothermal conditions (90 °C, in water) and in alkaline solution (at pH 12, for 1 week). The ordered mesostructure and gas permeation properties of the carbon membrane carbonized at 600 °C were maintained.

2.4 Control of mesostructure; effect of ethanol/water ratio

Table 1 summarizes the results of the pore structure analysis. The d-spacing, pore diameter, pore wall thickness, BET surface area, and pore volume change with carbonization temperature. Carbons carbonized at 400 °C have the lowest BET surface area, because the pore wall is still composed of an intermediate between a polymer and carbon. There is only a small difference in micropore volume of the samples carbonized at the same temperature but different ethanol/water molar ratios. The mesopore volume increases with increasing the ethanol/water molar ratio. Micropores are generated inside the carbonaceous framework by solidification and gasification of the polymer with a further increase in the carbonization temperature. Framework of ordered mesoporous carbon consists of imperfect graphenes of a very small size. A micropore is the space between the nanographenes. Table 1 shows that micropore volume increases with increasing pyrolysis temperature. On the other hand, the carbonization causes shrinkage of the mesostructure.

The pore size increases with increasing ethanol/water molar ratio. Control of pore size is one of the most important subjects in the study of mesoporous materials, and many methods for achieving control have been reported. The most common methods focus on the use of various swelling agents such as 1,3,5-trimethylbenzene (Beck et al., 1992), 1,3,5-triisopropylbenzene (Kimura et al., 1998), and decane (Blin et al., 2000). The strategy here is that a micellar array in which the core is composed of hydrophobic hydrocarbon chains participates in the solubilization of hydrophobic molecules. Incremental addition of the swelling agent results in an increasing pore size. This method has been shown to lead to pore expansion of up to 30%, usually accompanied by a loss of the long-range order of the mesostructure. In the organic-templating method for preparation of ordered mesoporous carbons, ethanol seems to play an important role in determining the characteristics of the porous structure. It is well known that the aqueous phase behavior of surfactants is influenced by the presence of short-chain alcohols ($n_c \geq 4$) (Ekwall et al., 1969). Unlike non-polar organics that are located at the hydrophobic core of surfactant assemblies, alcohol witha polar group (–OH group) is believed to be located at the hydrophilic–hydrophobic

EtOH/water molar ratio	T^a /°C	d^b /nm	$d_{\text{pore-to-pore}}^c$ /nm	D^d /nm	w^e /nm	S_{BET}^f /m² g⁻¹	V_{T}^g /cc g⁻¹	V_{meso}^h /cc g⁻¹	V_{micro}^i /cc g⁻¹
0.5	600	—	—	5.0	—	510	0.31	0.11	0.20
	800	—	—	4.0	—	490	0.24	0.04	0.20
0.75	600	—	—	5.3	—	540	0.34	0.13	0.21
	800	—	—	4.8	—	550	0.30	0.08	0.22
1.0	400	15.5	17.9	6.8	11.1	270	0.27	0.17	0.10
	600	—	—	5.8	—	530	0.40	0.19	0.21
	800	—	—	5.0	—	520	0.34	0.12	0.22
1.25	400	15.5	17.9	6.9	11.0	330	0.39	0.26	0.13
	600	—	—	6.0	—	520	0.40	0.20	0.20
	800	—	—	5.2	—	510	0.36	0.15	0.21
2.5	400	16.7	19.3	7.6	11.7	530	0.66	0.46	0.20
	600	15.5	17.9	7.2	10.7	650	0.72	0.50	0.22
	800	12.3	14.2	5.7	8.5	640	0.57	0.33	0.25

[a] Carbonization temperature. [b] d-spacing calculated from SAXS or Fourier diffractogram of TEM. [c] Distance between pores calculated by the formula $2d/\sqrt{3}$ assuming a hexagonal unit cell. [d] Pore diameter calculated by the BJH method using adsorption branches. [e] Pore wall thickness calculated by subtracting the pore size from the distance between pores. [f] BET surface area. [g] Total pore volume calculated as the amount of nitrogen adsorbed at a relative pressure of 0.95. [h] Mesopore volume calculated by subtracting the amount of nitrogen adsorbed at a relative pressure of 0.1 from that at a relative pressure of 0.95. [i] Micropore volume calculated from the amount of nitrogen adsorbed at a relative pressure of 0.1.

Table 1. Structure characteristics of mesoporous carbon powders prepared using different amounts of ethanol. (Tanaka et al., 2009)

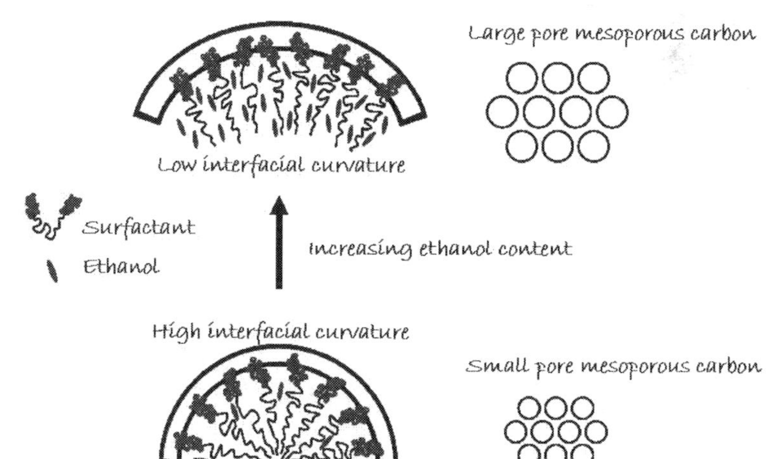

Fig. 13. Schematic representation of the role of ethanol as a swelling agent. (Tanaka et al., 2009)

interface to help stabilize liquid crystals and determine their surface curvatures. The method for controlling the mesostructure composed of silica using a ternary triblock copolymer–butanol–water system has been reported, as well as similar systems using pentanol and hexanol instead of butanol (Feng et al., 2000; Kleitz et al., 2004; Kim et al., 2005). The addition of these alcohols, which act as cosurfactants or swelling agents, results in not only increased pore size but also formation of a mesophase with a decreased curvature.

Fig. 13 shows that the mesophase may change with the micellar interfacial curvature, which varies with ethanol content. Ethanol swells the hydrophobic volume of the triblock copolymer micelles and interacts with both PPO and PEO segments because it is highly polar molecule. Thus, ethanol is located at the hydrophilic–hydrophobic interface (PEO/PPO) and stabilizes the interface, leading to the formation of micellar aggregates with decreased interfacial curvature.

3. Conclusion

As has been demonstrated in this chapter, the organic-templating approach is a very powerful method for the preparation of various types of ordered mesoporous carbons. The supramolecular templating technique opens an avenue for ordered mesoporous carbons and has advantages for controlling the morphology and configuration. The recent progress made in the development of organic-templating method was reviewed. Research efforts to produce ordered mesoporous carbons have focused on the use of phenolic polymer resins and thermally-decomposable Pluronic triblock copolymers. The choice of an appropriate set of thermosetting polymers and thermally-decomposable organic templates is the most important factor in controlling the mesophase topologies and morphology control of ordered mesostructured carbons. In powder preparation by simple precipitation process, the addition of ethanol expands both the pore size and d-spacing. An approach using a copolymer–alcohol–water system can be advantageous for tuning pore size even in the synthesis of mesoporous carbons. In film preparation, self-assembly of triblock copolymer–phenolic resin nanocomposites is affected by the substrate surface and mesostructure oriented parallel to the substrate. In nanofiber preparation using AAO membranes as a nanosized mold, the alumina walls may also assist the growth of the mesostructure of triblock copolymer–phenolic resin nanocomposites. The direct triblock-copolymer-templating method using an ethanol/water system provides a simple route to fabricating mesoporous carbon and carbon–polymer materials with controlled morphology. Because of their high surface area, large pore volume, and large pore size, the mesoporous carbons have potential application in capacitors, electrodes for batteries, fuel cells, chemical sensors, bioseparations, and as hosts for the immobilization of biomolecules. For many biotechnological applications, mesoporous carbons having large interconnected porous structures need to be fabricated. For the capacitors and electrodes of electrochemical devices, carbons with highly graphitic structures are needed. The synthesis and application of hierarchical porous carbons are expected in the future. These future works have been very challenging. Simple template synthetic procedures will expand the possibility of synthesizing a variety of ordered porous carbon and carbon–polymer materials.

4. Acknowledgment

This work was supported by the Kinki Invention Center, the Murata Science Foundation (A91153), and the Japan Society for the Promotion of Science (JSPS) (KAKENHI #19860074 and #21760562). The GISAXS patterns were collected at the NSF funded facility for In-situ X-ray Scattering from Nanomaterials and Catalysts (MRI program award 0321118-CTS). The authors thank Dr. M. P. Tate and Associate Prof. H. W. Hillhouse (Purdue University) for the GISAXS measurements.

5. References

Kroto, H. W.; Heath, J. R.; O'Brien, S. C.; Curl, R. F. & Smalley, R. E. (1985). C-60 – buckminsterfullerene, *Nature*, Vol. 318, No. 6042, pp. 162–163, ISSN 0028–0836

Iijima, S. (1991). Helical microtubules of graphitic carbon, *Nature*, Vol. 354, No. 6348, pp. 56–58, ISSN 0028–0836

Marsh, H. & Rand, B. (1971). The process of activation of carbons by gasification with CO_2-II. The role of catalytic impurities, *Carbon*, Vol. 9, No. 1, pp. 63–72, ISSN 0008–6223

Tamai, H.; Kakii, T.; Hirota, Y.; Kumamoto, T. & Yasuda, H. (1996). Synthesis of extremely large mesoporous activated carbon and its unique adsorption for giant molecules, *Chem. Mater.*, Vol. 8, No. 2, pp. 454–462, ISSN 0897–4756

Hu, Z.; Srinivasan, M. P. & Ni, Y. (2000). Preparation of mesoporous high-surface-area activated carbon, *Adv. Mater.*, Vol. 12, No. 1, pp. 62–65, ISSN 0935–9648

Kyotani, T.; Tsai, L. -F. & Tomita, A. (1995). Formation of ultrafine carbon tubes by using an anodic aluminum-oxide film as a template, *Chem. Mater.*, Vol. 7, No. 8, pp. 1427–1428, ISSN 0897–4756

Kyotani, T. (2006). Synthesis of various types of nano carbons using the template technique, *Bull. Chem. Soc. Jpn.*, Vol. 79, No. 9, pp. 1322–1337, ISSN 0009–2673

Kyotani, T.; Nagai, T.; Inoue, S. & Tomita, A. (1997). Formation of new type of porous carbon by carbonization in zeolite nanochannels, *Chem. Mater.*, Vol. 9, No. 2, pp. 609–615, ISSN 0897–4756

Ma, Z.; Kyotani, T. & Tomita, A. (2000). Preparation of a high surface area microporous carbon having the structural regularity of Y zeolite, *Chem. Commun.*, No. 23, pp. 2365–2366, ISSN 1359–7345

Nishihara, H; Yang, Q. -H.; Hou, P. -X.; Unno, M.; Yamauchi, S.; Saito, R.; Paredes, J. I.; Martinez-Alonso, A.; Tascon, J. M. D.; Sato, Y.; Terauchi, M. & Kyotani, T. (2009). A possible buckybowl-like structure of zeolite templated carbon, *Carbon*, Vol. 47, No. 5, pp. 1220–1230, ISSN 0008–6223

Zakhidov, A. A.; Baughman, R. H.; Iqbal, Z.; Cui, C.; Khayrullin, I.; Dantas, S. O.; Marti, J. & Ralchenko, V. G. (1998). Carbon structures with three-dimensional periodicity at optical wavelengths, *Science*, Vol. 282, No. 5390, pp. 897–901, ISSN 0036–8075

Yu, J. -S.; Kang, S.; Yoon, S. B. & Chai, G. (2002). Fabrication of ordered uniform porous carbon networks and their application to a catalyst supporter, *J. Am. Chem. Soc.*, Vol. 124, No. 32, pp. 9382–9383, ISSN 0002–7863

Ryoo, R.; Joo, S. H. & Jun, S. (1999). Synthesis of highly ordered carbon molecular sieves via template-mediated structural transformation, *J. Phys. Chem. B*, Vol. 103, No. 37, pp. 7743–7746, ISSN 1089–5647

Lee, J.; Yoon, S.; Hyeon, T.; Oh, S. M. & Kim, K. B. (1999). Synthesis of a new mesoporous carbon and its application to electrochemical double-layer capacitors, *Chem. Commun.*, No. 21, pp. 2177–2178, ISSN 1359–7345

Kaneda, M.; Tsubakiyama, T.; Carlsson, A.; Sakamoto, Y.; Ohsuna, T.; Terasaki, O.; Joo, S. H. & Ryoo, R. (2002). Structural study of mesoporous MCM-48 and carbon networks synthesized in the spaces of MCM-48 by electron crystallography, *J. Phys. Chem. B*, Vol. 106, No. 6, pp. 1256–1266, ISSN 1520–6106

Kleitz, F.; Choi, S. H. & Ryoo, R. (2003). Cubic *Ia3d* large mesoporous silica: synthesis and replication to platinum nanowires, carbon nanorods and carbon nanotubes, *Chem. Commun.*, No. 17, pp. 2136–2137, ISSN 1359–7345

Xia, Y.; Yang, Z. & Mokaya, R. (2006). Simultaneous control of morphology and porosity in nanoporous carbon: Graphitic mesoporous carbon nanorods and nanotubules with tunable pore size, *Chem. Mater.*, Vol. 18, No. 1, pp. 140–148, ISSN 0897–4756

Kresge, C. T.; Leonowicz, M. E.; Roth, W. J.; Vartuli, J. C. & Beck, J. S. (1992). Ordered mesoporous molecular-sieves synthesized by a liquid-crystal template mechanism, *Nature*, Vol. 359, No. 6397, pp. 710–712, ISSN 0028–0836

Beck, J. S.; Vartuli, J. C.; Roth, W. J.; Leonowicz, M. E.; Kresge, C. T.; Schmitt, K. D.; Chu, C. T. W.; Olson, D. H.; Sheppard, E. W.; McCullen, S. B.; Higgins, J. B. & Schlenker, J. L. (1992). A new family of mesoporous molecular-sieves prepared with liquid-crystal templates, *J. Am. Chem. Soc.*, Vol. 114, No. 27, pp. 10834–10843, ISSN 0002–7863

Zhao, D.; Huo, Q.; Feng, J.; Chmelka, B. F. & Stucky, G. D. (1998). Nonionic triblock and star diblock copolymer and oligomeric surfactant syntheses of highly ordered, hydrothermally stable, mesoporous silica structures, *J. Am. Chem. Soc.*, Vol. 120, No. 24, pp. 6024–6036, ISSN 0002–7863

Tanaka, S.; Nishiyama, N.; Egashira, Y. & Ueyama, K. (2005). Synthesis of ordered mesoporous carbons with channel structure from an organic-organic nanocomposite, *Chem. Commun.*, No. 16, pp. 2125–2127, ISSN 1359–7345

Tanaka, S.; Katayama, Y.; Tate, M. P.; Hillhouse, H. W. & Miyake, Y. (2007). Fabrication of continuous mesoporous carbon films with face-centered orthorhombic symmetry through a soft templating pathway, *J. Mater. Chem.*, Vol. 17, No. 34, pp. 3639–3645, ISSN 0959–9428

Tanaka, S.; Doi, A.; Nakatani, N.; Katayama, Y. & Miyake, Y. (2009). Synthesis of ordered mesoporous carbon films, powders, and fibers by direct triblock-copolymer-templating method using an ethanol/water system, *Carbon*, Vol. 47, No. 11, pp. 2688–2698, ISSN 0008–6223

Jin, J.; Tanaka, S.; Egashira, Y. & Nishiyam, N. (2010). KOH activation of ordered mesoporous carbons prepared by a soft-templating method and their enhanced electrochemical properties, *Carbon*, Vol. 48, No. 7, pp. 1985–1989, ISSN 0008–6223

Jin, J.; Nishiyama, N.; Egashira, Y. & Ueyama, K. (2009). Pore structure and pore size controls of ordered mesoporous carbons prepared from resorcinol/formaldehyde/triblock polymers, *Micropor. Mesopor. Mater.*, Vol. 118, No. 1–3, pp. 218–223, ISSN 1387–1811

Simanjuntak, F. H.; Jin, J.; Nishiyama, N.; Egashira, Y. & Ueyama, K. (2009). Ordered mesoporous carbon films prepared from 1,5-dihydroxynaphthalene/triblock copolymer composites, *Carbon*, Vol. 47, No. 10, pp. 2531–2533, ISSN 0008–6223

Meng, Y.; Gu, D.; Zhang, F. Q.; Shi, Y. F.; Yang, H. F.; Li, Z.; Yu, C.; Tu, B. & Zhao, D. (2005). Ordered mesoporous polymers and homologous carbon frameworks: Amphiphilic surfactant templating and direct transformation, *Angew. Chem., Int. Ed.*, Vol. 44, No. 43, pp. 7053-7059, ISSN 1433-7851

Zhang, F. Q.; Meng, Y.; Gu, D.; Yan, Y.; Yu, C. Z.; Tu, B. & Zhao, D. (2005). A facile aqueous route to synthesize highly ordered mesoporous polymers and carbon frameworks with *Ia3d* bicontinuous cubic structure, *J. Am. Chem. Soc.*, Vol. 127, No. 39, pp. 13508-13509, ISSN 0002-7863

Zhang, F. Q.; Meng, Y.; Gu, D.; Yan, Y.; Chen, Z. X.; Tu, B. & Zhao, D. (2006). An aqueous cooperative assembly route to synthesize ordered mesoporous carbons with controlled structures and morphology, *Chem. Mater.*, Vol. 18, No. 22, pp. 5279-5288, ISSN 0897-4756

Pekala, R. W. (1989). Organic aerogels from the polycondensation of resorcinol with formaldehyde, *J. Mater. Sci.*, Vol. 24, No. 9, pp. 3221-3227, ISSN 0022-2461

Yamaguchi, A.; Uejo, F.; Yoda, T.; Uchida, T.; Tanamura, Y.; Yamashita, T. & Teramae, N. (2004). Self-assembly of a silica-surfactant nanocomposite in a porous alumina membrane, *Nature Mater.*, Vol. 3, No. 5, pp. 337-341, ISSN 1476-1122

Hillhouse, H. W.; Egmond, J. W.; Tsapatsis, M.; Hanson, J. C. & Larese, J. Z. (2001). The interpretation of X-ray diffraction data for the determination of channel orientation in mesoporous films, *Micropor. Mesopor. Mater.*, Vol. 44-45, pp. 639-643, ISSN 1387-1811

Tanaka, S.; Tate, M. P.; Nishiyama, N.; Ueyama, K. & Hillhouse, H. W. (2006). Structure of mesoporous silica thin films prepared by contacting PEO_{106}-PPO_{70}-PEO_{106} films with vaporized TEOS, *Chem. Mater.*, Vol. 18, No. 23, pp. 5461-5466, ISSN 0897-4756

Tate, M. P.; Urade, V. N.; Kowalski, J. D.; Wei, T. C.; Hamilton, B. D.; Eggiman, B. W. & Hillhouse, H. W. (2006). Simulation and interpretation of 2D diffraction patterns from self-assembled nanostructured films at arbitrary angles of incidence: From grazing incidence (above the critical angle) to transmission perpendicular to the substrate, *J. Phys. Chem. B*, Vol. 110, No. 20, pp. 9882-9892, ISSN 1520-6106

Park, D. H.; Nishiyama, N.; Egashira, Y. & Ueyama, K. (2001). Enhancement of hydrothermal stability and hydrophobicity of a silica MCM-48 membrane by silylation, *Ind. Eng. Chem. Res.*, Vol. 40, No. 26, pp. 6105-6110, ISSN 0888-5885

Nishiyama, N.; Saputra, H.; Park, D. H.; Egashira, Y. & Ueyama, K. (2003). Zirconium-containing mesoporous silica Zr-MCM-48 for alkali resistant filtration membranes, *J. Membr. Sci.*, Vol. 218, No. 1-2, pp. 165-171, ISSN 0376-7388

Sakamoto, Y.; Nagata, K.; Yogo, K. & Yamada, K. (2007). Preparation and CO_2 separation properties of amine-modified mesoporous silica membranes, *Micropor. Mesopor. Mater.*, Vol. 101, No. 1-2, pp. 303-311, ISSN 1387-1811

Uhlhorn, R. J. R.; Keizer, K. & Burggraaf, A. J. (1989). Gas and surface-diffusion in modified gamma-alumina systems, *J. Membr. Sci.*, Vol. 46, No. 23, pp. 225-241, ISSN 0376-7388

Kimura, T.; Sugahara, Y. & Kuroda, K. (1998). Synthesis of mesoporous aluminophosphates using surfactants with long alkyl chain lengths and triisopropylbenzene as a solubilizing agent, *Chem. Commun.*, No. 5, pp. 559-560, ISSN 1359-7345

Blin, J. L. ; Otjacques, C.; Herrier, G. & Su, B. L. (2000). Pore size engineering of mesoporous silicas using decane as expander, *Langmuir*, Vol. 16, No. 9, pp. 4229-4236, ISSN 0743-7463

Ekwall, P.; Mandell, L. & Fontell, K. (1969). The cetyltrimethylammonium bromide-hexanol-water system, *J. Colloid. Interf. Sci.*, Vol. 29, No. 4, pp. 639–646, ISSN 0021–9797

Feng, P. Y.; Bu, X. H. & Pine, D. J. (2000). Control of pore sizes in mesoporous silica templated by liquid crystals in block copolymer-cosurfactant-water systems, *Langmuir*, Vol. 16, No. 12, pp. 5304–5310, ISSN 0743–7463

Kleitz, F.; Solvyov, L. A.; Anilkumar, G. M.; Choi, S. H. & Ryoo, R. (2004). Transformation of highly ordered large pore silica mesophases (*Fm3m*, *Im3m* and *p6mm*) in a ternary triblock copolymer-butanol-water system, *Chem. Commun.*, No. 13, pp. 1536–1537, ISSN 1359–7345

Kim, T. W.; Kleitz, F.; Paul, B. & Ryoo, R. (2005). MCM-48-like large mesoporous silicas with tailored pore structure: Facile synthesis domain in a ternary triblock copolymer-butanol-water system, *J. Am. Chem. Soc.*, Vol. 127, No. 20, pp. 7601–7610, ISSN 0002–7863

Part 4

Environmental Engineering: Modeling and Applications

25

Streambank Soil Bioengineering Approach to Erosion Control

Francisco Sandro Rodrigues Holanda and
Igor Pinheiro da Rocha
Universidade Federal de Sergipe
Brazil

1. Introduction

Rivers in tropical regions have been submitted to strong environmental impacts through changes in the hydrologic and sedimentological regime, and also to the ongoing destruction of their riparian vegetation, despite the important role of riparian vegetation in riverbank protection through root systems and plant cover, which improve soil particle aggregation in a low cohesion situation, reducing runoff and resulting in a lower erosion rate and sedimentation of the river channel. Rivers are in effect often referred to as dynamic systems which means they are in a constant state of change.

Techniques of stream bank and bed stabilization are needed and can be accomplished in several ways, such as the use of rockfill, which, though efficient, is quite expensive, precluding its use extensively along the river banks. In an attempt to solve the problem, riverine populations have resorted to various empirical solutions that, in addition to not producing the desired effect, cause problems for riparian vegetation recovery besides degrading the landscape (Holanda et al., 2010). The function of riverbank protection is to avoid bank erosion, that could cause movement of the river channel, which can be of vertical and horizontal direction, arise meandering, braiding, or moving and changing the river´s path.

As an alternative to the empirical practices of the riverines and to expensive bordering and rockfill techniques, the use of abundant raw material has been tested and used, providing a way of mitigating the problem that can be economically viable and with proven technical efficiency. This chapter intends to discuss soil bioengineering as a biotechnology that consists of the use of living materials or inert plant substances, biotextiles, associated or not with rocks, concrete, or metals that present themselves to be environmentally sustainable to riverbank erosion control at the various conditions of slope and soil texture along their water systems like reservoirs, irrigation canals, and rivers. Soil bioengineering can be applied in the mitigation of watershed disasters and protection and restoration of ecology. In soil bioengineering, plants assume an important ecological contribution (providing multiple ecological services), as well as an economic, and especially structural, contribution in contrast to other technologies in which plants are merely an aesthetic component of design. Also, a discussion will be developed on the vegetation component, which has a great importance in these biotechnologies, recognized not only for its landscaping qualities, but also for its beneficial hydromechanical effects and protection against soil erosion.

2. Basic concepts

2.1 Mechanisms of riverbank failure

The multiple demands for water resources show a typical picture of conflict for the use of waters required by the development´s policies and ecologic services. Watersheds around the world have been subjected to the installation of hydroelectric dams along river channels and surface water withdrawal to ensure water for agricultural, industrial, and domestic purposes; for hydroelectricity; or for flood protection. Based on the Mediterranean rivers, Salinas & Casas (2007) listed nine observed main categories of impacts, which can be applied to most of the rivers worldwide, as follows: 1) canalization, 2) substrate excavations and/or leveling of the channel floor, 3) traffic along the channel, 4) grazing by mixed flocks of sheep and goats, 5) fires, 6) up-stream water extraction for irrigation, 7) cutting of woody vegetation, 8) organic or inorganic rubbish dumps, and 9) farming activities in the riparian corridor.

When hydroelectric power plants are constructed they cause an irreversible modification in the morphology of the natural environment; the possibility of flooding over adjacent areas increases; new local climatic conditions are created, and there is a loss in water and sediments that should be given back to the river downstream (Carone et al., 2006).

The operation of reservoirs, centralized for the generation of electricity and the supply of water for irrigation, generally considered the attending of ecologic priorities to be marginal (Holanda et al., 2009), leading to a strong environmental debt, such as with bank erosion, river channel sedimentation, the growth of a large quantity of aquatic vegetation, and the decrease of sediments which harm the reproduction and preservation of fish and navigation. In addition as a result of the construction of these dams, land adjacent to floodplain is currently flooded and river flow regime has been altered. According to DeWine & Cooper (2007), the response of stream channels and riparian vegetation to river regulation is influenced by several factors including pre- and post-dam river flow regimes, channel type, and the species involved.

Serious disturbances in the major extension of riparian ecosystems along river margins have led to riverbank destabilization, increasing erosion, stream lateral migration, and sedimentation, which are reflected directly in the number and position of sand bars. Stream bank erosion is in effect a natural process that over time has resulted in the formation of the productive floodplains and alluvial terraces, and paradoxically, even stable river systems have some eroding banks.

These hydrological alterations change ecosystem structures and processes in running waters and associated environments the world over. Aquatic ecosystems have been strongly degraded, and many fish and other aquatic organisms are now threatened or endangered, particularly because of river development projects and artificial patterns of flow regulation (Fausch et al., 2002), compromising the traditional economic activities (waterlogged land farming and local fishing) (Holanda et al., 2005). With the decline of the population of fish, the majority of fishing communities have become impoverished and left with few alternatives for generating income for the subsistence of their families (Gutberlet et al., 2007). Another common downstream effect of large dams is that the flood peak, and hence the frequency of overbank flooding, is reduced and sometimes displaced in time.

According to Nilsson & Berggren (2000), hydroelectric power dams also change geomorphologic processes such as sediment cycling. The water released from a reservoir tends to restore its original load of sediment and nutrients, resulting in increased erosion downstream of the dam. This erosion leads to channel simplification and reduced geomorphologic activity in the river bed. Before the construction of a reservoir, bank erosion usually occurs in one local reach, while bank accretion also often happens in another local reach, which can maintain the dynamic balance of channel width. After the construction of a reservoir, the effects of the smoothing of flood peaks and decreasing of incoming sediment supply destroy the relative balance relationship between bank erosion and bank accretion, which often causes serious bank erosion (Xia et al., 2008).

The process of bank erosion is closely related to riverbank-soil composition and corresponding mechanical properties. Bank material may be cohesive or non-cohesive and may comprise numerous soil layers. Bank stability of cohesive riverbanks depends on numerous controlling variables such as soil properties and structure (Van Klaveren & McCool,1998), soil moisture conditions (Simon et al., 2000), and complex electrochemical forces between cohesive particles and flow and vegetation (Pizzuto et al., 2010; Wynn & Mostaghimi, 2006).

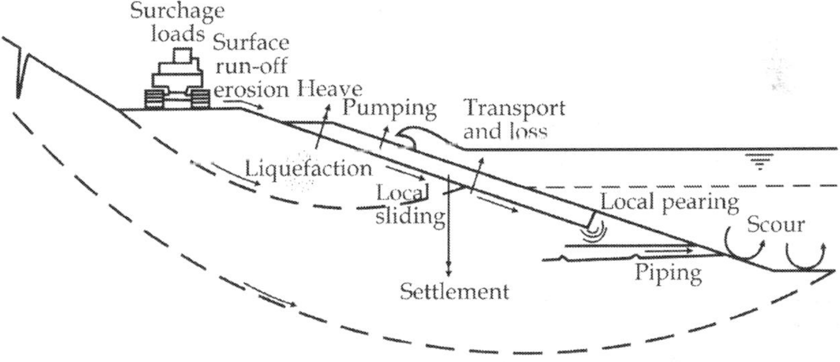

Fig. 1. Schematic representation of causes of geotechnical stability of riverbanks

A reasonable prediction of the bank ruptures can be provided by the qualitative evaluations of various elements influencing the river bank instability (Hunt, 1990). According to Queensland Government (2006) the various mechanisms of stream bank erosion generally fall into two main groups, bank scour and mass failure (Figure 1). In many cases of bank instability both will be evident, often with either scour or mass failure being dominant. Mass failure, which includes bank collapse and slumping, is where large chunks of bank material become unstable and topple into the stream or river in single events. Mass failure is often dominant in the lower reaches of large streams and often occurs in association with scouring of the lower banks. Landslides or mass failure occur when forces driving instability are greater than forces promoting slope stability (Conforth, 2005), that interact with river channel geometry and water flow, driving the sediment transport in the river. Bank scour is the direct removal of bank materials by the physical action of flowing water and the sediment that it carries. Piping is a subsurface form of erosion which involves the removal of subsurface soils in pipe-like erosional channels to a free or escape exit. As fluvial erosion at the bank toe takes place with the continuous removal of bank material, a change in the bank slope occurs with bank overdeepening and alteration of the bank angle (Bertrand 2010).

The likelihood of further mass movement will depend on the stability of the landslide itself and the surrounding soils. In addition, residual soils are likely to be unstable, subject to erosion and not readily colonized. Although many landslides occur naturally, humans are directly accelerating the frequency of landslides by land-use practices (e.g., roads, urbanization, agriculture, clear-cutting) and possibly through their indirect effects on weather patterns (e.g., increased storm frequencies) related to global climate change (Dale et al., 2000).

2.2 Fragmentation of riparian vegetation and restoration

Riparian vegetation on the riverbank has been seriously and continuously deforested because of roads, hydroelectric power dam's construction, urban occupation, adjacent land use, irrigated agriculture, livestock grazing, and the extraction of wood and minerals. Riparian ecosystems occupy the ecotone between upland and aquatic realms and more precisely, the riparian ecosystem can be defined as the stream channel between the low- and high-water marks plus the terrestrial landscape above the high-water mark, where vegetation may be influenced by elevated water tables or extreme flooding and by the ability of the soil to hold water (Naiman et al., 2002).

Natural riparian ecosystems include a variety of community types, with deciduous trees and shrubs on heterogeneous substrates, deltas with distinct plant zonation, and well-developed forests having diverse animal communities. Vegetation interacts with hydrological processes from the earliest stages of plant succession and can have significant impacts on hydraulic processes, particularly during periods of low flow, as well as at the beginning or at the end of flood periods (Tabacchi et al., 2005). Therefore, assessment of deforestation impacts on stream biodiversity and appropriate management practices for its conservation are urgently needed. There are strong evidences that past slash-and-burn agriculture exerted a "press disturbance", which reduced community diversity over a long period in the tropical streams.

An emphasis on the importance of promoting management practices that protect the diverse stream communities from poorly regulated land use in tropical rain forests (Iwata et al., 2003) is needed. Considering every social-ecological problem in river basins with small or large flows, it is necessary to deal with the effects of the impacts on seasonal flow, discharge influence from dams, and traditional knowledge to reach management practices that build resilience.

Because of a river basin's vulnerability to erosion and the unsustainable activities conducted there, flora has been the natural resource most rapidly and easily threatened. The spatial distribution and the structure and dynamics of the riparian vegetation are strongly influenced by the hydrological and sedimentological regime and by associated geomorphological and soil factors, which determine a certain degree of instability and heterogeneity of ecological parameters (Campos & Souza, 2002).

Following river damming and diversion, downstream aquatic and riparian ecosystems have collapsed along many streams.

2.3 Streams restoration

Many stream restoration efforts have targeted the reconstruction of small reaches through artificial measures such as boulder placement, vegetation planting, and fish stocking (Alpert et al., 1999). Live plants and other natural materials have been used for centuries to control erosion problems on slopes in different parts of the world.

According to Walker et al., (2009) once an initial vegetative cover is established on a landslide, many restoration projects end. The mechanism used includes the enhancement of soil shear strength using vegetation soil systems and limiting soil particle movements on slope via utilizing the effects of root systems on soil structure. In plant successional dynamics, great interest must be considered to explore further the influence of the engineering properties of the root system on slope stability and shallow landsliding.

The complex interactions of physical and biological processes in riverine ecosystems can complicate restoration efforts. An alternate approach is to restore more naturalized instream flow patterns to allow recovery through natural recruitment and growth processes (Molles et al., 1998; Richter & Richter, 2000; Rood et al., 2003).

According to Li & Eddleman (2002), traditional engineering methods for streambank stabilization that were once thought successful in the past are being re-evaluated in context of impacts resulting from excessive and rapid urbanization, and from the public awareness of these new environmental issues. These restoration strategies are very costly, may require perpetual effort, and often fail. Schiechtl (1985) mentioned that the stabilization of slopes through vegetation and soil treatment measures may be particularly appropriate in situations where an abundance of vegetative materials is present, and where manual labor, rather than machinery for installation, can be easily found.

The interest in natural techniques as biotechnical engineering has been raised, and the benefits and advantages of biotechnical engineering or ecological engineering have been gradually re-examined (Riley, 1998). It is necessary to understand the responses toward environmental changes for the management and sustainable use of resources, biological diversity, and ecosystems.

3. Defining soil bioengineering

In the last few decades, the seeking for ecologically correct technologies for environmental restoration has become very important. The new paradigm of economic development was built in order to create improvements in the livelihood of future generations, which incorporates a concept of agriculture production, and consequently less pollutants, associated to environmental techniques applied to restore natural systems and degraded agroecosystems. Researchers all around us have been pointing out signs that indicate that a paradigm shift is taking place both within and outside the engineering profession to accommodate ecological approaches to what was formerly done through rigid engineering and a general avoidance of any reliance on nature. Mitsch & Jørgensen (2003) brought the concept of ecological engineering that involves creating and restoring sustainable ecosystems that have value to both humans and nature. According to the authors, Ecological Engineering combines basic and applied science for the restoration, design, and construction of aquatic and terrestrial ecosystems.

Mitsch et al. (2002) provided example of ecological engineering techniques as a new field that has gained more and more importance, incorporating concepts that make it an increasingly attractive alternative to traditional engineering approaches, which are often much more expensive to construct and sustain. The merits of ecological engineering methods lie in the emphasis on comprehensive consideration of all aspects for soil and water conservation tasks (Wu & Feng, 2006). In addition to what has been said Mitsch & Jørgensen (2003) make prominent that Ecological engineering requires a more holistic viewpoint than in many ecosystem management strategies, with a strong emphasis, as does ecological modeling for systems ecologists, in the need to consider the entire ecosystem, not just species by species.

In this direction Pahl-Wostl (1995) argues that there are two ways that systems can be organized by rigid top-down control or external influence (imposed organization) or by self-organization (Table 1). Imposed organization, such as done in many conventional engineering approaches, results in rigid structures and little potential for adapting to change and desirable for engineering design where predictability of safe and reliable structures are necessary. Self-organization, like ecological engineering, develops flexible networks with a much higher potential for adaptation to new situations.

Characteristic	Imposed organization	Self-organization
Control	Externally imposed; centralized control	Endogenously imposed; distributed control
Rigidity	Rigid networks	Flexible networks
Potential for adaptation	Little potential	High potential
Application	Conventional engineering	Ecological engineering
Examples	Machine Fascist or Socialist society Agriculture	Organism Democratic society Natural ecosystem

Table 1. Systems categorized by types of organization (Pahl-Wostl, 1995).

Although practitioners have coined the terms ground (soil) bio and eco-engineering, confusion still exists as to the exact definition of each. It appears that the term bioengineering was first used as the translation from the German word 'Ingenieurbiologie', created in 1951 by V. Kruedener when referring to projects using both the physical laws of "hard" engineering and the biological attributes of living vegetation, which described the work that encompassed both engineering and biology (Schluter 1984; Stokes et al., 2010) that was considered in an "ecological engineering "context. In 1981, after many discussions with Dr. Schiechtl and other European practitioners, R. Sotir developed the new terminology 'soil bioengineering' for North America (Schiechtl, 1980). The differences between soil bioengineering and eco-engineering are largely due to their effectiveness over time and space. In soil bioengineering, from the first moment of installation no erosion should occur, as this would be considered part of the original criteria and may be alleviated by the angular arrangement and density of the installed measures. Still, Stokes et al. (2010) call to attention that in eco-engineering, civil engineering techniques are not used, although local organic material at the site, e.g. logs and stumps, may be positioned to prevent soil runoff.

Soil bioengineering, or biotechnical slope protection, has been defined variously as 'the use of mechanical elements (or structures) in combination with biological elements (or plants) to arrest and prevent slope failures and erosion' (Gray & Leiser, 1982), 'the use of living vegetation, either alone or in conjunction with non-living plant material and civil engineering structures, to stabilize slopes and/or reduce erosion', and 'the use of any form of vegetation, whether a single plant or collection of plants, as an engineering material (i.e. one that has quantifiable characteristics and behavior)' (Campbell et al., 2008). The biological and ecological concepts are to build based on the increase of the resistance of slopes to surface erosion by providing limited mechanical support to the soil, thereby reducing the potential for further surface erosion, gully formation, shallow failures, surface debris movement, and debris entrainment.

Soil bioengineering, in the context of upland slope protection and erosion reduction, combines mechanical, biological, and ecological concepts to arrest and prevent shallow slope failures and erosion (Gray & Sotir, 1992). Gray & Sotir (1996) describe soil bioengineering as a specific term that refers to 'the use of live plants and plant parts, in which live cuttings and stems are placed in the ground, or in earthen structures, where they provide additional mechanical support to soil, and act as hydraulic drains, barriers to earth movement, and hydraulic pumps or wicks'. Soil bioengineering systems commonly incorporate inert materials such as rock and wood, or geo-synthetics, geo-composites, and other manufactured products. Simplifying the concept, Sotir (2001) stated that soil bioengineering is the combined application of engineering practices and ecological principles to design and build systems that contain living plant materials. Thereby, bioengineering has as strategy to provide a sustainable ecosystem that benefits both human society and the natural environment (Zhai et al., 2010).

4. Bioengineering applications

The emphasis on ecosystem management, on improving fisheries, and on healthy watersheds has renewed interest in erosion control in the form of soil bioengineering. In these cases, what is focused on primarily is the erosion control that will start with a planted vegetation, and then establishment of a natural recovery by a "succession". According to Normaniza & Barakbah (2011), an understanding of these plant successional processes and pioneer vegetation will allow the development of effective strategies for revegetation of the slopes. Systems largely structured by a broad-scale physical process, such as riparian ecosystems worldwide, may be the most difficult to restore if the process is muted or extinct (Didham et al., 2005; Fremier & Talley, 2009). Managing plant communities that were created and maintained under extinct historic conditions, while not taking advantage of the impacted process (i.e., within site approaches), will lead to unexpected and often undesirable outcomes (Zedler, 2005). There are many biotechniques available to be applied in order to reduce bank erosion along rivers, pounds, and another water bodies.

As observed by Salix Applied Earthcare (2004), each one needs to focus on some elementary information about the site that will receive the bioengineering technique. Streambank soil bioengineering works are often useful on sensitive or steep sites, in areas with limited access, or where working space for heavy machinery is not feasible and its application involves the installation of woody plant materials, securely embedded in the ground and placed in specific planned configurations to create effective erosion control measures. They are intended to have an immediate effect and also to provide a foundation that will encourage colonization by the surrounding plants, thus ensuring long-term remediation and protection of slopes scarred by erosion, experiencing active soil erosion, and affected by shallow slope failures (Nilsson & Berggren, 2000). In addition, soil bioengineering measures are intended to both encourage and accelerate the processes of natural re-vegetation, thus enhancing natural diversity and sustaining the natural hillside ecosystems.

According to Stiles (1988), one of the benefits of these biotechniques compared to traditional engineering is their capacity to increase resistance over time. It is possible due to the strength increase that the plants provided the structures (as stakes, layering, etc.) as they grow and spread over the soil that they are holding. As we know, one of the main principles of soil bioengineering design for riverbank recovery is to provide support to forestation, especially to the native vegetation. Sometimes it is necessary to manage the vegetation so that it remains at the shrubby bush stage, without a main trunk, to reduce the risk of erosion. In fact, the development of trees is also to be avoided in order to maintain access for towing and other riverbank activities (Evette et al., 2009).

Soil bioengineering techniques to stabilize streambanks and shorelines are as effective, and sometimes more effective, than traditional engineering treatments (Li & Eddleman, 2002). Techniques to stabilize streambanks work by either reducing the force of the flowing water, by increasing the resistance of the bank to erosional forces, or by a combination of the two. They are generally appropriate for immediate protection of slopes against surface erosion, shallow mass wasting, cut and fill slope stabilization, earth embankment protection, and small gully repair treatment, also including dune stabilization, wetland buffers, reservoir drawdown areas where plants can be submerged for extended periods, and areas with highly toxic soils (Evette et al., 2009). Soil bioengineering for erosion control is not a method that imposes manmade structures on the site at the expense of existing native plant materials. Control of bank erosion can be accomplished in several ways, such as the use of rock-fill, which, though efficient, is quite expensive, precluding its use extensively along river banks.

In the Nineteenth Century, Defontaine (apud Evette et al., 2009) had suggested that traditional practices of engineering could be supplemented by soil bioengineering using stone and rock pavements (rip-rap) as shown in Figure 2.

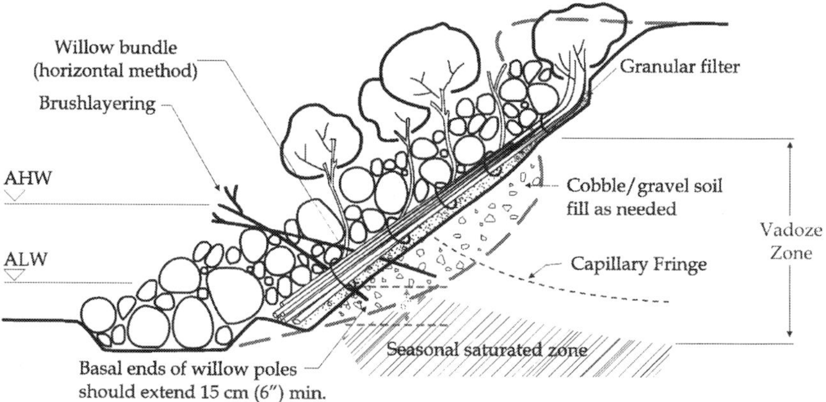

Fig. 2. Vegetated Riprap or Joint Planting composed live stakes, brushlayering and willow bundle, considering the average high or low water level. Adapted from Salix Applied Earthcare (2004).

Joint planting or Vegetated Riprap, in effect, involves tamping live cuttings of rootable plant material into soil between the joints or open spaces in rocks that have previously been placed on a slope (USDA–NRCS, 2007).

Petrone & Preti (2010) demonstrate that soil bioengineering for bank stabilization interventions regarding erosion occurrence is the most appropriate, because it is in accordance with the main concept of sustainable development, and also that soil bioengineering transfer provides users with an instrument that guarantees stability. This is essential to clearly demonstrate the objectives, risks, and reproducibility of the technology to local communities, certainly leading to a range of other innovative and sustainable technologies and a stimulating research environment. Like ecological engineering, soil bioengineering to provide riverbank restoration based on erosion control requires a more holistic viewpoint than what is common in many ecosystem management strategies; it considers all components of the riverine system simultaneously.

Another application for this technique is found to improve environmental factors. Its set can help protect environments that are still preserved and provide better conditions for the development of local fauna and flora. This has been widely used in public recreation areas, national parks, creeks, inlets, among others (Salix Applied Earth Care, 2004; Wu & Feng, 2006).

5. Planning of stream mitigation using soil bioengineering

Design and construction of specific soil bioengineering measures, selection of appropriate plant species, the maintenance requirements during the establishment period of the measures, and the subsequent monitoring and evaluation procedures are the procedures that guarantee the success of this technique (Campbell et al., 2008). Nevertheless, considering that soil bioengineering has unique attributes, and it is not appropriate for all sites and situations a list of factors and causes known to influence slope stability was chosen by Mickovski & Van Beek (2006) as part of a decision support system to implement eco-engineering practices, as shown in Table 2.

Once the decision is made the installation of the biotechniques plays a major structural role immediately or may become the major structural component over time. The effective installation of soil bioengineering measures requires careful planning and design, based upon the specific characteristics of each site. These include factors such as the site geology, soils, slope angle, slope aspect, hydrology, existing vegetation cover, etc., which should all be assessed before appropriate measures can be prescribed (Campbell et al., 2008).

Implementing projects in harmony with natural landscapes include the following considerations: careful selection of suitable construction machinery and tools matched to terrain characteristics; stable and correctly shaped banks; avoidance of steep gradients; use local building materials, e.g. stone, gravel, sand, soil, wood; use of local building materials that do not naturally occur at the construction site, e.g. rocks and boulders in fine grained alluvials, are best avoided; avoidance of artificial building materials, e.g. steel, concrete, plastics for surface cladding of grouting of river or stream beds; preferential use of live building materials; obtaining woody plants capable of vegetative propagation from the construction site, its environs or from similar nearby habitats; preservation of vegetation on the fringes of the construction or regulation area by the considerate use of moving machinery and equipment; removal, temporary storage and re-establishment (transplantation) of vegetation; restricted or, at best, total avoidance of cutting traces, fragmentation or clearing of alluvial woodland (Schiechtl & Stern, 1997).

Basic principles of soil bioengineering necessary for good planning are summarized as follows in Table 3.

Site characteristics	Slopes with high hazard of slope instability
Morphology	
Gradient	Moderately steep for landslides (>10_) to extremely steep for falls (>35_). Some flows can maintain momentum even on very gentle slopes.
Shape	Convergent or irregular in profile.
Height	Short steep slopes for rotational slides, long slopes for translational slides.
Material	
Slope material	Plastic soils, material sensitive to physical or chemical weathering or heavily fractured or jointed rock.
Stratigraphy	Alternation of weaker and stronger beds, of different permeability.
Hydrology	Signs of ponding and springs, presence of gleyic horizons indicating stagnating water in the soil.
Drainage	Heavily dissected by ephemeral or permanent streams with signs of undercutting at the base of the slope or signs of disrupted drainage.
Climate	Periods of intense or prolonged rainfall or rapid snowmelt; Strong diurnal and seasonal variations in temperature, e.g. freeze-thaw.
Seismicity	Evidence of moderately strong to strong earthquakes.
Past activity	Signs of previous slope movements (creep, sliding) and/or surface wash.
Vegetation	Irregular stands and/or deformed or underdeveloped vegetation; Exposure of roots in cracks or at the surface.
Human activity	Evidence of poor site management (leakage of sewer systems, blocked drains etc.) or extensive changes to the shape or composition of a slope. On a marginally stable slope, human intervention can easily upset the critical balance.

Table 2. Site characteristics and slopes with high hazard of slope instability.

In order to correctly plan and install a soil bioengineering project it must be considered that sites typically require some earthwork prior to the installation of soil bioengineering systems. A steep undercut or slumping bank, for example, requires grading to flatten the slope for stability.

1) The degree of flattening depends on the soil type, hydrologic conditions, geology, and other site factors; 2) Scheduling and timing planning and coordination are needed to achieve optimal timing and scheduling; 3) Vegetative damage to inert structures does not generally occur from roots. Plant roots tend to avoid porous, open-faced retaining structures because of excessive sunlight, moisture deficiencies, and the lack of a growing medium. 4) Moisture

1. Establishment of the cause of the damage if repair work is needed.
2. Establishment of the objective and final appearance of the project.
3. Evaluation of the hydro-engineering aspects of the project details.
4. Evaluation of the legal position (ownership, use, liability, etc).
5. Final selection of the bioengineering technique to be implemented.
6. Fit the soil bioengineering system to the site, which means that it has to consider information on site topography, geology, soils, vegetation, and hydrology. At a minimum, collect information on:
   i. Topography and exposure, related to the degree of slope in stable and unstable areas;
   ii. Geology and soils, related to geologic history and types of deposits (colluvium, glacial, alluvium, other), soil type and depth;
   iii. Hydrology, drainage area and the annual precipitation, and calculation of peak flows or mean discharge through the project area;
   iv. Site visit, alignment route, longitudinal and cross-section, (hydrological information);
   v. Evaluation of the soil analysis results of the bed material and watercourse bank stability;
   vi. Evaluation of the vegetation survey of the project area and its environment;
   vii. Evaluation of all available information on the hydro-ecology of the area;
In order to reach this information:
Obtain topographical maps, aerial photos, orthophotos and construction plans.
7. Selection of the construction method and type.
8. Selection of the live and dead vegetative material to be used.
9. Retain existing vegetation whenever possible - Limit removal of vegetation by the removal and storage of existing woody vegetation that may be used later in the project.
10. Stockpile and protect topsoil, related to the topsoil removal during clearing and grading operations that can be reused during planting operations.
11. Protect areas exposed during construction.
12. Divert, drain, or store excess water.

Table 3. Checklist for the planning of water bioengineering construction. Adapted from USDA–NRCS (1992) and Schiechtl & Stern (1997).

requirements and effects must consider that the backfill behind a stable retaining structure has certain specified mechanical and hydraulic properties. Ideally, the fill is coarse-grained, free-draining, granular material. Free drainage is essential to the mechanical integrity of an earth-retaining structure and also important to vegetation.

Soil bioengineering applications work directly with plants and live structures, so we must not forget that their basic science is ecology. Ecological knowledge is the fundamental scientific basis to planting and managing sustainable systems, and since holism and systems theory open up new perspectives and provide broader visions for planning, then it is highly desirable to have on staff people that are specialized in it. According to Leitão & Ahern (2002) sustainable planning represents a promising challenge for motivating and inspiring trans-disciplinary collaboration. Then, biologists, agronomists, engineers, geologists are part of the professionals necessary to develop plain environmentally and economically correct projects.

5.1 Suitable plant materials

The role of vegetation on slopes is increasingly being recognized and slope greening has become more important, as reflected in the number of landscaped slopes, government policies and business opportunities (Chong & Chu, 2007).

Although traditional erosion control practices have often focused on structures made from stone and other nonliving materials, interest in the use of plant materials, alone and in combination with nonliving materials ("soil bioengineering") for a range of applications, is increasing (Li et al., 2006). Some of the ecotechnological methods are not new and, in fact, some have been practiced for centuries, particularly in China (Stokes et al., 2010).

Vegetation helps to prevent erosion on slopes by: 1) Binding and restraining soil particles in place; 2) Reducing sediment transport; 3) Intercepting raindrops; 4) Retarding velocity of runoff; 5) Enhancing and maintaining infiltration capacity; 6) Minimizing freeze-thaw cycles of soils susceptible to frost Gray & Sotir (1992).

The selection of suitable plant species and species combinations in soil bioengineering measures must be based on careful vegetation surveys. The plants must tolerate thin, well-drained soils, steep slopes, and exposed sites. Native species, mainly shrubs, are preferred, once they are compatible with local ecosystems and are relatively inexpensive, because they can be harvested from areas adjacent to the site. Also, they are well suited to the local climate, soil, and moisture conditions. Exotic species may be considered in certain circumstances, to stabilize riverbanks generally has shown very good results, although native species are more suitable reducing the likelihood of erosion by mass failure due to reinforcement of riverbank soils by tree roots and this reduced likelihood of mass failure (Hubble et al., 2010). Despite that in some cases these techniques cannot resist in environments where the river's flow is continuous with high sediment transport. Live staking, live fascines, brush layers, and branchpacking have been current listed as soil bioengineering techniques that use stems or branch parts of living plants as initial and primary soil reinforcing and stabilizing material. Based in Li & Eddleman, (2002) and USDA–NRCS (2007), we listed some of the most important biotechnical streambank stabilization techniques in Table 4.

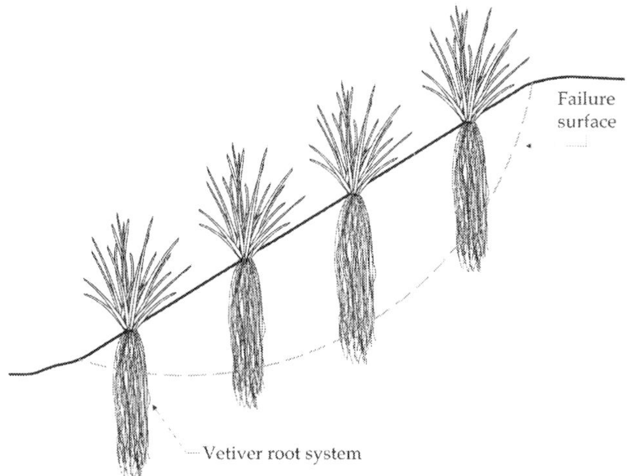

Fig. 3. Soil reinforcement by vetiver grass roots minimizing erosion risks.

Polser & Bio (2002) mentioned other biotechniques tipically applied to small streambanks or creeks such as wattle fences, live palisades, live gravel bar staking and live shade.

Live Fascine

Is a long bundle of live cuttings bound together into a rope or sausage-like bundles and their structure provides immediate protection for the toe. Since this is a surface treatment, it is important to avoid sites that will be toowet or too dry. The live cuttings eventually root and provide permanent reinforcement.

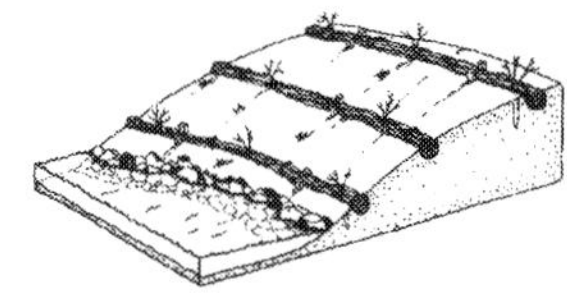

Live Stakes

Live pole cuttings are dormant stems, branches, or trunks of live, woody plant material inserted into the ground with the purpose of getting them to grow. Live stakes are generally shorter material that are also used as stakes to secure other soil bioengineering treatments such as fascines, brush mattresses, erosion control fabric, and coir fascines.

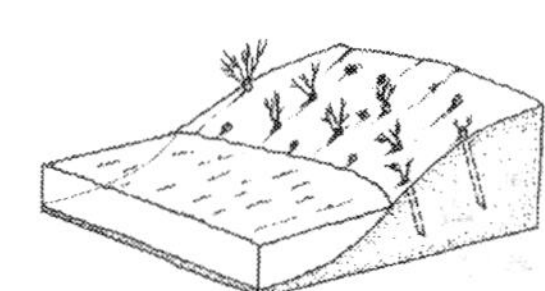

Brushlayering

Consists of alternating layers of live cuttings and soil. The cuttings protrude beyond the face of the slope approximately 6 to 18 inches. The installed live cuttings provide immediate frictional resistance to shallow slides, similar to conventional geotextile/geogrid reinforcement.

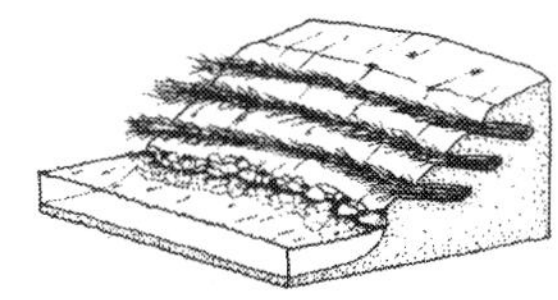

Branchpacking

Consists of alternating layers of live cuttings and soil to repair samalls slumps and holes in streambanks. The live cuttings reinforce the soil similar to conventional geotextile/geogrid reinforcements. The stems provide immediate frictional resistance to shallow slides.

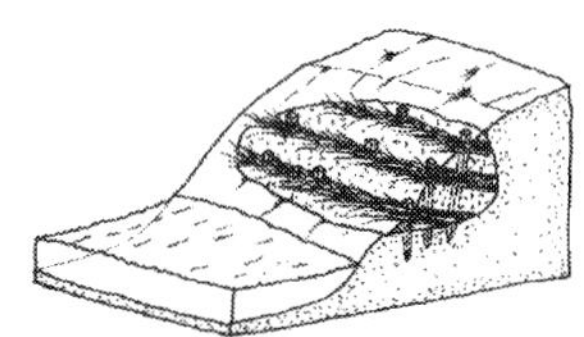

Live Cribwall

Is a hollow, boxlike structure of interlocking logs or timbers filled with rock, soil, and live cuttings, or rooted plants, that are intended to develop roots and top growth and take over some or all of the structural functions of the logs.

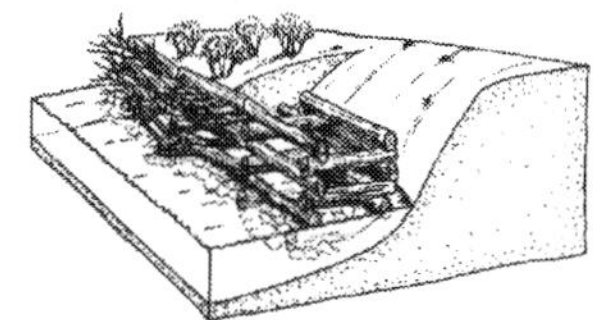

Brushmattress

Is a layer of live cuttings placed flat against the sloped face of the bank. Dead stout stakes and string are used to anchor the cutting material to the bank. This measure is often constructed using a fascine, joint planting, or riprap at the toe, with live cuttings in the upper mattress area.

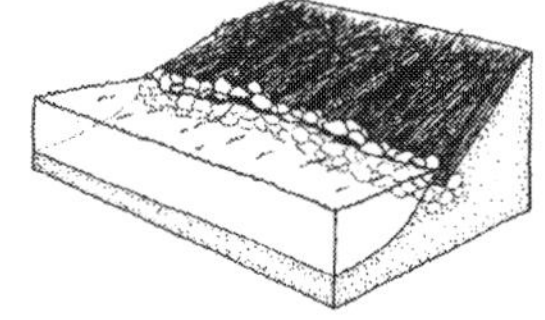

Coconut Fiber Rolls

Coconut fiber rolls or Coir fascines consist of coconut husk fibers bound together in a cylindrical bundle by natural or synthetic netting and are manufactured in a variety of standard lengths, diameters, and fill densities for different energy environments. They are flexible and can be fitted to the existing curvature of a streambank.

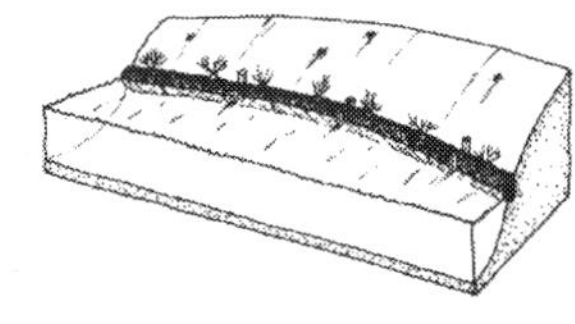

Erosion Control Blanket

They are produced from natural and synthetic materials such as straw, wood excelsior, woven coir, or combinations of these and turf reinforcement mats produced from nondegradable, synthetic, three-dimensional fibers. Jute mesh and coir mesh are the most used.

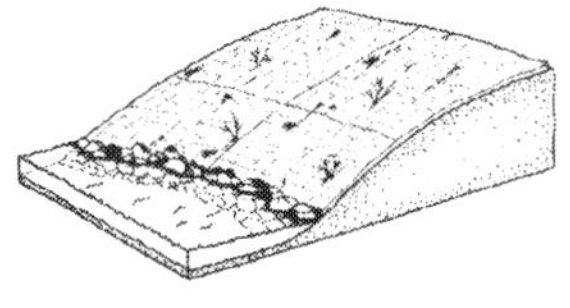

Table 4. Some biotechnical streambank stabilization´s techniques (Li & Eddleman, 2002; USDA–NRCS, 2007).

There is a certain influence of the root tensile strength on the increase in soil shear strength. The progress made during the past few years on the contribution of the root system in reinforcing mass-stability of slopes is an eye-opener. Soil cover with grass or herbaceous vegetation provides an efficient protection against surface erosion by reducing the impact of rainfall on bare soil (Davide et al., 2000), besides increased percolation of water, soil cohesion, and resistance on the banks, which are provided by the root systems (Burylo et al., 2009). Cazzuffi et al. (2006) mentioned that vetiver grass (*Vetiveria zizanioides* L. Nash) among other species is characterized by very resistant roots and confirms how they, could be successfully used with stabilizing effects on phenomena like shallow instability (Figure 3).

Vetiver grass have been used in practices of erosion control and slope stabilization (Mickovski & Van Beek, 2009; Mickovski et al., 2005; Truong, 2002), promoting a reduction by 50% and 70% of surface runoff and eroded soil (Phien & Tam, 2007). Being a very easy crop to grow, at various levels and fertility types of land, which very well resists both drought and immersion in water, Vetiver grass tolerates conditions of root asphyxia; it is easy to cultivate, almost without maintenance, and likes to be exposed to full sun; it is a long-lived crop, living more than 10 years, and for land conservation does not yield seed, and rhizomes or stolons (roots which can yield new crop), does not expand wildly outside the planned area, and consequently will not become an intruder upon other plants (Budinetro, 2004).

Fig. 4. Stages of installation of the biotechniques in Paramopama Creek in northeastern Brazil. a) Degraded river channel and riparian zone; b) Gabion at the toe and at the river bed, plus jute matting at the streambank; c) Development of the grass cover; d) Vegetation development six months after installation, composed by legume-shrubs mixture. Adapted from Holanda et al., (2009).

Other species are widely used in bank recovering projects, especially in the Northern Hemisphere. The genus Salix, also recognized as willow (*Salix* L.), has around 400 species between trees and shrubs, and the most used, generally, is found in soils with high moisture in temperate and arctic zones, but also can occur in subtropical and tropical zones; that is why it is highly desirable in this type of design. Among the range of agronomical, physiological and ecological characteristics of the genus Salix that are pertinent to ecological engineering, erosion control in order to protect slopes, streambanks and shorelines against water erosion, is very remarkable (Kuzovkina & Volk, 2009; Kuzovkina & Quigley, 2005 ;Wilkinson, 1999 ; Pezeshki et al., 2007 ; van Splunder et al., 1994; Shields Junior et al., 1995;), and if they are established successfully they alter the microclimate, improving soil conditions, control invasive species, and re-establish natural ecological complexity. Beside this, it became most useful due its fast growth rate allied to a dense root system that can rapidly stabilize the streambank and promote the secondary establishment of other vegetation (Figure 2). Willow (Salix spp) are analogous to annual or short-lived perennial grasses in a seed mixture (nurse or companion crop), and they provide a quick pioneer plant cover for soil protection. Their longevity depends on the region of the country and specific site conditions. In all cases, they prefer damp soils (USDA–NRCS, 2007).

Among the versatile leguminous trees, *Leucaena leucocephala* has been determined as a potential slope plant. Being a multipurpose tree that profusely produces propagules (beans) and has been used as an erosion control plant, Normaniza, et al (2008) identified a very important contribution of this species in terms of slope stability enhancement, showing that it plays a major mechanical role, as well as a hydrological role, in stabilizing slopes and protecting against soil erosion. It is suggested that the high capacity of root reinforcement and water absorption of *L. leucocephala* rank it as an outstanding future slope remedy for preventing slope failure.

5.2 Structural components

Structures can be built from natural or manufactured materials. Natural materials, such as earth, rock, stone, and timber, usually cost less, are environmentally more compatible, and are better suited to vegetative treatment or slight modifications than are manufactured materials. Natural materials may also be available onsite at no cost (USDA–NRCS, 1992). Live cribwalls, vegetated rock gabions, vegetated rock walls, and joint plantings are soil bioengineering techniques that use porous structures with openings through which vegetative cuttings are inserted and established (Figure 4). The inert structural elements provide immediate resistance to sliding, erosion, and washout, and as vegetation becomes established, roots invade and permeate the slope, binding it together into a unified, coherent mass.

6. Advantages and limitations of soil bioengineering practices

Several potential environmental benefits can be achieved by using soil bioengineering measures as opposed to conventional engineering methods. Notably, they generally require only minimal access provisions for equipment, materials and workers, and typically create only minor disturbances to the site during installation. In environmentally sensitive locations, where preservation of scenery or wildlife habitats may be critical, soil bioengineering measures can usually offer more environmentally compatible solutions. More importantly, for sensitive or remote sites, these measures do not require long-term maintenance, thereby creating fewer disturbances.

According to Schiechtl (1985) the use of natural building material requires spaces and it would be to attempt the implementing of vegetative methods in the construction of protection measures. Soil bioengineering systems generally require minimal access for equipment and workers and cause relatively minor site disturbance during installation, and cannot be installed where the site is in bedrock, on deep-seated failures with high back scars, or on steep slopes (over about 35-50°).

Soil bioengineering measures that combine mechanical, biological, and ecological principles and practices to protect and enhance slopes, repair erosion gullies, and remediate shallow mass movement scars are generally considered to be cost-effective techniques with desirable environmental and visual characteristics.

7. Site maintenance and monitoring

Designs for application of soil bioengineering techniques should also consider the periodic access of people, tools, supplies, and machinery (in some cases) to the site in order to guarantee the efficiency of the conjunct of elements involved in each situation. Commonly, when the site requires some kind of repair, this simple step in the planning can avoid unfortunate and expensive costs of material movement and replacement. The situation can be aggravated if the site was designed for experimental studies.

Recently, few techniques for evaluating streambank stability and the real ground geotechnical behavior are available, providing low-resolution monitoring and, in most cases, restricting visual comparison by photographic registries or invasive measurements in field, like topographic surveys. Nowadays, with an increasing need for high-resolution data in many areas involved with riverbank erosion (i.e. fluvial geomorphology and geotechnical engineering), the advance of remote technology, and the increment of electronic sensors, monitoring soil bioengineering sites has been becoming more accurate and trustful. For Lawler (2005), it also allows for collecting directly and routinely in the field, at event time scales, real-time high-resolution data.

Thus, it is possible to advance in knowledge and acquire data on the mechanism of riverbank erosion using high-resolution techniques, in addition to the meteorological and fluvial data that are already widely available.

Other tools have reached a great importance in erosion monitoring or in the effectiveness of the techniques toward its control, as they provide automated and continuous real-time bank erosion data. This information is of great importance to the field of geomorphology, as well as to numerical models such as the computer model Streambank Erosion CONCEPTS (Langendoen & Alonso, 2008), which simulates channel width adjustment by incorporating the two fundamental physical processes responsible for bank retreat: fluvial erosion or entrainment of bank-material particles by flow, and bank mass failure due to gravity.

Bertrand (2010), also studying erosion monitoring concluded that the Photo-Electric Erosion Pin, or PEEP, provided real-time monitoring of erosion events in terms of magnitude and frequency, which is not possible with manual instruments where only net changes from previous measurements are known. This real-time monitoring coupled with the automated nature of the instrument makes it ideal for certain sites that are not easy to access on a continuous basis.

8. Conclusions

There is a strong and urgent need to restoration the riparian zone with native or exotic plant species that have a fast vegetative development, in order to reduce riverbank erosion. Nevertheless, the preservation of riparian remnants is vital because they produce source of plant seeds, provide home for pollinators and dispersal agents, and contribute enormously to the recovery of the riparian zone. In the tropical region the riverine populations have tried their own solutions in order to control the riverbank´s erosion through the use of local low cost materials. At the same time public policies have focused on the Streambanks recovering mostly with the use of riprap to absorb the strong impact of rivers discharge regularization and its consequences. The use of soil bioengineering techniques have been motivated by practitioner's to promote immediate soil protection against erosion, by fast revegetation. It seems that it will take time and the participation of the public authorities, users and communities until these biotechniques will be recognized with its remarkable technical and environmental importance on the streambanks degraded recovery.

9. References

Alpert, P., F. T. Griggs, & D. R. Peterson. (1999). Riparian forest restoration along large rivers: Initial results from the Sacramento River Project. *Restoration Ecology*, Vol. 7, No.4, (date not avaliable), pp. 360–368, ISSN 1061-2971.

Bertrand, F. (2010). Fluvial erosion measurements of streambank using Photo-Electronic Erosion Pins (PEEP), In: *University of Iowa Website*, 06.04.2011, Avaliable from: <http://ir.uiowa.edu/etd/642/>

Budinetro, H. S. (2004). Low-cost treatment of river bank erosion, In: *Proceedings of APHW 2004 2nd International Conference on Hydrology and Water Resources in Asia Pacific Region*, 03.04.2011, Avaliable from: <http://www.wrrc.dpri.kyoto-u.ac.jp/~aphw/APHW2004/proceedings>

Burylo, M., Rey, F., Roumet, C., Buisson, E. & Dutoit, T. (2009). Linking plant morphological traits to uprooting resistance in eroded marly lands (Southern Alps, France). *Plant And Soil*, Vol.324, No.1-2, (March 2009), pp. 31-42, ISSN 0032-079X.

Carone, M. T., Greco, M., Molino, B. (2006). A sediment-filter ecosystem for reservoir rehabilitation. *Ecological Engineering*, Vol.26, No.2, (September 2005), pp. 182-189, ISSN 0925-8574.

Conforth, D. H. (2005). *Landslides in practice: investigations, analysis and remedial/preventive options in soils*, John Wiley & Sons, ISBN 0-471-67816-3, Hoboken, New Jersey.

Campbell, S. D. G., Shaw, R., Sewell, R. J. & Wong, J. C. F. (2008). *Guidelines for soil bioengineering applications on natural terrain landslide scars*, Civil Engineering and Development Department of the Government of Hong Kong Special Administrative Region, Homantin, Kowloon.

Cazzuffi, D., Corneo, A. & Crippa, E. (2006). Slope stabilization by perennial "gramineae" in Southern Italy: plant growth and temporal performance. *Geotechnical and Geological Engineering*, Vol.24, No.3, (June 2005), pp. 429–447, ISSN 1573-1529.

Campos, J. B. & Souza, M. C. (2002). Arboreous vegetation of an alluvial riparian forest and their soil relations: Porto Rico Island, Paraná river, Brazil. *Brazilian Archives of Biology and Technology.* Vol.45, No.2, (June 2002), pp. 137-149, ISSN 1516-8913.

Chong, C. W. & Chu, L. M. (2007). Growth of vetivergrass for cutslope landscaping: effects of container size and watering rate. *Urban Forestry & Urban Greening,* Vol.6, No.3, (date not available.), pp. 135–141, ISSN 1618-8667.

Dale, V. H., Joyce, L. A, McNulty, S. & Neilson, R. P. (2000). The interplay between climate change, forests and disturbances. *Science of the Total Environment,* Vol.262, No.3, (March 2000), pp. 201-204, ISSN 0048-9697.

Davide, A. C., Ferreira, R. A., Faria, J. M. R. & Botelho, S. A. (2000). Restauração de Matas Ciliares. *Informe Agropecuário,* Vol.21, No. 207, (date note avaliable), pp. 65-74, ISSN 0100-3364.

Didham, R. K., Tylianakis, J. M., Gemmell, N. J., Rand, T. A. & Ewers, R. M. (2005). Interactive effects of habitat modification and species invasion on native species decline. *Trends in ecology and evolution,* Vol.22, No.9, (date not available), ISSN 0169-5347.

DeWine, J. M. & Cooper, D. J. (2007). Effects of river regulation on riparian box elder (*Acer Negundo*) forests in canyons of the upper Colorado river basin, USA. *Wetlands,* Vol.27, No.2, (February 2007), pp. 278-289, ISSN 0277-5212.

Evette, A., Labonne, S., Rey, F., Liebault, F., Jancke, O. & Girel, J. (2009). History of bioengineering techniques for erosion control in rivers in western Europe. *Environmental Management,* Vol.43, No.6, (January 2009), pp. 972–984, ISSN 1432-1009.

Fremier, A. K. & Talley, T. S. (2009). Scaling riparian conservation with river hydrology: lessons from Blue Elderberry along four California rivers. *Wetlands,* Vol.29, No.1, (August 2008), pp. 150-162, ISSN 0277-5212.

Fausch, K. D., Torgersen, C. E., Baxter, C. V. & H. W. Li. (2002). Landscapes to riverscapes: bridging the gap between research and conservation of stream fishes. *BioScience* Vol.52, No., (date not avaliable), pp. 483–498, ISSN 0006-3568.

Gray, D. H. & Leiser, A. T. (1982). *Biotechnical slope protection and erosion control,* Van Nostrand Reinhold, ISBN 0-442-21222-4, New York, New York.

Gray, D. H., & Sotir, R. B. (1992). Biotechnical stabilization of highway cut slope. *Journal of Geotechnical Engineering,* Vol.118, No.9, (October 1990), pp. 1395-1409, ISSN 0733-9410.

Gray, D. H. & Sotir, R. B. (1996). *Biotechnical and soil bioengineering slope stabilization: a practical guide for erosion control,* John Wiley & Sons, ISBN 0-471-04978-6, New York, New York.

Gutberlet, J., Seixas, C. S., Glinfskoi, T. & Carolsfeld, J. (2007). Resource conflicts: challenges to fisheries management at the São Francisco river, Brazil. *Human Ecology,* Vol.35, No.5, (September 2007), pp. 623-638, ISSN 0300-7839.

Holanda, F. S. R, Santos, I. G. C., Santos, C. M. S., Casado, A. P. B., Pedrotti, A. & Ribeiro, G. T. (2005). Riparian vegetation affected by bank erosion in the Lower São Francisco river, Northeastern Brazil. *Revista Árvore,* Vol.29, No.2, (November 2004), pp. 327-336, ISSN 0100-6762.

Holanda, F. S. R., Ismerim, S. S., Rocha, I. P., Jesus, A. S., Araujo Filho, R. N. & Mello Junior, A. V. (2009). Environmental perception of the São Francisco riverine population in regards to flood impact. *Journal of Human Ecology*, Vol.28, No.1, (August 2008), pp. 37-46, ISSN 0970-9274.

Holanda, F. S. R., Gomes, L. G. N., Rocha, I. P, Santos, T. T., Araújo Filho, R. N., Vieira, T. R. S. & Mesquita, J. B. (2010). Initial development of forest species on riparian vegetation recovery at riverbanks under soil bioengineering technique. *Ciência Florestal*, Vol.20, No.1, (December 2009), pp. 157-167, ISSN 0103-9954.

Holanda, F. S. R., Bandeira, A. A, Rocha, I. P., Araújo Filho, R. N, Ribeiro, L. F. & Ennes, M. A. (2009). Riverbank erosion control at streams margin: from empiricism to soil bioengineering technique. *Ra'ega*, No.17, (date not avaliable), pp. 93-101, ISSN 1516-4136.

Hunt, E. (1990). Judgment assessment of slopes intropical climates. *Solos e Rochas - Revista Brasileira de Geotecnia*, Vol.13, (date not available), pp. 46-64, ISSN 0103-7021.

Hubble, T. C., Docker, B. B. & Rutherfurd, I. D. (2010). The role of riparian trees in maintaining river bank stability: a review on Australian experience and practice. *Ecological Engineering*, V.36, No.3, (April 2009), pp. 292–304, ISSN 0925-8574.

Iwata, T., Nakano, S. & Inoue, M. (2003). Impacts of past riparian deforestation on stream communities in a tropical rain forest in Borneo. *Ecological Applications*, Vol.13, No.2, (January 2002), pp. 461-473, ISSN 1051-0761.

Kuzovkina Y. A. & Quigley, M. F. (2005). Willows beyond wetlands: uses of *Salix* L, species for environmental projects. *Water, Air, and Soil Pollution*, Vol.162, No.1-4, (October 2004), pp. 183-204, ISSN 0049-6979.

Kuzovkina, Y. K. & Volk, T. A. (2009). The characterization of Willow (*Salix* L.) varieties for use in ecological engineering applications: co-ordination of structure, function and autecology. *Ecological Engineering*, Vol.35, No.8, (March 2009), pp. 1178-1189, ISSN 0925-8574

Lawler, D. M. (2005). The importance of high-resolution monitoring in erosion and deposition dynamics studies: examples from estuarine and fluvial systems. *Geomorphology*, Vol.64, No.1-2, (April 2004), pp. 1-23, ISSN 0169-555X.

Langendoen, E. J. & Alonso, C.V. (2008). Modeling the evolution of incised streams: I. Model formulation and validation of flow and streambed evolution components. *Journal of Hydraulic Engineering*, Vol.134, No.6, (September 2007), pp. 749-762, ISSN 0733-9429.

Leitão, A. B. & Ahern, J. (2002). Applying landscape ecological concept and metrics in sustainable landscape planning. *Landscape and Urban Planning*, Vol.59, No.2, (October 2001), pp. 65-93, ISSN 0169-2046.

Li, X., Zhang, L. & Zhang, Z. (2006). Soil bioengineering and the ecological restoration of riverbanks at the airport town, Shanghai, China. *Ecological Engineering*, Vol.26, No.3, (October 2005), pp. 304–314, ISSN 0925-8574.

Li, M. H. & Eddleman, K. E. (2002). Biotechnical engineering as an alternative to traditional engineering methods a biotechnical streambank stabilization design approach. *Landscape and Urban Planning*, Vol.60, No.4, (April 2002), pp. 225–242, ISSN 0169-2046.

Mickovski, S. B., Van Beek, L. P. H. & Salin, F. (2005). Uprooting of vetiver uprooting resistance of vetiver grass (Vetiveria zizaoindes). *Plant and Soil*, Vol.278, No.1-2, (February 2005), pp. 33-41, ISSN 0032-079X.

Mickovski, S. B. & Van Beek, L. P. H. (2006). A decision support system for the evaluation of eco-engineering strategies for slope protection. Geotechnical and Geological Engineering, Vol. 24, No.3, (June 2005), pp. 483-498, ISSN 0960-3182.

Mickovski, S. B. & Van Beek, L. P. H. (2009). Root morphology and effects on slope stability of young vetiver (*Vetiveria zizanioides*) plants grown in semi-arid climate. *Plant and Soil*, Vol.324, No.1-2, (July 2008), pp. 43-56, ISSN 0032-079X.

Mitsch, W. J., Lefeuvre, J. C. & Bouchard, V. (2002). Ecological engineering applied to river and wetland restoration. *Ecological Engineering*, Vol.18, No.5, (June 2002), pp. 529-541, ISSN 0925-8574.

Mitsch, W.J. & Jørgensen, S. E. (2003). Ecological engineering: a field whose time has come. *Ecological Engineering*, Vol.20, No.5, (May 2003), pp. 363–377, ISSN 0925-8574.

Molles Junior, M. C., Crawford, C. S., Ellis, L. M., Valett, H. M. & Dahm, C. N. (1998). Managed flooding for riparian ecosystem restoration. *BioScience*, Vol.48, No.9, (September 1998), pp. 749–756, ISSN 0006-3568.

Normaniza, O. & Barakbah, S. S. (2011). The effect of plant succession on slope stability. Ecological Engineering, Vol.37, No.2, (August 2010), pp. 139-147, ISSN 0925-8574.

Normaniza, O, Faisal, H. A. & Barakba, S. S. (2008). Engineering properties of *Leucaena leucocephala* for prevention of slope failure. *Ecological Engineering*, Vol.32, No.3, (October 2007), pp.215-221, ISSN 0925-8574.

Naiman, R. J., Bilby, R. E., Schindler, D. E. & Helfield, J. M. (2002). Pacific salmon, nutrients, and the dynamics of freshwater and riparian ecosystems. *Ecosystems*, Vol.5, No.4, (December 2001), pp. 399–417, ISSN 1435-0629.

Nilsson, C. & Berggren, K. (2000). Alterations of riparian ecosystems caused by river regulation. *Bioscience*, Vol.50, No. 9, (date not avaliable), pp. 783-792, ISSN 0006-3568.

Petrone, A & Preti, F. (2010). Soil bioengineering for risk mitigation and environmental restoration. *Hydrology and Earth System Sciences*, Vol.14, No.2, (January 2010), pp. 239-250, ISSN 1027-5606.

Polser, D. F. & Bio, R. P. (2002). Soil Bioengineering techniques for riparian restoration, In: *Proceeding of the 26th Annual British Columbia Mine Reclamation Symposium*, Dawson Creek, September 2002.

Pahl-Wostl, C., (1995). *The dynamic nature of ecosystems: chaos and order entwined*, John Wiley & Sons, ISBN 978-0-471-95570-2, New York, New York.

Pezeshki, S. R., Li, S., Shields Junior, F. D. & Martina, L. T. (2007). Factors governing survival of black willow (*Salix nigra*) cuttings in a streambank restoration project. *Ecological Engineering*, Vol.29, No.1, (July 2006), pp. 56–65, ISSN 0925-8574.

Phien, T. & Tam, T. T. (2007). Vetiver grass in hedgerow farming systems on sloping lands in Viet Nam, In: *The Vetiver network international website*, 01.04.2011, Avaliable from: <http://www.vetiver.org/VNN_Thai%20Phien.pdf>

Pizzuto, J., O'Neal, M. & Stotts, S. (2010). On the retreat of forested, cohesive riverbanks. *Geomorphology*, Vol.116, No.3-4, (August 2009), pp. 341-352, ISSN 0169-555X.

Queensland Government. (2006). What causes bank erosion?, In: *Queensland Department of Environment and Resourses Management Website*, 03.04.2011, Avaliable from: <http://www.derm.qld.gov.au/factsheets/pdf/river/r2.pdf>

Riley, A. L. (1998). *Restoring streams in cities: a guide for planners, policymakers, and citizens*, ISBN 1-55963-042-6, Island Press, Washington, District of Columbia.

Richter, B. D. & Richter, H. E. (2000). Prescribing flood regimes to sustain riparian ecosystems along meandering rivers. *Conservation Biology*, Vol.14, No.5, (January 2000), pp. 1467-1478, ISSN 0888-8892.

Rood, S. B., Gourley, C. R., Ammon, E. M., Heki, L. G., Klotz, J. R., Morrison, M. L., Mosley, D., Scoppettone, G. G., Swanson, S. & Wagner, P. L. (2003). Flows for floodplain forests: a successful riparian restoration. *Bioscience*, Vol.53, No.7, (July 2003), pp. 647-656, ISSN 0006-3568.

Salinas, M. J. & Casas, J. J. (2007). Riparian vegetation of two semi-arid mediterranean rivers: basin-scale responses of woody and herbaceous plants to environmental gradients. *Wetlands*, Vol.27, No.4, (date not avaliable), pp. 831–845, ISSN 0277-5212.

Salix Applied Earthcare. (2004). Environmentally-Sensitive Streambank Stabilization – EsenSS, version 2004, In: CD-ROM, 04.04.2011, Avaliable from: CD-ROM.

Stiles, R. (1988). Engineering with vegetation. *Landscape Design*, Vol.172, pp. 57-61, ISSN 0020-2908.

Schiechtl, H. M. (1980). *Bioengineering for land reclamation and conservation*, ISBN 0-888-64053-6, University of Alberta Press, Edmonton, Alta.

Schiechtl, H. M. (1985). *FAO Watershed Management Field: Vegetative and soil treatment measures*, ISBN 92-5-102310-7, Food and Agriculture Organization of the United Nations, Rome.

Schiechtl, H. M. & Stern, R. (1997). *Water bioengineering techniques: for watercourse bank and shoreline protection*, ISBN 0-632-04066-1, Blackwell Science, Malden, Massachusetts.

Shields Junior, F. D., Cooper, C. M. & Knight, S. S. (1995). Experiment in stream restoration. *Journal of Hydraulic Engineering*, Vol.121, No.6, (January 1994), pp. 494-502, ISSN 0733-9429.

Schluter, U. (1984). Zur Geschichte der Ingenieurbiologie. *Landschaft und Stadt*, Vol.16, No.1-2, (date not avaliable), pp. 2-9, ISSN: 0023-8058.

Simon, A., Curini, A., Darby, S. E. & Langenduen, E. J. (2000). Bank and near-bank processes in an incised channel. *Geomorphology*, Vol.35, No.3-4, (June 1999), pp. 193-217, ISSN 0169-555X.

Sotir, R. B. (2001). The value of vegetation - strategies for integrating soil bioengineering into civil engineering projects soil bioengineering - integrating ecology with engineering practice, sponsored by Maccaferri & Ground Engineering, pp. 6 - 9.

Stokes, A., Sotir. R., Chen, W. & Ghestem, M. (2010). Soil Bio- and Eco-Engineering in China: past experience and future priorities. *Ecological Engineering*, Vol.36, No.3, (June 2009), pp. 247-257, ISSN 0925-8574.

Sutili, F. J., Durlo, M. A. & Bressan, D. A. (2004). Bio-technical capability of "sarandi-branco" (*Phyllanthus sellowianus* Müll. Arg.) and vime (*Salix viminalis* L.) for re-vegetation water coarse edges. *Ciência Florestal*, Vol.14, No.1, (April 2004), pp.13–20, ISSN 0103-9954.

USDA–NRCS. (1992). Soil bioengineering for upland slope protection and erosion reduction, In: *Engineering Field Handbook*, R. W. Tuttle, (Ed.), pp. 18(33-52), United States Departmente of Agriculture – National Resources Concervation Service (USDA–NRCS), Retrieved from: <ftp://ftp-hq.sc.egov.usda.gov/NHQ/pub/outgoing/jbernard/CED-Directives/efh/EFH-Ch18.pdf.>

USDA–NRCS. (2007). Technical Supplement 14I Streambank Soil Bioengineering, In: Stream restoration design, pp. TS14I(1-76), United States Departmente of Agriculture – National Resources Concervation Service (USDA–NRCS), Retrieved from: <http://directives.sc.egov.usda.gov/viewerFS.aspx?id=3491>

Truong, P. (2002). Vetiver grass technology, In: *Vetiveria: The Genus Vetiveria*, M. MAFFEI, (Ed.), pp. 114-132, Taylor & Francis, ISBN 0-203-21873-6, New York, New York.

Tabacchi, E, Planty-Tabacchi, A. M., Roques, L. & Nadal, E. (2005). Seed inputs in riparian zones:implications for plant invasion. *River Res. Applic.* Vol. 21, (July 2004), pp. 299–313, ISSN 1535-1459.

Van Splunder, I., Coops, H. & Schoor, M. (1994). Tackling the bank erosion problem: reintroduction of willow on riverbanks. *Water Science Technology*, Vol.29, No.3, (date not available), pp. 379-381, ISSN 0273-1223.

Van Klaveren, R. W. & McCool, D. K. (1998). Erodibility and critical shear of a previously frozen soil. *Transactions of the ASABE*, Vol. 41, No.5, (date not avaliable), pp. 1315-1321, ISSN 0001-2351.

Wilkinson, A. G. (1999). Poplars and willows for soil erosion control in New Zealand. *Biomass and Bioenergy*, Vol.16, No.4, (December 1998), pp. 263–274, ISSN 0961-9534.

Walker, L. R., Velázquez, E. & Shiels, A. B. (2009). Applying lessons from ecological succession to the restoration of landslides. *Plant and Soil*, Vol.324, No.1-2, (July 2008), pp. 157-168, ISSN 0032-079X.

Wu, H. L. & Feng, Z. (2006). Ecological engineering methods for soil and water conservation in Taiwan. *Ecological Engineering*, Vol.28, No.4, (August 2006), pp. 333-344, ISSN 0925-8574.

Wynn, T. M. & Mostaghimi, S. (2006). Effects of riparian vegetation on streambank subaerial processes in Southwestern Virginia, USA. *Earth Surface Processes and Landforms*, Vol.31, No.4, (date not available), pp. 399-413, ISSN 1096-9837.

Xia J., Wu, B., Wang, Y. & Zhao, S. (2008). An analysis of soil composition and mechanical properties of riverbanks in a braided reach of the Lower Yellow River. *Chinese Science Bulletin*, Vol.53, No.15, (August 2008), pp. 2400-2409, ISSN 1861-9541.

Zedler, J. B. (2005). Restoring wetland plant diversity: a comparison of existing and adaptive approaches. *Wetlands Ecology and Management*, Vol.13, No.1, (May 2003), pp. 5-14, ISSN 0923-4861.

Zhai, H., Cui, B., Hu, B. & Zhang, K. (2010). Prediction of river ecological integrity after cascade hydropower dam construction on the mainstream of rivers in Longitudinal Range-Gorge Region (LRGR), China. *Ecological Engineering*, Vol.36, No.4, (July 2009), pp. 361-372, ISSN 0925-8574.

26

Improving Biosurfactant Recovery from *Pseudomonas aeruginosa* Fermentation

Salwa Mohd Salleh, Nur Asshifa Md Noh
and Ahmad Ramli Mohd Yahya
School of Biological Sciences, Universiti Sains Malaysia
Malaysia

1. Introduction

Surface active agents or surfactants or are amphiphilic compounds that are chemically synthesized which can greatly reduce the surface tension of a liquid. They are widely used industrially for various purposes such as detergents, wetting agents, foaming agents, emulsifiers, dispersants, lubricants and penetrants (Mulligan & Gibbs, 1993). To date, a large majority of surfactants used are chemically synthesized including alkylbenzene sulfonates (detergents) and lauryl sulfate (foaming agent) (Mukherjee *et al.*, 2006). These synthetic surfactants have been used in the oil industries to aid the clean up of oil spills, rapidly removing large amounts of oil from the ocean/soil surface (Banat, 1995). However, much like the oil they remove, these compounds exhibit poor biodegradability, are toxic to the environment and consequently, have limited applications.

In recent years, environmental compatibility and biodegradability have become increasingly important factors in the selection of industrial chemicals (Banat, 1995). For this reason, natural biosurfactants appear as a promising candidate to replace or reduce the usage of chemically synthesized surfactants. Biosurfactants are surface active biomolecules produced by a variety of microorganisms such as bacteria, yeast and fungi. As with surfactants, they too are amphipathic molecules, comprising hydrophilic and hydrophobic domains. The simultaneous existence of these domains provides biosurfactants the ability to partition themselves at the interphases between different fluid phases (Banat *et al.*, 2000). They show similar capability of reducing the surface and interfacial tensions using the same mechanisms as the synthetic surfactants. Owing to their unique characteristics which include lower toxicity, higher biodegradability, environmental compatibility and stable activity at extreme pH, salinity and temperature, biosurfactants have gained attention and importance in various fields (Maier, 2003; Mullican *et al.*, 2005).

1.1 Downstream processes

In general, biosurfactants are still unable to compete with the synthetic surfactants for commercial purposes due to their high production and recovery costs. As reported by Mukherjee *et al.*, (2006), three main factors that hinder the commercialization of biosurfactant are: i) the high cost of raw materials; ii) the high recovery and purification costs; and iii) the low yields in the production processes. Thus, in order to reduce the production cost of biosurfactants and to increase the efficiency of biosurfactant production, several techniques and approaches have been adopted worldwide. Inexpensive alternative substrates, optimized culture conditions in bioreactor operations, cost-effective recovery processes and strain improvements have been investigated to enhance biosurfactant yields (Chen *et al.*, 2006; Deleu and Paquot, 2004; Joshi *et al.*, 2008; Yeh *et al.*, 2006).

Efficient downstream processing techniques are required to minimize the overall production costs of any biotechnological products, including biosurfactants. Moreover, approximately 60% of the total biosurfactant production expenditure is from the downstream processes. Thus, it is prudent to recover and purify the biosurfactants in a cost-effective manner as this will contribute significantly in minimizing the total cost of production. In most reported studies, biosurfactants are recovered from the culture media using a combination of several techniques such as precipitation, centrifugation and solvent extraction (Desai and Desai, 1993). However, the solvents that are generally used for biosurfactant recovery, such as methanol, chloroform, acetone and dichloromethane are air-polluting, costly and toxic to environment.

1.2 Foam fractionation

Foam fractionation has drawn the most attention for the recovery of surface active molecules as this technique offers high effectiveness and requires a low cost of operation. It was originally proposed by Leonard and Lemlich in 1965 and has recently been practiced by a number of researchers (Davis *et al.*, 2001; Chen *et al.*, 2006a; Chen *et al.*, 2006b; Sarachat *et al.*, 2010). In this technique, foam is allowed to overflow from the bioreactor through a fractionation column, resulting in a highly concentrated product. To date, due to the outstanding features of this technique, such as high effectiveness, high purity of product, low space requirements and environmentally friendly, there are a number of reports that presented foam fractionation as one of the most proficient methods in biosurfactant recovery (Chen *et al.*, 2006a; Davis *et al.*, 2001; Noah *et al.*, 2002; Sarachat *et al.*, 2010).

The present study focused on the production and recovery of a biosurfactant produced by *Pseudomonas aeruginosa* isolated from local crude oil sample. This bacterium was found to produce rhamnolipid, a glycolipid-type of biosurfactant. The study was initiated with the cultivation of *Pseudomonas aeruginosa* USMAR-2 in a bioreactor, followed by the simultaneous recovery of rhamnolipid. This resulted in a process with a combined rhamnolipid production and recovery. An integrated foam recycler was employed to fractionate the foam produced and recycle the froth containing biosurfactant into the reactor. Several parameters were manipulated including aeration and agitation rate to improve the rhamnolipid recovery efficiency and productivity. The main objective in this study is to use the foaming problem as a key to purify and concentrate the rhamnolipid by recovering the overflowing foam from the modified bioreactor.

2. Methodology

2.1 Foam fractionation

Batch cultivation was carried out in 3.0 L bioreactor (Bioflo 115, New Brunswick, USA), integrated with a foam fractionation system for the primary rhamnolipid recovery. The foam fractionation system consisted of two main parts: i) a 3.0 L bioreactor and ii) a foam recycler system (Figure 1). The foam recycler system was equipped with a foam collector vessel (500 mL flask) and a foam recycler pump. The rapid stirring and aeration supplied during the fermentation ensured excessive foam and the overflowed foam was allowed to flow out of the bioreactor through an integrated foam tube. The resulting foamate was directed into the foam collector vessel and was continuously recycled into the bioreactor using the foam recycler pump until the rhamnolipid concentration in the foam remained constant. The foamate containing the concentrated product can be directly used for a suitable application or can be further purified should a higher concentration of rhamnolipid be required.

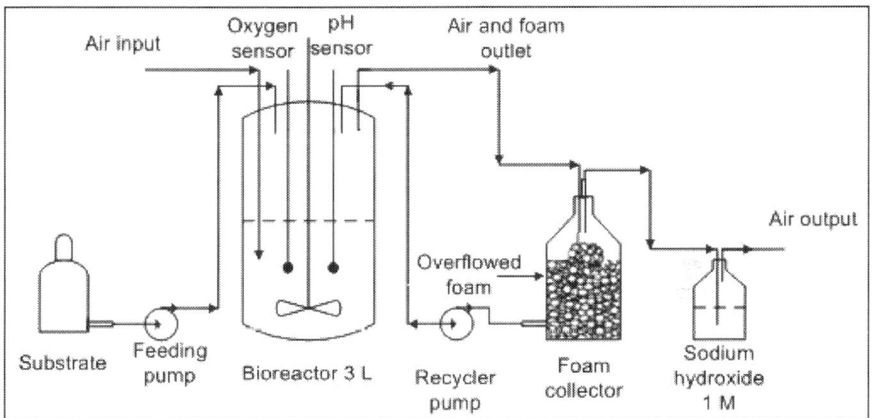

Fig. 1. Schematic figure of modified bioreactor with integrated foam recycler system

2.2 Aeration and agitation rate

The agitation speed and the aeration rate were manipulated throughout the study by setting the desired parameter values in the bioreactor controller, Biocommand OPC Version 1.30 (New Brunswick Scientific, USA).

2.3 Rhamnolipid recovery (%) and rhamnolipid enrichment

The formulas used to calculate the rhamnolipid recovery and enrichment in foam fractionation are as follows:

a. Rhamnolipid recovery, (%) =

$$\frac{\text{rhamnolipid concentration in foamate}}{\left(\begin{array}{c}\text{rhamnolipid concentration in foamate}\\+\\\text{rhamnolipid concentration in culture}\\\text{when foaming ceases}\end{array}\right)} \times 100$$

b. Rhamnolipid enrichment, E_R =

$$\frac{\text{rhamnolipid concentration in foamate}}{\text{rhamnolipid concentration remaining in the bioreactor}}$$

3. Results and discussion

In a normal bioprocess practice, foam formation is avoided at all costs as it will cause several problems, such as stripping of product, nutrients and cells into the foam. In most fermentation systems, the excessive foam can be controlled chemically or mechanically. The addition of chemical antifoams can often suppress a highly foaming culture. However, this technique will raise the cost of cultivation and lower its productivity as the antifoam (usually a surface tension lowering substance) can be costly and its presence may reduce the oxygen transfer rate and the nutrient uptake (Davis *et al.*, 2001). Moreover, antifoam often works upon addition, but the foam build-up will soon ensue as the fermentation progresses. On the other hand, mechanical foam breakers or any mechanical devices are only applicable for a moderately foaming culture as this technique can cause high energy consumption (Heinzle *et al.*, 2006).

The simultaneous rhamnolipid production and primary recovery using the foam fractionation system developed in this work was able to address the foaming problem associated with rhamnolipid production without the addition of an antifoam agent. As reported by Yeh *et al.*, (2006), the biosurfactant product preferentially distributed into the foam fraction, resulting in a higher concentration of biosurfactants in the foam. Consequently, this approach gave a concentrated biosurfactant product while at the same time, alleviated the foaming problem.

3.1 Effect of different aeration rate

Besides establishing a bioreactor design that can control foaming and recover the product, the other interest in this work is to study the effect of aeration and agitation speed towards rhamnolipid production and foam formation. These two parameters are highly correlated with the oxygen transfer efficiency in the bioreactor (David *et al.*, 2001, Yeh *et al.*, 2006). Thus, the effect of the aeration rate towards rhamnolipid production was investigated by fixing the stirrer speed at 400 rpm. As indicated in Figure 2, the cell dry weight of *P. aeruginosa* was 6.96 g/L when the aeration rate was set at 1.0 vvm. A similar fermentation run, aerated at 0.5 vvm gave a cell dry weight of 5.88 g/L. Interestingly, as depicted in Figure 3, the rhamnolipid concentrations obtained with both aeration rates (0.5 vvm and 1.0 vvm) were not significantly different, 1.15 g/L and 1.41 g/L, respectively. Thus, 0.5 vvm was used in subsequent experiments (Table 1). However, fixing the agitation rate at 400 rpm with either 0.5 vvm or 1.0 vvm resulted in no overflowing foam. It is tempting to speculate that a higher agitation speed is needed to enhance the foam formation. Thus, it is essential to find a suitable agitation rate, leading to a condition that is favorable for foam formation while maintaining a high rhamnolipid concentration.

3.2 Effect of different agitation rates with foam recycler system

Yeh and co-workers (2006) reported that efficient mass transfer and sufficient oxygen supply played major roles in rhamnolipid production. The influence of agitation rates in enhancing rhamnolipid yield and productivity was further investigated. The highest rhamnolipid concentration in the foamate (2.93 g/L) was achieved using the foam recycler system, with the bioreactor aerated at 0.5 vvm and agitated at 500 rpm. As shown in Figure

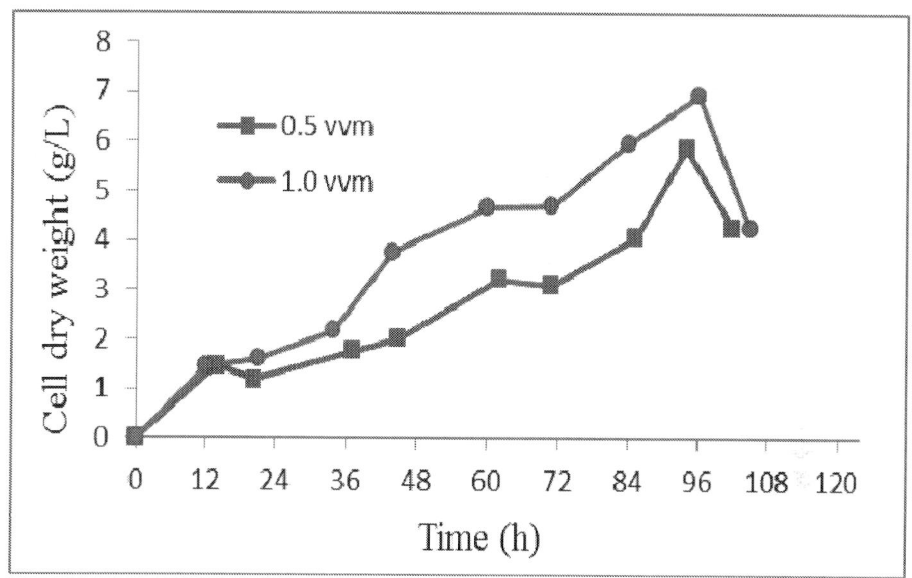

Fig. 2. Time profile of cell dry weight of *P. aeruginosa* USM AR-2 at different aeration rates

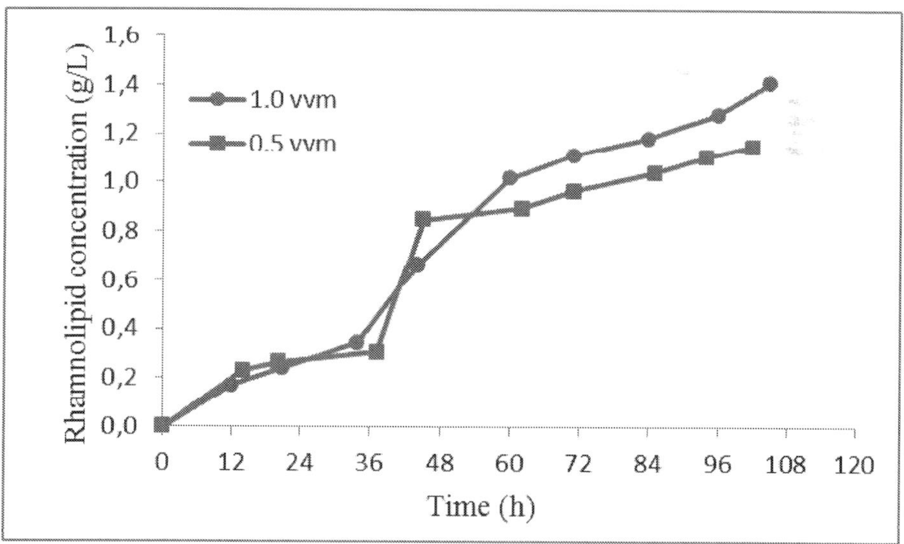

Fig. 3. Time profile of rhamnolipid concentration produced by *P. aeruginosa* USM AR-2 at different aeration rates

5, the highest concentration of rhamnolipid was detected from the recycled foam, not from the culture broth. This finding provides valuable information regarding the rhamnolipid enrichment and recovery process as summarized in Table 1, confirming that rhamnolipid was being concentrated in the foam.

Thus, this approach would be beneficial for subsequent downstream processes, providing an alternative source for rhamnolipid recovery as it was recovered from the froth rather than the spent broth. In addition, it presents a significant process cost reduction since the foamate is much smaller in volume relative to the spent broth. Rhamnolipid productivity in the foam recycler system was double the value of that in the conventional cultivation. In particular, the foam recycler system improved rhamnolipid production but did not contribute much to the growth of *P. aeruginosa* USM AR-2 in the cultivation system. The highest biomass densities for both conventional and foam recycler system remained similar (Figure 4). However, when the conventional cultivation was employed, approximately 50% of the culture broth was stripped from the vessel as a result of severe foaming during cultivation. It was assumed that this contributed to the lower concentration of rhamnolipid (1.15 g/L) as indicated in Figure 5. As summarized in Table 2, rhamnolipid production in the modified bioreactor from this work is highly competitive when compared with the findings from relevant studies done by several researchers (Table 2).

Cultivation condition			Yield [a]	Productivity [b]	Rhamnolipid enrichment	Rhamnolipid recovery (%)	Remarks
Aeration rate (vvm)	Agitation rate (rpm)	Foam recycler system					
1.0	400	√	0.203	0.015	-	-	No overflow foam
0.5	400	√	0.196	0.012	-	-	No overflow foam
0.5	500	-	0.144	0.014	-	-	50 % of culture broth was loss
0.5	500	√	0.233	0.028	1.89	65.40	Overflowed foam was recycled

[a] the yield of rhamnolipid on biomass (g g^{-1})
[b] volumetric production rate (g L^{-1} h^{-1})

Table 3. Summary of results obtained in an integrated bioreactor for the production of rhamnolipid produced by *P. aeruginosa* USM AR-2

Direct utilization of the foam produced can be applied in many fields, such as in bioremediation where the foam can be used to flush contaminated soils. For example, in a study conducted by Mulligan & Wang, 2006, the feasibility of rhamnolipid foam to enhance the remediation of heavy metals in contaminated soils was evaluated.

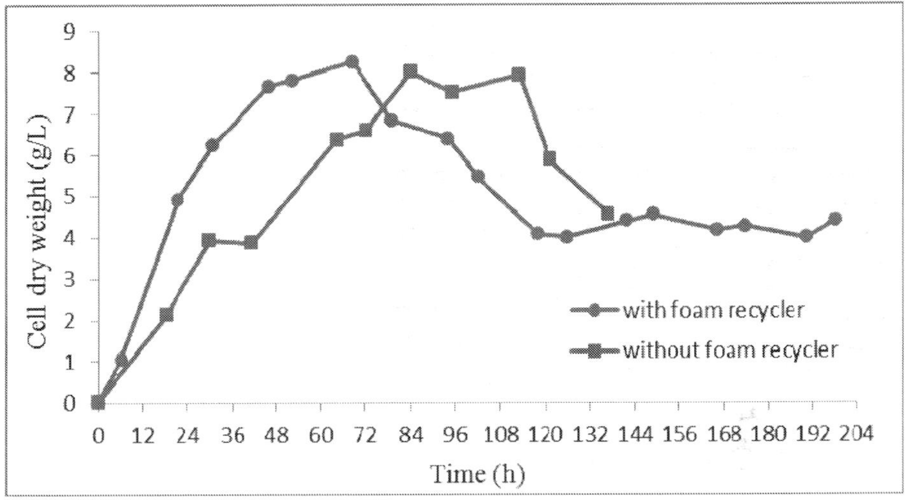

Fig. 4. Time profile of cell growth during batch fermentation agitated at 500 rpm under an aeration rate of 0.5 vvm

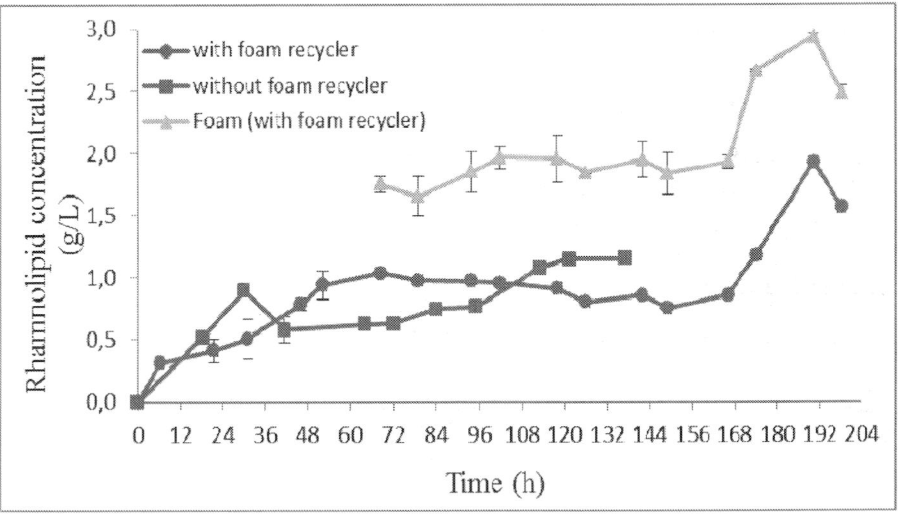

Fig. 5. Time profile of rhamnolipid concentration during batch fermentation agitated at 500 rpm under an aeration rate of 0.5 vvm

Biosurfactant	Microorganism	Strategy	Highest biosurfactant (g/L)	Biosurfactant recovery (%)	Carbon source	Remarks	References
Rhamnolipid	*Pseudomonas aeruginosa* USM-AR2	An integrated bioreactor with foam recycler system. The foam was continuously recycled back into the bioreactor.	2.93	65.40	Diesel oil	Batch	This work
	Pseudomonas aeruginosa	A foam fractionation column was integrated to bioreactor and sintered glasses were used to generate air bubbles with different pore size.	-	97.00	Palm oil	Batch	Sarachat *et al.*, (2010)
Surfactin	*Bacillus subtilis*	A foam collector and a cell recycler were integrated into the bioreactor. A solid carrier was added into the fermentation broth.	6.45	-	Glucose	Batch	Yeh *et al.*, (2006)
		A foam fractionation column was integrated to the bioreactor with its lower end penetrated the headspace of the fermenter. The overflowed foam was collected.	2.25	92.00	Glucose	Batch	Chen *et al.*, (2006a)
		Continuous cultivation with an integrated foam column. A mechanical rotor was used to break the foam and the resulting foamate was collected.	0.929	28.70	Glucose	Continuous cultivation	Chen *et al.*, (2006b)
		The foam was collected continuously whereby the foam collection was achieved using an inverted funnel.	3.74	-	Potato process effluent	Continuous cultivation using airlift bioreactor,	Noah *et al.*, (2002)
		Foam was allowed to flow out of the bioreactor and the foamate was collected in a separate vessel.	1.67	60	Glucose	Batch	Davis *et al.*, (2001)

Table 4. Comparison of foam fractionation for biosurfactant recovery from this work with other studies

4. Conclusion

At the end of this experiment, 2.93 g/L rhamnolipid was successfully obtained from the foam collected. The suitable agitation rate is very important to maintain the homogeneity of the medium and bacterial cells in the bioreactor. Using this combination (500 rpm and 0.5 vvm), a uniform distribution of the gas phase and sufficient gas-liquid mass transfer was achieved. With the foam recycler system, the foam was successfully fractionated and yielded a much higher rhamnolipid concentration. Severe foam formation was kept under control throughout the cultivation without the addition of an antifoam agent. Further research is needed to improve the system, namely to enhance the production where the other modes of cultivation (fed-batch or continuous) can be explored.

5. Acknowledgement

The authors gratefully acknowledge the financial support for this work in the form of the Postgraduate Incentive Grant, University Sains Malaysia (USM) and the National Science Fellowship, Ministry of Science, Technology and Innovation, Malaysia (MOSTI). All equipment and facilities are provided by USM.

6. References

Banat, I.M. (1995). A review: Biosurfactant production and possible uses in microbial enhanced oil recovery and oil pollution remediation. *Bioresource Technology,* Vol. 51, 1-12.

Banat, I.M., Makkar R.S. & Cameotra, S.S. (2000). Potential commercial application of microbial surfactants. *Applied Microbiology Biotechnology,* Vol. 53, 495-508.

Chen, C.Y., Baker, S.C., & Darton, R.C. (2006a). Continuous production of biosurfactant with foam fractionation. *Journal of Chemical Technology and Biotechnology,* Vol. 81, 1915–1922.

Chen, C.Y., Baker, S.C., & Darton, R.C. (2006b). Batch production of biosurfactant with foam fractionation. *Journal of Chemical Technology and Biotechnology,* Vol. 81, 1923–1931.

Davis, D.A., Lynch, H.C. & Varley, J. (2001). The application of foaming for the recovery of surfactin from *Bacillus subtilis* ATCC 21332 cultures. *Enzyme Technology,* Vol. 28, 346–354.

Deleu, M. & Paquot, M. (2004). From renewable vegetables resources to microorganisms: New Trends in Surfactants. *Comptes Rendus Chimie,* Vol. 7, 641-646.

Desai, J.D. & Desai, A.J., (1993). Production of biosurfactants. *In:* Kosaric N, (ed) Biosurfactants. New York: Marcel Dekker. pp. 65–97.

Heinzle, E.; Biwer, A. & Cooney, C. (2006). Development of Sustainable Bioprocess, In: *Development of Sustainable Bioprocess,* 32-57, John Wiley & Sons Ltd, ISBN-13: 978-0-470-01559-9, England.

Joshi, S., Bharucha, C., Jha, S., Yadav, S., Nerurkar, A. & Desai, A. J. (2008). *Bioresource Technology,* Vol. 99, 195–199.

Maier, R.M. (2003). Biosurfactant: evolution and diversity in bacteria. *Advanced Applied Microbiology,* Vol. 52, 101–121.

Mukherjee, S., Das, P. & Sen, R. (2006). Towards commercial production of microbial surfactants. *Trends in biotechnology.* Vol. 24, No. 11, 509–515.

Mullican, C.N. (2005) Environmental applications for biosurfactants. *Environmental Pollution.* Vol. 133, 183–198.

Mulligan, C.N. & Gibbs, B. F. (1993). Factors influencing the economics of biosurfactants. In: Kosaric, N. (ed.). Biosurfactants Production, Properties and Applications. New York: Marcel Dekker, Inc. 329-371.

Mulligan, C.N. & Wang, S. (2006). Remediation of a heavy metal-contaminated soil by rhamnolipid foam. *Engineering Geology,* Vol. 86, 75-81.

Noah, K.S., Fox S.L. & Bruhn, D.F. (2002). Development of continuous surfactin production from potato process effluent by *Bacillus subtilis* in an airlift reactor. *Applied Biochemistry Biotechnology.* Vol. 100, 803–813.

Sarachat, T., Pornsunthorntawee, O., Chadavej, S. & Rujiravanit, R. (2010). Purification and concentration of rhamnolipid biosurfactant produced by pseudomonas aeruginosa SP4 using foam fractionation. *Bioresource Technology.* Vol. 101, 324-330.

Yeh, M.S., Y. H. Wei & J. S. Chang (2006). Bioreactor design for enhanced carrier-assisted surfactin production with *Bacillus subtilis. Process Biochemistry,* Vol. 41, 1799-1805.

27

New Insight into Biodegradation of Poly (L-Lactide), Enzyme Production and Characterization

Sukhumaporn Sukkhum[1] and Vichien Kitpreechavanich[2]
[1]Faculty of Science, Srinakharinwirot University
[2]Faculty of Science, Kasetsart University
Thailand

1. Introduction

Owing to the global utillization of plastics in large quantities, their disposals as solid waste causes deleterious effect on the environment and global warming occurrence. The development of biodegradable plastics is considered to be a product innovation that can help to resolve the problems of plastic waste. The use of biodegradable plastic is one of method to resolve these problems. Biodegradable polymers were classified into four groups depend on the resources of monomers production such as agro-polymer (cellulose, chitin or starch), produced through fermentation by microorganisms (Polyhydroxyalkanoates, PHA), obtained by petrochemical products (Polycaprolactone, PCL) and conventional synthesis from bio-derived monomer (polylactic acid, PLA). In 2002, Cagill Dow was the global leader in commercialization of PLA production, launched a 300 million-USD effort to begin mass production of plastic based on PLA under the trade name Nature work™. The branded PLA is a compostable polymer used in a wide range of applications. Thus, PLA is expected to replace presently used plastic material synthesized from petrochemicals. Recently, the plastic compost by microbes has become a method of interest for plastic waste treatment. After worldwide use of PLA and disposal of PLA plastic waste, recycling of the PLA waste is necessary for utilizing materials efficiently. Biological processes by both microbial and enzymatic activities are currently considered to be sustainable recycling methods for PLA. Recently, several actinomycetes and thermophilic bacteria have been reported to exhibit PLA-degrading ability such as *Brevibacillus, Bacillus smithii, Geobacillus*, and *Bacillus licheniformis* (Tomita et al., 1999; Sakai et al., 2001; Tomita et al., 2004; Kim et al., 2007). Various reports on PLA-degrading enzyme were investigated such as protease, lipase or hydrolase. Proteinase K, a fungal serine protease of *Tritirachium album* has received attention since the early study by Williams (1981). At present study, a new incident on the production of enzyme by using statistical method was reported by Sukkhum et al. (2009b). The improvement of PLA-degrading enzyme production was successful and could be scale-up in 3L airlift fermenter. At this time, the available information on PLA-degrading microorganisms and enzymes are still less than that available other biodegradable plastics such as PCL or PHB. Thus, the study on PLA degradability and application of the enzyme for recycling of commercial PLA are very interesting in recent year. This article summarized and updated new insight into PLA-degrading microorganisms, development of enzyme production and characterization as well as demonstrated the biological recycling of PLA from various kinds of bacteria.

2. Poly(L-lactide), poly(Lactic acid), PLA

Poly(L-lactide) or PLA is one of biodegradable plastics, synthesized from lactic acid which can be produced from farm and agricultural product such as cassava, rice, corn and corncob by bacterial or fungal fermentation. Lactic acid, HOCH3CHCOOH, is exists as two enantiomers, L- and D- lactic acid that involves in the PLA processing and polymerization. Polymerization of lactic acid to high molecular weight PLA can be achieved by two ways: direct condensation and ring opening polymerization (Vink et al., 2002). Commercial PLA are copolymers of L- and D- lactides. Usually, L- isomer is the main product of L-lactic acid from biological sources such as bacterial and fungal fermentation. At present, Cargill Dow Polymers operates the world's largest PLA production from renewable resources. This company efforts at developing a comprehensive life cycle inventory for PLA pellet production span several years. Fig. 1 is a simplified flow and system boundary diagram for PLA production. The analysis depicted include impact associate with corn growing, transport of corn to the corn wet mill, processing of corn into dextrose by enzyme hydrolysis, conversion of dextrose into lactic acid by fermentation method, conversion of lactic acid into lactide and polymerization of lactide into polylactide by condensation and ring opening polymerization (Vink et al., 2002). PLA can be used for various applications. For example PLA make the fibre suitable for technical textile application especially for apparel and has inspired several studies on controlled drug delivery systems or surgical sutures are wound closure filaments bricated in various shapes (Wood, 1980; Laitinen et al., 1992) as well as Cargill Dow's Nature Works™ branded PLA is a compostable polymer used in a wide range of packaging (primarily for food), film, bottles and fibre applications such as short shelf life milk and oil packaging or cold drink cup.

3. Microbial degradation of PLA

The degradation of PLA by actinomycetes, Pranamuda et al. (1997) reported the ability of *Amycolatopsis* strain HT-32 formed clear zone on PLA plate. After their finding, several PLA-degrading microorganisms were recorded as show in the Table 1. Moreover, 15 strain of *Amycolatopsis* sp. formed clear zones on PLA agar plate, showing a large distribution of PLA-degrading actinomycetes within this genus (Pranamuda & Tokiwa, 1999). Ikura & Kudo (1999) isolated *Amycolatopsis* sp. strain 3118 from supernatant pond water or river water demonstrated PLA-degrading activity.

In the same time, *Amycolatopsis* sp. strain KT-5-9 was isolated and showed the degradability on PLA and silk fibroin (Tokiwa et al., 1999). Nakamura et al. (2001) also isolated and identified PLA degrading strain as *Amycolatopsis* sp. strain K104-1 and K104-2. Jarerat et al. (2002) investigated the distribution of PLA-degrading actinomycetes. Among 41 genera (105 strains) of tested actinomycetes found that PLA-degrading strain were limited to the family *Pseudonocardiaceae* and related genera such as *Amycolatopsis, Lentzea, Kibdelosporangium, Streptoalloteichus* and *Saccharothrix* (Jarerat et al., 2002). Interestingly, Sukkhum et al. (2009a) recently reported that PLA-degrading actinomycetes were not limited to this family. The results concluded that PLA-degrading strains were distributed to various families e.g.

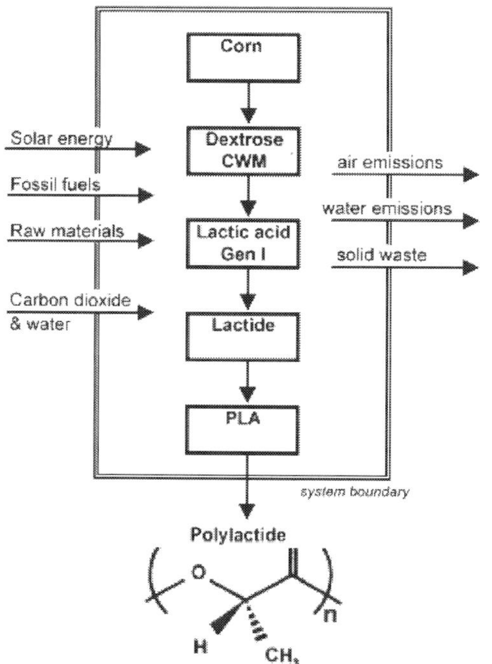

Fig. 1. PLA manufacturing overview of Cargill Dow under commercial name
NatureWorks™ (Vink et al., 2002)

Thermomonosporaceae, Micromonosporaceae, Streptosporangiaceae, Bacillaceae and
Thermoactinomycetaceae (Fig. 2). Among 13 isolates, strain T16-1 showed the highest PLA-
degrading activity at 50°C and identified as *Actinomadura keratinilytica* strain T16-1. This
finding draw a distinction to the report of Jarerat et al. (2002) that several actinomycetes
limited only in the family *Pseudonocardiaceae* and related genera were capable of degrading
PLA. Furthermore, strain T16-1 was further identified by Sukkhum et al. (2009a). It formed
cream-yellow substrate mycelium on ISP-2, 4 and 5. Aerial mycelium rarely forms. The color
of aerial mycelium was green on ISP-2 and ISP-4, and white on ISP-3. Oligosporic curved
chains of spinies spores were borne on aerial hyphae. The temperature ranges for growth
were 30-60 °C. The strain could degrade skim milk and tween 80. Cell hydrolysates
contained galactose, glucose, madurose, mannose and ribose. Phospholipids include
diphosphatidylglycerol, phosphatidylinositol, phosphatidylglycerol and
phosphatidylinositol mannoside. The major cellular fatty acids were iso $C_{16:0}$ and iso $C_{17:0}$.
The principal menaquinones were MK-9(H_4), MK-9(H_6) and MK-9(H_8). DNA G+C content of
the the strain was 72.2 mol%. Strain T16-1 showed 85% relatedness with *A. keratinilytica*
WCC-2265[T]. In agreement with phenotypic, chemotaxonomic and 16S rDNA sequencing, it
could be concluded that the strain was identified as *Actinomadura keratinilytica*, a novel
actinomycetes which produced high PLA-degrading activity. Several PLA-degrading
bacteria at high temperature (≥50 °C) have been reported. A thermophilic strain, *Brevibacillus*
sp. which degrades L-PLA film at 60°C was isolated from soil (Tomita et al., 1999).

Strain	Detection method of PLA degradation	Reference
Amycolatopsis sp. HT 32	Film-weight loss; monomer production	Pranamuda et al., 1997
Amycolatopsis sp. 3118	Film-weight loss; monomer production	Ikura & Kudo, 1999
Amycolatopsis sp. KT-s-9	Clear zone method	Tokiwa et al.,1999
Amycolatopsis sp. 41	Film-weight loss; monomer production	Pranamuda et al., 2001
Amycolatopsis sp. K104-1	Turbidity method	Nakamura et al., 2001
Lentzea waywayandensis	Film-weight loss; monomer production	Jarerat & Tokiwa, 2003
Kibdelosporangium aridum	Film-weight loss; monomer production	Jarerat et al., 2003
Tritirachium album ATCC 22563	Film-weight loss; monomer production	Jarerat & Tokiwa,2001
Brevibacillus	Change in molecular production and viscosity	Tomita et al., 1999
Bacillus stearothermophilus	Change in molecular production and viscosity	Tomita et al., 2003
Bacillus smithii PL 21	Change in molecular production and viscosity	Tomita et al., 2004
Bacillus licheniformis PLLA-2	Biodegradation test	Kim et al., 2007
Paenibacillus amylolyticus TB-13	Turbidity method	Shigeno et al., 2003
Bacillus clausii strain pLA-M4	Molecular technique	Mayumi et al., 2008
Bacillus cereus pLA-M7	Molecular technique	Mayumi et al., 2008
Treponema denticola pLA-M9	Molecular technique	Mayumi et al., 2008
Paecilomyces	Molecular technique	Sangwan & Wu, 2008
Thermomonospora	Molecular technique	Sangwan & Wu, 2008
Thermopolyspora	Molecular technique	Sangwan & Wu, 2008
Actinomadura keratinilytica T16-1	Clear zone and turbidity method	Sukkhum et al., 2009a
Micromonospora echinospora B12-1	Clear zone and turbidity method	Sukkhum et al., 2009a
Micromonospora viridifaciens B7-3	Clear zone and turbidity method	Sukkhum et al., 2009a

Strain	Detection method of PLA degradation	Reference
Nonomuraea terrinata L44-1	Clear zone and turbidity method	Sukkhum et al., 2009a
Nonomuraea fastidiosa T9-1	Clear zone and turbidity method	Sukkhum et al., 2009a
Bacillus licheniformis T6-1	Clear zone and turbidity method	Sukkhum et al., 2009a
Laceyella Sacchari T11-7	Clear zone and turbidity method	Sukkhum et al., 2009a
Thermoactinomyces vulgalis T7-1	Clear zone and turbidity method	Sukkhum et al., 2009a

Table 1. PLA-degrading microorganisms, type of enzyme and detection method for PLA degradation

Sakai et al. (2001) isolated thermophilic L-PLA-degrading bacteria from a garbage fermentor, identified as *Bacillus smithii*. The strain grew well in the medium containing 1% L-PLA and the molecular weight of L-PLA decreased by 35.6% after 3 days incubation with shaking at 60°C. Another isolation of L-PLA-degrading thermophile *Geobacillus* sp. strain 41 was reported by Tomita et al. (2004). The time course of L-PLA degradation was monitored at 60°C for 20 days and degradation was confirmed by the change in molecular weight and viscosity of the residual polymer. Newly thermophilic bacteria isolated from compost was isolated and identified as *Bacillus licheniformis*. It degraded not only low-molecular weight PLLA but also other PLLAs having higher molecular weight at 58°C (Kim et al., 2007). Recently, various molecular analyses such as 16S rDNA studies have confirmed that only less than 1% of microorganisms in the natural environment can be cultured by traditional culture-based methods (Bintrim et al., 1997; Rondon et al., 1999). Mayumi et al. (2008) suggested that some un-culturable microorganisms may also be associated with PLA degradation. Metagenomic library consisting of the DNA extraction from PLA disks buried in compost was constructed and identified three PLA-degrading genes which worked at high temperature (70°C). Afterward, screening of un-culturable microorganism's technique was also used. Gene sequences from genera *Paecilomyces*, *Thermomonospora*, and *Thermopolyspora* which play an important role in PLA degradation were most abundant in the compost sample containing PLA (Sangwan & Wu, 2008). Almost of reported PLA-degrading microorganisms are shown in the Table 1.

4. PLA-degrading enzyme purification and characterization

Usually, the enzymes play a significant role in degradation of polymers, even though the enzymes are not only responsible for the hydrolysis of polymers. The enzymatic degradation of aliphatic polyesters by hydrolysis is a two-step process. The first step is adsorption of the enzyme on the surface of the substrate through surface-binding domain and the second step is hydrolysis of the ester bond (Tokiwa & Calabia, 2006). Williams (1981) first reported the degradation of L-PLA by proteinase K from *T. album*. Naturally occurring amino acids as constituents of protein are L-isomers. As reported by Reeve et al. (1994) proteinase K was not able to cleave D-stereoisomer of PLA. It seems reasonable to conclude that PLA-degrading enzyme is protease-type enzyme which recognizes the repeated L-lactic acid unit of PLA as L-alanine unit of silk fibroin (protein). Currently, a few

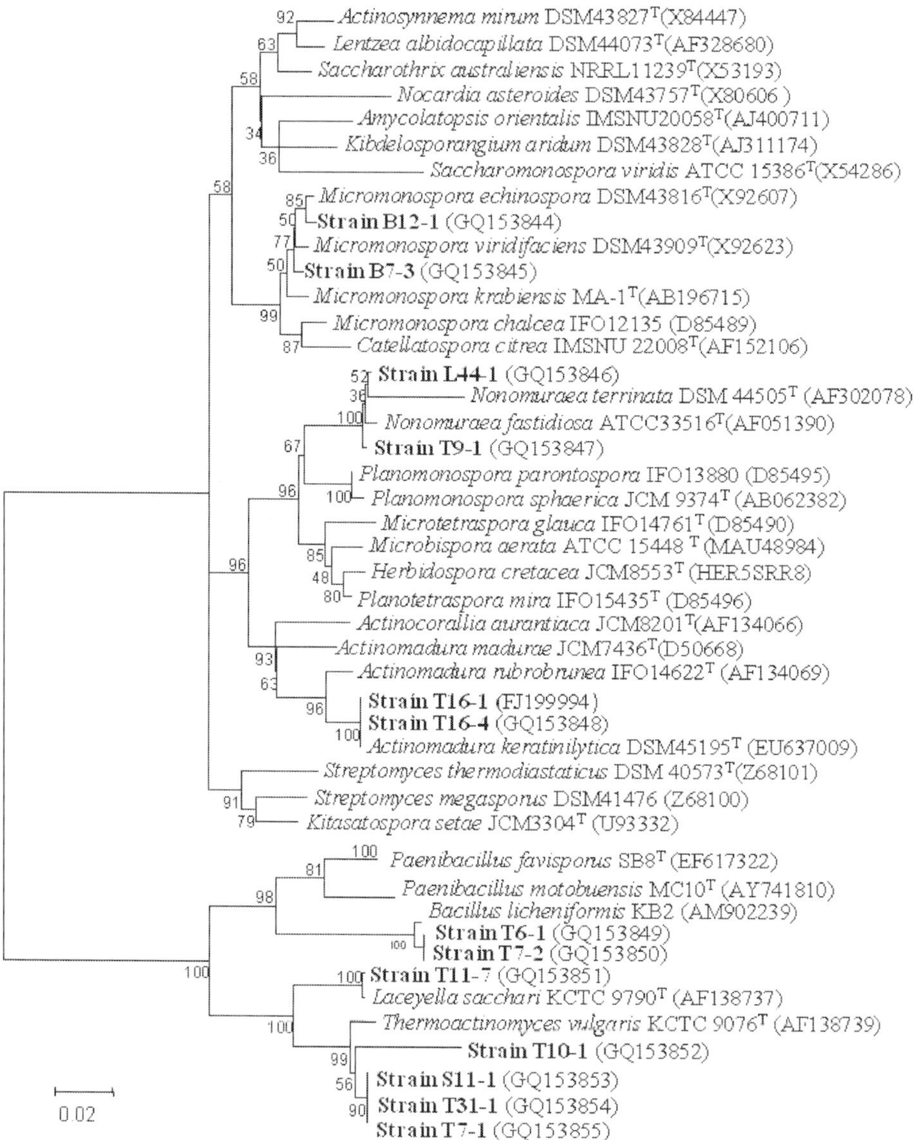

Fig. 2. Phylogenetic positions of PLA-degrading microorganisms. The phylogenetic tree was constructed by the neighbor-joining method. Bootstrap values at branching points are expressed as percentages from 1000 replications. The scale bar indicates 0.02 substitutions per nucleotide position. T = type strain. (Sukkhum et al., 2009a)

Strain	Mw (kDa)	Optimum pH and temperature	Substrate specificity	Enzyme type	Reference
Amycolatopsis sp. 41	40	pH 6.0 37-45°C	casein, silk powder, Suc-(Ala)$_3$-pNA	protease	Pranamuda et al., 2001
Amycolatopsis sp. K104-1	24	pH 9.5 55-60°C	Casein, fibroin	Serine protease	Nakamura et al., 2001
T. album	-	-	silk fibroin, elastin, (Suc-(Ala)$_3$-pNA)	protease	Jarerat & Tokiwa, 2001
B. smithii	62.5	pH 5.5 60°C	pNP-butyrate, capryrate, laurate, palmitate, stearate	acyltransferase	Sakai et al., 2001
Cryptococcus sp. S-2	20.9	-	L-PLA , PBS, PCL, PHB	Lipase	Masaki et al., 2005
Amycolatopsis orientalis ssp. *orientalis*	24, 19.5, 18	pH 9.5, pH 10. 5 pH 9.5 50-60°C	PLA, casein, C8 ester	Serine protease	Li et al., 2008
Actinomadura keratinilytica T16-1	30	pH 10.0 70°C	PLA, (Suc-(Ala)$_3$-pNA), gelatin	Serine protease	Sukkhum et al., 2009a

Table 2. The characteristics of purified PLA-degrading enzyme from various strains

reports on purification and characterization of PLA-degrading enzyme have been available (Table 2). Pranamuda et al. (2001) reported the purification of PLA-degrading enzyme from *Amycolatopsis* strain-41, with molecular weight of 43 kDa from size exclusion chromatography or 40 kDa from SDS-PAGE analysis. Optimum pH and temperature were 6.0 and 37-45°C, respectively.

Besides PLLA, the enzyme degrades casein, silk powder and Suc-(Ala)$_3$-pNA at an even lower level than Proteinase-K, but not Suc-(Gly)$_3$-pNA, poly (ε-caprolactone) and poly(β-hydroxybutyrate).The PLA-degrading enzyme of this genus with higher substrate specificity on PLA depolymerase produced by *Amycolatopsis* strain K104-1 was reported by Nakamura et al. (2001). The enzyme was classified as a serine type protease having molecular weight of 24 kDa, which exhibited degrading activity on casein and fibroin, but not collagen-type I, PCL and PHB. The optimum pH for enzyme activity is 9.5, and the optimum temperature is 55-60°C. In addition to this, the culture supernatant of *T. album* with 0.1% gelatin also showed hydrolytic activity on silk fibroin, elastin and (Suc-(Ala)$_3$-pNA) but not on PCL, PHB and PBS. From the results of a substrate specificity study, it was concluded that the enzyme of *T. album* might be protease rather than lipase or PHB depolymerase (Jarerat & Tokiwa, 2001). In addition, Sakai et al. (2001) purified an enzyme with a molecular weight of 62.5 kDa from a thermophilic *B. smithii*. This enzyme could be degraded various kinds of fatty acid esters and the molecular weight of L-PLA decreased at 60°C. However, control was not included in the experiment. A cutinase-like enzyme from the yeast *Cryptococcus* sp. strain S-2 was purified. The enzyme has a molecular weight of 20.9 kDa and could degrade L-PLA as well as other synthetic polymers such as PBS, PCL, and PHB (Masaki et al., 2005). Afterward, three novels purified PLA-degrading enzymes produced by *Amycolatopsis orientalis* ssp. *orientalis* named PLAase I, II and III, were purified. The molecular masses of these three PLAases as determined by SDS-PAGE were 24.0, 19.5 and 18.0 kDa, with the pH optima being 9.5, 10.5 and 9.5, respectively. The optimal temperature for the enzyme activities was 50–60°C (Li et al., 2008). Recently, Sukkhum et al., (2009a) reported that PLA-degrading enzyme was produced by *A. keratinilytica* strain T16-1 in liquid medium. Crude enzyme was purified by using DEAE-Toyopearl 650C and DEAE-Toyopearl 650S. The enzyme was purified to 13 folds with a recovery of 24% and a specific activity of 38.3 U/mg protein. The molecular weight of purified enzyme was approximately 30 kDa. The optimum pH and temperature were 10.0 and 70°C, respectively. The enzyme was stable at pH 11-12. Moreover, 70% of the enzyme activity remains when incubated at 70°C for 1h. The purified enzyme was inhibited by 5mM EDTA and 5mM Phenylmethyl sulfonyl fluoride as well as diisopropyl fluorophosphates, strongly hydrolyzed Suc-(Ala)$_3$-pNA, gelatin and PLA, but show low activity on casein. The n-terminal amino acid sequence of pure protein was determined for the initial 15 residues as follow: GYQNNPPSAGLDRAA which was different from that of other registered PLA-degrading enzymes but similar to serine protease from *Streptomyces avermitilis* (83% identity) and 81% identity with alkaline serine protease from *Streptomyces pristinaespiralis* ATCC 25486. Many PLA-degrading enzymes were identified as serine type protease. For example, n-terminal amino acid sequences of purified PLA-degrading enzyme from *Amycolatopsis orientalis* ssp. *orientalis*, PLAase III (YDVRGGDAYYINNSS) demonstrated the 86% identity with serine protease from *Streptomyces lividan* and a serine protease precursor of *Streptomyces coelicolor* A3 (Li et al., 2008). Pure PLA-degrading enzyme from *Amycolatopsis* sp. K104-1 was also classified by n-terminal amino acid sequence (IIGGSNATSGPYAARLF) as fibrinolytic serine proteases (F-I-1 and F-I-2) from the earthworm *Lumbricus rubellus* with 100% identity (Nakamura et al., 2001). In our work, suggested that the activity of purified enzyme was rapidly decreased when the enzyme was treated with EDTA, compared with the non treatment with EDTA as show in Fig. 3. This result strongly confirmed that EDTA affected the stability of PLA-degrading enzyme. These finding also previously reported by Hadj-Ali et al. (2007) mentioned that about 70% of serine protease activity was inhibited by EDTA, indicating the involvement of any metal ion such as Ca^{2+} in dimensional structure of protease from *Bacillus licheniformis*. Because of EDTA is chelating agent, due to its ability to sequester metal ions. Thus, the metals also affected the thermal stability of the enzyme through its binding mechanism on protein that prevented unfolding at high temperature in thermophilic bacteria.

According to the report on the effect of CaCl$_2$ on enzyme stability at high temperature, further experiment was investigated in our work. The pure PLA-degrading enzyme was incubated with various concentrations of CaCl$_2$ (1, 5, 10 mM) for 30 min at various temperatures (30-100°C). Fig. 4 indicates that after treatment of the enzyme with CaCl$_2$, the enzyme stability decreased when compared with the control (un-treated enzyme). Moreover, increasing the concentration of CaCl$_2$ resulted in decrease of the stability, indicating that CaCl$_2$ did not enhance the stability of enzyme but inhibited the activity of the enzyme.

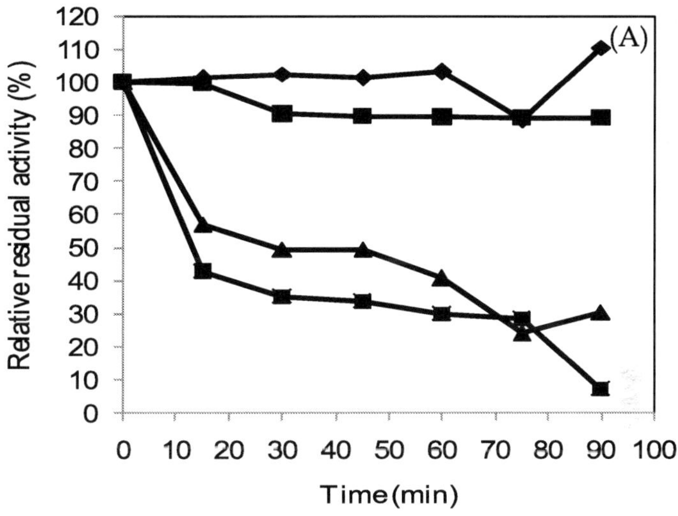

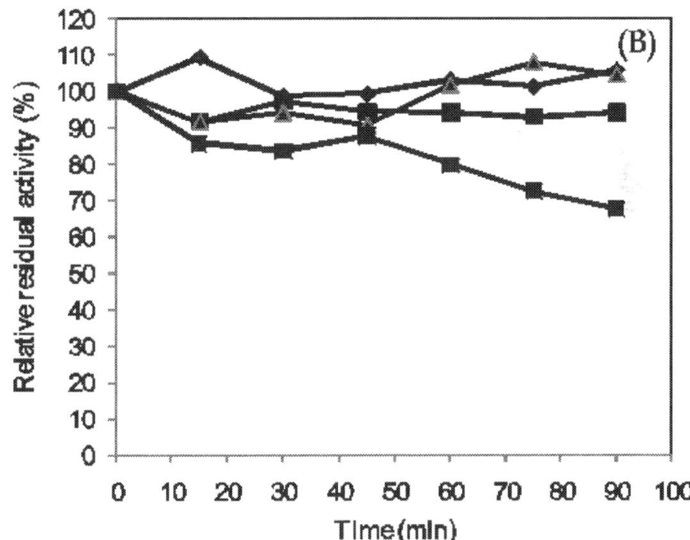

Fig. 3. Stability of purified PLA-degrading enzyme produced by *A. keratinilytica* strain T16-1 at various temperatures, 40°C (♦), 50°C (■), 60°C (▲) and 70°C (✗). (A) EDTA-treated enzyme and (B) EDTA-untreated enzyme.

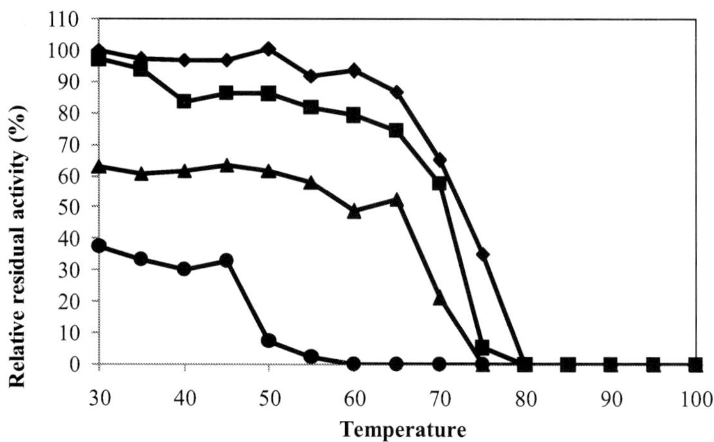

Fig. 4. The effect of CaCl$_2$ concentration on stability of purified PLA-degrading enzyme produced by *A. keratinilytica* strain T16-1. Without CaCl$_2$, (◆), 1 mM CaCl$_2$ (■), 5 mM CaCl$_2$ (▲), 10 mM CaCl$_2$ (●).

5. Development of fermentation process for PLA-degrading enzyme production

Factor affecting PLA-degrading enzyme production have not been studied so far. Mostly the researchers focused on the isolation and identification of new PLA-degrading microorganisms. *Amycolatopsis* sp. strain 3118 was isolated and identified by Ikura & Kudo (1999). The optimum condition for degradation of PLA film were 43°C at pH about 7.0 in a mineral salt medium with a low concentration of organic nutrients (0.002% yeast extract) and PLA film disappeared within 2 weeks. Jarerat & tokiwa (2001) demonstrated the degradation of PLA by using *T. album* ATCC 22563 found that in liquid basal medium containing PLA film, no film degradation was observed. However, by addition of 0.1% gelatin, about 76% of PLA film was degraded after 14 days of cultivation at 30°C. PLA-degrading enzyme production by *Tritirachium album*, *Lentzea waywayandensis* and *Amycolatopsis orientalis*, were inducible in the basal medium by 0.1% (w/v) poly-L-amino acids, peptides or amino acid. Silk fibroin was the best inducer for *A. orientalis* and that elastin was the best inducer for *L. waywayandensis* as well as *T. album* with the enzyme activity 150, 96 and 60 U/ml, respectively (Jarerat et al., 2004). In 2006, Jarerat et al. reported that silk fibroin was the best substrate for production of PLA-degrading enzyme in liquid medium by *Amycolatopsis orientalis* with 600 TOC formation:mg/l. The enzyme production was scaled-up in 5L jar fermenter with 550 TOC formation:mg/l after 3 days cultivation. Response surface method and central composite design were increasingly used for optimization of various phases in some fermentation process such as physical parameter and factors (temperature, pH, aeration rate, and agitation rate), fermentation medium (carbon and nitrogen sources, mineral salt and inducer) with various microorganisms. This method has been successfully applied to improve the production of many important enzyme for example α-amylase (Tanyidizi et al., 2005), xylanase (Li et al., 2007) and keratinase (Anbu et al., 2007). However, improvement of PLA-degrading enzyme production by using response surface method has not been studied. Sukkhum et al. (2009b) demonstrated that PLA film and gelatin were found to be major effects on the enzyme production by *A. keratinilytica* strain T16-1 based on the "one factor at a time method". The CCD experiments were designed to obtain the best condition for the maximum PLA-degrading enzyme production by the strain. The experimental design matrix and results obtained for enzymes activities were shown in Table 3. Treatment runs 9-11 were the center points in the design, which were repeated three times for estimation of error. By applying multiple regression analysis on the experimental data, the following second order polynomial equation (Eq. 1) was used to explain the enzyme production. The statistical optimal values of variables were obtained when moving along the major and minor axis of the contour, and the response at the centre point yielded maximum PLA-degrading enzyme production (Fig. 5). These observations were also verified from canonical analysis of the response surface. The canonical analysis revealed a minimum region for the model. The stationary point presenting a maximum PLA-degrading activity had the following critical values: PLA film: 0.035% (w/v) and gelatin: 0.238% (w/v). The predicted PLA-degrading activity for these conditions was 40.4 U/ml. The model was validated by repeating the experiment under the optimized conditions. The maximum experimental response for PLA-degrading enzyme production was 44.6 U/ml after 96 h cultivation with productivity of 0.46 U/ml/h. The enzyme activity obtained was 1.32 folds higher than the activity predicted by the optimized medium by one factor at a time method.

$$Y = -16.444 + \left(890.993X_1\right) + \left(347.718X_2\right) - \left(9004.569X_1^2\right)$$
$$- \left(650.037X_2^2\right) - \left(1108.419X_1X_2\right) \tag{1}$$

Where Y is the predicted response (PLA-degrading enzyme production); X_1, X_2 are coded values of PLA film and gelatin, respectively. The schematic diagrams of 3L airlift fermenter used throughout this study are shown in Fig.6. The fermentation was carried out in 3L airlift bioreactor with 2L working volume, which was 185 mm in diameter and 632 mm high. The bioreactor, which surrounded by a water jacket for temperature control, was made from glass. The air sparger was a multi porous plate (10 mm in diameter) located at the bottom of the bioreactor. The DO probe, pH probe and antifoam sensor were positioned at the top of the bioreactor. The antifoam sensor was located at 10 cm from the top of the upper broth surface. All the probes and sensor were interfaced with a control unit. The feasibility of the regression models was also carried out in a 3L airlift fermenter at aeration rate of 0.5 vvm, un-controlled pH 7.0 and 50°C. The maximum PLA-degrading enzyme production at 72 h cultivation was 150 U/ml with the productivity of 2.08 U/ml/h. A significant increase of 3.36 and 4.50 folds in PLA-degrading enzyme production and productivity, respectively, was observed in airlift fermenter.

Development of fermentation process of PLA-degrading enzyme production by strain T16-1 was summarized in Table 4. Yeast extract was used as organic nitrogen source for un-optimized medium. The enzyme activity 22 U/ml was obtained. The second step, gelatin was found to be a factor affecting the enzyme production which PLA-degrading activity 34 U/ml was obtained. The optimization of medium composition by using CCD in shake flasks was achieved which the concentration of 0.035% (w/v) PLA film and 0.24% (w/v) gelatin, with 45 U/ml of enzyme activity. At the last step of the enzyme production, the statistical model was validated in airlift fermenter using optimized medium under the condition: aeration rate of 0.5 vvm, initial pH 7.0 (un-controlled) and temperature 50°C. The enzyme activity increased up to 150 U/ml under this condition. In conclusion, PLA-degrading enzyme production by *A. keratinilytica* strain T16-1 was significantly increased about 7 folds. We suggest that experimental design by statistical method might be useful for improvement of PLA-degrading enzyme production from other strains.

Treatment number	Level		Actual level		PLA-degrading enzyme activity (U/ml)	
	X_1	X_2	X_1	X_2	Observed	Predicted
1	1	1	0.07	0.28	22.78	25.71
2	1	-1	0.07	0.12	19.78	24.16
3	-1	1	0.03	0.28	37.52	39.08
4	-1	-1	0.03	0.12	26.90	29.92
5	1.41	0	0.10	0.20	6.74	5.11
6	0	1.41	0.05	0.40	18.93	18.51
7	-1.41	0	0	0.20	27.57	27.10
8	0	-1.41	0.05	0	7.22	5.59
9	0	0	0.05	0.20	40.73	38.05
10	0	0	0.05	0.20	40.70	38.05
11	0	0	0.05	0.20	40.47	38.05

Table 3. Experimental design used in response surface methodology of 2 independent variables, (X_1) PLA film and (X_2) gelatin, with three center points, and the observed and predicted PLA-degrading activity.

Step	Method	Condition	Enzyme activity (U/ml)
1	Un-optimized medium/condition	0.1% PLA film 0.1% (w/v) yeast extract	22
2	Screening of factors affecting the enzyme production using one factor at a time method	0.05% (w/v) PLA film 0.2% (w/v) gelatin	34
3	Optimization of medium composition using CCD in shake flasks	0.035% (w/v) PLA film 0.24% (w/v) gelatin	45
4	Validation of the model in an airlift fermenter	0.035% (w/v) PLA film 0.24% (w/v) gelatin aeration rate of 0.5 vvm Initial pH of 7.0 (un-controlled) temperature at 50°C	150

Table 4. Development of fermentation process of PLA-degrading enzyme production by *A. keratinilytica* strain T16-1

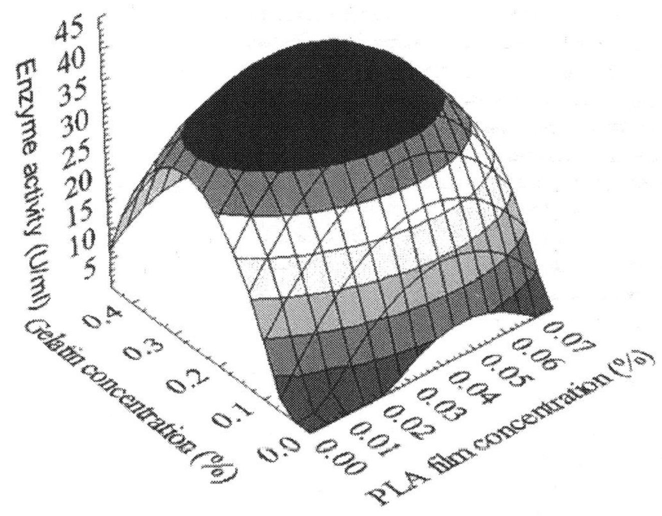

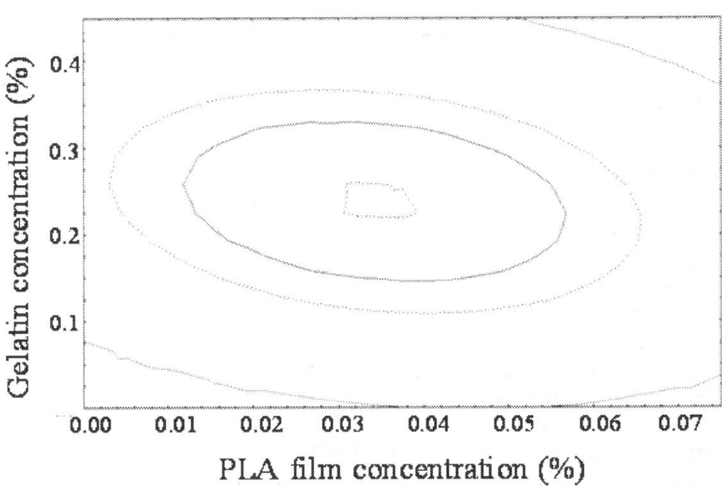

Fig. 5. Response surface described by the model, representing PLA-degrading enzyme activity (U/ml) as a function of PLA film and gelatin concentrations (Sukkhum et al., 2009b).

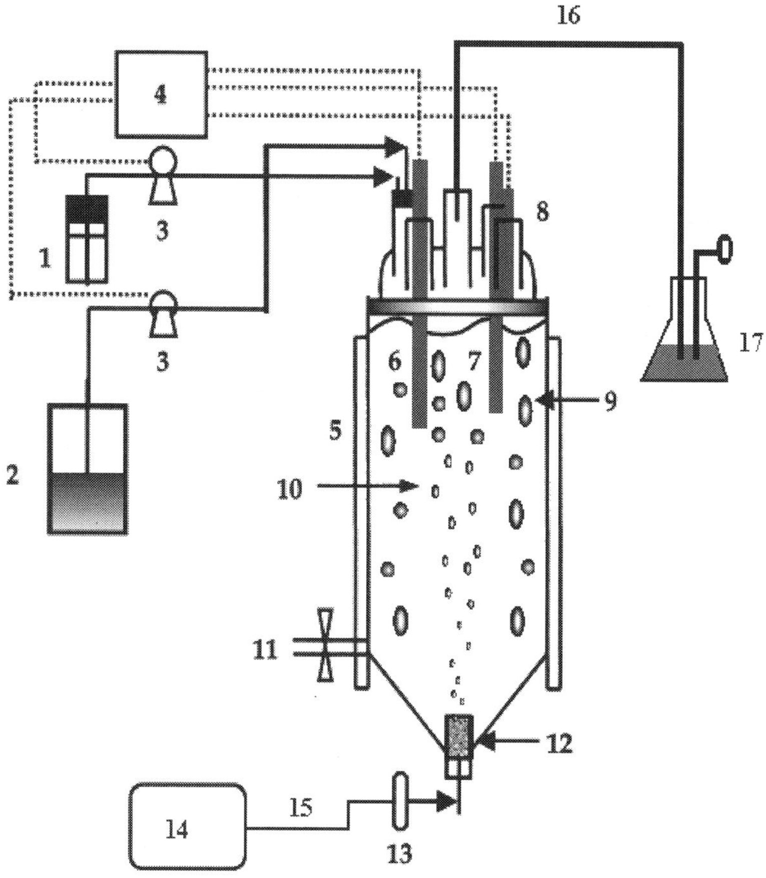

Fig. 6. Schematic diagrams of 3L airlift bioreactor used throughout this study. Antifoam, 1; Alkaline/acid reservoir, 2; Pump, 3; Controller, 4; Water jacket, 5; pH probe, 6; DO probe, 7; Antifoam sensor, 8; Dispersed bubble, 9; Bubble, 10; Sampling line, 11; Air filters (sparger), 12; Air flow meter, 13; Air pump, 14; Air inlet line, 15; Air outlet line, 16; 4% $CuSO_4.5H_2O$ solution, 17.

6. Biological recycling of PLA

Currently, recycling of polymers is necessary for utilizing materials efficiently and can help to reduce the plastics wastes which effect to global worming. Methods for PLA recycling, e.g. pyrolysis (Fan et al., 2003) and chemical hydrolysis (Tsuji & Nakahara, 2002), have been reported. Poly (L-lactide) with calcium salt end structure (PLLA-Ca) is a promising material for PLLA recycling because of the ease of lactide recovery through the unzipping depolymerization process. However, the pyrolysis of PLLA-Ca also causes to form DL-lactide at high temperature about 250°C (Fan et al, 2003). Amorphous and crystallized poly(L-lactide) (PLLA) films were prepared by quenching and annealing at 140°C for 600 min, respectively, from the melt. Their hydrolysis is investigated at pH 2.0 in HCl and DL-lactic acid (DLLA) solutions (37°C) for up to 300 days to obtain hydronium ions and the lactic acid oligomers and monomers. Biological processes by both microbial and enzymatic activities are currently considered as the sustainable recycling method for polyesters. Biological recycling of PLA is one application of PLA-degrading enzyme to recycle of the plastic wastes containing PLA. The treatment process by the activity of enzyme is under mild condition, regarding as a clean process, and does not contain any undesirable by-products such as racemic of PLA after degradation. Poly(lactic acid) such as poly(DL-lactic acid), poly(D-lactic acid) and poly(L-lactic acid) were degraded by the activity of lipase in an organic solvent to produce cyclic oligomer with several enzyme for example, lipase RM (Lipozyme RM IM) and lipase CA (Novozym 435) at 60-100°C (Takahashi et al., 2004). This report suggested that the cyclic oligomer might be suitable for repolymerization and recycling of PLA. The lipase-catalyzed transformation will open a novel route for sustainable chemical recycling of polymers (Fig. 7). Jarerat et al. (2006) demonstrated the biological recycling of PLA using enzyme activity by a process that is mild (at relatively low in temperature, 40°C) and clean (without organic solvent). In addition, a recycling process using the stereospecific enzyme activity of *Amycolatopsis orientalis* at low temperature prevented formation of a mixture of D- and L-lactic acids, which generally occurs in conventional hydrolysis in water at a temperature over 300°C. The obtained degradation products, monomer and oligomers, can be readily polymerized to high molecular weight PLA by one-step condensation polymerization after the removal of impurities.

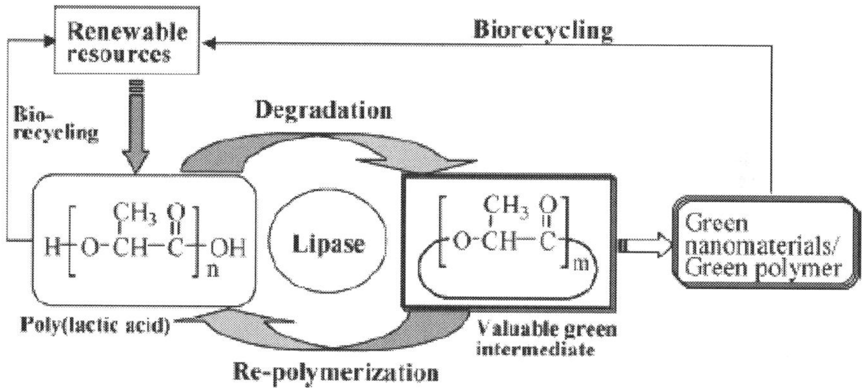

Fig. 7. Concept of poly (lactic acid) recycling (Takahashi et al., 2004)

7. Conclusion

Since 1997, many researchers have been isolated and identified PLA-degrading microorganisms. Several PLA-degrading microorganisms have been reported such as thermophilic bacteria, actinomycetes and fungi. Furthermore, PLA-degrading actinomycetes were found to distribute into various families for example *Pseudonocardiaceae*

Thermomonosporaceae, Micromonosporaceae, Streptosporangiaceae, Bacillaceae and *Thermoactinomycetaceae*. Molecular techniques like metagenomic library used to study the other unculturable PLA-degrading microorganisms in ecosystem. However, many other PLA-degrading microorganisms have not been isolated from the natural environment. So, further study on isolation and identification of a potent PLA-degrading strain which produced high activity should be investigated. Normally, purified PLA-degrading enzyme were characterized into two groups such as protease and lipase which showed the substrate specificity with PLA, protein, peptide and some of synthesized fatty acid at high temperature (37-70°C) and alkaline pH (9.5-10). However, the mechanisms of PLA degradation by microorganisms and enzymes should be more clearly understood by studying the enzyme purification from other microorganisms. Response surface method was successful to improve PLA-degrading enzyme production by *A. keratinilytica* strain T16-1. The optimum concentration of both PLA and gelatin as carbon and nitrogen sources, 0.035% and 0.24% (w/v) respectively, showed the maximum PLA-degrading activity was obtained by using this statistical method. The maximum PLA-degrading activity in 3L airlift fermenter with statistical optimized medium was 150 U/ml under the condition: pH 7.0 (uncontrolled), aeration rate of 0.5 vvm and 50°C. We suggest that this experimental design might be useful for improvement of PLA-degrading enzyme production from other strains. Beside, the recycling method of PLA by using enzymatic degradation was available and use mild condition without undesirable products. Recycling of biopolymer, e.g. PLA by using microbial enzyme should be achieved more details, especially in the process of biodegradation; bio-recycling and re-polymerization to open new technology for reduce the plastic wastes in the future.

8. References

Anbu, P., Gopinath, S.C.B., Hilda, A., Lakshmipriya, T & Annadurai, G. (2007). Optimization of extracellular keratinase production by poultry farm isolate Scopulariopsis brevicaulis. Bioresource Technology. Vol. 98, pp. 1298-1303, ISSN 0960-8524

Bintrim, S., Donohue, T. J., Handelsman, J., Roberts, G. P. & Goodman, R. M. (1997). Molecular phylogeny of Archaea from soil. Proceedings of the National Academy of Science. Vol. 94, pp. 277-282, ISSN 1091-6490

Fan, Y., Nishida, H. Hoshihara, S. Shirai, Y. Tokiwa Y. & Endo, T. (2003). Pyrolysis kinetics of poly(L-lactide) with carboxyl and calcium salt end structures. Polymer Degradation and Stability, Vol. 79, pp. 547–562, ISSN 0141-3910

Hadj-Ali, N.E., Agrebi, R., Ghorbel-Frikha, B., Sellami-Kamoun, A., Kanoun, S. & Nasri, M. (2007). Biochemical and molecular characterization of a detergent stable alkaline serine protease from a newly isolated Bacillus licheniformis NH1. Enzyme and Microbial Technology. Vol. 40, pp. 515-523, ISSN 0141-0229

Ikura, Y. & Kudo, T. (1999). Isolation of a microorganism capable of degrading poly(L-lactide). Journal of General and Applied Microbiology. Vol. 45, pp. 247–251, ISSN 0022-1260

Jarerat, A., Pranamuda H. & Tokiwa, Y. (2002). Poly(L-lactide)-degrading activity in various actinomycetes. Macromolecular Bioscience. Vol. 2, pp. 420–428, ISSN 1616-5187

Jarerat, A. & Tokiwa, Y. (2001). Degradation of poly(L-lactide) by a fungus. Macromolecular Bioscience. Vol. 1, pp. 136–140, ISSN 1616-5187

Jarerat, A. & Tokiwa, Y. (2003). Degradation of poly(L-lactide) by Saccharothrix waywayandensis. Biotechnology Letters. Vol. 25, pp. 401–404, ISSN 0141-5492

Jarerat, A., Tokiwa, Y. & Tanaka, H. (2003). Poly(L-lactide) degradation by Kibdelosporangium aridum. Biotechnology Letters. Vol. 25, pp. 2035–2038, ISSN 0141-5492

Jarerat, A., Tokiwa, Y. & Tanaka, H. (2004). Microbial poly(L-lactdie)-degrading enzyme induced by amino acids, peptides, and poly(L-amino acids). Journal of the Polymers and the Environment. Vol. 12, No. 3, pp.139-146, ISSN 1566-2543

Jarerat, A., Tokiwa, Y. & Tanaka, H. (2006). Production of poly(L-lactide)-degrading enzyme by Amycolatopsis orientalis for biological recycling of poly(L-lactide). Applied Microbiology and Biotechnology. Vol. 72, pp. 726-731, ISSN 0175-7598

Kim, M. N., Kim, W. G., Weon, H. Y. & Lee, S. H. (2007). Poly(L-lactide)-degrading activity of a newly isolated bacterium. Journal of Applied Polymer Science. Vol. 109, pp. 234–239, ISSN 0021-8995

Laitinen, O., Tormala, P., Taurio, R., Skutnabb, K., Saarelainen, K. & Iivonen, T. (1992). Mechanical properties of biodegradable ligament augmentation device of poly(L-lactide) in vitro and in vivo. Biomaterials. Vol. 13, pp. 1012-1016, ISSN 0142-9612

Li, Y., Cui, F., Liu, Z., Xu, Y. & Zhao, H. (2007). Improvement of xylanase production by Penicillium oxalicum ZH-30 using response surface methodology. Enzyme and Microbial Technology. Vol. 40, pp. 1381-1388, ISSN 0141-0229

Li, F., Wang, S., Liu, W. & Chen, G. (2008). Purification and characterization of poly(L-lactic acid)-degrading enzymes from Amycolatopsis orientalis ssp.orientalis. FEMS Microbiology Letters. Vol. 282, pp. 52–58, ISSN 0378-1097

Masaki, K., Kamini, N.R., Ikeda, H. & Iefuji, H. (2005). Cutinase-like enzyme from the yeast Cryptococcus sp. strain S-2 hydrolyses polylactic acid and other biodegradable plastics. Applied and Environmental Microbiology. Vol. 7, pp. 7548–7550, ISSN 0099-2240

Mayumi, D., Shigeno, Y.A., Uchiyama, H., Nomura, N. & Kambe, T.N. (2008). Identification and characterization of novel poly(DL-lactic acid) depolymerases from metagenome. Applied Microbiology and Biotechnology. Vo. 79, pp. 743–750, ISSN 0175-7598

Nakamura, K., Tomita, T., Abe, N. & Kamio, Y. (2001). Purification and characterization of an extracellular poly(L-lactic acid) depolymerase from a soil isolate, Amycolatopsis sp. strain K104-1. Applied and Environmental Microbiology. Vol. 67, pp. 345–353, ISSN 0099-2240

Pranamuda, H. & Tokiwa, Y. (1999). Degradation of poly (L-lactide) by strains belonging to genus Amycolatopsis. Biotechnology Letters. Vol. 21, pp. 901–905, ISSN 0141-5492

Pranamuda, H., Tokiwa, Y. & Tanaka, H. (1997). Polylactide degradation by an Amycolatopsis sp. Applied and Environmental Microbiology. Vol. 63, pp. 1637–1640, ISSN 0099-2240

Pranamuda, H., Tsuchii, A. & Tokiwa, Y. (2001). Poly (L-lactide)-degrading enzyme produced by Amycolatopsis sp. Macromolecular Bioscience. Vol. 1, pp. 25–29, ISSN 1616-5187

Reeve, M. S., McCarthy, S. P., Downey, M. J. & Gross, R. A. (1994). Polylactide stereochemistry: effect on enzymatic degradability. Macromolecules. Vol. 27, pp. 825–831, ISSN 0024-9297

Rondon, M.R., Robert, M., Goodman & Handelsman, J. (1999). The Earth's bounty: assessing and accessing soil microbial diversity. Trends Biotechnology. Vol. 17, No. 10, pp. 403-409, ISSN 0167-7799

Sakai, K., Kawano, H., Iwami, A., Nakamura, M. & Moriguchi, M. (2001). Isolation of a thermophilic poly-L-lactide degrading bacterium from compost and its enzymatic characterization. Journal of Bioscience and Bioengineering. Vol. 92, pp. 298–300, ISSN 1389-1723

Sangwan, P. & Wu, D.Y. (2008). New insights into polylactide biodegradation from molecular ecological techniques. Macromolecular Bioscience. Vol. 8, pp. 304–315, ISSN 1616-5187

Shigeno, Y.A., Teeraphatpornchai, T., Teamtisong, K., Nomura, N., Uchiyama, H., Nakahara, T. & Kambe, T.N. (2003). Cloning and sequencing of a poly(DL-Lactic Acid) depolymerase gene from Paenibacillus amylolyticus strain TB-13 and its functional expression in Escherichia coli. Applied and Environmental Microbiology. Vol. 69, No. 5, pp. 2498–2504, ISSN 0099-2240

Sukkhum, S., Tokuyama, S., Tamura, T. & Kitpreechavanich, V. (2009a). A novel poly(L-lactide) degrading actinomycetes isolated from Thai forest soils, phylogenic relationship and enzyme characterization. Journal of General and Applied Microbiology. Vol. 55, pp. 459-467, ISSN 0022-1260

Sukkhum, S., Tokuyama, S. & Kitpreechavanich, V. (2009b). Development of fermentation process for PLA-degrading enzyme production by a new thermophilic Actinomadura sp. T16-1. Biotechnology and Bioprocess Engineering. Vol. 14, pp. 302-306, ISSN 1226-8372

Takahashi, Y., Okajima, S., Toshima, K. & Matsumura, S. (2004). Lipase-catalyzed transformation of poly(lactic acid into cyclic oligomers. Macromolecular Bioscience. Vol. 4, pp. 346–353, ISSN 1616-5187

Tanyildizi, M. S., Ozer, D. & Elibol, M. (2005). Optimization of α-amylase production by Bacillus sp. using RSM. Process Biochemistry. Vol. 40, pp. 2291-2296, ISSN 1359-5113

Tokiwa, Y. & Calabia, B. P. (2006). Biodegradability and biodegradation of poly(lactide). Applied Microbiology and Biotechnology. Vol. 72, pp. 244–251, ISSN 0175-7598

Tokiwa, Y., Konno, M. & Nishida, H. (1999). Isolation of silk degrading microorganisms and its poly(L-lactide) degradability. Chemistry Letters. Vol. Vol. 28, No. 4, pp. 355–356, ISSN 0366-7022

Tomita, K., Kuroki, Y. & Nakai, K. (1999) Isolation of thermophiles degrading poly(L-lactic acid). Journal of Bioscience and Bioengineering. Vol. 87, pp. 752–755, ISSN 1389-1723

Tomita, K., Nakajima, T., Kikuchi, Y. & Miwa, N. (2004). Degradation of poly(L-lactic acid) by a newly isolated thermophile. Polymer Degradation and Stability. Vol. 84, pp. 433-438, ISSN 0141-3910

Tomita, K., Tsuji, H., Nakajima, T., Kikuchi, Y., Ikarashi, K. & Ikeda, N. (2003). Degradation of poly(D-lactic acid) by a thermophile. Polymer Degradation and Stability. Vol. 81, pp. 167-171, ISSN 0141-3910

Tsuji, H. & Nakahara,K. (2002). Poly(L-lactide) IX. Hydrolysis in acid media. Journal of Applied Polymer Science. Vol. 86, pp.186–194. ISSN 0021-8995

Vink, E. T. H, Rabago, K. R., Glassner, D. A. & Gruber, P. R. (2002). Application of life cycle assessment to Nature Works™ polylactide (PLA) production. Polymer Degradation and Stability. Vol. 80, pp. 403-419, ISSN 0141-3910

Williams, D.F. (1981). Enzymatic hydrolysis of polylactic acid. Engineering in Medicine. Vol. 10, No. 1, pp. 5-7, ISSN 0046-2039

Wood, A.D. (1980). Biodegradable drug delivery systems. International of journal Pharmaceutics. Vol. 7, pp. 1–18, ISSN 0378-5173

28

Engineering Bacteria for Bioremediation

Elen Aquino Perpetuo, Cleide Barbieri Souza
and Claudio Augusto Oller Nascimento
CEPEMA-University of São Paulo
Brazil

1. Introduction

Bioremediation is a process that uses microorganisms or their enzymes to promote degradation and/or removal of contaminants from the environment. The use of microbial metabolic ability for degradation/removal of environmental pollutants provides an economic and safe alternative compared to other physicochemical methodologies. However, although highly diverse and specialized microbial communities present in the environment do efficiently remove many pollutants, this process is usually quite slow, which leads to a tendency for pollutants to accumulate in the environment and this accumulation can potentially be hazardous. This is especially true for heavy metals. Heavy metal contamination is one of the most significant environmental issues, since metals are highly toxic to biota, as they decrease metabolic activity and diversity, and they affect the qualitative and quantitative structure of microbial communities. For treating heavy metal contaminated tailings and soils, bioremediation is still the most cost-effective method, although various heavy metals are beyond the bioaccumulation capabilities of microorganisms. Perhaps, because of the toxicity of these compounds, microorganisms have not evolved appropriate pathways to bioaccumulate them; populations of microorganisms responsible for this bioaccumulation are not large or active enough to remove these compounds completely, or complex mixtures of pollutants resist removal by existing pathways. The pathway used to accumulate these compounds is adsorption, where metals are taken up by microbial cells (biosorption). Biosorption mechanisms are numerous and are not yet fully understood. However, biosorption capacity often varies with test conditions, such as initial metal concentration, solution pH, contact time, biomass dosage, processing method, and so on. Accordingly, populations of microorganisms that are able to promote metal adsorption and accumulate them are not large or active enough to support these compounds by existing pathways. Furthermore, there are several strategies that optimize the bioremediation process of pollutants. One approach to enhance populations of microorganisms capable of pollutant removal is the addition of exogenous microorganisms in order to expand indigenous populations. This process, commonly known as bioaugmentation, can be performed either by adding microorganisms that naturally contain catabolic genes or those that have been genetically modified (GMOs). This strategy can also result in the transfer of plasmids containing the necessary genetic material between the different populations. Recent advances in the molecular biology field have been applied to microorganisms in order to produce novel strains with desirable properties for the bioremediation processes. These include the construction or adaptation of catabolic pathways; redirection of carbon flux to prevent the formation of harmful intermediates; modification of catabolic enzyme affinity and specificity; improvement of the genetic stability of catabolic activities; increasing the bioavailability of pollutants; and enhancement of the monitoring, yield, control, and efficiency of processes. Despite the many advantages of GMOs with regards to bioremediation, their use is still limited in the environment because of the instability of the inserted genetic material. There are two major reasons for this: first, the efficiency of GMOs is dependent on their ability to carry the genetic material in a stable manner; second, the transfer of genetic material to the indigenous organisms is perceived to be a negative attribute, despite the fact that this transfer is a common phenomenon among native organisms. These factors have incentivized the study of survival, competition and persistence of GMOs in the environment, as well as the potential risks involved in their use. Besides the significant advances that have already been made with regards to the development and utilization of GMOs for bioremediation of contaminants in the environment, many more challenges still remain. In this chapter, we will detail how genetic engineering may improve bioremediation through the engineering of bacteria. Several genetic approaches have been developed and used to optimize enzymes, metabolic pathways and organisms that are relevant to biodegradation. New information on metabolic routes is still being accumulated, thus the available toolbox is continuously being expanded. With molecular methods enabling the characterization of microbial community structure, metabolic pathways and enzyme activities, the performance of microorganisms under *in situ* conditions can be improved by making heavy metal bioremediation a much more efficient process. The present review also highlights the current situation pertaining to biosorbents, their mechanisms and advantages and disadvantages. Thus, this article reviews the achievements and current status of biosorption technology, which exploits natural biodiversity and molecular tools, in order to engineer microorganisms and provide new information about this research frontier.

1.1 Heavy metals and toxicicity

Heavy metals are considered to be chemical elements with an atomic mass greater than 22 and a density greater than 5g/mL. This definition includes 69 elements, of which 16 are synthetic. Some of these elements are extremely toxic to human beings, even at very low concentrations (Roane & Pepper, 2000; Wang & Chen, 2006). The main heavy metals associated with environmental contamination, and which offer potential danger to the ecosystem, are copper (Cu), zinc (Zn), silver (Ag), lead (Pb), mercury (Hg), arsenic (As), cadmium (Cd), chromium (Cr), strontium (Sr), cesium (Cs), cobalt (Co), nickel (Ni), thallium (Tl), tin (Sn) and vanadium (V) (Wang & Chen, 2006). In general, metal ions can be classified as: 1) Essential and important for metabolism (Na, K, Mg, Ca, V, Mn, Fe, Co, Ni, Cu, Zn, Mo and W); 2) Toxic heavy metals (Hg, Cr, Pb, Cd, As, Sr, Ag, Si, Al, Tl), which have no biological function (in ecotoxicology terms, hexavalent forms of Hg, Cr, Pb and Cd ions are the most dangerous); 3) Radionuclides (U, Rn, Th, Ra, Am, Tc), which are radioactive isotopes and, although toxic to cells, they are nonetheless important in nuclear medicine procedures; 4) Semi-metals or metalloids (B, Si, Ge, As, Sb, Te, Po, At, Se), which exert distinct biological effects on metals. However, metals are predominantly present in the environment in cationic and anionic forms in semimetals, and As is often classified as heavy metal (Roane & Pepper, 2000; Ahluwalia & Goyal, 2007). In the environment, metals can be divided into two categories: 1) bioavailable (soluble, non-absorbed, mobile); and 2) non-bioavailable (precipited, complexed, sorbed, non-mobile). The ionic form (speciation) of a metal determines its bioavailability and its destination. Most heavy metals are cations and this determines their sorption to negatively charged functional groups that are present in: cell surfaces, which are generally anionic at a pH of between 4 and 8; surfaces with residual hydroxides (OH^-) or thiol (SH^-) and anionic salts, such as PO^{4-} and SO^{4-}, humic acid, and clay minerals (Roane & Pepper, 2000). Heavy metal ions possess great electrostatic attraction and high binding affinities to the same sites that essential metal ions normally bind to various cellular structures, causing destabilization of the structures and biomolecules (cell-wall enzymes, DNA and RNA), thus inducing replication defects and consequent mutagenesis, hereditary genetic disorders and cancer. This occurs, for example, with arsenate, which competes with phosphate, and cadmium, which competes with zinc. By employing microarray technology, Kawata *et. al.* (2007), found that six heavy metals (arsenic, cadmium, nickel, antimony, mercury and chromium) induce gene expression patterns that are very similar to the pattern induced by DMNQ (2.3-dimethoxy-1, 4-naphthoquinone), the reactive oxygen species (ROS) chemical generating agent, which causes "oxidative stress", leading to deleterious effects (membrane damage or other cellular lipid structures, modification of proteins, fragmentation and cross-links, changes in DNA that can induce mutations or be repaired by repair mechanisms). Therefore, the ions of heavy metals cause oxidative damage, both directly, by producing ROS, and indirectly, by inactivating the cellular antioxidant system, thus leading to cell damage (Mannazzu *et. al.*, 2000; Liu *et. al.*, 2005).

1.2 Heavy metals and the environment

Among the different contaminants, heavy metals have received special attention due to their strength and persistence in accumulating in ecosystems, where they cause damage by moving up the food chain to finally accrue in human beings, who are at the top of this chain (Figure 1) (Volesky, 2001; Ahluwalia & Goyal, 2007; Machado *et. al.*, 2008).

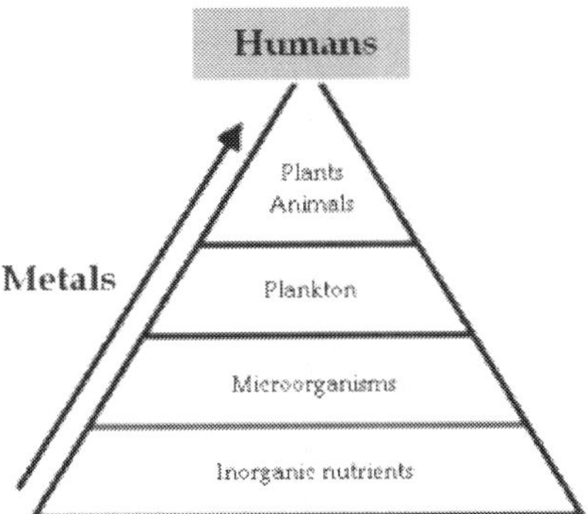

Fig. 1. The destiny of heavy metals released into the environment and their accumulation throughout the chain food. Adapted from Volesky (2001).

The toxic potential of heavy metals with regards to the human body is diverse and, because of their toxicity and persistence in nature, the levels of heavy metals in the environment need to be controlled by mandating waste treatment at the sources of pollution. The development of new treatment technologies is required at these sources; however, even though there is awareness of this problem, sustainable solutions are not easily accessible. In general, the conventional treatment methods used to remove metals from wastewater are inefficient and cost-prohibitive.

1.3 Conventional technologies for treating environments contaminated by heavy metals

Environments contaminated by heavy metals are treated by means of conventional technologies based on physicochemical principles, which are considered inefficient and uneconomic. One method often employed to remove metals from aqueous solutions involves the addition of reagents that increase pH, thus converting metals from a soluble form into an insoluble form (hydroxides), which results in their precipitation. This procedure produces large quantities of mud in the final wastewater with concentrations of metals in the order of mg/L, which is difficult to dispose of. More complex processes of this type can involve single or multiple steps: 1) precipitation with hydroxides, carbonates or sulfides; 2) redox chemistry; 3) sorption (adsorption with activated carbon/ion exchange); 4) use of membranes (ultrafiltration, electrodialysis and reverse osmosis-RO); 5) electrolytic recovery; 6) evaporation; 7) liquid-liquid extraction; 8) electrodeposition. On the other hand, bioremediation is increasingly gaining importance as an alternative technology, due the advantages it offers: simplicity, efficiency and low cost (Goyal *et. al.*, 2003; Tabak *et. al.*, 2005; Hameed, 2006; Machado *et. al.*, 2008; Wang & Chen, 2009).

1.4 Types of bioremediation

Bioremediation involves the use of plants or microorganisms, viable or not, natural or genetically engineered to treat environments contaminated with organic molecules that are difficult to break down (xenobiotics) and to mitigate toxic heavy metals, by transforming them into substances with little or no toxicity, hence forming innocuous products (Dobson & Burgess, 2007; Li & Li, 2011). With the objective of improving the process of bioremediation, different strategies can be employed, depending of the state of the contaminated environment. One of these strategies, biostimulation, involves promoting the growth of autochthonous microorganisms at the contaminated site, in order to introduce pH-correction substances, nutrients, surfactants and oxygen. As a consequence, the rate of biodegradation/bioremediation can be increased. Another strategy, bioaugmentation or bioaddition, is the addition of microbial populations to indigenous, alien or genetically modified organisms (GMO), in places where there is an insufficiency of indigenous microorganisms with ecophysiological characteristics compatible with the habitat conditions that are conducive to the promotion of bioremediation (Vidali, 2001; Silva *et. al.*, 2004; Gaylard *et. al.*, 2005; Li & Li, 2011). Bioremediation is a versatile process that can be applied *in-situ*, at the contaminated site, or *ex-situ*, involving removal of contaminated material to be treated elsewhere. *In-situ* bioremediation technologies are more economical and release fewer pollutants into the environment; however, they may require longer treatment timeframes than the *ex-situ* techniques (Vidali, 2001; Tabak *et. al.*, 2005; Costa *et. al.*, 2007). Currently, there is wide variety of microorganisms (bacteria, fungi, yeasts and algae) that are being studied for use in bioremediation processes, and some of these have already been employed as biosorbents of heavy metals (Roane & Pepper, 2000; Machado *et. al.*, 2008;

Bogacka, 2011). The main advantages of biosorption over conventional treatment methods include: low cost; high efficiency; minimization of chemical and biological sludge; selectivity to specific metals; no additional nutrient requirement; regeneration of the biosorbent; and the possibility of metal recovery (Kratochvil & Volesky, 1998). Most studies on the bioremediation of heavy metals present the following as a biphasic biosorption process: 1) rapid initial phase (adsorption), not dependent on metabolism or temperature, which is generally reversible; and, 2) slower phase (intracellular accumulation) metabolism-dependent, influenced by environmental factors such: temperature, which decreases the biosorption capacity due to damage to the target sites (in general, the temperature considered "ideal" is between 25-35°C); and metabolic inhibitors (Roane & Pepper, 2000; Malik, 2004; Tabak, 2005). Desorption is a very important process in the success of the applicability of bioremediation for the treatment of wastewater, because it allows for the recovery of adsorbed metal ions, as well as the recycling and reuse of biomass for a new cycle of metal recovery. There is obviously great interest in the development of a procedure that enables the recovery of removed metal ions, as well as for the cellular integrity of the biosorbents to be maintained, thus allowing for their regeneration and reuse in successive cycles of sorption-desorption. This results in the simultaneous acquisition of two valuable products: treated water and low-cost recovery of metal (Aldor *et. al.*, 1995; Volesky, 2001; Yu & Kaewsarn, 2001).

2. Exploitation of the natural biodiversity

2.1 Microorganisms as biosorbents of heavy metals

Several studies have shown that many organisms, prokaryotes and eukaryotes, have different natural capacities to biosorb toxic heavy metal ions (Table 1), giving them different degrees of intrinsic resistance, particularly in diluted solutions (between 10 to 20 mg/L^{-1}), due to their mobility, as well as the solubility and bioavailability capacities of these metal ions (Malik, 2004; Tabak *et. al.*, 2005; Kim *et. al.*, 2007; Chen & Wang, 2008).

The manner in which microorganisms interact with heavy metal ions depends partly on whether they are eukaryotes or prokaryotes. Eukaryotes are more sensitive to metal toxicity than bacteria. In the presence of toxic concentrations, several resistance mechanisms are activated, for example: the production of peptides of the family of metal binding proteins, such as metallothioneins (MTs); the regulation of the intracellular concentration of metals, with expression of protein transporters of ligand-metal complexes from the cytoplasm to the inside of vacuoles, and efflux of metal ions by ion channels present in the cell wall. In bacteria, these tolerance mechanisms are often encoded by plasmids, which facilitate their dispersion from cell to cell (Valls & Lorenzo, 2002). For the bioremediation of heavy metals on an industrial scale, it is important to use low-cost biomaterial, which can be a byproduct or waste material with high removal capacity, since the low cost of this biomass is crucial for the process to be economically viable (Volesky & Holan, 1995; Zouboulis *et. al.*, 2001). Several studies have been conducted with the purpose of improving the resistance and/or the ability of microorganisms to accumulate heavy metal ions, including a number of studies that follow parameters: pH (Naeem *et. al.*, 2006); temperature; different metal concentrations and biomass (Soares *et. al.*, 2003; Kim *et. al.*, 2005), competitiveness of ions of different elements; microorganism-metal contact time (Kotrba & Ruml, 2000), composition of the culture medium (Ghosh *et. al.*, 2006); bioaugmentation/biostimulation (Roane &

Pepper, 2001; Silva *et. al.*, 2004), resistance to toxicity of heavy metals of Gram positive/Gram-negative bacteria (Samuelson *et. al.*, 2002); intracellular/extracellular bioaccumulation; viable /non-viable cells, free /immobilized cells, and biological processes by aerobic/anaerobic microorganisms (Dias *et. al.*, 2002; Liu *et. al.*, 2005; Tabak *et. al.*, 2005; Wang & Chen, 2006).

Organisms	Genus/species	Reference
Bacteria	*Arthrobacter*	Roanne & Pepper, 2001
	Bacillus sp	Gupta *et. al.*, 2000; Dias *et. al.*, 2002 ; Kim *et. al.*, 2007
	Citrobacter	Renninger *et. al.*, 2001
	Cupriavidus metallidurans	Roanne & Pepper, 2001 ; Grass *et. al.*, 2005
	Cyanobacteria	Gupta *et. al.*, 2000
	Enterobacter cloacae	Hernandes *et. al.*, 1998; Gupta *et. al.*, 2000
	Pseudomonas aeruginosa	Dias *et. al.*, 2002; Zhang *et. al.*, 2005
	Streptomyces sp	Dias *et. al.*, 2002
	Zoogloea ramigera	Gupta *et. al.*, 2000
Archea	Filo *Crenarchaeota*	Sandaa *et. al.*, 1999
	Phanerochaete chrysosporium	Wu & Yu, 2007
Fungi	*Aspergillus tereus*	Kumar *et. al.*, 2008
	Penicillium chrysogenum	Dias *et. al.*, 2002
Yeast	*Candida utilis*	Kujan *et. al.*, 2006
	Hansenula anomala	Breierová *et. al.*, 2002
	Rhodotorula mucilaginosa	Dias *et. al.*, 2002
	Rhodotorula rubra GVa5	Ghosh *et. al.*, 2006
	Saccharomyces cerevisiae	Gupta *et. al.*, 2000; Dias *et. al.*, 2002; Ghosh *et. al.*, 2006

Table 1. Examples of microorganisms studied and strategically treated for bioremediation of heavy metals.

The search for new technologies for the removal of toxic metals from contaminated sites has focused on biosorption, which is based on the metal binding capacities of various biological materials. Biosorption can be defined as the ability of biological materials to accumulate heavy metals from wastewater through metabolically mediated or physicochemical uptake pathways (Fourest & Roux, 1992). Algae, bacteria, fungi and yeasts have all proven to be potential metal biosorbents (Volesky, 1987). Many indigenous organisms isolated from sites contaminated with heavy metals have tolerance to heavy metal toxicity and these microbial activities have always been the natural starting point for all biotechnological applications. It is therefore necessary to isolate bacterial strains with novel metabolic capabilities and to establish degradation pathways both biochemically and genetically. Potent metal biosorbents in the bacteria class include the genus *Bacillus* (Nakajima & Tsuruta, 2004; Tunali *et. al.*, 2006), *Pseudomonas* (Chang *et. al.*, 1997; Uslu & Tanyol, 2006) and *Streptomyces* (Mameri *et. al.*, 1999; Selatnia *et. al.*, 2004a). The biosorption process involves a solid phase (sorbent or biosorbent; biological material) and a liquid phase (solvent, normally water) containing a dissolved species to be sorbed (sorbate, metal ions). Due to higher affinity of the sorbent for the sorbate species, the latter is attracted and bound there by different mechanisms. The process continues until equilibrium is established between the amount of solid-bound sorbate species and its portion remaining in the solution. The degree of sorbent affinity to the sorbate determines its distribution between solid and liquid phases.

Biosorbent material: Strong biosorbent behaviour of certain microorganisms towards metallic ions is a function of the chemical make-up of the microbial cells. This type of biosorbent consists of dead and metabolically inactive cells. Some types of biosorbents probably have a broad range, binding and collecting the majority of heavy metals with no specific preference, while others are specific to certain metals. Some laboratories have used easily available biomass whereas others have isolated specific strains of microorganisms and some have also processed existing raw biomass to a certain degree to improve its biosorption properties. Recent biosorption experiments have focused attention on waste materials, which are by-products or residues from large-scale industrial operations. E.g., mycelia waste available from fermentation processes, solid waste from olive oil manufacturing plants (Pagnanelli *et. al.,* 2002), activated sludge from sewage treatment plants (Hammaini *et. al.,* 2003), biosolids (Norton *et. al.,* 2004) and aquatic macrophytes (Keskinkan *et. al.,* 2003), etc. The biosorption mechanism is complex, including mainly ion exchange, chelation, adsorption by physical forces, entrapment in intra and interfibrilliar capillaries and spaces of the structural polysaccharide network as a result of the concentration gradient and diffusion through cell walls and membranes. There are several chemical groups that are thought to attract and sequester metals in biomass: acetamido groups of chitin, structural polysaccharides of fungi, amino and phosphate groups in nucleic acids, amido, amino, sulphydryl and carboxyl groups in proteins, hydroxyls in polysaccharide and mainly carboxyls and sulphates in polysaccharides of marine algae that belong to the divisions Phaeophyta, Rhodophyta and Chlorophyta. However, this does not necessarily mean that the presence of a particular functional group guarantees biosorption, perhaps due to steric, conformational or other barriers. The choice of metal for biosorption process: The appropriate selection of metals for biosorption studies is dependent on the angle of interest and the impact of different metals. Accordingly, they can be divided into four major categories: (i) toxic heavy metals (ii) strategic metals (iii) precious metals and (iv) radionuclides. In terms of environmental threats, categories (i) and (iv) are the ones of most concern with regards to their removal from the environment and/or from point source effluent discharges. Apart from toxicological criteria, the interest in specific metals may also be based on how representative their behavior is in terms of an eventual generalization of results regarding their biosorbent uptake. The toxicity and interesting solution chemistry of elements such as chromium, arsenic and selenium make them good candidates for study. Although not environmentally threatening, strategic and precious metals are important to consider from the standpoint of their recovery value. Research activities related to biosorption were initiated in the 1980s (Volesky & Holan, 1995; Volesky, 2001). Previously, research on this topic involved the use of biorremediation involving microorganisms alone to degrade organic compounds (Lovley & Coates, 1997). Since then, there has been an intensive effort to investigate the binding properties of heavy metals to different types of biomass (Chen & Wang, 2008). In general, biosorption is physicochemical interactions of metal ions with specific groups of the cell surface of microorganisms, which may enhance or inhibit intracellular transport or influence the processes of transformation and extracellular precipitation (biomineralization) (Davis *et. al.*, 2003; Gavrilescu, 2004). Biosorption is based on passive (metabolism-independent) or active (metabolism-dependent) accumulation processes (Wang & Chen, 2006), in combinations that differ qualitatively and quantitatively, depending on the type of biomass, its origin, feasibility, and type of processing (Veglio & Beolchini, 1997). A classification of the mechanisms of metal accumulation can be made based on the dependence of cellular metabolism or according to the location of the metal within the cell.

2.2 Biosorption mechanisms

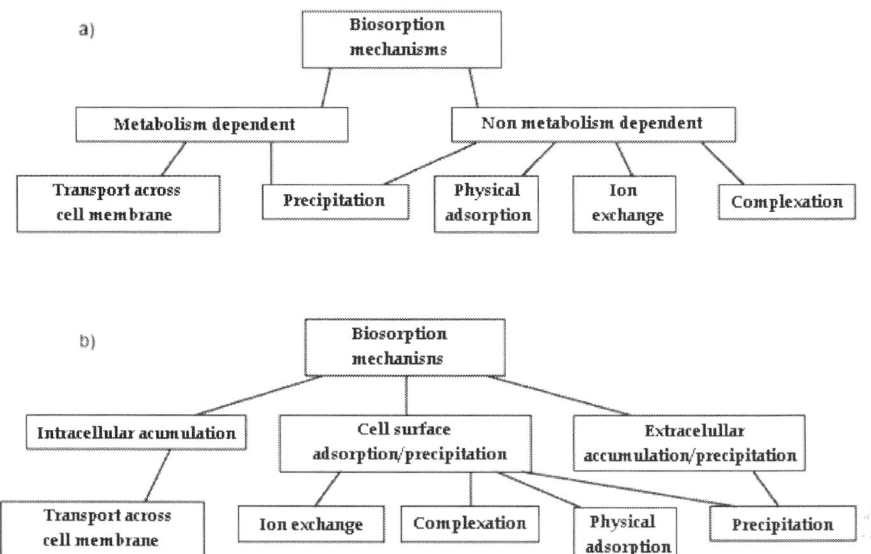

Fig. 2. Mechanisms of biosorption, a) classification according to dependence on cell metabolism, b) classification according to the location within the cell and the metal removed (Veglio & Beolchini, 1997).

The complex structure of microorganisms implies that there are many ways for the metal to be taken up by the microbial cell. There are various biosorption mechanisms and they are not yet fully understood. They may be classified according to various criteria (Figure 2).
The transportation of metal across the cell membrane yields intracellular accumulation, which is dependent on the metabolism of the cell. This means that this kind of biosorption may only take place within viable cells. It is often associated with an active defense system of the microorganism, which reacts in the presence of a toxic metal. During non-metabolism dependent biosorption, metal uptake occurs by means of physicochemical interaction between the metal and the functional groups present on the microbial cell surface. This is based on physical adsorption, ion exchange and chemical sorption, which is not dependent on cell metabolism. Cell walls of microbial biomass, which consists mostly of polysaccharides, proteins and lipids, have abundant metal binding groups such as carboxyl,

sulphate, phosphate and amino groups. This type of biosorption, i.e., non-metabolism dependent, is relatively rapid and can be reversible (Kuyucak & Volesky, 1988).

In the case of precipitation, metal uptake may take place both in the solution and on the cell surface (Ercole *et. al.*, 1994). Furthermore, it may be dependent on cell metabolism if, in the presence of toxic metals, the microorganism produces compounds that favor the precipitation process. Precipitation may not be dependent on cell metabolism, if it occurs after a chemical interaction between the metal and the cell surface.

Due to the complexity of biomaterial, it is possible for several of these mechanisms to occur simultaneously, in degrees that depend on the biosorbent and environmental conditions (Kefala *et. al.*, 1999; Volesky, 2001; Gavrilescu, 2004). Studies on biosorption by Wang & Chen (2006) have focused on selecting the most promising types of biomass and its future success by sorption will depend on interaction between three knowledge areas: biology, chemistry and engineering (Kefala *et. al.*, 1999). Biosorption capacity is affected mainly by three factors: 1) characteristics of the metal ion (atomic weight, ionic ray, valence), 2) environmental conditions (pH, temperature, ionic strength, contact time, biomass concentration), and 3) the nature of the biosorbent, which may determine differences in selectivity and affinity to metal ions, since the type and species of microorganisms, conditions of growth, physiological state and cell age may all affect the binding mechanism to heavy metals (Wang & Chen, 2006; Chen & Wang, 2008).

Different strategies have been developed using recombinant DNA technology to produce genetically improved strains for use in the biosorption process, and many of these strategies "equip" the bacterial cell wall with metal ion-binding polypeptides to act as anchors. One of these studies was on the fusion protein: the β-domain of IgA protease of N. *gonorrhoeae* with metallothionein (MT) from rats (Valls *et. al.*, 2000), and lpp-ompA-various sizes of peptides (EC20) (Bae *et. al.*, 2000). The following items are the main biosorption mechanisms, which can also be considered mechanisms of resistance or tolerance to heavy metals developed by microorganisms.

Transport across cell membranes: Heavy metal transportation across microbial cell membranes may be mediated by the same mechanism used to convey metabolically important ions such as potassium, magnesium and sodium. The metal transport systems may become confused by the presence of heavy metal ions of the same charge and ionic radius associated with essential ions. This kind of mechanism is not associated with metabolic activity. Basically, biosorption by living organisms comprises of two steps: first, an independent binding metabolism where metals are bound to the cell walls; and second, metabolism-dependent intracellular uptake, whereby metal ions are transported across the cell membrane (Costa *et. al.*, 1990; Gadd *et. al.*, 1988; Huang *et. al.*, 1990; Nourbaksh *et. al.*, 1994).

Physical adsorption: In this category, physical adsorption takes place with the help of van der Waals forces. Kuyucak & Volesky (1988), hypothesized that uranium, cadmium, zinc, copper and cobalt biosorption by dead biomasses of algae, fungi and yeasts takes place through electrostatic interactions between the metal ions in solutions and the cell walls of microbial cells. Electrostatic interactions have been proven to be responsible for copper biosorption by the *Zoogloea ramigera* bacterium and the *Chiarella vulgaris* alga (Aksu *et. al.*, 1992), and for chromium biosorption by the *Ganoderma lucidum* and *Aspergillus niger* fungi.

Ion Exchange: Cell walls of microorganisms contain polysaccharides and bivalent metal ions exchange with the counter ions of the polysaccharides. For example, the alginates of marine algae occur as salts of K^+, Na^+, Ca^{2+}, and Mg^{2+}. These ions can exchange with counter ions such as CO^{2+}, Cu^{2+}, Cd^{2+} and Zn^{2+}, resulting in the biosorptive uptake of heavy metals (Kuyucak & Volesky, 1988). The biosorption of copper by *Ganoderma lucidium* (Muraleedharan & Venkobachr, 1990) and *Aspergillus niger* fungi was also taken up by the ion exchange mechanism.

Complexation: An extracellular complexation or coordination is the result of electrostatic attraction between a metallic ion chelating agent and a polymer that can be excreted by a microorganism that is viable or not. It can be caused by: biosurfactants, polysaccharides, proteins and nucleic acids. These chelating agents contain pairs of electrons that present electrostatic attraction and if they cling to the metallic ions, there is no electron transfer. The final structure has the electric charge of the sum of individual charges of the participants of the complex (Veglio & Beolchini, 1997; Davis *et. al.*, 2003). When this detoxification/complexation system is overloaded, "oxidative stress" of the cell occurs (Gavrilescu, 2004). Metal removal from the solution may also take place by complex formation on the cell surface after interaction between the metal and the active groups. Aksu *et. al.* (1992) hypothesized that biosorption of copper by C. *vulgaris* and Z. *ramigera* takes place through both adsorption and the formation of coordination bonds between metals and amino and carboxyl groups of cell wall polysaccharides. Complexation was found to be the only mechanism responsible for the accumulation of calcium, magnesium, cadmium, zinc, copper and mercury by *Pseudomonas syringae*. Microorganisms may also produce organic acids (e.g., citric, oxalic, gluonic, fumaric, lactic and malic acids), which may chelate toxic metals, thus resulting in the formation of metallo-organic molecules. These organic acids help in the solubilization of metal compounds and leaching from their surfaces. Metals may be biosorbed or complexed by carboxyl groups found in microbial polysaccharides and other polymers.

Precipitation: Precipitation may be either dependent on the cellular metabolism or independent of it. In the former case, metal removal from solutions is often associated with the active defense system of the microorganisms. They react in the presence of toxic metal-producing compounds, which favor the precipitation process. In the case of precipitation that is not dependent on cellular metabolism, it may be a consequence of the chemical interaction between the metal and the cell surface. The various biosorption mechanisms mentioned above can take place simultaneously.

Adsorption: Physical adsorption is a process where the metal ion in solutions binds onto polyelectrolytes present in microbial cell wall through electrostatic interactions, Van der Waals forces, covalent bonding, redox interaction and biomineralization to achieve electroneutrality. This process is independent of metabolism, and it is reversible and very promising, as it presents many advantages, especially for treating large volumes of wastewater with low concentrations of contaminants (Ahluwalia & Goyal, 2007; Kuroda & Ueda, 2010; Nishitani *et. al.*, 2010). In physical adsorption, the metal ions are attracted by the potential negative of the cell wall, and both are dependent on pH. The influence of pH on metal accumulation by yeasts, algae and bacteria are very similar; for example, in yeast, pH < 2 the accumulation of metals is practically zero, because in low pH the active sites of the cell wall are associated with protons, restricting the approach of metal cations, and thus resulting in a repulsive force. Therefore, as the pH increases, an increasing number of sites (acetamide chitin, structural polysaccharides of fungi, phosphate and amino groups of nucleic acids, amino and carboxyl groups of proteins and hydroxyl groups of polysaccharides) are replaced by negative charges, increasing the attraction of metallic cations and adsorption on the cell surface. Accordingly, the solubility of metallic ions decreases and, consequently, their bioavailability is reduced, and precipitation occurs (Esposito *et. al.*, 2002; Chen & Wang, 2008; Nishitani *et. al.*, 2010; Kuroda & Ueda, 2011). Thus, a pH range of between 4 and 8 is generally accepted as "good" for the biosorption of heavy metals for almost all types of biomass (Borroka & Fein, 200; Wang & Chen, 2006; Machado *et. al.*, 2010). Also, has been studied the role of extracellular polymeric substances (EPS) excreted by bacteria in the removal of heavy metal ions by adsorption process (Gupta *et. al.*, 2004).

Siderophores: Like heavy metal chelating agents, some types of microorganisms have low molecular weight, and these are called siderophores. When microorganisms are grown in an iron deficient medium, they produce specific iron chelators, so-called siderophores, in the medium. They play an important role in the complexation of toxic metals and radionuclides, by increasing their solubility. Siderophores have low molecular weights, and have compounds of the catecholate, phenolate or hydroxamate as their binding groups. Many bacteria, such as *Actinomycetes Azotobacter* and those of the genus *Pseudomonas*, synthesize these substances to capture the iron ions they require for their metabolic activity, or for biosorption (Pattus & Abdallah, 2000; Gazsó, 2001).

Biosurfactants: Most surfactants used in bioremediation are produced industrially from petroleum, but they can also be synthesized by microorganisms. Biosurfactants are natural surfactants synthesized by plants: (saponins), microorganisms (glycolipids) and the bodies of organisms (bile salts), and they have several advantages over industrially-produced surfactants, such as lower toxicity to degrading microorganisms and less recalcitrance in the environment, greater diversity of chemical structures; performance over a broader range of conditions at different temperatures and pHs (Bognolo, 1999). Biosurfactants of microbial origin (mainly aerobic) are the result of the metabolic byproducts of bacteria, fungi and yeasts, which are released into the medium. The hydrophilic portion may consist of amino acids, peptides or saccharides, and the hydrophobic portion is usually formed by saturated or unsaturated fatty acids. Biosurfactants are able to form various structures, such as micelles, vesicles and spherical or irregular lamellar structures, among others (Figure 3) (Champion *et. al.*, 1995; Mulligan, 2005; Li & Li, 2011).

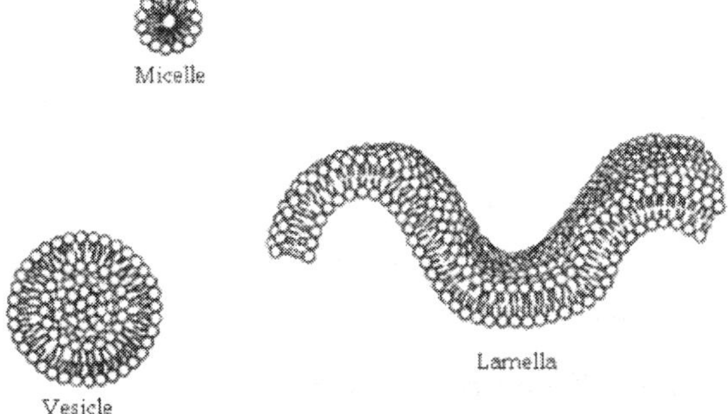

Fig. 3. Basic structures formed by biosurfactants (Champion *et. al.*, 1995).

Microorganisms naturally overcome their limitations when faced with insoluble organic and inorganic pollutants (hydrocarbons, oil, pesticides, heavy metals - uranium, cadmium, lead, etc.), performing the excretion of these structures to the culture medium, associated their cell walls. This facilitates the transport and translocation of insoluble substrates, in turn facilitating biosorption. In the process of heavy metal ion bioremediation, a complex non-toxic biosurfactant/metallic ion structure is formed, resulting in emulsification and solubilization of the ions, and, consequently, their physical sequestration. One of the most widely used natural biosurfactants in bioremediation is rhamnolipids, produced by *Pseudomonas aeruginosa* (Bognolo, 1999; Champion *et. al.*, 1995; Mulligan, 2005; Tabak *et. al.*, 2005; Zhang *et. al.*, 2005).

Oxidation-reduction (redox): Microorganisms can mobilize or immobilize metal ions, metalloid and organometal compounds, thus promoting redox processes. Only prokaryotes are capable of oxidizing metals such as Mn^{2+}, Fe^{2+}, Co^{2+}, Cu, AsO^{2-}, Se^0 or SeO_3^{2-}, or reducing Mn^{4+}, Fe^{3+}, Co^{3+}, AsO_4^{2-}, SeO_4^{2-} or SeO_3^2, whilst obtaining energy from these reactions (Gavrilescu, 2004). When, for example, the Fe^{3+} ion is reduced to Fe^{2+}, or the Mn^{4+} ion is reduced to Mn^{2+}, there is an increase in the solubility of these metals. Microbial species can efficiently immobilize heavy metals through their ability to reduce heavy metal ions, reducing them to a lower oxidation state, and giving rise to metallic elements (load zero) which are less bioactive (Valls & Lorenzo, 2002; Gadd, 2004).

Biomethylation: Microorganisms can transform metal ions from a more toxic to a less toxic form through the biomethylation mechanism. Hg, As, Cd, Se, Sn, Te and Pb ions can be methylated by a variety of bacteria, filamentous fungi and yeasts, under both aerobic and anaerobic conditions, which results in increased mobility, and also in their suitability for involvement in processes that result in a decrease in their toxicities. This enzymatic mechanism involves a transfer from the methyl group (CH_3) to metals and metalloids. The methylated compounds formed differ in their solubility, volatility and toxicity (Roane & Pepper, 2000; Gadd, 2004). For example, methyl and dimethyl mercury are more toxic than inorganic Hg ions, however, these are intermediate forms of processing for Hg^0. Inorganic forms of As are more toxic than methylated species (acids and methyl-As dimethyl-As), the methylated and inorganic forms of Se and Cd are highly toxic (Roane & Pepper, 2000; Tabak, 2005).

Metal-binding cysteine-rich peptides: When cells are exposed to heavy metals in toxic concentrations an induction of expression of peptides rich in cysteine residues, metallothioneins (MTs), glutathione (GSH) or phytochelatin (PCs) occurs. These are non-enzymatic molecules, with low molecular weight, which are resistant to thermo-coagulation and acid precipitation. The main feature of these peptides is to form complexes with divalent metals and metal-thiols, which consist of important metabolites to combat ROS (Bae *et. al.*, 2000; 2001).

Metallothioneins (MTs): Metallothioneins (MTs) are a group of well-preserved structures of proteins that act as antioxidants, and they are distributed among all living organisms. They have low molecular weight and they are rich in cysteine residues. The presence of thiol groups (SH) in the cysteine chemical structure enables the capture of metal ions such as Cd^{2+}, Fe^{2+}, Hg^{2+}, Cu^{2+} and Zn^{2+}. MTs are composed of two separate domains, the β domain in the N-terminal region, and the α domain in the C-terminal region. The β domain possesses nine cysteine residues linking three divalent ions, the α domain already has eleven cysteine residues and binds four ions, thus a total of seven ions per molecule are bound (Cobbett & Goldsbrough, 2002; Thirumoorthy *et. al.*, 2007). The MTs have several functions, including the detoxification of heavy metals and protection against the presence of ROS. Thus, MTs are responsible for reducing the effect of oxidative stress caused by these ions, but they are also responsible for maintaining homeostatic cellular redox balance.

According to these characteristics, protein synthesis is only induced by metals. (Cobbett & Goldsbrough, 2002; Smith *et. al.*, 2007).

Glutathione (GSH): Glutathione (GSH), L-Glutamyl-L-cysteinyl-glycine (Figure 4), is a soluble antioxidant, recognized as the most important non-protein thiol present in all living organisms. It consists of three amino acids (Glu-Cys-Gly), and it is the cysteine thiol group of the active site that is responsible for its biochemical properties (Bae & Mehra, 1997; Penninckx, 2000; 2002; Mendonza-cózatl *et. al.*, 2005).

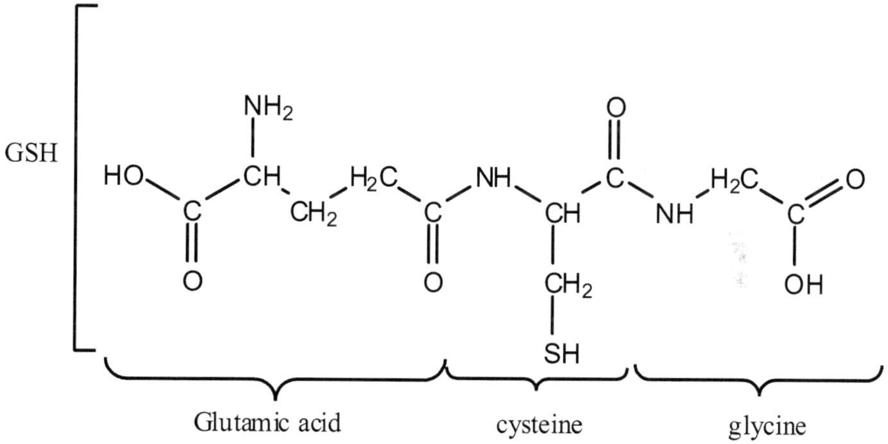

Fig. 4. Chemical structure of glutathione (GSH) (Bae & Mehra, 1997).

The GSH controls its own synthesis and participates in multiple processes: regulation of the intracellular redox state, inactivation of ROS, transport of GSH-conjugated amino acid and other molecules, stock of the sulfur and cysteine. In mammals, it is present in greater quantities in the liver. Its biosynthesis is similar in plants, yeast and protists. The mitochondria and the nucleus have their own GSH reservation, which is critical or instrumental in protecting these structures against ROS action (Penninckx, 2002; Inouhe, 2005; Mendoza -Cózatl *et. al.*, 2005). In yeast, the cell protection system against the toxicity of Cd^{2+} occurs through the formation of a GSH- Cd^{2+} complex, because this causes a decrease in the levels of lipid peroxidation of the cell membrane and allows for the transportation of GSH-Cd^{2+} conjugate to the vacuole. This results in a decreased concentration of toxic metals in the cytosol, thus promoting a reduction in the levels of oxidative stress (Penninckx, 2000; Adamis *et. al.*, 2004; Kobayashi *et. al.*, 2005; Preveral *et. al.*, 2006).

Natural phytochelatins (PCs) and synthetic phytochelatin (EC20): PCs are small cysteine-rich peptides with the general structure (Glu-Cys) nGly (n = 2-11) (Figure 5-A) (Grill et. al., 1985; Cobbett, 2000). PCs are synthesized from glutathione (GSH) in steps catalyzed by PC synthase (Grill et. al., 1985; Gupta et. al., 2004). They enable ions to bind to various heavy metal ions through thiol residues and carboxyl (Kobayashi et. al., 2005; Inouhe, 2005). These PCs are present in plants, fungi, nematodes, parasites and algae, including cyanobacteria. Despite being classified as MT-III, the PCs have a greater capacity for binding heavy metal ions (1 atom per cysteine basis) than MTs. The pioneering work that targeted the expression of recombinant PCs in *E.coli* were faced with the difficulty imposed by the chemical bonds of type γ present between Glu-Cys units, which are the result of multi-enzyme processes. These bindings are different for type α among the present amino acid chains of all proteins (Bae *et. al.*, 2001; Cobbett, 2000; Penninckx, 2000; Gupta *et. al.*, 2004; Inouhe, 2005; Hirata, 2005; Mendoza-Cózatl, 2005; Wu *et. al.*, 2006). An alternative was to synthesize an *in vitro* gene that codes for proteins similar to PCs, whose general structure corresponded to (Glu-Cys) nGly (ECs) with all amino acids linked chemically by type α (Figure 5-B). Thus, the synthetic phytochelatin EC20 contains 20 units of repeated Glu-Cys (EC). The synthetic phytochelatin EC20 has greater capacity for binding to heavy metal ions than natural PCs. Recombinant strains of bacteria are currently being constructed (Bae *et. al.*, 2000; 2001; Xu *et. al.*, 2002; Lee *et. al.*, 2006; Wu *et. al.*, 2006), and though few studies have been developed for yeast, we can cite (Schmitt *et. al.*, 2006) expressing EC20 with the aim of identifying microorganisms with increased capacity for binding heavy metal ions for use in bioremediation processes.

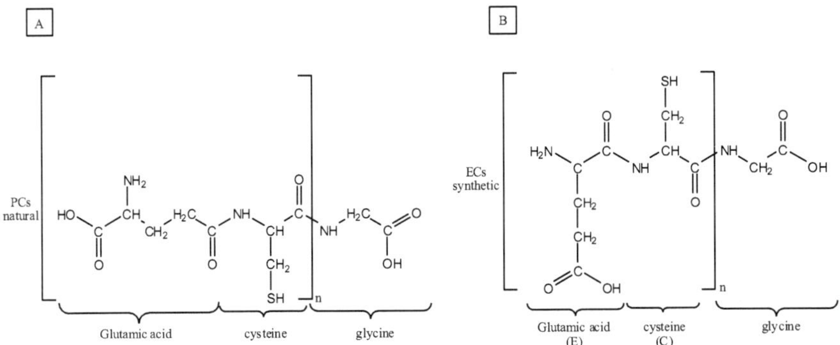

Fig. 5. Chemical structures of molecules binding heavy metals: A) natural phytochelatin (PC) and B) synthetic phytochelatin (EC) - glutamic acid (glu-E) and cysteine (Cys-C) (Bae *et. al.*, 2000).

Among these genetic engineering techniques for the construction of new recombinant microbial strains (Deng *et. al.*, 2003; Merle *et. al.*, 2003; Kim *et. al.*, 2005; Nishitani *et al.*, 2010), MTS, PCs and ECs were also explored so that these proteins remain anchored to the outer surface cell ("cell-surface display"). The aim was to increase the capacity these special microorganisms in adsorption processes involving heavy metal ions in comparison with non-recombinant microorganisms (Bae *et. al.*, 2000; 2001; Kuroda *et. al.*, 2002; Kuroda & Ueda, 2003; Jiang *et. al.*, 2007).

The "cell-surface display" system: Cell surface proteins constitute an important class of biomolecules because they are situated at the interface between the cell and the environment. The cells have systems for anchoring specific proteins to the surface and confining them to certain areas. A great many systems are being used in bacteria and in *S. cerevisiae* (Kondo & Ueda, 2004). The expression of heterologous peptides on the cell surface ("cell-surface display ") is a powerful technique widely used in the biotechnology area in the following processes: production of recombinant vaccines, antigens, antibodies, enzymes and library peptides (Kuroda *et. al.*, 2001; Chen & Georgiou, 2002; Samuelson *et. al.*, 2002; Rutherford & Moures, 2006; Wang *et. al.*, 2007; Kuroda & Ueda, 2010; Nishitani *et. al.*, 2010; Kuroda & Ueda, 2011).

2.3 Factors affecting biosorption

The investigation of the efficacy of metal uptake by microbial biomass is essential for the industrial application of biosorption, as it yields information about the equilibrium of the process that is necessary to design the equipment to be employed.

Metal uptake is usually measured by the parameter 'q', which indicates the milligrams of metal accumulated per gram of biosorbent material, while 'qH' is reported as a function of the metal accumulated, the sorbent material used and the operating conditions.

The following factors affect the biosorption process:

1) temperature does not seem to influence biosorption performance in the 20-35 °C range (Aksu *et. al.*, 1992); 2) pH seems to be the most important parameter in the biosorption process: it affects the solution chemistry of the metals, the activity of the functional groups in the biomass and competition between metallic ions (Machado *et. al.*, 2010); 3) biomass concentration in solution seems to influence specific uptake: for lower values of biomass concentrations, there is an increase in the specific uptake (Fourest & Roux, 1992; Gadd *et. al.*, 1988). Gadd *et. al.* (1988) suggested that an increase in biomass concentration leads to interference among the binding sites. Fourest & Roux (1992) invalidated this hypothesis by attributing the responsibility of the decrease of specific uptake to a shortage of metal concentration in the solution. Hence this factor needs to be taken into consideration in any application of microbial biomass as a biosorbent; 4) biosorption is mainly used to treat wastewater where more than one type of metal ion are probably present; the removal of one metal ion may be influenced by the presence of other metal ions. For example: Uranium uptake by bacterium, fungus and yeast biomass was not affected by the presence of manganese, cobalt, copper, cadmium, mercury and lead in solution (Sakaguchi & Nakajima, 1991). In contrast, the presence of Fe^{2+} and Zn^{2+} was found to influence uranium uptake by *Rhizopus arrhizus* (Tsezos & Volesky, 1982) and cobalt uptake by different microorganisms seemed to be completely inhibited by the presence of uranium, lead, mercury and copper (Sakaguchi & Nakajima, 1991).

3. New developments in organisms capable of enhanced bioremediation

3.1 The use of recombinant bacteria for metal removal

The use of recombinant bacteria to remove specific metals from contaminated water is currently being investigated. For example a genetically engineered *E.coli*, which expresses the Hg^{2+} transport system and metallothionein (a metal binding protein), was able to selectively accumulate 8 μmole Hg^{2+}/g cell dry weight. The presence of the chelating agents Na^+, Mg^{2+} and Ca^{2+} did not affect bioaccumulation.

3.2 Genetically modified biosorbents

Genetic engineering has the potential to improve or redesign microorganisms, where biological metal-sequestering systems will have a higher intrinsic capability as well as specificity and greater resistance to environmental conditions (Bae *et. al.*, 2000; Majare & Bulow, 2001). It is well known that virgin biosorbents usually lack specificity in metal-binding, which may cause difficulties in the recovery and recycling of the desired metal(s). Genetic modification is a potential solution for enhancing selectivity as well as the accumulating potential of cells (Pazirandeh *et. al.*, 1995). Genetic modification would be feasible, especially when the microbial biomass is produced from fermentation processes where genetically engineered microorganisms are used. Currently, many kinds of amino acids and nucleic acids are being produced on an industrial scale by using genetically engineered microbial cells. Higher organisms respond to the presence of metals, with the production of cysteine-rich peptides, such as glutathione (GSH) (Singhal *et. al.*, 1997), phytochelatins (PCs) and metallothioneins (MTs) (Mehra & Winge, 1991), which can bind and sequester metal ions in biologically inactive forms (Hamer, 1986; Bae *et. al.*, 2000). The overexpression of MTs in bacterial cells will result in enhanced metal accumulation, thus offering a promising strategy for the development of microbial-based biosorbents for the remediation of metal contamination (Pazirandeh *et. al.*, 1995). In addition to the high selectivity and accumulation capacity, Pazirandeh *et. al.* (1995) demonstrated that uptake by recombinant *E. coli* (expressing the *Neurospora crassa* metallothionein gene within the periplasmic space) was rapid. Greater than 75% Cd uptake occurred within the first 20 min, with maximum uptake achieved in less than 1 h. However, the expression of such cysteine-rich proteins is not devoid of problems, due to the predicted interference with redox pathways in cytosol. More importantly, the intracellular expression of MTs may prevent the recycling of biosorbents, as the accumulated metals cannot easily be released (Gadd & White, 1993). Chen & Georgiou (2002) suggested a solution to bypass this transport problem by expressing MTs on the cell surface. Sousa *et. al.* (1996) demonstrated the possibility of inserting MTs into permissive site 153 of the *LamB* sequence. The expression of the hybrid proteins on the cell surface dramatically increased the whole-cell accumulation of cadmium. Also, the expression of proteins on the surface offers an inexpensive alternative for the preparation of affinity adsorbents (Georgiou *et. al.*, 1993). The use of PCs in a similar manner to MTs has also been suggested (Bae *et. al.*, 2000). PCs are short, cysteine-rich peptides, with the general structure (γGlu-Cys)nGly (n=2–11) (Zenk, 1996). PCs offer many advantages over MTs, due to their unique structural characteristics, particularly the continuously-repeating γGlu-Cys units. Also, PCs have been found to exhibit higher metal-binding capacity (on a per cysteine basis) than MTs (Mehra & Mulchandani, 1995). However, the development of overexpressing PC organisms requires a thorough knowledge of the mechanisms involved in the synthesis and chain elongation of these peptides. Several biosorbents, displaying metal-binding peptides on the cell surface, have been successfully engineered. A typical example includes creating a repetitive metal-binding motif, consisting of (Glu-Cys)nGly (Bae *et. al.*, 2000). These peptides emulate the structure of PCs; however, they differ in the fact that the peptide bond between the glutamic acid and cysteine is a standard α peptide bond. Phytochelatin analogs were found to be present on the bacterial surface, which enhanced the accumulation of Cd^{2+} and Hg^{2+} by 12- (Bae *et. al.*, 2000) and 20-fold (Bae *et. al.*, 2001), respectively. Attempts to create recombinant bacteria with improved metal binding capacity have so far been restricted mostly to *E. coli*. This is because *E. coli* greatly facilitates genetic engineering experiments and it is found to have more surface area per unit of cell mass, which potentially should result in higher rates of metal removal from solutions (Chen & Wilson, 1997). Nevertheless, a Gram-positive surface display system also possesses its own merits, compared to Gram-negative bacteria (Malik *et. al.*, 1998; Samuelson *et. al.*, 2000): (a) translocation through only one membrane is required; and (b) Gram-positive bacteria have been shown to be more rigid and, therefore, less sensitive to shear forces (Kelemen & Sharpe, 1979) due to the thick cell wall surrounding the cells, which potentially make them more suitable for field applications, such as bioadsorption. Samuelson *et. al.* (2000) generated recombinant *Staphylococcus xylosus* and *Staphylococcus carnosus* strains, with surface-exposed chimeric proteins containing polyhistidyl peptides.

Both strains of Staphylococcaceae gained improved nickel-binding capacities due to the introduction of the H1 or H2 peptide into their surface proteins. Owing to their high rate of selectivity, genetically engineered biosorbents may prove very competitive in the separation of toxins and other pollutants from diluted contaminated solutions.

3.3 Survivability and stability of GMOs

Although the utilization of GMOs in the field has been limited due to possible risks involved in the horizontal transfer of genetic material, the results that have been obtained are nevertheless important in assessing the benefits and obstacles associated with their applications in bioremediation. Such knowledge is necessary in view of the future possibility of releasing GEMs into contained environments for bioremediation. To be of practical use in the field, a bacterial GMO must be able to survive and grow in such environments. Important parameters in this regard are growth rate, inoculum size, environmental conditions, including spatial distribution, and the presence of competing microorganisms. The spatial distribution of a GMO introduced into the environment is important because it helps define its interactions with the members of the indigenous bacterial community and other components of the ecosystem (Dechesne *et. al.*, 2005). In general, a bacterium that has been recently isolated from a natural environment is more likely to survive when released back into that same environment. A crucial consideration regarding the introduction of engineered bacteria into field sites is their effect on the structure and function of natural ecosystems.

3.4 Natural horizontal transference of DNA in bacteria

It is perhaps most useful to consider horizontal transfer of recombinant DNA in the overall context of horizontal gene transfer among bacteria, which is a natural and presumably widespread phenomenon. The role of horizontal gene transfer in bacterial evolution has been demonstrated in many studies (Rensing *et. al.*, 2002; Dennis, 2005). It has been suggested that it is an important process contributing to the development of novel biodegradation capacities of microbial communities when they are exposed to organic pollutants (Rittmann *et. al.*, 1990; Dennis, 2005). The transfer of genes encoding biodegradation functions appears to occur through the action of conjugative plasmids, transposable elements, and "integrative and conjugative transposons" (also known as "genomic islands") (Springael & Top, 2004; Top *et. al.*, 2002; Van Der Meer & Sentchilo, 2003). There is evidence to suggest that the genes in at least some of these elements were assembled in stepwise processes (Springael & Top, 2004). That such horizontal transfer is apparently quite common is suggested by the variety of specific examples. Direct measurement of horizontal transfer has been carried out under both well-defined conditions and in microcosms, the latter serving as models for *in situ* situations. An interesting example of horizontal gene transfer in a completely natural environment is provided by the *pheBA* operon, which originated from the strain EST1001 of *Pseudomonas* sp. It encodes two enzymes involved in phenol catabolism and, like several examples described above, it is carried on a conjugative plasmid with transposable element characteristics. It was transferred by conjugation to *P. putida* PaW85 and this strain has already been released into the field for the large-scale bioremediation of river water contaminated with phenolics, originating from a fire in an oil shale mine (Peters *et. al.*, 1997). Six years later, despite the absence of the PaW85 strain, the operon was once again detected in the watershed, apparently having been transferred to nine *Pseudomonas* strains belonging to four different species (*P. corrugata*, *P. fragi*, *P. stutzeri* and *P. fluorescens* biotypes B, C and F). In eight cases the operon was plasmid-borne, and in one case it had integrated into the host chromosome. The environment in which these bacteria were found was subject to continual pollution with phenolics. This presumably provided positive selection for their perseverance, which appears to have had continual beneficial effects regarding bioremediation in this location. The examples discussed above can be viewed as evidence that horizontal gene transfer of recombinant genes could occur rapidly by the same mechanisms. On the other hand, it may be that any predicted horizontal transfer of DNA from GMOs may occur naturally with the same genes in non-recombinant organisms at rates which make any contribution from GMOs insignificant. As discussed above, there is concern that GMOs introduced into polluted sites to enhance bioremediation may have adverse environmental effects because of horizontal transfer of recombinant DNA (Davison, 1999). In many cases, released GMOs do not survive long, and disappear before they have any effect on biodegradation (Davison, 1999). In other cases, however, transfer of plasmid from an introduced GMO to an indigenous microorganism may occur even though the introduced strain does not survive (Davison, 1999; Peters *et. al.*, 1997). Thus, in general it appears that the potential impact of introduction of GMOs on native microbial populations is not uniform and therefore must be evaluated on a case-by-case basis.

3.5 Containment strategies to diminish horizontal transfer of DNA from GMOs

As mentioned above, one important concern with the use of GMOs bearing recombinant genes on plasmids arises from the instability of the plasmids, which can lead to loss of the desired phenotype. On the other hand, the relative chemical and physical stability of plasmids can contribute to their spreading, along with both the recombinant genes and selective markers (e.g., antibiotic resistance), to other bacteria (Pieper & Reineke, 2000). One solution is to replace antibiotic resistance selective markers with selective markers that do not center on antibiotic resistance (Sanchez-Romero *et. al.*, 1998; Herrero *et. al.*, 1990). One way to obviate the problem of plasmid transfer is to use "mini transposons" for the stable integration of genes into the chromosome of recipient strains (De Lorenzo *et. al.*, 1998; Herrero *et. al.*, 1990). If the chosen GMO recipient proves not to be able to support plasmid maintenance, the transformant bearing gene(s) of interest must undergo transposition of the cloned gene into the host chromosome, where it will remain integrated in a stable manner. Some mini-transposon vectors have also been engineered for use on heavy metals or herbicides, rather than antibiotic resistance, as a selectable marker, to eliminate the problem of horizontal transfer of this characteristic altogether (Herrero *et. al.*, 1990). Other "biological containment strategies" that can minimize this problem are suicide systems, where the GMO dies after it has completed its required task (De Lorenzo *et. al.*, 1998; Molin *et. al.*, 1993; Molina *et. al.*, 1998; Pandey *et. al.*, 2005; Paul *et. al.*, 2005; Ronchel *et. al.*, 1998; Ronchel & Ramos, 2000). Such suicide mechanisms are based on the controlled expression in the gene host that encodes proteins that are lethal to it. A common system induces the "suicide gene(s)" when the pollutant that the host degrades is absent. Expression of these genes, which cause holes to form in the cell membrane, leads to cell death.

Composting is another containment strategy for using GMOs in field applications; this process has recently been reviewed by Singh *et. al.* (2006). The conditions during the composting process include elevated temperatures (as high as 80–90°C), decreases in pH due to organic acid production, and the production of toxic metabolites that can greatly decrease microbial populations. The subsequent lysis of dead microbes releases their DNA into the environment, where it is subject to degradation. These effects presumably minimize horizontal gene transfer from the GMOs, suggesting that composting could be a safe solution for their disposal after they have completed their required functions.

4. Conclusion

The use of genetic engineering to produce microorganisms capable of degrading specific contaminants or to enhance such processes in native organisms with such capabilities has become a popular way of increasing the efficiency of bioremediation in laboratory studies. Techniques used can include engineering with single genes, pathway construction, and alteration of the sequences of existing genes (both coding and controlling sequences). But, before releasing a GMO into the environment, the researchers should emphasize the ethical responsibilities to be considered before using such novel strategies for bioremediation. In this context, several points should be taken into account. The stability of GMOs and their horizontal transfer of engineered DNA are crucial issues regarding the potential impact of their release into the field for bioremediation. In order to determine how released GMOs are affecting the environment, it is necessary to be able to detect and enumerate them in complex samples. Important parameters in this context are survival, number, activity, and dispersion of released GMOs (Widada *et. al.*, 2002). Ideally such methods should be applicable in the field and in real time, and should be simple and inexpensive while also being accurate (Wu *et. al.*, 2001). In addition to the GMO itself, it is useful to track the recombinant DNA with which the GMO has been engineered so as to monitor potential loss of these genes and their possible horizontal transfer to other microorganisms.

In conclusion, a number of important molecular tools have been developed for genetic and metabolic engineering of microorganisms for the degradation of environmental contaminants. These new tools will make the construction of new or improved strains much easier and quicker than in the past. However, these genetic modifications should beunderstood in full and any research must always determine the actual risks and benefits involved.

5. Acknowledgments

We thank Ingrid Regina Avanzi and Louise Hase Gracioso for helping with the references and Stanley Vasconcelos for drawing the chemical diagrams. INCT-EMA-CNPq (Brasília, DF, Brazil) and FAPESP (São Paulo, SP, Brazil) provided financial support for this study.

6. References

Adamis PD, Gomes DS, Pinto ML, Panek AD, Eleutherio EC (2004). The role of glutathione transferases in cadmium stress. Toxicol. Lett., Vol.154, n.1, pp. 281-288.

Adamis PDB, Panek AD, Leite SGF, Eleutherio ECA (2003) Factors involved with cadmium absorption by a wild-type strain of Saccharomyces cerevisiae. J. Microbiol., Vol. 34, pp. 55-60.

Ahluwalia SS & Goyal D (2007). Microbial and plant derived biomass for removal of heavy metals from wastewater. Bioresour. Technol. Vol.98, n.12, pp. 2243-2257.

Asku Z (1992). The biosorption of Cu (II) by C. vulgaris and Z. ramigera. Environ. Tech. Vol. 13, No1, pp.579-586.

Bae W, Chen W, Mulchandani A, Mehra RK (2000). Enhanced bioaccumulation of heavy metals by bacterial cells displaying synthetic phytochelatins. Biotechnol Bioeng. Vol. 70, pp.518–24.

Bae W, Chen W, Mulchandani A, Mehra RK (2001). Genetic engineering of Escherichia coli for enhance uptake and bioaccumulation of mercury. Appl. Environ Microbiol., Vol.67, pp. 5335-5338.

Bae W, Mehra RK (1997). Metal-Binding characteristics of a phytochelatin analog (Glu-Cys)2Gly. Elsevier Sci., pp. 201-210.

Bogacka E K (2011). Surface properties of yeast cells during heavy metal biosorption[+]. Cent. Eur. J. Chem., Vol. 9, n.2, pp. 348-351.

Bognolo G (1999) Biosurfactants as emulsifying agents for hydrocarbons Colloids and surfaces A: Physicochemical and Engineering Aspects Vol.152, pp. 41–52.

Borro, DM, Fein JB (2005). The impact of ionic strength on the adsorption of protons, Pb, Cd, and Sr onto the surfaces of gram negative bacteria: testing non-eletrostatic, diffuse, and triple-layer models. J. Colloid Interf. Sci., Vol. 286, pp. 110-126.

Breierová E, Vajcziková V, Sasinková V, Styratilová E, Fisera M, Gregor T, Sajbidor J (2002). Biosorption of cadmium ions by different yeast species. Inst. Chem. Z. Naturforch, Vol. 57, pp. 634-639.

Bruins MR, Kapil S, Oehme FW (2000). Microbial resistance to metals in the environment. Ecotoxicol and Environ Safety 45: 198-207.

Champion, J. T, Gilkey, C.J, Lamparski, H, Retterer, J, Miller, R.M. (1995) Electrom microscopy of rhamnolipid (biosurfactant) morfphology: Effects of pH, cadmium, and octadecane. J. Colloid and interface science, Vol. 170, pp. 569-574.

Chang L, Meier J, Smikth M (1997). Application of plant and earthworm bioassays to evaluate remediation of a lead-contaminated soil. Arch. Environ. Contam. Toxicol, v.32, pp.166-171.

Chen C, Wang J (2008). Removal of Pb2+, Ag +, Cs+, and Sr 2+ from aqueous solution by brewery s waste biomass. J. Hazard. Mater., Vol. 151, pp. 65-70.

Chen S & Wilson DB (1997). Genetic engineering of bacteria and their potential for Hg^{2+} bioremediation. Biodegradation. Vol.8, n.2, pp. 97–103.

Chen W, Bruhlmann F, Richnis RD, Mulchandani A (1999). Engineering of improved microbes and enzymes for bioremediation. Curr Opin Biotechnol, Vol. 10, pp.137–141.

Chen W, Georgiou G (2002). Cell-surface display of heterologous proteins: from high-throughput screening to environmental applications. Biotechnol Bioeng., Vol.79, pp.496–503.

Cobbett C, Goldsbrough P (2002). Phytochelatins and Metallothioneins: Roles in heavy metal detoxification and homeostasis. Annu. ReVol. Plant Biol., Vol. 53, pp. 159-82.

Cobbett CS (2000). Phytochelatins and their roles in heavy metal detoxification. Plant Physiol., Vol. 123, pp. 825-832, 2000.

Costa ACA & Leite SGF (1991). Metals biosorption by sodium alginate immobilized Chiarella homosphaera cells. Biotechnol Left, Vol.13, pp.555-562.

Davis TA, Volesky B, Mucci AA (2003). Review of the biochemistry of heavy metalbiosorption by brown algae, Water Res., Vol. 37, n. 18, pp. 4311- 4330.

Dejonghe W, Goris J, El Fantroussi S, Hofte M, De Vos P, Verstraete W & Top EM (2000). Effect of dissemination of 2,4 –D degradation plasmids on 2,4-D degradation and on bacterial community structure in two different soil horizons. Appl Environ Microbiol., Vol. 66, pp. 3297-3304.

Deng X, Li QB, Lu YH, Sun DH, Huang YL, Chen XR (2003). Bioaccumulation of nickel from aqueous solutions by genetically engineered E. coli. Water Res., Vol. 37, pp. 2505-2511.

Dias, MA, Lacerda, ICA, Pimentel, PF, Castro HF, Rosa CA (2002) Removal of heavy metals by an Aspergillus terreus strain immobilized in a polyurethane matrix. Lett Appl. Microbiol., Vol. 34, pp. 46-50.

Dobson RS, Burgess, JE (2007). Biological treatment of precious metal refinery wastewater: A review. Miner. Eng. Vol. 20, pp. 519-532.

Ercole C, Veglio F, Toro, L, Ficara G and Lepidi A. (1994). Immobilisation of microbial cells for metal adsorption and desorption. In: Mineral Bioprocessing II. Snowboard. Utah.

Esposito A, PAgnanelli FE, Veglio F (2002). pH-related equilibria models for biosorption in single metal systems. Chem. Eng. Sci., Vol. 57, n. 3, pp. 307-313.

Fourest E & Roux JC (1992). Heavy metal adsorption by fungal mycelial by-products: influence of pH. Appl. Microbiol. Biotechnol, v.37, pp. 399-403.

Gadd GM and White C (1993). Microbial treatment of metal pollution: a working biotechnology? TIBTECH. Vol.11, pp.353–9.

Gadd G M (1988). Heavy metal and radionuclide by fungi and yeasts. In: P.R. Norris and D.P. Kelly (Editors), Biohydrometallurgy. A. Rowe, Chippenham, Wilts , U.K.

Gadd GM (2004). Microbial influence on metal mobility and application for bioremediation. Geoderma, Vol. 122, n. 2-4, pp. 109-119.

Gadd GM & De Rome L (1988). Biosorption of copper by fungal melanine. Appl. Microbiol. Biotech. Vol. 29 N.6, pp. 610–617.

Gavrilescu M (2004). Removal of heavy metals from the environment by biosorption, Eng. Life Sci., Vol. 4, n. 3, pp. 219-232.

Gazsó LG (2001). The key microbial processes in the removal of toxic metals and radionuclides from the environment. Mini ReVol. CEJOEM, Vol.7, pp. 178-185.

Gentry TJ, Rensing C, Pepper IL (2004). New approaches for bioaugmentation as a remediation technology. Crit. Rev. Environ Sci Techn., Vol. 34, n. 5, pp. 447-494.

Georgiou G, Poetschke HL, Stathopoulos C, Francisco JA (1993). Practical applications of engineering Gram-negative bacterial cell surfaces. TIBTECH. Vol.11, pp. 6–10.

Ghosh SK, Chaudhuri R, Gachhui R, Mandal A, Ghosh S (2006). Effect of mercury and organomercurials on cellular glucose utilization: a study using resting mercury-resistant yeast cells. J. Appl. Microbiol., Vol. 102, pp. 375-383.

Goyal N, Jain SC, Banerjee UC (2003). Comparative studies on the microbial adsorption of heavy metals. AdVol. Environ. Res., Vol. 7, n. 2, pp. 311-319.

Grill E, Winnacker EL, Zenk MH (1985). Phytochelatins: the principal heavy-metal complexing peptides of higher plants. Science, Vol.230, pp. 674-676.

Gupta DK, Tohoyama H, Joho M, Inouhe M (2005). Changes in the levels of phytochelatins and related metal-binding peptides in chickpea seedlings exposed to arsenic and different heavy metal ions. J. Plant Res., Vol. 17, pp. 253-256.

Gupta R, Ahuja P, Khan S, Saxena RK, Mohapatra H (2000) Microbial biosorbents: Meeting challenges of heavy metal pollution in aqueous solutions. Curr. Sci., Vol.78, n.8, pp. 967-973.

Hameed MSA (2006). Continuous removal and recovery of lead by alginate beads free and alginate-immobilized Chlorella vulgaris. Afr. J. Biotechnol., Vol. 5, n. 19, pp. 1819-1823.

Hamer DH. (1986). Metallothionein. Annu Rev Biochem. Vol.55, pp. 913–51.

Hammaini A (2003). Simultaneous uptake of metals by activated sludge. Minerals Engineering. Vol. 16, pp. 723-729.

Hernandez A, Mellado RP, Martinez JL, (1998) Metal Accumulation and Vanadium-Induced Multidrug Resistance by Environmental Isolates of Escherichia hermannii and Enterobacter cloacae. American. Society for Microbiology Applied and Environmental Microbiology. Vol. 64, n. 11, pp. 4317–4320.

Hirata K, Tsuji N, Miyamoto K (2005). Biosynthetic regulation of phytochelatins, heavy metal-binding peptides. J. Biosc. Bioeng. ReVol., Vol. 100, n. 6, pp. 593-599.

Huang C, Huang C, Morehart AL (1990). The removal of copper from dilute aqueous solutions by Saccharomyces cerevisiae. Water Res., v.24, pp.433– 439.

Inouhe M (2005). Phytochelatins. Braz. J. Plant Physiol., Vol. 17, n. 1, pp. 650-678.

Jiang, ZB, Song HT, Gupta N, Ma LX, Wu ZB (2007). Cell surface display of functionally active lipases from Yarrowia lipolytica in Pichia pastoris. Protein Expr. Purif., Vol. 56, pp. 35-39.

Kawata K, Yokoo H, Shimazaki R, Okabe S (2007). Classification of heavy-metal toxicity by human DNA microarray analysis. Environ Sci Technol., Vol. 15, n. 41, pp. 3769-377.

Kefala MI, Zouboulis AIE, Matis KA (1999). Biosorption of cadmium ions by Actinomycetes and separation by flotation. Environ. Pollut., Vol.104, n. 2, pp. 283-293.

Kelemen MV, Sharpe JE (1979). Controlled cell disruption: a comparison of the forces required to disrupt different micro-organisms. J Cell Sci., Vol. 35, pp. 431–441.

Keskinkan O, Goksu MZL, Yuceer A, Basibuyuk M, Forster C.F (2003). Heavy metal adsorption characteristics of a submerged aquatic plant (Myriophyllum spicatum). Process Biochemistry, Vol.39, n.2, pp.179-183.

Kim SU, Cheong YH, Seo DC, Hu JS, Heo JS, Cho JS (2007). Characterisation of heavy metal tolerance and biosorption capacity of bacterium strain CPB4 (Bacillus spp.). Water Sci. Tech., Vol. 55, n. 1-2, pp. 105-111.

Kim, SK, Lee, BS, Wilson, DB, Kim, EK (2005). Selective cadmium accumulation using recombinant E. coli. J. Biosci. Bioeng., Vol. 99, n. 2, pp. 109-114.

Kobayashi I, Fujiwara S, Saegusa H, Inohe M, Metsumoto H, Tsuzuki M (2006). Relief of arsenate toxicity by Cd-stimulated phytochelatin synthesis in the green alga Chlamydomonas reinhardtii. Mar. Biotechnol., Vol. 8, pp. 94-101.

Kondo A, Ueda M (2004). Yeast cell-surface display-applications of molecular display. Appl. Microbiol. Biotechnol., Vol. 64, pp. 28-40.

Kotrba P, Ruml T (2000). Bioremediation of heavy metal pollution exploiting constituents, metabolites and metabolic pathway of livings. Appl. Environ. Microbiol., Vol. 65, n. 8, pp. 1205-1247.

Kratochvil D & Volesky B (1998). Advances in the biosorption of heavy metals. Trends Biotechnol., Vol. 16, pp. 291-300.

Kujan P, Prell A, Safar H, Sobotka M, Rezanka T, Holler PP. (2006). Use of the industrial yeast Candida utilis for cadmium sorption. Folia Microbiol., Vol. 51, pp. 257 -260.

Kumar R, Bishnoi NR, Bishnoi GK (2008). Biosorption of chromium (VI) from aqueous solution and electroplating wastewater using fungal biomass. J. Eng. Chem., Vol.135, n. 3, pp. 202-208.

Kuroda K, Shibasaki S, Ueda M, Tanaka A (2001). Cell surface engineered yeast displaying a histidine oligopeptide (hexa-His) has enhanced adsorption of and tolerance to heavy metal ions. Appl. Microbiol. Biotechnol., Vol. 57, pp. 697-701.

Kuroda K, Ueda M, Shibasaki A, Tanaka A (2002). Cell surface-engineered yeast with ability to bind, and self-aggregate in response tom copper ion. Appl. Microbiol. Biotechnol., Vol. 59, pp. 259-264.

Kuroda K, Ueda M (2003). Bioadsorption of cadmium ion by cell surface-engineered yeasts displaying metallothionein and hexa-His. Appl. Microbiol. Biotechnol., Vol. 63, pp. 182-186.

Kuroda K, Ueda M (2010). Engineering of microorganisms towards recovery of rare metal ions: Mini Review. Appl Microbiol Biotechnol., Vol. 87, PP. 53-60.

Kuroda K, Ueda M (2011). Yeast Biosorption and Recycling of Metal Ions by Cell Surface Engineering. Microbial Biosorption of Metals. n. 10, pp. 235-247, [DOI: 10.1007/978-94-007-0443/5-10].

Kuyucak N, Volesky B. Desorption of cobalt-laden algal biosorbent (1988). Biotechnol Bioeng., Vol. 33, pp. 815-822.

Lee W, Wood T, Chen W (2002). Engineering TCE-degrading Rhizobacteria for heavy metal accumutation and enhanced TCE degradation. Biotechnol. Bioeng., pp. 399-403. doi: 10.1002/bit. 20950.

Li Y, Li B (2011). Study on fungi-bacteria consortium bioremediation of petroleum contaminated mangrove sediments amended with mixed biosurfactants. Adv.Mat. Res., Vols. 183-185, pp. 1163-1167.

Liu J, Zhang Y, Huang D, Song G (2005). Cadmium induced MTs synthesis via oxidative stress in yeast Saccharomyces cerevisiae. Mol. Cell Biochem., Vol. 280, n. 1-2, pp. 139-145.

Machado MD, Santos MSF, Gouveia C, Soares HMVM, Soares, EV (2008). Removal of heavy metal using a brewer's yeast strain of Saccharomyces cerevisiae: The flocculation as a separation process. Bioresour. Technol., Vol. 99, pp. 2107-2115.

Machado MD, Soares EV, Helena MVM, Soares HMVM (2010). Removal of heavy metals using a brewer's yeast strain of Saccharomyces cerevisiae: Chemical speciation as a tool in the prediction and improving of treatment efficiency of real electroplating effluents. J. Hazard. Mat., Vol. 180, pp. 347-353.

Majáre M, Bülow L (2001). Metal-binding proteins and peptides in bioremediation and phytoremediation of heavy metals. TIBTECH. Vol.19, pp.67–73.

Malik, A. Metal bioremediation through growing cells. Enkviron. Int., Vol. 30, n. 2, pp. 261-278.

Malik P, Terry TD, Bellintani F, Perham RN (1998). Factors limiting display of foreign peptides on the major coat protein of filamentous bacteriophage capsids and a potential role for leader peptidase, FEBS Lett., Vol. 436, pp. 263–266.

Mameri, N., N. Boudries, L. Addour, D. Belhocine, H. Lounici, H. Grib and A. Pauss (1999). Batch zinc biosorption by a bacterial nonliving Streptomyces rimosus biomass. Water Res., Vol. 33, pp. 1347-1354.

Mannazzu I, Guerra E, Ferretti R, Pediconi D, Fatichenti F (2000). Vanadate and copper induce overlapping oxidative stress responses in the vanadate-tolerant yeast Hansenula polymorpha. Biochim. Biophys. Acta, Vol. 1475, pp. 151-156.

Mehra RK & Mulchandani P (1995). Glutathione-mediated transferof Cu(I) into phytochelatins. Biochem. J., Vol. 307, pp. 697-705.

Mehra RK,Winge DR (1991). Metal ion resistance in fungi-molecular mechanisms and their regulated expression. J Cell Biochem., Vol. 45, pp. 30–40.

Mendoza-Cózatl D, Tavera HL,Navarro, AH, Sánchez RM (2005). Sulfur assimilation and glutathione metabolism under cadmium stress in yeast, protists and plants. FEMS Microbiol., Vol. 29, n. 4, pp. 653-671.

Merle SS, Cuiné S, Carrier P, Pradines CL, Luu DT, Peltier G (2003). Enhanced toxic metal accumulation in engineered bacterial cells expressing Arabidopsis thaliana phytochelatin synthase. Appl. Environ. Microbiol., Vol. 69, n. 1, pp. 490-494.

Mulligan CN (2005). Environmental applications for biosurfactants. Environ. Pollut., Vol. 133, n. 2, pp. 183-198.

Murai T, Ueda M, Yamamura M, Atomi H, Shibasaki Y, Kamasawa N, Osumi M, Amachi T, Tanaka A (1997). Construction of a starch-utilizing yeast by cell surface engineering. Appl. Environ. Microbiol., Vol. 63, n. 4, pp. 1362-1366.

Muraleedharan TR & Venkobachar C (1990). Mechanism of biosorption of copper (II) by Ganoderma lucidum. Biotechnol. Bioeng., Vol. 35, pp. 320–325

Naeem A, Woertz JR, Fein JB (2006). Experimental measurement of proton, Cd, Pb, Sr, and Zn adsorption onto the fungal species Saccharomyces cerevisiae. Environ. Sci. Technol., Vol. 40, n.18, pp. 5724-5729.

Nakajima A & Tsuruta T (2004). Competitive biosorption of thorium and uranium by Micrococcus luteus. J. Radioanal Nucl. Chem., Vol. 260, pp. 13–8.

Nishitani T, Shimada M, Kuroda K, Ueda M (2010). Molecular design of yeast cell surface for adsorption and recovery of molybdenum, one of rare metals. Appl Microbiol Biotechnol., Vol. 86, pp. 641–648.

Norton L, Baskaran K, McKenzie T (2004). Biosorption of zinc from aqueous solutions using biosolids. Adv. Environ. Res., Vol. 8, n. 3-4, pp.629-635.

Nourbakhsh, M., Y. Sag, D. Ozer, Z. Aksu, T. Kustal and A. Caglar (1994). A comparative study of various biosorbents for removal of chromium (VI) ions from industrial waste waters. Process Biochem., Vol. 29, pp. 1-5.

Pagnanelli F, Esposito A, Vegliò F (2002). Multi-metallic modelling for biosorption of binary systems. Water Research, Vol. 36, pp. 4095–105.

Pagnanelli F, Toro L & Veglio F (2002). Olive mild solid residues as heavy metal sorbent material: a preliminary study. Waste Management, Vol. 22, pp.901-907.

Pattus F & Abdallah M (2000). Siderophores and iron-transport in microorganisms: Review. J. Chin. Chem. Soc., Vol. 47, pp. 1-20.

Paul D, Pandey G, Pandey J& Jain R K (2005). Acessing microbial diversity for bioremediation and environmental restoration. Trends in Biotechnol, Vol. 23, n.3, pp. 135-142.

Pazirandeh M, Chrisey LA, Mauro JM, Campbell JR, Gaber BP (1995). Expression of the Neurospora crassa metallothionein gene in Escherichia coli and its effects on heavy-metal uptake. Appl Microbiol Biotechnol., Vol.43, pp.1112–1117.

Penninckx MA (2000). A short review on the role of glutathione in the response of yeasts to nutritional, environment, and oxidative stresses. Enzyme Microbiol. Technol., Vol. 26, pp. 737-742.

Penninckx MJ (2002). An overview on glutathione in Saccharomyces versus non-conventional yeasts. FEMS Yeast Res., Vol. 2, pp. 295-305.

Pieper DH & Reineke W (2000). Engineering bacteria for bioremediation. Curr. Opin. Biotechnol., Vol. 11, pp. 262–270.

Preveral S, Ansoborlo E, Mari S, Vavasseur A, Forestier C (2006). Metal(loid)s and radionuclides cytotoxicity in Saccharomyces cerevisiae. Role of YCF1, glutathione and effect of buthionine sulfoximine. Biochimie, Vol. 88, n. 11, pp. 1651-1663.

Renninger, N, McMahon, KD, Knop R, Nitsche H, Clark DS, Keasling JD (2001). Uranyl precipitation by biomass from an enhanced biological phosphorus removal reactor. Biodegradation, Vol. 12, pp. 401-410.

Ríos VG, Pelegrín YF, Robledo D, Mendoza-Cozáti D, Sánchez RM, Bouchot GG (2007). Cell wall composition effects cadmium accumulation and intracellular thiol peptides in marine red algae. Aquat. Toxicol., Vol. 81, n. 1, pp. 65-72.

Rittmann BE, Seagren E, Wrenn BA, Valocchi AJ, Ray C & Raskin L (1994). In situ bioremediation. Park Ridge, NJ, Noyes Publications.

Roane TM, Pepper IL (2001). Environmental Microbiolology In: ROANE, T. M, PEPPER, I. L. (Ed.). Microorganisms and Metal Pollutants. Academic Press, Vol.17, pp.403-423.

Rutherford N, Mourez M (2006). Surface display of proteins by Gram-negative bacterial autotransporters. Microbiol. Cell Factores, Vol. 5, n. 22.

Sakaguchi, T & Nakajima, A (1991). Accumulation of heavy metals such as uranium and thorium by microorganisms. Smith RW, Misra M. (Eds.), Mineral Bioprocessing. The Minerals, Metals and Materials Society.

Samuelson H, Wernérus M, Svedberg , Ståhl S (2000). Staphylococcal surface display of metal-binding polyhistidyl peptides, Appl Environ Microbiol., Vol.66, pp. 1243–1248.

Samuelson P, Gunneriusson E, Nygren PPA, Stahl S (2002). Display of proteins on bacteria. J. Biotechnol., Vol. 96, n. 2, pp. 129-154.

Sandaa RA, Enger O, Torsvik V (1999). Abundance and Diversity of Archaea in Heavy-Metal-Contaminated Soils. Appl. Environ. Microbiol., Vol. 65, n. 8, pp. 3293-3297.

Sayler GS & Ripp S (2000) Field application of genetically engineered microorganisms for bioremediation processes. Curr. Opin. Biotechnol., Vol. 11, pp. 286–289.

Schmitt M, Schwanewilm P, Ludwig J, Fraté LH (2006). Use of PMA1 as a housekeeping Biomarker for Assessment of Toxicant- induced Stress in Saccharomyces cerevisiae. Appl. Environ. Microbiol., Vol. 72, n. 2, pp. 1515-1522.

Selatnia A, Boukazoula A, Kechid BN, Bakhti MZ, Chergui A, Kerchich Y (2004). Biosorption of lead (II) from aqueous solution by a bacterial dead Streptomyces rimosus biomass. J. Eng. Biochem., Vol. 19, n.2, pp.127-135.

Silva E, Fialho AM, Sá-Correia I, Burns RG, Shaw LJ (2004). Combined bioaugmentation and biostimulation to cleanup soil contamined with high concentrations of atrazine. Environ. Sci. Technol., Vol. 15-38, n. 2, pp. 632-637.

Singhal RK, Andersen ME, Meister A (1997). Glutathione, a first line of defense against cadmium toxicity. FASEB J. Vol.1, pp. 220–223.

Smets BF, Rittmann BE, Stahl DA (1990). Genetic capabilities of biological processes, Part II. Environ. Sci. Technol., Vol. 24, pp.162-170.

Smith MC, Sumner ER, Avery SV (2007). Gluthatione and Gts1p drive beneficial variability in the cadmium resistances of individual yeast cells. Mol. Microbiol., Vol. 66, n. 3, pp. 699-712.

Soares EV, Hebbelinck K, Soares HM (2003). Toxic effects caused by heavy metals in the yeast Saccharomyces cerevisiae: a comparative study. J. Microbiol., Vol. 49, n. 5, pp. 336-343.

Sousa C, Cebolla A, De Lorenzo V (1996). Enhanced metalloadsorption of bacterial cells displaying poly-His peptides. Nat. Biotechnol., Vol. 14, pp. 1017–1020.

Sze, KF, Lu, YJ, Wong PK (1996). Removal and recovery of copper ion (Cu^{2+}) from electroplating effluent by a bioreactor containing magnetite-immobilized cells of Pseudomonas putida 5X. Resour. Conserv. Recy., Vol. 18, n. 1-4, pp. 175-193.

Tabak HH, Lens P, Hullebusch EDV, Dejonghe W (2005). Developments in bioremediation of soil and sediments polluted with metals and radionuclides–1. Microbiolal processes and mechanisms affecting bioremediation of metal contamination and influencing meal toxicity. Rev. Environ. Sci. Biotechnol., Vol. 4, pp. 115-156.

Thirumoorthy N, Kumar KTM, Sundar AS, Panayappan L, Chatterjee M (2007). Metallothionein: An overview. World J. Gastroenterol., Vol. 13, n. 7, pp. 993-996.

Timmis KN & Piper DH (1999). Bacteria designed for bioremediation. Trends Biotechnol., Vol. 17, pp. 201–204.

Travieso L, Cañizares RO, Borja R, Benítez F, Domínguez AR, Dupeyrón R, Valiente V (1999). Heavy metal removal by microalgas. B. Environ. Contam. Tox., Vol. 62, n. 2, pp. 144-151.

Tsezos, M & Volesky, B (1982). The mechanism of uranium biosorption by Rhizopus arrhizus. Biotech. Bioeng., Vol. 24, n.2, pp.385-401.

Tunali S, Çabuk A, Akar T (2006). Removal of lead and copper ions from aqueous solutions by bacterial strain isolated from soil. J. Eng. Chem., Vol. 115, pp. 203-211.

Uslu G & Tanyol M (2006) Equilibrium and thermodynamic parameters of single and binary mixture biosorption of lead(II) and copper(II) ions onto Pseudomonas putida: effect of temperature. J. Hazard. Mater., Vol. 135, pp. 87-93.

Valls M, Atrian S, Lorenzo V, Fernadéz LA (2000). Engineering a mouse metallothionein on the surface of Ralstonia eutropha CH34 for immobilization of heavy metals in soil. Nat. Biotechnol., Vol. 18, pp. 661-665.

Valls M, Lorenzo V (2002). Explointing the genetic and biochemical capacities of bacteria for the remediation of heavy metal pollution. FEMS Microbiol. Rev., Vol. 26, n. 4, pp. 327-338.

Veglio F, Beolchini F (1997). Removal of metals by biosorption: a review. Chem. Hydrometallurgy, Vol. 44, pp. 301-316.

Velkov VV (2001). Stress-induced evolution and the biosafety of genetically modified microorganisms released into the environment. J. Biosci., Vol. 26, pp. 667-683.

Vidali M (2001). Bioremediation. Overview. Pure Appl. Chem., Vol. 73, n. 7, pp. 1163-1172.

Volesky B (2001). Detoxification of metal-bearing effluents: biosorption for the next century. Hydrometallurgy, Vol. 59, n. 2-3, pp. 203-216.

Volesky B, Holan ZR (1995). Biosorption of heavy metal. Biotechnol. Prog., Vol. 11, pp. 235-250.

VoleskyB (1987). Biosorbents for metal recovery. Trends in Biotechnology, Vol. 5, n. 4, pp. 96-101.

Wang J & Chen C (2009). Biosorbents for heavy metals removal and their future. Biotechnol. Adv., Vol. 27, pp. 195-226.

Wang J, Chen C (2006). Biosorption of heavy metal by Saccharomyces cerevisiae. Biotechnol. Adv., Vol. 24, pp. 427-451.

Wang Q, Li L, Chen M, Qi Q, Wang PG (2007). Construction of a novel system for cell surface display of heterologous proteins on Pichia pastoris. J. Biotechnol. Lett., Vol. 29, n. 10, pp. 1561-1566.

Wu CH, Wood TK, Mulchandani A, Chen W (2006). Engineering Plant-microbe symbiosis for rhizoremediation of heavy metal. Appl. Environ. Microbiol., Vol. 72, n. 2, pp. 1129-1134.

Wu J & Yu HQ (2007). Biosorption of 2,4-dichlorophenol by immobilized white-rot fungus Phanerochaete chrysosporiyum from aqueous solutions. Bioresour. Technol., Vol. 98, n. 2, pp. 253-259.

Xu Z, Bae W, Nulchandani A, Mehra RK, Chen W (2002). Heavy Metal Removal by Novel CBD-EC20 Sorbents Immobilized on Cellulose. Biomacromolecules, Vol. 3, pp. 462-465.

Yu QZ, Kaewsarn P (2001). Desorption of Cu^{2+} from a biosorbent derived from the marine alga Durvillaea potatorum. Separ. Sci. Technol., Vol. 36, n. 7, pp. 1495-1507.

Zenk MH (1996). Heavy metal detoxification in higher plants: a review. Gene. Vol. 179, pp. 21-30.

Zhang GL, Wu YZ, Qian XP, Meng Q (2005). Biodegradation of crude oil by Pseudomonas aeruginosa in the presence of rhamnolopids. J. Zhejiang Univ. Sci., Vol. 6, n. 8, pp. 725-730.

Zouboulis AI, Matis KA, Lazaridis NK (2001). Removal of metal ions from simulated wastewater by Saccharomyces yeast biomass: Combining biosorption and flotation processes. Separ. Sci. Technol., Vol. 36, n. 3, pp. 349-365.

29

Construction and Characterization of Novel Chimeric β -Glucosidases with *Cellvibrio gilvus* (CG) and *Thermotoga maritima* (TM) by Overlapping PCR

Kim Jong Deog[1] and Hayashi Kiyoshi[2]
[1]Department of Biotechnology, Research Center on Anti-Obesity and Health Care (RCAOHC), Chonnam National University, Yosu
[2]National Food Research Institute, Tskuba
[1]Korea
[2]Japan

1. Introduction

It is known that β-glucosidase is very important for obtaining clean energy from cellulolytic materials such as plants and wood. For most bioconversion processes with cellulose, endo(1-4)-β-D-glucan glucanhydrolases and exo(1-4)-β-D-glucan cellobiohydrolases are necessary for catalyzing the random hydrolysis of cellulose to produce cellobiose. On the basis of substrate specificity, β-glucosidases can be classified into aryl-β-glucosidases, cellobioses hydrolysing only oligosaccharides, and those hydrolysing both aryl-β-glucosides and oligosaccharides [1, 2]. β-glucosidases are important components of the cellulase enzyme complex required for the hydrolysis of cellulose into glucose by catalysing the final step, which is the conversion of cellobiose into glucose [3]. From the basis of sequence homology β-glucosidases have been divided into two sub-families, namely BGA (β-gucosidases and phospho-β-glucosidases from bacteria to mammals) and BGB (β-glucosidases from yeast, mold and rumen bacteria) [4]. The study of these enzymes has been facilitated by the use of recombinant DNA technology [5,6,7,8]. A number of cellulase genes including several forms of β-glucosidases, have been cloned and expressed in both *E.coli* and *S.cereviciae* [9,10]. In recent years, protein engineering has become an increasingly important tool in the development of novel hybrid enzymes with useful catalytic functions [11,12,13]. The catalytic activities and thermal stabilities of enzymes can be improved by the construction of chimeric enzymes with gene-shuffling from different species of genes [14]. For this purpose, chimeric genes are normally constructed by the application of one of two methods: using restriction enzymes or by overlapping polymerase chain reaction (PCR) [15]. Although the use of restriction enzymes is the easier method, a high level of identity is required between the parental DNA sequences to obtain common restriction enzyme sites in both genes. Often, the availability of common restriction enzyme sites limits the use of this strategy in the construction of chimeric genes. In contrast, there are no such limitations in defining shuffling sites for chimeric genes constructed via overlapping PCR technique.

In order to construct new types of the chimeric β-glucosidases, *Thermotoga maritima* (TM) gene and *Celvibrio gilvus* (CG) gene were used for gene-shuffling. *Celvibrio gilvus* [16], a cellulose-metabolizing bacterium, has the unique property of producing cellobiose in high yields from acidic swollen cellulose. *Thermotoga maritima* is a fermentative, marine, hyperthermophilic eubacterium that can be grown in temperatures of up to 90°C with an optimal temperature of around 80°C [17,18]. Two novel active β -glucosidase were successfully constructed from the pool of chimeric genes and their properties were characterized. These results proved the chimeric gene construction is promising approach for searching for the better β-glucosidase

2. Materials and methods

Bacterial strain and plasmids
The plasmids pET-CG and pET-TM were constructed by introducing the CG gene and the TM gene for β-glucosidase into the pET28a (+) vector (Novagen, Germany) carrying an N-terminal His tag and a thrombin cleavage site. Topo-XL TOP 10 (Invitrogen, USA) was used for cloning of chimeric genes, and the pET28 (a) vector was used for introducing the Topo-XL-chimeric gene. The BL21 (DE3) (NextGen Sciences, Inc., USA) strain was used for expression of the chimeric protein produced through the pET28a (+) vector.

Overlapping PCR for chimeric gene
For the construction of 8 kinds of chimeric genes from CG and TM, three steps of PCR were applied. The first step was the amplification of gene fragments for domain-shuffling using the CG and TM gene as templates for overlapping PCR. The outside primer was prepared from the shuffling region of both templates and contained the restriction site, while the inside primer contained the opposite gene region. The second step involved targeting the chimeric gene with overlapped PCR from the templates prepared in the first step. At this stage the primers were not used as this step was exclusively focused on annealing of the two templates of CG and AT genes. PCR was initially performed at 98°C for 3 min, in order to convert the single DNA into a template. Polymerase was not added in this step. PCR was repeated at 25°C for 1 hr for purposes of overlapping both genes. DNA polymerase was then added and 15 PCR cycles were performed at 25°C. This completed the construction of the chimeric gene. The third step involved the amplification of the newly constructed chimeric gene using the primers from the first step, allowing for an exponential increase in the amount of chimeric genes produced.

Primers
Each chimeric gene required 4 primers to be constructed, and thus, 20 kinds of primers (forward and reverse) were designed as Figure 1 and Table 1. Figure 1 showed the position of forward and reverse primers and Table 1 explained the structure of primers for each chimeric gene, respectively.

Cloning and colony PCR
All PCR products were purified after agarose gel electrophoresis, using the QIAquick PCR purification kit (QIAGEN, Germany). The amplified chimeric genes were cloned using the pCR-TOPO cloning kit (Invitrogen, USA). The resulting recombinant pCR-TOPO plasmid

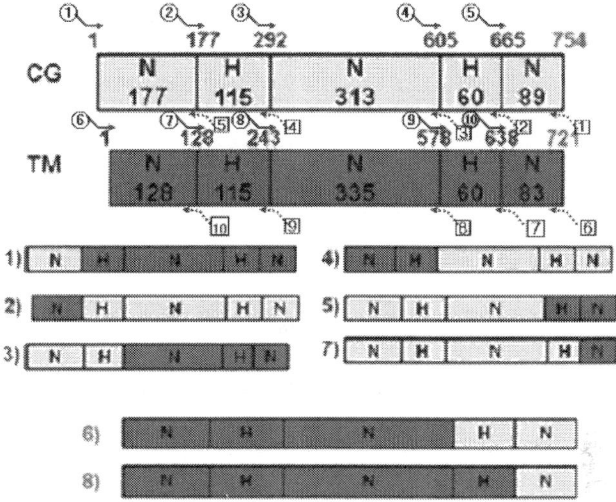

Fig. 1. Design of 8 kinds of chimeric β-glucosidase gene from CG and TM with higher
homology region. Numerics in bold gothic are designated as numbers of amino acids, and
numerics in circle are designated as forward primers, in square are reverse primers for
constructing chimeric enzymes. N: non-homology region, H: higher homology region.

containing chimeric β-glucosidase genes was confirmed by colony PCR, and then extracted
using a QIAminiprep kit (QIAGEN, Germany), and sequenced with a DNA sequencer
(Model 373A, Applied Biosystem, USA). The sequence data was analyzed using the
GENETYX program (Software Development Co., Tokyo, Japan).

Selection of active clonies

The TOPO-cloned chimeric β-glucosidase genes were transferred into the pET28a (+) vector
with hydrolysing *NdeI* and *HindIII* restriction enzymes. The host *E.coli* BL21 (DE3)
competent cells were tranformed with pET28a (+) vector carrying the chimeric β-
glucosidase genes to check its activity. Active colonies were selected with solid medium
including pNP-glucopyranoside.

Purification of enzymes

BL21 (DE3) cells carrying the chimeric gene were cultivated in Luria-Bertani broth (LB)
medium (1000 ml) containing kanamycin (30 mg/ml) at 37°C. On reaching an optimum
density of 0.2 at 600 nm, protein production was induced by the addition of 1 mM
isopropyl-β-D-thiogalactopyranoside (IPTG) into the culture media and incubating it
overnight. When the absorbance of the broth at 660 nm reached values between 0.4-0.5, the
cells were harvested by centrifugation (8,000 rpm, 10 min at 4°C) and suspended in 50 mM
MOPS buffer pH 6.5. The cells were then disrupted by sonication (Branson sonifier 250D).
The intact cells and debris were removed by centrifugation (12,000 rpm, 10 min at 4°C).
Purification was achieved by binding of the cleared lysate with Ni-NTA resin, which bound
to the target protein. This resin was then packed into a column to facilitate the washing and

| Chimera | 1st PCR primer | | Overlapping PCR primer | |
No	FWD	REV	FWD	REV
1)	① ccatgggcagcagc catc	⑤ ttccttccgttgcgc ggctcg	② gccgcgcaacggaag gaatttcgagtacta	⑥ aagcttatggttt gaatctcttct
2)	⑥ ccatatggaaaggat cgatg	⑩ agttgcggccacaa agagggtttctgtgaat	⑦ ccctctttgtggccgcaa cttcgaatac	① caagctttcagcgtg cgc
3)	①	④ ctcccgcgtaccagt cggacatgacataac	③ gtccgactggtacg cgggagacaac	⑥
4)	⑥	⑨ gggtggcgccccag tcgctcatcacgaaa	⑧ gagcgactggggcg ccaccca	①
5)	①	⑧ acctgtatccgaccgc cgcacc	④ tgcggcggtcgcatacagg tactacgacacc	⑥
6)	⑥	③ acttgtagcccacgtagatgtcttcct	⑨ cattcacgtgggctacaa gtggttcgacct	①
7)	①	⑦ gagctttgatgtacac ctgcggcacgtc	⑤ gcaggtgtacatcaa agctccaaaa	⑥
8)	⑥	② tcggcgcggcgtaga cctgtgagacttcctt	⑩ acaggtctacgccgcg ccga	①

Table 1. Forward and reverse primers for constructing each chimeric gene.

elution steps. Washing and elution were performed by the batch procedure described in TheQIA expressionist TM (QAIGEN, Germany) with a step gradient of 250 mM imidazol in 50 mM Na-phosphate buffer, pH 8.0. The fractions of 1 ml were collected and assayed for chimeric β-glucosidase activity. The active fractions were combined and dialyzed against 5 mM MOPS buffer, pH 6.5 overnight at 4°C. The dialysed enzyme solution was futher purified on a Mono-S HR 5/5 column (Pharmacia, Sweden) equilibrated with 20 mM acetate buffer (pH 5.0). Chromatography was performed with a linear gradient of NaCl (20 ml, 0-0.5 M) at a flow rate of 1.0 ml/min using a Pharmacia FPLC system (Sweden). The removal of salt from the pooled active fraction was performed by dialyzing against 5 mM MPOS buffer, pH 6.5.

Enzyme assay

The enzyme activity of chimeric β-glucosidase was determined by measuring the absorbance at 405 nm, which was related to the amount of pNP (p-nitrophenol) released from the pNP-β-D-glucopyranoside by the enzyme at 30°C. The assay mixture, consisting of 5 mM substrate (pNP-β-D-glc) and 50 mM MOPS buffer (pH 6.5) was incubated with the enzyme, in a total volume of 0.5 ml. The reaction was stopped by the addition of 0.5 ml of

0.2 M glycine-NaOH (pH 10.5). One unit of β-glucosidase was defined as the amount of enzyme releasing 1 μmole of p-nitrophenol per minute under the above-mentioned conditions.

3. SDS-page

Sodium dodecyl sulfate polyacrylamide gel electrophoresis (SDS-PAGE) was conducted on 1 mm thick, 12% acrylamide slab gels. The samples were dissolved in Tris/HCl loading buffer containing 1% (w/v) SDS, 20% (v/v) glycerol and 2% (v/v) 2-mercaptoethanol and then heated at 95°C in heating block for 5 min. Proteins were stained with 1% (w/v) Coomassie Brilliant Blue R 250 in methanol/acetic acid/water (30:10:60), and then destained with the electric-range.

Stability and optimization of pH and temperature

To determine the effect of pH, 50 mM concentrations of various buffers namely sodium phosphate (pH 1.1-3.1), sodium formate (pH 2.8-4.45), sodium succinate (pH 3.23-6.4), 3-[N-morpholino]propanesulfonic acid (MOPS; pH 6.2-8.18), N-[2-hydroxyethyl] piperazine-N'-[2-ethanesulfinic acid] (HEPES; pH 5.93-9.13), piperidine (pH 10.0-12.0), 2-[N-cyclohexylamino]ethanesulfonic acid (CHES; pH8.15-10.2), McIlvaine buffer (pH 2.6-7.6) were used at 30°C. A test of the pH stability of the enzyme at 30°C was performed by pre-incubating the enzyme with each of the above buffers for 30 min and determining the remaining activity using a standard procedure. For optimum pH, the enzyme was incubated with substrates in each of the above buffers at 30°C for 10 min, then reaction was stopped with 0.5M glycine(pH 10.3) solution, and followed enzyme assay. The optimum temperature of the enzyme was determined by incubating with substrates in a temperature range from 20°C to 100°C for 10 min, at three different pH levels, pH 3.0, 4.5 and 5.0, and analysed the enzyme activity. Similarly, the thermal stability of the enzyme was also determind by incubating the purified protein for 30 min at temperatures ranging from 20 to 100°C. After cooling the sample on ice for 10 min, the remaining activity was determined using standard procedures.

Kinetic parameters

Michaelis-Menten constants were determined from Line Weaver-Burk plots. The data used were obtained by measuring the initial rate of pNP-glucose hydrolysis by incubating the enzyme with appropriate concentrations of the substrate in 25 mM MOPS at 30°C. The reaction was monitored at 405 nm on a Beckman spectrophotometer (model DU 640, USA) equipped with a temperature-controlled cell holder. The initial rate was determined at six different concentrations ranging from approximately 0.5 to 2.0 times the Km. The catalytic constant K_{cat} was deduced using the molecular mass of 80 kDa, while the kinetic parameters (KM and Vmax) with their standard error was determined using the linear regression analysis.

4. Results and discussion

Structure of chimeric β-glucosidases gene

The characterization of β-glucosidase from CG and TM, as well as the cloning and analysis of the gene coding for these enzymes have previously been investigated by the National

Food Research Institute (NFRI). Based on the amino acid alignment between TM and CG enzymes, an N-terminal domain, a C-terminal region, and a non-homologus region were reported(5,16). The amino acid homology between TM and CG was 30.7%. The homology of N-terminal region and C-terminal region were found to be 32% and 36%, respectively, both of which fall within the highly homologus regions of the amino acid sequence. From these data, 8 kinds of chimeric β-glucosidases were designed from the TM gene and CG gene as shown in Fig. 1, and these chimeras were constructed by overlapping PCR.

Confirmation of introduced chimeric genes

After conducting overlapping PCR for construction of 8 chimeric genes, the PCR products were introduced into the pCR-TOPO plasmid. For confirmation of the inserted chimeric genes, colony PCR was performed using expressed colonies as a template. 8 kinds of chimeric β-glucosidase genes were introduced into the pCR-TOPO plasmid and these were confirmed with colony PCR. This confirmation was also shown positively in the other 7 chimeric genes (data not shown). From the colony PCR, some successfully inserted chimeric genes for 8 different chimeric β-glucosidase types were obtained. Colony PCR was found to be a fast and convenient method for confirmation of inserted target DNA fragments.

Selection of active chimeric β-glucosidase colonies

For the selection of active chimeric β-glucosidase colonies, TOPO plasmids were transferred into the pET28a (+) vector, and colony PCR was performed to confirm the presence of the inserted chimeric gene. BL21 (DE3) cells were then transformed with the plasmid and spread on a solid plate containing pNP-β-D-glucopyranoside. We obtained some successfully inserted chimeric gene colonies for 8 designed chimeric types from colony PCR. However, only 2 kinds of active chimeric β-glucosidases were obtained from No.6 and No.8 chimeric types. Activities of other types of chimeric β-glucosidases were not expressed. Inclusion bodies were produced due to the over-production by the pET28a (+) vector [19,20,21]. In some instances, the chimeric proteins were unfolded in structure [22,23].

Purification of enzymes

The pET vector is known to be a powerful expression system [24,25], and over-produced chimeric β-glucosidases can accumulate in an insoluble form. Harvested cells were suspended in 25 mM MOPS buffer (pH 6.5) and then disrupted by sonication. The soluble fraction was obtained by centrifugation of the sonically treated cell suspension at 15,000 rpm for 15 min. The resultant pellet was solublized in 8 M urea solution in the above-mentioned buffer, and after centrifugation (15,000 rpm for 15 min), the supernatant was used as the insoluble fraction [15,26,27]. The proteins in both the soluble and insoluble fractions of No.6 and No.8 were analyzed by 7.5% SDS-PAGE followed by Coomassie Brilliant Blue staining and the molecular mass of both protein constructs were found to be around 80 kDa (Fig. 2). The activities of the chimeric β-glucosidases were monitored at each step of the purification process and these activities are summarized in Table 2. Enzyme activities of both chimeric β-glucosidases were not very different from each other based on the specific activity that was calculated.

Purification step	Protein (mg)		Total activity (unit)		Specific activity (unit/mg)		Purification factor(fold)		Recovery (%)	
	No.6	No.8	No.6	No.8	No.6	No.8	No.6	No.8	No.6	No.8
Crude extract	335	350	55.3	49.5	0.17	0.14	1.0	1.0	100	100
Ni-NTA	16.75	15.05	24.8	18.8	1.48	1.25	8.71	8.93	44.8	38.0
Mono-S	2.43	2.03	13.5	9.2	5.55	4.53	32.65	32.35	24.4	18.6

Table 2. Purification step for chimeric β-glucosidases.

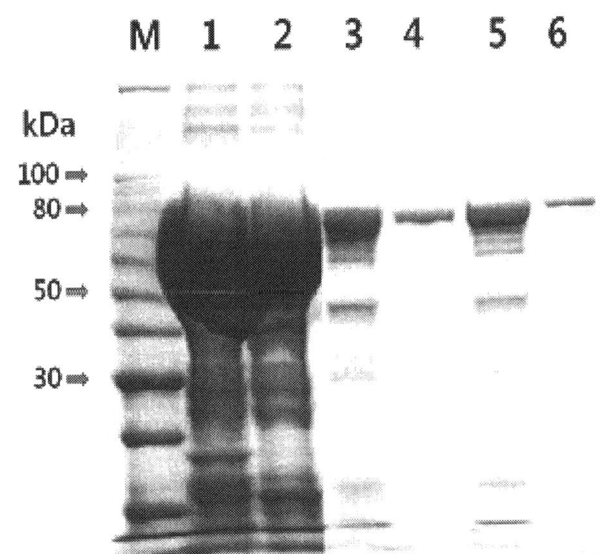

Fig. 2. SDS-PAGE of soluble and insoluble fractions of chimeric β-glucosidase from No.6 and No. 8.
Lane M, Marker; lane 1 and 2, insoluble fraction of pET-chimeric β-glucosidase after IPTG induction of No.6 and No.8, respectively; lane 3 and 5, soluble fraction of pET-chimeric β-glucosidase after IPTG induction of No.6 and No.8, respectively; lane 4 and 6, purification on Mono-S column after Ni-NTA-agarose slurry.

pH stability and optimum pH

The influence of pH on the chimeric β-glucosidase activity was determined using a series of various buffers at 30°C. The stability test [28] of the purified enzyme at different pH levels indicated that it was stable in the pH range of 3.0 to 5.0 for No.6 chimeric enzyme and 3.0 to 7.0 for No.8 chimeric enzyme at 30°C. The pH stability for the parents of these chimeric enzymes was followed the pervious paper(5,15), CG and Tm were pH 4.0 to 8.0 and pH 4.0 to 12.0, respectively (Fig. 3). For pH stability, chimeric enzymes generally correlate with the parental enzymes. The shuffling regions in the C-terminal domain may have played a very important role in enzyme-folding and stability because both chimeric genes were replaced

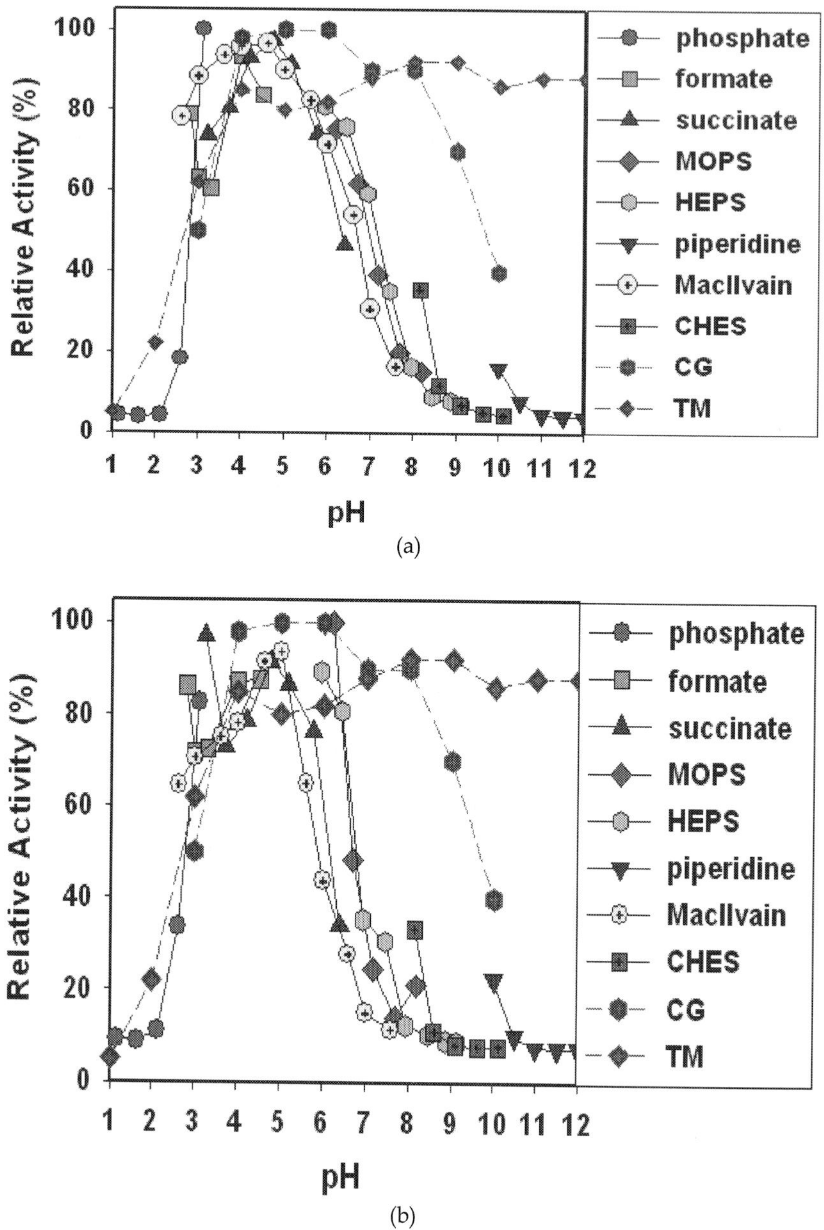

(a)

(b)

Fig. 3. pH stability range of No.6 (a) and No.8 (b) chimeric enzyme. CG and Tm as parents of these chimeric enzymes.

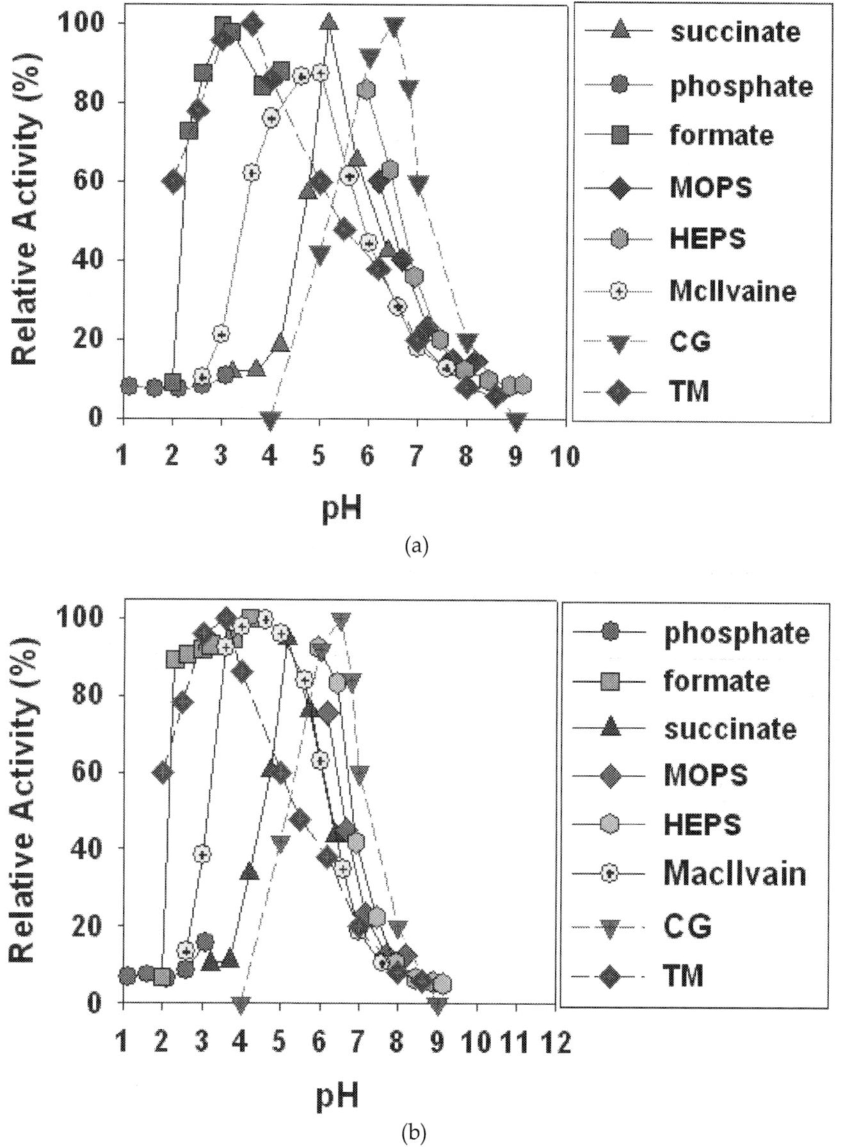

Fig. 4. Optimum pH of No.6 (a) and No.8 (b) chimeric enzyme. CG and Tm as parents of these chimeric enzymes.

with CG terminal domain by 20% (No.6-chimeric enzyme) and 12% (No.8-chimeric enzyme) on the whole TM gene. For determining the optimum pH of the chimeric enzymes, the pH of the parent enzymes was investigated. The optimum pH of CG and Tm was reported to be 6.4 and 3.4, respectively (Fig. 4). The optimum pH of No.6 chimeric enzymes was found to be 3.0 and 5.0, which showed different maximum activity according to the buffer used. No.8 chimeric enzyme showed an optimum pH of 4.5. The optimum pH profile of both chimeric enzymes is closer to that of TM than that of CG. The optimum pH of No.6 and No.8 was shifted between both the optimum pH properties of CG and TM, which is an indication of inherited pH properties.

Thermal stability and optimum temperature

The enzyme activity correlated with the pH unit. The optimum pH was reported at two different positions that varied according to the buffer used. The optimum temperature therefore, also varied between the two different pH points, pH 3.0 (formate buffer) and pH 5.0 (succinate buffer) for No.6 chimeric enzyme, and pH 4.5 (McIlvain buffer) for No.8 chimeric enzyme (Fig. 5). The optimum temperature of CG and TM was found to be 30°C and 70°C, respectively. The optimum temperature of the chimeric enzymes of No.6 (pH 5.0) and No.8 (pH 4.5) was reported to be at 60°C, and No.6 (pH 3.0) at 50°C, even though the activity was very low(Fig. 6). The optimum temperature of both chimeric β-glucosidases were found to lie between that of CG and TM. A temperature stability test was carried out for the parental and chimeric enzymes. For estimation of the heat stability of each β-glucosidase, residual activity was determined using a standard assay after a 30 min pre-incubation at various temperatures. The heat stability of CG and TM was at 41°C and 87°C

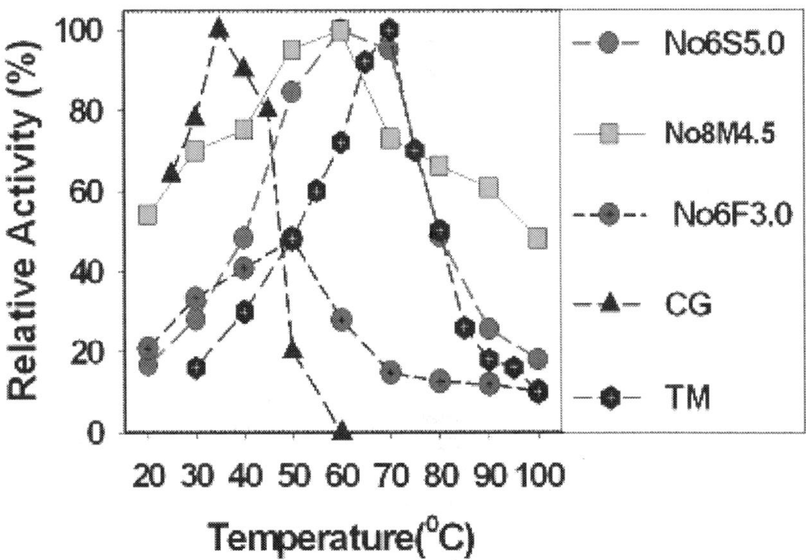

Fig. 5. Optimum temperature of No. 6 and No. 8 chimeric β-glucosidase with the buffer used, CG and TM as parents β-glucosidase.
S: Succinate buffer, M: McIlvain buffer, F: Formate buffer

respectively, while those of the chimeric enzymes of No.6 and No.8 were 70°C and 74°C respectively. The heat stability of the chimeric enzymes increased by 29-33°C relative to CG. himeric β-glucosidases were more stable than CG, but less stable than TM. These results demonstrated that the temperature stability of the chimeric enzymes generally correlated well with the thermal stability of the parental enzymes. In addition, the optimal temperature for the enzymatic reaction correlated successfully with the heat stability of the enzyme.

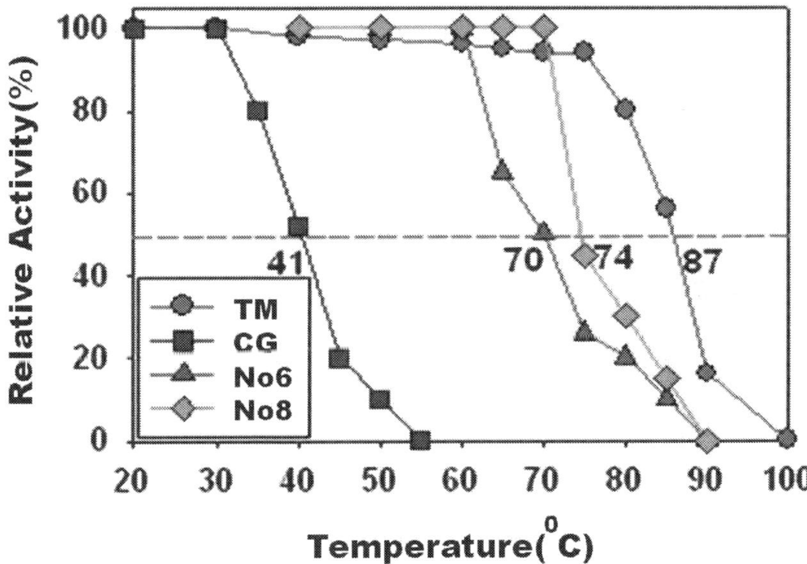

Fig. 6. Effect of temperature on the thermal stability of the chimeric β-glucosidase, CG and TM as parents of β-glucosidase. For estimation of the heat stability of the each β-glucosidase, residual activity was determined using a standard assay after a 30 min pre-incubation at various temperatures.

Kinetic parameters

The kinetic parameters of the two parental and chimeric β-glucosidases were investigated using pNP-β-D-glucopyranoside as the substrate(Table 3). The observed Km values for the chimeric No.6 and No.8 enzymes toward pNP-β-D-glucopyranoside were calculated to be 0.012 and 0.0082 mM, respectively. These Km values were lower than that of CG (0.44 mM), but are more or less similar to that obtained for TM (0.0039 mM). However, the Kcat values observed for the chimeric No.6 and No.8 enzymes were 5.62/sec and 3.38/sec respectively, which is lower than that obtained for both prental enzymes, i.e. TM (6.4/sec) and CG (42.2/sec). This indicated that the substrate specificities of the chimeric enzymes No.6 and No.8 were similar to each other and were closer to that of TM than CG. Also, No.8 chimeric β-glucosidase showed greater affinity toward pNP-β-D-glucopyranoside than that of No.6 chimeric enzyme. Thus the shuffling regions in the C-terminal domain produced a slight effect on the enzyme's substrate specificity and catalytic efficiency.

Enzymes	Km (mM)	Kcat (/sec)	Kcat/Km (mM/sec)
TM	0.0039	6.4	1640
No.6	0.012	5.62	468
No.8	0.0082	3.84	468
CG	0.44	42.2	95.9

Table 3. Kinetic parameters of the parental and chimeric β-glucosidases toward pNP-β-D-glucopyranoside

5. Concluding remarks

Chimeric β-glucosidases with improved enzymatic properties can be prepared in a convenient and effective way manipulating homology region of parental enzymes with overlapping PCR . The characteristics of the daughter enzymes suach as stability and optimization of pH and temperature, kinetic parameters were inherited with inter-mediate properties of parents. Also, the shuffling regions in the C-terminal domain may have played a very important role in determining enzyme characteristics, and changes in enzymatic properties.

6. References

[1] Béguin, P.(1990) Molecular biology of cellulose degradation. *Annu. Rev. Microbiol.* 44; 219-248.

[2] Paavilainen, S., J. Hellman, and T. Korpela(1993) Purification, characterization, gene cloning, and sequencing of a new beta-glucosidase from *Bacillus circulans subsp. alkalophilus. Appl. Environ. Microbiol.* 59;927-932.

[3] Shewale, J.G.(1982) Beta-Glucosidase: its role in cellulase synthesis and hydrolysis of cellulose. *Int. J. Biochem.*14; 435-443.

[4] Singh, A., K. and Hayashi(1995) Construction of chimeric β-glucosidases with improved enzymatic properties. *JBC.* 270; 21928-21935.

[5] Glick, B.R. and J. J. Pasternak(1989) Isolation, characterization and manipulation of cellulase genes. *Biotechnol. Adv.* 7; 361-386.

[6] Guo, R., M. Ding,S. L. Zhang, G. J. Xu , and F.K. Zhao(2008) Molecular cloning and characterization of two novel cellulase genes from the mollusc Ampullaria crossean. *J. Comp. Physiol.*178; 209-215.

[7] Roy, P., S. Mishra, and T. K. Chaudhuri(2005) Cloning, sequence analysis, and characterization of a novel beta-glucosidase-like activity from *Pichia etchellsii. Biochem. Biophys. Res. Commun.*336; 299-308.

[8] YKim , Y.M, S.H. Jung , Y.H. Chung , C. B. Yu , and I. K. Rhee(2008) Cloning and Characterization of a Cyclohexanone Monooxygenase Gene from *Arthrobacter* sp. L661. *Biotechnol. Bioprocess Eng.* 13; 40-47.

[9] Singh, A. and K. Hayashi(1995) Microbial celluoases, protein architecture, molecuar properties and biosynthesis. *Adv. Appl. Microbiol.*40;38-44.

[10] Tao, A. L. and S. H. He(2004) Bridging PCR and partially overlapping primers for novel allergen gene cloning and expression insert decoration. *World J. Gastroenterol.* 10; 2103-2108.

[11] Crameri, A., S.A.Raillard, E. Bermudez, and W.P.Stemmer(1998) DNA shuffling of a family of genes from diverse species accelerates directed evolution. *Nature*. 391; 288-291.

[12] Lehmann, M., L. Pasamontes, S. F. Lassen, and M. Wyss(2000) The consensus concept for thermostability engineering of proteins. *Biochim.Biophys. Acta*. 1543; 408-15

[13] Singh, S.P., J. D. Kim, S. Machida, and K. Hayashi(2002) Overexpression and protein folding of a chimeric β-glucosidase constructed from *Agrobacterium tumefaciens* and *Cellvibrio gilvus, Ind. J. Biochem. Biophys*. 39; 235-239.

[14] Lisy, O., B.K. Huntley, D.J. McCormick, P.A. Kurlansky, and J.C. Burnett Jr(2008) Design, synthesis, and actions of a novel chimeric natriuretic peptide: CD-NP. *J. Am. Coll. Cardiol*.52; 60-68.

[15] Shibuya, H., S. Kaneko, and K. Hayashi(2000) Enhancement of the thermostability and hydrolytic activity of xylanase by random gene shuffling. *Biochem. J.* 349;651-656.

[16] Singh, A., K. Hayashi,T.T. HOA, Y. Kashiwagi, and K. Tokuyasu(1955) Construction and characterization of chimeric β-glucosidase. *Biochem. J.* 305; 715-719.

[17] Bhat, K.M.,J.S. Gaikwad, and R. Maheshwari(1993) Purification and characterization of an extracellular β-glucosidase from thermophiic fungus *S.thermophile* and its influence on cellulae activity. *J. Gen. Microbiol.* 139; 2825-2832.

[18] D'Auria, S., R. Nucci, M. Rossi, E. Bertoli, F.Tanfani, I.Gryczynski, H. Malak, and J. R. Lakowicz (1999) β-glucosidase from the hyperthermophilic Archaeon Sulfoobus Solfataricus: structure and activity in presence of alcohols. *J. Biochem.* 126;545-52.

[19] Di Lorenzo, M.,A. Hidalgo, M. Haas, and U. T. Bornscheuer(2005) Heterologous production of functional forms of *Rhizopus oryzae* lipase in Escherichia coli. *Appl. Environ. Microbiol.* 71; 8974-8977.

[20] Gasparian, M.E.,V.G. Ostapchenko, A.V. Yagolovich, I.N. Tsygannik,B.V. Chernyak, D.A. Dolgikh, M.P. Kirpichnikov (2007) Overexpression and refolding of thioredoxin/ TRAIL fusion from inclusion bodies and further purification of TRAIL after cleavage by enteropeptidase. *Biotechnol. Lett.* 29; 1567-1573.

[21] Wakarchuk, W.W., N.M. Greenberg, D.G. Kilburn, R.C. Miller Jr, and R.A. Warren(1988) Structure and transcription analysis of the gene encoding a cellobiase from *Agrobacterium* sp. strain ATCC 21400. *J. Bacteriol*.170; 301-307.

[22] Iwahashi, J., S. Furuya, K. Mihara, and T. Omura(1992) Characterization of adrenodoxin precursor expressed in *Escherichia coli. J. Biochem*.111; 451-455.

[23] Parakhnevitch, N.M., A.A. Malygin, and G.G. Karpova(2005) Recombinant human ribosomal protein S16: expression, purification, refolding, and structural stability. *Biochemistry*(Mosc).70; 777-781.

[24] Pan, S.H. and B.A. Malcolm(2000) Reduced background expression and improved plasmid stability with pET vectors in BL21 (DE3). *Biotechniques*. 29;1234-1238.

[25] Stemmer, W.P.(1994) Rapid evolution of a protein in vitro by DNA shuffling. *Nature*. 370; 389-91.

[26] Machida, S., Y. YU, S.P. SINGH, J.D. Kim, K. Hayashi, and Y. Kawata(1998) Overproduction of β-glucosidase as active form by *Eschericha coli* system coexpressing the chaperonin GroELS at 25. *FEMS. MICROBIOL. LETT.* 159; 41-46.

[27] Wang ,X., M. Fu, J. Ren, and X. Qu(2007) Evaluation of different culture conditions for high-level soluble expression of human cyclin A2 with pET vector in BL21 (DE3) and spectroscopic characterization of its inclusion body structure. *Protein Expr. Purif.* 56; 27-34.

[28] Lien , T. S., S. T. Yu, S. T. Wu, and J. R. Too(2007)Induction and Purification of a Thermophilic Chitinase Produced by *Aeromonas* sp. DYU-Too7 Using Glucosamine. *Biotechnol. Bioprocess Eng.* 12 ; 610-617.